T0306019

# Amorphous and Polycrystalline Thin-Film Silicon Science and Technology—2008

MATERIALS RESEARCH SOCIETY
SYMPOSIUM PROCEEDINGS VOLUME 1066

# Amorphous and Polycrystalline Thin-Film Silicon Science and Technology—2008

Symposium held March 25–28, 2008, San Francisco, California, U.S.A.

**EDITORS:**

## Arokia Nathan
University College London
London, United Kingdom

## Andrew Flewitt
University of Cambridge
Cambridge, United Kingdom

## Jack Hou
General LED
San Luis Obispo, California, U.S.A.

## Seiichi Miyazaki
Hiroshima University
Hiroshima, Japan

## Jeffrey Yang
United Solar Ovonic LLC
Troy, Michigan, U.S.A.

Materials Research Society
Warrendale, Pennsylvania

# CAMBRIDGE
## UNIVERSITY PRESS

University Printing House, Cambridge CB2 8BS, United Kingdom

One Liberty Plaza, 20th Floor, New York, NY 10006, USA

477 Williamstown Road, Port Melbourne, VIC 3207, Australia

314-321, 3rd Floor, Plot 3, Splendor Forum, Jasola District Centre, New Delhi - 110025, India

79 Anson Road, #06-04/06, Singapore 079906

Cambridge University Press is part of the University of Cambridge.

It furthers the University's mission by disseminating knowledge in the pursuit of education, learning and research at the highest international levels of excellence.

www.cambridge.org
Information on this title: www.cambridge.org/9781605110363

Materials Research Society
506 Keystone Drive, Warrendale, PA 15086
http://www.mrs.org

© Materials Research Society 2008

First published 2008
First paperback edition 2012

Single article reprints from this publication are available through University Microfilms Inc., 300 North Zeeb Road, Ann Arbor, MI 48106

CODEN: MRSPDH

*A catalogue record for this publication is available from the British Library*

ISBN 978-1-605-11036-3 Hardback
ISBN 978-1-107-40857-9 Paperback

# CONTENTS

Preface ..........................................................................................................xvii

Materials Research Society Symposium Proceedings.............................xviii

## *FILM GROWTH*

\* Crystallinity Uniformity of Microcrystalline Silicon
Thin Films Deposited in Large Area Radio Frequency
Capacitively-Coupled Reactors ...................................................................3
  Benjamin Strahm, Alan A. Howling, and
  Christoph Hollenstein

Cone Kinetics Model: Insights Into the Morphologies
of Mixed-Phase Silicon Film Growth .........................................................15
  Howard M. Branz, Paul Stradins, and
  Charles W. Teplin

Evolution of Film Crystalline Structure During the
Ultrafast Deposition of Crystalline Si Films................................................21
  Haijun Jia, Hiroshi Kuraseko, Hiroyuki Fujiwara,
  and Michio Kondo

## *DEFECTS AND TRANSPORT*

Electronic Transport in Co-Deposited Hydrogenated Amorphous/
Nanocrystalline Thin Films.........................................................................29
  Y. Adjallah, C. Blackwell, C. Anderson, U. Kortshagen,
  and J. Kakalios

Metastable Defects in Light Soaked Amorphous Silicon
at 77 K..........................................................................................................35
  Tong Ju, Paul Stradins, and P. Craig Taylor

Improved Passivation of a-Si:H /c-Si Interfaces Through Film
Restructuring ...............................................................................................41
  M.Z. Burrows, U.K. Das, S. Bowden, S.S. Hegedus,
  R.L. Opila, and R.W. Birkmire

*Invited Paper

## SOLAR CELLS I

\* Understanding of Passivation Mechanism in
Heterojunction c-Si Solar Cells..................................................................................49
    Michio Kondo, Stefaan De Wolf, and
    Hiroyuki Fujiwara

Correlation of Hydrogen Dilution Profiling to
Material Structure and Device Performance of
Hydrogenated Nanocrystalline Silicon Solar Cells ......................................................61
    Baojie Yan, Guozhen Yue, Yanfa Yan,
    Chun-Sheng Jiang, Charles W. Teplin,
    Jeffrey Yang, and Subhendu Guha

MW Plasma Enhanced CVD of Intrinsic Si for Thin-Film
Solar Cells.............................................................................................................67
    Bas B. Van Aken, Hans Leegwater,
    Maarten Dorenkamper, Camile Devilee,
    Jochen Loffler, Maurits C.R. Heijna, and
    Wim J. Soppe

## CHARACTERIZATION

\* Characterization of Amorphous/Crystalline Silicon
Interfaces From Electrical Measurements .................................................................75
    J.P. Kleider and A.S. Gudovskikh

Probing Carrier Depletions on Grain Boundaries in
Polycrystalline Si Thin Films by Scanning Capacitance
Microscopy ..........................................................................................................87
    C.-S. Jiang, H.R. Moutinho, B. To, P. Dippo,
    M.J. Romero, and M.M. Al-Jassim

Characterization of the Mobility Gap in μc-Si:H Pin Devices ...................................93
    Bart Elger Pieters, Sandra Schicho, and Helmut Stiebig

Characterization of Gap Defect States in Hydrogenated
Amorphous Silicon Materials .................................................................................99
    Lihong (Heidi) Jiao and C.R. Wronski

*Invited Paper

*POSTER SESSION:*
*THIN-FILM GROWTH*

**Polysilazane Precursor Used for Formation of Oxidized Insulator** .................................................................................107
    Yuji Urabe and Toshiyuki Sameshima

**Boron Incorporation and Its Effect on Electronic Properties of Ge:H Films Deposited by LF Plasma** ...............................113
    Andrey Kosarev, Alfonso J. Torres, Nery D. Checa,
    Yurii Kudriavtsev, Rene Asomoza, and
    Salvador G. Hernandez

**Low Temperature Deposition of Si-Based Thin Films on Plastic Films Using Pulsed-Discharge PECVD Under Near Atmospheric Pressure** ..........................................................119
    Mitsutaka Matsumoto, Yohei Inayoshi, Maki Suemitsu,
    Setsuo Nakajima, Tsuyoshi Uehara, and
    Yasutake Toyoshima

**Pulsed Laser Heating-Induced Surface Rapid Cooling and Amorphization** .................................................................125
    Longzhang Tian and Xinwei Wang

**Studies on the Surface Reactions of Substituted Disilanes with Silica Surface** ..........................................................131
    Tom Blomberg, Raija Matero, Suvi Haukka, and
    Andrew Root

*POSTER SESSION:*
*AMORPHOUS, MICRO-, NANO-*
*AND POLYCRYSTALLINE SILICON*

**Polycrystalline Silicon Thin-Film Solar Cells on ZnO:Al Coated Glass** .................................................................139
    Christiane Becker, Pinar Dogan, Benjamin Gorka,
    Florian Ruske, Tobias Hänel, Jan Behrends,
    Frank Fenske, Klaus Lips, Stefan Gall, and Bernd Rech

**Photoluminescence of Different Phase Si Nanoclusters in Amorphous Hydrogenated Silicon** ....................................145
    Tatyana V. Torchynska

**Electronic Properties of Nanocrystalline Silicon
Deposited with Different Crystallite Fractions and
Growth Rates** .................................................................................................................149
    P.G. Hugger, J. David Cohen, Baojie Yan,
    Guozhen Yue, Xixiang Xu, Jeffrey Yang,
    and Subhendu Guha

**Doping Effects in Co-Deposited Mixed Phase Films
of Hydrogenated Amorphous Silicon Containing
Nanocrystalline Inclusions** ...........................................................................................155
    C. Blackwell, Xiaodong Pi, U. Kortshagen,
    and J. Kakalios

**Nanocrystalline Silicon Diodes for Rectifiers on
Flexible RFID Tags**........................................................................................................161
    Ian Chi Yan Kwong, Hyun Jung Lee, and
    Andrei Sazonov

**Low Temperature Synthesis of Nanocrystalline Silicon
and Silicon Oxide Films by Plasma Chemical Vapor
Deposition** .....................................................................................................................167
    Atsushi Tomyo, Hirokazu Kaki, Eiji Takahashi,
    Tsukasa Hayashi, Kiyoshi Ogata, and
    Yukiharu Uraoka

**Cyclically Varying Hydrogen Dilution for the Growth
of Very Thin and Doped Nanocrystalline Silicon Films
by Hot-Wire CVD** ..........................................................................................................173
    Fernando Villar, Aldrin Antony, Delfina Muñoz,
    Fredy Rojas, Jordi Escarré, Marco Stella,
    José Miguel Asensi, Joan Bertomeu, and
    Jordi Andreu

**Microstructure Effects in Hot-Wire Deposited
Undoped Microcrystalline Silicon Films**.......................................................................179
    Wolfhard Beyer, Reinhard Carius,
    Dorothea Lennartz, Lars Niessen, and
    Frank Pennartz

**Seeding Solid Phase Crystallization of Amorphous
Silicon Films with Embedded Nanocrystals** ..................................................................185
    Curtis Anderson and Uwe Kortshagen

## POSTER SESSION:
## ALLOYS, STRUCTURAL PROPERTIES
## AND SOLAR CELLS

**Nanostructures with Group IV Nanocrystals Obtained by LPCVD and Thermal Annealing of SiGeO Layers** .................................................... 193
Bruno Morana, Andrés Rodríguez, Jesús Sangrador,
Tomás Rodríguez, Óscar Martínez, Juan Jiménez,
and Andreas Kling

**On Determination of Properties of Ultrathin and Very Thin Silicon Oxide Layers by FTIR and X-Ray Reflectivity** ............................ 199
Martin Kopani, Matej Jergel, Hikaru Kobayashi,
Masao Takahashi, Robert Brunner, Milan Mikula,
Kentarou Imamura, Stanislav Jurecka, and Emil Pincik

**Structural and Opto-Electronic Properties of a-Si:H/a-SiN$_x$:H Superlattices** ...................................................................................................... 205
Stefan L. Luxembourg, Frans D. Tichelaar, Peter Kúš, and
Miro Zeman

**Transient Photoconductivity Study of the Distribution of Gap States in 100°C VHF-Deposited Hydrogenated Silicon Layers** ............................................................................................................... 211
Monica Brinza, Guy J. Adriaenssens, Jatindra K. Rath,
and Ruud E.I. Schropp

**Characterization and Light Emission Properties of Osmium Silicides Synthesized by Low Energy Ion Implantation** ............................................................................................................ 217
P.R. Poudel, K. Hossain, J. Li, B. Gorman, A. Neogi,
B. Rout, J.L. Duggan, and F.D. McDaniel

## NOVEL APPLICATIONS

**Multilayered a-SiC:H Device for Wavelength-Division (de)Multiplexing Applications in the Visible Spectrum** ............................ 225
Manuela Vieira, Miguel Fernandes, Paula Louro,
Manuel Augusto Vieira, Manuel Barata, and
Alessandro Fantoni

**\* Floating-Gate a-Si:H TFT Nonvolatile Memories** .................................... 231
Yue Kuo and Helinda Nominanda

\*Invited Paper

# THIN-FILM TRANSISTORS I

\* Micro Crystalline Silicon TFT by the Metal Capped Diode
Laser Thermal Annealing Method ..................................................................... 243
 Toshiaki Arai, Narihiro Morosawa, Yoshio Inagaki,
 Koichi Tatsuki, and Tetsuo Urabe

# ALLOYS: MICROCRYSTALLINE SILICON

Analysis of Compositionally and Structurally Graded Si:H
and $Si_{1-x}Ge_x$:H Thin Films by Real Time Spectroscopic
Ellipsometry ......................................................................................................... 253
 Nikolas J. Podraza, Jing Li, Christopher R. Wronski,
 Mark W. Horn, Elizabeth C. Dickey, and
 Robert W. Collins

Two-Step Capacitance Transients From an Oxygen Impurity
Defect ................................................................................................................... 259
 Shouvik Datta, J. David Cohen, Yueqin Xu, and
 Howard M. Branz

Micro Photovoltaic Modules for Micro Systems ......................................... 265
 Nicolas Wyrsch, Sylavain Dunand, and
 Christophe Ballif

# FILM GROWTH AND CHARACTERIZATION I

Magnetic Resonance in Hydrogenated Nanocrystalline
Silicon Thin Films ............................................................................................. 273
 Tining Su, Tong Ju, Baojie Yan, Jeffrey Yang,
 Subhendu Guha, and P. Craig Taylor

Voids in Hydrogenated Amorphous Silicon: A Comparison
of ab initio Simulations and Proton NMR Studies ...................................... 279
 Sudeshna Chakraborty, David C. Bobela, P.C. Taylor,
 and D.A. Drabold

Quality and Growth Rate of Hot-Wire Chemical Vapor
Deposition Epitaxial Si Layers ....................................................................... 285
 Charles W. Teplin, Ina T. Martin, Kim M. Jones,
 David Young, Manuel J. Romero, Robert C. Reedy,
 Howard M. Branz, and Paul Stradins

\*Invited Paper

## CRYSTALLIZATION TECHNIQUES

Influence of the Structural Properties of Microcrystalline
Silicon on the Performance of High Mobility Thin-Film
Transistors ................................................................................................293
    Kah-Yoong Chan, Dietmar Knipp, Reinhard Carius,
    and Helmut Stiebig

## THIN-FILM TRANSISTORS II

The Positive Gate Bias Annealing Method for the
Suppression of a Leakage Current in the SPC-Si TFT
on a Glass Substrate ................................................................................301
    Sang-Geun Park, Joong-Hyun Park, Seung-Hee Kuk,
    Dong-Won Kang, and Min-Koo Han

High Performance Bottom Gate μc-Si TFT Fabricated by
Microwave Plasma CVD ............................................................................307
    Akihiko Hiroe, Akinobu Teramoto, and Tadahiro Ohmi

## SOLAR CELLS II

* Production Technology of Large-Area, Light-Weight,
Flexible Solar Cell and Module ................................................................315
    Makoto Shimosawa, Shinichi Kawano,
    Takamasa Ishikawa, Tetsuro Nakamura,
    Yasushi Sakakibara, Shinji Kiyofuji,
    Hirofumi Enomoto, Hironori Nishihara,
    Tomoyoshi Kamoshita, Masahide Miyagi,
    Junichiro Saito, and Akihiro Takano

Study of Large Area a-Si:H and nc-Si:H Based
Multijunction Solar Cells and Materials ....................................................325
    Xixiang Xu, Baojie Yan, Dave Beglau, Yang Li,
    Greg DeMaggio, Guozhen Yue, Arindam Banerjee,
    Jeff Yang, Subhendu Guha, Peter G. Hugger, and
    J. David Cohen

Improved Photon Absorption in a-Si:H Solar Cells
Using Photonic Crystal Architectures ......................................................331
    Rana Biswas and Dayu Zhou

*Invited Paper

# FILM GROWTH AND CHARACTERIZATION II

**Mechanical Properties and Reliability of Amorphous vs. Polycrystalline Silicon Thin Films** ..................................................339
Joao Gaspar, Oliver Paul, Virginia Chu, and
Joao Pedro Conde

**In Situ Transmission Electron Microscopy Investigation of Aluminum Induced Crystallization of Amorphous Silicon** .......................................345
Ram Kishore, Renu Sharma, Satoshi Hata,
Noriyuki Kuwano, Yoshitsuga Tomokiyo,
Hameed Naseem, and W.D. Brown

**Blue and Yellow Electroluminescence of MOSLED Made on Si-Rich SiO$_x$ Film with Detuning Buried Si Nanoclusters Size** ...........................353
Chung-Hsiang Chang, Chi-Wee Liu, Chin-Hua Hsieh,
Li-Jen Chou, and Gong-Ru Lin

## POSTER SESSION: THIN-FILM TRANSISTORS

**The Effect of Electrical Stress on the New Top Gate N-type Depletion Mode Polycrystalline Thin-Film Transistors Fabricated by Alternating Magnetic Field Enhanced Rapid Thermal Annealing** ..................................................361
Won-Kyu Lee, Sang-Myeon Han, Sang-Geun Park,
Sung-Hwan Choi, Joonhoo Choi, and Min-Koo Han

**Negative Bias Temperature Instability for P-Channel of LTPS Thin-Film Transistors with Fluorine Implantation** .......................................367
Chyuan-Haur Kao and W.H. Sung

**Hysteresis Phenomenon in Sequential Lateral Solidification Poly-Si Thin-Film Transistor at Low Temperature (213K)** ..................................373
Sung-Hwan Choi, Sang-Geun Park, Won-Kyu Lee,
Tae-Jun Ha, and Min-Koo Han

**Hydrogenated Nanocrystalline Silicon Thin-Film Transistor Array for X-Ray Detector Application** ..................................................379
Kyung-Wook Shin, Mohammad R. Esmaeili-Rad,
Andrei Sazonov, and Arokia Nathan

Temperature and Humidity Effects on the Stability of
On-Plastic a-Si:H Thin-Film Transistors with Various
Conduction Channel Layer Thicknesses .....................................................385
    Jian Z. Chen and I-Chun Cheng

POSTER SESSION:
CRYSTALLIZATION TECHNIQUES

Laser Fabrication of Sharp Conical Microstructures on Si
Thin Films by Nd:YAG Laser Single Pulse Irradiation ............................393
    Joe Moening and Daniel Georgiev

Impurities and Grain Size Modeling in Recrystallized
Silicon...........................................................................................................399
    Valeri V. Kalinin, Alexandre M. Myasnikov, and
    Vladislav E. Zyryanov

POSTER SESSION:
IMAGERS, SENSORS AND
NOVEL APPLICATIONS

Improvement in pinpi'n' Device Architectures for Imaging
Applications...................................................................................................407
    P. Louro, A. Fantoni, M. Fernandes, G. Lavareda,
    N. Carvalho, and M. Vieira

Noise Analysis of Image Sensor Arrays for Large-Area
Biomedical Imaging ......................................................................................413
    Jackson Lai, Denis Striakhilev, Yuri Vygranenko,
    Gregory Heiler, Arokia Nathan, and Timothy Tredwell

Noise in Different Micro-Bolometer Configurations with
Silicon-Germanium Thermo-Sensing Layer .............................................419
    Mario M. Moreno, Andrey Kosarev, Alfonso J. Torres,
    and Ismael Cosme

Transient Current in a-Si:H-Based MIS Photosensors .............................425
    Miguel Fernandes, Yuriy Vygranenko, Manuela Vieira,
    Gregory Heiler, Timothy Tredwell, and Arokia Nathan

**Physically Based Compact Model for Segmented a-Si:H
n-i-p Photodiodes** ...................................................................................................**431**
    Jeff Hsin Chang, Timothy Tredwell, Gregory Heiler,
    Yuri Vygranenko, Denis Striakhilev, Kyung Ho Kim,
    and Arokia Nathan

**Luminescent Colloidal Silicon Nanocrystals Prepared by
Nanoseconds Laser Fragmentation and Laser Ablation in
Water** ...........................................................................................................................**437**
    Vladimir Svrcek, Davide Mariotti, Richard Hailstone,
    Hiroyuki Fujiwara, and Michio Kondo

**Optimized O/Si Composition Ratio for Enhancing
Si Nanocrystal Based Luminescence in Si-Rich SiO$_x$
Grown by PECVD with Argon Diluted SiH$_4$** .............................................................**443**
    Chung-Hsiang Chang, Chin-Hua Hsieh,
    Li-Jen Chou, and Gong-Ru Lin

*SENSORS, TRANSISTORS AND
ACTIVE MATRIX ARRAYS I*

**Fluorescence Detection of DNA Hybridization Using
an Integrated Thin-Film Amorphous Silicon n-i-p
Photodiode** ..................................................................................................................**451**
    A.C. Pimentel, R. Cabeça, M. Rodrigues,
    D.M.F. Prazeres, V. Chu, and J.P. Conde

**Noise Characterization of Polycrystalline Silicon
Thin-Film Transistors for X-Ray Imagers Based on
Active Pixel Architectures** ........................................................................................**457**
    L.E. Antonuk, M. Koniczek, J. McDonald,
    Y. El-Mohri, Q. Zhao, and M. Behravan

**High Fill Factor a-Si:H Sensor Arrays with Reduced
Pixel Crosstalk** ...........................................................................................................**463**
    Y. Vygranenko, A. Sazonov, D. Striakhilev,
    J.H. Chang, G. Heiler, J. Lai, T. Tredwell, and
    A. Nathan

## SENSORS, TRANSISTORS AND ACTIVE MATRIX ARRAYS II

**Self-Aligned Amorphous Silicon Thin-Film Transistors with Mobility Above 1 $cm^2V^{-1}s^{-1}$ Fabricated at 300°C on Clear Plastic Substrates**................................................................471
Kunigunde H. Cherenack, Alex Z. Kattamis,
Bahman Hekmatshoar, James C. Sturm, and
Sigurd Wagner

**Aligned-Crystalline Si Films on Glass**.........................................................477
Alp T. Findikoglu, Ozan Ugurlu, and
Terry G. Holesinger

**Monolithic 3D Integration of Single-Grain Si TFTs**...............................483
Mohammad Reza Tajari Mofrad, Ryoichi Ishihara,
Jaber Derakhshandeh, Alessandro Baiano,
Johan van der Cingel, and Cees Beenakker

**Author Index**...............................................................................................491

**Subject Index**.............................................................................................495

# PREFACE

Amorphous, nano-, micro-, and poly-crystalline silicon thin films, and associated alloys are used in a plethora of applications ranging from active matrix displays and imaging arrays to solar panels. These applications make large-area electronics the fastest growing semiconductor technology today, pushing material requirements and device performance to new limits. As one of the longest running MRS symposia, Symposium A, "Amorphous and Polycrystalline Thin-Film Silicon Science and Technology," held March 25–28 at the 2008 MRS Spring Meeting in San Francisco, California, continued its long-standing tradition to provide scientists and engineers an excellent forum to discuss issues ranging from film deposition and associated electronic and optical properties to design, fabrication and analysis of devices, and their integration into systems. Materials addressed include amorphous, nano, micro, and poly-crystalline silicon, and their alloys with germanium, carbon, and other elements.

The current challenges in thin-film silicon technology were addressed with an opening full-day tutorial given by Friedhelm Finger and Michio Kondo, followed by 10 invited talks, 61 contributed oral presentations and 56 poster presentations. Topics addressed included the understanding of growth processes; producing high-quality films at high growth rates or low temperatures; *in-situ* characterization techniques for monitoring growth; understanding amorphous, mixed-phase and crystalline structures, along with the principles for augmenting crystallinity; developing post-deposition processes, such as thermal or laser annealing; identifying fundamental issues in electronic structure and carrier transport in three-, two-, and one-dimensions; understanding metastability and the role of hydrogen; integrating photovoltaic devices and thin-film electronics into systems on glass, flexible polymeric, and other non-conventional substrates; and designing, fabricating, and testing new and improved devices and applications.

The success of Symposium A would not have been possible if not for the high-quality invited and contributed oral and poster presentations, including the contributors and reviewers of this proceedings volume. Our thanks also goes to Mary Ann Woolf for managing the tireless manuscript reviewing process and for the timely production of this volume.

Last but not least, we are grateful for the generous financial support of our corporate sponsors: AU Optronics Corp., CYTEK Taiwan Ltd., Fuji Electric Advanced Technology Co. Ltd., ITRI, Merck Chemicals UK., Sharp Labs of America, and United Solar Ovonic LLC.

Arokia Nathan
Andrew Flewitt
Jack Hou
Seiichi Miyazaki
Jeffrey Yang

August 2008

# MATERIALS RESEARCH SOCIETY SYMPOSIUM PROCEEDINGS

Volume 1066 — Amorphous and Polycrystalline Thin-Film Silicon Science and Technology—2008, A. Nathan, J. Yang, S. Miyazaki, H. Hou, A. Flewitt, 2008, ISBN 978-1-60511-036-3

Volume 1067E —Materials and Devices for "Beyond CMOS" Scaling, S. Ramanathan, 2008, ISBN 978-1-60511-037-0

Volume 1068 — Advances in GaN, GaAs, SiC and Related Alloys on Silicon Substrates, T. Li, J. Redwing, M. Mastro, E.L. Piner, A. Dadgar, 2008, ISBN 978-1-60511-038-7

Volume 1069 — Silicon Carbide 2008—Materials, Processing and Devices, A. Powell, M. Dudley, C.M. Johnson, S-H. Ryu, 2008, ISBN 978-1-60511-039-4

Volume 1070 — Doping Engineering for Front-End Processing, B.J. Pawlak, M. Law, K. Suguro, M.L. Pelaz, 2008, ISBN 978-1-60511-040-0

Volume 1071 — Materials Science and Technology for Nonvolatile Memories, O. Auciello, D. Wouters, S. Soss, S. Hong, 2008, ISBN 978-1-60511-041-7

Volume 1072E —Phase-Change Materials for Reconfigurable Electronics and Memory Applications, S. Raoux, A.H. Edwards, M. Wuttig, P.J. Fons, P.C. Taylor, 2008, ISBN 978-1-60511-042-4

Volume 1073E —Materials Science of High-k Dielectric Stacks—From Fundamentals to Technology, L. Pantisano, E. Gusev, M. Green, M. Niwa, 2008, ISBN 978-1-60511-043-1

Volume 1074E —Synthesis and Metrology of Nanoscale Oxides and Thin Films, V. Craciun, D. Kumar, S.J. Pennycook, K.K. Singh, 2008, ISBN 978-1-60511-044-8

Volume 1075E —Passive and Electromechanical Materials and Integration, Y.S. Cho, H.A.C. Tilmans, T. Tsurumi, G.K. Fedder, 2008, ISBN 978-1-60511-045-5

Volume 1076 — Materials and Devices for Laser Remote Sensing and Optical Communication, A. Aksnes, F. Amzajerdian, 2008, ISBN 978-1-60511-046-2

Volume 1077E —Functional Plasmonics and Nanophotonics, S. Maier, 2008, ISBN 978-1-60511-047-9

Volume 1078E —Materials and Technology for Flexible, Conformable and Stretchable Sensors and Transistors, 2008, ISBN 978-1-60511-048-6

Volume 1079E —Materials and Processes for Advanced Interconnects for Microelectronics, J. Gambino, S. Ogawa, C.L. Gan, Z. Tokei, 2008, ISBN 978-1-60511-049-3

Volume 1080E —Semiconductor Nanowires—Growth, Physics, Devices and Applications, H. Riel, T. Kamins, H. Fan, S. Fischer, C. Thelander, 2008, ISBN 978-1-60511-050-9

Volume 1081E —Carbon Nanotubes and Related Low-Dimensional Materials, L-C. Chen, J. Robertson, Z.L. Wang, D.B. Geohegan, 2008, ISBN 978-1-60511-051-6

Volume 1082E —Ionic Liquids in Materials Synthesis and Application, H. Yang, G.A. Baker, J.S Wilkes, 2008, ISBN 978-1-60511-052-3

Volume 1083E —Coupled Mechanical, Electrical and Thermal Behaviors of Nanomaterials, L. Shi, M. Zhou, M-F. Yu, V. Tomar, 2008, ISBN 978-1-60511-053-0

Volume 1084E —Weak Interaction Phenomena—Modeling and Simulation from First Principles, E. Schwegler, 2008, ISBN 978-1-60511-054-7

Volume 1085E —Nanoscale Tribology—Impact for Materials and Devices, Y. Ando, R.W. Carpick, R. Bennewitz, W.G. Sawyer, 2008, ISBN 978-1-60511-055-4

Volume 1086E —Mechanics of Nanoscale Materials, C. Friesen, R.C. Cammarata, A. Hodge, O.L. Warren, 2008, ISBN 978-1-60511-056-1

# MATERIALS RESEARCH SOCIETY SYMPOSIUM PROCEEDINGS

Volume 1087E —Crystal-Shape Control and Shape-Dependent Properties—Methods, Mechanism, Theory and Simulation, K-S. Choi, A.S. Barnard, D.J. Srolovitz, H. Xu, 2008, ISBN 978-1-60511-057-8

Volume 1088E —Advances and Applications of Surface Electron Microscopy, D.L. Adler, E. Bauer, G.L. Kellogg, A. Scholl, 2008, ISBN 978-1-60511-058-5

Volume 1089E —Focused Ion Beams for Materials Characterization and Micromachining, L. Holzer, M.D. Uchic, C. Volkert, A. Minor, 2008, ISBN 978-1-60511-059-2

Volume 1090E —Materials Structures—The Nabarro Legacy, P. Müllner, S. Sant, 2008, ISBN 978-1-60511-060-8

Volume 1091E —Conjugated Organic Materials—Synthesis, Structure, Device and Applications, Z. Bao, J. Locklin, W. You, J. Li, 2008, ISBN 978-1-60511-061-5

Volume 1092E —Signal Transduction Across the Biology-Technology Interface, K. Plaxco, T. Tarasow, M. Berggren, A. Dodabalapur, 2008, ISBN 978-1-60511-062-2

Volume 1093E —Designer Biointerfaces, E. Chaikof, A. Chilkoti, J. Elisseeff, J. Lahann, 2008, ISBN 978-1-60511-063-9

Volume 1094E —From Biological Materials to Biomimetic Material Synthesis, N. Kröger, R. Qiu, R. Naik, D. Kaplan, 2008, ISBN 978-1-60511-064-6

Volume 1095E —Responsive Biomaterials for Biomedical Applications, J. Cheng, A. Khademhosseini, H-Q. Mao, M. Stevens, C. Wang, 2008, ISBN 978-1-60511-065-3

Volume 1096E —Molecular Motors, Nanomachines and Active Nanostructures, H. Hess, A. Flood, H. Linke, A.J. Turberfield, 2008, ISBN 978-1-60511-066-0

Volume 1097E —Mechanical Behavior of Biological Materials and Biomaterials, J. Zhou, A.G. Checa, O.O. Popoola, E.D. Rekow, 2008, ISBN 978-1-60511-067-7

Volume 1098E —The Hydrogen Economy, A. Dillon, C. Moen, B. Choudhury, J. Keller, 2008, ISBN 978-1-60511-068-4

Volume 1099E —Heterostructures, Functionalization and Nanoscale Optimization in Superconductivity, T. Aytug, V. Maroni, B. Holzapfel, T. Kiss, X. Li, 2008, ISBN 978-1-60511-069-1

Volume 1100E —Materials Research for Electrical Energy Storage, J.B. Goodenough, H.D. Abruña, M.V. Buchanan, 2008, ISBN 978-1-60511-070-7

Volume 1101E —Light Management in Photovoltaic Devices—Theory and Practice, C. Ballif, R. Ellingson, M. Topic, M. Zeman, 2008, ISBN 978-1-60511-071-4

Volume 1102E —Energy Harvesting—From Fundamentals to Devices, H. Radousky, J. Holbery, B. O'Handley, N. Kioussis, 2008, ISBN 978-1-60511-072-1

Volume 1103E —Health and Environmental Impacts of Nanoscale Materials—Safety by Design, S. Tinkle, 2008, ISBN 978-1-60511-073-8

Volume 1104 — Actinides 2008—Basic Science, Applications and Technology, B. Chung, J. Thompson, D. Shuh, T. Albrecht-Schmitt, T. Gouder, 2008, ISBN 978-1-60511-074-5

Volume 1105E —The Role of Lifelong Education in Nanoscience and Engineering, D. Palma, L. Bell, R. Chang, R. Tomellini, 2008, ISBN 978-1-60511-075-2

Volume 1106E —The Business of Nanotechnology, L. Merhari, A. Gandhi, S. Giordani, L. Tsakalakos, C. Tsamis, 2008, ISBN 978-1-60511-076-9

Volume 1107 — Scientific Basis for Nuclear Waste Management XXXI, W.E. Lee, J.W. Roberts, N.C. Hyatt, R.W. Grimes, 2008, ISBN 978-1-60511-079-0

**Prior Materials Research Society Symposium Proceedings available by contacting Materials Research Society**

# Film Growth

Mater. Res. Soc. Symp. Proc. Vol. 1066 © 2008 Materials Research Society          1066-A01-01

# Crystallinity Uniformity of Microcrystalline Silicon Thin Films Deposited in Large Area Radio Frequency Capacitively-coupled Reactors

Benjamin Strahm, Alan A. Howling, and Christoph Hollenstein
Centre de Recherches en Physique des Plasmas, Ecole Polytechnique Fédérale de Lausanne (EPFL), Lausanne, CH-1015, Switzerland

## ABSTRACT

The microcrystalline silicon ($\mu$c-Si:H) intrinsic layer for application in micromorph tandem photovoltaic solar cells has to be optimized in order to achieve cost-effective mass production of solar cells in large area, radio frequency, capacitively-coupled PECVD reactors. The optimization has to be performed with regard to the deposition rate as well as to the crystallinity uniformity over the substrate area. The latter condition is difficult to achieve since the optimal solar grade $\mu$c-Si:H is deposited at the limit between a-Si:H and $\mu$c-Si:H material, where the film crystallinity is very sensitive to the plasma process. In this work, a controlled RF power nonuniformity was generated in a large area industrial reactor. The resulting film uniformity was studied as a function of the deposition regimes. Results show that the higher the input silane concentration, the more the uniformity of the crystallinity is sensitive to the RF power nonuniformity for films deposited at the limit between a-Si:H and $\mu$c-Si:H. The effect of the input silane concentration on the microstructure uniformity could be explained on the basis of an analytical plasma chemistry model. This result is important for reactor design. In reactors generating nonuniform plasma the input silane concentration has to be limited to low values in order to deposit films with uniform microstructure. To benefit from the high silane flow rate utilization fraction encountered only for higher input silane concentration, the RF power distribution has to be as uniform as possible over the whole substrate area.

## INTRODUCTION

Plasma enhanced chemical vapour deposition of silicon for large area microelectronics has to fulfil various objectives to achieve a successful production of devices such as TFT displays or low cost thin film photovoltaic solar cells. High deposition rate ($\approx 10$ Å/s) of silicon film, especially of microcrystalline silicon ($\mu$c-Si:H), is one of these objectives which for many years has been the centre of interest of research groups all around the world. The quality of the deposited films is of course also one of the major concerns to reach cost effective production of high-efficiency solar cells. Therefore, properties such as the crystalline fraction or the hydrogen content of the film have to be as uniform as possible (less than ± 10 %) over the whole substrate area, similarly for the film thickness uniformity. The film uniformity becomes a crucial issue when processing large area devices that nowadays can reach sizes of several square meters [1, 2]. Up to now, studies on the uniformity are mainly performed by industrial companies due to the scarcity of large area equipment in academic research laboratories. Moreover, these generally empirical studies are mainly focused on the uniformity of the deposition rate, i.e. uniformity of the film thickness, because generally it is assumed that to achieve a film with uniform properties one has to have a uniform deposition rate over, i.e. a uniform thickness.

The present work is a study of the effect of an RF voltage perturbation in a large area radio-frequency PECVD reactor on both the film thickness and the film crystallinity. Two typical conditions for the deposition of microcrystalline silicon are compared. One is from a low silane input concentration (1 %) with a low RF power density, and the other is from a high silane input concentration (7 %) with a high RF power density. These two experiments show that the film thickness uniformity alone is not sufficient to obtain a uniform microstructure and that the crystallinity uniformity depends strongly on the deposition parameters. These results are interpreted on the basis of a simple analytical plasma chemistry model introduced in earlier work [3-5], and it is demonstrated that the higher the input silane concentration, the more the microstructure uniformity is sensitive to RF voltage nonuniformity.

## EXPERIMENTAL ARRANGEMENT

Experiments presented in this work were performed in a modified version of a KAI-S (47 x 57 cm$^2$), radio-frequency (40.68 MHz) capacitively-coupled parallel plate reactor manufactured by Oerlikon AG. This reactor is not as large as actual production devices which are generally larger than 1 m$^2$. However, this reactor is much larger than parallel plate reactors conventionally used in research laboratories which are often cylindrical with diameter about 20 cm. The use of a large area reactor permits the study of film nonuniformities caused by sources other than edge effects such as the telegraph effect [6, 7] or the silane back-diffusion from the surrounding vacuum chamber [5, 8]. In the case of small open laboratory reactors, these effects are dominant and most of the plasma volume is dominated by edge effects, which makes it difficult to study the deposition uniformity in a controlled manner.

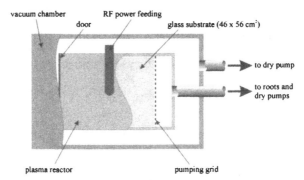

**Figure 1**: Top view of the KAI-S reactor based on the Plasma-Box$^{TM}$ concept with a unilateral pumping of the plasma reactor.

The reactor consists of a closed, grounded box with unilateral pumping as shown in Fig. 1. The 40.68 MHz RF power feeding is connected to a RF electrode suspended in the box (see Fig. 2b). This RF electrode acts also as a gas showerhead providing a uniform gas (SiH$_4$ and H$_2$) density distribution over the whole deposition area even with unilateral pumping [9].

In order to have a well-defined and controlled plasma nonuniformity, a non constant interelectrode distance was set across the width of the reactor, as shown in Fig. 2. The resulting variation in interelectrode distance is 3 mm (from 19 (left) to 22 mm (right)). The 46 x 56 cm² 3 mm thick glass substrate, which covers almost the whole electrode area, was placed horizontally by using an insulator strip as shown in Fig. 2b. This precaution was taken in order to guarantee a constant gas residence time (constant plasma height of 16 mm) across the width of the reactor. This configuration results in a capacitive division of the interelectrode voltage as schematized in Fig. 2a, making the RF voltage amplitude across the plasma, $U_p$, higher for narrower interelectrode distance. The relative variation of $U_p$ across the width of the reactor (Fig. 2b) can be estimated by using the model presented in Fig. 2a assuming two constant capacitive sheaths (width=3 mm) in series with a variable capacitance due to the vacuum under the glass substrate. Assuming a relation $P_{RF} \propto (U_p)^2$ between the RF power, $P_{RF}$, and $U_p$, the distribution of the RF power across the width of the reactor is almost linear with a total relative variation of $\approx 70$ % as shown in Fig. 2b.

Two depositions were performed during 60 minutes with the following parameters:
- 500 W, 1 mbar, 1 % of silane, 20 sccm of silane, 1980 sccm of $H_2$, 230 °C and,
- 1000 W, 1 mbar, 7 % of silane, 20 sccm of silane, 265 sccm of $H_2$, 230 °C.

The crystalline volume fraction of the resulting films was estimated by micro-Raman spectroscopy. The Raman crystallinity was defined as the ratio of the area of the peaks associated with crystalline phase at 518 and 510 cm⁻¹ to the sum of all Raman peaks including the peak associated with amorphous phase at 480 cm⁻¹ [10]:

$$\phi_c = \frac{A_{518} + A_{510}}{A_{518} + A_{510} + A_{480}}.$$

This method is described in more detail elsewhere [4].

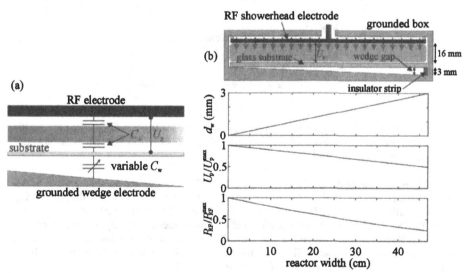

**Figure 2:** Lateral view of the wedge electrode design: (a) corresponding electrical circuit used to determine (b) the RF power relative distribution across the width of the reactor.

The thickness uniformity was determined by light interferometry techniques. Two-dimensional interferograms of the whole substrate area were obtained using a diffuse white light source behind the coated glass substrate. The resulting light was then filtered at 720 nm before the interferogram acquisition with a highly sensitive camera. The thickness of the film was determined by white light reflectometry using a commercial reflectometer (NanoCalc2000).

## RESULTS AND DISCUSSION

The two sets of deposition parameters were chosen in order to deposit films with microstructure at the transition from amorphous to microcrystalline silicon. This was chosen because the best PV cells are obtained when films are deposited in such conditions [11]. The film deposited with a 1 % input silane concentration was performed at low power (500 W) and high total flow rate (2000 sccm) in order to have a low silane depletion. On the other hand, for the film deposited from a high input silane concentration (7 %), the RF power has been increased to 1000 W and the total flow rate decreased to 285 sccm in order to obtain a high fraction of depleted silane necessary to achieve microcrystalline silicon. These two parameter sets correspond to two different regimes, named Regime 1 and Regime 2 in previous work [4].

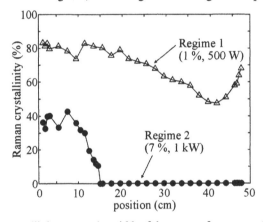

**Figure 3**: Raman crystallinity across the width of the reactor for two conditions: Regime 1 with a silane concentration of 1 % and a RF power of 500 W, and Regime 2 with a silane concentration of 7 % and a RF power of 1 kW.

Figure 3 presents the results of the Raman crystallinity measured across the width of the reactor in the central part of the substrate for the two different regimes. It shows that, depending on the deposition regime, the effect of the given RF power distribution is very different. In the case of Regime 1 at low silane concentration and RF power, the crystallinity is everywhere higher than 50 % and decreases only a little from the high plasma density side to the low plasma density side of the wedge reactor. However, the variation of the Raman crystallinity is small compared to the 70 % variation of the RF power across the reactor.

On the other hand, for the film deposited in Regime 2 at high silane concentration and RF power, the Raman crystallinity falls abruptly from 40 % to 0 % at 10 cm from the reactor edge. The conclusion is that the RF power nonuniformity created by the wedge electrode has a very different effect depending on the deposition regime.

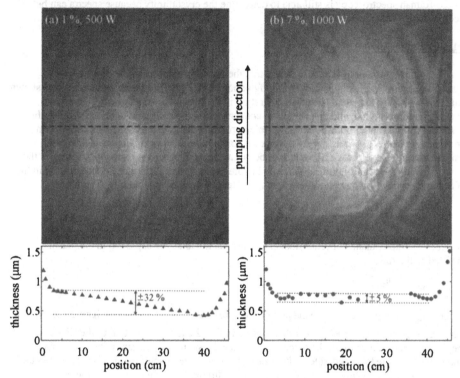

**Figure 4:** Thickness uniformity determined by (top) 2D monochromatic (720 nm) light interferometry and (bottom) white light interferometry across the width of the substrate following the dashed lines on the interferograms.

The transition from microcrystalline to amorphous silicon cannot be attributed to a difference in deposition rate as shown by thickness uniformity measurements presented in Fig. 4. Indeed, the zone where the microstructure transition occurs is almost flat (± 5 %). Neither can it be attributed to the amorphous incubation layer thickness generally observed before the growth of microcrystals, because the films are everywhere thicker than 400 nm, which is larger than amorphous incubation layer thickness generally observed. Moreover, in Plasma-Box[TM] type reactors, i.e. a closed and directly pumped reactor, the plasma chemistry transient at ignition is only about 1 second and cannot be at the origin of a thick incubation layer [5]. The thickness profiles presented in Fig. 4 also confirm the fact that nonuniformities due to edge effects are restricted to a 5 cm strip from the edge of the reactor. There, the increase of thickness on the substrate at the wall has two different sources: (i) the telegraph effect induced by the asymmetry between the RF and the grounded electrodes, and (ii) more intense plasma at the wall due to the geometrical edge of the RF electrode.

The absence of thickness measurements between 25 and 35 cm in the case of Regime 2 is due to film peeling off the substrate making the optical measurement inapplicable. However, micro-Raman spectroscopy is still able to determine the crystallinity because spectra can be acquired even from peeled films.

### Plasma composition

In previous work [4], it was shown that the plasma composition, i.e. the silane concentration in the plasma, is a determining parameter for the film microstructure. It was shown that it is necessary to have a silane concentration in the plasma lower than 0.5 % to obtain microcrystalline silicon, and that for silane concentration in the plasma higher than 1.2 % the resulting films are amorphous. These two microstructure zones are separated by a transition zone in which parameters such as the pressure or the ion bombardment may determine the final microstructure. Nevertheless, the silane concentration in the plasma principally determines the two different crystallinity profiles presented in the present work. The silane concentration in the plasma, $c_p$, is determined not only by the input silane concentration, but by the relation

$$c_p = c \cdot (1 - D),$$

between the input silane concentration, $c$, and the fractional silane depletion in the plasma, $D$, defined as the ratio

$$D = \frac{p^0_{SiH_4} - p_{SiH_4}}{p^0_{SiH_4}},$$

where $p^0_{SiH_4}$ and $p_{SiH_4}$ are the silane partial pressure in the gas before plasma ignition and in the plasma, respectively. The RF power distribution in Fig. 2 causes a variation of the silane depletion across the width of the reactor, and hence the silane concentration in the plasma increases from the left-hand-side to the right-hand-side of the reactor (neglecting edge effects). However, this alone is not sufficient to explain the different behaviour between the two experiments presented here.

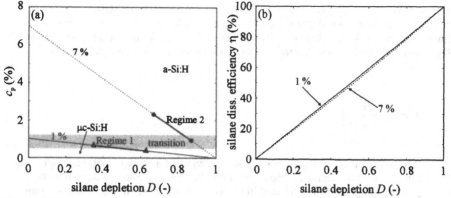

**Figure 5:** (a) Silane concentration in the plasma, $c_p$, and (b) silane dissociation efficiency, $\eta$, as a function of the silane depletion for the two experimental conditions Regimes 1 and 2, with corresponding estimated silane depletion range of 0.35-0.63 and 0.67-0.87, respectively.

Figure 5a presents the silane concentration in the plasma as a function of the silane depletion fraction for the two present cases: Regime 1 (1 %) and Regime 2 (7 %). The difference in slope of the two regimes shows clearly that they have a different impact on the plasma composition for a nonuniform silane depletion as in the present case. With an input silane concentration of 1 %, the deposition is everywhere – i.e. independently of the silane depletion - in the microcrystalline zone ($c_p <$ 0.5 % in Fig. 5a) or at least in the transition zone between μc-Si:H and a-Si:H (0.5 % < $c_p$ < 1.2 % in Fig. 5a). For higher concentration, this is no longer the case. The strong gradient of the silane concentration in the plasma as a function of the silane depletion makes the microstructure of the deposited film nonuniform for film deposited at the limit between a-Si:H and μc-Si:H, where the PV cells show the best performances [11]. This explains why in Fig. 3 the microstructure uniformity between the two Regimes is so different for the used parameters, i.e. at the transition between a-Si:H and μc-Si:H.

However, even if the uniformity of the plasma composition varies strongly from one case to the other, the silane dissociation efficiency, $\eta$, defined as the ratio between the silane dissociation rate ($kn_e n_{SiH_4}$) to the silane flow rate ($\Phi_{SiH_4}$) [3, 4]

$$\eta = \frac{kn_e n_{SiH_4}}{\Phi_{SiH_4}} = \frac{D}{1 + (1 - D)c}$$

is almost independent of the input silane concentration, as shown in Fig. 5b. The silane dissociation efficiency is mostly determined by the silane depletion fraction and depends only little on the input silane concentration. This explains why even with a strong plasma composition nonuniformity, the thickness profiles of the two Regimes are similar in Fig. 4. Moreover, it shows that having a uniform thickness does not guarantee a uniform microstructure.

However, for the case of Regime 2 (7 %) the thickness profile is more uniform (± 5 %) in the central region which is not affected by edge effects than the case at lower silane concentration (± 32 %). This effect may be attributed to the different range in silane depletion fraction between the two experiments. In the case of Regime 1, the parameters are a low RF power (500 W) and a high total flow rate (2000 sccm). This corresponds respectively to a low electron density, $n_e$, and a high pumping rate, $a$. Therefore, the dissociation rate over the pumping rate ratio is low, making the silane depletion fraction also low, i.e. about 0.35 to 0.63 ($\pm$ 28 %), since it depends on $kn_e/a$, where $k$ is the silane dissociation constant, following the expression [4]

$$D = \left(1 + \frac{a\big/kn_e}{(1+c)}\right)^{-1} .$$

This dependence is plotted in Fig. 6, and it shows that for low $kn_e/a$ ratio corresponding to the Regime 1, a given variation of the electron density such as the one resulting from the wedge electrode design of the reactor used in the present study (here 70 %), induces a strong variation of the silane depletion fraction. On the other hand, for deposition in the microstructure transition zone with higher input silane concentration, i.e. higher $kn_e/a$ ratio, the silane depletion fraction is less sensitive to electron density variation because of the asymptotic behaviour of $D=D(kn_e/a)$. The silane depletion variation induced by the 70 % variation of $n_e$ is estimated from Fig. 6 to be lower than 0.2 ($\pm$ 12 %) in the Regime 2. Consequently, the silane dissociation efficiency is more uniform across the reactor, making the film thickness more uniform as shown in the central part of Fig. 4 where edge effects such as intense plasma due to the proximity of ground and RF electrodes have no influence on the plasma. This is not in contradiction with the strong nonuniformity in microstructure observed in Fig. 3 since for films deposited at the limit between a-Si:H and $\mu$c-Si:H, even a silane depletion fraction variation as low as 0.1 may induce a change in microstructure.

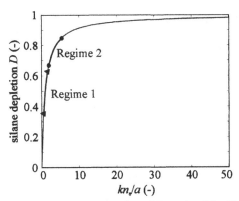

**Figure 6**: Silane depletion fraction as a function of the ratio of the dissociation rate ($kn_e$) to the pumping rate ($a$). The two regimes corresponds to a 70 % variation of the electron density $n_e$.

## Impact on large area production reactor design and manufacturing

Up to now, only an intentional nonuniform voltage across the plasma has been discussed. However, even in parallel plate reactors, i.e. without wedge electrode, nonuniform voltage distribution may occur. First of all, the reactor edges are the origin of many complications and their design is not trivial. The so-called telegraph effect can be eliminated by using symmetrical electrodes [6, 7], but the intense electric fields due to the RF and ground proximity that produces locally very nonuniform films as shown in Fig. 4 may be much more complicated to eliminate. However, these are edge effects which can be ignored simply by only considering the central part of large area reactors where the influence of edges has vanished.

Nevertheless, there also exists sources of complication within the uniform plasma of large area reactors. Imperfect contact between the glass substrate and the bottom electrode may lead to a capacitive division of the interelectrode voltage, exactly as shown in Fig. 2, resulting in local nonuniformity as shown by Meiling et al [12]. Another example of nonuniformity source is the standing-wave effect, which has electromagnetic origins as opposed to the mechanical origins of the previous example. The standing-wave effect is observed in large area capacitively-coupled parallel-plate reactors when very high excitation frequencies (VHF) are used. The standing-wave effect is observed when the reactor size is larger than about a tenth of the vacuum wavelength of the excitation frequency [13]. Therefore, the correction of the standing-wave effect is of great importance when processing $\mu$c-Si:H for the PV industry, since the dimension of the reactors is nowadays about 1 $m^2$ to reduce the production costs and the excitation frequencies are in the range of 40 to 100 MHz to improve the deposition rate [14-17] and to reduce the bombardment by energetic ions [14, 17].

The standing-wave effect, as opposed to the telegraph effect, is only weakly damped, therefore it influences the whole deposition area. Moreover, it does not affect only the plasma potential, as for the case of the telegraph effect, but it affects the RF voltage amplitude across the plasma. The vertical electric field in the reactor is then perturbed as shown by vacuum measurements, i.e. without plasma, performed by Schmidt et al [18] and it results in a variation of the plasma intensity across the area of the reactor. This plasma nonuniformity affects the gas dissociation rate uniformity over the substrates, hence films deposited in the presence of the standing-wave nonuniformity have a nonuniform thickness profile [19] which is related to the profile of the RF voltage.

Sansonnens et al [20] have shown that shaping one or both electrodes can compensate the higher RF voltage amplitude at the center of the reactor due to the standing-wave. The shaping has to be filled or screened by a dielectric in order to permit a capacitive division of the interelectrode RF voltage amplitude, hence reducing the potential drop across the plasma. The profile of the shaping can be calculated analytically to compensate the interelectrode RF voltage amplitude in a cylindrical reactor [18]. But numerical calculations are required for a shaped electrode which takes into account the plasma in large area rectangular reactors [19].

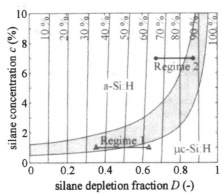

**Figure 7**: Silane dissociation efficiency in the $(D, c)$ plane, with contours of constant silane dissociation efficiency from 10 % to 100 %. Efficient deposition of silicon at the limit between a-Si:H and μc-Si:H can be achieved only in the Regime 2, i.e. at high input silane concentration and depletion.

These perturbations of the silane depletion fraction may only have a marginal effect if we consider only deposition of microcrystalline silicon from very low input silane concentration as shown in Figs. 3 and 4. However, Regime 1 is not interesting if the objective is to reduce the cost of PV cells production. Indeed, microcrystalline silicon at high deposition rates can be achieved within this Regime 1, but as shown in Fig. 7 the silane dissociation efficiency is so low ($\eta \approx 30\text{-}40$ %) that a significant amount of expensive raw material (silane and hydrogen) is lost and adds to the final PV cells cost. In order to achieve the deposition of solar grade μc-Si:H, i.e. at the transition between a-Si:H and μc-Si:H [11], with high deposition and high gas utilization efficiencies, the silane concentration and depletion have to be increased in order to reach the deposition Regime 2. However, in this case, as opposed to the Regime 1, the uniformity of the silane depletion fraction is crucial to achieve a uniform microstructure over the whole substrate area. Therefore, all imperfections of the voltage drop across the plasma have to be cured. This means that the substrate has to have a perfect contact with the bottom electrode, the interelectrode distance has to be maintained constant by using stiffener to avoid top-electrode bending under its own weight, and, if VHF is used, the standing-wave effect has to be compensated by adequate electrode design. Moreover, the precision of these adjustments to the reactor design has to be increased as the working input silane concentration increases. This is because the higher the input silane concentration, the more the uniformity of the microstructure is sensitive to silane depletion nonuniformity for film deposited at the limit between a-Si:H and μc-Si:H.

**CONCLUSION**

In this work, two deposition conditions of silicon at the limit between amorphous silicon and microcrystalline silicon were performed in a large area radio-frequency reactor modified to create a nonuniform RF power distribution across its width. The two sets of parameters were (i) an input silane concentration of 1 % with a low RF power density, and (ii) an input silane concentration of 7 % with a high RF power density. This resulted in two different thickness and crystallinity profiles. On one hand, at low input silane concentration, the film presented a crystallinity above 50 % across the whole substrate. On the other hand, at high input silane concentration, the microstructure of the film dropped from microcrystalline to fully amorphous over the deposition area. Even though the crystallinity profiles of the two films were very different, the thickness profiles were almost similar and uniform, showing that a good thickness uniformity does not guarantee a good microstructure uniformity.

These results were interpreted on the basis of an analytical plasma chemistry model and it was shown that the crystallinity nonuniformity was due to a nonuniform silane concentration in the plasma across the width of the reactor. Moreover, it was shown that the higher the input silane concentration, the more the crystallinity uniformity is sensitive to RF voltage variation for films deposited at the limit between amorphous and microcrystalline silicon. The thickness uniformity does not follow the same tendency, and a given RF voltage variation has less influence on the thickness uniformity for higher input silane concentration as shown by the model as well as by the experiments.

Therefore, to benefit from the high gas utilization efficiency associated with input silane concentration higher than 5 %, in order to reduce the PV cells processing costs, the design and the manufacturing of the reactor has to be carefully executed in order to achieve uniform films, with regard to film thickness and microstructure.

## ACKNOWLEDGMENTS

We wish to acknowledge Dr Christoph Ellert from Oerlikon-Solar AG for fruitful discussions about thickness and microstructure film uniformity that are the origin of this work.

## REFERENCES

[1] J. Meier, U. Kroll, J. Spitznagel, S. Benagli, T. Roschek, G. Pfanner, C. Ellert, G. Androutsopoulos, A. Hügli, M. Nagel, C. Bucher, L. Feitknecht, G. Büchel, and A. Buechel, *Proc. 31st IEEE Photovoltaic Specialists Conference, Orlando (USA)*, 1464 (2005).

[2] H. Takatsuka, Y. Yamauchi, K. Kawamura, H. Mashima and Y.Takeuchi, *Thin Solid Films*, **506-507**, 13-16 (2006).

[3] A.A. Howling, L. Sansonnens, J. Ballutaud, F. Grangeon, T. Delachaux, Ch. Hollenstein, V. Daudrix, and U. Kroll, *16th European Photovoltaic Solar Energy Conference, Glasgow (UK)*, 375-379 (2000).

[4] B. Strahm, A.A. Howling, L. Sansonnens, and Ch. Hollenstein, *Plasma Sources Sci. Technol.*, **16**, 80-89 (2007).

[5] A.A. Howling, B. Strahm, P. Colsters, L. Sansonnens and Ch. Hollenstein, *Plasma Sources Sci. Technol.*, **16**, 679-696 (2007).

[6] A.A. Howling, L. Derendinger, L. Sansonnens, H. Schmidt, Ch. Hollenstein E. Sakanaka and J.P.M. Schmitt, *J. Appl. Phys.*, **97**, 123308 (2005).

[7] L. Sansonnens, B. Strahm, L. Derendinger, A.A. Howling, Ch. Hollenstein, C. Ellert and J.P.M. Schmitt, *J. Vac. Sci. Technol. A*, **23**, 922-926 (2005)

[8]    M.N. van den Donker, B. Rech, F. Finger, W.M.M. Kessels and M.C.M. van de Sanden, *Appl. Phys. Lett.*, **87**, 263503 (2005).

[9]    L. Sansonnens, A.A. Howling and Ch. Hollenstein, *Plasma Sources Sci. Technol.*, **9**, 205-209 (2000).

[10]   C. Droz, E. Vallat-Sauvain, J. Bailat, L. Feitknecht, J. Meier and A. Shah, *Solar Energy Mater. Solar Cells*, **81**, 61-71 (2004).

[11]   O. Vetterl, F. Finger, R. Carius, P. Hapke, L. Houben, O. Kluth, A. Lambertz, A. Mück, B. Rech and H. Wagner, *Solar Energy Mater. Solar Cells*, **62**, 97-108 (2000).

[12]   H. Meiling, W.G.J.H.M. van Sark, J. Bezemer and W.F. van der Weg, *J. Appl. Phys.*, **80**, 3546-3551 (1996).

[13]   A.A. Howling, L. Sansonnens and Ch. Hollenstein, *Thin Solid Films*, **515**, 5059-5064 (2007).

[14]   A.A. Howling, J.-L. Dorier, Ch. Hollenstein, U. Kroll and F. Finger, *J. Vac. Sci. Technol. A*, **10**, 1080-1085 (1992).

[15]   L. Sansonnens, A.A. Howling and Ch. Hollenstein, *Plasma Sources Sci. Technol.*, **7**, 114-118 (1998).

[16]   E. Amanatides, D. Mataras and D.E. Rapakoulias, *Thin Solid Films*, **383**, 15-18 (2001).

[17]   F. Finger, P. Hapke, M. Luysberg, R. Carius, H. Wagner and M. Scheib, *Appl. Phys. Lett.*, **65**, 2588-2590 (1994).

[18]   H. Schmidt, L. Sansonnens, A.A. Howling, Ch. Hollenstein, M. Elyaakoubi and J.P.M Schmitt, *J. Appl. Phys.*, **95**, 4559-4564 (2004).

[19]   L. Sansonnens, H. Schmidt, A.A. Howling, Ch. Hollenstein, C. Ellert and A. Buechel, *J. Vac. Sci. Technol. A*, **24**, 1425-1430 (2006).

[20]   L. Sansonnens and J.P.M. Schmitt, *Appl. Phys. Lett.*, **82**, 182-184 (2003).

Mater. Res. Soc. Symp. Proc. Vol. 1066 © 2008 Materials Research Society      1066-A01-03

# Cone Kinetics Model: Insights into the Morphologies of Mixed-phase Silicon Film Growth

Howard M. Branz, Paul Stradins, and Charles W. Teplin
National Center for Photovoltaics, National Renewable Energy Laboratory, 1617 Cole Blvd, Golden, CO, 80401

## ABSTRACT

The 'cone kinetics' model, which explains development of cone-shaped inclusions during nanocrystalline silicon film growth and during low-temperature silicon epitaxy breakdown, is extended to protocrystalline ('edge') amorphous silicon, Si heterojunctions and other Si film morpholologies. We generalize the physics underlying cone formation and present diagrams that delineate the deposition regimes giving rise to the different film morphologies; these regimes are determined by the nucleation rate of the second phase and the relative growth rate of the phases present. The model predicts cone growth during thin-film deposition by plasma-enhanced and other chemical vapor deposition techniques when isotropic growth is coupled with the isolated nucleation of a second phase with a higher growth rate. Protocrystalline amorphous silicon and other embedded crystallite phases are formed when a second phase successfully nucleates but has a smaller growth rate than the surrounding amorphous film. The cone kinetics phase diagram provides a simple explanation of the various nanocrystalline film morphologies observed when silane precursors are diluted with different concentrations of hydrogen gas.

## INTRODUCTION

Several types of film silicon layers undergo spontaneous phase transition during chemical vapor deposition (CVD) and develop regular morphologies of the second material phase. The exact morphologies depend on the particular CVD technique used, the deposition conditions and the substrate materials [1-9]. For example, nanocrystalline (nc-Si:H) and protocrystalline Si films deposited by plasma-enhanced (PE) and hot-wire (HW) CVD begin growth as a hydrogenated amorphous silicon (a-Si:H) incubation layer, but later develop crystalline inclusions, cones and other structures [3-6]. During epitaxy below about 500°C, there is often a breakdown from near-perfect c-Si to conical inclusions of a-Si:H [7] or nc-Si:H [8, 9]. The morphologies of the crystalline and amorphous phases in all these layers are determined by the substrate material and the deposition conditions.

Improved understanding of the mechanisms that lead to complex morphologies during CVD could allow better control of materials properties and even development of new functional materials based on the combinations of the incubation thickness and second phase morphology. In this paper, we review the cone kinetics model that reveals the underlying mechanisms responsible for cone growth in nc-Si:H and epitaxy breakdown and use the model to place various materials systems of interest into a qualitative morphology phase diagram. A quantitative understanding of the phase space would allow improved morphology control during silicon film growth.

## CONE MORPHOLOGY AND THE CONE KINETICS MODEL

Our cone kinetics model [1, 2] predicts the formation of included cones of a new phase when three conditions are satisfied: 1) deposition is isotropic due to low surface diffusion and impinging atoms are directionally-randomized in the gas phase, 2) a second material phase nucleates sporadically at isolated locations on the growing surface, and 3) the nuclei of the second phase grow at a higher rate than the initial phase. When growth is isotropic, point nuclei naturally grow into spheres and planar surfaces grow as planes. If the nucleated phase grows faster than the planar phase, the point nucleus will spread as a conical section of the sphere (i.e., a spherical cone). The angle of the cone is determined by the relative growth rates of the initial and nucleated phases [1, 2, 5].

The cone kinetics model was originally developed to describe low-temperature HWCVD epitaxy breakdown to hydrogenated a-Si:H cones below 500°C [2]. It was later applied to PECVD nanocrystalline silicon development of conical structures from an a-Si:H nucleation layer [1]. In both cases, analysis of cone morphology with atomic force microscopy and transmission electron microscopy reveal that the cones have the particular geometry of spherical cones which protrude from the planar surface as a sphere centered on the cone apex. The cones have uniform cone angles throughout the film. Usually, these cones nucleate after a well-defined incubation period whose origin remains unexplained. In the case of epitaxial breakdown cones, Thiesen et al. [10] proposed that a buildup of H in the film causes the a-Si:H cone nucleation while others suggest that roughness is the cause [11]. In nanocrystalline silicon films, Fujiwara et al. [12] implicated stress in the crystalline cone formation. The cone kinetics model explains the morphological development of the second phase but does not attempt to provide a mechanism for nucleation of the second phase.

## MORPHOLOGY DIAGRAM FOR NANO- AND PROTO-CRYSTALLINE SILICON

Figure 1 shows a highly schematic morphology phase diagram for thin silicon films deposited on amorphous substrates such as glass. The morphology of a film will depend upon the nucleation rate of the second phase (horizontal axis) and the relative growth rates of the two phases (vertical axis). Qualitative demarcation among different growth regimes are indicated with dashed lines; dotted lines indicate zones that favor specific materials within a given growth regime. If the a-Si:H growth rate is significantly faster than the nc-Si:H rate (below horizontal dashed line) and the nucleation rate of nc-Si:H is very low, stable amorphous silicon will grow as indicated in the lower left region of Figure 1. However, as the nucleation rate and relative growth rate of nanocrystalline material increases, critical-sized nuclei will form and expand into embedded crystallites before they are overgrown by a-Si:H (lower right of Fig. 1). The smallest embedded crystallites, with sizes at nanometer scale, are often referred to as protocrystalline or 'edge' a-Si:H [13, 14]. When the crystalline phase grows more rapidly than a-Si:H (above the horizontal dashed line), crystalline nuclei will expand as spherical cones with cone angles that increase with the relative crystalline growth rate [2, 5]. Extremely rapid breakdown to the nanocrystalline phase occurs when the nucleation rate is also very high (upper right of Fig. 1).

Tsai et al. [15] proposed in 1989 that nc-Si:H forms during growth from silane/hydrogen gas mixtures because of preferential etching of the amorphous silicon phase under these H-diluted growth conditions. We interpret 'preferential etching' to mean that introduction of $H_2$ in the reactant gases increases the c-Si growth rate relative to the a-Si:H growth rate -- both phases will still grow. The higher c-Si growth rate with increasing H-dilution will favor large and more rapidly growing crystallite forms, as observed by Vallat-Sauvain et al.[3] and others. In Fig. 1, we have reproduced the sketches from Vallat-Sauvain et al.[3] of the morphologies they deposited and measured. Morphologies obtained with larger H-dilutions are positioned near the top of Figure 1 and the H-dilution *decreases* as the sketches move *clockwise* around the boxed morphology phase diagram. The large dot at the end of each arrow indicates approximately where in our phase diagram each morphology will appear. The main effect indicated by our placement of these dots is an increase of crystalline-to-amorphous growth rate ratio with H-dilution. However, measurements by Fujiwara et al.[16] suggest that the nucleation rate also increases with H-dilution. We indicate this change in nucleation rate by our placement of the various mixed-phase morphologies in Fig. 1.

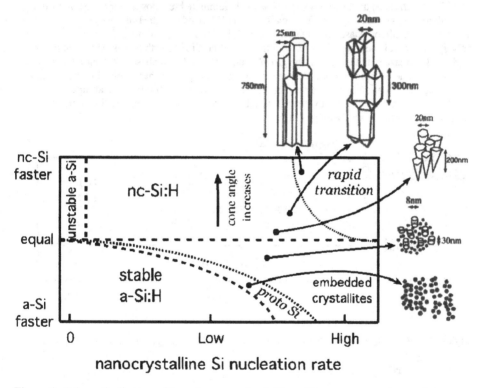

**Figure 1.** Schematic diagram of two-phase growth of silicon films on an amorphous substrate. Arrows point to sketches of crystallite morphologies observed by Vallat-Sauvain et al. [3]. H-dilution during deposition increases counterclockwise from the bottom sketch to the top. Large dots at arrow ends indicate approximate placement of these phases on the diagram; the primary role of H-dilution is to increase the c-Si growth rate relative to the a-Si:H growth rate.

We note that deviations from the ideal spherical cone geometry have been observed [5, 17], usually with initial cone growth evolving to more columnar growth (decreasing cone angle) capped by a spherical surface. Such morphologies may result from changes in the growth rate ratio due to either changing gas conditions or a change of the stress in the film as it thickens.

## MORPHOLOGY DIAGRAM FOR SILICON EPITAXY

Figure 2 shows a highly schematic morphology phase diagram for breakdown of epitaxial growth on a silicon wafer [1,2]. Below the dashed horizontal line, the epitaxial c-Si phase grows faster than any breakdown phase and epitaxy is stable; above this horizontal line, a breakdown phase grows faster. Cones will develop in this breakdown region when they have a sufficient probability of nucleation. The breakdown phase will be determined by the relative a-Si:H and nc-Si:H growth rates; the faster growing phase will dominate breakdown. For deposition from pure silane, the breakdown phase is normally a-Si:H [7]. Under high H-dilution, the polycrystalline silicon phase grows faster than a-Si:H (see Fig. 1) and breakdown to polycrystalline Si is observed [9]. As with the case of nc-Si:H growth discussed above, the cone angle will increase with the ratio of the growth rate of the included phase to the initial phase.

The zone of rapid epitaxial failure (upper right of Fig. 2) is characterized by both a high second phase nucleation rate and a smaller relative epitaxial growth rate. This abrupt development of a-Si:H on a c-Si wafer is crucial to deposition of silicon heterojunction solar cells with high open-circuit voltage [18].

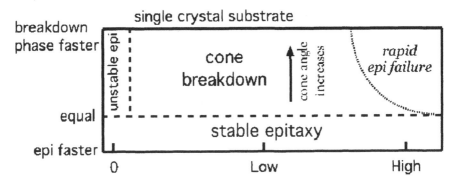

**Figure 2.** Schematic diagram of two-phase growth of silicon films on a crystalline Si substrate.

## CONCLUSIONS

Our quantitative cone kinetics model has been extended qualitatively to understand the wide variety of morphologies observed in silicon layer growth. These morphologies on amorphous substrates include nc-Si:H cones and abrupt nc-Si:H, protocrystalline ('edge') silicon and larger embedded crystallites in amorphous silicon. The morphologies deposited on c-Si substrates include epitaxial breakdown cones of either a-Si:H or nc-Si:H and abrupt a-Si:H heterojuction breakdown layers. The morphology phase diagrams depend upon the substrate material and the deposition chemistry. The deposition chemistry controls the ratio of the growth rates of the initial and included phases, and also the nucleation rate of the included phase. It will be important to define the morphology zone boundaries of Figures 1 and 2 more quantitatively, both by experiment and through quantitative modeling.

In the technologically-important case of growth on amorphous substrates, H-dilution appears to govern the ratio of the growth rates of the phases, and also influences the nucleation rate of the nanocrystallites and nc-Si:H cones. This complication makes morphology understanding and control more difficult. If we are to develop full control of these mixed phase film morphologies it will be essential to discover additional growth parameters that afford independent control of the nucleation rate and the phase deposition rate ratios.

## ACKNOWLEDGMENTS

We thank B. Yan and J. Yang of United Solar Ovonics LLC for providing samples and C.-S. Jiang, B. To and K. Jones for measurements that vitally contributed to development of the cone kinetics model, as detailed in References 1 and 2. We gratefully acknowledge support from the U.S. DOE under Contract DE-AC36-99GO10337.

## REFERENCES

[1]    C.W. Teplin, C. Jiang, P. Stradins, and H.M. Branz, *Appl. Phys. Lett.* **92**, 093114 (2008).
[2]    C.W. Teplin, E. Iwaniczko, B. To, H. Moutinho, P. Stradins, and H.M. Branz, *Phys. Rev. B* **74**, 235428 (2006).
[3]    E. Vallat-Sauvain, U. Kroll, J. Meier, A. Shah, and J. Pohl, *J. Appl. Phys.* **87**, 3137 (2000).
[4]    H. Fujiwara, M. Kondo, and A. Matsuda, *Phys. Rev. B* **63**, 115306 (2001).
[5]    A. Fejfar, T. Mates, O. Certik, B. Rezek, J. Stuchlik, I. Pelant, and J. Kocka, *J. Non-Cryst. Solids* **338-40**, 303 (2004).
[6]    R.W. Collins, A.S. Ferlauto, G.M. Ferreira, C. Chen, J. Koh, R.J. Koval, Y. Lee, J.M. Pearce, and C.R. Wronski, *Solar Energy Mats. and Solar Cells*, **78**, 143 (2003).
[7]    C.W. Teplin, Q. Wang, E. Iwaniczko, K.M. Jones, M. Al-Jassim, R.C. Reedy, and H.M. Branz, *J. Cryst. Growth* **287**, 414 (2006).
[8]    B. Rau, I. Sieber, B. Selle, S. Brehme, U. Knipper, S. Gall, and W. Fuhs, *Thin Sol. Films* **451-52**, 644 (2004).
[9]    C. Richardson, M. Mason, and H.A. Atwater, *Thin Solid Films* **501**, 332 (2006).
[10]   J. Thiesen, H.M. Branz, and R.S. Crandall, *Appl. Phys. Lett.* **77**, 3589 (2000).
[11]   O.P. Karpenko, S.M. Yalisove, and D.J. Eaglesham, *J. Appl. Phys.* **82**, 1157 (1997).
[12]   H. Fujiwara, M. Kondo, and A. Matsuda, *J. Non-Cryst. Solids* **338-340**, 97 (2004).
[13]   D.V. Tsu, B. S. Chao, S.R. Ovshinsky, S. Guha, and J. Yang, *Appl. Phys. Lett.* **71**, 1317 (1997).
[14]   R.W. Collins and A.S. Ferlauto, *Curr. Op. Sol. St. & Mat. Sci.* **6**, 425 (2002).
[15]   C.C. Tsai, G.B. Anderson, R. Thompson, and B. Wacker, *J. Non-Cryst. Solids* **114**, 151 (1989).

[16]   H. Fujiwara, M. Kondo and A. Matsuda, *Phys. Rev. B*, **63**, 115306 (2001).
[17]   P.C.P. Bronsveld, J.K. Rath, R.E.I. Schropp, T. Mates, A. Fejfar, B. Rezek, and J. Kocka, *Appl. Phys. Lett.* **89**, 51922 (2006).
[18]   T.H. Wang, E. Iwaniczko, M.R. Page, D.H. Levi, Y. Yan, H.M. Branz, and Q. Wang, *Thin Solid Films* **501**, 284 (2006).

Mater. Res. Soc. Symp. Proc. Vol. 1066 © 2008 Materials Research Society    1066-A01-04

# Evolution of Film Crystalline Structure During the Ultrafast Deposition of Crystalline Si Films

Haijun Jia[1], Hiroshi Kuraseko[2], Hiroyuki Fujiwara[1], and Michio Kondo[1]

[1]Research Center for Photovoltaics, National Institute of Advanced Industrial Science and Technology (AIST), Central 2, 1-1-1 Umezono, Tsukuba, Ibaraki, 305-8568, Japan

[2]The Furukawa Electric Co., Ltd., Yawata-Kaigandori,Chiba, 290-8555, Japan

## ABSTRACT

By using a high density microwave plasma source, an ultrafast deposition rate over 1000 nm/s has been achieved for polycrystalline silicon (poly-Si) film deposition. We find that crystalline structure of the deposited film evolves along the film growth direction, i.e. large grains in surface region while small grains in the bottom region of the film. Systematic study of the deposition process has been performed as a function of the deposition duration. Based on the observed results, a possible mechanism, the annealing-assisted plasma-enhanced chemical vapor deposition, is proposed to describe the film growth process.

## INTRODUCTION

Crystalline silicon thin films are attracting considerable attention as a promising material for cost-effective thin film solar cell fabrications due to their strong stability and wide-range spectral sensitivity [1,2]. Preparation of the crystalline Si films can be realized by using plasma enhanced chemical vapor deposition (PECVD) method [3]. Due to a large thickness of the Si film needed for sufficient sunlight absorption, high rate deposition of high quality Si films becomes a critical issue [4,5]. For the film growth process, it is usually considered that the film deposition proceeds through several specific growth regimes including incubation, nucleation and stable bulk growth [6]. In addition, atomic H plays an essential role for film crystallization [7]. In this letter, by using a high density microwave plasma source, an ultrafast rate over 1000 nm/s is achieved for polycrystalline Si (poly-Si) film deposition. We find that the film crystalline structure evolves during the deposition process. Annealing-assisted plasma enhanced chemical vapor deposition is proposed to describe the evolution.

## EXPERIMENTS

The microwave-induced plasma source used in this study employs a hybrid-mode resonator cavity. Proper design of the cavity based on a mode-switching technique suppresses the impedance variation before and after plasma ignition. Plasmas are generated inside a hollow quartz tube with an inner diameter of 10 mm positioned in the center of the cavity. By using the plasma source, He and/or Ar plasmas can be generated in a wide pressure range between 1 Torr and atmospheric pressure. Details about the plasma source were described elsewhere [8].

Si film depositions using the microwave plasma source were performed by using SiH$_4$, He and H$_2$ as source gases. A microwave power of ~100 W was supplied during the depositions under a working pressure of 2 Torr. The depositions were continued for 60 seconds and no intentional substrate heating was used. A quartz fiber used as a substrate (0.3 mm in diameter) was placed in the center of the quartz tube. The use of fiber, a thermal-stable and flexible substrate, is expected to provide a fiber-type thin film solar cell with advantages of light weight, flexibility and possibility for roll-to-roll fabrication process [9]. Deposited Si films were characterized using Raman spectroscopy, scanning electron microscope (SEM) and transition electron microscope (TEM).

## RESULTS

Cross sectional SEM image of a deposited Si film is shown in Fig. 1 (a). The deposition was performed using a SiH$_4$ flow rate of 50 sccm, a He flow rate of 400 sccm and a H$_2$ flow rate of 400 sccm. A dense Si film with high uniformity can be deposited on the fiber surface. A film thickness of ~ 60 μm is achieved during the 60 s deposition with a deposition rate of 1000 nm/s. Its Raman spectrum demonstrates a sharp peak centered at 520 cm$^{-1}$ corresponding to the crystalline Si phase with negligible contribution of the amorphous Si peak at 480 cm$^{-1}$, confirming the very high film crystallinity.

Microstructure of the deposited Si film was investigated by using the TEM. Figure 1 (b) shows the cross sectional TEM micrographs and the corresponding selective area diffraction (SAD) patterns for different parts of the deposited Si film at near film surface region, middle region and near substrate region. A dense grain arrangement coalesced with oriented crystallites extends over the entire surface region. The final size of the grains extends over 10 μm along the film growth direction. The corresponding SAD pattern of this part that shows an obvious spot image confirms the large grain. The middle region of the deposited film is composed of crystallites with size of several hundreds nanometers. A transition region from granular crystallites to columnar grains can be identified in the image as denoted by a letter R in Fig. 1 (b). In addition, this part shows an almost identical SAD pattern with that of the surface region. For the region

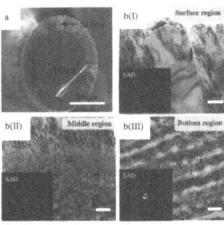

Fig. 1 (a) Cross sectional SEM image of the deposited Si film together with its Raman spectrum. (b) Cross sectional TEM micrographs and corresponding SAD patterns for different parts of the deposited Si film at near surface region (part I), middle region (part II) and near substrate region (part III).

near the substrate, no a-Si component can be detected. The TEM image and SAD pattern show a fully crystallized microstructure with small grains and random crystallographic orientation. These TEM observations clarify two important results: i) complete crystallization is realized across the entire film; ii) there is an evolution of film crystalline structure along the growth direction with oriented large grains in the upper region and random small grains in the bottom region.

To explore insight into the film growth process, Si film depositions were carried out further as a function of the deposition duration. The depositions were performed using a $SiH_4$ flow rate of 5 sccm and a He flow rate of 400 sccm. To exclude the effect of H, $H_2$ dilution was not used. In fact, the absence of $H_2$ does not affect the structural evolution. Other deposition parameters were same as those of the samples in Fig. 1. Fig. 2 shows the Raman crystallinity $I_c/I_a$ [an intensity ratio of the crystalline peak at ~520 cm$^{-1}$ ($I_c$) to the amorphous peak at ~ 480 cm$^{-1}$ ($I_a$)] of the deposited films versus the deposition duration. The $I_c/I_a$ increases from $I_c/I_a<5$ for the film deposited for 5 s to $I_c/I_a>25$ when the deposition was continued for 40 s. This result shows the presence of amorphous phase near the substrate region and is apparently different with the results shown in Fig. 1. To determine the film crystallinity along the growth direction and confirm the discrepancy, reactive ion etching (RIE) was performed for the film deposited for 40 s together with Raman measurement after each etching turn. Fig. 3 shows the Raman spectra corresponding to (a) surface region, (b) middle region and (c) bottom region of the film obtained after the RIE etching. For the RIE, $SF_6$ and $O_2$ mixture was used with a RF power of 100 W and a working pressure of 20 Pa. For comparison, spectrum of the fiber substrate is also shown in Fig. 3. We can see that the spectrum of bottom region can be reproduced simply by combining the spectrum of surface region and spectrum of substrate. These results clearly demonstrate that after 40 s deposition, even the initially deposited film shows rather high crystallinity, which is obviously different from the film deposited for 5 s.

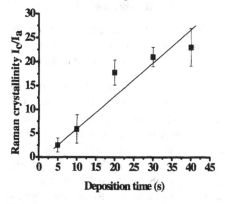

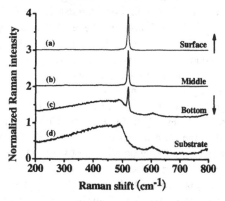

Fig. 2 Raman crystallinity $I_c/I_a$ of deposited Si films as a function of the deposition time. The line is a guide to the eye.

Fig. 3 Raman spectra of the Si film after etching by RIE process for a short time (a), for a medium time (b) and for a long time (c). For comparison, Raman spectrum of the fiber substrate is also shown in (d).

Figure 4 shows the film deposition rate as a function of deposition duration and the time evolution of H concentration in the deposited films measured by thermal desorption spectrometer (TDS). The $R_d$ increases from ~100 nm/s to ~300 nm/s when the deposition duration extends from 5 s to 40 s. On the other hand, the H concentration markedly decreases with increasing the deposition duration. The rather low H concentration in the film deposited for 40 s suggests the increase in the process temperature during the deposition. In fact, the increasing process temperature during the deposition was confirmed by the measurement using a thermal couple placed in the position of substrate in the plasma, and the temperature was confirmed to increase rapidly up to ~650 ℃ within 60 s.

## DISCUSSION

As shown in Figs. 2 and 4, the film crystallinity increases while the H concentration decreases with increasing the deposition time. These behaviors are also generally observed in the crystallization process of a-Si:H films by solid phase crystallization. The results in Fig. 3 demonstrate that after long time deposition (~40s), complete crystallization is realized even for the initially deposited Si film which exhibits amorphous component. These findings imply that during the film deposition, a thermal annealing process promotes crystallization throughout the entire film and leads to the evolution of crystalline structure along the growth direction. In conventional solid phase crystallization process, the annealing effect

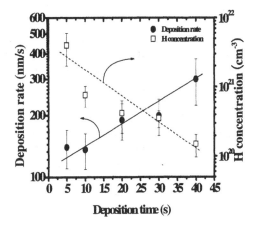

Fig. 4 Film deposition rate and H concentration in the deposited Si films plotted versus the deposition time. The line is a guide to the eye.

usually occurs when temperature is higher than 600℃ [10]. In this work, as metioned earlier, the gradual increase in temperature up to 650℃ has been confirmed experimently. In addition, as shown in Fig. 4, $R_d$ increases with time even though growth conditions are the exactly same. Thus the main reason responsible for the increase in $R_d$ with the deposition time can be the enhanced sticking probability of the precursors on film surface. It is reasonable to assume that the high surface temperature induces the H effusion, and the film surface is covered with dangling bonds which drastically increases the sticking probability. According to the above analyses, a possible mechanism for structural evolution, the annealing-assisted plasma-enhanced chemical vapor deposition (AA-PECVD), can be proposed. Since the deposition also involves the plasma process, other factors such as ion bombardment may also contribute to the crystallization process [11]. Further investigation, however, is needed to understand the phenomenon more clearly.

## CONCLUSION

In summary, TEM observations of a poly-Si film deposited at an ultrafast rate of 1000 nm/s revealed that the film crystalline structure evolved along the growth direction. Systematic investigation of the film deposition process was performed as a function of the deposition duration. A possible mechanism, the annealing-assisted plasma enhanced chemical vapor deposition, was proposed to describe the evolution.

## ACKNOWLEDGEMENT

This work was supported by the New Energy and Industrial Technology Development Organization (NEDO) of Japan.

## REFERENCES

[1] M. Kondo, Solar Energy Material & Solar Cell 78, 543 (2003)
[2] B. Rech, T. Repmann, M. N. van den Donker, M. Berginski, T. Kilper, J. Hupkes, S. Calnan, H. Stiebig, S. Wieder, Thin Solid Films 511-512, 548 (2006)
[3] A. Matsuda, Jpn. J. Appl. Phys. 43, 7909 (2004)
[4] H. Jia, J. K. Saha, H. Shirai, Jpn. J. Appl. Phys. 45, 666 (2006)
[5] Y. Mai, S. Klein, X. Geng and F. Finger, Appl. Phys. Lett. 85, 2839 (2004)
[6] H. Fujiwara, M. Kondo and A. Matsuda, J. Appl. Phys. 93, 2400 (2003)
[7] H. Fujiwara, M. Kondo and A. Matsuda, Jpn. J. Appl. Phys. 41, 2821 (2002)
[8] H. Jia, H. Kuraseko, M. Kondo, J. Appl. Phys. 103, 024904 (2008)
[9] H. Kuraseko, T. Nakamura, S. Toda, H. Koaizawa, H. Jia and M. Kondo, Proceeding of the 4th World Conference on Photovoltaic Energy Conversion, 2006, p 1380
[10] A. T. Voutsas, Applied Surface Science 208-209, 250 (2003)
[11] G. Holmen, J. Linnros and B. Svensson, Appl. Phys. Lett. 45, 1116 (1984)

# Defects and Transport

Mater. Res. Soc. Symp. Proc. Vol. 1066 © 2008 Materials Research Society    1066-A02-03

# Electronic Transport in Co-Deposited Hydrogenated Amorphous/Nanocrystalline Thin Films

Y. Adjallah[1], C. Blackwell[1], C. Anderson[2], U. Kortshagen[2], and J. Kakalios[1]

[1]Physics and Astronomy, University of Minnesota, 116 Church Street S.E., Minneapolis, MN, 55455

[2]Mechanical Engineering, University of Minnesota, 111 Church Street S.E., Minneapolis, MN, 55455

## ABSTRACT

Mixed-phase hydrogenated amorphous silicon thin films containing nanocrystalline silicon inclusions have been synthesized in a dual chamber co-deposition system. A PECVD deposition system produces small crystalline silicon particles (3-5 nm diameter) in a flow-through reactor, and injects these particles into a separate capacitively-coupled plasma chamber in which hydrogenated amorphous silicon is deposited. Raman spectroscopy is used to determine the volume fraction of nanocrystals in the mixed phase thin films, while infra-red spectroscopy characterizes the hydrogen bonding structure as a function of nanocrystalline concentration. At a moderate concentration of 5 nm silicon crystallites, the dark conductivity and photoconductivity are consistently found to be higher than in mixed phase films with either lower or higher densities of nanocrystalline inclusions.

## INTRODUCTION

There has been considerable recent interest in the properties of mixed phase materials consisting of nanocrystallites embedded within an amorphous matrix, owing to their potentially superior properties for a wide range of technological applications. Hydrogenated amorphous silicon thin films containing silicon nanocrystalline inclusions (a/nc-Si:H) have been investigated as potential materials for photovoltaic devices [1,2], while silicon nanocrystals within an insulating matrix are employed in non-volatile memory and electroluminescent devices [3,4]. These materials are typically synthesized in a Plasma Enhanced Chemical Vapor Deposition (PECVD) system operated at high silane gas chamber pressures, where silicon cluster formation is known to occur [5,6]. However, the plasma conditions that yield silicon nanocrystals are far from those that are known to produce high electronic quality a-Si:H, and the concentration of silicon particles embedded within the a-Si:H matrix is not easily controlled. Moreover, when synthesized using in a single capacitively-coupled plasma chamber, one is limited to growing silicon nanocrystalline particles in a-Si:H. We have consequently constructed a dual chamber co-deposition system, where the silicon nanocrystals are formed in one plasma deposition system, and are then entrained in a carrier gas and injected into a second PECVD system, where hydrogenated amorphous silicon is deposited [7]. A preliminary report on this dual chamber system has been published previously [7]. This paper reports the structural and electronic properties of undoped a/nc-Si:H films containing silicon nanocrystals of 5-6 nm in diameter, as a function of nanocrystalline concentration.

## MATERIALS PREPARATION

The undoped a/nc-Si:H films reported here are synthesized in a dual-chamber co-deposition system, as sketched in Fig. 1, described previously [7]. The Particle Synthesis Reactor consists of a 3/8 inch diameter quartz tube with two fitted ring electrodes, connected to a 13.56 MHz power supply and matching network. The plasma conditions in this tube (pressure of 1.5 Torr, RF power of 15 W, 5/95% mixture of silane/helium with additional argon, total flow rate 60 sccm) are selected for the formation of silicon nanocrystals of diameter of approximately 5 nm. Previous studies have confirmed that the particles generated in the first feeder tube are indeed crystalline, with a monodisperse size distribution set by the detailed plasma conditions [8]. The nanoparticles are entrained by the inert carrier gas and injected into the second capacitively coupled plasma deposition system. Substrates are mounted onto the lower, heated electrode in this second chamber, while the upper RF (13.56MHz) electrode is not heated. When the silicon particles enter the second chamber, gas convection drives them towards the substrates labeled A, C and E in figure 1. The a/nc-Si:H films are deposited onto Corning 7059 glass or crystalline silicon substrates at temperatures of 250°C. The thicknesses of the thin films measured using a profilometer vary from 200 to 700 nm for plasma deposition times of 60 min (deposition rates of ~ 1 – 2 Å/sec). Tapping mode Atomic Force Microscopy studies indicate that the largest concentration of deposited nanoparticles is found in the sample labeled E in Figure 1, while the lowest density of particles is observed in sample A (AFM confirms that nanocrystals are deposited even in the outer electrode region) [7].

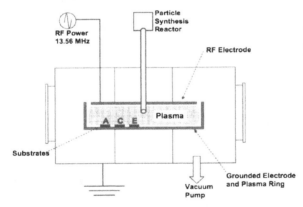

**Figure 1:** Schematic of the dual-chamber co-deposition system. The concentration of silicon nanocrystals in the a/nc-Si:H films depends on the substrates location relative to the injection tube from the Particle Synthesis Reactor.

The a/nc-Si:H films are deposited onto Corning 7059 glass or crystalline silicon substrates at temperatures of 250°C. The thickness of the hydrogenated silicon thin films range from 200 to 700 nm as measured using a profilometer. For the measurements of the electronic properties, chromium electrodes typically 1 cm long and separated by a gap either 2mm or 4 mm wide are evaporated onto the films.

## STRUCTURAL AND ELECTRONIC CHARACTERIZATION

Transmission electron microscopy studies verify that silicon nanocrystals are indeed introduced into the amorphous silicon matrix, and that these crystallites are synthesized in the particle synthesis reactor, rather than forming within the silane plasma in the CCP chamber [7]. The particle size distribution and the particle crystallinity were studied by transmission electron microscopy (TEM) analysis [8]. TEM imaging is performed on 10 nm thick films deposited onto crystalline silicon wafers, which are then cross-sectioned via using a small angle cleaving technique (SACT) [9], in order to avoid structural damage induced by ion milling. The TEM imaging confirms that the nanocrystallites are present throughout the thickness of the a/nc-Si:H films, and that the average diameter of the nanocrystals is approximately 3 nm. Focal series imaging confirms that the lattice fringes are not microscope artifacts [10].

The Raman spectra of the a/nc-Si:H films from the three different substrate positions in the CCP chamber (Fig. 1), corresponding to low, medium and high nanocrystalline concentrations, measured using a Witec Alpha 300 R confocal Raman microscope equipped with a UHTS 200 spectrometer using an Ar ion laser (wavelength 514.5 nm) are shown in Figure 2. The mixed phase film deposited in the A position (Fig. 1) corresponding to a low nanocrystalline concentration, displays a Raman spectrum with a broad peak reflecting the TO mode at 480 cm$^{-1}$, essentially indistinguishable from the Raman spectrum observed in homogenous a-Si:H [11] deposited in the same system with the Particle Synthesis Reactor turned off. The mixed-phase film deposited in the C position, that has a higher density of nanocrystallites, shows a small but distinct second peak at approximately 512 cm$^{-1}$. This 512 cm$^{-1}$ peak has a greater intensity for the film deposited at the E position (Fig. 1), which has the highest concentration of nanocrystalline inclusions.

In bulk crystalline silicon one observes a sharp peak in the Raman spectrum centered at 520 cm$^{-1}$, but quantum confinement effects can shift the location of the peak to lower wavenumbers [12]. A downward shift of peak location of 8 cm$^{-1}$ is consistent with a nanocrystalline diameter of 3 nm [12] while TEM studies indicate an average diameter for the nanocrystalline inclusions of 5 – 6 nm [8]. The Raman spectra in Fig. 2 are measured with a low incident laser power of 5 mW in order to avoid any shifts in peak position from heating of the sample. There is also evidence for a Raman peak at 500 cm$^{-1}$ in the film with a high nanocrystalline concentration, which has been interpreted as arising from the grain boundary region surrounding the nanocrystalline inclusions [13]. If the atoms at the boundaries of the nanocrystal are contributing to this peak at 500 cm$^{-1}$, then a smaller core may give rise to the peak at 512 cm$^{-1}$. Comparing the area under the 512 cm$^{-1}$ peak, arising from the silicon in the crystalline phase, to the area of the 480 cm$^{-1}$ peak, corresponding to the amorphous silicon matrix (and assuming a ratio of Raman backscattering cross-sections of the crystalline and amorphous phases of 1.0) one finds that for the films shown in Fig. 2 that the crystalline fraction $X_c$ is 10, 2 and less than 0.4 at. % for the films grown in the E, C and A locations, respectively. Assuming a crystallite size of 5 - 6 nm, a crystalline fraction of 10% for the film grown closest to the injection tube from the Particle Synthesis Reactor (E in Fig. 1) corresponds to an average spacing between nanoparticles of 13 - 14 nm, while for the film with the lowest crystalline fraction (A in Fig. 1) the spacing between inclusions is > 37 nm.

The infrared absorption spectra for a/nc-Si:H films deposited onto crystalline silicon substrates were measured using a Nicolet Magna 750 FTIR spectrometer, as shown in Fig. 3. The data are plotted in the range of 1800 to 2200 cm$^{-1}$, showing the presence of the vibrational bond-stretching modes of Si-H at 2000 cm$^{-1}$ and Si=H$_2$ at 2090 cm$^{-1}$. The latter has been associated with higher amounts of structural and electronic disorder [11, 14, 15], and in a/nc-Si:H films has been ascribed to the bonded hydrogen in grain boundary regions surrounding the nanocrystalline inclusions [16]. The microstructure fraction, defined as R = I(2090cm$^{-1}$) / [I(2000cm$^{-1}$) + I(2090cm$^{-1}$)] increases with nanocrystalline density, from 0.45, to 0.60 to 0.78, for the films in the A, C, and E positions in Fig. 1, respectively.

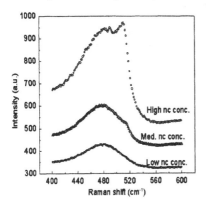 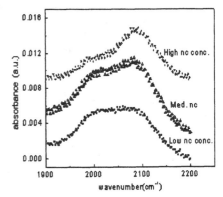

**Figure 2:** Plot of the Raman absorption spectrum for a/nc-Si:H films as a function of nanocrystalline concentration. The curves are offset vertically for clarity.

**Figure 3:** Infrared absorption spectra of a/nc-Si:H films as a function of nanocrystalline concentration. The curves are offset vertically for clarity.

Arrhenius plots of co-planar dark conductivity of the a/nc-Si:H thin films are shown in Fig. 4. The film with the lowest nanocrystalline density has essentially the same conductance and activation energies as pure a-Si:H films deposited with the Particle Synthesis Reactor off. Surprisingly, the highest conductivity, and lowest activation energy are found for films containing a moderate density of nanocrystals (approximately 2 – 4% crystal fraction as determined by Raman spectroscopy). Films with the highest concentration of nanoparticles have a dark conductivity intermediate between the A and C films. The films shown in Fig. 4 correspond to a single deposition run. Comparing data for five separate deposition runs, while the exact values of the conductivities vary slightly from run to run, we consistently observe that the C films have the highest conductivity, and the E film's conductance is lower than C but higher or roughly equal to the A film. The conductivity values in Fig. 4 are for state A, measured following annealing under vacuum at 450 K for 120 mins. The photoconductivity is measured using 100 mW/cm$^2$ heat-filtered W-Ha lamp at room temperature, as well as to determine the extent of light-induced degradation [17] in these a/nc-Si:H films. The state B dark conductivity, measured following two hours of light-soaking, is also shown in Fig. 4 (open symbols). The electronic properties of the films shown in Fig. 4 are summarized in Table I. The photosensitivity, defined as the ratio of the initial photoconductivity to the annealed state A dark conductivity, indicates that the low nanocrystal concentration films may be good candidates for solar cells.

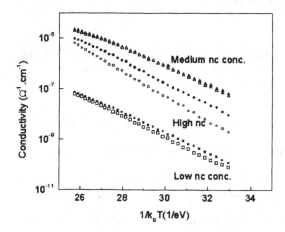

**Figure 4:** Arrhenius plot of the dark conductivity for the a/nc-Si:H films as a function of nanocrystal concentration. The conductivity measured after light-soaking (opened symbols) is not far from the annealed state A conductivity 450K (filled symbols).

**Table I:** Summary of the material and electronic proprieties for the a/nc-Si:H films.

| Sample | $X_c$ % from Raman | Dark $\sigma$ ($\Omega^{-1}$ cm$^{-1}$) at 350K | Dark $\sigma$ Activation energy (eV) | Photo-conductivity $\sigma_{dark}(\Omega^{-1}$ cm$^{-1})$ at 320K | Photosensitivity $\sigma_{ph}$ / $\sigma_{dark}$ at 320K |
|---|---|---|---|---|---|
| A (low nc conc.) | 0.4 | 1.2 x10$^{-10}$ | 0.77 | 6x10$^{-5}$ | 10$^4$ |
| C (med. nc conc.) | 2 | 6.4 x10$^{-8}$ | 0.59 | 10$^{-6}$ | 10$^2$ |
| E (high nc conc.) | 10 | 9.2 x10$^{-9}$ | 0.84 | 6.4x10$^{-6}$ | 10$^2$ |

The presence of the nanocrystalline inclusions influences electronic transport in a/nc-Si:H films by at least two separate mechanisms. At lower particle concentrations the material is essentially an amorphous tissue, and the addition of the inclusions acts to dope the material. Higher nanocrystal densities provide additional long-ranged disorder at the mobility edges, increasing the activation energy [15] and decreasing the dark conductivity. Measurements are underway of doped a/nc-Si:H films, comparing the temperature dependence of the thermopower to that of the dark conductivity. The difference in activation energies observed between these two measurements is interpreted as reflecting the influence of long-range disorder on electronic transport [15]. These

studies will help elucidate the role that the silicon nanocrystalline inclusions play in electronic transport in these mixed-phase films.

## ACKNOWLEDGEMENTS

This work was partially supported by NSF grants NER-DMI-0403887, DMR-0705675 and the University of Minnesota Center for Nanostructure Applications.

## REFERENCES

1.  J. Yang, K. Lord, S. Guha, S. R. Ovshinski, *Mater. Res. Soc. Proc.* **609**, A15.4.1 (2000).
2.  R. W. Collins, A. S. Ferlauto, G. M. Ferreira, K. Joohyun, *Mater. Res. Soc. Proc.* **762**, A. 10.1 (2001).
3.  M. L. Ostraat, J. W. DeBlauwe, M. L. Green, L. D. Bell, M. L. Brongersma, J. Caasperson, R.C. Flagen and H. Atwater, *Appl. Phys. Lett.*, **79**, 433 (2001).
4.  N.-M. Park, T. -S. Kim and S. -J. Park, *Appl. Phys. Lett.* **78**, 2575 (2001).
5.  P. Roca i Cabarrocas, A. Fortcuberta i Morral and Y. Poissant, *Thin Solid Films*, **403-404**, 39 (2002).
6.  S. Thompson, C. R. Perrey, C. B. Carter, T. J. Belich, J. Kakalios, U. Kortshagen, *J. Appl. Phys.* **97**, 34,310 (2005).
7.  C. Anderson, C. Blackwell, J. Deneen, C. Barry Carter, J. Kakalios and U. Kortshagen, *Mater. Res. Soc. Proc.* **910**, 79 (2006).
8.  L. Mangolini, E. Thimsen, U. Kortshagen, *Nano Lett.* **5**, 655 (2005).
9.  S.D. Walck and J.P. Mcaffey, *Mater. Res. Soc. Proc.*, **480**, 149 (1997).
10. D. B. Williams and C. B. Carter, *Transmission Electron Microscopy* (Plenum, New York, 1996).
11. G. Lucovsky, R. J. Nemanich, and J. C. Knights, *Phys. Rev. B* **19**, 2064 (1979).
12. G. Viera, S. Huet and L. Boufendi, *J. Appl. Phys.* **90**, 4175 (2001); C. Min, Z. Weijia, W. Tianmin, J. Fei, L. Guohua and D. Kun, *Vacuum* **81**, 126 (2006).
13. R. Saleh and N.H.Nickel, *Thin Solid Films* **427**, 266 (2003).
14. E. Bhattacharya and A. H. Mahan, *Appl. Phys. Lett.* **52**, 1587 (1988).
15. D. Quicker and J. Kakalios, *Phys. Rev. B* **60**, 2449 (1999).
16. D. C. Marra, E. A. Edelberg, R. L. Naone and E. S. Aydil, *J. Vac. Sci. Tech. A* **16**, 3199 (1998).
17. D. L .Staebler and C.R.Wronski, *Appl. Phys. Lett.* **31**, 292 (1976); *J. Appl. Phys.* **51**, 3262 (1980).

Mater. Res. Soc. Symp. Proc. Vol. 1066 © 2008 Materials Research Society       1066-A02-04

## Metastable Defects in Light Soaked Amorphous Silicon at 77 K

Tong Ju[1], Paul Stradins[2], and P. Craig Taylor[3]

[1]University of Utah, Salt Lake City, UT, 84112

[2]National Renewable Energy Lab, Golden, CO, 80401

[3]Physics, Colorado School of Mines, Golden, CO, 80401

## ABSTRACT

We have observed the growth of defects caused by optical illumination in liquid nitrogen. We kept the sample in liquid nitrogen for over one year. After a year and a half the ESR signal reached $\sim 10^{18}$ cm$^{-3}$ with no evidence of saturation. After that, we step-wise annealed the sample isochronally up to room temperature, where two thirds of original defects were annealed out. After annealing at room temperature, the sample was annealed isothermally around 300 K for several months. At this temperature, the defects slowly annealed. After a hundred hours at 295 K, the defect density decreased by a factor of 10 from its original value at 77 K.

## INTRODUCTION

The appearance of optically or electrically induced defects in hydrogenated amorphous silicon (a-Si: H), especially those that contribute to the Staebler-Wronski(S-W) effect, has been the topic of numerous studies, yet the mechanism of defect creation and annealing is far from clarified. In particular, understanding of the kinetics for the production of silicon dangling bonds at different temperatures, T, is very important. The recombination processes that govern defect creation are strongly affected by T.

We report the observation of the S-W effect in hydrogenated amorphous silicon at 77 K up to very long (1 year) exposure times. The sample was irradiated in-situ at 77 K with white-light and the light-induced defects were measured intermittently by electron spin resonance (ESR) at the same temperature. The defect density grows sublinearly with $N_d \propto t^{0.7}$ at 77 K, which is different from the $\propto t^{0.33}$ behavior at room temperature.

## EXPERIMENTAL

The amorphous silicon sample was made by plasma enhanced chemical vapor deposition (PECVD). It was deposited at glass substrate temperatures of 250 °C. The size of sample was 0.5 cm wide, 2 cm long, 1μm thick. We used an ELH lamp as the optical source.

All ESR measurements were made using a Bruker ESR spectrometer operating at 9.5 GHz with 4 gauss magnetic field modulation amplitude. ESR measures only the paramagnetic defects, such as neutral silicon dangling bonds. ESR does not measure the charged dangling bonds.

We first annealed the sample at 180 °C for several hours to remove any bulk defects. Then, we immersed the sample in liquid nitrogen in a glass dewar. We used a small liquid nitrogen dewar to keep the sample cold during transfer to the ESR cavity, which was maintained at 77 K. The light intensity was approximately 100 mw/cm$^2$ at the sample. During the time the sample was kept at 77 K, we used ESR periodically to track the defect densities. After over a year, the defect density reached approximately 10$^{18}$ cm$^{-3}$ with no evidence of saturation. At this time, we annealed the sample isochronally up to room temperature and isothermally at room temperature after that.

## RESULTS

After annealing the films, the defects began to accumulate in the dark at 77 K. The spin densities as functions of time stored at 77 K are shown in Fig. 1

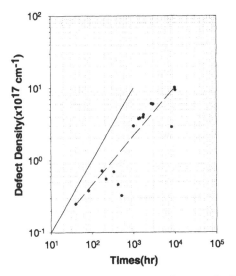

Figure 1.   Increase of the defect density after irradiation at 77 K.

The initial defect density is probably due to surface or interface defects. The defects grow measurably after about 40 hrs after which they continue growing in liquid nitrogen without saturation. The three points around 300 hours are lower because of inaccurate sample positioning in the cavity. The highest value is about 1.0x10$^{18}$/cm$^3$ after 10,000 hours of irradiating the sample. The growth is slightly sub-linear. The power-law exponent is 0. 7.

After about 10,000 hours of irradiation the sample was annealed for 30 minutes isochronally at successive temperatures of 130, 180, 230, and 300 K.   All measurements were performed by returning to 77 K. Each measurement involved checking for saturation due to the microwave power incident on the sample. Figure 2 shows data for stepwise annealing of the sample up to 300 K.

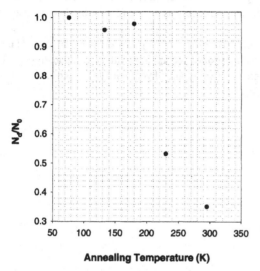

Figure 2 Isochronal annealing of the a-Si:H sample light-soaked at 77 K.

From Fig.2, it is clear that the defect density has decreased by about a factor of two from its original value after annealing at 230 K. At room temperature approximately 30% of original defects remain. The final defect spin density is about $3.7 \times 10^{17}$ cm$^{-3}$. These results are similar to those obtained by Schultz and Taylor [1]. After isochronal annealing of the a-Si:H sample at room temperature, we isothermally annealed the sample at room temperature. We kept the sample at room temperature and checked the defect density approximately every day to track the decay process. These results are shown in Fig 3.

After 300 hours, the defect density is around $10^{17}$/cm$^3$, which is at least an order of magnitude larger than the defect density after annealing at 180 C before we started irradiating the sample at 77 K. After annealing for several months (not shown), the defect density finally dropped to about $3 \times 10^{16}$/cm$^3$.

## DISCUSSION

The microscopic mechanism for the creation and annealing of metastable defects in a light-soaked sample is still an important problem that remains unsolved after 30 years from the initial discovery of degradation by Stabler and Wronski in 1977. There are several models to explain the Stabler-Wronski effect, two of which are widely used. The first model was proposed by Stutzmann, Jackson, and Tsai (SJT) in 1985 [2]. This model supposes that the recombination of excited photocarriers sometimes breaks a weak Si-Si bond and generates a defect. Bonding breaking is accompanied by H motion to stabilize two light-induced dangling bonds. This model predicts that at intermediate times $N_D (t) = (const)(G^{2/3}t^{1/3})$ where $N_D$ is the defect density, G is the generation rate of carriers, and t is the time. At 300K the defect density initially increases roughly as a cube root of the time and saturates after long time illumination.

The second model due to Branz in 1997 [3], invokes the light-induced production of "mobile" hydrogen that most often recombines with silicon dangling bond defects, but rarely the "collision" of two mobile hydrogen atoms produces a metastably trapped hydrogen pair. [4].

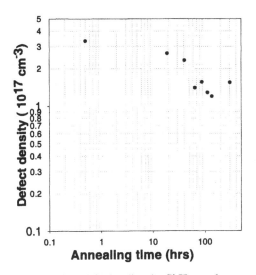

Figure 3 Isothermal annealing of the irradiated a-Si:H sample at room temperature.

This unlikely reaction produces a metastable, paired hydrogen complex and two silicon dangling bonds. Those are the two most often cited models to explain how the dangling bonds are created in light soaking experiments. However in the 77 K light soaking experiment, the time dependence is different from the room temperature experiment.

In 1993 Stradins and Fritzsche [5] performed light-soaking experiments at several low temperatures, such as 4.2, 80, and 180 K, by measuring the optical absorption coefficient, $\alpha$, using the constant photocurrent method (CPM). CPM can measure the optical absorption, which is proportional to the defect density, by changing the light intensity to keep the photocurrent constant. In this experiment, photo-induced defects were created by $h\nu = 2.1$ eV heat-filtered light of intensity 700 mW/cm$^2$. The sample was exposed to the light at 4.2 K, 80 K and 300 K for up to 1000 minutes.

At 4.2 K the sub gap absorption as measured by CPM obeys the following kinetic equation: $\triangle N_D \propto G^{0.42} t^m$ (m=0.35 ± 0.01) [5]. In the present experiments, our time dependence is also sublinear but with a much greater power law exponent of m=0.67.

There are several differences between the two experiments that might explain this discrepancy. First, the CPM experiments used a Xe lamp of different light intensity with a short-wavelength cutoff filter that passes wavelengths longer than about 560 -- 600 nm. Second, the CPM experiments employed a heat filter that blocks infrared light. Most of these differences should not affect kinetics too much, but the difference in excitation intensity may be important. In addition, the very long irradiation times discussed in this work as compared to the $\approx 10^3$ s employed in the CPM experiments may also contribute to the differences.

## SUMMARY

We have shown that the defect densities of light soaked amorphous silicon samples at 77 K increase sublinearly. After a year and a half, the final densities reach $\approx 10^{18}$ cm$^{-3}$. There is no evidence of saturation at 77 K. The power–law growth exponent is about 0.7 instead of 0.3 for illumination at room temperature. We cannot tell whether the creation mechanism results from the breaking of weak silicon-silicon bonds or silicon-hydrogen bonds, but in either case the kinetics must be different from those invoked to explain the room temperature experiments. The present results provide further information on the role of hydrogen in creating and /or stabilizing defects in light soaked samples at low temperature [2, 6, 7].

## ACKNOWLEDGEMENTS

Research at the University of Utah was supported by NREL under subcontract No. XXL-5-44205-09 and by NSF under grant No. DMR 0073004. Work at NREL was supported DOE contract #DE-AC36-99G010337.

## REFERENCES

1. N.A. Schultz, P.C. Taylor, Phys. Rev. B 65, 235207 (2002).
2. M. Stutzmann, W.B. Jackson, C. C. Tsai, Phys. Rev. B 32, 5510 (1985)
3. H. M. Branz, Solid State Communications. 105, 6 (1998).
4. H. M. Branz, Solar Energy Materials and Solar Cells 78 (2003).
5. P. Stradins and H. Fritzsche, Philosophical Magazine Part B, 69, 121(1994).
6. R. A Street, Hydrogenated Amorphous Silicon (Cambridge Univ. Press, Cambridge, 1991).
7. H. M. Branz, Phys. Rev. B 60, 7725 (1999).

Mater. Res. Soc. Symp. Proc. Vol. 1066 © 2008 Materials Research Society       1066-A02-05

# Improved Passivation of a-Si:H / c-Si Interfaces Through Film Restructuring

M. Z. Burrows[1,2], U. K. Das[1], S. Bowden[1], S. S. Hegedus[1], R. L. Opila[2], and R. W. Birkmire[1]
[1]Institute of Energy Conversion, University of Delaware, 451 Wyoming Rd., Newark, DE, 19713
[2]Materials Science and Engineering, University of Delaware, 201 Dupont Hall, Newark, DE, 19716

## ABSTRACT

The as-deposited passivation quality of amorphous silicon films on crystalline silicon surfaces is dependent on deposition conditions and resulting hydrogen bonding structure. However the initial surface passivation can be significantly improved by low temperature post-deposition anneal. For example an improvement in effective lifetime from 780 μsec as-deposited to 2080 μsec post-anneal is reported in the present work. This work probes the hydrogen bonding environment using monolayer resolution Brewster angle transmission Fourier transform infrared spectroscopy of 100 Å thick films. It is found that there is significant restructuring at the a-Si:H / c-Si interface upon annealing and a gain of mono-hydride bonding at the c-Si surface is detected. Calculations show an additional $3.56 - 4.50 \times 10^{14}$ cm$^{-2}$ mono-hydride bonding at c-Si surface due to annealing. The estimation of the surface hydride oscillator strength in transmission mode is reported for the first time to be $7.2 \times 10^{-18}$ cm on Si (100) surface and $7.5 \times 10^{-18}$ cm on Si (111).

## INTRODUCTION

Silicon heterojunction technology continues to be a topic of research interest due to continual advances in device performance and understanding. A key feature in maximizing silicon wafer based photovoltaic conversion efficiency is the minimization of saturation current. An excellent example of this application is the heterojunction-with-intrinsic-thin-layer cell design [1]. This report focuses on minimizing the saturation current through the passivation of crystalline silicon (c-Si) surface with hydrogenated intrinsic amorphous silicon ((i) a-Si:H). Specifically the aim is to display the relative insensitivity of minority carrier lifetime to bulk (i) a-Si:H film material properties and highlight the importance of the (i) a-Si:H / c-Si interface. The bulk and interface hydrogen bonding environment are studied predominantly by Fourier transform infrared absorption spectroscopy (FTIR). Short, low temperature anneals have been performed causing a large improvement in effective lifetime assuming an epitaxial growth regime is avoided. The change in the hydrogen bonding structure as a result of this anneal has been characterized and leads to an improved understanding of the importance of the hydrogen bonding structure at the c-Si interface.

## EXPERIMENT

The experiment consists of two (n) c-Si substrate orientations, 30 Ω cm Si (111) and 1.5 Ω cm Si (100). DC plasma enhanced chemical vapor deposition is used for deposition of the a-Si:H films. A simple yet rigorous substrate cleaning and H-termination procedure was applied independent of substrate orientation. The main steps include ultrasonic detergent clean, wet-chemical oxidation in self-heated 4:1 $H_2SO_4$ and $H_2O_2$, and H-termination with a 60 sec 10 %

HF room temperature dip. The material properties of the a-Si:H were controlled through variation of the hydrogen to silane gas flow ratio ($R_H$) during deposition. An annealing step was performed which consisting of a post-deposition HF dip immediately followed by a 25 min anneal at 285 °C in $10^{-4}$ Torr vacuum. It has been shown elsewhere that a HF dip effectively removes native oxide from an (i) a-Si:H film [2]. Samples are of symmetric structure and films are expected to be $100 \pm 20$ Å thick. For more complete information regarding substrate preparation and film deposition conditions please refer to U.K. Das *et al.* [3].

Surface passivation quality was evaluated by photoconductive decay measured with a Sinton WCT silicon-wafer minority carrier lifetime tester. Effective lifetime ($\tau_{eff}$) is calculated using the generalized analysis of transient decay mode at excess carrier concentration $10^{15}$ cm$^{-3}$. It is expected that due to the high quality and moderate resistivity of the float zone c-Si substrates the effective lifetime measured will be limited by surface recombination. The material properties of the deposited (i) a-Si:H films are characterized by FTIR. Due to the thinness of the films and large relative error in thickness estimation, neither absorption coefficient nor derived H concentration after the BCC method [4] are reported. To enable IR signal comparison the microstructure factor ($R_{MF}$) as introduced by Bhattacharya and Mahan will be employed [5]. This factor is the ratio of integrated di-hydride stretch peak area centered at 2080 – 2090 cm$^{-1}$ to total hydride and therefore not dependent on accurate thickness determination. The spectra are collected in Brewster angle transmission mode at a resolution of 4 cm$^{-1}$ on a Bruker Tensor 27 fitted with a liquid $N_2$ cooled HgCdTe detector and analyzed with Opus software. Excellent sensitivity down to $10^{-5}$ absorbance units, sufficient for monolayer (ML) sensitivity, is demonstrated by this set-up.

## RESULTS AND DISCUSSION

### Influence of Bulk Film Material Properties

Two Shockley-Reed-Hall recombination paths exists at an a-Si:H / c-Si interface region. The first will be considered to occur through interface defect states directly at the c-Si surface, commonly referred to as surface states. This defect has been characterized by photoemission and photoelectron yield spectroscopy as an unbound electron [6]. The second path will be through the unbound electron state or dangling bond defect known to exist around midgap of a-Si:H [7] in the film bulk itself. While the precise recombination dynamics and cross sections of these defects are beyond the scope of the current work; the importance of the latter defect at the c-Si surface will be demonstrated.

Figure 1 depicts the relationship between passivation quality and $R_{MF}$ for the DC (i) a-Si:H on Si (111) (n) c-Si case. Note that over several depositions at $R_H = 6$ it is difficult to replicate an identical passivation quality, with values spanning from 800 – 1800 µsec. However the bulk material has little variation with respect to $R_{MF}$ and simple monotonic trends are clear in the FTIR signal. It has been shown that films with high $R_{MF}$ are less ordered or more defective [5] and thus would be expected to provide a lower passivation quality. However maxima in the lifetime data can be seen at $R_H = 6$ while the bulk $R_{MF}$ continues to decline beyond $R_H = 10$. Finally, as will be further discussed below, there is a large post-anneal lifetime improvement with very little *relative* change in hydrogen bonding and no measurable change in $R_{MF}$. A closer inspection of H bonding environment change with annealing is therefore warranted.

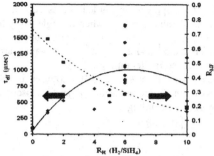

Figure 1. Effective lifetime and microstructure factor as a function of hydrogen dilution ratio of 10nm DC (i) a-Si:H film on Si (111).

## Oscillator Strength of Si–H$_{c-Si}$

Historically, IR absorption measurement of hydride bonding on c-Si surface has been achieved in the multiple-internal-reflection (MIR) mode [8,9]. Surface hydride peak identification has made full use of the high resolution and signal to noise levels achievable. Figure 2 displays the absorbance intensity of a monolayer (ML) coverage of hydride over Si (111) and Si (100) surfaces obtained by HF dip collected in single pass Brewster angle transmission. This is the first time such a measurement is reported to the best of author's knowledge.

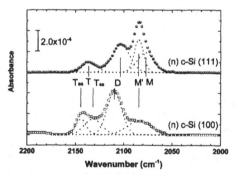

Figure 2. FTIR absorption of ML hydride coverage post HF dip on (n) c-Si (111) and (100), sum of front and rear surfaces.

Peak identification follows that established by Chabal et al. [8] and Shinohara et al [9]. On the Si (111) surface it is important to note ideal mono-hydride termination is not achieved. This is to be expected as it has been shown by several groups that HF pH buffered with NH$_4$F [10] or just NH$_4$F itself [9] is required for ideal termination and that dilute HF solutions lead to a roughening of the Si (111) surface [11]. Mode $M'$ is uncoupled mono-hydride on Si (111) at 2084 cm$^{-1}$ bonded normal to the surface, $M$ is the coupled mono-hydride mode at 2077 cm$^{-1}$ whose bond is tilted with respect to the surface normal [8]. $D$ is di-hydride at 2103 cm$^{-1}$ on Si (111) and 2110 cm$^{-1}$ on Si (100). The current experimental set-up does not resolve the di-hydride into symmetric and asymmetric stretch expected at 2103 cm$^{-1}$ and 2113 cm$^{-1}$ [9] and likely leads to a higher FWHM and peak position variation between the substrate orientations. The tri-hydride peak $T$ on Si (111) appears uncoupled and is centered at 1036 cm$^{-1}$ as expected [8]. Interestingly a very poor fit quality occurs when a single tri-hydride mode is applied for the Si (100) spectrum. In this case the tri-hydride mode appears split into the symmetric and asymmetric modes at 2132 cm$^{-1}$ and 2144 cm$^{-1}$. These values are in acceptable agreement with 2129 cm$^{-1}$ and 2143 cm$^{-1}$ listed in literature [8,9].

The absorbance signal from surface ML coverage is related to a *surface* density and therefore the thickness is an ill-defined value. A surface oscillator strength is calculated from integrated absorbance, $A_i(v)$, according to

$$\Gamma_i = \frac{\int A_i(v)dv}{N_i} \tag{1}.$$

The absorbance integral, $\int A_i$, has been determined experimentally and is shown in Table I. However $N_{i,surf}$ (cm$^{-2}$), the H atomic surface density is yet unknown for each mode shown in Figure 2. For a first approximation however, if one assumes the IR absorption cross-section of each mode is nearly equivalent and there is complete hydride surface coverage, the total H atomic surface density can be calculated. Given the fact the surfaces are atomically rough on the order of 5 Å [11], and similar assumptions regarding oscillators strength of different stretch modes in a-Si:H bulk [12] it is assumed this is an acceptable starting point that is open to refinement in future work. Taking the ideal Si (100) surface density at $SS_{100} = 6.78 \times 10^{14}$ cm$^{-2}$ [13] and Si (111) at $SS_{111} = 7.83 \times 10^{14}$ cm$^{-2}$ [14] and considering a tri-hydride adds three states to the H atomic surface density per Si back bond ($n_{3,j} = 3$), di- adds two ($n_{2,j} = 2$) and mono-hydride just one ($n_{1,j} = 1$), calculation of $N_{i,j}$ follows Eqn. (2)

$$N_{i,j} = n_{i,j} \frac{\left( \dfrac{\int A_{i,j}(v)dv}{n_{i,j}} \right)}{\left( \sum_i \dfrac{\int A_{i,j}(v)dv}{n_{i,j}} \right)} SS_j \tag{2}.$$

Where i varies over the number of hydrogen bonded to the surface Si state, and j denotes the different surface orientations. Table I shows the results of these calculations and the resulting oscillator strength. Also from Table I it can be calculated that 1.75 ML of hydride is present on the Si (100) surface, close to the saturation coverage of 1.85 ML expected for Si (100) - 2 × 1 [13]. The 1.37 ML on the Si (111) surface is larger than the ideal mono-hydride coverage of 1 ML but reasonable considering the large fractions of di- and tri-hydride observed.

Table I. Summary of measured peak position and absorbance integral per surface. The H atomic surface density and oscillator strength are calculated assuming mono-, di-, and tri-hydride modes have equivalent IR absorption cross-sections.

| | Si (100) | | | | | Si (111) | | | | |
|---|---|---|---|---|---|---|---|---|---|---|
| Oscillator strength (cm) | 7.23x10$^{-18}$ | | | | | 7.47x10$^{-18}$ | | | | |
| | M | D | T$_{ss}$ | T$_{as}$ | Total | M | M' | D | T | Total |
| Peak position (cm$^{-1}$) | 2084 | 2110 | 2132 | 2144 | | 2077 | 2084 | 2103 | 2136 | |
| Absorbance integral (cm$^{-1}$) | 0.0021 | 0.0038 | 0.0017 | 0.0010 | 0.0086 | 0.0021 | 0.0019 | 0.0030 | 0.0010 | 0.0080 |
| H atomic surface density (x10$^{14}$ cm$^{-2}$) | 2.90 | 5.30 | 2.32 | 1.37 | 11.89 | 2.78 | 2.58 | 4.06 | 1.33 | 10.74 |

## Influence of Hydrogen Bonding at Interface

Improvements in effective lifetime from a few hundreds of microseconds to greater than a millisecond are achievable with a short low temperature anneal [15]. Yet the H bonding environment of the bulk FTIR, i.e. $R_{MF}$ and absorption integral, remain relatively static. In a separate report this observation was analyzed in detail for three specific cases [16]. It was concluded that there exists a restructuring in the di-hydride interface layer [17] that leads to an increase in mono-hydride bonding to the c-Si surface. This would be expected to reduce surface defect density and lead to the measured increase in effective lifetime. Here, the above defined oscillator strength is invoked to calculate the actual increase in mono-hydride bonding on the c-Si surface and compare this to the expected amount of interface states removed.

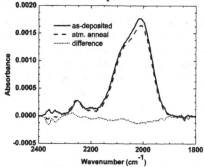 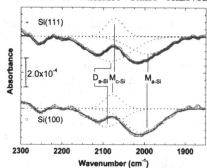

Figure 3(a). FTIR absorption Si–H stretch region of 2 × 100Å (front and rear surface) $R_H = 2$, DC (i) a-Si:H on Si (100).

Figure 3(b). Difference in FTIR absorption due to 30 min 285 °C, $10^{-4}$ Torr, annealing of 2 × 100Å $R_H = 2$, DC (i) a-Si:H on Si (100) and Si (111).

Table II. Effective lifetime pre- and post- 30 min, 285 °C anneal, absorbance integral per surface, $\int A_{Mc\text{-}Si}$, and increase in $M_{c\text{-}Si}$ surface bonding density.

|          | Lifetime (μsec) | | $\int A_{Mc\text{-}Si}$ | $\Delta N_{Mc\text{-}Si}$ |
|----------|-------------|--------------|--------------------|-------------------------------------|
|          | Pre-Anneal  | Post-Anneal  | (cm$^{-1}$)        | ($\times 10^{14}$ cm$^{-2}$)        |
| Si (100) | 570         | 1630         | 0.00257            | 3.6                                 |
| Si (111) | 780         | 2080         | 0.00336            | 4.5                                 |

Figure 3(a) illustrates the bulk Brewster angle FTIR signal from the sum of two 100 Å films (front and rear surface). The H bonding change due to annealing is simply the subtraction of the absorbance curves, labeled as the difference curve. Figure 3(b) shows two difference curves from identically processed samples with Si (111) and Si (100) surface orientations. The two negative peaks indicate a small loss of mono-hydride ($M_{a\text{-}Si}$) and di-hydride ($D_{a\text{-}Si}$) bonding in an amorphous Si matrix at their characteristic frequencies of 2000 cm$^{-1}$ and 2090 cm$^{-1}$ [17]. The positive peak indicates a gain of mono-hydride ($M_{c\text{-}Si}$) bonded to a crystalline surface at peak frequency ~2075 cm$^{-1}$. This frequency is in the expected range for mono-hydride bonded to a crystalline surface as can be seen in Table I and references [9,14]. The $M_{c\text{-}Si}$ mode is 2-3 times broader than on HF treated Si (111) surfaces but nearly equal to that on Si (100). Broadening is likely due to additional roughening of the substrate surface during initial plasma growth and the localized fluctuation in the surrounding medium dielectric. Also note that two negative peaks centered at 2030 cm$^{-1}$ and 2115 cm$^{-1}$ do not provide an acceptable fit due to the asymmetry of the intersection region. Table II reports the effective lifetime improvement as well as the absorbance integral of the peak $M_{c\text{-}Si}$ per surface and finally the calculated increase in mono-hydride bonding to the c-Si surface using Eqn. (2).

For comparison we look at the modeling work done by Garin et al. [18] and surmise the density of interface defects, $D_{it}$, would need to be reduced from approximately $5 \times 10^{12}$ to $1 \times 10^{12}$ cm$^{-2}$ for an improvement in lifetime from 500 to 2000 µsec. That is, approximately two orders of magnitude more surface hydride states are gained to each defect state removed. Hence a large degree of interfacial H bonding environment restructuring occurs relative to defect density reduction.

## CONCLUSIONS

FTIR spectroscopy has been applied to probe the relationship between hydrogen bonding environment in a-Si:H / c-Si and surface passivation. Effective lifetime dependency on bulk film material property, as indicated by $R_{MF}$, is mitigated beyond a certain hydrogen dilution ratio. $R_{MF}$ is also insensitive to the material changes that occur with annealing which lead to substantial effective lifetime improvements. Upon annealing effective lifetime improves from 780 to 2080 µsec in the $R_H = 2$, Si (111) case. A concomitant increase in $M_{c\text{-}Si}$ mode is detected and estimated to be the result of a gain in $4.5 \times 10^{14}$ cm$^{-2}$ mono-hydride at the c-Si surface. This is approximately 100 times the number of surface defect states that would be removed for the measured improvement in effective lifetime upon annealing.

## REFERENCES

1 M. Taguchi, A. Terakawa, E. Maruyama et al., Progress in Photovoltaics 13 (6), 481 (2005).
2 M. Burrows, U. Das, M. Lu, S. Bowden, R. Opila, and R. Birkmire, Mater. Res. Soc. Symp. Proc. **989**, San Francisco, 431-436 (2007)
3 U.K. Das, M. Burrows, M. Lu, S. Hegedus, S. Bowden, and R. Birkmire, 22nd European Solar Energy Conference, Milan, Italy (2007).
4 M. H. Brodsky, M. Cardona, and J. J. Cuomo, Physical Review B 16 (8), 3556 (1977).
5 E. Bhattacharya and A. H. Mahan, Applied Physics Letters 52, 1587-1589 (1988).
6 S. Miyazaki, J. Schafer, J. Ristein et al., Applied Physics Letters 68 (9), 1247 (1996).
7 R. A. Street, Hydrogenated Amorphous Silicon, Cambridge University Press, 1991, Cambridge.
8 Y. J. Chabal, G. S. Higashi, K. Raghavachari, and V. A. Burrows, J. Vac. Sci. Technol. A 7 (3), 2104-2109 (1998).
9 M. Shinohara, T. Kuwano, Y. Akama, Y. Kimura, M. Niwano, H. Ishida, and R. Hatakeyama, J. Vac. Sci Technol. A 21 (1), 25-31 (2003).
10 G. S. Higashi, Y. J. Chabal, G. W. Trucks, and K. Raghawachari, Appl. Phys. Lett. 56 (7), 656-658 (1989).
11 W. Henrion, M. Rebien, H. Angermann, and A. Roeseler, App. Surf. Sci. 202, 199-205 (2002).
12 A. H. M. Smets, W. M. M. Kessels, M. C. M. de Sanden, Appl. Phys. Lett. 82(10), 1547-1549 (2003).
13 K. Oura, J. Yamane, K. Umezawa, M. Naitoh, F. Shoji, and T. Hanawa, Phys. Rev. B 41 (2), 1200-1203 (1990).
14 D. C. Marra, E. A. Edelberg, R. L. Naone, and E. S. Aydil, J. Vac. Sci Technol. A 16(6) 3199-3210 (1998).
15 U.K. Das, M.Z. Burrows, M. Lu, S. Bowden, and R. W. Birkmire, Applied Physics Letters 92 (6), 063504 (2008).
16 M. Z. Burrows, U. K. Das, R. L. Opila, S. De Wolf, and R. W. Birkmire, J. Vac. Sci Technol. A 4 (2008) (accepted).
17 H. Fujiwara, Y. Toyoshima, M. Kondo, and A. Matsuda, Physical Review B 60, 13598-13604 (1999).
18 M. Garin, U. Rau, W. Brendle, I. Martin, and R. Alcubilla, J. of App. Phys. 98, 093711 (2005).

# Solar Cells I

Mater. Res. Soc. Symp. Proc. Vol. 1066 © 2008 Materials Research Society      1066-A03-01

# Understanding of Passivation Mechanism in Heterojunction c-Si Solar Cells

Michio Kondo[1], Stefaan De Wolf[1,2], and Hiroyuki Fujiwara[1]

[1]RCPV, AIST, Umezono, Tsukuba, 305-8568, Japan

[2]Institute of Microtechnique, University of Neuchâtel, Rue Breguet 2, Neuchâtel, CH-2000, Switzerland

Intrinsic hydrogenated amorphous silicon ($a$-Si:H) films can yield in outstanding electronic surface passivation of crystalline silicon ($c$-Si) wafers as utilized in the HIT (heterojunction with intrinsic thin layer) solar cells. We have studied the correlation between the passivation quality and the interface nature between thin amorphous layers and an underlying c-Si substrate for understanding the passivation mechanism. We found that a thin (~5nm) intrinsic layer is inhomogeneous along the growth direction with the presence of a hydrogen rich layer at the interface and that completely amorphous films result in better passivation quality and device performance than an epitaxial layer. Post annealing improves carrier lifetime for the amorphous layer, whereas the annealing is detrimental for the epitaxial layer. We have also found that the passivation quality of intrinsic $a$-Si:H($i$) film deteriorates severely by the presence of a boron-doped $a$-Si:H($p^+$) overlayer due to Si-H rupture in the $a$-Si:H($i$) film. Finally, for a passivation layer in the hetero-junction structure, $a$-Si$_{1-x}$O$_x$ will be demonstrated in comparison with $a$-Si:H.

## INTRODUCTION

Hydrogenated amorphous silicon ($a$-Si:H) films deposited on crystalline silicon ($c$-Si) surfaces have increasingly attracted attention over the past years. Initially it was discovered that abrupt electronic heterojunctions can be created with such structure[1], followed by an application to solar cells[2]. Structures featuring a wider bandgap emitter have in the past already been suggested to allow for maximal photovoltaic device performance[3]. For $a$-Si:H / $c$-Si heterostructures, it was found that the output parameters benefit substantially from inserting a few nm thin intrinsic $a$-Si:H($i$) film between the doped amorphous emitter and $c$-Si substrate. Direct deposition of doped $a$-Si:H films on $c$-Si surfaces has been speculated to result in poor interface properties because of a defective property of doped $a$-Si:H. Presently, for solar cells that feature a heterostructure emitter and back surface field, impressive energy conversion efficiencies exceeding 22% have been reported [4]. Nevertheless, despite these results, a physical understanding of the $a$-Si:H / $c$-Si interface is not yet complete.

In this paper, we summarize our study on understanding the mechanism of the electronic passivation and the interface nature between thin amorphous layers and an underlying $c$-Si substrate. The article is organized as follows: First we discuss results obtained by real-time spectroscopic ellipsometry (SE) and attenuated total reflection (ATR) measurements. This yields insight in the growth mechanism of hydrogenated $a$-Si:H layers on $c$-Si surfaces.

In a second part, an annealing study of thin stacked doped $a$-Si:H films is presented where electronic passivation results are linked to material properties. These studies are useful for two reasons: Firstly, postdeposition annealing offers a single parameter to vary both electronic and material properties of the samples under study, from which additional physical insights may be obtained. Secondly, such treatments may be beneficial for the structures under study, and hence are also relevant from device processing point of view.

Lastly, interpretation of the deposition condition dependence of the solar cell performance is discussed in conjunction with different amorphous materials for heterojunction.

## EXPERIMENTAL

### Sample preparation

For the experiments, 300-320 μm thick relatively low resistivity boron-doped (~3.0 Ω.cm) as well as phosphorus-doped (~0.7 Ω.cm) high quality float zone (100) FZ-Si wafers have been used. Both surfaces of the substrates were mirror polished to eliminate the influence of substrate surface roughness on the passivation properties [5], and to allow for SE measurements.

For pre-deposition surface cleaning, the samples were first immersed in a ($H_2SO_4$:$H_2O_2$) (4:1) solution for 10 min to grow a chemical oxide, which was followed by a rinse in de-ionized water. The oxide was then stripped off in a dilute HF solution (5%) for 30 sec. After this the samples were immediately transferred to the load lock of the deposition system.

For a-Si:H film deposition, a parallel plate direct-plasma enhanced chemical vapor deposition (PECVD) systems were used. During film deposition, all chambers were operated at radio frequency (rf) (13.56 MHz) power and a pressure between 0.05 and 0.5 Torr. For soft film deposition, the power desity was consistently the minimum required to maintain stable plasmas, typically ~10 mW/ $cm^2$. The value for $T_{depo}$ was varied from 105 °C to 255 °C. For intrinsic films 20 SCCM $SiH_4$ was used, for boron doped films this was 10 SCCM, mixed in 30 SCCM $B_2H_6$ (4660 ppm in $H_2$).

The a-Si:H/c-Si heterojunction solar cells in this study consist of Ag grid/$In_2O_3$:Sn/a-Si:H(p)/a-Si:H(i)/c-Si(n)/Al. The solar cell efficiencies were determined from solar cells having active area of 0.21 $cm^2$ under AM 1.5 illumination (100 mW/$cm^2$), and solar cell efficiencies designated in this study represent active area efficiencies.

### In-situ measurements

For real-time monitoring of the film deposition processes, real-time spectroscopic ellipsometry (SE) and infrared attenuated total reflection spectroscopy (ATR) have been performed using the system shown in Fig. 1 [6-9]. During the a-Si:H p-i layer growth, ellipsometry spectra (ψ,Δ) were collected using a rotating-compensator instrument (J. A. Woollam, M-2000). The SE analysis was performed in real time using a two-layer model for the a-Si:H film consisting of ambient/surface roughness layer/bulk layer/substrate. The dielectric function of the surface roughness layer was modeled as a 50/50 vol.% mixture of the bulk layer material and voids. We obtained the dielectric function of the a-Si:H bulk layer in advance from an a-Si:H i-layer (200 Å) using a global error minimization scheme [8]. The a-Si:H p-layer was analyzed by applying the dielectric function of the i-layer, as we find that the dielectric function of the p-layer is rather similar to that of the i-layer when the p-layer thickness is thin (~50 Å). On the other hand, the real-time ATR spectra were measured by employing a Fourier-transform infrared instrument (Nicolet, Magna 560). For the ATR measurement, we used trapezoidal non-dope c-Si substrates (~1000 Ωcm). Thus, only the SE measurement was carried out when the a-Si:H layers were deposited on the n-type substrate to fabricate the solar cell. Details of the measurements and data analysis procedures for real-time SE [8,9] and ATR [6,9] have been reported elsewhere.

**Real-time ATR system**

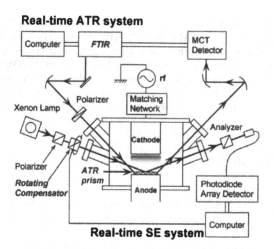

**Fig. 1.** Schematic of the plasma-enhanced CVD system and the real-time measurements by SE and ATR.

## Ex-situ measurements

To gain knowledge about the electronic surface passivation properties of these interfaces, the most straightforward technique is to measure the effective carrier lifetime, $\tau_{eff}$, of the samples. The value for $\tau_{eff}$ of the samples was measured with a Sinton Consulting WCT-100 quasi steady state photoconductance system [10], operated in the so-called *generalized* mode. Since high quality FZ-Si wafers have been used throughout the experiments, the contribution of the bulk to the total recombination expressed by $\tau_{eff}$ can be neglected. In such a case, the effective surface recombination velocity, $S_{eff}$, which value can be regarded as a direct measure for the passivation quality of the films present at the surfaces, may be approximated by $S_{eff} = d*(2\tau_{eff})^{-1}$, with $d$ being the wafer thickness. All reported values for $\tau_{eff}$ and $S_{eff}$ are evaluated at a constant minority carrier injection density, $\Delta n$, of $1.0 \times 10^{15}$ cm$^{-3}$. The thickness of the deposited films was *ex-situ* determined by measuring ellipsometry spectra ($\psi,\Delta$) using a Woollam M-2000 rotating-compensator instrument.

We have studied the effects of low temperature (up to 260 °C) post-deposition annealing upon the surface passivation quality of a-Si:H($i$) films with or without a boron-doped a-Si:H($p^+$) over-layer on top of c-Si / a-Si:H($i$). The samples were consecutively annealed in a vacuum furnace (30 min, with annealing temperatures, $T_{ann}$, ranging from 120 °C to 260 °C and a 20 °C increment per step).

For bulk characterization of the films, thermal desorption spectroscopy (TDS) measurements were taken. For this an ESCO EMD-WA1000S system operated at ultra high vacuum (< $1.0 \times 10^{-9}$ Torr) is used in which the samples are lamp-heated up to 1000 °C, with a linear temperature ramp of 20 K.min$^{-1}$. During the annealing, a Balzers AG QMG 421 quadrupole mass spectrometer was used to determine the $H_2$ effusion rate from the a-Si:H films. Throughout this article the following shorthand notations are used: $i$, $p^+$ and $i/p^+$ for respectively the c-Si / a-Si:H($i$), c-Si / a-Si:H($p^+$) and c-Si / a-Si:H($i$) / a-Si:H($p^+$) structures.

## RESULTS AND DISCUSSION

### Thickness dependence of the i-layer

Figure 2 shows (a) depth profiles of hydrogen contents determined from SE-ATR results and (b) solar cell efficiency, plotted as functions of the thickness of a-Si:H i-layer deposited at 130 °C. The result shown in Fig. 2(a) was obtained from the growth of a thick a-Si:H i-layer, and the formation of a ~2 nm-thick interface layer with a maximum SiH$_2$-hydrogen content of 27 at.% can be seen at the a-Si:H/c-Si interface. The large SiH$_2$ content in the interface layer indicates poor network formation in this layer. From the SE result obtained simultaneously, we concluded that the formation of the SiH$_2$ interface layer is induced by the island growth of the a-Si:H i-layer on the substrate [9]. As confirmed from Fig. 2, the best solar cell efficiency coinsides with a thickness where the a-Si:H growth reaches a steady state after the interface layer formation.

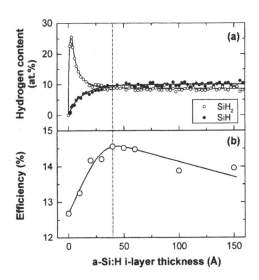

**Fig. 2.** (a) depth profiles of hydrogen contents and (b) solar cell efficiency, plotted as functions of the a-Si:H i-layer thickness [7].

In general, the creation of SiH$_2$ bonds in amorphous network is accompanied by the generation of defects. Accordingly, the improvements in the solar cell characteristics up to 40

A are likely caused by reduction in defect density with increasing the i-layer thickness; in other words, solar cell characteristics degrades seriously when the defect-rich interface layer maintains extensive contact with the defective p-layer at the p/i interface. The reduction in $J_{sc}$ observed at 0<i-layer<4 nm could also be explained by the presence of the interface layer. When the i-layer thickness is increased further (>4 nm), on the other hand, the solar cell efficiency reduces due to the decrease in $J_{sc}$ [7].

So far, a number of studies have reported the improvement of a-Si:H/c-Si cell performance by the incorporation of a-Si:H i-layer at the a-Si:H/c-Si heterointerface [2,11,12]. This effect has been attributed to better properties of an a-Si:H(i)/c-Si interface, compared with an a-Si:H(doped)/c-Si interface [2,11,12].

We have also found that the a-Si:H p-layer deposited directly on H-terminated c-Si becomes partially epitaxial due to the presence of $H_2$ gas during the growth and that the conversion efficiency of this solar cell shows a low value of 12.7 %. This is the case for the a-Si:H i-layer grown epitaxially on the crystalline substrate. The epitaxial interface formed at low temperatures is so defective as to deteriorate the interface quality.

Based on these results, we propose a passivation mechanism where the defective interface formed by the doped a-Si:H p-layer is spatially separated by the a-Si:H i-layer at a proper distance ( ~ 4nm), suggesting high quality of the passivation at a-Si:H i-layer and c-Si substrate interface even though the Si-$H_2$ rich layer is involved.

**Deposition temperature dependence**

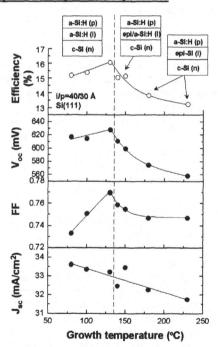

Fig. 3. Characteristics of the a-Si:H/c-Si heterojunction solar cell, plotted as a function of the growth temperature of the p-i layers. In the solar cells, the thicknesses were fixed to i/p=4/3 nm. The inset represents the phase structures of the p-i layers deduced from SE.

Figure 3 shows the solar cell performance, plotted as a function of the growth temperature $T_G$ of the p-i layers. In this figure, the same temperature was used for the growth of the p-i layers. Accordingly, the result shown in Fig. 3 includes the temperature effects for both p and i-layers. The inset of Fig. 3 shows the phase structures of the p-i layers deduced from SE. One of our important finding is that the epitaxial growth of the i-layer occurs at $T_G \geq 140$ °C even without the presence of $H_2$ gas, as illustrated in the inset. In particular, the thickness of the p-i layers becomes zero when the epitaxial Si layer (epi-Si) is characterized by SE, since the complex refractive index of the epi-Si is essentially the same as that of the substrate. In other words, SE does not see the presence of the homoepitaxial layer optically. Thus, the epitaxial/ amorphous growth can be distinguished rather easily from the thickness or instantaneous growth rate during the p-i layer growth using SE.

As shown in Fig. 3, the SE analysis revealed that the i-layer is partially epitaxial at $140 \leq T_G < 180$ °C. At $T_G \geq 180$ °C, however, the i-layer becomes completely epitaxial, although the p-layer is still amorphous. It should be emphasized that transmission electron microscopy (TEM) also confirmed the heterointerface structures shown in Fig. 3 [14]. It can be seen that the epi-Si growth degrades the solar cell characteristics severely even when the thickness of epi-Si is quite thin (~2 nm). In particular, the $V_{oc}$ shows a low value of ~560 mV when there is no a-Si:H i-layer ($T_G \geq 180$ °C). Thus, the a-Si:H i-layer is quite important for obtaining high $V_{oc}$. At $T_G \leq 130$ °C, on the other hand, the p-i layers are amorphous. At lower process temperatures, however, the efficiency reduces due to the decrease in FF. We attributed this to the low activation of the boron dopant in the a-Si:H p-layer owing to the low process temperatures. The result in Fig. 3 shows clearly that the optimum growth temperature for the a-Si:H/c-Si solar cells is just below the onset temperature of the Si epitaxial growth. Furthermore, the above results support the validity of real-time SE in characterizing the complicated structural evolution. The evaluation of such structures is of significant importance, since the epitaxial growth seriously degrades the solar cell performance [14].

## Annealing studies

In the series of the annealing studies, we have adopted the lifetime measurements for evaluating the passivation quality. Figure 4 shows how the surface passivation quality of a single intrinsic a-Si:H film of about 50 nm thin (present on both wafer surfaces), for different values of $T_{depo}$, changes as a function of the annealing temperature, $T_{ann}$. The substrates here are 3.0 $\Omega$.cm p-type FZ wafers. For the films deposited at the lowest temperature ($T_{depo} = 105$ °C), initially the passivation is poor, but improves to a remarkable extent by annealing. By increasing the deposition temperature, $T_{depo}$, up to 180 °C, the surface passivation quality of as-deposited films improves. Here, post annealing up to 260 °C increases the values for $\tau_{eff}$ further, well in excess of 1 ms. This situation is different for films deposited at 205 °C. Annealing does not give rise anymore now to an improvement. For even higher values of $T_{depo}$, the figure shows that the passivation quality actually goes down dramatically.

Although the data in Fig. 4 was for still relatively thick (50nm) intrinsic a-Si:H films with variable values for $T_{depo}$, in Fig. 5 it is shown how the electronic passivation quality of much thinner (few nm) i-, $p^+$- and $i/p^+$- structures changes by low temperature ($\leq 260$ °C) post-deposition annealing. Here, $T_{depo}$ was constant at 155°C. The table in the inset gives the values of the layer thickness $d_{bulk}$ and surface roughness $d_{rough}$ for the respective layers, as extracted again from SE measurements. For the i-case, the applied low temperature annealing treatment is again seen to have a beneficial influence on the passivation. This is different however for the $p^+$-case: here, annealing rapidly leads to passivation degradation. The latter situation is seen to be only slightly improved when an intrinsic buffer layer (of similar thickness as in the i-case) has been inserted underneath the $p^+$-film ($i/p^+$-case): Starting with values similar as for the $p^+$-case, initially the passivation quality benefits from annealing. Nevertheless, from about $T_{ann} = 220$ °C on also here degradation sets in. Such trends have in the past been found to be irrespective of the dopant-type of low resistivity wafers [15]. We will now argue that the degradation mechanisms as seen in these two figures are of fundamentally different nature.

First we discuss the case of passivation by intrinsic $a$-Si:H films. From the slopes of the exponential fits in the shown Arrhenius plot in Fig. 4, an activation energy, $E_A$, can be extracted. These values are given in figure 6 (b) as function of $T_{depo}$. The area showing

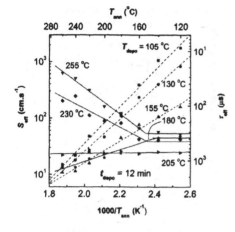 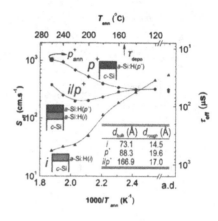

FIG. 4 Influence of $T_{ann}$ on the surface passivation quality for PECVD $a$-Si:H($i$) films deposited on mirror polished (100) FZ-Si($p$) surfaces at different temperatures, $T_{depo}$. All deposition times were 12 min, whereas annealing times were 30 min. Starting from the $T_{ann}$ onset of annealing induced passivation changes, the shown lines are exponential fits of the data.

FIG. 5 Influence of $T_{ann}$ on the electronic passivation quality for thin intrinsic and doped $a$-Si:H layers deposited on mirror polished (100) FZ-Si($n$) surfaces. For reference, values for as-deposited films are given as well (label a.d. in the abscissa). The lines here are guides for the eye. The table in the inset gives the values of $d_{bulk}$, and $d_{rough}$ for the respective films, determined from SE measurements.

negative values for $E_A$ corresponds to films for which post annealing is harmful for the passivation quality. This has been cross-hatched in the figure. Part (a) of the same figure gives the $a$-Si:H($i$) film thickness, $d_{bulk}$, during the initial deposition stages, as function of $T_{depo}$. This data is obtained from fitting the measured SE data of the different samples to a two-layer model, taking the 50% void surface roughness on top of the $a$-Si:H film into account. For all films deposited at $T_{depo} \geq 205$ °C, hardly any film growth can be observed with SE during at least the first 72 sec. The difficulty to fit the initial film thickness to the SE data suggests that for the complete surface the deposited material is crystalline [14]. This area is cross-hatched in the figure too: A good correspondence exists between the onset where crystalline material is grown at the full interface during film deposition, and the point where the value of $E_A$ becomes negative[16].

The value for $E_A$ is extracted from $\tau_{eff}$ measurements, hence its value will mainly be determined by the density of electronically active defects of the film. These are the defects that are within reach of the $c$-Si minority carrier wave function. The impact of these defects is determined by their energetic position within the bandgap and their electron and hole capture cross sections. This contrasts with, e.g., electron spin resonance (ESR) measurements of $a$-Si:H($i$) layers, which rather reveal paramagnetic defect densities in the bulk of relatively thick (typically a few µm) films deposited on quartz substrates. Nevertheless, for 4 µm thick films deposited at 25 °C, Biegelsen et al. observed in their ESR studies activation energies of about 0.5 eV for the decrease of the Si dangling bond density by post deposition annealing at values for $T_{ann}$ up to 250 °C [17]. Considering the differences in measurement techniques, a good correspondence exists between this value for $E_A$ and the ones obtained from $\tau_{eff}$ measurements for the films deposited at the lowest values $T_{depo}$ as presented in figure 6.

For higher values of $T_{depo}$, the value of $E_A$ lowers and eventually passes through zero. Likely, this is because by increasing $T_{depo}$ the deposited material contains a lower dangling

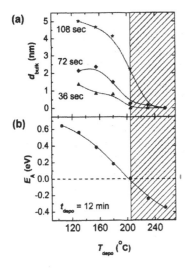

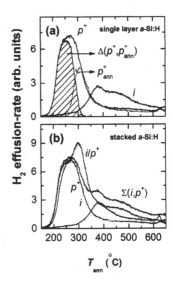

FIG. 6 (a) Calculated film-thickness, $d_{bulk}$, from *ex-situ* SE measurements as function of $T_{depo}$, given for several deposition times, $t_{depo}$. Cross-hatched area shows films with epitaxially grown interface. (b) Extracted activation energy $E_A$ as function of $T_{depo}$ for the films shown in figure 5. Cross-hatched area shows films for which the surface passivation degrades by annealing. All films are intrinsic. The lines are guides for the eye.

FIG. 7 Influence of $T_{ann}$ on $H_2$ effusion rate of *a*-Si:H films as given in Fig. 1: (a) data for few nm thin single layer *a*-Si:H films, as deposited and after the low-temperature annealing cycle described in Fig. 4 and 5. (b) data for as-deposited stacked doped films. Measurements by TDS.

bond density. When $E_A$ equals to zero there is no net decrease in dangling bond density any more. For negative values of $E_A$ annealing increases the dangling bond density. These results suggest that at all values for $T_{depo}$ two competing mechanisms may be at work during annealing, resulting in an increase and decrease of the dangling bond density, respectively. The net effect of these two mechanisms likely depends on how much hydrogen is available in the film for dangling bond passivation. This may explain why negative values of $E_A$ correspond to epitaxially grown films, where it can be expected that the amount of available hydrogen is limited, and where it is known that such growth during *a*-Si:H(*i*) deposition is detrimental for heterojunction device performance [18]. This result also shows that annealing combined with carrier lifetime measurements offers an alternative means, compared to transmission electron microscopy (TEM) or SE [19], to determine the abruptness of the interface.

To explain the experienced degradation in case of boron-doped films, first we show in figures 7 (a) and (b) TDS data of exactly the same structures as displayed in Fig. 5. Part (a) shows that for such thin films, the boron doping ($p^+$-case) leads to $H_2$ effusion at significant lower temperatures shown compared to intrinsic films, occurs. This phenomenon has been reported earlier in literature for thicker films [20]. The same figure also shows similar data (label $p_{ann}^+$) for the doped film case after the stepwise annealing cycle, as described in Fig. 5. The crosshatched area in this figure (label $\Delta(p^+, p_{ann}^+)$) represents the difference between these two signals and clearly demonstrates that during the latter cycle already significant $H_2$ effusion takes place. Figure 7(b) compares $H_2$ effusion rate data of the $i/p^+$-structure with that of the summed $i$- and $p^+$-case (label $\Sigma(i, p^+)$). It is seen that at low temperatures more hydrogen effuses out for the $i/p^+$-case than for the combined $i$- and $p^+$-cases. Note that the combined $i$- and $p^+$-layer thickness practically equals that of the stacked $i/p^+$ structure (see inset in Fig. 5).

For intrinsic $a$-Si:H material annealed at higher temperatures (> 300 °C), a correspondence between the $H_2$ effusion rate and defect-generation in the film has been demonstrated in the past by comparing TDS, IR absorption and electron spin resonance measurements[ ] Consequently, for boron-doped $a$-Si:H($p^+$) material, the effusion data suggests that hydrogen likely is already transferred at much lower temperatures from a Si-H to a $H_2$ state, creating defects in the material. The Si-H bond rupture energy has been argued to depend on the Fermi energy (rather than on the actual dopants) in such material [21,22] . As discussed, in the present case, annealing rapidly results in electronic passivation losses for the $p^+$-case. Also here, the origin of this phenomenon likely is Fermi-level dependent defect-generation, occurring close to the $a$-Si:H / $c$-Si interface.

For the $i/p^+$-structure, the initial improvement by annealing in passivation is again most probably due to out-annealing of defects in the intrinsic layer. At higher values for $T_{ann}$, the passivation degradation of this stacked structure may be caused by two related phenomena: Firstly, dangling bonds are created already at low temperatures by annealing in the doped layer. As a consequence, the wavefunction associated with the minority carriers close to the interface in the $c$-Si material may probe through the intrinsic film, when sufficiently thin, to collapse in the defect rich doped overlayer [23]. Secondly, in addition to this, due to the presence of the $p^+$-layer on top of the $i$-layer, also in the latter layer the Fermi-level will be shifted towards the valence band of the material. As a result, also here already at moderate annealing temperatures, Si-H bond-rupture can be expected to take place in the buffer layer too [24]. This is evidenced in Fig. 7 (b), where it can be seen that at lower temperatures more $H_2$ effuses out of the stack than for the two films measured separately. Conversely, at higher temperatures a smaller amount effuses out. Consequently, it must be concluded that the presence of a $p^+$-type overlayer likely enhances Si-H bond rupture in the intrinsic buffer layer. As a result, also for the $c$-Si / $a$-Si:H($i$) / $a$-Si:H($p^+$) structure, high dangling bond densities may be generated not only in the doped over layer, but also close to the $c$-Si / $a$-Si:H interface, already at relatively low values for $T_{ann}$.

## Application of a-SiO:H p-i layers

Thus, a-Si:H has been employed for the passivation of c-Si, whereas it is still an open question that a-Si:H is the best material for heterojunction. A disadvantage of a-Si:H is the absorption of short wavelength light and the partially epitaxial growth on the c-Si substrate. In order to overcome these problems, we have developed a-SiO:H/c-Si heterojunction structures. For the deposition of the a-SiO:H p-i layers, a rather high growth temperature of 180 °C was used to improve the film quality [25]. Figure 8 shows the variation of the solar cell performance with $CO_2$ flow rate in the a-SiO:H i layer. In Fig. 8, the oxygen contents in the a-SiO:H i layers were estimated to be 4 at.% ([$CO_2$]=1 SCCM) and 7 at.% ([$CO_2$]=2 SCCM). For the deposition of the a-SiO:H p layer, [$CO_2$]=2 SCCM was also used.

At [$CO_2$]=0 SCCM, formation of the epi-Si layer occurs, since the growth temperature is 180 °C in Fig. 4. Quite interestingly, we find that a small flow rate of [$CO_2$]=0.5 SCCM ([$CO_2$]/[$SiH_4$]=10 %) is sufficient to suppress the Si epitaxial growth completely. As shown in

Fig. 8, we obtained the best efficiency of 16.0 % when the oxygen content in the a-SiO:H i layer is 4 at.% ([CO$_2$]=1 SCCM). This efficiency can be compared with 15.8 % obtained from the optimized a-Si:H p-i layers. The greatest advantage of the a-SiO:H p-i layers, however, is suppression of the epi-Si growth on c-Si substrates by the a-SiO:H i layer. In particular, the a-SiO:H i layer enables us to change the deposition conditions freely without forming epi-Si layers. The application of the a-SiO:H i layer further improves the reproducibility for solar cell fabrication, since the optimum deposition conditions of a-Si:H i layers lie near a-Si:H/epi-Si phase boundaries [14]. Thus, by applying a-SiO:H p-i layers, high-efficiency solar cells can be fabricated more easily, compared with conventional a-Si:H/c-Si solar cells.

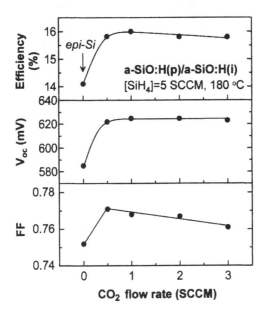

**Fig. 8** Variation of the solar cell performance with CO$_2$ flow rate in the a-SiO:H i layer [25].

## SUMMARY

In summary, we discussed how interface properties of $a$-Si:H/$c$-Si structures may determine the electronic passivation behavior in conjunction with the solar cell performance. The beneficial combination of a-Si:H($p,n$) and a-Si:H(i) for passivation is explained in terms of the defective interface with a doped layer and the excellent interface with an intrinsic layer. Post annealing treatments have been found to be a good tool to unravel the physical mechanism of passivation. The annealing studies demonstrate the need for a careful assessment of process temperature during $c$-Si/$a$-Si:H heterostructure device fabrication. For intrinsic film deposition, the deposition temperature should be sufficiently low to prevent epi-Si growth. For doped layer deposition, care has to be taken not to generate harmful defects already at moderate temperatures. The novel material a-SiO:H demonstrates the advantage of the prevention of harmful epitaxial growth and less optical absorption.

## Acknowledgements

The authors would like to thank Drs. M. Tanaka, M. Taguchi and A. Terakawa (Sanyo Electric) for their fruitful discussion. This work is partly supported by NEDO.

## REFERENCES

[1] H. Matsuura, T. Okuno, H. Okushi, and K. Tanaka, J. Appl. Phys. 55, 1012 (1984).

[2] M. Tanaka, M. Taguchi, T. Matsuyama, T. Sawada, S. Tsuda, S. Nakano, H. Hanafusa, and Y. Kuwano, Jpn. J. Appl. Phys. 31, 3518 (1992).

[3] E. Yablonovitch, T. Gmitter, R.M. Swanson, and Y.H. Kwark, Appl. Phys. Lett. 47, 1211 (1985).

[4] M. Taguchi, A. Terakawa, E. Maruyama, and M. Tanaka, Prog. Photovolt: Res. Appl. 13, 481 (2005).

[5] S. De Wolf, G. Agostinelli, G. Beaucarne, and P. Vitanov, J. Appl. Phys. 97, 063303 (2005).

[6] H. Fujiwara, M. Kondo, and A. Matsuda, J. Appl. Phys. 91 (2002) 4181.

[7] H. Fujiwara and M. Kondo, J. Appl. Phys. 101 (2007) 054516.

[8] H. Fujiwara, J. Koh, P. I. Rovira, and R. W. Collins, Phys. Rev. B 61 (2000) 10832.

[9] H. Fujiwara, J. Koh, P. I. Rovira, and R. W. Collins, Phys. Rev. B 61 (2000) 10832.

[10] R.A. Sinton and A. Cuevas, Appl. Phys. Lett. 69, 2510 (1996).

[11] M. Taguchi, K. Kawamoto, S. Tsuge, T. Baba, H. Sakata, M. Morizane, K. Uchihashi, N. Nakamura, S. Kiyama, and O. Oota, Prog. Photovolt: Res. Appl. 8 (2000) 503.

[12] M. W. M. van Cleef, J. K. Rath, F. A. Rubinelli, C. H. M. van der Werf, R. E. I. Schropp, and W. F. van der Weg, J. Appl. Phys. 82 (1997) 6089.

[14] H. Fujiwara and M. Kondo, Appl. Phys. Lett. 90 (2007) 013503.

[15] S. De Wolf and M. Kondo, in *Proceedings of the 4th World Conference on Photovoltaic Energy Conversion*, Waikoloa, Hawaii (IEEE, Piscataway, NJ, 2006), p. 1469.

[16] S. De Wolf and M. Kondo, Appl. Phys. Lett. 90, 042111 (2007).

[17] D.K. Biegelsen, R.A. Street, C.C. Tsai, and J.C. Knights, Phys. Rev. B 20, 4839 (1979).

[18] T.H. Wang, E. Iwaniczko, M.R. Page, D.H. Levi, Y. Yan, H.M. Branz, and Q. Wang, Thin Solid Films 501, 284 (2006).

[19] D.H. Levi, C.W. Teplin, E. Iwaniczko, Y. Yan, T.H. Wang, and H.M. Branz, J. Vac. Sci. Technol. A 24, 1676 (2006).

[20] W. Beyer, H. Wagner, and H. Mell, Solid State Comm. 39, 375 (1981).

[21] R.A. Street, C.C. Tsai, J. Kakalios, and W.B. Jackson, Philos. Mag. B 56, 305 (1987).

[22] W. Beyer, Physica B 170, 105 (1991).

[23] S. De Wolf and G. Beaucarne, Appl. Phys. Lett. 88, 022104 (2006).

[24] S. De Wolf and M. Kondo, Appl. Phys. Lett. 91, 112109 (2007).

[25] H. Fujiwara, T. Kaneko and M. Kondo, Appl. Phys. Lett. 91 (2007) 133508.

Mater. Res. Soc. Symp. Proc. Vol. 1066 © 2008 Materials Research Society 1066-A03-03

# Correlation of Hydrogen Dilution Profiling to Material Structure and Device Performance of Hydrogenated Nanocrystalline Silicon Solar Cells

Baojie Yan[1], Guozhen Yue[1], Yanfa Yan[2], Chun-Sheng Jiang[2], Charles W. Teplin[2], Jeffrey Yang[1], and Subhendu Guha[1]

[1]United Solar Ovonic LLC, 1100 West Maple Road, Troy, MI, 48084

[2]National Renewable Energy Laboratory, 1617 Cole Blvd, Golden, CO, 80401

## ABSTRACT

We present a systematic study on the correlation of hydrogen dilution profiles to structural properties materials and solar cell performance in nc-Si:H solar cells. We deposited nc-Si:H single-junction solar cells using a modified very high frequency (VHF) glow discharge technique on stainless steel substrates with various profiles of hydrogen dilution in the gas mixture during deposition. The material properties were characterized using Raman spectroscopy, X-TEM, AFM, and C-AFM. The solar cell performance correlates well with the material structures. Three major conclusions are made based on the characterization results. First, the optimized nc-Si:H material does not show an incubation layer, indicating that the seeding layer is well optimized and works as per design. Second, the nanocrystalline evolution is well controlled by hydrogen dilution profiling in which the hydrogen dilution ratio is dynamically reduced during the intrinsic layer deposition. Third, the best nc-Si:H single-junction solar cell was made using a proper hydrogen dilution profile, which caused a nanocrystalline distribution close to uniform throughout the thickness, but with a slightly inverse nanocrystalline evolution. We have used the optimized hydrogen dilution profiling and improved the nc-Si:H solar cell performance significantly. As a result, we have achieved an initial active-area cell efficiency of 9.2% with a nc-Si:H single-junction structure, and 15.4% with an a-Si:H/a-SiGe:H/nc-Si:H triple-junction solar cell structure.

## INTRODUCTION

Hydrogen dilution is a key parameter for controlling the transition from a-Si:H to nc-Si:H. The structure analyses showed that nc-Si:H films deposited with constant hydrogen dilution normally exhibit an initial amorphous incubation layer, followed by nucleation and growth of cone-like nanocrystalline clusters. During the growth of the nanocrystalline cone-structures, the crystalline volume fraction and average grain-size increase with the film thickness [1-4], a phenomenon called nanocrystalline evolution. It was also found that nc-Si:H solar cells made with a hydrogen dilution close to the amorphous/nanocrystalline transition showed a better performance than those with a higher crystalline volume fraction [5]. The nc-Si:H materials with a high crystalline volume fraction normally have a high density of micro-voids and cracks, which results in ambient degradation caused by impurity diffusion into the material [6]. Obviously, with a constant hydrogen dilution, it is difficult to maintain the material structure close to the transition regime throughout the sample thickness. In order to solve the nanocrystalline evolution problem, we have previously reported a hydrogen dilution profiling method, where the hydrogen dilution ratio is dynamically reduced during the deposition [7]. We also showed that this method is an effective way to increase the solar cell efficiency. In this paper, we present a systematic study of material structure and device performance for nc-Si:H solar cells made with various hydrogen dilution profiles.

## EXPERIMENT

nc-Si:H *n-i-p* solar cells were made on bare stainless steel (SS) and Ag/ZnO back reflector coated SS substrates. The doped layers as well as *n/i* and *i/p* buffer layers were deposited using RF glow discharge under the same conditions for all samples. The nc-Si:H intrinsic layer was made using a modified very high frequency (MVHF) glow discharge technique at high rates with various hydrogen dilution profiles. Before the nc-Si:H deposition, an optimized seeding layer was made with very high hydrogen dilution to reduce the amorphous incubation layer. The solar cell performance was characterized using current density versus voltage (J-V) measurements under an AM1.5 solar simulator at 25°C and quantum efficiency (QE) measurements under the short-circuit condition for short-circuit current density ($J_{sc}$) calibration. The material structure properties were characterized using Raman spectroscopy, cross-sectional transmission electron microscopy (X-TEM), atomic force microscopy (AFM), and conductive AFM (C-AFM).

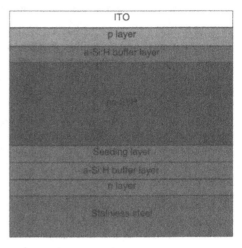

**Figure 1.** schematic of sample structure.

## RESULTS AND DISCUSSION

We made four nc-Si:H solar cells on bare SS to compare the effect of hydrogen dilution profiling on solar cell performance, electronic structure, and microstructure. The sample structure is illustrated in Fig. 1. On top of the *n* layer, an a-Si:H buffer layer was used to reduce P-incorporation into the nc-Si:H intrinsic layer. Then, a seeding layer made with very high hydrogen dilution was deposited to promote the nucleation in the nc-Si:H intrinsic layer. An a-Si:H buffer layer was inserted between the nc-Si:H intrinsic layer and the *p* layer to reduce the shunt current and improve the open circuit voltage ($V_{oc}$) [8,9]. During the intrinsic layer growth, the hydrogen flow rate was kept constant and the same value was used for all of the samples, whereas the $Si_2H_6$ flow rate decreased as different functions of time, which produced different degrees of hydrogen dilution profiling. The total amount of $Si_2H_6$ flow during the intrinsic layer deposition was adjusted to be the same for all of the samples, and therefore the average hydrogen dilution ratio was the same.

Figure 2 shows the Raman spectra of the four nc-Si:H samples excited by (a) a green laser with a wavelength of 532 nm and by (b) a red laser with a wavelength of 633 nm. Normally, the Si Raman TO mode in nc-Si:H can be de-convoluted into three components: a broad amorphous peak at ~480 cm$^{-1}$, an intermediate peak at ~510 cm$^{-1}$, and a crystalline peak at ~520 cm$^{-1}$. The intermediate peak could be assigned to the scattering from distorted Si-Si bonds in the grain boundaries or in the grains with a size smaller than 3 nm. The crystalline volume

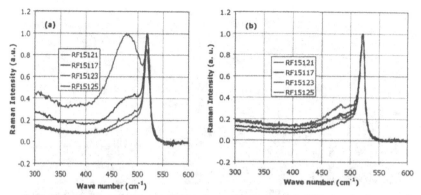

**Figure 2.** Raman spectra of four nc-Si:H solar cells made with different hydrogen dilution profiles. The measurements were made with (a) a green laser of 532-nm wavelength and (b) a red laser with 632.8-nm wavelength.

fraction $(f_c)$ is calculated from the ratio of the component areas obtained by de-convolution of Raman spectra. Table I lists the results of the Raman spectrum de-convolutions and the crystalline volume fraction for the four samples with the two lasers. Normally, one believes that the green laser probes the material structure in the top surface region and the red laser measures the material structure in the bulk of the film. Because the a-Si:H buffer layer made a large contribution to the Raman spectra measured with the green laser, the overall crystalline volume fractions obtained from the green laser Raman spectra are lower than those from the red laser Raman spectra. However, a clear trend was observed that with the increase of the degree of hydrogen dilution profiling, the nanocrystalline volume fraction from the green laser Raman spectra was decreased, especially the one with the highest degree of hydrogen dilution profiling showed a large amorphous component. The Raman spectra excited by the red laser is supposed to detect the bulk structure in the intrinsic layer; however, the contribution from the top layer was still greater than from the bottom layer, therefore the crystalline volume fractions measured with the red laser also deceased with the increase of the degree of hydrogen dilution profiling. From the Raman measurement, we conclude that the hydrogen dilution profiling indeed controls the nanocrystalline volume fraction as reported previously [7].

**Table I.** Raman spectrum de-composition parameters and crystalline volume fractions of nc-Si:H samples made with different hydrogen dilution profiles, where $a$, $i$, and $c$ represent the amorphous, intermediate, and crystalline peaks.

| Run No. | $\lambda_{ex.}$ (nm) | Area | | | Peak (cm$^{-1}$) | | | $f_c$ (%) | Dilution profiling |
|---|---|---|---|---|---|---|---|---|---|
| | | $a$ | $i$ | $c$ | $a$ | $i$ | $c$ | | |
| 15125 | 532 | 9.1 | 4.9 | 7.9 | 487.4 | 513.0 | 520.5 | **58.4** | None |
| | 632.8 | 7.3 | 5.1 | 6.5 | 487.2 | 514.9 | 521.4 | **61.4** | |
| 15123 | 532 | 12.7 | 5.3 | 8.3 | 481.4 | 510.0 | 519.0 | **51.7** | Low |
| | 632.8 | 8.5 | 4.8 | 6.2 | 483.5 | 514.9 | 521.4 | **56.3** | |
| 15117 | 532 | 21.8 | 5.0 | 7.9 | 479.9 | 508.5 | 519.0 | **37.2** | Middle |
| | 632.8 | 11.1 | 4.8 | 7.1 | 482.5 | 513.1 | 520.5 | **51.8** | |
| 15121 | 532 | 43.0 | 14.4 | 5.6 | 470.8 | 499.5 | 519.0 | **31.7** | High |
| | 632.8 | 15.1 | 4.9 | 6.5 | 482.5 | 514.0 | 521.4 | **43.0** | |

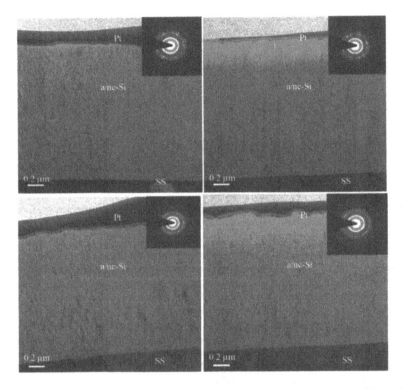

**Figure 3.** X-TEM images of the four samples made with different hydrogen profiles, where the upper left picture is from the sample made with no hydrogen dilution profiling, the upper right from the low hydrogen dilution profiling, the lower left from the middle hydrogen dilution profiling, and the lower right from the high hydrogen dilution profiling.

In order to confirm the Raman results, the samples were characterized using X-TEM. Figure 3 shows the medium resolution X-TEM pictures of the four samples. It appears that all of the samples show some elongated structures perpendicular to the substrates. Based on the analyses in the literature [10], these structures are assigned to large nanocrystalline clusters called large grains. If this assignment is correct, the following observations are made. First, there is no incubation layer in the $n/i$ interface region. The 70-nm thick smooth region near the substrate is the $n$ layer and the a-Si:H buffer layer. It indicates that the seeding layer effectively removed the amorphous incubation layer, which is normally observed in nc-Si:H deposition. Second, the sample made with no hydrogen dilution profiling showed some degree of increase in the nanocrystalline volume fraction as confirmed by the high resolution X-TEM, where extra elongated structures were mostly found in the top half of the sample than the bottom half of the sample. Second, the sample made with the low hydrogen dilution profiling showed a nanocrystalline distribution near uniform throughout the film thickness with a tendency of decrease in the nanocrystalline volume fraction the film thickness. Fourth, the inversed nanocrystalline evolution becomes clear for the samples with the middle hydrogen dilution profiling and the high hydrogen dilution profiling. In particular, the sample made with the high hydrogen dilution profiling has a 200-300 nm thick layer in the top region with no indication of nanocrystallite inclusions. High resolution images reveal that the top layer is indeed amorphous silicon.

We took the AFM and C-AFM images on the top surfaces of the nc-Si:H solar cell made with constant hydrogen dilution. The AFM topographies are very similar for the four samples, with large hill-like structures, corresponding to the clusters of nanograins. While, the C-AFM images showed large current spikes in the samples made with the constant hydrogen dilution and such current spikes were significantly reduced in the samples with hydrogen dilution profiles, especially in the sample with high hydrogen dilution profile. Our previous results showed that a thick a-Si:H $p/i$ buffer layer effectively reduces the shunt current [8, 9]. In the current situation, the extra amorphous layer caused by the hydrogen dilution profile acts as a thick a-Si:H buffer layer.

We used hydrogen dilution profiles similar to those listed in Table I for nc-Si:H single-junction solar cells on Ag/ZnO coated SS substrates. We found that that the cell with a no hydrogen dilution profiling shows the lowest efficiency due to a lower $V_{oc}$ and poorer FF, which could arise from a high crystallinity near the $i/p$ interface and a large shunt current. The cell made with the high hydrogen dilution profiling also shows a poor performance. The FF is low due to a high series resistance caused by the amorphous component near the $i/p$ interface, as shown by the Raman and X-TEM results. A moderate degree of hydrogen dilution profiling improves the cell performance remarkably; in particular the cell with the low hydrogen dilution profiling results in the highest cell efficiency. We further optimized the average hydrogen dilution and improved the efficiency to 9.2% in nc-Si:H single-junction solar cells deposited on Ag/ZnO back reflectors. We have used the optimized nc-Si:H as the bottom cell in an a-Si:H/a-SiGe:H/nc-Si:H triple-junction structure and achieved an initial active-area efficiency of 15.4% with J-V characteristics and QE curves shown in Fig. 4.

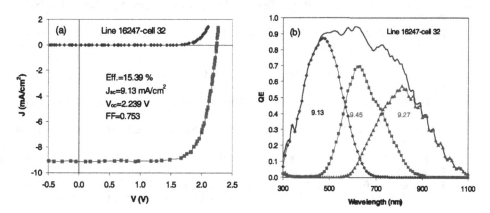

**Figure 4.** (a) J-V characteristics and (b) quantum efficiency curves of an a-Si:H/a-SiGe:H/nc-Si:H triple-junction cell with an initial active-area efficiency of 15.4%.

## SUMMARY

We have systematically studied the structural properties of nc-Si:H materials made with various hydrogen dilution profiles using Raman spectroscopy, X-TEM, AFM, and C-AFM. The solar cell performance correlates well with the material structure. Based on the experimental results, we make the following conclusions. The nc-Si:H materials do not have an amorphous incubation layer, indicating that the seeding layer is well optimized and works as per design. Even with the optimized seeding layer, some degree of nanocrystalline evolution was still observed when no hydrogen dilution profiling was used. Hydrogen dilution profiling can control the microstructure growth. The best nc-Si:H single-junction solar cell was made using an optimized hydrogen dilution profiling. We have used the optimized hydrogen dilution profiling and improved the nc-Si:H solar cell performance significantly. As a result, we have achieved an initial active-area cell efficiency of 9.2% with a nc-Si:H single-junction structure, and 15.4% with an a-Si:H/a-SiGe:H/nc-Si:H triple-junction structure.

## ACKNOWLEDGEMENTS

This work was partially supported by NREL under the Thin Film Partnership Program Subcontract No. ZXL-6-44205-14 and DOE under the Solar America Initiative Program Contract No. DE-FC36-07 GO 17053 at United Solar and by DOE under Contract No. DE-AC36-99GO10337 at NREL.

## REFERENCE:

[1]  E. Vallat-Sauvain, U. Kroll, J. Meier, A. Shah, and J. Pohl, J. Appl. Phys. 87, 3137 (2000).
[2]  Y. Nasuno, M. Kondo, and A. Matsuda, Proc. of 28th IEEE Photovoltaic Specialists Conference (IEEE, Anchorage, Alaska, 2000), p.142.
[3]  F. Finger, S. Klein, T. Dylla, A. L. Baia Neto, O. Vetterl, and R. Carius, Mater. Res. Soc. Symp. Proc. 715, 123 (2002).
[4]  C. W. Teplin, C.-S. Jiang, P. Stradins, and H. M. Branz, Appl. Phy. Lett. 92, 093114 (2008).
[5]  T. Roschek, T. Repmann, J. Müller, B. Rech, H. Wagner, Proc. of 28th IEEE Photovoltaic Specialists Conf., Anchorage, AK, September 15-22, 2000, (IEEE New York, 2000), p. 150.
[6]  B. Yan, K. Lord, J. Yang, S. Guha, J. Smeets, and J.-M. Jacquet, Mater. Res. Soc. Symp. Proc. 715, 629 (2002).
[7]  B. Yan, G. Yue, J. Yang, S. Guha, D. L. Williamson, D. Han, and C.-S. Jiang, Appl. Phys. Lett. 85, 1955 (2004).
[8]  G. Yue, B. Yan, C. W. Teplin, J. Yang, and S. Guha, J. Non.-Crystal. Solids, (2008), in press.
[9]  B. Yan, C.-S. Jiang, C.W. Teplin, H.R. Moutinho, M.M. Al-Jassim, J. Yang, and S. Guha, J. Appl. Phys. 101, 033711 (2007).
[10] J. Kočka, A. Fejfar, H. Stuchlíková, J. Stuchlík, P. Fojtík, T. Mates, B. Rezek, K. Luterová, V. Švrček, and I. Pelant, Sol. Energy Mater. & Sol. Cells 78, 493 (2003).

Mater. Res. Soc. Symp. Proc. Vol. 1066 © 2008 Materials Research Society
1066-A03-05

# MW plasma enhanced CVD of intrinsic Si for thin film solar cells

Bas B. Van Aken, Hans Leegwater, Maarten Dorenkamper, Camile Devilee, Jochen Loffler, Maurits C.R. Heijna, and Wim J. Soppe
ECN - Solar Energy, P.O. box 1, Petten, 1755 ZG, Netherlands

## ABSTRACT

The aim of the thin film silicon PV research program at ECN is the development of high-throughput production technology for high efficiency, microcrystalline and amorphous thin film silicon photovoltaics (PV) on flexible substrates. For this purpose, a roll-to-roll system has been designed and constructed, consisting of three deposition chambers for the continuous deposition of n-type, intrinsic and p-type Si layer. In this paper, we will present optical and electrical characterisation of device quality intrinsic Si layers, deposited with Microwave (MW) plasma enhanced chemical vapour deposition (PECVD), with a special focus on UV-reflection spectroscopy (UVRS). UVRS can be used to determine the crystallinity in very thin silicon layer and is interesting as a possible inline tool for layer quality assessment and crystallinity control.

## INTRODUCTION

Roll-to-roll (R2R) production of thin film Si solar cells combines the advantages of flexible substrates and high-throughput fabrication. Advantages include reduced handling costs, clean (closed) environment processing and the absence of pump-down cycles leading to potential significant cost reduction compared to batch-type processing using glass superstrates. However, in contrast to equipment adapted from display technology for the fabrication of thin Si layers on glass, equipment for roll-to-roll production is not commercially available. Therefore, ECN is developing high-throughput, roll-to-roll production technology for high efficiency thin film silicon solar cells. In our concept, steel foil, coated with an insulating barrier layer, is used as substrate. A sputtered back-contact and –reflector is applied, before the active Si layers are deposited by PECVD. On top of the Si layers, a transparent conductive oxide is sputtered. In order to achieve monolithical series connection, cells are defined and isolated by laser scribing followed by printing of insulating and conducting lines. In principle, the same concept can be used for plastic substrates. For the continuous deposition of amorphous and microcrystalline silicon n-i-p layers and thin film solar cells by PECVD, ECN and Roth&Rau AG developed a roll-to-roll system, the FLEXICOAT300.

In this contribution, we will report on intrinsic Si layers grown with MW-PECVD. We will pay particular attention to UVRS as characterization tool for determination of the crystalline fraction in thin silicon layers. The standard method for analysis of crystallinity of silicon layers is Raman spectroscopy, which, however, is less suitable for analysis of very thin layers (on top of other silicon layers), unless a UV laser is used for excitation. Further, Raman spectroscopy is rather difficult to implement as an inline monitoring tool in a PECVD system. UVRS is suited for measurement on very thin films and is easier to implement as inline monitoring tool. However, there is no standard evaluation method yet to determine the crystalline fraction in thin silicon layers from UVRS. We will present UVRS-analyses of crystalline fraction in microcrystalline silicon, based on an approach developed by Harbeke and Jastrzebski [1], and show that this method is indeed a potentially well suited tool to determine crystalline fractions.

## EXPERIMENT

The FLEXICOAT300 is designed for the continuous deposition of silicon on 300 mm wide (stainless steel) moving foil. The sources are all linear and the entire system including plasma sources can be scaled up to a width of 1500 mm. Plasma boxes shield both plasma and substrate from the water-cooled walls, in order to reduce oxygen contamination of the deposited layers due to water desorption. Gas locks are installed allowing to use different pressures in each of the deposition chambers and to prevent the contamination of the intrinsic chamber with dopant gases such as $PH_3$ and $B_2H_6$. All chambers are pumped with roots pumps allowing a base pressure in the range of $10^{-5}$ mbar

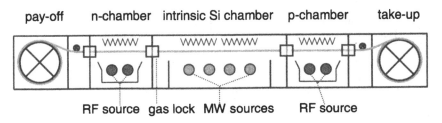

**Figure 1.** Cross-section through the FLEXICOAT300 roll-to-roll coater.

The intrinsic layer is deposited with MW-PECVD as this plasma method combines high deposition rates, uniform deposition on large areas and good layer quality [2]. High deposition rates, in the order of 1 nm/s, are needed for the deposition of the relatively thick intrinsic Si layers, ~300 nm for a-Si and 1-2 µm for µc-Si, in order to achieve high process throughput and thus low cost of ownership of the equipment. The linear microwave source consists of a linear antenna, which is placed inside a quartz tube to shield it from the plasma. Pulsed microwaves are fed in from MW power supplies on both ends of the antenna. Each MW source contains two linear arrays of magnets, along each side of the MW antenna. The magnet configuration influences the shape and extent of the plasma confinement around the source. This greatly affects the plasma physics and chemistry, and thus the layer quality and deposition rate.

While the MW sources are excellent for deposition of intrinsic silicon, they are not suitable for doped layer deposition, since parasitic deposition of conductive layers on the quartz tube will decrease the microwave power that can be coupled into the plasma. For that reason, we had to resort to another plasma source for the growth of n-type Si and p-type Si. The linear RF source that has been implemented in the Flexicoat300 has been developed by Roth&Rau AG and consists of two aluminium rods which are symmetrically connected to the 13.56 MHz RF generator, with a maximum power of 600 W [3]. In contrast to conventional asymmetric RF sources, here the substrate does not form part of the RF network and therefore does not need to be grounded. The absence of the need for grounded substrates is a significant advantage in roll-to-roll processing.

The ion energy distribution of the symmetric RF source leads to "soft" deposition conditions on the surface of the substrate. With this RF source, device quality n-type a-Si:P and µc-Si:P have been deposited at economically feasible deposition rates of 0.2-0.4 Å/s [4].

We use a wide range of characterization techniques to determine the optoelectronic quality of the deposited silicon layers, and most of them are extensively described in previous papers [4]. Raman spectra have been obtained with a Renishaw inVia Raman Microscope, using laser excitation wavelengths of 514 and 633 nm.

UV-reflection spectra are measured with a DH-2000 Deuterium Halogen light source from Top Sensor Systems, which provides a deuterium and a tungsten halogen light source in a single optical path with a stable wavelength range of 210 – 1700 nm. The light is guided through Oceanic Optics XSR optical fibers to a 30 mm Mikropack integrating sphere, which has a reflectivity of 98 % in the visible and infrared region and 95 % in the UV region. For detection of the reflected UV light an Avantes AvaSpec 2048 spectrometer, with a wavelength range of 200 – 1100 nm, is used. All measured spectra are scaled to a 99%-reflectance sample, however, the reflectivity of this sample in the UV-region is somewhat lower and less accurately known.

## RESULTS and DISCUSSION

### Si layers

Both a-Si and μc-Si layers have been deposited with MW-PECVD. Systematic variation of the MW pulse parameters, i.e., the pulse width, repetition rate and peak power, showed that shorter times between the MW pulses lead to dense, low defect a-Si layers. For instance, in continuous MW mode, a-Si layers with refractive index n ~ 3.35 and a microstructure factor R* ~ 0.12 could be grown. The properties of these layers are summarised in Table I. For reference, device quality criteria, as stated by Schropp and Zeman [5], are given as well. The layers deposited with MW-PECVD are, almost, of device quality, only the absorption coefficient at 600 nm, $\alpha(600)$, is somewhat too low.

Table I. MW-PECVD Si material parameters compared with device quality criteria [5].

| | Amorphous silicon | | Microcrystalline silicon | |
|---|---|---|---|---|
| | MW-PECVD | Device quality | MW-PECVD | Device quality |
| n (@0.5 eV) | 3.35 | - | 3.0 | ~3.4 |
| R* | 0.12 | <0.10 | - | - |
| $E_{urb}$ [meV] | 57 | <60 | 72 | <60 |
| $\alpha(600)$ [cm$^{-1}$] | $1.8\times10^4$ | $\geq 3.5\times10^4$ | - | - |
| $\sigma_d$ [S/cm] | $2.8\times10^{-11}$ | $<10^{-10}$ | $4.2\times10^{-8}$ | $<1.5\times10^{-7}$ |
| $E_{act}$ [eV] | 0.88 | ~0.8 | 0.53 | 0.53-0.57 |
| $\sigma_{ph}/\sigma_d$ | $1.5\times10^6$ | $>10^5$ | 89 | >100 |

Table I also gives the data for microcrystalline Si. Although the electrical properties are nearly of device quality, the optical parameters still need some improvement before these μc-Si layers can be incorporated in high efficiency μc-Si solar cells.

Figure 2 shows the absorption curve for a device quality intrinsic a-Si layer, by combining regular reflection/transmission measurements and FTPS experiments [6]. Note that the absorption values cover over 7 orders of magnitude. The defect density in the band tail states is quantified by the Urbach tail energy $E_{urb}$. $E_{urb}$ can be found by calculating the inverse slope below the band gap at ~1.80 eV as indicated in the graph.

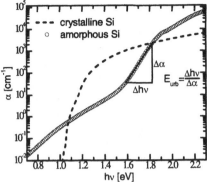

**Figure 2.** Absorption $\alpha$ as a function of the energy h$\nu$ for a device quality thin film a-Si sample. For reference, the absorption coefficient for single crystalline Si with $E_g = 1.12$ eV is shown.

## Raman and UV-reflection

Figure 3a shows typical UV-reflection spectra of a c-Si wafer, an intrinsic a-Si sample and a µc-Si sample. The c-Si spectrum shows, respectively at 281 nm and 365 nm, the $X_4$-$X_1$ and $\Gamma_{25'}$-$\Gamma_{15}$ optical interband transitions [7]. The first of these transitions is also clearly visible in the microcrystalline sample (solid line in Figure 3b), but the second is very hard to identify. As expected, the a-Si spectrum (dashed line in Figure 3b) does not show these optical transitions.

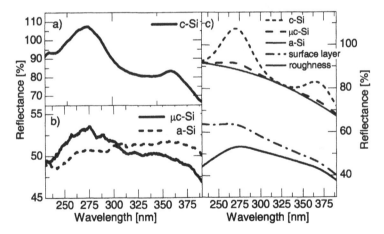

**Figure 3.** UV reflection data of different Si surfaces, scaled to the reflectance of a 99%-reflectance reference sample, with a lower reflectance in the UV. a) c-Si UV-reflectance spectrum. b) Smoothed UV-reflectance spectra for µc-Si (blue line) and a-Si (green dashes). c) Modelled effect of decreasing crystallinity, surface contamination and surface roughness on the UV-spectra. See text for an elaborate explanation of the model.

The UV-reflection spectra of thin Si films are not only influenced by the crystalline fraction, but also by the surface roughness and the presence of contamination of the surface or a *foreign* surface layer, like $SiO_x$ [1]. Roughness of the surface leads to a wavelength-dependent decrease of the reflection, whereas surface layer or contamination will decrease the reflection uniformly. A decrease in crystallinity is reflected by a reduction of the height of the two peaks. This is shown schematically in Figure 3b. The topmost curve is a numerical fit to the UV-reflection spectrum of c-Si. The next-lower curve takes the decrease in crystallinity into account for a typical μc-Si sample. For reference, the solid, black line shows the same curve for a completely amorphous sample. Next, the decrease as a result of a surface layer is added for the μc-Si curve. And finally, the Rayleigh-like scattering effect of surface roughness is included. Of course, using the reference c-Si sample and a measured UV-spectrum of a μc-Si or a-Si layer, we can calculate three parameters that quantify the three mentioned effects.

Using the above described model by Harbeke and Jastrzebski [1], we extracted a UV crystallinity parameter for a range of a-Si and μc-Si layers, deposited with MW-PECVD as shown in Figure 4a. There is a clear linear relationship between the UV crystallinity parameter and the Raman crystalline volume fraction. UV-reflection spectroscopy can thus be used to monitor the crystallinity during continuous, roll-to-roll fabrication of Si layers.

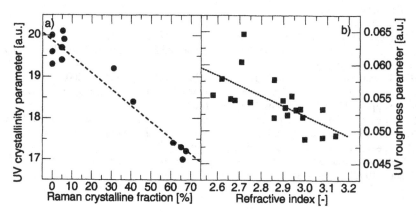

**Figure 4.** Fit parameters obtained from UV-reflection data. a) UV crystallinity parameter as a function of Raman crystalline volume fraction. b) UV roughness parameter as a function of the refractive index @ 0.5 eV for μc-Si samples with Raman crystal fraction of ~50-60%. Dotted and dashed lines are linear fits of the data, to guide the eye.

For a range of μc-Si samples with crystalline volume fractions of 50-60%, determined by Raman spectroscopy, the refractive index has been determined from FTIR spectra. Low refractive index is indicative of porous layer growth. We plotted the UV roughness parameter as a function of the refractive index. With decreasing refractive index, i.e. increasing porosity, the UV roughness parameter increases. This agrees with the expectation that porous, microcrystalline layers will also have rough surfaces.

## CONCLUSIONS

To conclude: we developed, in collaboration with Roth&Rau AG, a roll-to-roll PECVD system, the FLEXICOAT300, for the continuous deposition of silicon n-i-p layers. Intrinsic layers with sufficient quality for solar cells can be grown with MW-PECVD. We will extend the research to the fabrication of nip and pin thin film Si solar cells on steel substrates by roll-to-roll processing. To facilitate continuous roll-to-roll production of Si layers, UV-reflection spectroscopy is an easy, fast and robust tool to control the crystallinity and quality of those layers.

## ACKNOWLEDGMENTS

This work has been financially supported by the Dutch Ministry of Economic Affairs (Project No. TSIN3043) and by the European Commission under contract no. FP6-2006-Energy 3-019948.

## REFERENCES

1. G. Harbeke and L. Jastrzebski, J. Electrochem. Soc. **137**, 696 (1990).
2. J. Löffler, C. Devilee, M. Geusebroek, W.J. Soppe and H.-J. Muffler, Proc. 21st Europ. PV Sol. Energy Conf. 1597 (2006).
3. H. Schlemm, M. Fritzsche and D. Roth, Surf. Coat. Technol. **200**, 958 (2005).
4. B.B. Van Aken, C. Devilee, M. Dörenkämper, M. Geusebroek, M.C.R. Heijna, J. Löffler and W.J. Soppe, J. Non-Cryst. Solids, **354**, 2392 (2008).
5. R.E.I. Schropp and M. Zeman, *Amorphous and microcrystalline silicon solar cells*, 1st ed. (Kluwer Academic Publishers, Dordrecht, the Netherlands, 1998) p. 47.
6. M. Vanecek and A. Poruba, Appl. Phys. Lett. **80**, 719 (2002).
7. R.J. Turton, "Band structure of Si: Overview", *Properties of crystalline silicon*, ed. R. Hull (Inspec, 1999) pp. 381-382.

# Characterization

Mater. Res. Soc. Symp. Proc. Vol. 1066 © 2008 Materials Research Society     1066-A04-01

# Characterization of amorphous/crystalline silicon interfaces from electrical measurements

J. P. Kleider[1], and A. S. Gudovskikh[2]

[1]Laboratoire de Génie Electrique de Paris, CNRS UMR8507; SUPELEC; Univ Paris-Sud; UPMC Univ Paris 06, 11 rue Joliot-Curie, Gif sur Yvette, F-91192, France
[2]St.-Petersburg Physics and Technology Centre for Research and Education of the RAS, Hlopina str. 8/3, St.-Petersburg, 194021, Russian Federation

## ABSTRACT

Electrical techniques based on capacitance and conductance measurements are powerful tools for interface characterization in semiconductor heterostructures. Here we detail their application to the study of the heterointerface between hydrogenated amorphous silicon (a-Si:H) and crystalline silicon (c-Si). The main parameters governing the device applications are the conduction and valence band mismatch, and the density of interface states. The presence of a high interface states density can be revealed by capacitance versus temperature and frequency measurements. However, for very high quality interfaces that are required for instance to reach high conversion efficiencies in solar cells, the usual measurements performed in the dark and at zero or reverse bias are not sensitive enough. We show that the sensivity to interface states can be enhanced by performing capacitance measurements under illumination and at a forward bias close or equal to the open-circuit voltage. In this case, the measured capacitance is determined by the diffusion of free carriers in c-Si and limited by recombination at the interface. Regarding the determination of band offsets, the method using a plot of the inverse square capacitance as a function of bias to determine the diffusion potential from the intercept of the extrapolated linear region is shown to lead to errors even in the absence of any interface charge. This is due to the presence of a strong inversion layer in c-Si at the interface, the effect of which has been ignored so far in the literature. The presence of this strong inversion layer is evidenced from planar conductance measurements on (n) a-Si:H/(p) c-Si structures. We emphasize that these measurements are very sensitive to details of the band structure profile. In particular, it is shown that the temperature dependence of the sheet electron density allows the determination of the conduction band offset between a-Si:H and c-Si with a good precision: $\Delta E_C = E_C^{\text{a-Si:H}} - E_C^{\text{c-Si}} = 0.15 \pm 0.04$ eV.

## INTRODUCTION

Silicon heterojunction solar cells combining crystalline and thin film silicon technologies are attracting a lot of attention. This is because very high efficiencies (above 20%) have been demonstrated, while using low temperature deposition of doped silicon thin films to form the junctions and the back surface field reduces the cost and energy consumption during fabrication [1, 2]. As far as the physics is concerned, however, the precise description and characterization of the heterojunctions is still a matter of debate. In particular, there is still a lack of knowledge on the interface defect density and on the conduction and valence band offsets between hydrogenated amorphous silicon (a-Si:H) and crystalline silicon (c-Si). These are key parameters in the solar cell performance, and their knowledge is required for reliable device modelling.

Many techniques can be used to gain insight into the heterojunction physics and the corresponding relevant parameters. These can be roughly separated into optical and electrical techniques [3]. Among optical techniques one can emphasize spectroscopic ellipsometry and infrared spectroscopy which can be used for process control and for learning about the crystallinity of the deposited thin films [4], photoluminescence that is sensitive to the recombination both in the bulk and at interfaces [5], and photoyield spectroscopy [6]. The latter technique has been successfully used to determine the valence band offset between a-Si:H and c-Si [7, 8]. The conduction band offset can thus be deduced from the knowledge of the a-Si:H band gap energy. However, there is still an uncertainty arising from the possible difference between the optical gap that determines optical transitions and the mobility gap that determines electronic transport properties in amorphous semiconductors [9, 10]. This is why electrical techniques are also mandatory to characterize interface properties. Moreover, values reported in the literature for the band offsets at the a-Si:H/c-Si heterojunction are spreading over a large range [11], so there is a need for a critical review of the various techniques. Here we review electrical techniques based on capacitance and conductance measurements. Trapping and de-trapping of carriers at defect levels due to ac bias modulation is the basis of defect characterization in semiconductors using space charge spectroscopy [12, 13], either in quasi steady-state or transient (DLTS or ICTS) regimes [14, 15]. This has been widely used to characterize bulk defects in pn and Schottky junctions made on crystalline or amorphous semiconductors [16, 17], and also to characterize surface defects at the interface between crystalline silicon and insulators [18]. Here we show that space charge spectroscopy can also detect defects at the a-Si:H/c-Si interface, but with a rather poor sensitivity. We explain how and why this can be improved by using capacitance measurements at forward bias close to the open-circuit voltage and under illumination. Another aspect of capacitance is developed in the so-called $C$-$V$ method. In this method, one probes the dc bias dependence of the space charge capacitance near a depleted semiconductor interface, rather than checking for the response of gap states. This technique has been used for the characterization of band offsets at either isotype or anisotype heterojunctions [19, 20]. We discuss here the limit of the $C$-$V$ method in the case of (n) a-Si:H/(p) c-Si heterojunctions, due to the presence of a strong inversion layer at the interface. Finally, we present measurements of the planar conductance of such structures. We emphasize that these measurements are very sensitive to details of the band structure profile and show how the temperature dependence allows the unambiguous determination of the conduction band offset between a-Si:H and c-Si with a very good precision.

## SAMPLES AND TOOLS

Two types of samples are being considered here: solar cells based on (n) a-Si:H/ (p) c-Si front interfaces, and samples with parallel top coplanar electrodes, as described in Figure 1. Experimental details on the fabrication of the structures can be found elsewhere [21, 22]. In addition to the single heterojunction solar cell (SHJ) shown in Fig.1, some experimental data were also obtained on double heterojunction (DHJ) cells, where the aluminum back surface field (BSF) was replaced by a (p) c-Si/ (p) a-Si:H heterojunction. Capacitance measurements and DC current measurements were performed on the solar cells and coplanar samples, respectivley, in a liquid nitrogen cryostat pumped down to $10^{-5}$ mbar and in a wide range of temperatures.

Numerical modelling was performed using AFORS-HET v2.2 software, which calculates the current in both dc and ac modes in the drift-diffusion theory [23]. Unless otherwise specified the following parameters were used. We introduced the density of states (DOS) typical for n-type a-Si:H consisting of two exponential band tails with characteristic energies, $k_B T_C$, and $k_B T_V$, of 0.055 eV and 0.12 eV for the conduction and valence band, respectively, and with a pre-exponential factor of $2\times10^{21}$ cm$^{-3}$ eV$^{-1}$, and two Gaussian deep defect distributions of donor and acceptor nature being located at 0.58 eV and 0.78 eV above the top of the valence band, respectively, with a maximum value of $8.7\times10^{19}$ cm$^{-3}$ eV$^{-1}$ and a standard deviation of 0.23 eV. A donor doping density of $5.34\times10^{19}$ cm$^{-3}$ was introduced to reproduce the Fermi level position in a-Si:H. The influence of these parameters on the capacitance is quite weak. On the contrary, interface states play a significant role. These states have been taken into account by inserting a thin ($d = 1$ nm) defective c-Si layer at the interface. In this interface layer, the DOS, $g_{it}$, has been taken independent of energy, with states of the upper (lower) part of the gap being of acceptor (donor) type, and with capture cross sections for electrons and holes equal to $10^{-14}$ cm$^{-2}$. The interface density of states was then defined as $D_{it} = g_{it} \times d$.

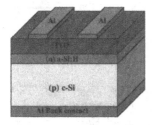 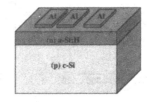

**Figure 1.** Schematic view of the two types of samples: solar cells and coplanar structures.


**RESULTS AND DISCUSSION**

<u>**Capacitance versus frequency and temperature in the dark, at zero or reverse bias**</u>

Free carrier densities can be modulated by an ac bias in the space charge region of a Schottky, pn or MIS structure, which in turn can produce a modulation of gap states occupancy that contributes to the device capacitance. This is due to the capture and release processes that can occur if their time constant is smaller than the period of the ac bias. As an example, in a Schottky structure made on an n-type semiconductor, bulk gap states can exchange electrons with the conduction band provided the angular frequency $\omega$ ($=2\pi f$) is smaller than $2e_n$, $e_n$ being the electron emission frequency given by $e_n(T) = \nu_n \exp[(E-E_C)/k_B T]$, where $\nu_n$ is the attempt-to-escape frequency (proportional to the electron capture cross section), $E$ the gap state energy, $E_C$ the bottom of the conduction band, $k_B$ Boltzmann's constant and $T$ the temperature. Thus, there is an onset temperature, $T_0$, and an onset angular frequency, $\omega_0$, related by $e_n(T_0)=\omega_0$, that mark the response of gap states to the ac modulation resulting in a capacitance ($C$) step as either the temperature is increased or the frequency is decreased. Since $e_n$ is an increasing function of temperature, increasing the frequency leads to an increase of $T_0$ resulting in a shift of the capacitance step to higher temperatures in a $C(T)$. Similarly, in a $C(\omega)$ plot, the onset is shifted to higher frequencies when the temperature is increased. It is worth stressing that, while the position of the capacitance step depends on the capture cross section of the carrier being exchanged, the amplitude of the step is directly related to the quantity of defects regardless of the capture cross section. Turning now to the a-Si:H/c-Si solar cells, one could also expect the capacitance measurement to be sensitive to interface defects that may exchange carriers with the conduction or valence band. Indeed, Figure 2 reveals capacitance steps that follow the above described dependence upon frequency and temperature. However, such steps were not observed in our best solar cells, where the capacitance is found frequency independent and only exhibits a small monotonous increase with temperature (cell#2 in Fig. 2a).

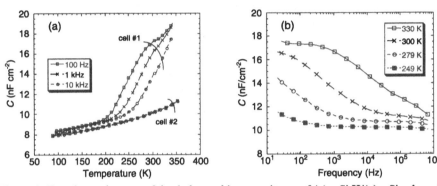

**Figure 2.** Experimental curves of the dark zero bias capacitance of (n) a-Si:H/(p) c-Si solar cells as a function of temperature (a) or frequency (b). In (a) results are shown for two different solar cells. For the best solar cell (cell#2), no capacitance step is observed.

Indeed, when the interface states density is low enough, their contribution to the total capacitance becomes negligible compared to that of the c-Si space charge capacitance, which is frequency independent. There are two other reasons for the absence of interface defect contribution in the capacitance of high quality solar cells. Indeed, defects must not only be present, they must be modulated by the ac bias. Here, the situation is different from that of an MIS structure, where capacitance measurements can reveal even very low defect densities. Indeed, while the interface states of an MIS structure are governed by the single semiconductor Fermi level and can exchange carriers only with the single semiconductor, that of a pn solar cell can be modulated by either the electron or the hole quasi Fermi level and carriers can be exchanged with either a-Si:H or c-Si. It then depends on how a modulation of the bias applied to the junction is reflected in changes in the electron and hole concentrations at the interface. As illustrated in Fig. 3, there is a strong band bending at the (n) a-Si:H/ (p) c-Si interface. We have shown that electrons can be exchanged with a-Si:H and give only a small contribution to the capacitance at low temperature. On the other hand, due to their low emission rate, holes can be exchanged with the c-Si only at much higher temperature (that may be outside the explored range). However, in case of a high interface defect density, the band bending in c-Si is reduced which increases the hole concentration, reduces their emission time at the hole quasi Fermi level, and makes the response of holes visible in the explored temperature range [24]. As can be seen in Figure 4a, numerical modelling of the solar cell capacitance shows that for values of $D_{it}$ below $5\times10^{12}$ cm$^{-2}$ eV$^{-1}$, no step in the capacitance is observed, as for our best solar cells. For higher values, we do indeed observe a capacitance step both in the $C(T)$ and in the $C(\omega)$ curves (Figures 4a and 4b, respectively), as for lower quality solar cells.

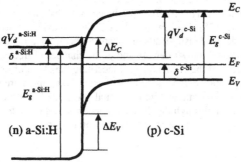

**Figure 3.** Schematic illustration of the (n) a-Si:H/ (p) c-Si band diagram at equilibrium.

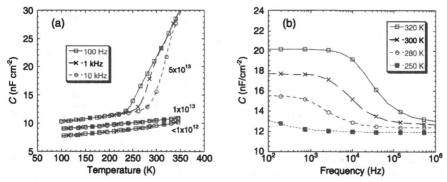

**Figure 4.** Calculated curves of the dark zero bias capacitance of (n) a-Si:H/(p) c-Si solar cells as a function of temperature (a) or frequency (b). In (a) the three sets of curves correspond to the indicated values of $D_{it}$ (in cm$^{-2}$ eV$^{-1}$). In (b) $D_{it}= 5\times10^{13}$ cm$^{-2}$ eV$^{-1}$.

## Capacitance versus frequency under illumination, at forward bias close or equal to open-circuit voltage

The sensitivity to interface states of the usual capacitance technique being rather poor, we recently suggested to measure the capacitance under strong forward bias, close to the AM1.5 open-circuit voltage of the cell and under illumination [25]. Here we want to stress that interface states are detected in these measurements in a completely different way compared to the previously discussed regime, where measurements are performed at zero or reverse bias in the dark. Indeed, in the proposed conditions, the dominant contribution to the capacitance comes from the modulation of injected free carriers, known as the diffusion capacitance. If the capacitance is measured as a function of frequency at a given temperature, one obtains a curve that may resemble that obtained in the preceding regime, namely a low frequency plateau, $C_{LF}$, followed by a drop above a turn-on frequency. However, the different behaviour compared to the preceding regime is emphasized by the dependence of $C_{LF}$ on the cell quality and by the temperature dependence as seen in Fig. 5. In the lower quality cell (#1 in Fig. 5a), $C_{LF}$ is about two orders of magnitude lower. This is also reproduced in the calculations, where $C_{LF}$ is found to decrease when $D_{it}$ increases [25]. The reason is that $C_{LF}$ is limited by interface recombination. An increase in interface defect density leads to enhanced recombination at the interface. This produces a decrease of the free carrier density and of the changes of free carrier densities with ac voltage, and thus a decrease of the capacitance. Concerning the effect of temperature, almost no frequency shift of the capacitance transition is observed if the temperature is changed. This is because the transition is not determined by the thermally activated process of gap states response to the ac modulation. Instead, through the diffusion mechanism it is related to the mobility and recombination of free carriers that depend only weakly on temperature.

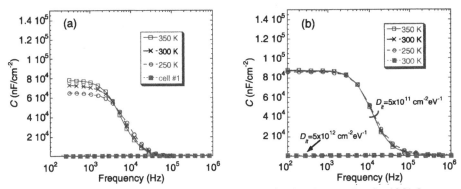

**Figure 5.** Capacitance curves versus frequency at open-circuit voltage and under AM1.5 illumination: (a) experimental, and (b) calculated. Only the curve at 300 K is shown for the lower quality cell (experimental cell#1, and $D_{it}=5\times10^{12}$ cm$^{-2}$ eV$^{-1}$ in the calculation).

An important consequence of the dependence of $C_{LF}$ on recombination is that $C_{LF}$ depends also on the interface states capture cross sections and on band discontinuities. This is illustrated in Fig. 6a. At low $D_{it}$ values (where interface recombination is negligible) $C_{LF}$ is almost independent of $D_{it}$, which indicates the detection limit. However, beyond a given value of $D_{it}$, $C_{LF}$ exhibits a sharp drop. This transition is shifted to lower $D_{it}$ values if the states have higher capture cross sections because a lower $D_{it}$ is then needed to obtain the same recombination rate. Also, if the conduction band discontinuity, $\Delta E_C$, between both semiconductors is increased, this transition is shifted to higher $D_{it}$. This is because a higher $\Delta E_C$ produces a stronger band bending, which acts as a better front surface field reducing recombination at the front interface. Higher $D_{it}$ values are then needed to reproduce the same interface recombination rate. From the Shockley-Read-Hall theory, the recombination rate through interface states can be written

$$U = \int_{E_V}^{E_C} D_{it}(E) \frac{c_n c_p \left[ np - n_i^2 \right]}{c_n n + e_n(E) + c_p p + e_p(E)} dE, \tag{1}$$

where $n$ and $p$ are the electron and hole concentrations at the interface, $n_i$ is the intrinsic electron concentration, $c_n$ ($c_p$) is the electron (hole) capture coefficient, $e_n$ ($e_p$) is the electron (hole) emission rate. Owing to the band diagram at the (n) a-Si:H/(p) c-Si interface, $c_p p$ can be neglected with respect to $c_n n$, so that:

$$U \approx c_p p \int_{E_1}^{E_2} D_{it}(E) dE, \tag{2}$$

where $E_2 = E_F^e$, $E_1 = E_i - (E_F^e - E_i) - k_B T \ln(c_n/c_p)$, $E_i$ and $E_F^e$ being the intrinsic and electron quasi-Fermi levels, respectively. This expression clearly shows that the recombination rate depends on the hole capture cross section and is independent of the electron capture cross section. The dependence on the conduction band offset is hidden in the hole concentration and in the integration energy range, $E_2 - E_1$. Indeed, if we consider that the electron and hole quasi-Fermi levels at the interface correspond to the a-Si:H and c-Si Fermi levels, respectively, an approximate expression is: $E_2 - E_1 \approx 2(E_g^{c-Si}/2 - \delta^{a-Si:H} - qV_d^{a-Si:H} + \Delta E_C)$, where notations of Fig. 3 have been used.

The demonstration that $C_{LF}$ depends on the recombination at the interface is given in Fig. 6b. We indeed observe that the various curves of Fig. 6a gather into a single curve, whatever the values of defect density, capture cross section and conduction band discontinuity.
Another important issue is that $C_{LF}$ is also sensitive to recombination at the back surface, which allows the technique to be used for the optimization of the back interface in heterojunction solar cells [25]. Finally, a last evidence for the sensitivity of $C_{LF}$ to recombination was provided by the good correlation with the band-to-band recombination photoluminescence (PL) signal [26].

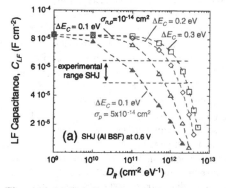 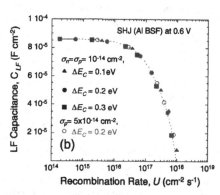

**Figure 6.** (a) Calculated low frequency capacitance as a function of interface states density, for various values of conduction band offset and capture cross sections; (b) the same data are plotted against the recombination rate at the interface, where all data align on a single curve.

## Capacitance versus DC voltage

The capacitance-voltage technique (C-V) was proposed as one of the easiest electronic technique to gain insight into band offset parameters in heterojunction structures [20]. Based on the band diagram shown in Fig. 3, the total diffusion potential across the junction, $V_d$, can be expressed as:

$$qV_d = qV_d^{\text{a-Si:H}} + qV_d^{\text{c-Si}} = E_g^{\text{c-Si}} + \Delta E_C - \delta^{\text{a-Si:H}} - \delta^{\text{c-Si}} . \qquad (3)$$

Since $E_g^{\text{c-Si}}$, $\delta^{\text{c-Si}}$, and $\delta^{\text{a-Si:H}}$ are known or can be estimated from other measurements, $\Delta E_C$ can be obtained from this equation if one can determine $V_d$. It is widely believed and assumed that $V_d$ can be found from C-V measurements, and, more precisely, from the intercept of the linear extrapolation of $1/C^2$ versus voltage (the capacitance being measured at reverse bias) with the voltage axis, $V_{int}$. Although, in principle, this method is a simple tool for the determination of energy band offsets, there are several sources of errors related to non-ohmic contacts, bulk or interface traps, and non-uniform bulk free-carrier concentrations [20]. In the case of (n) a-Si:H/ (p) c-Si heterojunctions, there is another source of error which has been ignored so far in the literature, and that we will discuss here, namely the presence of an electron rich inversion layer. In Fig. 7a we present an example of experimental plot of $1/C^2$ versus applied bias measured at 1 kHz. It has a well-defined linear behavior at reverse bias and the slope allows one to deduce the doping level in c-Si, $N_a = 7.5 \times 10^{14}$ cm$^{-3}$, in good agreement with the nominal resistivity of the silicon wafer. The intercept of the linear extrapolation with the bias axis, $V_{int}$, is equal to 0.65 V. $\delta^{\text{a-Si:H}}$ being estimated at 0.2 eV from the a-Si:H conductivity, application of Eq. (3) using $V_{int}$ for $V_d$ would yield $\Delta E_C \approx 0$. In the simulations, we always obtained a quite good linear dependence of $1/C^2$ versus applied bias, and the error in the determination of $V_{int}$ can be estimated at less than 0.05 V. An example of calculated curve (for $N_a = 7.5 \times 10^{14}$ cm$^{-3}$) is also presented in Fig. 7a. It perfectly reproduces the experimental data. However, this curve was calculated with $\Delta E_C = 0.15$ eV, which is significantly larger than the value deduced from Eq. (3). The results of simulations made for $N_a = 10^{16}$ cm$^{-3}$ and $N_a = 10^{15}$ cm$^{-3}$ and different values of $\delta^{\text{a-Si:H}}$ are presented in Fig. 7b, where the calculated values of $V_{int}$ are plotted versus $\Delta E_C$. According to Eq. (3), if $V_{int} = V_d$, $V_{int}$ should increase linearly with $\Delta E_C$. In contrast, we observe that $V_{int}$ increases only for low values of $\Delta E_C$ but it saturates and becomes independent of $\Delta E_C$ at a value $V_{int}^{\text{sat}}$ above a critical value $\Delta E_C^{\text{sat}}$. $V_{int}^{\text{sat}}$ is lower for the lower doping level in c-Si and curves are shifted towards higher values of $\Delta E_C$ with increasing $\delta^{\text{a-Si:H}}$. The observed saturation of $V_{int}$ is caused by the existence of a strong inversion layer in c-Si at the (n) a-Si:H/ (p) c-Si interface meaning that the electron concentration, $n$, at the interface becomes higher than the doping level in c-Si, $N_a$. For low values of $\Delta E_C$ where no inversion layer is present, the usual theory of p-n junctions based on the depletion layer approximation applies. The measured capacitance being then inversely proportional to the depletion layer in c-Si, the C-V method is working properly and we find $V_{int} \approx V_d^{\text{c-Si}} \approx V_d$. In this regime, an increase of $\Delta E_C$ will lead to the same increase in $V_d^{\text{c-Si}}$ and in $V_{int}$. On the other hand, when a strong inversion layer at the c-Si interface is present at equilibrium, the potential drop in the depletion region of c-Si is fixed and does not depend any more on $\Delta E_C$. Indeed, when increasing $\Delta E_C$ further, the band bending increases within the inversion layer and within a-Si:H, while it does not change in the c-Si depletion region. In this strong inversion regime, the C-V method will allow us to determine only a part of the diffusion potential in c-Si since the other part is across the strong inversion layer and the a-Si:H depletion region [27]. Using $V_{int}$ for $V_d$ in Eq. (4) will thus lead to underestimate $\Delta E_C$.

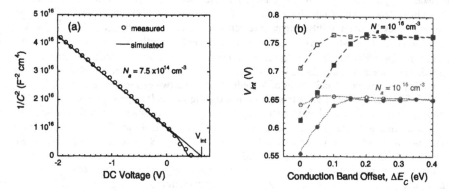

**Figure 7.** (a) Inverse square capacitance in the dark as a function of the DC voltage applied to the solar cell at 300 K and 1 kHz. Symbols show experimental data, while the full line shows the calculated curve. (b) Calculated intercept voltage of the linear extrapolation of the $1/C^2$ vs voltage plot as a function of conduction band offset, for two values of doping density of c-Si. Open and full symbols are for $\delta^{a\text{-}Si:H} = 0.2$ eV and $\delta^{a\text{-}Si:H} = 0.3$ eV, respectively.

## Planar conductance measurements

We now turn to the measurements and modelling of the planar conductance of the second type of structures of Fig. 1. We have compared the conductance measured on (n) a-Si:H/(p) c-Si samples with that of (n) a-Si:H/glass samples. The latter exhibits values and a temperature dependence typical for (n) a-Si:H films, with an activation energy of 0.17 eV. The coplanar conductance of (n) a-Si:H/(p) c-Si samples is several orders of magnitude higher, and the temperature dependence is much weaker (activation energy of 0.018 eV). This is not caused by the conduction of holes through the (p) c-Si substrate which cannot be revealed because the current flowing from one electrode to the other through the c-Si wafer has to cross one reverse biased (n) a-Si:H/ (p) c-Si junction. It is due to a high planar electron conductance at the c-Si surface. This was proved by the drastic decrease of conductance and increase of its activation energy observed after the a-Si:H layer had been removed by dry etching [22]. Thus, the presence of an electron rich inversion layer at the c-Si surface suggested in the previous section was experimentally demonstrated by our coplanar conductance measurements. This inversion layer occurs due to the strong band bending (see Fig. 3), and it is related to the conduction band offset. Its value is determined by the sheet electron density, $N_s$, being the integral of the electron concentration over the c-Si wafer thickness, $L$, the main contribution to this integral coming from the strong inversion layer. Simulations of (n) a-Si:H/(p) c-Si band diagrams for various temperatures allowed us to calculate the temperature dependence of $N_s$. In Fig. 8a the Arrhenius plots of the calculated $N_s$ for different values of $\Delta E_C$ are presented. $N_s$ increases with increasing $\Delta E_C$ and its temperature dependence becomes weaker with a lower activation energy, $E_a$. This is because $E_a$ is determined by the difference $E_C - E_F$ in c-Si near the interface, which becomes smaller with increasing $\Delta E_C$ due to a stronger band bending. Experimental values of $N_s$ can be easily deduced from the measured conductance, $G$, using $N_s = L\,G/(q\,\mu_n\,h)$, $h$ being the length of the electrode, $q$ the unit electron charge and $\mu_n$ the electron mobility. The electron mobility may be quite different in a strong inversion layer as compared to the bulk material. We thus used two extreme cases for the mobility temperature dependence: (i) $\mu_n(T) = \mu_{n,300K}\,(T/300)^{-2.42}$, with $\mu_{n,300K} = 1500$ cm$^2$ V$^{-1}$s$^{-1}$ which is characteristic of high quality, lightly doped bulk c-Si [28], and (ii) $\mu_n(T) = \mu_{n,300K} = 500$ cm$^2$ V$^{-1}$s$^{-1}$ without any temperature dependence, which accounts for the lowest value and weakest temperature dependence reported for a MOS inversion layer in the literature [29]. The experimental values of $N_s$ deduced in these two extreme cases are also presented in Fig. 8a, and we observe that the corresponding values for $E_a$ are 0.07 and 0.018 eV, respectively. The true value and temperature dependence in our case should be within these two extreme cases. In Fig. 8b the calculated dependence of $E_a$ on $\Delta E_C$ is shown. This allows a precise determination of the conduction band offset between a-Si:H and c-Si. Indeed, from the estimated range for the experimental values of $E_a$ we obtain $\Delta E_C = 0.15$ eV $\pm\,0.04$ eV. This is consistent with previous determinations from photoelectron spectroscopy [7, 8]. However, our determination is more precise and does not suffer from uncertainties due to the possible difference between the optical and mobility gaps in a-Si:H, because it is based only on electrical measurements in the dark.

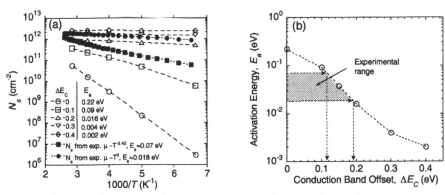

**Figure 8.** (a) Dependence of the electron sheet concentration upon inverse temperature calculated for different values of $\Delta E_C$ (open symbols) and deduced from the measured planar conductance for two extreme cases of the mobility temperature dependence (full symbols); the activation energies are also indicated; (b) Calculated dependence of the electron sheet concentration activation energy on conduction band offset $\Delta E_C$, showing also the range deduced from the measurements.


## CONCLUSIONS

Quasi-stationnary measurements of the capacitance in the dark at zero dc bias can be used for a rough estimation of the interface quality. Indeed, a step observed in $C(T)$ or $C(\omega)$ curves that shifts when changing the frequency or the temperature is a signature of trapping and release of carriers at defects that reveals a rather high interface defect density (higher than $5\times10^{12}$ cm$^{-2}$ eV$^{-1}$). The sensitivity to interface defects is improved when the capacitance is measured at strong forward bias (close to or at open-circuit voltage) and under AM1.5 illumination. In these conditions the capacitance originates from injected free carriers that can diffuse in the c-Si wafer. For high quality c-Si wafers (with minority carrier lifetimes above a few hundreds of microseconds), the value of the low-frequency capacitance is then determined by recombination at both interfaces. This is interesting since the technique can be used to perform a comparative study of the quality of either the front (n) a-Si:H/(p) c-Si anisotype interface or the back (p) c-Si/(p) a-Si:H isotype interface in a-Si:H/c-Si double heterojunction solar cells. However, absolute values of defect densities cannot be obtained because capture cross sections and band discontinuities also determine the recombination rate at the interface.

We have shown that the current analysis of the conventional $C$-$V$ method should be reconsidered in (n) a-Si:H/(p) c-Si heterojunction diodes due to the presence of a strong inversion layer at the c-Si interface. Indeed, when a strong inversion layer is present, the potential corresponding to the intercept of the linear part of $1/C^2$ versus bias is limited to the potential drop in the c-Si depletion region, which only depends on the doping level of c-Si and is independent of $\Delta E_C$. This can lead to underestimate the conduction band offset from the $C$-$V$ method.

The presence of an electron rich inversion surface layer at the (n) a-Si:H/(p) c-Si interface was experimentally demonstrated by coplanar conductance measurements. This inversion layer occurs at the c-Si surface due to the strong band bending, and is related to the conduction band offset. The temperature dependence of the sheet electron density allows a precise determination of the conduction band offset between a-Si:H and c-Si, which is a key parameter for the physics of the a-Si:H/c-Si interface. Our measurements yield: $\Delta E_C = 0.15 \pm 0.04$ eV. Being based only on electrical measurements in the dark, the proposed method does not suffer from uncertainties due to the possible difference between the optical and mobility gaps in a-Si:H. It can thus be reliably used for modelling a-Si:H/c-Si based devices.


## ACKNOWLEDGMENTS

This work was partly supported by ANR (Agence Nationale de la Recherche) and ADEME (Agence de l'Environnement et de la Maîtrise de l'Énergie) in the framework of several french photovoltaic projects, as well as by CNRS and the Russian Foundation for Basic Research in the framework of a joint Russian-French project.

## REFERENCES

1. M. Taguchi, A. Terakawa, E. Maruyama and M. Tanaka, *Prog. Photovolt. Res. Appl.* **13**, 481 (2005).
2. M. Tanaka, M. Taguchi, T. Matsuyama, T. Sawada, S. Tsuda, S. Nakano, H. Hanafusa and Y. Kuwano, *Jap. J. Appl. Phys.* **31**, 3518 (1992).
3. F. Capasso and G. Margaritondo (Eds.), *Heterojunction Band Discontinuities – Physics and Device Applications* (North-Holland, Amsterdam, 1987).
4. H. Fujiwara and M. Kondo, *Appl. Phys. Lett* **86**, 032112 (2005).

5. S. Tardon, M. Rösch, R. Brüggemann, T. Unold and G. H. Bauer, *J. Non-Cryst. Solids* **338-340**, 444 (2004).
6. L. Ley, *J. Non-Cryst. Solids* **114**, 238 (1989).
7. M. Sebastiani, L. Di Gaspare, G. Capellini, C. Bittencourt and F. Evangelisti, *Phys. Rev. Lett.* **75**, 3352 (1995).
8. M. Schmidt, L. Korte, A. Laades, R. Stangl, Ch. Schubert, H. Angermann, E. Conrad and K.V. Maydell, *Thin Solid Films* **515**, 7475 (2007).
9. C.R. Wronski, S. Lee, M. Hicks and S. Kumar, *Phys. Rev. Lett.* **63**, 1420 (1989).
10. M. Vanecek, J. Stuchlik, J. Kocka and A. Triska, *J. Non-Cryst. Solids* **77-78**, 299 (1985).
11. M. Rösch, R. Brüggemann and G.H. Bauer in *Proc. of the 2$^{nd}$ World Conference and Exhibition on Photovoltaic Solar Energy Conversion*, edited by J. Schmid, H.A. Ossenbrink, P. Helm, H. Ehmann and E.D. Dunlop (1998), pp. 964-967.
12 D. V. Lang in *Thermally Stimulated Relaxation in Solids*, edited by P. Braunlich, Topics in Applied Physics Vol. 37 (Springer-Verlag, Berlin, 1979), p. 93.
13. L. Losee, *J. Appl. Phys.* **46**, 2204 (1975).
14. D. V. Lang, *J. Appl. Phys.* **45**, 3023 (1974).
15. H. Okushi and Y. Tokumaru, *Jap. J. Appl. Phys.* **19**, L335 (1980).
16. D. V. Lang, J. D. Cohen and J. P. Harbison, *Phys. Rev. B* **25**, 5285 (1982).
17. J. P. Kleider, *Thin Solid Films* **427**, 127 (2003).
18. E. H. Nicollian and J. R. Brews, *MOS (Metal Oxide Semiconductor) Physics and Technology* (Wiley-Interscience, 1982).
19. H. Kroemer, Wu-Yi Chien, J. S. Harris and D. D. Edwall, *Appl. Phys. Lett.* **36**, 295 (1980).
20. S. R. Forrest, in Ref. [3], pp. 311-375.
21. Y. Veschetti, J.-C. Muller, J. Damon-Lacoste, P. Roca i Cabarrocas, A.S. Gudovskikh, J. P. Kleider, P.-J. Ribeyron and E. Rolland, *Thin Solid Films* **511-512**, 543 (2006).
22. J. P. Kleider, M. Soro, R. Chouffot, A. S. Gudovskikh, P. Roca i Cabarrocas, J. Damon-Lacoste, D. Eon and P.-J. Ribeyron, *J. Non-Cryst. Solids*, in press.
23. R. Stangl, M. Kriegel and M. Schmidt, in *Conference Record of the 2006 IEEE 4th World Conf. on Photovoltaic Energy Conversion* (IEEE, Piscataway, NJ, USA, 2006), pp. 1350-1353.
24. A. S. Gudovskikh, J. P. Kleider, J. Damon-Lacoste, P. Roca i Cabarrocas, Y. Veschetti, J.-C. Muller, P.-J. Ribeyron and E. Rolland, *Thin Solid Films* **511-512**, 385 (2006).
25. A. S. Gudovskikh and J. P. Kleider, *Appl. Phys. Lett.* **90**, 034104 (2007).
26. R. Chouffot, S. Ibrahim, R. Brüggemann, A. S. Gudovskikh, J. P. Kleider, M. Scherff, W.R. Fahrner, P. Roca i Cabarrocas, D. Eon and P.-J. Ribeyron, *J. Non-Cryst. Solids*, in press.
27. A. S. Gudovskikh, S. Ibrahim, J. P. Kleider, J. Damòn-Lacoste, P. Roca i Cabarrocas, Y. Veschetti and P.-J. Ribeyron, *Thin Solid Films* **515**, 7481 (2007).
28. C. Jacoboni, C. Canali, G. Ottaviani and A. Quaranta, *Solid-State Electron.* **20**, 77 (1977).
29. J. P. Kleider, A. S Gudovskikh and P. Roca i Cabarrocas, *Appl. Phys. Lett.*, in press.

Mater. Res. Soc. Symp. Proc. Vol. 1066 © 2008 Materials Research Society  1066-A04-02

## Probing Carrier Depletions on Grain Boundaries in Polycrystalline Si Thin Films by Scanning Capacitance Microscopy

C.-S. Jiang, H.R. Moutinho, B. To, P. Dippo, M.J. Romero, and M.M. Al-Jassim
National Renewable Energy Laboratory, Golden, CO, 80401

### ABSTRACT

Grain boundaries (GBs) in polycrystalline Si thin-film solar cells are believed to limit the photovoltaic efficiencies. In this paper, we report on a nanometer-resolution measurement of the carrier depletion at the GBs, using scanning capacitance microscopy (SCM). The SCM images exhibit the following features: (1) Carrier concentrations are lower at locations around the GBs than on center regions of the grains; (2) The depletion width at the GBs varies considerably, between 0 and 100 nm, depending on individual GBs; (3) Intra-grain carrier depletion was also observed at point and line defects; and (4) The faceted features that were observed on the topography of the as-grown film surface appeared on the SCM images even after the film surface was polished flat. The direct measurement of the carrier depletion on the GBs demonstrates that the GBs in Si thin films indeed create charged gap states. The nonuniformity of the carrier depletions suggests that the gap states depend on specific GB structures, which should relate directly to the grain orientations and grain facets adjacent to the GB. The depletion around the intragrain defects indicates that the defects are charged and can be recombination centers, and thus, harmful to device performance. This paper reports the first step of our studies toward understanding the relationships between the electronic and structural properties on specific GBs.

### INTRODUCTION

Polycrystalline Si thin film is a promising candidate in next-generation thin-film solar cells because of its advantages in material saving, low cost, and potential high performance. In recent years, great efforts have been made to improve the quality of this material, and energy conversion efficiency of the devices steadily improved [1,2]. Enlargement of grain size and passivation of grain boundaries (GBs) in the polycrystalline material are considered effective ways to stimulate research and further improve efficiency [2–5]. Most of the literature reports that GBs in Si films harm the efficiency because GBs act as recombination centers for photo-excited carriers, and the carrier transports are scattered at the GBs by electrostatic potential barriers [2–4,6–8]. The GBs negatively impact all device performance parameters, which include short-circuit current density ($J_{sc}$), open-circuit voltage ($V_{oc}$), and fill factor (FF) [3,6–8]. On the other hand, positive impacts of GBs were proposed by two-dimensional device simulation and by quantum efficiency analysis, suggesting that the efficiency would be enhanced if the GBs are doped to a considerable length-scale by diffusion of impurities in the emitter [9,10]. In this case, the minority carriers become majority carriers at the type-inversed GB, and such minority-carrier collections can suppress the negative impact such as potential fluctuation at the GBs.

Numerous studies have been reported on the electronic properties of polycrystalline Si thin films at GBs, such as energy position in the bandgap, density of the gap states, minority-carrier recombination velocity, electrostatic potential barriers, and carrier scattering at the GBs [3,4,11–18]. However, most of the measurements and characterizations were carried out on a large section of sample that contained a huge number of grains on the order of micrometers. Therefore, the measurements resulted from collective effects from the huge number of GBs, and cannot resolve the electronic property at specific GBs, which depends on the specific GB structures. In this paper, we tackle this problem by reporting on our recent nanometer-resolution characterizations of the electronic properties at the GBs. These nanometer-resolution measurements allow us to examine the electronic properties on individual GBs. Our ultimate goal is to understand the relationships between the electronic and atomic structures on individual GBs. As a first step, we observed the carrier depletion behavior and resolved various carrier depletions at individual GBs.

## EXPERIMENT

Si thin films 500 nm thick were deposited by hot-wire chemical vapor deposition (HWCVD) from pure $SiH_4$ gas onto ~100-nm-thick silicon seed layers formed by aluminum-induced crystallization (AIC) on glass substrates [19] (seed layers prepared at the Hahn-Meitner-Institute in Germany). The deposition process and epitaxial alignment of the HWCVD-grown layer have been described previously [20,21]. Typically, the individual grains are ~10 μm in size. The film is unintentionally doped by residual phosphorus in the growth chamber and aluminum from the AIC process. Secondary ion mass spectrometry measurement shows that Phosphorus is the dominant dopant in the $n$-type layer grown by HWCVD, and aluminum in the $p$-type seed layer.

Scanning capacitance microscopy (SCM) measures two-dimensional distributions of carrier concentrations with spatial resolutions on the nanometer scale or about equal to the size of the atomic force microscopy (AFM) tip apex [22,23]. The SCM measurement is based on the contact mode of AFM. Simultaneously with the AFM topographic image, the metal-oxide-semiconductor (MOS) capacitance was measured using an ultra-high-frequency (~1 GHz) resonant capacitance sensor. The MOS structure consists of the AFM tip, oxide layer on top of the Si film, and the film. The high-frequency capacitance-voltage (C-V) curve describes the capacitance either increasing or decreasing monotonously with the voltage for $n$- or $p$-type semiconductors. With this C-V relationship, changes in the MOS capacitance, driven by an ac voltage (0.5-1.0 V) with an intermediate frequency (10–100 kHz) is measured by a lock-in amplifier at the frequency. The absolute value of the SCM signal dC/dV is smaller with a higher carrier concentration because of its smaller rate of change on the C-V curve, and the dC/dV has opposite signs for $n$- and $p$-type samples. A negative 0.5-1.0 V dc voltage was also applied to the sample to drive the MOS to a flat band condition, which is for the optimization of dC/dV signal. A uniform and less-defected oxide layer on top of the film is necessary to obtain reliable SCM signal. A conventional way to achieve this layer is to polish the film surface chemical-mechanically by using silica colloid with fine ~50-nm particles. After polishing, ultraviolet irradiation treatment while heating the sample at ~300°C was expected to improve the quality of the oxide layer, as reported in ref. [23].

## RESULTS AND DISCUSSIONS

The AFM and SCM measurements were done on an as-grown Si film, as shown in Figs. 1(a) and 1(b). The AFM topographic image shows that the surface morphology varies from grain to grain, which is considered to depend on the grain orientations. The AFM images show several types of surface morphologies, which may correspond to the low-index orientations in the film growth direction. However, the simultaneously obtained SCM image does not show clear

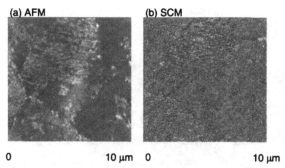

(a) AFM　　　　　　　　(b) SCM

0　　　　　　10 μm　0　　　　　　　10 μm

Fig. 1. (a) An AFM topographic image and (b) the corresponding SCM image taken on an as-grown polycrystalline Si thin-film.

features except for the topographic effect; the image shows large contrasts on regions with rough topographies.

To get a high-quality oxide layer and minimize the topographic effect, the film surface was chemical-mechanically polished. An AFM image and the corresponding SCM image on a polished sample are shown in Figs. 2(a) and 2(b). From the AFM image, we know that the film surface is indeed polished flat; the surface corrugation was reduced from ~50 nm before to less than 2 nm after the polishing. On the SCM image, carrier depletions on the GBs clearly appeared as the dark lines corresponding to larger absolute dC/dV values. All the dC/dV values in this image are negative and a larger absolute dC/dV value represents a lower carrier concentration or a carrier depletion. Although changing phase in the lock-in amplifier inverts the sign of SCM dC/dV signal, the signs at both close and away the GBs are always the same, indicating carrier depletions but no carrier inversions. Because the film surface is flat, the GBs cannot be identified from the AFM image. However, the GBs can be recognized from the high-quality SCM image. They can be identified from the depletion on the GBs or from the subtle topography-like textures on the grains. We will explain the topography-like textures later. An identification of the grains is shown in Fig. 2(c), as identified from Fig. 2(b).

The SCM image shows that the GBs indeed exhibit carrier depletions and that the carrier depletion behavior is not uniform among the GBs. For example, the GBs between the grains A/C, B/C, and A/B in Figs. 2(b) and 2(c) exhibit clear carrier depletions, and the depletions on the GBs between grains A/C and B/C are deeper than that of A/B. However, the GBs between grains C/D and C/E do not exhibit significant depletions. A deeper depletion represents a larger GB charge density and a wider depletion width. This nonuniformity of the GB depletions demonstrates that the electronic properties of the GBs vary, possibly due to specific GB structures that are decided by grain orientations and the facets adjacent to the GB. A zoomed-in SCM image is shown in Fig. 3(a), and a profile across the line in Fig. 3(a) is shown in Fig. 3(b). From the line profile, we know that the depletion width at each side of the GB is ~100 nm. This is the maximum depletion width among the GBs measured on this sample.

Most research indicates that the electrically active GBs are one of the main causes for losses in the conversion efficiency of Si thin-film solar cells, because they negatively $J_{sc}$, $V_{oc}$, and FF [3,6–8]. Recombination of photo-excited carriers with significantly large recombination velocity ($>\sim10^4$ cm/s) at the GBs makes the minority-carrier diffusion length ($<\sim$ μm) smaller than the

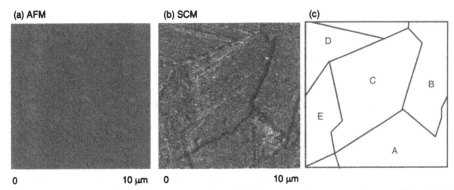

Fig. 2. (a) An AFM and (b) the corresponding SCM images taken on a polycrystalline Si thin-film after the film surface was polished. (c) illustrates the grains and grain boundaries as recognized from Fig. 2(b).

film thickness [3,4,6–8]. However, Beaucarne et al. reported an enhanced minority-carrier collection by the preferentially doped GBs [9]. Similar effects of the GBs on the minority-carrier collection was also reported on polycrystalline Cu(In,Ga)Se$_2$ solar cells [24]. In this positive role of the GBs, preferential doping on the columnar GBs is the key factor [9,10]. On the other hand, it was reported that GB passsivation by post-hydrogenation effectively improved J$_{sc}$ [3,4]. In addition, control of grain orientations and subsequently the GB structures, as well as increasing the grain size, are expected to relieve negative impacts of the GBs. Small-angle or low-index GBs with fewer defects and dangling bonds are expected to reduce the recombination at the GBs. Our SCM measurement on the GBs suggests that the electrical properties depend on detailed GB structures. Smaller recombination velocities are expected in GBs without detectable carrier depletion resulting from fewer defects and lower charges.

V$_{oc}$ values of the devices are negatively impacted by the GBs through the increase in dark saturation current density (J$_0$) [3,4]. Most studies believe that the carrier transport mechanism in the $p$-$n$ junction is governed by carrier recombination at the GBs, where the defect states act as recombination centers [4]. A few studies reported that it is governed by the recombination at intragrain defects [25]. In addition, band-edge fluctuations at both conduction and valence bands increase J$_0$ by reducing the excitation energy [26]. Both the electrostatic potential fluctuation induced by the charges at the GBs and the bandgap fluctuation at the GBs due to local defect configurations result in the band-edge fluctuations at the conduction and valence bands [26]. Similar to the effects on J$_{sc}$, hydrogenation passivation of the GBs significantly improved V$_{oc}$ [2,3], and our SCM measurement suggests a possible improvement of V$_{oc}$ by controlling the GB structure.

In addition to the depletion on the GBs, the SCM images also exhibit intragrain defects and texture-like features on grains. An example of the SCM images is shown in Fig. 4. Four examples of the intragrain defects are indicated in the image: two of them are line defects, labeled LD, and two others are point defects, labeled PD. We note that the intragrain defects are not observed on every SCM image. This is because only the defects at the interface of oxide-layer/Si-film or in the region close to the interface with a distance less than the depletion width can be detected by SCM. If the defects are deeper in the film bulk than the depletion width, they are screened by carriers and cannot be detected. Similar to the GBs, these charged intragrain

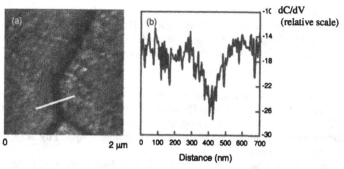

Fig. 3. (a) A zoomed-in SCM image taken on a polished polycrystalline Si thin film and (b) a line profile of SCM dC/dV along the white line in (a).

defects can be active recombination centers for minority carriers and can result in potential fluctuations. Therefore, the intragrain defects are considered harmful to photovoltaic performance.

The texture-like features on the grains (Fig. 4) originate from nonuniformity of the oxide layer on top of the film. These features are similar to the topographic features on the film surface (Fig. 1(a)) before the surface was polished. For example, fine stripes on grain A in Fig. 4 trend in the direction from top left to bottom right, and they trend from top right to bottom left on grain B. Although the surface was polished flat, during the polishing, the oxide layer may have been grown nonuniformly, following the facet features of the grain orientations. This nonuniformity of the oxide layer may include layer thickness and charged defects at the interface between the oxide layer and the film, which are sensitive to the SCM measurements.

## SUMMARY

Carrier depletions on individual GBs in the polycrystalline Si thin films were measured and resolved using the SCM technique. The measurement demonstrates that the carrier-depletion behavior is highly nonuniform among the GBs, suggesting various electronic properties of the GBs that possibly relate to the specific GB structures. This paper reports the first step of our studies toward understanding the relationships between the electronic and structural properties on specific GBs

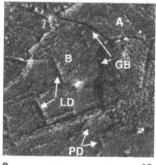

Fig. 4. An SCM image showing that, in addition to the GBs, there are intragrain defects of line defects (LD) and point defects (PD); also shown are the surface topography-like textures on the grains.

ACKNOWLEDGMENTS

This work was supported by DOE under Contract No. DE-AC36-99GO10337. The authors thank Charles W. Teplin for growing the epitaxial HWCVD layers and Stefan Gall at Hahn-Meitner-Institute in Germany for providing the AIC silicon seed layer on glass.

REFERENCES

[1]   A.G. Aberle, Proc. 4$^{th}$ World Conf. Photovoltaic Energy Conversion, Hawaii, 2006, p.1481.
[2]   P.A. Basore, Proc. 19$^{th}$ European Photovoltaic Solar Energy Conf., Paris, 2004, p.455.
[3]   T. Yamazaki, Y. Uraoka, and T. Fuyuki, Thin Solid Films 487, 26 (2005).
[4]   G. Beaucarne, S. Bourdais, A. Slaoui, and J. Poortmans, Appl. Phy. A79, 469 (2004), and the references therein.
[5]   N. Kawamoto, A. Matsuda, N. Matsuo, Y. Seri, T. Nishimori, Y. Kitamon, H. Matsumura, H. Hamada, and T. Miyoshi, Jpn. J. Appl. Phys. 45, 2726 (2006).
[6]   K. Kurobe, Y. Ishikawa, Y. Yamamoto, T. Fuyuki, and H. Matsunami, Solar Energy Material & Solar Cells 65, 201 (2001).
[7]   T. Matsui, T. Yamazaki, A. Nagatani, K. Kino, H. Takakura, and Y. Hamakawa, Solar Energy Materials & Solar Cells 65, 87 (2001).
[8]   Y. Ishikawa, Y. Yamamoto, T. Hatayama, Y. Uraoka, and T. Fuyuki, Jpn. J. Appl. Phys. 40, 6783 (2001).
[9]   G. Beaucarne, S. Bourdais, A. Slaoui, and J. Poortmans, Proc. 28$^{th}$ IEEE PVSC, Alaska, 2000, p.128.
[10]  E. Christoffel, M. Rusu, A. Zerga, S. Bourdais, S. Noël, and A. Slaoui, Thin Solid Films 403-403, 258 (2002).
[11]  J.Y.W. Seto, J. Appl. Phys. 46, 5247 (1975).
[12]  B. Warren, N.M. Jackson, and D.K. Biegelsen, Appl. Phys. Lett. 43, 195 (1983).
[13]  J. Werner and M. Peisl, Phys. Rev. B31, 6881 (1985).
[14]  Y. Alpern and J. Shappir, J. Appl. Phys. 63, 2694 (1988).
[15]  H. Hasegawa, M. Arai, and Y. Kurata, J. Appl. Phys. 71, 1462 (1992).
[16]  F. Cleri, P. Keblinski, L. Colombo, S.R. Phillpot, and D. Wolf, Phys. Rev. B57, 6247 (1998).
[17]  S. Ostapenko, Applied Physics A69, 225 (1999).
[18]  W. Choi, V. Matias, J.-K. Lee, and A.T. Findikoglu, Appl. Phys. Lett. 87, 152104 (2005).
[19]  S. Gall, J. Schneider, J. Klein, K. Hübener, M. Muske, B. Rau, E. Conrad, I. Sieber, K. Petter, K. Lips, M. Stöger-Pollach, P. Schattschneider, and W. Fuhs, Thin Solid Film 511-512, 7 (2006).
[20]  Q. Wang, C.W. Teplin, P. Stradins, B. To, K.M. Jones, and H.M. Branz, J. Appl. Phys. 100, 093520 (2006).
[21]  C.W. Teplin, H.M. Branz, K.M. Jones, M.J. Romero, P. Stradins, and S. Gall, Mat. Res. Soc. Symp. Proc. 989, 133 (2006).
[22]  C.C. Williams, W.P. Hough, and S.A. Rishton, Appl. Phys. Lett. 55, 203 (1989).
[23]  V.V. Zavyalov, J.S. McMurray, and C.C. Williams, Rev. Sci. Instruments 70, 158 (1999).
[24]  C.-S. Jiang, R. Noufi, K. Ramanathan, J.A. AbuShama, H.R. Moutinho, and M.M. Al-Jassim, Appl. Phys. Lett. 85, 2625 (2004).
[25]  R. Brendel, R.B. Bergmann, B. Fischer, J. Krinke, R. Plieninger, U. Rau, J. Reiss, H.P. Strunk, H. Wanka, and J. Werner, Proc. 26$^{th}$ IEEE PVSC, California, 1997, p.635.
[26]  J.H. Werner, J. Mattheis, and U. Rau, Thin Solid Films 480-481, 399 (2005).

Mater. Res. Soc. Symp. Proc. Vol. 1066 © 2008 Materials Research Society 1066-A04-03

## Characterization of the Mobility Gap in μc-Si:H Pin Devices

Bart Elger Pieters, Sandra Schicho, and Helmut Stiebig
Institut für Energieforschung - Photovoltaik, Forschungszentrum Jülich, Leo-Brandt-Straße,
Jülich, 52428, Germany

## ABSTRACT

For the mobility gap of hydrogenated micro-crystalline silicon (μc-Si:H) a value near 1.1 eV is commonly found, similar to the bandgap of crystalline silicon. However, in other studies mobility gap values have been reported to be in the range of 1.48-1.59 eV. Indeed, for accurate modeling of μc-Si:H solar cells it is paramount that key parameters like the mobility gap are accurately determined. In this work we will discuss a method to determine the mobility gap of μc-Si:H using the dark current activation energy of μc-Si:H *pin* devices, and apply this method to μc-Si:H solar cells with varying crystalline volume fraction. We found the mobility gap is around 1.2 eV to 1.26 eV for μc-Si:H solar cells with a crystalline volume fraction between 50 % and 70 %. For a highly crystalline solar cell we found a mobility gap of 1.07 eV.

## INTRODUCTION

For the application in multi-junction thin-film solar cells μc-Si:H is a promising material. In comparison to hydrogenated amorphous silicon (*a*-Si:H), μc-Si:H is more stable to light exposure and has a high spectral response in the red wavelength region. In 1994 IMT Neuchâtel presented the "micromorph" concept consisting of an *a*-Si:H top cell and a μc-Si:H bottom cell [1].

Modeling and characterization of μc-Si:H silicon is complicated by its complex micro structural properties involving a mixed phase of crystalline and amorphous tissue, grain boundaries, inhomogeneity in the growth direction and columnar growth of the grains. The complex structural properties give rise to complex properties of electronic transport in the material, impeding a detailed analysis of its optoelectronic properties and make it sometimes difficult to determine proper values for model parameters. In particular the value of the mobility gap is controversial. Commonly a value of 1.1 eV is found, similar to the bandgap of crystalline silicon [2]. However, in other studies mobility gap values have been reported to be in the range of 1.48-1.59 eV [3,4], depending on crystalline volume fraction. We will present a method to determine the mobility gap of μc-Si:H using the dark current activation energy of μc-Si:H *pin* devices, and apply the method to a series of μc-Si:H *pin* solar cells with varying crystalline volume fraction.

In the next section we will derive an analytical expression for the thermal activation energy of μc-Si:H *pin* diodes. The derivation presented here is similar to the derivation of the dark J-V characteristics of *a*-Si:H *pin* devices from Berkel et al. [5]. However, in our derivation we include several additional temperature effects. As the derivation includes many approximations, we compared the derived analytical expression with numerical simulations to verify the results are accurate. Furthermore the numerical simulations allow for a detailed comparison between simulations and experimental results. Finally we determine and discuss the mobility gap of various μc-Si:H solar cells with varying crystalline volume fraction.

## THEORY

The dark current can be written as the integral of the recombination through the device including surface recombination at the contacts. Under low forward bias conditions recombination in $a$-Si:H and $\mu c$-Si:H pin devices will take place primarily in the intrinsic layer. In the doped layers the low minority carrier concentrations limit the recombination rate and surface recombination does not play a major role due to the many defect states that are always present in $a$-Si:H and $\mu c$-Si:H materials. The dark current of $a$-Si:H and $\mu c$-Si:H pin devices will therefore be primarily determined by the intrinsic layer and therefore we can neglect recombination in the doped layers and surface recombination.

For simplicity we assume that the capture cross-section of trap states, $\sigma_R$, is the same for electrons and holes, and we apply the Tailor and Simmons 0 K approximation [6] to describe the Shockley-Read-Hall (SRH) recombination. We write the dark current through the *pin* device as the integral of the recombination rate in the intrinsic layer:

$$J = q\int_0^W Rdx \approx qv_{th}\sigma_R \int_0^W \frac{np}{n+p} \int_{E_{fpt}}^{E_{fnt}} g(E)dEdx,$$ (1)

where $q$ is the electronic charge, $W$ is the width of the intrinsic layer, $R$ is the recombination rate as a function of the position, $x$, $v_{th}$ is the thermal velocity, $n$ and $p$ are the electron and hole concentrations in the device, respectively, $E_{fnt}$ and $E_{fpt}$ are the quasi-Fermi levels for trapped electrons and holes, respectively, and $g(E)$ is the one-electron density of states distribution in the device.

To solve the integral in Equation 1 analytically we assume a uniform electric field and constant quasi-Fermi levels in the device with a separation equal to the applied voltage. As the electric field and the quasi-Fermi levels are constant, the separation between the quasi-Fermi levels and their respective bands become linear functions of position. We can write for the carrier concentrations in the intrinsic layer:

$$n = n_0 \exp\left(-\frac{E_0(x-x_0)}{kT}\right),$$ (2)

$$p = p_0 \exp\left(\frac{E_0(x-x_0)}{kT}\right),$$ (3)

where, $E_0$ is the uniform electric field in the intrinsic layer, $n_0$ and $p_0$ are the electron and hole concentration respectively at position $x_0$ in the intrinsic layer.

For convenience we define $x_0$ such that $n_0 = p_0$. The recombination rate scales with the term $np/(n+p)$ (see Equation 1). By substituting Equation 2 and 3 in the term $np/(n+p)$ it can be seen that the recombination rate peaks at $x_0$, where the carrier concentrations are equal, and decays exponentially in either direction away from $x_0$. Therefore the integral of the recombination over the intrinsic layer is dominated by the recombination in the region where the electron and hole concentrations are approximately equal. We introduce the term "density of active recombination centers", $N_R(x)$, defined as the integral of $g(E,x)$ between the quasi-Fermi levels for trapped charge. For simplicity we assume that the density of active recombination centers does not vary much in the region where the recombination is high, and thus we take a position-independent density of active recombination centers in Equation 1. Using Equation 2 and Equation 3, and substituting $n_0 = p_0$ in Equation 1 gives:

$$J = q v_{th} \sigma_R N_R \int_0^W \frac{n_0}{\exp\left(-\dfrac{E_0(x-x_0)}{kT}\right) + \exp\left(\dfrac{E_0(x-x_0)}{kT}\right)} dx \approx q v_{th} \sigma_R N_R \frac{n_0}{E_0} kT\pi \tag{4}$$

In Equation 4, the thermal velocity and the effective density of states in the valence and conduction band are temperature dependent, following the expressions [7]:

$$v_{th}(T) = \sqrt{\frac{3kT}{m_e^*}}, \text{ and} \tag{5}$$

$$N_{c,v}(T) = N_{c,v}^{T_0} \left(\frac{T}{T_0}\right)^{\frac{3}{2}}, \tag{6}$$

where, $m_e^*$ is the effective electron mass and $N_{c,v}^{T_0}$ is the effective density of states in the conduction or valence band at temperature $T_0$. Substituting these temperature dependent relations in Equation 4, and substituting $n_0 = N_c(T)\exp\left(\dfrac{V - E_\mu}{2kT}\right)$ we obtain for the current density:

$$J \propto N_R T^3 \exp\left(\frac{V - E_\mu}{2kT}\right). \tag{7}$$

Equation 7 is essentially the diode equation with an ideality factor of 2. Note, however, that the density of active recombination centers around $x_0$, $N_R$, is also voltage dependent, and thus the ideality factor of $\mu c$-Si:H diodes will in general be less than two. As $N_R$ is expected to show little temperature dependence, the term will not affect the dark current activation energy much (i.e. the slope of the Arrhenius plot). From the Arrhenius plot of Equation 7 we obtain the activation energy as:

$$E_a = \frac{E_\mu - V}{2} + 3kT \tag{8}$$

Note that the denominator in Equation 8 is equal to the ideality factor of Equation 7 without the influence of $N_R$. This "thermal ideality factor" is therefore a good indicator whether the previous derivation holds, or whether other effects such as surface recombination or series resistance influence the current.

## RESULTS AND DISCUSSION

### Determining the mobility gap in μc-Si:H *pin* solar cells

In the derivation of Equation 8 numerous approximations have been made. In order to verify that Equation 8 is indeed accurate, we compared Equation 8 with numerical simulations and the experimentally obtained dark current activation energy of a μc-Si:H solar cell.

We prepared a μc-Si:H solar cell that consisted of a glass substrate covered with ZnO, a 15-20 nm μc-Si:H p-layer, a 1.15 μm intrinsic layer, and a 15-20 nm a-Si:H n-layer and an Al back contact. In order to accurately measure the current at low applied voltages through a μc-Si:H solar cell, it is important to prevent lateral currents through the n-layer [8]. To this end we removed the silicon in the surrounding of the contacts, using Reactive Ion Etching (RIE), where the aluminum back contacts serve as a mask. This way we obtain isolated solar cells on a ZnO covered glass substrate.

We measured the dark current at 4 temperatures, 289 K, 298 K, 308 K, and 323 K. Figure 1.a shows the dark current at 298 K. From the Arrhenius plot at each voltage we determined the thermal activation energy of the dark current, shown in Figure 1.b. From Figure 1.b it can be seen that the dark current activation energy has the predicted slope of -0.5 between 0 V and 0.4 V. Above 0.4 V other effects such as series resistance start to influence the dark current activation energy (see also Figure 1.a). Using Equation 8 we determined the mobility gap as 1.19 eV, which corresponds with the dashed line in Figure 1.b. We fitted the simple diode equation to the dark J-V characteristics from 0 V up to 0.3 V and obtained an ideality factor of 1.6 (dashed line in Figure 1.a). The thermal ideality factor in the same voltage range is 2 (see Figure 1.b). The J-V characteristics under AM 1.5 illumination are shown in Figure 1.c.

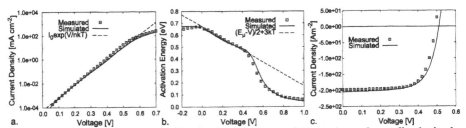

**Figure 1.** Comparison between simulated solar cell characteristics and experimentally obtained characteristics. (a) Measured and simulated dark current density. The diode equation with an ideality factor of 1.6 is indicated. (b) Measured, simulated and theoretical (see Equation 8) dark current activation energy as a function of voltage. (c) The simulated and measured J-V characteristics under AM1.5 illumination.

Using the device simulator ASA [9], we simulated the dark current activation energy, the dark current density and the current density under AM1.5 illumination, using the determined mobility gap of 1.19 eV. It can be seen from Figure 1, that a good match is obtained between simulation, experiments and the analytical expression of Equation 8. The close match between the numerical simulations and Equation 8, indicates that the used approximations in the derivation of Equation 8 do not affect the obtained activation energy much between 0 V and 0.35 V.

## The crystalline volume fraction and the mobility gap

In order to investigate the influence of the crystalline volume fraction on the mobility gap we prepared a series of solar cells, similar to the solar cells described above. However, in order to vary the crystalline volume fraction we varied the silane concentrations during the deposition of the intrinsic layer. To evaluate material properties using Raman spectroscopy, we prepared a second sample in parallel, for which we omitted the n-layer and back contact. A third sample consisted of a crystalline wafer where we deposited the intrinsic $\mu c$-Si:H layer in parallel to the other two samples, which was used to evaluate the hydrogen content using Fourier Transform Infra Red (FTIR) spectroscopy. The hydrogen concentration is of interest as in a-Si:H films the optical bandgap is related to the hydrogen concentration [10]

The crystalline volume fraction was determined from Raman measurements where we used a wavelength of 647 nm ($X_c^{red}$) and a wavelength of 488 nm ($X_c^{blue}$) in order to evaluate the average crystallinity and the crystallinity at the surface, respectively. The hydrogen concentration in the deposited intrinsic $\mu c$-Si:H layers was evaluated from FTIR spectroscopy measurements by integrating the peak at 640 cm$^{-1}$. Using the same procedure as discussed above we determined the mobility gap of the $\mu c$-Si:H solar cells from the dark current activation energy of the solar cells with aluminum back contacts.

Figure 2.a shows the obtained mobility gaps versus the crystalline volume fractions. The determined mobility gaps vary from 1.07 eV to 1.26 eV. The mobility gap of the sample with the highest crystalline volume fraction (sample 6 with $X_c^{blue}$ is 82 % and $X_c^{red}$ is 74 %) is 1.07 eV, which is significantly lower than the mobility gap of all other samples. The mobility gap of the other samples varied between 1.23 eV to 1.26 eV, without a clear correlation with the crystalline volume fraction. The crystalline volume fraction for these samples varied between 50 % and 70 %. Figure 2.b shows the determined mobility gap versus the hydrogen concentration in the film. The hydrogen concentration of the most crystalline sample could not be determined as the $\mu c$-Si:H pealed off before a FTIR measurement could be carried out. Figure 2.b shows no clear correlation between the hydrogen concentration and the mobility gap of the samples.

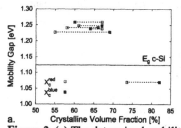
a.
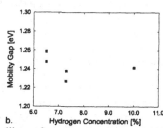
b.

**Figure 2.** (a) The determined mobility gap versus the crystalline volume fraction determined by Raman measurements with a red laser, open squares, and a blue laser, closed squares. For each sample the two corresponding crystalline volume fractions are connected with a dotted line. (b) The mobility gap versus the hydrogen concentration in the film.

As it is well known that the mobility gap of a-Si:H depends strongly on the hydrogen concentration this is an indication the mobility gap in $\mu c$-Si:H is primarily determined by the crystalline phase rather than the amorphous phase, which is in line with the percolation theory of conduction in $\mu c$-Si:H [11] which states that for sufficiently crystalline films conduction primarily takes place through the crystalline phase.

We find values for the mobility gap which are significantly lower than the values reported by Xu *et al.* [3] (1.59 eV) and Hamma *et al.* [4] (1.48 eV and 1.55 eV, for material with a crystalline fraction of 70 % and 30 %, respectively). The found values for the mobility gap in the range of 1.2-1.26 eV are, however, larger than the bandgap of crystalline silicon (1.12 eV). Xu *et al.* [3] explain the difference between the mobility gap of $\mu c$-Si:H and the bandgap of crystalline silicon with quantum-size effects. However, Carius *et al.* [12] report the optical bandgap $\mu c$-Si:H does not show a large influence of quantum-size effects and is close to the bandgap of crystalline silicon [12]. A large difference between mobility gap and optical gap implies a large number of localized states and thus would render the material quite useless for device applications. The reported mobility gaps in the range of 1.48-1.59 eV are therefore contradicting with the observed optical properties by Carius *et al.* [12]. The here reported values for the mobility gap are more in line with the observed optical properties, although a significant difference exists between the observed mobility gap values of 1.2-1.26 eV and the bandgap of crystalline silicon. The reported value of 1.07 eV for the mobility gap of a highly crystalline sample indicates that for highly crystalline samples transport may take place through the band-tail states.

## CONCLUSIONS

We derived that the dark current activation energy provides an accurate method to determine the mobility gap of $\mu c$-Si:H *pin* diodes. Using the dark current activation energy we determined the mobility gap of $\mu c$-Si:H solar cells with varying crystalline volume fraction. We found the mobility gap is around 1.2 eV to 1.26 eV for $\mu c$-Si:H solar cells with a crystalline volume fraction between 50 % and 70 %. For a highly crystalline solar cell we found a mobility gap of 1.07 eV.

## ACKNOWLEDGMENTS

The authors wish to thank Kah-Yoong Chan, C. Sellmer, R. van Aubel and F. Birmans.

## REFERENCES

1. J. Meier, *et al.*, *proc. IEEE 1st. WCPEC*, pp. 409-412, 1994.
2. T. Brammer, *et al.*, *Mater. Res. Soc. Symp. Proc. 644*, pp. A19.1-6, 2001
3. X. Xz, *et al.*, *Appl. Phy. Lett.*, vol. 67, pp 2323-2325, 1995.
4. S. Hamma, *et al.*, *Appl. Phys. Lett.*, vol. 74, pp. 3218-3220, 1999.
5. C. van Berkel, *et al.*, *J. App. Phys.*, vol 73, no 10, pp. 5264-5268, 1993.
6. J.G. Simmons, *et al.*, *Phys. Rev. B*, vol. 4, no. 2 pp.502-511, 1971.
7. S. M. Sze, *et al.*, *Physics of Semiconductor Devices.* Wiley-Interscience, 3 ed.,2006.
8. J.A. Willemen, PhD thesis, Delft University of Technology, 1998.
9. M. Zeman, *et al.*, *Sol. En. Mat. Sol.Cells*, vol. 46, pp. 81-91, 1997.
10. C.C. Tsai, *et al.*, *Solar Energy Mater.*, vol. 1, no.1-2, pp. 29-42, 1979.
11. H. Overhof, *et al.*, J. of Non-Cryst. Solids, vol. 227-230, p. 992, 1998.
12. R. Carius, *et al.*, *J. Optoelectron. Adv. Mater.*, Vol. 7, No. 1, pp. 485-489, 2005.

Mater. Res. Soc. Symp. Proc. Vol. 1066 © 2008 Materials Research Society          1066-A04-05

## Characterization of Gap Defect States in Hydrogenated Amorphous Silicon Materials

Lihong (Heidi) Jiao[1], and C. R. Wronski[2]
[1]School of Engineering, Grand Valley State University, Grand Rapids, MI, 49504
[2]Electrical Engineering, Pennsylvania State University, University Park, PA, 16802

### ABSTRACT

An enhanced simulation model based on the carrier recombination through these states was developed to characterize the gap defect states in hydrogenated amorphous silicon materials (a-Si:H). The energy dependent density of electron occupied gap states, kN(E), was derived directly from Dual Beam Photoconductivity (DBP) measurements at different bias currents. Through Gaussian de-convolution of kN(E), the energy peaks of the multiple defect states, including both neutral and charged states, were obtained. These energy levels, together with the information on the capture cross sections, were used as known input parameters to self-consistently fit the subgap absorption spectra, the electron mobility-lifetime products over a wide range of generation rates, as well as the energy dependent density of electron occupied gap state spectra. Accurate gap state information was obtained and the nature of the defect states was studied. Simulation results on light degraded hydrogen diluted, protocrystalline a-Si:H show that the density of charged states is 4.5 times that of neutral states. The two states close to the midgap act as effective recombination centers at low generation rates and play key roles in photoconductivity studies.

### INTRODUCTION

Hydrogenated amorphous silicon (a-Si:H) materials have been an interest in many device applications and it is important to understand the nature of the gap defect states in these materials. Attempts have been made in the past to characterize the gap defect states in both as-grown and light degraded states. Extensive analysis has been carried out on the subgap optical absorption spectra obtained from Constant Photocurrent Method (CPM) or Photothermal Deflection Spectroscopy (PDS) [1,2]. However, it is difficult to obtain reliable gap defect state parameters from just the subgap absorption data. Dual Beam Photoconductivity (DBP) measurements at different generation rates along with the mobility-lifetime products, $\mu\tau$, measurements have allowed detailed analysis to be carried out [3]. Due to the large number of parameters involved in the simulation, it still remained a challenge to obtain accurate information on the gap defect states. Recently, Pearce et al introduced the energy dependent density of electron occupied gap states derived directly from the DBP measurements [4]. In their approach, by first-order approximation, the densities of states in the conduction band were treated as a constant. Taking the derivative of subgap absorption, $\alpha(h\upsilon)$, the densities of electron occupied gap states were obtained from the equation $kN_{gap}(E) = (h\upsilon)(d[\alpha(h\upsilon)]/dE) + \alpha(h\upsilon)$, where $kN_{gap}$ represents the energy dependent density of electron occupied gap states that pertains to the distribution of the gap defect states. Through de-convolution of this spectrum, the energy peaks and half-widths of the multiple defect states were derived [4]. In this study, an enhanced simulation program was developed to characterize the gap defect states. The model takes into account the carrier recombination processes and the capture cross sections for each gap defect state. It uses the information from the de-convolution of the kN(E) spectra, such as the peaks and half-widths of the defect states, and self-consistently simulates the density of electron occupied gap state spectra, the subgap absorption, and the mobility-lifetime ($\mu\tau$) products at different generation rates to obtain accurate gap state parameters and to provide insights into the distribution of the gap defect states.

In this paper, an enhanced simulation model and the simulation results on protocrystalline a-Si:H thin films are presented. The nature of the gap defect states is discussed.

## EXPERIMENTAL DETAILS

The 0.8-$\mu$m-thick protocrystalline a-Si:H thin films studied here were prepared with RF PECVD. The deposition conditions were the same as reported previously [5]. The subgap absorptions were measured using the DBP technique. The DBP measurements cover the photon energies ranging from 0.5 eV to 2.0 eV to allow gap defect states well above midgap to be characterized. The volume absorbed bias light was used to establish the steady state generation rates or the splitting of the quasi-Fermi levels. The ac photocurrents generated by the probing photon energy less than the bandgap were collected using the lock-in amplifier. Different gap defect states were probed by adjusting the bias light intensities. The energy dependent density of electron occupied gap states, kN(E), was derived from the derivative of subgap absorption spectrum. The absolute magnitude of kN(E) was obtained by normalizing to the densities of states at the mobility edge. The mobility gap, $E_\mu$, of the film was determined from the Internal Photoemission measurements and the optical gap, $E_{opt}$, was determined from the Transmission and Reflection measurements. $E_\mu = 1.85$eV and $E_{opt} = 1.75$eV for the protocrystalline a-Si:H thin films studied here. The electron mobility-lifetime products were measured with the red light to ensure uniform absorption and carrier generation. Neutral density filters were used to change the carrier generation rates. The thin films were annealed for four hours at 170°C and light soaked for 30 minutes with 100mW/cm$^2$ (AM1.5) of white light from a 300W ELH lamp.

## GAP DEFECT STATES MODELING

The enhanced simulation model is based on the carrier recombination and generation processes under steady state condition. In this model, two different types of gap defect states are included. One is associated with the mono-vacancy or the dangling bond, $B^0$, located near midgap. The other is associated with the di-vacancy, which exhibits three distinctive defect levels: one close to the conduction band called $D^+$, one close to the midgap called $A^-$, and one close to the valence band called $C^-$ [6,7]. Gaussian distributions are assumed for all these states. The key parameters are the peak, the strength, and the half-width of the Gaussian. The peaks and half-widths of the Gaussians, $A^-$, $B^0$ and $C^-$, are derived from de-convolution of the density of electron occupied gap states. For DBP measurements, the bias light is much stronger than the probing light and the sample is at steady state, i.e. the generation rate equals the recombination rate. The generation rate is determined by the bias light intensity and the rate of emission of electrons/holes from the traps. On the other hand the recombination rate is determined by the rate of capture of electrons/holes from the conduction/valence band by the traps. From the rate equations of electrons and holes, the occupation function for an arbitrary distribution of traps can be derived [8, 9]. As shown in equation 1, the occupation function $f(E)$ of the gap state i is independent of the energy distribution of gap states.

$$f^i = \frac{\bar{n} + e_p}{e_n + \bar{n} + \bar{p} + e_p} \qquad (1)$$

Where $\bar{n} = \upsilon\sigma_n^{\,i}n$, $\bar{p} = \upsilon\sigma_p^{\,i}p$, $e_n = \upsilon\sigma_n^{\,i}N_c e^{\frac{E_t - E_c}{kT}}$, and $e_p = \upsilon\sigma_p^{\,i}N_v e^{\frac{E_v - E_t}{kT}}$. The above equation also shows that the probability of occupation of a state is determined by the ratio of capture cross sections for electrons and holes and free carrier concentrations. Gap states with different ratios of the capture cross sections will have different occupation functions. For the defect states located between the trapped electron and hole quasi-Fermi levels, the occupation function is constant [8],

$$f^i = \frac{Rn}{Rn + p} \qquad (2)$$

where $R = \sigma_n^{\,i}/\sigma_p^{\,i}$.

The calculation of subgap absorptions takes into account all the optical transitions from electron occupied gap states to the parabolic extended states. The capture cross section for electrons is determined by the $\mu\tau$ products and the ratio of the capture cross sections by the occupation function – the density of electron occupied gap states. By self-consistently simulating the subgap absorption spectra, the $\mu\tau$ products at different generation rates and the density of electron occupied gap states spectra, the gap state parameters are derived.

**RESULTS AND ANALYSIS**

The energy dependent density of electron occupied gap states of protocrystaline a-Si:H thin films after 30 minutes of AM1.5 illumination was de-convoluted into three Gaussian distributions A, B, and C. The peak energy values for the three Gaussians were 1.01 eV, 0.84 eV, and 0.57 eV above the valence band edge $E_v$, respectively. The corresponding half-widths of respective states were 0.19 eV, 0.13 eV, and 0.19 eV. Due to the experimental limitation, the information of the gap states below 0.65 eV cannot be obtained from kN(E) de-convolution. Figure 1 shows the normalized kN(E) spectra with de-convoluted three Gaussian distributions. With the increase in bias currents, the peak energies and half-widths for A and B Gaussians

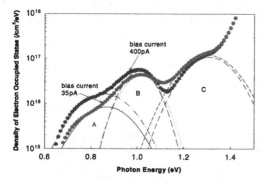

**Figure 1** The density of electron occupied gap states kN(E) at two different bias currents for protocrystalline a-Si:H after 30 minutes AM1.5 illumination. The symbols represent kN(E) derived from the subgap absorption and the dotted lines are A, B, and C states derived from Gaussian de-convolution.

remained the same. There are changes for C states due to the sweep of hole quasi-Fermi level through these states. The changes in the parameters of A, B, and C Gaussians with bias currents can be used to characterize the electron occupation and the ratio of the capture cross sections for electrons and holes. States A, B, and C are associated with the actual gap defect states $A^-$, $B^0$, and $C^-$ through the occupation function.

The peak energies and half-widths of $A^-$, $B^0$, and $C^-$ states used in the simulation were taken as those of A, B, and C states. It can be seen from figure 1 that there is a big change in the occupation of A states with the bias light, which indicates that the ratio of the capture cross sections for electrons and holes is much smaller than 1. All three states have different ratios of the capture cross sections. For states $A^-$ and $C^-$, the capture cross sections of holes are smaller than that of electrons. These states will be negatively charged when occupied by electrons and neutral when empty. The densities and capture cross sections of these three states, and gap state parameters of $D^+$ were derived from self-consistent fitting to the density of electron occupied gap states, the subgap absorption, and the $\mu\tau$ products at different generation rates. The total density of gap states derived from the simulation for protocrystalline a-Si:H after 30 minutes illumination is shown in figure 2. It can be seen that the density of charged states is about 4.5 times that of neutral states. The dark Fermi level $E_F$ is located at 0.08eV above the midgap.

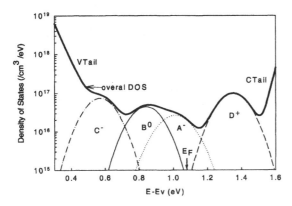

**Figure 2** Density of gap states for protocrystalline a-Si:H after 30 minutes of AM1.5 illumination. The solid line represents the resultant densities of states, the dotted lines the Gaussian distributions.

The simulation results on the density of electron occupied gap states at two different generation rates, $2 \times 10^{15}$ /cm$^3$s and $7 \times 10^{16}$ /cm$^3$s, are illustrated in figure 3a. Those on the $\mu\tau$ products as a function of generation rates are illustrated in figure 3b. In both figures, the

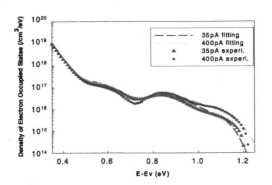

**Figure 3a** The density of electron occupied gap states at two generation rates for protocrystalline a-Si:H after 30 minutes of AM1.5 illumination. The symbols are the experimental results and the dotted lines the simulation results.

experimental results are shown as symbols and the simulation results are shown as dotted lines. There is a good agreement between the experimental and simulation results. The quasi-Fermi levels of free electrons, $E_{fn}$, at different generation rates were determined by the electron $\mu\tau$ products. $E_{fn}$ is 1.16 eV above $E_v$ at generation rate of 2 x 10$^{15}$ /cm$^3$s and is 1.23 eV above $E_v$ at generation rate of 7 x 10$^{16}$ /cm$^3$s.

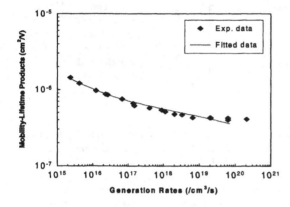

**Figure 3b** Electron $\mu\tau$ products as a function of generation rates for protocrystalline a-Si:H after 30 minutes of AM1.5 illumination. The symbols are the experimental results and the solid line the simulation results.

At very low generation rates, most of B$^0$ states and part of A$^-$ states are between the two quasi-Fermi Levels. These states act as effective recombination centers. The electron $\mu\tau$ products are sensitive to the capture cross sections of these states. At higher generation rates, the effects of D$^+$ and C$^-$ states on the $\mu\tau$ products become evident. The reiterative procedure was used to obtain the other gap state parameters. The densities of gap states D$^+$, A$^-$, B$^0$, and C$^-$ are 2 x 10$^{16}$ /cm$^3$, 6 x 10$^{15}$ /cm$^3$, 9 x 10$^{15}$ /cm$^3$, and 1.5 x 10$^{16}$ /cm$^3$, respectively. The capture cross sections, $\sigma_n/\sigma_p$, for respective states are 1x10$^{-15}$/1x10$^{-16}$ cm$^2$, 5x10$^{-16}$/1x10$^{-14}$ cm$^2$, 1x10$^{-15}$/1x10$^{-16}$ cm$^2$, and 5x10$^{-18}$/1x10$^{-14}$ cm$^2$. The peak energy of D$^+$ states is located at 1.35eV above $E_v$, derived after taking into account the charge neutrality. Good agreement between experimental results and simulation results was also obtained on the subgap absorption at two generation rates.

The above simulation results indicate that at low generation rates the electron occupied A and B states act as effective recombination centers and play key roles in the photoconductivity studies. This can be further seen from the light degradation studies. Figure 4 shows the inverse of the mobility-lifetime products, 1/$\mu\tau$, as a function of light soaking time for protocrystalline a-Si:H thin films. Also shown in figure 4 is the combined defect densities of electron occupied A and B states. These are obtained after subtracting the initial density of states and represent just the light induced defects. The kinetics are the same for both results. In addition, it shows that the combined density of states is created with a t$^{1/2}$ time dependence as previously reported on the solar cell results [10].

**CONCLUSIONS**

An enhanced simulation model was developed to study gap defect states in a-Si:H materials. The model takes into account the recombination and generation processes and the optical transitions between localized states and extended states. Through self-consistent simulation to

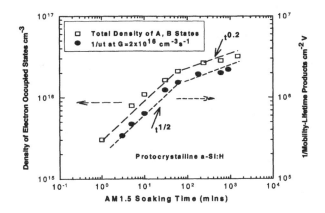

**Figure 4** Evolution of $1/\mu\tau$ and the combined density of just the light induced states under AM1.5 illumination. The circles represent $1/\mu\tau$ at generation rate of $2 \times 10^{16}$ /cm$^3$s and the squares the combined density of the two states at the same generation rate.

the density of electron occupied gap defect states, the subgap absorption, and the $\mu\tau$ products at different generation rates, reliable gap defect state parameters including the distribution of gap states and the capture cross sections for electrons and holes were derived. States B from deconvolution of kN(E) spectra were associated with the mono-vacancy B$^0$. States A and C were associated with the di-vacancy defect levels A$^-$ and C$^-$. States D$^+$ were introduced and associated with the top di-vacancy defect level. The energy peak of positively charged states D$^+$ was determined by taking into account the charge neutrality. At low generation rates, A$^-$ and B$^0$ states act as effective recombination centers. States C$^-$ and D$^+$ play key roles at higher generation rates.

## ACKNOWLEDGEMENTS

The authors would like to thank Professor Thomas Jackson for the valuable discussion and Xinwei Niu for providing the experimental data.

## REFERENCES:

1. W. Pickin, J.C. Alonso, and D. Mendoza, J. Phys. C, vol. 20, p341, (1987)
2. Z.A. Yasa, W.B. Jackson, and N.M. Amer, Appl. Opt., vol. 21, p21 (1982)
3. L. Jiao, H. Liu, S. Semoushikiana, Y. Lee and C.R. Wronski, 9th International Photovoltaic Science and Engineering Conf., p641, (1996)
4. J. M. Pearce, J. Deng, V. Vlahos, R. W. Collins, C. R. Wronski, 3$^{rd}$ World Conference on Photovoltaic Energy Conversion, 2, p1588 (2003)
5. Y. Lee, L. Jiao, H. Liu, Z. Lu, R.W. Collins, and C. R. Wronski, Conf. Record of 25th IEEE PVSEC (IEEE, 1996), 1165 (1996)
6. A.H. Kalma et al., Ed. F. L. Vook (Plenum Press, New York – London) p153, (1968)
7. K. Matsui and P. Baruch, University of Tokyo Press, Toyko, p282, (1968)
8. J.G. Simmons and G.W. Taylor, Phys. Rev. B, 4, p502 (1971)
9. A. Rose, Phys. Rev. 97, p322 (1955)
10. J. Deng, B Ross, M. Albert, R. W. Collins, and C. R. Wronski, Mater. Res. Soc. Symp. Proc., 0910-A02-02, (2006 )

# Poster Session:
# Thin-Film Growth

Mater. Res. Soc. Symp. Proc. Vol. 1066 © 2008 Materials Research Society          1066-A05-02

## Polysilazane Precursor Used for Formation of Oxidized Insulator

Yuji Urabe, and Toshiyuki Sameshima
Tokyo University of Agriculture and Technology, 2-24-16, nakacho, koganei-shi, Japan

## ABSTRACT

We report on $SiO_2$ film formation using Polysilazane precursor treated with remote oxygen plasma and high-pressure $H_2O$ vapor heating. Polysilazane precursor films with a thickness of 130 nm were spin coated on silicon substrates. They were annealed at 350°C in remote oxygen plasma at a pressure of $2.0x10^{-2}$ Pa, frequency of 13.56 MHz and power of 300 W, and then followed by 13-atmospheric-pressure-water vapor heating at 260°C for 3 hrs. It was found that the films made by Polysilazane precursor were entirely oxidized by high-pressure $H_2O$ vapor thermal treatment, and the densities of Si-N and Si-H bonds inside those films diminished by the combination of double oxidized treatment. While Metal-Oxide-Semiconductor (MOS) capacitors fabricated only by high-pressure $H_2O$ vapor heat treatment had a high specific dielectric constant of 6.1, an oxide charge density of $1.3x10^{12}$ $cm^{-2}$ and a density of interface trap of $5.4x10^{11}$ $cm^{-2}eV^{-1}$, the combination of oxygen-plasma-then-water-vapor-thermal oxidized treatment allowed us to reduce them to 4.1, $1.6x10^{11}$ $cm^{-2}$ and $4.2x10^{10}$ $cm^{-2}eV^{-1}$, respectively, indicating better $SiO_2$ dielectric material and $SiO_2$/Si interface.

## INTRODUCTION

The formation of good-quality $SiO_2$ films and $SiO_2$/Si interface at low temperature is important to fabricate high-quality and low-cost electronic devices, such as liquid crystal and organic electroluminescent displays. A low temperature process for $SiO_2$ formation allow us to fabricate thin film transistors on inexpensive glass substrates and flexible substrates, as well reduce the production cost of solar cells. The $SiO_2$ films from Polysilazane precursor, made by post-annealing process at 450°C in air, have been commercially applied for passivation films in LSI process. Moreover, the improvement in $SiO_2$ films and $SiO_2$/Si interface by high-pressure $H_2O$ vapor thermal treatment has been reported by us [1-8], which is ascribed to the better thermal relaxation inside a network of Si-O bonding. In this study, we report $SiO_2$ film formation using Polysilazane precursor treated with remote oxygen plasma and high-pressure $H_2O$ vapor heating. The structural properties of Polysilazane and $SiO_2$ films were characterized by Fourier transform infrared spectrometry (FTIR). The electrical properties of $SiO_2$ films and $SiO_2$/Si interface were characterized by Capacitance-Voltage (CV) measurement. It was demonstrated that the Polysilazane precursor were completely oxidized to $SiO_2$ films with better electrical properties for gate dielectric material by our oxygen-plasma-then-water-vapor-thermal oxidized treatment.

## EXPERIMENT

Polysilazane films with a thickness of 130 nm were spin coated on P-type single crystalline silicon substrates. 13.56 MHz radio frequency (RF) remote oxygen plasma treatment was carried out in $O_2$ gas with a flow rate of 2 sccm at a pressure of $2.0x10^{-2}$ Pa, and a power of 300 W for 3 hrs at 130, 260 and 350°C, respectively. The FTIR absorption spectra of the films were measured to determine the absorption ratio of Si-N, Si-O and Si-H bonds. MOS capacitors were also fabricated with an active area of 0.01 cm$^2$ after the remote oxygen plasma treatment. The samples were subsequently heated at 260°C with water vapor at 1.3x10$^6$ Pa for 3 hrs. The CV at 1 MHz was measured to compare the bulk and interfacial electrical properties.

RESULT AND DISCUSSION

Figure 1 shows optical absorption spectra measured for the samples of Polysilazane precursor film as-spin-coated (1), remote oxygen plasma treatment at 350°C (2), remote oxygen plasma treatment at 350°C followed by high-pressure H$_2$O vapor thermal treatment (3). Optical absorption peaks corresponding to the Si-N vibration mode around 875 cm$^{-1}$ and the Si-H vibration mode around 2100 cm$^{-1}$ were observed in the spectrum of as-spin-coated [9]. They are clearly the compositions of Polysilazane precursor films. In order to investigate the oxidization of Polysilazane precursor films, we focused on these two peaks and those optical absorption peaks corresponding to the Si-O vibration mode around 1080 cm$^{-1}$. The intensities of Si-N and Si-H peaks were reduced by remote oxygen plasma treatment. On the other hand, the intensity of Si-O peak was slightly increased by remote oxygen plasma treatment. It means that Polysilazane precursor films were slightly oxidized by remote oxygen plasma treatment. The intensities of Si-N and Si-H peaks were diminished significantly, while the intensity of the Si-O peak was markedly increased by high-pressure H$_2$O vapor heat treatment after remote oxygen plasma treatment and its line shape became similar to that of thermally grown SiO$_2$ films at high temperature. These results indicate that Polysilazane precursor films were completely oxidized by high-pressure H$_2$O vapor heat treatment. Next step, a semi-quantitative analysis is performed by integrating the peaks' intensities at 750 - 900 cm$^{-1}$, 1000 - 1300 cm$^{-1}$ and 2000 - 2300 cm$^{-1}$, which is corresponding to Si-N, Si-O and Si-H bonds, respectively.

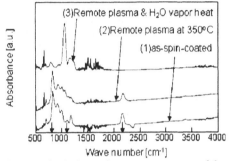

Fig.1: optical absorption spectra measured by FTIR for as-spin-coated (1), remote oxygen plasma treatment at 350°C with the RF power of 300 W for 3 h (2), remote oxygen plasma treatment at 350°C with the RF power of 300 W for 3 h followed by 1.3x10$^6$-Pa-H$_2$O vapor heat treatment at 260°C for 3h (3). The arrows indicate optical absorption peak corresponding to Si-N (875cm$^{-1}$), Si-O (1080cm$^{-1}$) and Si-H (2100cm$^{-1}$) vibration mode.

Figure 2 shows total absorbance optical absorption peaks corresponding to Si-N bonding (a), Si-O bonding (b) and Si-H bonding (c) as a function of heating temperature for remote oxygen plasma treatment. Total absorbance of optical absorption peaks corresponding to Si-O bondings treated by remote oxygen plasma treatment at 350°C followed by water-vapor-thermal oxidized treatment was determined to be 100. The intensity of Si-N peak was 47 for As-spin-coated sample, and it was 51 by remote oxygen plasma treatment at 350°C, as shown in Figure.2 (a). This is because the nitrogen atoms are composition of Polysilazane. On the other hand, it was markedly reduced to 7 by remote oxygen plasma treatment at 350°C followed by water-vapor-thermal oxidized treatment. From this result, the Si-N bondings in Polysilazane precursor were effectively dissociated by water-vapor-thermal oxidized treatment. The as-spin-coated sample had a Si-O peak with an intensity of 37 as shown in Figure.2 (b). Although Polysilazane intrinsically has no oxygen atoms, the high intensity of Si-O peak was observed. This result indicates that Polysilazane precursor was slightly oxidized by air. Total absorbance of optical absorption peaks corresponding to Si-O bonding was increased to 45 by remote oxygen plasma treatment at 350°C. This means that oxygen radicals slightly oxidized Polysilazane precursor. It was markedly increased to 100 by remote oxygen plasma treatment at 350°C followed by water-vapor-thermal oxidized treatment. This result indicates that water-vapor-thermal oxidized treatment entirely oxidized Polysilazane precursor. Total absorbance of optical absorption peaks corresponding to Si-H bonding was 18 for the as-spin-coated sample, as shown in Fig.2 (c). It was reduced to 13 by remote oxygen plasma treatment at 350°C, and markedly reduced to 0.9 by remote oxygen plasma treatment at 350°C followed by water-vapor-thermal oxidized treatment. These results indicate that the Si-H bondings in Polysilazane precursor were effectively dissociated by the combination of double oxidized treatment.

Figure 3 shows capacitance responses as a function of bias voltage for water-vapor-thermal oxidized treatment alone and remote oxygen plasma treatment at 350°C followed by water-vapor-thermal oxidized treatment. We could not measure the capacitance responses for the sample fabricated by remote oxygen plasma treatement alone. We thought that this resulted from insufficient oxidation of the Polysilazane precursor, as seen from the fact that the sample had a small intensity of Si-O peak in FTIR optical absorption spectrum. Although there was no change in the thickness of $SiO_2$ films after each

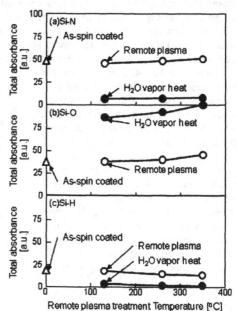

Fig.2: Total absorption corresponding to the intensity of Si-N peak (a), the intensity of Si-O peak (b) and the intensity of Si-H peak (c) as a function of heating temperature for remote oxygen plasma treatment.

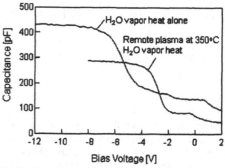

Fig.3: Capacitance responses as a function of bias voltage at a frequency of 1 MHz.

fabrication process, the maximum capacitance of capacitance response was reduced by remote oxygen plasma at 350°C followed by water-vapor-thermal oxidized treatment. Moreover, the overall capacitance response was shifted to positive bias voltage direction. These results indicates that Polysilazane precursor was oxidized well by remote oxygen plasma at 350°C followed by water-vapor-thermal oxidized treatment.

Figure 4 shows specific dielectric constant (a), fixed oxide charge density (b) and density of interface trap (c) as a function of remote plasma treatment temperature. The MOS capacitor treated with water-vapor-thermal oxidized treatment alone had a specific dielectric constant of 6.4, as shown in Fig. 4(a). Remote oxygen plasma treatment at 350°C followed by water-vapor-thermal oxidized treatment reduced to 4.1. A high specific dielectric constant for the sample treated with water-vapor-thermal oxidized treatment alone is caused by bonding distortion of Si-O associated with the lack of oxygen atoms in the $SiO_2$ films [2]. Remote oxygen plasma treatment followed by water-vapor-thermal oxidized treatment oxidized polysilazane precursor well and reduced the maximum capacitance associated with reduction of the specific dielectric constant. The MOS capacitor treated with water-vapor-thermal oxidized treatment alone had a high fixed oxide charge of $1.3 \times 10^{12}$ cm$^{-2}$ and a high interface trap states of $5.4 \times 10^{11}$ cm$^{-2}$eV$^{-1}$, as shown in figure 4(b) and (c). Remote oxygen plasma treatment at 350°C followed by water-vapor-thermal oxidized treatment reduced them to $1.6 \times 10^{11}$ cm$^{-2}$, $4.2 \times 10^{10}$ cm$^{-2}$eV$^{-1}$, respectively. Remote oxygen plasma treatment reduced the specific dielectric constant, the fixed oxide charge density and the interface trap states with increasing the heating temperature. These results indicate that remote oxygen plasma treatment at 350°C effectively reduced the defects of $SiO_2$ films and $SiO_2$/Si interface.

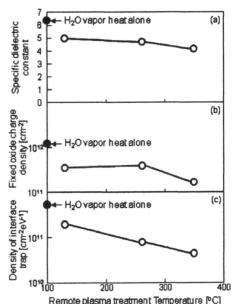

Fig.4: The specific dielectric constant (a), the fixed oxide charge density (b) and the density of interface trap (c) as a function of heating temperature for remote oxygen plasma treatment.

## CONCLUSIONS

We investigated $SiO_2$ films formation using Polysilazane precursor treated by remote oxygen plasma and water-vapor-thermal oxidized treatment. From the result of FTIR optical absorption spectrum, we confirmed that the as-spin-coated sample has the optical absorption peaks corresponding to Si-N and Si-H. On the other hand, the high-pressure $H_2O$ vapor heat treatment after remote oxygen plasma treatment significantly diminished the intensities of Si-N and Si-H peaks, and markedly increased the intensity of the Si-O peak. The shape of the Si-O peak became similar to that of thermally grown $SiO_2$ films at high temperature. The combination of oxygen-plasma-then-water-vapor-thermal oxidized treatment entirely oxidized Polysilazane precursor. The MOS capacitor fabricated by high-pressure $H_2O$ vapor heat treatment had a high specific dielectric constant of 6.4, a high fixed oxide charge density of $1.3 \times 10^{12}$ $cm^{-2}$ and a density of interface trap of $5.4 \times 10^{11}$ $cm^{-2}eV^{-1}$. On the other hand, remote oxygen plasma treatment at $350^\circ C$ followed by water-vapor-thermal oxidized treatment reduced them to 4.1, $1.6 \times 10^{11}$ $cm^{-2}$ and $4.2 \times 10^{10}$ $cm^{-2}eV^{-1}$, respectively. From these results, the combination of oxygen-plasma-then-water-vapor-thermal oxidized treatment oxidized Polysilazane precursor well and improve in electrical properties of $SiO_2$ films and $SiO_2/Si$ interface.

**ACKNOWLEDGMENTS**

Author would like to appreciate "Human Resource Development Program for Scientific Powerhouse" for their support.

**REFERENCES**

1. T. Sameshima, A. Kohno, M. Sekiya, M. Hara, and N. Sano, Appl. Phys. Lett. 64 8 (1994)
2. K. Sakamoto and T. Sameshima, Jpn. J. Appl. Phys. 39 (2000) pp.2492-2496
3. T. Sameshima and M. Satoh, Jpn. J. Appl. Phys. 36 (1997) pp L687-L689
4. T. Sameshima, K. Sakamoto, T. Tsunoda and M. Saitoh, Jpn. J. Appl. Phys. 37, L1452, (2000)
5. T. Sameshima, M. Satoh and K. Sakamoto, Thin Solid Films 335 (1998) 138
6. H. Watakabe and T. Sameshima, Jpn. J. Appl. Phys. 41 L974 (2002)
7. H. Watakabe and T. Sameshima, Proc in 1st Thin Film Material & Devices Meeting (Nara, 2004)
8. T. Sameshima, M. Satoh and K. Sakamoto, K. Ozaki and K. Saitoh, Jpn. J. Appl. Phys. 37 (1998) 4254
9. G. Lucovsky, J. Yang, S. S. Chao, J. E. Tyler, and W. Czubatyj, Phys. Rev. B 28 (1983) 3234

Mater. Res. Soc. Symp. Proc. Vol. 1066 © 2008 Materials Research Society          1066-A05-04

# Boron Incorporation and Its Effect on Electronic Properties of Ge:H Films Deposited by LF Plasma

Andrey Kosarev[1], Alfonso J Torres[1], Nery D Checa[1], Yurii Kudriavtsev[2], Rene Asomoza[2], and Salvador G Hernandez[2]

[1]Electronics, National Institute for Astrophysics, Optics and Electronics, L.E.Erro No.1, col. Tonantzintla, Puebla, 72840, Mexico
[2]CINVESTAV Institute Poltechnical National, Mexico DF, 71000, Mexico

## ABSTRACT

Previously the deposition conditions that provided low absorption related to both band tail and deep localized states have been found. In this work boron doping of Ge:H films have been systematically investigated. The films were deposited by low frequency plasma under such conditions that resulted in a low density of localized states. The boron incorporation in solid phase was observed to increase linearly with the increase of the doping in the gas phase. The hydrogen concentration in the films was determined from FTIR and SIMS measurements. In the entire region of boron concentrations here studied, the hydrogen content changed non-monotonously by a factor of 1.5 as it was determined from both stretching mode absorption at k $\approx 1870$ cm$^{-1}$ and SIMS data. The activation energy of conductivity increased in the range of $[B]_{sol} = 0$ to $0.05\%$ suggesting a compensation of electron conductivity, reaching maximum value $E_a = 0.5$ eV  (corresponding approximately to $E_g/2$) at $[B]_{sol} = 0.05\%$. Then with further incorporation of boron reduced the value of the activation energy to a minimum value $E_a = 0.27$ eV (corresponding Fermi energy $E_F(RT) = 0.15$ eV  at $[B]_{sol} = 0.12$ %. After this point, a behaviour showing a trend to saturation with further boron increase is observed. This behaviour is related to the change of charge transport from electron to intrinsic at $[B]_{sol} = 0.05\%$ and beyond this point to hole transport. A significant reduction of both band tail and deep localized states were observed at $[B]_{sol} = 0.004\%$. The latter is presumably related to the improvement of the lattice structure.

## INTRODUCTION

Hydrogenated Germanium film (Ge:H) deposited by plasma are of much interest because of its possible applications in devices such as low band gap solar cells (or the long wavelength part of the tandem of a photovoltaic structure), photo-detectors, which are well suited for fibre optic communications, thermo-photovoltaic devices, etc. A Ge:H p-i-n detector has been reported [1], but with the p-n layers made of silicon films. Usually, the plasma deposited Ge:H films demonstrate worse electronic properties in comparison with those of silicon films. The deposition conditions providing the best electronic properties for silicon films did not allow to prepare good quality Ge:H films. Dalal et al [2] have reported the important role of ion bombardment during growth in the plasma deposition of high quality Ge:H films. In this respect low frequency (LF) PECVD with its inherent higher ion bombardment than standard RF plasma is very attractive. In our previous studies [3], we have demonstrated a LF PECVD fabrication of Ge:H films with both low tail and low deep localized states absorption. Boron doping of plasma deposited silicon films have been widely studied and applied in many devices, while only a few papers have been reported on B-doping of Ge:H films [ 4-6].

This work is devoted to the study of boron incorporation and its effect on the electronic properties of Ge:H films deposited by LF PECVD.

## EXPERIMENT

The films were obtained by low frequency (LF) plasma enhanced chemical vapour deposition (PECVD) from a mixture of $GeH_4$ +$B_2H_6$ diluted with hydrogen. The deposition parameters were as follows: substrate temperature $T_s$= 300 °C, the discharge frequency f= 110 kHz, pressure P= 0.6 Torr, power W= 300 W, germane flow $Q_{GeH4}$= 50 sccm, hydrogen flow $Q_{H2}$=3500 sccm. The Diborane ($B_2H_6$) flow was varied in the range of $Q_{B2H6}$=0 to 20 sccm providing boron concentration in gas phase in the range of $[B]_{gas}$ = 0 to 4%. The composition of the films was determined by SIMS profiling. Hydrogen bonding was studied by FTIR. The measurement -in a vacuum thermostat- of the temperature dependence of conductivity in DC regime was employed to study carrier transport. From transmittance and reflectance measurements were determined the optical gap, sub-gap absorption and refraction index of the deposited films.

## DISCUSSION

Before to starting the study of doping we have investigated various deposition regimes for intrinsic films, and have found some optimal conditions for the deposition of intrinsic films with the both the lowest deep and tail density of localized states (see ref. [3]). These depositing conditions were employed in this study for fabrication of both reference and doped films.

The deposition rate of the deposited films changed non -monotonously with the atomic concentration of boron (from diborane as a stock gas) in gas phase as shown in Figure 1. The variation of the deposition rate in the entire range of the boron concentrations here studied, was from $V_d$= 3.3 Å/s to a minimum 2.5 Å/s at $[B]_{gas}$= 3 %. Nevertheless, the deposition rate here obtained is higher than that of the typical in RF discharge ($V_d \approx 1$Å/s).

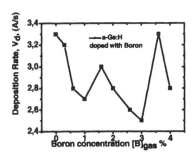

Figure 1. Deposition rate as function of boron doped in gas phase for a-Ge:H films.

The composition of the films and the boron incorporation were determined by SIMS. An example of a SIMS profile is shown in Figure 2 a). In order to analyze the boron incorporation and also to control such components as hydrogen and oxygen, the concentrations of the elements at approximately the middle part of sample thickness were considered. Figure 2 b) presents the boron concentration in solid phase versus that in gas phase. The solid line is the result of the best linear fit described as $[B]_{sol} \% \approx k [B]_{gas}\%$ ($k = 0.035 \pm 0.001$).

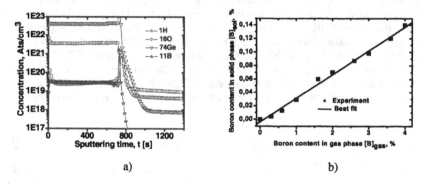

a)                                                    b)

Figure 2. Composition SIMS profile of the sample # 595    (a) and boron gas into solid phase incorporation (b).

The hydrogen incorporation was studied through FTIR and SIMS. In IR spectra two modes are attributed to Ge-H; deformation ($k \approx 560$ cm$^{-1}$) and stretching ($k \approx 1870$ cm$^{-1}$). The corresponding areas of each IR spectra of the absorption lines versus boron concentration in solid phase are shown in Figure 3. The hydrogen concentration obtained from SIMS measurements are also presented in this figure and there is observed a close correlation with the FTIR measurements.

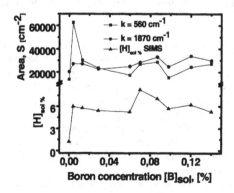

Figure 3. IR absorption related to Ge- H bonding versus boron concentration in the films.

In the entire region of boron concentrations here studied, the hydrogen content changed non- monotonously by a factor of 1.5 as it was determined from both stretching mode absorption at $k \approx 1870 \, cm^{-1}$ and SIMS data. Absorption at $k \approx 560 \, cm^{-1}$ demonstrated similar trend with the only exception at very low boron concentration $[B]_{sol} = 0.004\%$, where the hydrogen concentration showed a value 3 times larger than the other values.

From measurements of temperature dependence of conductivity $\sigma(T)$ for the different B concentrations in solid phase, have shown changes on electrical parameters such as, room temperature conductivity $\sigma_{RT}$, activation and Fermi energies $E_a$, $E_F(RT)$ respectively, as shown in Figure 4. $\sigma_{RT}$ reduces with $[B]_{sol}$ in the range of $[B]_{sol}$ from 0 to 0.03% and in a linear plot it practically remains constant. On the other hand, $E_a$ increases from 0.25 eV at $[B]_{sol}=0$ to a maximum value of $E_a = 0.45$ eV for $[B]_{sol}=0.06\%$, then it reduces to minimum of 0.27 eV for $[B]_{sol}=0.12\%$, followed by a slight increase at $[B]_{sol}=0.13\%$. The behavior of $E_F(RT) =f([B]_{sol})$ is similar to that of $E_a([B]_{sol})$. The remarkable difference of $E_a$ and $E_F(RT)$ observed at $[B]_{sol}= 0.12\%$ suggests that these films have a less rigid lattice in comparison the rest of the deposited films.

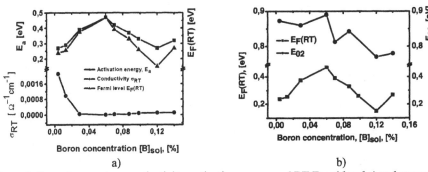

Figure 4. Room temperature conductivity, activation energy and RT Fermi level a) and energy $E_{02}$ corresponding photon energy with observed absorption $\alpha =10^2 \, cm^{-1}$ b) versus boron concentration in the films.

The increase of $E_a$ with $[B]_{sol}$ at low boron concentration is thought to be due to compensation of electronic conductivity with respect to the un-doped reference sample. In fact at $[B]_{sol}= 0.06\%$ we have compensated material with a Fermi energy near the center of the mobility gap. Further increase of boron concentration results in p- type material with further reduction on the activation energy. In Figure 4 b) the position of the Fermi level referred to the conductivity band edge is shown in comparison with energy $E_{02}$, this energy corresponds to the photon energy observed for the absorption value $\alpha =10^2 \, cm^{-1}$. The energy $E_{02}$ practically does not changes for $[B]_{sol}$ in the range of $[B]_{sol}$ from 0 to 0.05% and then decreases with further boron incorporation. This behavior in general agree with the behavior of the Urbach energy and deep states versus boron content as is shown in Figure 5 b).

The Figure 5 a) shows the spectral dependence of the optical absorption $\alpha(h\nu)$ for the samples with different boron concentrations. It can be seen that the boron incorporation changes both the absorption related to band tail states and that related to deep defect states. The tail state absorption is characterized by the Urbach energy $E_U$. For a quantitative characterization of the deep states, we have employed the absorption coefficient $\alpha_D$ measured at the photon energy hv = 0.85 eV. Both $E_U$ and $\alpha_D$ are shown in Figure 5 b) as a function of the boron concentration in the films. It is interesting to note that the incorporation of a very small boron quantity reduced both tail and deep states (both $E_U$ and $\alpha_D$ have minimum values at [B]$_{sol}$= 0.004%).

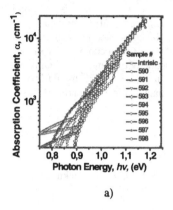
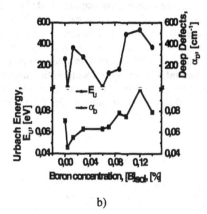

a)                                                    b)

Figure 5. Spectral dependence of absorption coefficient α(hv) a), and Urbach energy $E_U$ and deep defect absorption $\alpha_D$ b) in the films with different boron concentration.

Table I. Optical characteristics and electrical properties of the films
with different boron concentration.

| [B/Ge]$_{sol}$ % | $\sigma_o$ [$\Omega^{-1}$ cm$^{-1}$] | $E_a$ [eV] | $\sigma_{RT}$ [$\Omega^{-1}$ cm$^{-1}$] | $\gamma \times 10^4$ [eV/k] | $E_F$ (RT) [eV] | $E_{03}$ [eV] | $E_{04}$ [eV] | $\Delta E$ [eV] | $n_{(\infty)}$ | $E_U$ [meV] | Defect absorption at hv=0.85 eV, [cm$^{-1}$] |
|---|---|---|---|---|---|---|---|---|---|---|---|
| 0.0 | 17.6 | .22 | 3.5E-3 | -2.14 | 0.20 | 0.96 | 1.12 | 0.16 | 4.21 | 71 | 260 |
| 0.004 | 59.7 | 0.27 | 2.1E-3 | -1.06 | 0.24 | 0.96 | 1.12 | 0.15 | 4.28 | 46 | 0 |
| 0.013 | 56.8 | 0.29 | 9.7E-4 | -1.11 | 0.26 | 0.95 | 1.12 | 0.16 | 4.29 | 55 | 370 |
| 0.029 | 114.4 | 0.39 | 4.9E-5 | -0.49 | 0.38 | 0.94 | 1.11 | 0.17 | 4.27 | 63 | 280 |
| 0.060 | 200.3 | 0.47 | 3.5E-6 | -.0015 | 0.47 | 0.96 | 1.10 | 0.14 | 4.21 | 63 | 0 |
| 0.070 | 86.5 | 0.42 | 1.9E-5 | -0.74 | 0.39 | 0.94 | 1.10 | 0.16 | 4.21 | 65 | 130 |
| 0.087 | 44.7 | 0.37 | 4.5E-5 | -1.32 | 0.33 | 0.91 | 1.11 | 0.19 | 4.21 | 78 | 170 |
| 0.098 | 20.0 | 0.33 | 8.7E-5 | -2.04 | 0.26 | 0.93 | 1.10 | 0.17 | 4.25 | 75 | 490 |
| 0.12 | 2.69 | 0.27 | 1.0E-4 | -3.8 | 0.15 | 0.91 | 1.10 | 0.19 | 4.22 | 100 | 540 |
| 0.14 | 21.1 | 0.32 | 1.2E-4 | -1.98 | 0.27 | 0.95 | 1.11 | 0.16 | 4.25 | 79 | 370 |

It is worth to note that a small amount of boron incorporation [B]$_{sol}$ = 0.004 to 0.013% resulted in practically no change on the conductivity of the films, but it is evident that changes have occurred in the film structure i.e. a reduction in the Urbach energy $E_U$ and an increase in the refraction index $n_\infty$ as is shown in Table I.

## CONCLUSIONS

The LF PE CVD a-Ge:H(B) films doped with boron in the range of $[B]_{sol}$= 0 to 0.14% using the deposition conditions corresponding to the reference intrinsic film with the lowest density of localized states have been fabricated and studied. It was observed the following: a linear incorporation of boron from gas source into the solid phase; no significant effect of boron incorporation on the hydrogen concentration; a small boron amount in the film, significantly reduced the density of both band tail and deep localized states; the boron incorporation in the range of $[B]_{sol}$ = 0.004% to 0.06% resulted in compensation of the electron conductivity in the intrinsic material; further boron incorporation increased the conductivity to $\sigma_{RT}$ = $10^{-4}$ Ohm$^{-1}$cm$^{-1}$ and reduced the Fermi level to a minimum $E_F(RT)$ = 0.15 eV at $[B]_{sol}$ = 0.12%; these films were deposited at deposition rate $V_d$ = 3.3 Å/s which is higher than that in a standard RF discharge.

## ACKNOWLEDGMENTS

This work was supported by CONACyT project No. 48454F. N.D.Checa acknowledges CONACyT for the scholarship No 171132.

## REFERENCES

1. M.Krause, H.Stiebig, R.Carius, H.Wagner, *Mat.Res.Symp.Proc.*, **664**, A26.5 (2001).
2. V.L. Dalal. *Current opinion in Solid State&Material Science*, **6**, 455 (2002).
3. L.Sanchez, A.Kosarev, A.Torres, A.Ilinskii, Y.Kudriavtsev, R.Asomoza, P.Roca i Cabarrocas, A.Abramov. *Thin Solid Films*, **515**, 7603 (2007).
4. M.Stutzman, D.K. Biegelsen, R.A.Street. *Phys.Rev.* **B 35**, 5666 (1987),
5. B.Ebersberger, W.Krueller, W.Fuhs, H.Mell. *Appl. Phys.Lett.,* **65** (13), 26 Sept, 1683 (1994).
6. W.B.Jordan, S.Wagner. *Mat.Res. Soc.Symp.Proc.*, **762**, A5.7.1-6 (2003).

Mater. Res. Soc. Symp. Proc. Vol. 1066 © 2008 Materials Research Society 1066-A05-05

# Low Temperature Deposition of Si-based Thin Films on Plastic Films Using Pulsed-Discharge PECVD under Near Atmospheric Pressure

Mitsutaka Matsumoto[1], Yohei Inayoshi[1], Maki Suemitsu[1], Setsuo Nakajima[2], Tsuyoshi Uehara[2], and Yasutake Toyoshima[3]

[1]Research Institute of Electrical Communication, Tohoku University, 2-1-1 Katahira, Aoba-ku, Sendai, 980-8577, Japan

[2]Sekisui Chemicals Co. Ltd, 2-3-17 Toranomon, Minato-ku, Tokyo, 105-8450, Japan

[3]Energy Technology Research Institute, AIST, 1-1-1 Umezono, Tsukuba, Tukuba, 305-8568, Japan

## ABSTRACT

Low temperature (150 °C) deposition of doped and undoped polycrystalline Si (poly-Si) as well as $SiN_x$ films on polyethylene terephthalate (PET) films has been achieved with practical deposition rates by using pulsed-plasma CVD under near-atmospheric pressure. The precursor is $SiH_4$ diluted in $H_2$ for poly-Si while $N_2$ has been additionally used for $SiN_x$. No inert gases such as He was used. A short-pulse based power system has been employed to maintain a stable discharge in the near-atmospheric pressures. With this technique, deposition of poly-Si thin film with virtually no incubation layer is possible, which in the case of P-doped poly-Si shows a Hall mobility ($\mu_H$) of 1.5 cm$^2$/V·s.

## INTRODUCTION

Recently, low-temperature depositions of silicon-based thin films, such as amorphous Si [1], polycrystalline Si [2] and $SiN_x$ [3], have been reported on plastic films, which have attracted much attention because they form an essential part of the fabrication process in flexible electronics such as for flexible solar cells and flexible displays. Despite the ever increasing needs for flexible electronics, however, conventional low-temperature deposition techniques such as represented by plasma enhanced chemical vapor deposition (PECVD), still require temperatures that are significantly higher than the melting points of most plastic films (>300 °C). Here, we demonstrate that a pulsed-discharge (PD)-PECVD under near atmospheric pressure provides a quite feasible method to form various Si-based thin films on plastic substrates. Plasma operation under Near atmospheric pressure is advantageous since the amount of the density of the gas molecules or radicals is considerably increased from the conventional PECVD operated at lower pressures, leading to an increased in the deposition rate and a reduction in the process cost. Actually, a high rate deposition (~60 nm/min) of poly-Si thin film has been achieved on glass substrates [4] using the same plasma. In this report, we show that a high rate deposition of poly-Si films, its n-type doping with phosphine ($PH_3$), and a low-temperature (~100 °C) deposition of $SiN_x$ films can be realized on polyethylene terephthalate (PET) substrates using PD-PECVD under near atmospheric pressure.

## EXPERIMENT

Figure 1 shows the schematic diagram of the PECVD apparatus. The discharge plasma is generated by applying a pulsed electric bias on the hot electrode pair ($20\times20$ mm$^2$ each) located opposite to the substrate on the grounded electrode. The $H_2/SiH_4/N_2/PH_3$ mixture gas flows through the gap between the hot electrodes and then between the hot electrodes and the substrate. The distance between the hot electrodes and the substrate is 1 mm. Pulsed discharge is operated by applying 30 kHz bipolar pulses of about 12 kV peak height, with a single pulse discharge duration of about 5 $\mu$s and the intermission between the pulse of about 33 $\mu$s. A flexible PET substrate of $0.2\times50\times50$ mm$^3$ was used in every experiment. After being rinsed with pure water for ten minutes, the PET substrate was introduced into the chamber, and its surface impurities were etched off using a $H_2$ plasma treatment for five minutes. The deposited films were characterized using Raman scattering spectroscopy and X-Ray Photoelectron spectroscopy (XPS). Cross sections of the films were observed by transmission electron microscopy (X-TEM) and scanning electron microscopy (X-SEM). The electrical properties of the n-type doped Si films are determined by Hall Effect measurement.

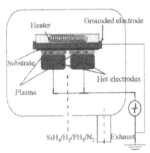

**Figure 1.** A schematic diagram of the apparatus. The discharge plasma is generated by applying a pulsed electric bias on the hot electrode pair, located opposite to the substrate on the grounded electrode.

## RESULTS AND DISCUSSION

### Polycrystalline Si thin film deposition

Figure 2 shows the Raman scattering spectra from the films deposited using the $H_2$-diluted ratio ($R_{H2}=H_2/SiH_4$) of 1000, 750 and 400. The pressure of 500 Torr, the $H_2$ flow rate of 1500 ml/min, and the substrate temperature of 150 °C are common. These Raman spectra are peak-separated into three Gaussian components, which are comprised of the crystalline component at 520 cm$^{-1}$, the nanocrystalline (<10 nm) component at 510 cm$^{-1}$, and the amorphous component at 480 cm$^{-1}$ [5,6]. Houben et al. [7] have reported that the area-based ratio of the crystalline components at 520 and 510 cm$^{-1}$ to the total band intensity, described by equation (1), provides a lower limit of the crystalline volume fraction of the film. Thus, the crystalline fractions obtained are 80, 70 and 0 % for the $R_{H2}$ of 1000, 750 and 400, respectively.

$$R_c \geq \frac{A_{c-Si} + A_{nc-Si}}{A_{c-Si} + A_{nc-Si} + A_{a-Si}} \tag{1}$$

Figure 3 shows the temperature dependence of the Raman scattering spectra. The $R_{H2}$ and the pressure of deposited Si films were fixed at 1000 and 500 Torr, respectively. Above 100 °C "poly-Si" films are obtained with no significant temperature dependence.

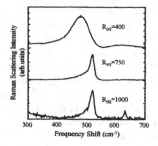

**Figure 2.** H$_2$-diluted ratio (R$_{H2}$) dependence of the Raman scattering spectra for growth at 150 °C and 500 Torr.

**Figure 3.** Temperature dependence of the Raman scattering spectra for growth at R$_{H2}$=1000 and 500 Torr.

Meanwhile, no film deposition was observed for temperatures below 100 °C. It is known that hydrogen atoms generated in H$_2$ plasma can etch Si with negative temperature dependence (etch rate is higher at lower temperature) [8]. No film deposition below 100 °C is explained by this negative temperature dependence of etch rate, assuming a competition of deposition by silane-related radicals to etching by hydrogen atoms in our reaction conditions.

Figure 4 shows the typical X-TEM bright and dark field image of the film deposited at 500 Torr, 150 °C and the R$_{H2}$=1000. From these images, the thickness of the film is estimated to be about 200 nm, which corresponds to a deposition rate of 40 nm/min. This deposition rate is generally higher than those obtained in conventional PECVD [9]. The crystallites in the dark field image are almost uniformly distributed in the entire Si film, indicating that initiation of the crystalline deposition is almost immediate just on the polymer substrate. This is one of the major advantages in our deposition technique since it is generally noticed that incubation layer exists in conventional PECVD deposited Si films in its initial part before poly-Si starts to deposit.

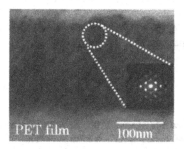
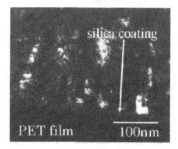

**Figure 4.** X-TEM bright (left) with diffraction pattern (inset) and dark (right) field images of the film grown at 150 °C. The bright parts seen in dark field image are crystallites aligned to the <111> directions.

## n-type doping Si thin films deposition

It is well known that under $H_2$ large dilution conditions, PECVD can provide the poly-Si deposition. However, Brogueira et al. [10] reported that $PH_3$ doping often caused the suppression of the crystallization. To investigate the suppression of crystallization by $PH_3$, n-doped Si films were prepared using a gas mixture of $SiH_4$, $PH_3$ and $H_2$. Figure 5 shows the Raman scattering spectra from the films deposited using the $PH_3$ flow rate ($R_{PH3}$= $PH_3/SiH_4 \times 100$ [%]) of 0.05, 0.10, and 1.0. The substrate temperature of 150 °C, the $R_{H2}$ of 1000, and the pressure of 500 Torr are common. With increasing $R_{PH3}$, the peak of the amorphous Si tends to increase. Hence, this implies that $PH_3$ inclusion suppresses crystallization even for the case of large $H_2$ dilution conditions ($R_{H2}$=1000). Moreover, the LO-TO phonon line shifts to 517.68 cm$^{-1}$, 517.28 cm$^{-1}$ and 515.73cm$^{-1}$ for the $R_{PH3}$ of 0.05, 0.10 and 1.0, respectively. Nickel et al. [11] reported that as the P concentrations increase, the LO-TO phonon line shifts to low wave number.

Figure 6 shows the effect of the $R_{PH3}$ on the conductivity of the n-doped Si film at 150 °C, 500 Torr and $R_{H2}$=1000. This figure shows that the conductivity increases with decreasing $[SiH_4]/[PH_3]$, and its maximum is about 0.1S/ cm. This result is due to the carrier scattering by $PH_3$. The hall mobility ($\mu_H$) was determined from as the ratio between the Hall coefficient and the sheet resistivity. As a result, the $\mu_H$ of the n-doped Si films at 150 °C, 500 Torr, $R_{H2}$=1000 and $R_{PH3}$=0.05 was 1.5 cm$^2$/V·s. This $\mu_H$ is generally higher than those obtained in amorphous Si at low temperature (=150 °C) [12].

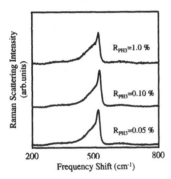

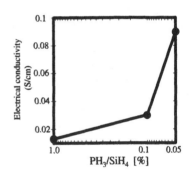

**Figure 5.** $PH_3$ flow rate ($R_{PH3}$) dependence of the Raman scattering spectra for growth at 150 °C, 500 Torr and $R_{H2}$=1000.

**Figure 6.** $PH_3$ flow rate ($R_{PH3}$) dependence of the conductivity for growth at 150 °C, 500 Torr and $R_{H2}$=1000.

## SiN$_X$ thin film deposition

The N/Si ratio of $SiN_X$ films prepared at the $N_2$ flow rate of 500, 1500, 3000 ml/min, at 100 °C, which is obtained from the Si2p and N1s core-level bands in XPS spectra shown in

Figure 7, at 0.7 0.9 and 1.0, respectively. Si2p peak position is calibrated to C1s (284.5 eV) originating from casual surface contaminant. The Si2p spectra are deconvoluted into four Gaussian components: the Si bonding feature at 99.6 eV, the $Si_3N_2$ one at 100.8 eV, the $Si_3N_4$ one at 101.7 eV and the $SiO_2$ one at 103 eV. With increasing N/Si ratio, both the Si and $Si_3N_2$ peaks decrease their intensity, leaving the $Si_3N_4$ peak dominant.

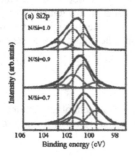 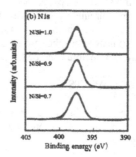

**Figure 7.** Si 2p (left) and N 1s (right) spectra of a-SiN$_x$ films growth at 100 °C and 500 Torr with different N/Si rate, as indicated.

In Figure 7, however, the $SiO_2$ peak intensity also increased with increasing $N_2$ flow rate, in which the origin has not been identified yet. One possibility is the increase of defects in the film, which act as an oxidation site during exposure to air after depositing. This is suggested because the plasma discharge becomes unstable with the increase of the $N_2$ flow rate as a result of the higher dissociation energy of $N_2$ molecule than that of $H_2$ molecule. The N1s spectra consist of a single band, indicating that all N atoms are in the same chemical environment (highly likely to be fully coordinated to three Si atoms) in agreement with Refs. [13] and [14].

Figure 8 shows the typical X-SEM image of the film deposited at 100 °C with $N_2$ flow rate of 3000 ml/min. From the image, the film thickness is estimated to be about 1450 nm, which corresponds to a deposition rate of 290 nm/min. This deposition rate is generally higher than those obtained in low pressure PECVD [15, 16].

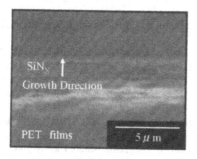

**Figure 8.** Cross section of the film deposition at 100 °C and 500 Torr with N/Si rate of 1.0 observed by scanning electron microscope (SEM).

## CONCLUSIONS

Low temperature deposition of Poly-Si, n-type doping Si and $SiN_X$ films on PET films have been achieved by employing a pulsed discharge based PECVD operated at near atmospheric pressures. Highly crystallized features of silicon-based thin films are shown by Raman scattering spectroscopy, XPS and X-TEM, and the $\mu_H$ of the n-doped Si films measured by Hall Effect measurement was 1.5 $cm^2$/V·s for the film deposited at low temperature (<150 °C). The authors believe that the present result may suffice to indicate the high potentiality of this pulsed-discharge near-atmospheric-pressure PECVD in flexible electronic devices such as flexible display and flexible solar cells.

## ACKNOWLEDGMENTS

This research has been supported by the Japan Science and Technology Agency and the Tohoku University Global COE program "Center of Education and Research for Information Electronics Systems".

## REFERENCES

1. C. S. McCormick, C. E. Weber, J. R. Abelsona, and S. M. Gates, Appl. Phys. Lett. **70**, 13 (1997).
2. N. D. Young, G. Harkin, R. M. Bunn, D. J. McCulloch, R. W. Wilks, and A. G. Knapp, IEEE Electron. Device Lett. **18**, 19 (1997).
3. J.K. Holt, D.G. Goodwin, A.M. Gabor, F. Jiang, M. Stavola, and H.A. Atwater, Thin Solid Films **430**, 37 (2003).
4. H. Kitabatake, M. Suemitsu, H. Kitahata, S. Nakajima, T. Uehara, and Y. Toyoshima, Jpn. J. Appl. Phys. **44**, L683 (2005).
5. Z. Iqbal, and S. Veprec, J. Phys. C. **15**, 377 (1982).
6. Hua Xia, Y. L. He, L. C. Wang, W. Zhang, X. N. Liu, X. K. Zhang, and D. Feng, J. Appl. Phys. **78**, 6705 (1995).
7. L. Houben, M. Luysberg, P. Hapke, R. Carius, and F. Finger, Philos. Mag. 77, 1447 (1998)
8. Y. Toyoshima, K. Arai, and A. Matsuda, J. Non-Cryst. Solids **114**, 819 (1989).
9. A. Matsuda, J. Non-Cryst. Solids **338**, 1 (2004).
10. P. Brogueira, V. Chu, A. C. Ferro, and J. P. Conde, J. Vac. Sci. & Technol. A**15**, 2968 (1997).
11. N. H. Nickel, P. Lengsfeld, and I. Sieber, Phys. Rev. B**61**, 15561 (2000).
12. R. Martins, A. Mac̦arico, I. Ferreira, R. Nunes, A. Bicho, and E. Fortunato, Thin Solid Films **317**, 144 (1998).
13. C.H.F. Peden, J.W. Rogers, N.D. Shinn, K.B. Kidd, and K.L. Tsang, Phys. Rev. B **47**, 15622 (1993).
14. L.G. Jacobsohn, R.K. Schulze, L.L. Daemen, I.V. Afanasyev-Charkin, and M. Nastasi, Thin Solid Films **494**, 219 (2006).
15. K.M. Chang, C.C Cheng, and C.C Lang, Solid-State Electron. **46**, 1399 (2002).
16. Y.T. Kim, D.S. Kim, and D.H. Yoon, Mater. Sci. Eng. B **118**, 242 (2005).

Mater. Res. Soc. Symp. Proc. Vol. 1066 © 2008 Materials Research Society      1066-A05-06

## Pulsed Laser Heating-induced Surface Rapid Cooling and Amorphization

Longzhang Tian[1], and Xinwei Wang[2]
[1]Department of Mechanical Engineering, University of Nebraska-Lincoln, Lincoln, NE, 68588
[2]Department of Mechanical Engineering, Iowa State University, Ames, IA, 50011

## ABSTRACT

In this work, hybrid atomistic-macroscale simulation is conducted to explore the crystallization process of Si surface in the situation of fast melting and solidification induced by ultrafast laser heating and heat conduction. Using the environment-dependent interatomic potential, samples containing 2,880 and 11,520 Si atoms are modeled to provide accurate details for the relationship between the final crystal structure and the parameters of laser pulses. An empirical correlation $E_c = 448.76 \times \left(t_g\right)^{0.56}$ is obtained to relate the critical fluence for amorphization to the laser pulse width. It is found that the final thickness of amorphous layer is related to the fluence of the laser pulse with the same full width at half maximum (FWHM). Employing laser pulses with FWHM = 6.67 ns, the formation and recrystallization processes of a 12 nm thick amorphous layer is further investigated, which may have great potential in laser manufacture techniques for Si-associated structures.

## I. INTRODUCTION

Pulsed-laser annealing techniques are widely used in the fabrication of amorphous semiconductors, which have various applications in solar cells, xerography, and flat-panel displays [1-6]. To understand the microscopics of new melting and solidification characteristics induced by the pulsed laser, much numerical and experimental work has been done in this field. Using nanosecond lasers, Gullis et al. demonstrated that the maximum growth velocities of the principal low index Si surfaces lie in the order of (001) > (011) > (112) > (111) [7]. When the crystal growing rate is beyond the maximum velocity, the crystal growth breaks down and a final amorphous solid phase is produced. This order of growth velocities was also predicted by molecular dynamics (MD) simulations conducted by Landman [8]. The accrual amorphization threshold of interface velocity has been measured by transient conductivity experiments and shown to be 15.8 m/s for (001) Si and 14.6 m/s for (111) Si [9-11].

The MD simulations done in this area mostly focused on the crystallization process of liquid Si driven by a constant cooling. This neither provided the relationship between the laser properties and final structures of Si nor studied the solidification process under the actual cooling rate caused by heat conduction through the solid Si substrate. In this work, hybrid atomistic-macroscale modeling is conducted to explore the solidification/crystallization characteristics of (001) Si under ultrafast laser heating and natural cooling situation. The structure change is captured during laser heating and post-laser solidification. The critical fluence for amorphization, the thickness and the recrystallization process of the final amorphous layer are explored as well.

## II. METHODOLOGIES OF SIMULATION

### A. Computational domain construction and laser absorption in the material

The potential used for MD simulation of Si is the EDIP which was developed by Bazant *et al.* and Justo *et al.* for bulk Si [12, 13]. Figure 1 illustrates the design of the computational domain for hybrid atomistic-macroscale modeling. Materials in Domain I is subjected to picosecond (ps) and nanosecond (ns) laser heating and experience intense structural change. MD simulation is conducted to model the material behavior in Domain I. Materials in Domain II experiences heat conduction. Therefore, the FD method is used to model the heat transfer in Domain II. The one-dimensional heat conduction equation is solved in Domain II for heat transfer. The thermophysical properties (specific heat and thermal conductivity) of the material are temperature dependent and updated in the FD simulation every time step. Periodical boundary conditions are used along the $x$ and $y$ directions and free boundary conditions are used in the $z$ direction shown in Figure 1.

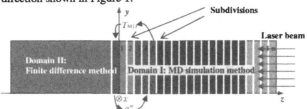

**Figure 1.** Schematic of the computational domain

In our study, two types of Domain I with different sizes are used. A small Domain I with a size of $1.629 \times 1.629 \times 21.72$ nm$^3$ ($x \times y \times z$) with 2,880 atoms is used for studying surface melting and solidification (section III.A) and critical laser fluence for amorphization (section III.B). A large Domain I of size $1.629 \times 1.629 \times 86.88$ nm$^3$ ($x \times y \times z$) with 11,520 atoms is used for the effect of laser fluence on the amorphous layer thickness (section III.C) and the recrystallization of the amorphous layer (section III.D) simulation. The size of Domain II is $1.629 \times 1.629 \times 12,591.304$ nm$^3$ ($x \times y \times z$) which is large enough to capture the heat conduction in the $z$ direction.

The laser pulses used in our simulation has a wavelength of 248 nm, a uniform spatial energy distribution in the $x$-$y$ plane (in Figure 1) and a temporal Gaussian distribution, $I(t) = I_0 \exp[-(t - t_0)^2 / t_g^2]$ where $I_0 = E / (t_g \times \pi^{1/2})$, $I$ is the laser intensity, $E$ the pulse energy, $t_g$ is equal to 0.6×FWHM (full width at half maximum), $t_0$ is the peak time of the laser pulse. In crystal Si (c-Si), the laser is absorbed exponentially with an optical absorption depth ($\tau$) of 5.54 nm following the formula of $dI / dz = I / \tau$ (details could be found in Ref. 14, 15 and 16).

### B. Procedure of MD and FD simulation

The time step chosen for the MD and FD simulation is 2 femtoseconds (fs). During the simulation, the initial equilibrium temperature of the system is uniform and set to room temperature (300 K). The program takes 500 ps to adjust the system to 300 K. The following 40 ps are used to equilibrate the system and calculate the first average temperature of the three bottom layers (dark red region in Domain I). Then laser heating and heat conduction begin, which are calculated by MD and FD simulations simultaneously. The heating time is set to $2t_0$ for different laser pulses, which is large enough to cover the whole laser pulse. After the laser heating process finishes, the MD simulation for the sample and FD simulation for the substrate continue for a sufficiently long time to simulate the cooling process driven by heat conduction.

## III. RESULTS AND DISCUSSION

### A. Melting and solidification

As the results of simulations using large Domain I (detailed in section II.A) and high laser fluences, Figure 2 provides general pictures of the melting and solidification process with amorphization caused by pulsed laser heating and natural heat conduction.

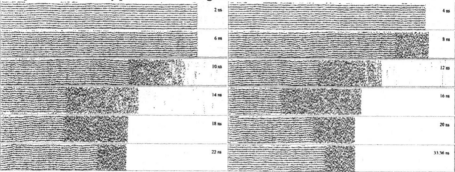

**Figure 2.** Snapshots of atomic positions in a $z$-$x$ plane of Domain I for laser amorphization of Si (horizontal: $z$ coordinate, 0~100 nm, vertical: $x$ coordinate, 0~1.7 nm). The laser fluence is 1300 J/m$^2$, $t_g$ = 4 ns. Size of Domain I: 1.629×1.629×86.88 nm$^3$ ($x$×$y$×$z$).

In Figure 2, snapshots of the Si film (Domain I in Figure 1) at different times illustrate its structural change during the laser heating and natural cooling process. Heated by a laser pulse with $t_g$ = 4 ns and fluence of 1300 J/m$^2$ from 0 to 16 ns (4$t_g$), the Si surface (Domain I in Figure 1) melts after 6 ns and the top layer of Domain I reaches the boiling point at about 10 ns when the top layer begins to vaporize. The vaporization process ends and Domain I begins to solidify/recrystallize after 14 ns. The solidification concludes at around 22 ns leaving a 12 nm-thick amorphous layer on the top of Domain I.

Using the crystallinity function, Figure 3 shows the evolution of the crystal-liquid/amorphous interface in the fast melting and solidification process [17].

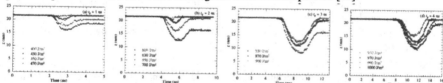

**Figure 3.** Crystal-liquid/amorphous interface evolution during laser heating and subsequent cooling: (a) $t_g$ = 1 ns, (b) $t_g$ = 2 ns, (c) $t_g$ = 3 ns, and (d) $t_g$ = 4 ns.

For laser pulses with the same $t_g$, the sample recrystallizes to a single crystal perfectly when the laser fluence is smaller than a critical value $E_c$ (detailed in section III.B). Above this value, amorphous layers with different thicknesses are formed on top of the sample. The thickness of the amorphous layer is greater when higher laser fluences being used and when the temperature of the top layer of the sample not reaching the boiling point of Si. Further discussion about the thickness of the amorphous layer and the laser fluence is provided in Section III.C. The velocity of the interface movement is around 3~5 m/s. This velocity is less than the critical value (15.8

m/s) measured by transient conductivity experiments [9-11].

## B. Critical laser fluence for amorphization

As discussed in Section III.A, amorphization happens when the laser fluence exceeds a certain critical value. In this work, laser pulses with different $t_g$ are used to identify the critical laser fluence above which an amorphous layer will form during post-laser solidification. The $t_g$ takes the range of 0.05~4 ns. Small Domain I is used for simulations in this section. Figure 4 shows the critical fluences of lasers with different FWHMs, which do not exhibit a linear relationship with $t_g$.

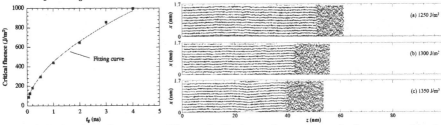

**Figure 4.** Critical laser fluences with different $t_g$ for amorphization to occur.

**Figure 5.** Final sample structures for laser amorphization process ($t_g$ = 4 ns). The laser fluences are (a) 1250 J/m²; (b) 1300 J/m²; and (c) 1350 J/m².

Using power function curve fitting, an empirical formula (fitting curve in Figure 4) is given for the $t_g$ and critical laser fluence $E_c = 448.76 \times (t_g)^{0.56}$. This relationship could be explained approximately by the heat conduction from the melting region (in Domain I) to the solid part (in Domain II). The formation of the amorphous layer is induced by the high cooling rate. The critical cooling rate is caused by the critical heat flux between Domains I and II (shown in Figure 1), which means a unique critical heat flux does exist above which the epitaxial re-growth cannot be sustained. Consider the heat conduction from Domain I to the layer at the thermal diffusion length position of Domain II (at room temperature). The thickness of Domain I could be neglected for the thermal diffusion length ($\Delta l$) is much longer than it. Noting that the temperature difference is caused by critical laser fluence ($E_c$) when the heating laser induces the critical heat flux and neglecting the temperature change caused by heat conduction during the fast laser heating, the critical heat flux ($q_c''$) is $q_c'' \propto (T - T_{room}) / \Delta l \propto E_c / \Delta l$, where $T$ is the average temperature of Domain I at the time of laser pulse stops. In our simulation, we find that superheating happens and $T$ is proportional to the overall laser energy input, which leads to the above equation. Then $E_c$ can be estimated as $E_c \propto q_c'' \times \Delta l \propto q_c'' \times (t_g \times \alpha)^{1/2} \propto t_g^{1/2}$ where $\alpha$ is the thermal diffusivity of Si in Domain I. This first order estimation is very close to the formula concluded from the simulation results.

## C. Effect of laser fluence on the amorphous layer thickness

As discussed in Section III.A, the laser fluence is one of the factors affecting the thickness of the amorphous layer besides FWHM. It is shown in Figure 3 that the thickness of the amorphous layer is proportional to the laser fluence when it exceeds $E_c$ but not too high to cause the top of Domain I to reach the boiling point. The thickness of the amorphous layer created by the laser pulse is usually several nanometers. To achieve a thicker amorphous layer, e.g. 10 nm, high laser fluences are needed. But this may make the situation complicated because high laser fluences may make Domain I (Si surface) become shorter than its initial thickness by vaporizing its top part. In this section, large Domain I and laser pulses with $t_g = 4$ ns (FWHM = 6.67 ns) are used to study the amorphous layer thickness controlled by the laser fluence when the fluence is large enough to cause the Domain I to experience evaporation (as shown in Figure 2).

With $t_g = 4$ ns, laser pulses with different fluences result in different amorphous layer thickness. Figure 5 shows the final structure of Domain I after amorphization. When increasing the laser fluence, the sample becomes shorter after solidification because more atoms reach the boiling point of Si and vaporize. Another reason is that a-Si has a higher density than c-Si. Therefore, when the laser fluence is higher, the amorphous layer is thicker, and the overall thickness of the sample looks shorter. When the laser fluence varies from 1250 to 1350 J/m$^2$, the resulting thicknesses of amorphous layer changes from 10 to 14.5 nm.

## D. Recrystallization of the amorphous layer by a second laser pulse

It is reported that the melting point of a-Si is 200K - 600K lower than that of c-Si [18, 19]. When heated by a second laser pulse of low/moderate fluence, the amorphous layer melts and recrystallizes to crystal structure for the low laser pulse provides a small heating and cooling rate. Figure 6 shows the crystal-liquid/amorphous interface and temperature evolution of the sample with a 12-nm-thick amorphous layer created by 1300 J/m$^2$ laser ($t_g = 4$ ns, discussed in Section III.A, Figure 2).

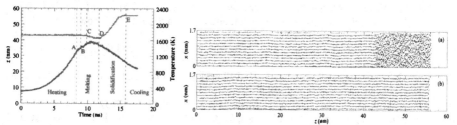

**Figure 6.** Crystal-liquid/amorphous interface (red) and average temperature (blue) evolution of the sample during recrystallization.

**Figure 7.** Snapshots of atomic positions: (a) before laser recrystallization (0 ns); (b) after laser recrystallization (17 ns).

Heated by a laser pulse of 900 J/m$^2$ ($t_g = 4$ ns, pulse width is $4t_g$), the amorphous layer begins to melt at point A and the crystal domain begins to melt at time point C. The heating process between time point B (8.9 ns) and C (9.8 ns) heats the following crystal domain to the melting point of c-Si. After point D (11.5 ns), the solidification process begins, and finally the sample recrystallizes to a single crystal at time point E (15.8 ns). Figure 7 shows the snapshots of atomic positions of the sample before and after laser recrystallization using pulsed laser with $t_g = 4$ ns and 900 J/m$^2$ in fluence. It is shown in Figure 7 that the amorphous layer crystallizes to crystal structure with a low energy laser pulse. The crystallinity function of the whole crystal part shown in Figure 7 (a) and (b) is also calculated and is 0.91 for (a) and 0.90 for (b). These values also prove the final crystal structure after the recrystallization process.

## IV. CONCLUSIONS

With the combined MD/FD simulation, the structural evolution of Si surface heated by pulsed lasers was studied. When the laser fluence was above a critical value ($E_c$), amorphous layers formed after laser heating and natural conduction cooling. The critical fluence seemed to satisfy a power function relationship with $t_g$. When the fluence of a $t_g = 4$ ns pulsed laser is increased from 1250 to 1350 J/m$^2$, the thickness of the amorphous layer formed after solidification varies from 12 to 14.5 nm. Using a lower laser fluence, the amorphous layer could be recrystallized completely. These amorphization and crystallization characteristics of Si may have great potentials in the fabrication of Si material.

## ACKNOWLEDGEMENTS

Support for this work from NSF (CMS: 0457471), Nebraska Research Initiative, Air Force Office for Scientific Research, and MURI from ONR is gratefully acknowledged.

## REFERENCES

1.  D. E. Carlson, and C.R. Wronski, Appl. Phys. Lett. **28**, 671-673 (1976).
2.  B. K. Nayak, B. Eaton, J. A. A. Selvan, J. Mcleskey, M. C. Gupta, R. Romero, and G. Ganguly, Appl. Phys. A **80**, 1077 (2005).
3.  T. Suzuki, and S. Adachi, Jpn. J. Appl. Phys. **32**, 4900 (1993).
4.  J. S. Im, and H. J. Kim, Appl. Phys. Lett. **63**, 1969 (1993).
5.  M. Miyasaka, and J. Stoemenos, J. Appl. Phys. **86**, 5556 (1999).
6.  Y. -C. Wang, J. -M. Shieh, H. -W. Zan, and C. -L. Pan, Opt. Express **15**, 6982 (2007).
7.  A. G. Cullis, N. G. Chew, H. C. Webber, and D. J. Smith, J. Cryst. Growth **68**, 624 (1984).
8.  U. Landman, W. D. Luedtke, M. W. Ribarsky, R. N. Barnett, and C. L. Cleveland, Phys. Rev. B **37**, 4637 (1988).
9.  M. O. Thompson, J. W. Mayer, A. G. Cullis, H. C. Webber, N. G. Chew, J. M. Poate, and D. C. Jacobson, Phys. Rev. Lett. **50**, 896 (1983).
10. P. A. Stolk, A. Polman, and W. C. SInke, Phys. Rev. B **47**, 5 (1993).
11. A. Polman, P. A. Stolk, D. J. W. Mous, W. C. SInke, C. W. T. Bulle-Lieuwma and D. E. W. Vandenhoudt, J. Appl. Phys. **67**, 4024 (1990).
12. J. F. Justo, M. Z. Bazant, E. Kaxiras, V. V. Bulatov, and S. Yip, Phys. Rev. B **58**, 2539 (1998).
13. M. Z. Bazant, E. Kaxiras, and J. F. Justo , Phys. Rev. B **56**, 8542 (1997).
14. D. R. Lide, *CRC Handbook of Chemistry and Physics*, 86th Edition, 12 (2005).
15. Wang, X., J. Phys.: D Applied Physics **38**, 1805 (2005).
16. X. Wang and X. Xu, ASME Journal of Heat Transfer **124**, 265 (2002).
17. X. Wang and Y. Lu, J. Appl. Phys. **98**, 114304: 1-10 (2005).
18. P. Baeri, G. Foti, J. M. Poate and A. G. Cullis, Phys. Rev. Lett. **45**, 2036 (1980).
19. Ya. V. Fattakhov, M. F. Galyautdinov, T. N. L'vova and I. B. Khaibullin, Tech. Phys. **42** (12), 1457 (1997).

Mater. Res. Soc. Symp. Proc. Vol. 1066 © 2008 Materials Research Society          1066-A05-07

# Studies on the Surface Reactions of Substituted Disilanes with Silica Surface

Tom Blomberg[1], Raija Matero[1], Suvi Haukka[1], and Andrew Root[2]

[1]R&D, ASM Microchemistry Ltd., Väinö Auerin katu 12 A, Helsinki, 00560, Finland

[2]MagSol, Tuhkanummenkuja 2, Helsinki, 00970, Finland

## ABSTRACT

Both CVD and ALD deposition techniques benefit from a detailed understanding of the reaction mechanisms of the precursor molecules with the surface. In this paper, the reactions of hexakis ethylaminodisilane (AHEAD™), hexamethoxydisilane and hexamethyldisilane were studied on high surface area silica granules at 200-375 °C. Silica was heat treated at 200-820 °C to control the number of surface Si-OH groups. The samples were characterized by FTIR and solid state NMR spectroscopy. After the chemisorption of the precursors with silica, Si-H bonds, not originally present in the molecules, were identified for AHEAD and hexamethoxydisilane, but not for hexamethyldisilane. It is suggested that with AHEAD and hexamethoxydisilane, cleavage of the Si-Si bond takes place during the chemisorption with Si-OH sites. Since no reaction for hexamethyldisilane at the studied temperatures was observed, a prerequisite for the reaction with Si-OH groups seems to be the presence of electronegative O or N atoms in the ligands. In the paper, possible reaction mechanisms with the various surface species are discussed.

## INTRODUCTION

In the semiconductor industry, $SiO_2$ is typically deposited by CVD at elevated temperatures. For instance from silane and oxygen between 300 and 500 °C, from dichlorosilane and $N_2O$ around 900 °C, and from TEOS ($Si(OC_2H_5)_4$) and $O_2$ between 650 and 750 °C. The requests for lower thermal budgets in the device processing flows call for new silicon precursors and deposition methods that work at lower temperatures. ALD is one of the most promising advanced deposition methods that can provide good quality films with excellent step coverage in deep vias and trenches.

In CVD the growth is based on the continuous introduction and decomposition of the precursors to the growing surface, while in ALD the precursors are pulsed alternately to the surface and the growth relies on the chemisorption or reaction of the precursors with the reactive sites. For instance $Al(CH_3)_3$ reacts with Al-OH groups forming $Al-O-Al(CH)_2$ surface species and $CH_4$ gas. Thus the ligands stay intact until the ligand removal agent is introduced. Both CVD and ALD deposition techniques benefit from a detailed understanding of the reaction mechanisms of the precursor molecules with the surface. For ALD in particular, it is important to know whether the precursor is able to chemisorb and how it actually chemisorbs, in other words, which type of reactive sites it uses. In the growth of $SiO_2$ the possible reactive sites are Si-OH or Si-O-Si groups. In this paper the reactivity and reaction mechanisms of three different disilanes, namely hexakis ethylaminodisilane (AHEAD™), hexamethoxydisilane and hexamethyldisilane were studied on high surface area silica granules at 200-375 °C. It is shown that AHEAD™ and hexamethoxydisilane are suitable ALD precursors, whereas hexamethyldisilane does not chemisorb on the surface of silica and therefore cannot be applied as an ALD precursor. Also the reason for this difference in reactivity is proposed.

## EXPERIMENT

The reaction mechanisms of hexakis-ethylaminodisilane (AHEAD™), hexamethoxydisilane and hexamethyldisilane with high surface area silica granules (EP10) were studied at 200-375 °C. Silica granules were heat treated at 200-820 °C before the exposures to the disilanes in order to control the number of surface Si-OH groups. The exposure tests were done in F-120 ALD reactor from ASM Microchemistry Ltd equipped with a sample holder for porous silica particles. After the exposures, the samples were characterized by FTIR (Nicolet) and solid state NMR spectroscopy (MagSol company, Finland).

## RESULTS

Figure 1 compares the IR spectra of the pure precursors and pure silica with those, after the surface reactions. Clearly the spectrum of hexamethyldisilane after the surface reaction does not show any significant variation from the spectrum of pure silica, thus confirming that no reaction had taken place between the two. For hexakis-ethylaminodisilane (AHEAD™) and hexamet-hoxydisilane the IR spectra after the surface reactions show a new intensive absorption peak at approximately 2200-2300 cm⁻¹. In the literature, this peak position has been assigned to be indicative of the Si-H bond [1]. Further confirmation of the Si-H bond was achieved with $^1$H MAS NMR measurements shown in Figure 2. The Si-H peak at 3.7 ppm is clear for hexakis-ethylaminodisilane (AHEAD™), but unfortunately for hexamethoxydisilane the Si-H peak position is very close to the O-CH₃ peak and thus cannot be clearly separated.

Figure 3 shows the $^{29}$Si MAS NMR measurements of the samples. In addition to the typical peak or peaks due to the Si in silica at -100-120 ppm, the spectra show four peaks at -65,-75,-85 and -95 ppm for hexamethoxydisilane and a number of peaks, not clearly separable at -10-70 ppm for hexakis-ethylaminodisilane (AHEAD™). These peaks can be assigned to Si with many different ligand configurations. For example, the four peaks in Figure 3b can be assigned for the following surface species:

1)  2)  3)  4)

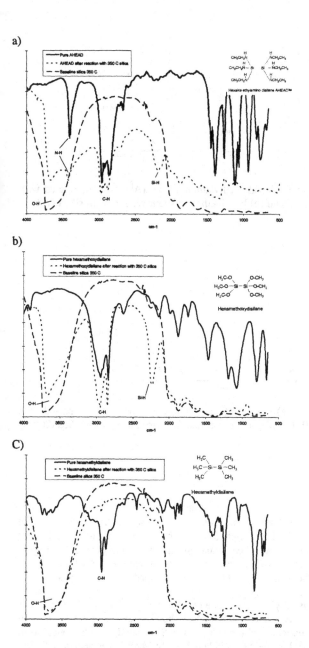

**Figure 1.** IR spectra of (a) hexakis-ethylaminodisilane(AHEAD™), (b) hexamethoxydisilane and (c) hexamethyldisilane before and after the surface reaction with silica granules.

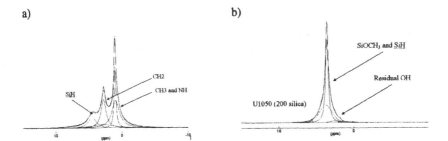

**Figure 2.** 1H MAS NMR spectra of (a) hexakis-ethylaminodisilane(AHEAD™), reacted with silica (450 °C heat treatment) at 250 °C and (b) hexamethoxydisilane reacted with silica (200 °C heat treatment) at 200 °C.

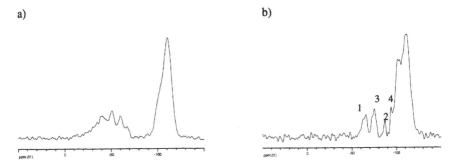

**Figure 3.** $^{29}$Si MAS NMR spectra of (a) hexakis-ethylaminodisilane(AHEAD™), reacted with silica (200 °C heat treatment) at 200 °C and (b) hexamethoxydisilane reacted with silica (200 °C heat treatment) at 200 °C.

## DISCUSSION

The IR and NMR experiments indicate the formation of Si-H bonds after the reaction of hexakis-ethylaminodisilane (AHEAD™) and hexamethoxydisilane with silica surface in the studied temperature range, 200 - 375 °C. The presence of Si-H bonds in the samples must result from the cleavage of the Si-Si bond, because these bonds are not originally present in the precursors and a direct ligand release reaction would not lead to the formation of Si-H bonds. The Si-H bonds seem to be located in the adspecies and not directly bonded to the surface silicon of the silica. This can be concluded from the absence of the $O_3$–Si-H peak (-84.5 ppm [2]) in the $^{29}$Si MAS NMR spectrum of hexakis-ethylaminodisilane (AHEAD™) and the splitting of the IR absorption peak of the Si-H bond to two maximums (a Si-H bond at the surface silicon atom would lead to only one type of chemical environment for the Si-H bond, which would mean singlet peaks in the IR spectrum). For hexamethoxydisilane the $^{29}$Si MAS NMR peak at -85 ppm was not assigned for surface Si-H bond, because the O-Si-$(OCH_3)_3$ peak coincidences with it and was assumed to be the more likely species responsible for this peak.

The Si-Si bond was not directly identified, because it is very weakly IR active (Si-Si stretching at 500-600 cm$^{-1}$[3], but weak and outside the wavelength range of the spectrometer we used) and because it cannot be separated in the NMR spectra. It is still assumed however, that also the ligand release reaction scheme leaving the Si-Si bond intact in the adspecies takes place. This has been indirectly suggested by the high ALD growth rates observed in our SiO$_2$ growth studies [4]. In future studies the confirmation of the Si-Si bonds is tempted with Raman spectroscopy. Figure 4 shows a schematic presentation of the proposed reaction routes for the precursors studied in this study.

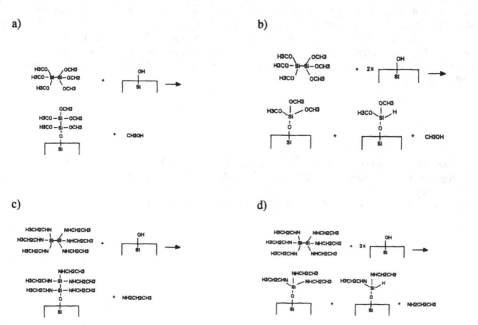

**Figure 4.** Proposed reaction routes for the surface reactions of hexamethoxydisilane (a, b) and hexakis-ethylaminodisilane (AHEAD™) (c, d) with the hydroxyl groups on the silica surface.

## CONCLUSIONS

Hexamethyldisilane does not chemisorb on silica surfaces. Hexakis-ethylaminodisilane (AHEAD™) and Hexamethoxydisilane chemisorb following two possible reaction schemes: 1) Reaction with the surface hydroxyl groups by cleavage of the Si-Si bond followed by a ligand release reaction of the formed monosilane fraction leaving two monosilane adspecies attached to the surface. 2) Ligand release reaction with the surface hydroxyl groups leaving the Si-Si bond intact in the adspecies. Doubly and possibly also triply bonded adspecies are also possible with both mechanisms.

Chemisorption of substituted disilanes with surface hydroxyl groups seems to require electronegative substituents surrounding the silicon-silicon bond. Indeed, the electronegative ligands most likely change the polarization of the Si-Si bond making it more reactive to electrofilic attack by the oxygen atom in the surface hydroxyl group.

Further tests will focus on the determination of the Si-Si bond in the adspecies. Additionally, the influence of different types of hydroxyl groups on the relative importance of the two reaction mechanisms, and the possible effect it has on the ALD growth rates (Å/cycle) will be studied.

## REFERENCES

[1] Lucovsky, G., Nemanich, R. J., Knights, J. C., Phys. Rev. B 19, 2064 - 2073 (1979)
[2] Agaskar, P. A., Klemperer, W. G., Inorg. Chim. Acta, 229, 355-364 (1995)
[3] Hinchley et al., J. Chem. Soc. Dalton Trans., 2916-2925 (2001)
[4] Matero et al., Proceedings of the 213th ECS meeting 2008, Phoenix, USA.

# Poster Session:
## Amorphous, Micro-, Nano-
## and Polycrystalline Silicon

Mater. Res. Soc. Symp. Proc. Vol. 1066 © 2008 Materials Research Society

## Polycrystalline Silicon Thin-film Solar Cells on ZnO:Al Coated Glass

Christiane Becker, Pinar Dogan, Benjamin Gorka, Florian Ruske, Tobias Hänel, Jan Behrends, Frank Fenske, Klaus Lips, Stefan Gall, and Bernd Rech
Dep. Silicon Photovoltaics, Helmholtz-Zentrum Berlin für Materialien und Energie (formerly Hahn-Meitner-Institut Berlin), Kekulestr. 5, Berlin, 12489, Germany

## ABSTRACT

Thin film solar cells based on polycrystalline silicon are an appealing option combining the advantages of thin film technologies, namely low cost, and the superior electrical properties of crystalline silicon. The specific structure aimed at in this work uses the relatively simple contacting scheme used in amorphous silicon thin film solar cells which relies on a front contact consisting of a transparent conducting oxide (TCO). Electron-beam evaporation is applied as preparation method for the silicon films with high deposition rate. Solid phase crystallization (SPC) is used to crystallize the silicon films after deposition.

In this work the properties of as-deposited and crystallized silicon films produced by e-beam evaporation are investigated as a function of deposition temperature. It is shown that the largest crystallites are obtained for deposition around 300°C and subsequent treatment at 600°C, while deposition at higher temperatures leads to small crystal grains which are not changed significantly during SPC.

Solar cells have been prepared on TCO-coated glass and were studied by measurements of the external quantum efficiency.

## INTRODUCTION

Thin film solar cells based on silicon usually use amorphous and/or microcrystalline silicon layers as absorbers. Tandem solar cells consisting of two stacked thin film solar cells currently reach stabilized efficiencies of about 10 %. An appealing alternative is the use of thin crystalline films as absorbers as, unlike amorphous silicon, they do not suffer from light-induced degradation. The main challenge is the formation of high quality films with a thickness of a few micrometers on a low cost substrate at high deposition rate and low temperatures.

The desired substrate for this type of solar cells is glass, which limits the feasible process temperatures to around 600°C. At these temperatures, films of amorphous silicon are transformed into polycrystalline silicon by solid phase crystallization (SPC). This concept has already been used over ten years ago by Sanyo to produce a single junction solar cell with a conversion efficiency of almost 10% [1]. CSG Solar has recently shown mini-modules of 10 x 10 cm$^2$ with an efficiency of 10.4% [2].

A key technological challenge is the formation of suitable contacts to the device. While CSG uses a sophisticated scheme of point contacts, thin film technology based on amorphous silicon uses a simple design, which is based on a TCO (transparent conductive oxide) front contact and laser scribing for integrated series connection of solar cells.

This work aims at the formation of poly-Si thin film solar cells on TCO-coated glass substrates by SPC in superstrate configuration. Recently, it was found that ZnO:Al films are thermally stable upon SPC-temperatures[3] and can be used as TCO for our solar cell approach.

In order to achieve higher deposition rates as compared to the usually used PECVD process, e-beam evaporation is used to deposit the silicon films with deposition rates around 200 nm/min. E-beam evaporation of silicon for SPC is also investigated by other research groups [4,5].

This paper describes the influence of the substrate temperature during e-beam deposition on the properties of silicon films before and after SPC. First solar cell results based on silicon films grown at high deposition temperature without further thermal treatment are discussed.

## EXPERIMENT

### Sample preparation

The silicon films under investigation with a thickness of around 1 μm were grown by e-beam evaporation onto bare Corning 1737 glass. Doping of the absorber layer was obtained by co-evaporation of boron using a high-temperature effusion cell. The substrate temperature was varied from 200 to 600°C leading to silicon films with different morphologies from amorphous to fine-crystalline phase. These layers were solid phase crystallized after deposition by tempering up to 20 h at 600°C in nitrogen atmosphere.

For preparation of solar cells, glass substrates with a 700 nm thick ZnO:Al layer were used produced by RF magnetron sputtering at Forschungszentrum Jülich (Germany) [6]. The ZnO:Al layer serves as transparent front contact, on which an $n^+/p^-/p^+$ solar cell structure is subsequently deposited. The highly doped $n^+$ and $p^+$ layers are produced by PECVD, the $p^-$ absorber layer is deposited by e-beam evaporation as described above. SPC is carried out after $p^-$ deposition followed by a hydrogen passivation treatment in a special plasma tool which allows for substrate temperatures of up to 650°C. Details on the solar cell preparation can also be found in [7].

### Characterization methods

The structural properties of the poly-Si films were investigated by Raman spectroscopy with an excitation wavelength of $\lambda = 632.8$ nm and scanning electron microscope (SEM).

Standard room temperature electron spin resonance (ESR) was utilized for spin density measurements of the Si dangling bonds. For this purpose the samples were cut into an appropriate size (4 x 10 mm$^2$) and placed in a continuous-wave (cw) ER4104OR resonator of a Bruker ESP300 X-Band (9.5 GHz) spectrometer. A magnetic-field modulation of 0.4 mT at a frequency of 100 kHz was used. The microwave power was set to 2 mW such that saturation effects did not occur. The spin density calibration was achieved by comparing the ESR response of the samples to a known spin standard.

The external quantum efficiency (EQE) of the solar cells was measured under short circuit conditions.

## RESULTS AND DISCUSSION

The solar cell performance is expected to strongly depend on the properties of the absorber layer. Due to the highly doped TCO, $n^+$ and $p^+$ layers many characterization techniques could not be carried out on complete solar cell stacks, so that p⁻ absorber layers deposited onto bare glass were studied first before and after SPC. Solar cell results in the second part refer to the complete solar cell structure as shown in Fig. 1.

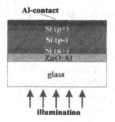

**Figure 1.** Schematic drawing of a poly-Si thin-film solar cell on ZnO:Al-coated glass in superstrate configuration. The structure consists of a glass substrate, a ZnO:Al layer as transparent conductive oxide (TCO), an $n^+$-type poly-Si emitter, a p-type poly-Si absorber, a $p^+$- type a-Si:H back surface field (BSF), and an aluminum contact.

## Solid phase crystallized silicon thin films

During deposition of the p⁻ absorber layers the substrate temperature was varied from 200°C and 600°C. Raman spectra of these samples (thickness 1 µm) can be seen in Fig. 2a). For low substrate temperatures a broad peak around 480 cm⁻¹ can be seen, indicating that the morphology is amorphous. At a deposition temperature of 400°C a small sharp peak at about 520 cm⁻¹ becomes visible resulting from the optical phonon of the crystalline fraction inside the material. For even higher deposition temperatures (T ≥ 500°C) the layers are grown directly crystalline.

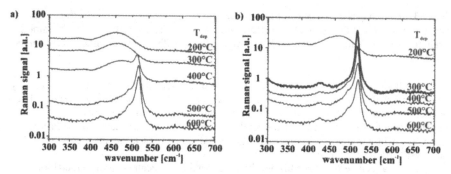

**Figure 2.** Raman spectra of silicon layers on glass deposited by electron-beam evaporation at various deposition temperatures $T_{dep}$ a) as-grown and b) after annealing for 8 h at 600°C in N₂ atmosphere. The curve with the sharpest peak (FWHM = 4.5 cm⁻¹) at the Raman shift of about 520 cm⁻¹ is highlighted by bold line. All curves are vertically displaced for better clarity.

After annealing of the samples for 8 h at 600°C, the Raman spectra were measured again (Fig. 2b). No change could be observed for samples that have already been crystalline before SPC ($T_{dep}$ = 500°C and 600°C). The amorphous fraction of the 400°C-sample vanished during annealing, indicating SPC of the amorphous parts. The sharpest peak (FWHM = 5.06 cm$^{-1}$) in the Raman spectrum was found for the film deposited at $T_{dep}$ = 300°C after annealing. The sample at $T_{dep}$ = 200°C was not crystallized after 8 h heat treatment at 600°C.

The structural properties of the layers were investigated by SEM for $T_{dep}$ = 300°C-600°C. Figure 3 shows the observed surface morphologies before and after SPC. As expected for an amorphous film the sample grown at 300°C shows no distinctive morphology. After SPC no change is observed in the surface morphology, but crystal grains of about 1 µm size can be distinguished from the brightness contrast. This shows that rather flat films with large grains can be obtained by SPC.

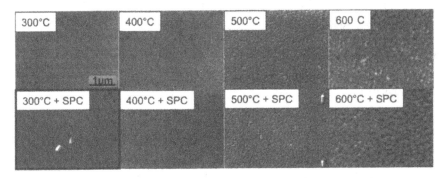

**Figure 3**. SEM images of the silicon film surface directly grown on glass by electron-beam evaporation at various deposition temperatures (upper row). The lower row shows the same set of samples after SPC for 8 h at 600°C. The sample marked with the black frame is also marked in Fig. 2b.

For higher deposition temperatures ($T_{dep}$ > 400°C) small grains can be observed already in the as deposited samples. Their grain size strongly increases with substrate temperature but does not exceed a few 10 nm. Annealing does not cause a remarkable difference for these samples. These findings fit well with the results of the Raman measurements above: For the silicon layer with the sharpest Raman peak ($T_{dep}$ = 300°C after annealing) the biggest crystal grains were observed in SEM.

For solar cell performance the density of recombination centers plays a crucial role which, if they are paramagnetic, can be identified by ESR measurements. In ESR a single line is observed in all studied films with a g-value that ranges between 2.0050(5) and 2.0060(5) which is attributed to Si dangling-bonds [8]. As shown in Figure 4, the spin density, $N_S$, of the Si dangling bonds clearly decreases in the as-grown state with increasing deposition temperature. Most surprisingly, after SPC those films deposited at $T_{dep}$ = 300°C show the lowest dangling bond density ($4 \times 10^{17}$ cm$^{-3}$). For all samples no hydrogen-passivation was applied. These results correlate well with the outcomes from Raman and SEM. The sample with the sharpest Raman peak and the biggest crystal grains, observed by SEM, also has the lowest spin density. SPC seems to have no influence on the samples with already crystalline morphology ($T_{dep} \geq 500°C$) which was already seen in Raman and SEM.

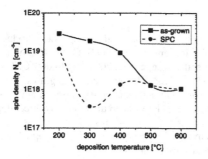

**Figure 4.** Spin density of Si dangling bonds measured by ESR at T = 300 K of 1 µm thick Si deposited by e-beam evaporation at different temperatures before (as-grown) and after SPC for 8 h at 600°C (solid line).

### Silicon thin-film solar cells on ZnO

First solar cell structures were processed as described in the first part of this paper. The best solar cell so far has an absorber layer deposited at 600°C and therefore fine-crystalline morphology. The total silicon thickness in the solar cells was about 1.2 µm. No light trapping, i.e. texturing of the layers, was implemented. Figure 5 shows EQE measurement on this sample without bias light background. The maximum EQE value is about 52%. The short circuit density deduced out of the EQE-measurement by integration is 9.4 mA/cm². The $V_{oc}$ values measured under 1 Sun illumination is about 380 mV.

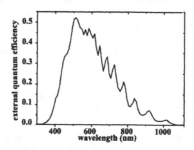

**Figure 5.** External quantum efficiency (EQE) versus wavelength of a poly-Si thin-film solar cells (thickness about 1.2 µm, no texture) on ZnO:Al-coated glass measured without bias light. The cell was prepared by direct deposition of the absorber at 600°C. The corresponding short circuit current density is 9.4 mA/cm².

As a next step, the knowledge gained by the investigations of the absorber material will be transferred to the preparation of poly-Si thin-film solar cells. Furthermore, other fabrication steps, e.g. thermal defect annealing, hydrogen-passivation, light trapping and contacting, have to be optimized.

## CONCLUSIONS

In conclusion, a poly-Si thin-film solar cell concept was introduced using TCO-coated glass as substrate. The poly-Si films are obtained by SPC at 600°C of layers deposited by e-beam evaporation in the temperature range from 200°C up to 600°C. The layers were characterized by Raman, SEM and ESR. The biggest crystal grains, the sharpest Raman peak and the lowest dangling bond density were obtained by deposition at 300°C and subsequent SPC. Best solar cell results, however, could be reached up to now by direct deposition at 600°C and showed a short circuit current density $J_{SC}$ of 9.4 mA/cm$^2$ and an open circuit voltage $V_{OC}$ of about 380 mV. Further steps in solar cell preparation will be adapted to silicon films obtained by low temperature deposition and subsequent SPC.

## ACKNOWLEDGMENTS

The authors would like to thank Jürgen Hüpkes from Forschungszentrum Jülich (Germany) for the preparation of the ZnO:Al films and S. Common, K. Jacob, C. Klimm and A. Scheu from HMI for their assistance during sample preparation.
The work has been supported by the FP6 research project ATHLET (Contract No. 019670-FP6-IST-IP) and BMU project (Contract No. 0327581).

## REFERENCES

1. T. Matsuyama, N. Terada, T. Baba, T. Sawada, S. Tsuge, K. Wakisaka and S. Tsuda, *J. Non-Cryst. Solids* **198-200**, 940 (1996).
2. M.J. Keevers, T.L. Young, U. Schubert, M.A. Green, *Proc. of the 22nd European Photovoltaic Solar Energy Conference*, Milan, Italy (2007) , p. 1783
3. K. Y. Lee, C. Becker, M. Muske, F. Ruske, S. Gall, B. Rech, M. Berginski and J. Hüpkes, *Appl. Phys. Lett.* **91**, 241911 (2007).
4. D. Song, D. Inns, A. Straub, M. L. Terry, P. Campbell and A. G. Aberle, *Thin Solid Films* **513**, 356 (2006).
5. C. Secouard, C. Ducros, P. Roca i Cabarrocas, T. Duffar and F. Sanchette, *Proc. of the 22nd European Photovoltaic Solar Energy Conference*, Milan, Italy (2007), p. 2036
6. C. Agashe, O. Kluth, J. Hüpkes, U. Zastrow, B. Rech, and M. Wuttig, J. Appl. Phys. **95**, 1911 (2004).
7. C. Becker, E. Conrad, P. Dogan, F. Fenske, B. Gorka, T. Hänel, K. Y. Lee, B. Rau, F. Ruske, T. Weber, M. Berginski, J. Hüpkes, S. Gall and B. Rech, *submitted for publication in Solar Energy Materials and Solar Cells (Proceedings of the 17th Photovoltaic Solar Energy Conference, Fukuoka, Japan (2007))*.
8. K. Lips, P. Kanschat and W. Fuhs, *Solar Energy Materials and Solar Cells* **78**, 513 (2003).

Mater. Res. Soc. Symp. Proc. Vol. 1066 © 2008 Materials Research Society 1066-A06-05

# Photoluminescence of Different Phase Si Nanoclusters in Amorphous Hydrogenated Silicon

Tatyana V. Torchynska

Material Science, National Polytechnic Institute, av. IPN, U.P.A.L.M., Ed.9, Mexico D.F., -
7738, Mexico

## ABSTRACT

This paper presents the results of XRD and PL spectrum studies for Si nano-clusters embedded in the amorphous silicon matrix. Investigated layers were deposited by the hot-wire CVD method on glass substrates at the wafer temperature 300°C and different filament temperatures from the range 1735-1885°C. It was shown that variation of filament temperatures allows producing the films with desirable parameters. Using of X-ray diffraction and photoluminescence methods the correlation between the intensity of some photoluminescence bands and the concentrations of Si nanocrystals and amorphous Si nanoclusters has been shown. The nature of light emission is discussed.

## INTRODUCTION

In the field of silicon based materials emitted in visible spectral range at room temperature a great number of films fabricated by sputtering, gas evaporation, chemical vapor deposition and so on has been reported last decade. These films are of two phase's structures, such as nano-crystallites embedded in amorphous matrix. The last could be the silicon oxide, or silicon nitride, or amorphous silicon. It was shown that these films are more stable then the porous silicon in both mechanical and luminescent properties. Many models were proposed to explain the nature of efficient emission of Si-based nanostructures. However the mechanism of light emission is a subject of a debate. Usually, silicon nanocrystals (NCs) were considered as the main source of luminescence [1]. At the same time the defects located in silicon oxide or at the $Si/SiO_2$ interface could be alternative sources of photoluminescence (PL) [2-5]. From this point of view it is clear that the substitution of $SiO_2$ matrix by amorphous silicon is promising for PL nature understanding.

Silicon crystallites embedded in the matrix of amorphous silicon could be prepared by different methods. The first one is based on the re-crystallization of amorphous film using a laser, flash lamp or conventional furnace annealing. Another one, such as chemical vapor deposition (CVD) or rf sputtering, with plasma of pure hydrogen allows deposit directly the NC layers. As it is shown earlier Si NCs embedded in the matrix of hydrogenated amorphous silicon prepared by HW-CVD method show the bright emission not only in the visible range but also in the near-infrared [6]. This new paper presents the results of the investigation of structural and luminescent properties of Si NCs embedded in hydrogenated amorphous Si films prepared by HW-CVD method.

## EXPERIMENT

The films were grown by HW-CVD technique in a high vacuum deposition system with a background pressure of $\sim 8 \times 10^{-7}$ Torr. All films were deposited on Corning 50 glass substrates at 0.1 Torr at the constant ratio of silane and hydrogen gases ([$SiH_4$]:[$H_2$]=5:20) and the substrate temperature of (300 °C). The gases flowed through the 0.75 mm diameter and 13 cm long coiled tungsten filament which a temperature was kept at 1735, 1785, 1835 and 1885 °C. As a result, deposited α-Si films contained a substantial concentration of hydrogen. PL setup was presented earlier in [7]. The X-ray diffraction experiments have been done using Siemens D5000 model XRD spectrometer with copper X-ray beam (1.54A). The X-ray diffraction (XRD) spectra of the

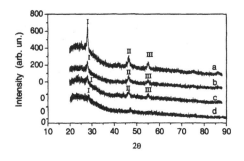

**Fig. 1** X-ray diffraction spectra for amorphous silicon films with Si NCs, prepared at the filament $T_f$: 1885 (a), 1835 (b), 1785 (c) and 1735 (d) °C.

**Fig.2.** Decomposition of the XRD peak I for the film prepared at $T_f$=1885 °C.

samples prepared at different filament temperatures $T_f$ are presented in Fig. 1.

As one can see (Fig.1), all XRD spectra demonstrate the broad peak at $2\Theta$=22.86°-23.39° deals with amorphous Si. Besides this broad peak, XRD spectra of the films prepared at $T_f$=1785-1885 °C demonstrate the peaks at $2\Theta$=28.40°, 47.38° and 56.08° that corresponds to (111), (220) and (311) silicon crystal planes, respectively (Table1).

Table1. Results based on the XRD peak I analysis

| Fig.1 curve | Filament temperature | Peak | Peaks | FWHM | Average D(nm) | Area | Angle |
|---|---|---|---|---|---|---|---|
| a) | $T_F$=1885 °C | I | 1 | 16 | a-Si | 8552 | 25.3° |
| | | | 2 | 0.4 | 21.38 | 175 | 28.4° |
| b) | $T_F$=1835 °C | I | 1 | 16 | a-Si | 6000 | 25.3° |
| | | | 2 | 0.56 | 15.26 | 58 | 28.4° |
| | | | 3 | 1.5 | 5.70 | 20 | 27.9° |
| c) | $T_F$=1785 °C | I | 1 | 16 | a-Si | 6340 | 25.3° |
| | | | 2 | 0.60 | 14.36 | 38 | 28.4° |
| | | | 3 | 2.0 | 4.69 | 13 | 27.9° |
| d) | $T_F$=1735 °C | I | 1 | 16 | a-Si | 5640 | 25.3° |
| | | | 2 | | | | |
| | | | 3 | 2.5 | 3.62 | 8 | 27.9° |

The intensities of all peaks deal with NCs increase with $T_f$ increasing and the highest is observed for the films prepared at $T_f$=1885 °C. The XRD spectrum of the film prepared at $T_f$=1735 °C shows the a-Si peak at 2Θ=25.3° and small NC peak at 2Θ=27.9°. The estimation of the size of Si NCs using Scherrer's formula [8] and decomposition of XRD spectra (Fig.2) show that there are the Si NCs of small and big sizes in the films. The smallest NC size is 3.62-5.70 nm, while the biggest one is in the range of 14.36-21.38 nm (Table 1).

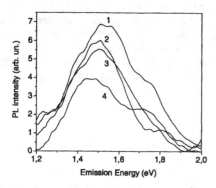 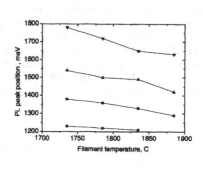

Fig. 3 PL spectra of amorphous Si films prepared at different filament temperatures: 1-1735 °C, 2-1785 °C, 3-1835 °C, 4-1885 °C.

Fig.4 The shift of PL peak positions versus filament temperature

PL spectra measured at 300 K are shown in Fig.3. As one can see all PL spectra are complex and contain the PL elementary bands peaked at ~1.22, 1.36, 1.45-1.52 and ~1.75 eV (Fig.3). The increase of $T_f$ leads to the decrease of contribution of the 1.45-1.52 and 1.7eV PL bands and to the relatively increasing of the contribution of the 1.2 and 1.36eV PL bands (curve 2 and 3). At the same time the crystalline part enlarges in the amorphous phase (Table 2). The increase of $T_f$ leads to gradual low-energy shift of all PL components (Fig. 4).

## DISCUSSION

The XRD peaks corresponding to silicon crystal planes demonstrate the presence of Si NCs in amorphous films. Their number increase with the increase of $T_f$. At the same time the amorphous silicon phase decreases with the rise of filament temperature that is confirmed by decreasing of the intensity (XRD peak area) of a-Si broad peak at 2Θ=22.86°-23.39° (Table 2)

Table 2. Dependence of amorphous and crystal phases on temperature $T_f$

| Fig.1 curves | Filament temperature | Peak | Amorphous phase, % | Crystal phase, % |
|---|---|---|---|---|
| a) | $T_F$=1885 °C | I | 97.99 | 2.01 |
| b) | $T_F$=1835 °C | I | 98.72 | 1.28 |
| c) | $T_F$=1785 °C | I | 99.21 | 0.79 |
| d) | $T_F$=1735 °C | I | 99.85 | 0.15 |

The integrated PL intensity for all PL bands decreases with the growth of big size Si NCs and corresponding PL peak positions shift to the low energy spectral side. Moreover the temperature dependence of PL peaks shows the shift of PL peaks versus temperature by the same way as the Si band gap shrinkages with temperatures [9]. The latter confirm that all PL bands can be attributed to quantum confinement PL mechanism in Si clusters of different sizes and phases.

It is well known that amorphous phase has the short ordering and does not have the long range ordering. It is possible to consider the amorphous phase as the mixture of a-Si:H clusters of smallest sizes in comparison with Si NCs. These small Si clusters can be responsible for the quantum confinement PL mechanism of high energy PL bans (1.45-1.52eV and 1.75 eV). The Si NCs with the highest sizes (18-21nm) do not participate in the quantum confinement light emission process. Nevertheless they could be responsible for the absorption of excitation light. The Si NCs of the middle size (3.62-5.7 nm) can be responsible for quantum confinement PL emission as well and, apparently, connected with low energy PL bands (1.22 and 1.36eV). Note that these PL bands appear in the PL spectrum when the XRD study (Fig.1, curve d) permits the identification of the small size (3.62-5.7 nm) Si NCs (Table 1).

CONCLUSIONS

We investigated the interrelation of photoluminescence and structural properties of hydrogenated amorphous Si films with embedded Si NCs. Four PL bands in the red and IR spectral ranges with the maxima at 1.22, 1.36, 1.45-1.52 and 1.75 eV have been revealed. All PL bands can be explained as a radiative transition between quantum confined levels within Si QDs deal with amorphous Si nanoclusters (1.45-1.52 and 1.75 eV) and small size Si NCs (1.22 and 1.36eV).

ACKNOWLEDGMENTS

The author thanks Dr. Yosuhiro Matsumoto from CINVESTAV-IPN, Mexico for growing the studied structures by the HW-CVD method. This work was partially supported by CONACYT Mexico project (N58358) as well as by SIP-IPN, Mexico.

REFERENCES
1. A. G. Cullis, L.T. Canham, P. D. J. Calcott, J. Appl. Phys. **82**, 909 (1997).
2. T. Shimizu-Iwayama, S. Nakao, K, Saitoh, Appl. Phys.Lett. **65**, 1814 (1994).
3. G. Z. Ran, J. S. Fu, W. C. Qin, B. R. Zhang, Y.P. Qioa, G. G. Qin, Thin Solid Films **388**, 213 (2001).
4. G. Z. Ran, Y. Chen, F. C. Yuan, Y. P. Qiao, J. S. Fu, Z. C. Ma, W. H. Zong, G. G. Qin, Sol. St. Commun. **118**, 599 (2001).
5. L. Khomenkova, N. Korsunska, M. Sheinkman, T. Stara, T.V. Torchynska and A. Vivas Hernandez, J. Lumin., **115**, 117 (2005).
6. T. V. Torchynska, A. Vivas Hernandez, M. Dybiac, Yu. Emirov, I. Tarasov, S. Ostapenko, Yusuhiro Matsumoto, phys. stat. sol. (c), **2**, 1832 (2005).
7. A. L. Quintos Vasques, T.V. Torchynska, G. Polupan, Y. Matsumoto, L. Khomenkova, L.V. Shcherbyna, Solis State Phenomena, **131-133**, 71 (2008).
8. H. P. Klug and L. E. Alexander, X-ray diffraction procedures for polycrystalline and amorphous materials, Willey and Sons, 1954.
9. T. V. Torchynska, J. Non-Crystaline Solids, **352**, 2484 (2006).

Mater. Res. Soc. Symp. Proc. Vol. 1066 © 2008 Materials Research Society    1066-A06-06

# Electronic Properties of Nanocrystalline Silicon Deposited With Different Crystallite Fractions and Growth Rates

P. G. Hugger[1], J. David Cohen[1], Baojie Yan[2], Guozhen Yue[2], Xixiang Xu[2], Jeffrey Yang[2], and Subhendu Guha[2]

[1]Department of Physics, University of Oregon, Eugene, OR, 97405
[2]United Solar Ovonic, LLC, Troy, MI, 48084

## ABSTRACT

Junction capacitance measurements were used to characterize the properties of nanocrystalline silicon (nc-Si:H) solar cells. These methods included drive-level capacitance profiling (DLCP) to obtain spatially-resolved defect densities, as well as transient photocapacitance (TPC) and transient photocurrent (TPI) spectra to reveal optically responsive states in the band-gap, and to estimate minority carrier behavior before and after prolonged light exposure (lightsoaking). Crystalline volume fractions were estimated using Raman spectroscopy. Previously we had identified at least two types of distinct behaviors in such nc-Si:H materials that correlated with the crystalline volume fraction. Here, in one case, we report results indicating that both types of behavior can occur in a single sample, possibly indicating that the structural properties of that sample have evolved during growth.

## INTRODUCTION

Hydrogenated nanocrystalline silicon (nc-Si:H) has shown promise as a thin film photovoltaic material for it's long wavelength optical response, high growth rates, and it's high resilience to optical degradation. The mixed-phase nature of nc-Si:H accounts for these benefits but also contributes to the unique and complex electrical behavior of this material.

In this study, junction capacitance measurements were used in conjunction with Raman spectroscopy to probe both the bulk electrical properties and the structural characteristics of *n-i-p* nc-Si:H solar cell devices. The samples consisted of several series of solar cells deposited on stainless steel substrates (SS/$n^+$/i nc-Si:H/$p^+$/ITO). Intrinsic layers were deposited using both RF and MVHF glow-discharge methods and were grown under conditions producing samples with different average crystalline volume fractions, thicknesses, and nanocrystallite distributions. We typically examined samples both in their annealed and light-soaked state. The latter was produced by exposing the samples to tungsten-halogen light through a 610nm long-pass filter at an intensity of 200mW/cm$^2$ for 100 hours.

## EXPERIMENTAL METHODS

Densities of states in the bandgap were investigated using drive-level capacitance profiling (DLCP) [1]. Similarly to C-V profiling methods, the DLCP measurement takes advantage of the fact that the depletion region width, i.e. the junction capacitance, is intimately related to the density of states in the bandgap. It can be shown [2] that under AC biasing conditions ($C = C_0 + C_1 dV + C_2 dV^2 + \ldots$) the DLCP density can be written as an integral over the band-gap density of states:

$$N_{DLCP} = -\frac{C_0{}^3}{2q\varepsilon A^2 C_1} = n + \int_{E_C-E_e}^{E_F} g(E,x)dE \qquad (1)$$

Here $E_e$ represents a thermal emission cutoff energy: $E_e = k_B T \ln(\nu/2\pi f)$, where T is the measurement temperature, f is the measurement frequency, and $\nu$ is the thermal emission prefactor. $N_{DLCP}$ is then typically plotted against a distance parameter $\langle x \rangle = \varepsilon A / C_0$ where $\langle x \rangle$ is the first moment of charge response, and A is the area of the device. Determining $N_{DLCP}$ for a number of DC bias values produces a defect density profile as a function of distance from the junction.

While DLCP is a powerful method for investigating band-gap states, care should be taken when interpreting DLC profiles. When $\langle x \rangle$ does not correspond closely with the actual depletion region width the DLC profile may exhibit useful, but intuitively misleading features. Such features will be presented in the Discussion section below.

Optical methods were also used to investigate bandgap states. Transient Photocapacitance (TPC) and Transient Photocurrent (TPI) [3] are absorption-like spectra that are obtained by observing the recovery of junction capacitance (TPC) or current (TPI) following a voltage pulse. The behavior of these recovery transients is then measured over a suitable time window in the presence of monochromatic sub-bandgap light. The resulting spectra closely resemble an integral over the density of states in the band-gap. In addition to yielding information regarding densities of states, taking the ratio of TPC signal to TPI signal produces an estimation of the faction of carriers collected during the time window that were *minority carriers* [4]. That is:

$$\frac{Signal_{TPC}}{Signal_{TPI}} \propto \frac{n-p}{n+p} \qquad (2)$$

This ability to resolve charge type results directly from junction capacitance dependence on the total amount of charge stored in the depletion region.

Finally, in addition to DLCP and TPC/I, Raman spectra were also obtained for these samples for an excitation wavelength of 785nm. In order to estimate nanocrystalline volume fractions, the procedures of Bustarret [5] and Han [6] were followed. The data were baseline adjusted between 400 and 550 $cm^{-1}$ and the resulting spectra were deconvolved into three peaks: two Gaussian peaks centered at 480 $\pm$ 5 $cm^{-1}$ and 510 $\pm$ 5 $cm^{-1}$, and one final peak at 520 $\pm$ 3 $cm^{-1}$ whose Gaussian/Lorenzian mixing fraction was allowed to float as an extra fitting parameter. The integrated intensities of these peaks were then used to estimate crystalline volume fractions.

RESULTS

In general, we have seen that nanocrystalline devices with higher crystalline volume fractions yield higher DLCP densities [7]. In addition, the samples with high crystalline volume fractions tend to give DLC profiles with strong temperature dependence. This behavior, illustrated in Figure 1, is for a nanocrystalline device of crystalline volume fraction $X_c \cong 0.6$ that shows a temperature dependent series of DLC profiles in the fully light soaked state. This behavior signifies a deep state response in DLCP and is attributed to the temperature dependence of the integration limits of Equation (1).

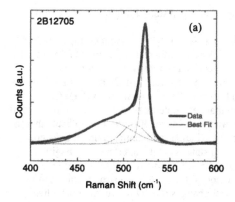 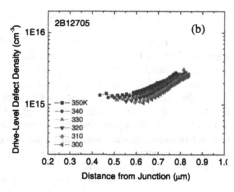

FIG 1. (a) Raman spectrum and (b) DLC profile of a nc-Si:H intrinsic layer with a high crystalline volume fraction. Deconvolution of the Raman spectrum yielded an estimated crystalline volume fraction of $X_c = 0.57$. The DLC defect density shows temperature dependence, signifying deep state response.

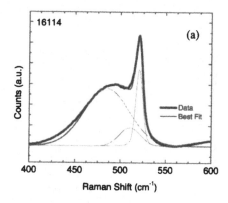 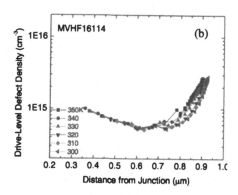

FIG 2. (a) Raman spectrum and (b) DLC profile of a nc-Si:H intrinsic layer with a low crystalline volume fraction. Deconvolution of the Raman spectrum yielded an estimated crystalline volume fraction was $X_c = 0.30$. (b) DLC defect density shows little temperature dependence, indicating a low density of deep defects. The slight temperature variation of DLCP near <x>=0.9μm may signify aberrant behavior near the back (i-n) junction.

On the other hand, primarily amorphous devices in general do not show strongly temperature dependant DLC profiles. Therefore, the devices with more amorphous component have the DLC profiles that exhibit a weak dependence on temperature. Figure 2 shows an amorphous device of crystalline volume fraction $X_c \cong 0.3$. Note that even in the fully light soaked state, the device with the greater amorphous component does not show a significant temperature dependent defect response from occupied mid-gap states lying below the Fermi level.

Until recently either one behavior or another was seen in DLCP: samples appeared highly intrinsic, showing little evidence of deep states, or yielded a temperature dependent DLC profile. However Figure 3(b) shows a DLC profile that exhibits a sharp transition from a more "amorphous-like" intrinsic DLCP response (temperature independent DLCP signal signifying no deep defect density response to the measurement) to a more "nanocrystalline like" DLCP response (temperature dependent DLCP signifying that deep defects are attributing to the integral in Equation 1) in the region near the p-i interface. Although Figure 3(b) seems to indicate that the distance from the interface where this transition occurs changes with temperature, it does not. Instead, this behavior is attributed to the fact that deep states near the interface only begin to cross the Fermi level (thus begin to contribute to the integral in Equation 1) for a particular reverse bias. As these near-interface states begin to contribute to $N_{DLCP}$, a strong temperature dependence is seen in the profile and the value of <x> changes rapidly to accommodate this near-interface charge response.

To further characterize the device showing this interesting structural transition, we applied the optical methods of TPC and TPI to probe the mid-gap density of states and to invest (a) ie degree of hole collection shown in these measurements both before and after lights Figure 4 demonstrates that the effect of 200h of 610nm long-pass filtered light

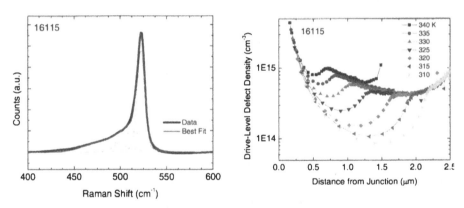

FIG 3. (a) Raman spectrum and (b) DLC profile of a nc-Si:H sample with a thick nc-Si:H i layer. The estimated average crystalline volume fraction was $X_C = 0.77$. The DLCP shows evidence of both amorphous and nanocrystalline-type behavior in the annealed state: Mostly shallow states responded far from the junction and deep states, roughly 0.5eV below the conduction band, responded in the region within 0.3μm of the p-i interface. This transition of behavior may correspond to changing structural characteristics in the i layer.

exposure on these measurements. Unlike the TPC spectra of higher amorphous fraction nc-Si:H samples reported previously [8], these spectra exhibited little evidence of any a-Si:H component. The most significant effect of lightsoaking on this sample was to suppress hole collection. This effect is seen as an increase in the magnitude of the ratio $Signal_{TPC} / Signal_{TPI}$ after light soaking for photon energies higher than ~ 0.9 eV.

Note that the primary effect of lightsoaking has been to effect this degree of hole collection, and not to effect the defect distribution near 0.7 eV. These changes in optical spectra correlate with

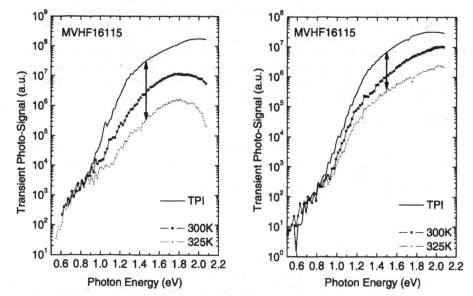

**FIG 4.** (a) TPC spectra compared with a TPI spectrum in the fully annealed state; (b) a similar collection of spectra for the light-degraded state. The TPC signal has a larger magnitude relative to the TPI signal in the degraded state, which corresponds to a decrease in hole collection. Regardless of lightsoaking, the density of states corresponding to an absorption below ~ 1.0 eV is unchanged.

changes in device performance parameters after light soaking, where the J-V characteristic parameter changes were $\Delta J_{SC} = -0.31$ mA/cm$^2$, $\Delta V_{OC} = -5$ mV, and $\Delta FF = -0.026$.

## DISCUSSION AND CONCLUSIONS

We modeled the DLCP results shown in Figure 3 using a custom numerical simulation in which the AC charge response of a *p-n* junction is calculated using Poisson's equation. Using this model, theoretical DLCP curves were obtained for user-defined shallow and deep defect distributions. Preliminary simulation results have shown that a relatively specific defect distribution is responsible for the DLCP behavior seen in Figure 3. In our model, material properties undergo a transition at 0.3 μm from the *p-i* interface. Far from the barrier, mid-gap states consist of shallow band-tail states and a low (~ $10^{14}$ cm$^{-3}$) density of mid-gap defects (0.7eV below $E_C$). Near the barrier these states transition to a distribution consisting of a relatively large (~$8 \cdot 10^{15}$ cm$^{-3}$) density of shallower (~0.5 eV) defects. It was verified experimentally that these defects share a thermal prefactor value of ~$10^{13}$ $s^{-1}$. It has not yet been verified through modeling that this interpretation is consistent with modeling results for devices showing no such material transition.

We do not withhold the possibility that the transition behavior seen in Figure 3(b) is indicative of a rapid transition from one dominant structural phase to another. Experiments are currently underway verify this assumption. We also can not yet verify the origins of the near-barrier 0.5 eV defect, but possibilities remain grain boundary states, or mid-gap "amorphous-like" dangling bond defects in the presence of a local potential fluctuation [9]. However, more important to the characterization of device performance may still be quantification of the decrease in minority carrier collection in these materials after lightsoaking, as this behavior may relate directly to metastable decreases in hole mobility.

## ACKNOWLEDGMENTS

This work was partially supported by NREL under the Thin Film Partnership Program subcontract No. ZXL-6-44205-14 at United Solar and No. ZXL-5-44205-11 at the University of Oregon, and by US DOE under the Solar America Initiative Program Contract No. DE-FC36-07 GO 17053 at both United Solar and the University of Oregon.

## REFERENCES

1.  C.E. Michelson, A.V. Gelatos, and J.D. Cohen, *Applied Physics Letters* **47**, 4 (1985).
2.  J.T. Heath, J.D. Cohen, and W.N. Shafarman, *Journal of Applied Physics* **95**, 3 (2004).
3.  A.V. Gelatos, *et al.*, *Applied Physics Letters* **53**, 5 (1988).
4.  J.D. Cohen, J.T. Heath, and W.N. Shafarman, "Photocapacitance Spectroscopy in Copper Indium Diselenide Alloys", in *Wide-Gap Chalcopyrites*, ed. S. Siebentritt and U. Rau. 2006, Springer: Berlin. p. 69-87.
5.  E. Bustarret, M.A. Hachicha, and M. Brunel, *Applied Physics Letters* **52**, 20 (1988).
6.  D. Han and J.D. Lorentzen, *Journal of Applied Physics* **94**, 5 (2003).
7.  P.G. Hugger, *et al.*, *Journal of Non-Crystalline Solids* doi:10.1016/j.jnoncrysol.2007.09.088 (2008). (in press)
8.  A. Halverson, J.J. Gutierrez, and J.D. Cohen, *Appl. Phys. Lett.* **88**, 071920 (2006).
9.  P.G. Hugger, *et al. Electronic Characterization and Light-Induced Degradation in nc-Si:H Solar Cells.* in *Amorphous and Polycrystalline Thin-Film Silicon Science and Technology. - 2006*, edited by S. Wagner, V. Chu, J. Harry A. Atwater, K. Yamamoto and H.-W. Zan (Mater. Res. Soc. Symp. Proc. **910**, San Francisco, CA, 2006), 0910-A01-05.

Mater. Res. Soc. Symp. Proc. Vol. 1066 © 2008 Materials Research Society          1066-A06-08

# Doping Effects in Co-deposited Mixed Phase Films of Hydrogenated Amorphous Silicon Containing Nanocrystalline Inclusions

C. Blackwell[1], Xiaodong Pi[2], U. Kortshagen[2], and J. Kakalios[1]

[1]School of Physics and Astronomy, University of Minnesota, 116 Church Street S.E., Minneapolis, MN, 55455

[2]Mechanical Engineering, University of Minnesota, 111 Church Street S.E., Minneapolis, MN, 55455

## ABSTRACT

Hydrogenated amorphous silicon films containing silicon nanocrystalline inclusions (a/nc-Si:H) that have been n-type doped have been synthesized using a dual-plasma co-deposition system. We report the structural and electronic properties of n-type doped a/nc-Si:H as a function of phosphine doping level and nanocrystalline concentration. The volume fraction of nanocrystals in the doped a/nc-Si:H thin films is measured using Raman spectroscopy, and the hydrogen binding configurations are characterized using infra-red absorption spectroscopy. In undoped a/nc-Si:H, the inclusion of low and moderate nanocrystalline concentrations results in an increase in the dark conductivity, compared to a-Si:H films grown without nanocrystalline inclusions. In contrast, the addition of even a low concentration of silicon nanoparticles in doped a/nc-Si:H thin films leads to a decrease in the dark conductivity and photoconductivity, compared to pure a-Si:H films.

## INTRODUCTION

Interest in hydrogenated amorphous silicon with nanocrystalline silicon inclusions (a/nc-Si:H)-based photovoltaic devices stems from reports of an enhanced resistance to light-induced degradation, coupled with high solar conversion efficiencies and high deposition rates.[1-4] There has been considerable effort in determining the deposition processes and conditions that would yield optimal electronic properties of undoped a/nc-Si:H mixed-phase thin films.[1-7] However, there has not been to date comparable effort in synthesizing and studying the properties of doped a/nc-Si:H films. An elucidation of the opto-electronic properties of doped a/nc-Si:H films is a necessary step in the development of a mixed-phase *pin* a/nc-Si:H-based solar cell.

Mixed phase a/nc-Si:H thin films are typically synthesized in a single PECVD reactor chamber, using high gas pressures and a high hydrogen concentration.[1,2,5] The deposition parameters necessary for silicon nanocrystalline formation in the silane plasma are very far from those known to yield high electronic quality a-Si:H. While one can control the concentration of nanocrystals embedded within the a-Si:H matrix through a thermophoretic force in a single chamber plasma deposition system, the electronic conductivity of the resulting a/nc-Si:H has been found to be decreased for thermal gradients imposed across the silane plasma during film growth that enhance nanoparticle incorporation.[8] These results have motivated the construction of a dual chamber co-deposition to better control the quality of the a-Si:H film.[9,10] In this system silicon nanocrystals are synthesized in one plasma deposition system, and are then entrained in a carrier gas and injected into a second PECVD system, where hydrogenated amorphous silicon is deposited. The growth conditions in the second chamber can be independently optimized to yield a high quality a-Si:H matrix. In this paper we report the first studies of n-type doped a/nc-Si:H films as a function of silicon nanocrystalline concentration.

## MATERIALS PREPARATION

The doped a/nc-Si:H films reported here were synthesized in a dual chamber co-deposition system shown schematically in Figure 1. Reactive gases consisting of silane ($SiH_4$) and phosphine ($PH_3$), diluted with a carrier gas of argon, are first passed through the Particle Synthesis Reactor, which consists of a 3/8 inch diameter quartz tube with two fitted ring electrodes, connected to a 13.56 MHz power supply and matching network. Silane and phosphine are dynamically mixed, and diluted with an inert carrier gas of argon; a total gas flow rate of 90 sccm was maintained through both systems. The plasma conditions in the Particle Synthesis Reactor (1.7 Torr, 70 W RF power) were selected to generate silicon nanocrystals, that were then carried by the flowing argon into the second PECVD chamber (700 mTorr, RF power 4 W). The flow rates were adjusted to control both the nanoparticle and doping concentration. The silane flow rates varied between 2.5 sccm and 5.5 sccm corresponding to high nanocrystalline particle concentration to low nanocrystalline particle concentration, where the conditions governing the nanocrystalline diameter are fixed. The doping concentration was changed by changing the flow rates of dopant gas to produce $[PH_3]/[SiH_4] = 6 \times 10^{-5}$ (low doping) and $[PH_3]/[SiH_4] = 6 \times 10^{-3}$ (high doping). Mixed phase a/nc-Si:H films are deposited on substrates held at $250^0$C placed on the lower, grounded electrode in the second PECVD chamber.

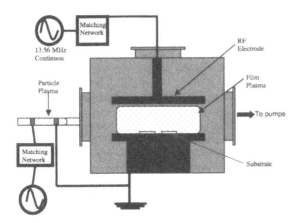

**Figure 1:** Sketch of the dual chamber co-deposition system. Dopant gases and silane pass through the particle deposition chamber (left) and then flow into the second chamber where amorphous silicon film is deposited.

The samples were deposited onto Corning 7059 glass substrates and crystalline silicon substrates for infra-red absorption measurements. Film thicknesses are measured using a profilometer, thicknesses and deposition rates are summarized in Table 1. Deposition rates can be found by dividing the film thickness (Table 1) by the deposition time of 62 minutes. Coplanar chromium electrodes (0.2 cm wide, 1 cm long) for electronic measurements are deposited using a shadow mask via e-beam evaporation.

## STRUCTURAL CHARACTERIZATION

The structural composition of the doped a/nc-Si:H films were determined using Raman and Fourier Transform Infra-red (FTIR) spectroscopy. Raman spectra are recorded using a Witec Alpha 300 R confocal Raman microscope equipped with a UHTS 200 spectrometer using an Ar ion laser (wavelength 514.5 nm). Figure 2 shows the Raman spectra for the doped a/nc-Si:H films grown under the low doping condition ($[PH_3]/[SiH_4] = 6 \times 10^{-5}$) when no nanocrystals are generated in the Particle synthesis Reactor (that is, the film is pure, doped a-Si:H) and for medium and high nanocrystalline concentrations, as determined by the growth conditions. A broad peak at 480 cm$^{-1}$ is observed for all three films, reflecting the TO mode of the a-Si:H material.[11] Bulk crystalline silicon has a narrow peak at 520 cm$^{-1}$ while the mixed phase doped nanocrystalline silicon exhibits a Raman peak downshifted to 518 cm$^{-1}$.[12] As the particle density is increased from zero, a shoulder forms and develops into a larger peak. The crystalline fraction was determined by fitting each spectrum using Gaussian curves and relating the areas under each curve. The crystalline fraction is defined as $X_c = I_{518}/(I_{480}+I_{518})$ where $I_{480}$ is the area under the peak at 480 cm$^{-1}$ and $I_{518}$ is the area under the peak at 518 cm$^{-1}$. The curves in Fig. 2 are representative of measurements taken at several locations on the a/nc-Si:H films. The resulting crystalline fractions for these films, and for the films deposited at the higher doping level, are listed in Table 1.

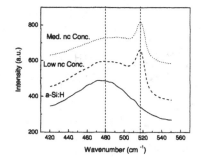 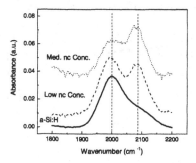

**Figure 2:** Plot of the Raman absorption spectra for $6 \times 10^{-5}$ $[PH_3]/[SiH_4]$ doped a/nc-Si:H films as a function of nanocrystalline concentration. The curves are offset vertically for clarity.

**Figure 3:** Infrared absorption spectra for $6 \times 10^{-5}$ $[PH_3]/[SiH_4]$ doped a/nc-Si:H films as a function of nanocrystalline concentration. The curves are offset vertically for clarity.

The infrared absorption spectra for a/nc-Si:H films deposited onto crystalline silicon substrates were measured using a Nicolet Magna 750 FTIR spectrometer, are shown in fig. 3

(these are from the same deposition runs as in Fig. 2). The data are plotted in the range of 1800 to 2200 cm, showing the presence of the vibrational bond-stretching modes of Si-H at 2000 cm$^{-1}$ and Si=H$_2$ at 2090 cm$^{-1}$.[11] The microstructure fraction, defined as $R = I_{2090}/(I_{2090}+I_{2000})$ where $I_{2000}$ is the area under the peak at 2000 cm$^{-1}$ and $I_{2090}$ is the area under the peak at 2090 cm$^{-1}$ is found to increase with nanocrystalline density.[11] The latter has been associated with higher amounts of structural and electronic disorder,[13,14] and in a/nc-Si:H films has been ascribed to the bonded hydrogen in grain boundary regions surrounding the nanocrystalline inclusions.[15] The microstructure fractions as determined by FTIR increases in the doped a/nc-Si:H films with crystalline fraction $X_c$ as measured using Raman spectroscopy, as summarized in Table 1, in support of the interpretation that the absorption at 2090 cm$^{-1}$ in the a/nc-Si:H films results from hydrogenated regions surrounding the nanocrystalline inclusions.[15] Details of the curve fitting procedure used in Figs. 2 and 3, as well as correlation between the infra-red absorption at 2090 cm$^{-1}$ and the nanoparticle density, will be described in a later publication.

## ELECTRONIC CHARACTERIZATION

The temperature dependence of the dark conductivity for these doped a/nc-Si:H films as a function of nanocrystalline concentration in the annealed state and following extended light exposure (the Staebler-Wronski effect)[16] are shown in Figure 4. The conductivity of the films with the lowest nanocrystalline concentration overlapped those with a moderate density of nanoparticle inclusions, these data have been omitted from Fig. 4 for clarity. The films are first annealed at 450 K for 120 minutes under vacuum and then slowly cooled to 320 K. The conductance in the annealed state A is measured upon warming, at a rate of 3°/min (solid symbols in Fig. 4). The sample is re-cooled to 320 K and then light soaked (using a heat-filtered W-Ha lamp intensity ~ 100 mW/cm$^2$) for 22 hours.

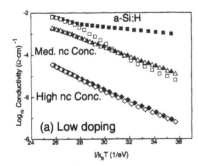 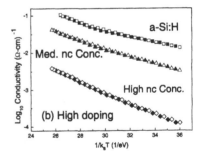

**Figure 4**: Arrhenius plots of the dark conductivity of doped a/nc-Si:H films as a function of increasing nanocrystalline concentration, for (a) low doping (6 x 10$^{-5}$ [PH$_3$]/[SiH$_4$]) and (b) high doping levels (6 x 10$^{-3}$ [PH$_3$]/[SiH$_4$]). The filled symbols represent the dark conductivity in the annealed state A while the conductivity following extended light-soaking (state B) is represented by the open symbols.

**Table1:** Summary of the material and electronic proprieties for n-type doped a/nc-Si:H films. Films labels as Low doping had a doping level of 6 x $10^{-5}$ [PH$_3$/SiH$_4$] while the films labeled as High doping had a doping level of 6 x $10^{-3}$ [PH$_3$/SiH$_4$].

| Doping level | Thickness (nm) | $X_c$ % from RAMAN | Microstructure Fraction from FTIR | Activation Energy (eV) | Photoconductivity $\sigma_{ph}$ ($\Omega$cm)$^{-1}$ | Photosensitivity $\sigma_{ph}/\sigma_a$ |
|---|---|---|---|---|---|---|
| Low | 1210.0 | 0.0 | 0.30 | 0.181 | 1.7 x $10^{-03}$ | 2.09 |
| Low | 590.9 | 9.6 | 0.42 | 0.349 | 1.7 x $10^{-04}$ | 6.66 |
| Low | 240.2 | 11.9 | 0.54 | 0.500 | 5.6 x $10^{-05}$ | 7.29 |
| Low | 73.3 | 29.7 | 0.70 | 0.623 | 2.6 x $10^{-06}$ | 52.94 |
| High | 1186.7 | 0.0 | 0.37 | 0.204 | 1.6 x $10^{-02}$ | 1.04 |
| High | 1114.0 | 0.4 | 0.52 | 0.235 | 4.0 x $10^{-03}$ | 1.02 |
| High | 421.4 | 1.0 | 0.40 | 0.242 | 4.1 x $10^{-03}$ | 1.12 |
| High | 117.8 | 3.2 | 1.00 | 0.347 | 2.6 x $10^{-04}$ | 2.68 |

The State B conductivity is then measured upon warming (open symbols in Fig. 4). The mixed phase doped a/nc-Si:H show no significant light induced degradation. The dark conductivity measured at 325K and activation energy obtained from the Arrhenius plots, as well as the photoconductivity and photosensitivity (defined at the ratio of the photoconductivity when illumination begins to the annealed state A dark conductivity) are summarized in Table 1. The addition of nanocrystalline inclusions leads to a more pronounced decrease in the dark conductivity than the photoconductivity, such that the photosensitivity increases slightly for these films.

## CONCLUSION

We have reported the dual-chamber co-deposition and characterization of n-type doped a-Si:H films containing silicon nanocrystalline inclusions. Studies of undoped a/nc-Si:H films synthesized in a dual chamber system similar to, but not the same as the one employed in this work have found that the addition of nanocrystalline inclusions leads to a nonmontonic increase in the dark conductivity, compared to pure a-Si:H films. That is, the highest conductance values are consistently observed in mixed phase thin films with a crystalline fraction of 2 – 4 %, as determined by Raman spectroscopy as in Fig. 2, and even when the crystalline fraction is 10% or higher, the dark conductivity is greater than in films for which no nanocrystals are present.[10] In contrast, for the doped a/nc-Si:H films reported here, the addition of nanocrystals always leads to a reduction on the dark conductivity. The increase in activation energy observed, particularly for the lightly doped films with the highest concentration of nanocrystals, is striking, as typically in doped a-Si:H the Fermi energy will not reside deeper in the mobility gap than the minimum in the density of states formed between the negatively charged dangling bond states and the conduction band tail, approximately 0.4 eV from the conduction band edge.[17] It is likely that the larger activation energies observed here result from increased long-range disorder introduced by the nanocrystalline inclusions, which would have the effect of shifting the electronic transport level to higher energies, resulting in a larger conductivity activation energy.[14] Experiments are underway, comparing the conductivity and thermopower activation energies, to ascertain the influence of increasing disorder on charge transport in these mixed phase thin films.

## ACKNOWLEDGEMENTS

This work was partially supported by NSF grants DMR-0705675, NER-DMI-0403887, and the University of Minnesota Center for Nanostructured Applications. Parts of this work were carried out at in the University of Minnesota I.T. Characterization Facility, which receives partial funding from the NSF through the NINN program.

## REFERENCES

1   P. Roca i Cabarrocas, A. Fontcuberta i Morral, and Y. Poissant, *Thin Solid Films* **403-404**, 39 (2002).
2   C. R. Wronski, J. M. Pearce, R. J. Koval, X. Niu, A. S. Ferlauto, J. Koh, and R. W. Collins, *Mc Res. Soc. Symp. Proc.* **715**, 459 (2002).
3   Y. Lubianiker, J. D. Cohen, H.-C. Jin, J.R. Abelson, *Phys. Rev. B* **60**, 4434 (1999); D. Kwon, C.-C. Chen, J. D. Cohen, H.-C. Jin, E. Hollar, I. Robertson, J. R. Abelson, *Phys. Rev. B* **60**, 4442 (1999).
4   C. R. Wronski, J. M. Pearce, R. J. Koval, X. Niu, A. S. Ferlauto, J. Koh, and R. W. Collins, *Mater. Res. Soc. Symp. Proc.***715**, A13.4.1 (2002).
5   J. Yang, K. Lord, S. Guha, S. R. Ovshinski, *Mater. Res. Soc. Proc.* **609**, A15.4.1 (2000).
6.  R. W. Collins, A. S. Ferlauto, G. M. Ferreira, K. Joohyun, *Mater. Res. Soc. Proc.* **762**, A. 10.1 (2001).
7.  S. Thompson, C. R. Perrey, C. B. Carter, T. J. Belich, J. Kakalios, U. Kortshagen, *J. Appl. Phys.* **97**, 34,310 (2005).
8.  C. Blackwell, C. Anderson, J. Deneen, C. B. Carter, U. Kortshagen and J. Kakalios, *Mater. Res. Soc. Proc.* **910**, 181 (2006).
9.  C. Anderson, C. Blackwell, J. Deneen, C. Barry Carter, J. Kakalios and U. Kortshagen, *Mater. Res. Soc. Proc.* **910**, 79 (2006).
10. Y. Adjallah, C. Blackwell, C. Anderson, U. Kortshagen and J. Kakalios, *Mater. Res. Soc. Proc.*, this volume (2008).
11. G. Lucovsky, R. J. Nemanich, and J. C. Knights, *Phys. Rev. B* **19**, 2064 (1979).
12. G. Viera, S. Huet and L. Boufendi, *J. Appl. Phys.* **90**, 4175 (2001); R. Saleh and N. H. Nickel, *Thin Solid Films* **427**, 266 (2003).
13. E. Bhattacharya and A. H. Mahan, *Appl. Phys. Lett.* **52**, 1587 (1988).
14. D. Quicker and J. Kakalios, *Phys. Rev. B* **60**, 2449 (1999).
15. D. C. Marra, E. A. Edelberg, R. L. Naone and E. S. Aydil, *J. Vac. Sci. and Tech.* **16**, 3199 (1998).
16. D. L .Staebler and C.R.Wronski, *Appl. Phys. Lett.* **31**, 292 (1976); *J. Appl. Phys.* **51**, 3262 (1980).
17. J. Kakalios and R. A. Street, *Phys Rev. B* **34**, 6014 (1986).

Mater. Res. Soc. Symp. Proc. Vol. 1066 © 2008 Materials Research Society          1066-A06-09

# Nanocrystalline Silicon Diodes for Rectifiers on Flexible RFID Tags

Ian Chi Yan Kwong, Hyun Jung Lee, and Andrei Sazonov
Department of Electrical and Computer Engineering, University of Waterloo, 200 University
Avenue West, Waterloo, Ontario, N2L 3G1, Canada

## ABSTRACT

There has been an on-going effort to produce low cost radio frequency identification (RFID) tags as a replacement for traditional barcodes. One method to achieve low cost production is to integrate the manufacturing of the substrate, antenna and active devices into one single continuous process. Hydrogenated nanocrystalline silicon (nc-Si:H) is a suitable material for manufacturing the active devices in such a process.

We present a nc-Si:H diode suitable for use in rectifiers on RFID tags. It consists of a Cr bottom contact, an undoped layer of nc-Si:H, an n-doped nc-Si:H and an Al top contact. We demonstrate the current-voltage characteristics of the nc-Si:H diode are much improved over a-Si:H diodes. Current density of 10 $A/cm^2$ and ON/OFF ratio greater than $10^6$ was measured at 2 V forward bias. Output DC voltage of 2.6 V was achieved using four nc-Si:H diodes in a full-wave bridge rectifier. The input AC signal was a sine wave at 14 MHz and 2 $V_{RMS}$ amplitude.

## INTRODUCTION

The use of RFID technology on the commercial market is being lauded as the next big revolution in enhancing business intelligent. However, the deployment of RFID by business has been slow and one major factor is the cost associated with deploying large number of RFID tags. However, advanced RFID tags that provide data processing and writable storage capability requires many transistors. Currently, these advanced RFID tags uses traditional CMOS chips and are still too expensive for wide deployment [1]. Recent research in RFID tags manufacturing has been centered on different technology in reducing the cost of manufacturing tags with many transistors [2-4].

One method to reduce the manufacturing cost of RFID tags is to switch to using large area electronics manufacturing methods and technology. Current RFID tags are manufactured using a packaged CMOS chip made on crystalline silicon which is then bonded to an antenna and substrate of the tag. The packaging cost of the CMOS chip, the bonding cost and the subsequent testing cost of the final tag adds to the total cost of the finished RFID tag. By switching to large area electronics technology, where the antenna and active circuitry are manufactured in one integrated process, there is much potential for cost savings.

Research has been done on using amorphous silicon and organic semiconductors in making the active components for RFID tags [2-4]. Our approach is to use hydrogenated nanocrystalline silicon (nc-Si:H) in an integrated process for manufacturing the same active component. Previous research has demonstrated that nc-Si:H possesses higher electron and hole mobilities than a-Si:H and organic semiconductors, which will lead to superior performing devices, necessary for RFID tags operating at 13.56 MHz and above. We demonstrate here a diode which can be used for rectification purpose, to convert an input AC voltage to a DC voltage, which is an essential requirement to power RFID tags.

## EXPERIMENT

The nc-Si:H diode is made on Corning 1737 glass substrate in a vertical structure and the cross-section is shown in Figure 1. The bottom contact is 300 nm thick layer of chromium (Cr). On top of the bottom contact is a bilayer of nc-Si:H, consisting a 300 nm thick layer of intrinsic nc-Si:H and a 100 nm thick layer of n+ nc-Si:H. A layer of silicon dioxide ($SiO_2$) is used to passivate the nc-Si:H. The top metal contact consists of 300 nm thick layer of aluminum (Al).

Diodes of different areas were fabricated, including 50*50 $\mu m^2$, 100*100 $\mu m^2$, 150*150 $\mu m^2$, 200*200 $\mu m^2$ and 250*250 $\mu m^2$. The Cr and Al films used in the diode were RF sputtered. The nc-Si:H was deposited using a 13.56 MHz PECVD system at 260 °C. The intrinsic nc-Si:H was deposited using silane ($SiH_4$) and hydrogen source gases only, with a $[H_2]/[SiH_4+ H_2]$ dilution of 99%. The n+ nc-Si:H was doped with phosphine ($PH_3$) at a $[PH_3]/[SiH_4]$ ratio of 2/100. The $SiO_2$ was deposited using the same PECVD system and at the same temperature of 260 °C. The diodes were patterned using standard lithographic technique. Etching of chromium, aluminum and $SiO_2$ was done using $Ce(NH_4)_2(NO_3)_6$ plus acetic acid, PAN and buffered HF respectively. Etching of the nc-Si:H bilayer was done using reactive ion etching (RIE) with $SF_6$ and $O_2$ gas mixture.

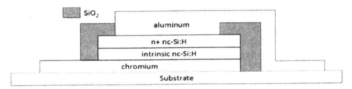

**Figure 1.** Cross section of the experimental nc-Si:H diode.

## RESULTS

### Diode behavior and characteristics

Figure 2 below shows the IV characteristic for a sample 150*150 $\mu m^2$ diode measured at room temperature between 2 V forward and reverse bias.

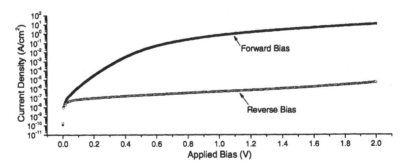

**Figure 2.** Diode forward and reverse IV characteristics.

The forward current at low forward applied bias, from 0.1 V – 0.4 V, displays an exponential growth relationship as expected of a diode. Starting from 0.5 V onwards, forward current growth decays possibly due to high series resistance effect or space charge limited current (SCLC). The ON/OFF ratio of the diodes is around $3*10^5$ at 1 V bias. The ON/OFF ratio increases as voltage increases with the ratio exceeding $10^6$ at 2 V.

To elucidate the nature of the nc-Si:H diode, whether the diode is a Cr/nc-Si:H Schottky diode or p-n nc-Si:H diode, we compare the experimental results with the theoretical models. Electrical current for a Schottky diode can be attributed to one of four different sources: thermionic emission (TE), tunneling, generation-recombination and leakage [5]. The total current is described by equation 1.

$$I = I_{TEO}\left[\exp\left(\frac{q(V - IR_s)}{kT}\right) - 1\right] + I_t \exp\left(\frac{q(V - IR_s)}{E_0}\right) + I_{gr}\left[\exp\left(\frac{q(V - IR_s)}{2kT}\right) - 1\right] + \frac{V - IR_s}{R_L} \qquad (1)$$

Where $I_{TEO}$, $I_t$ and $I_{gr}$ are saturation current components due to thermionic, tunneling and generation-recombination, $E_0$ is a junction-dependent factor, $R_S$ is the series resistance and $R_L$ is a leakage current parameter. Equation 1 is often simplified by using a non-ideality factor "n" and the resulting equation is equation 2.

$$I = I_s \exp\left(\frac{q(V - IR_s)}{nkT}\right) \qquad (2)$$

Where $I_S$ is the general saturation current, n is the non-ideality factor, $R_S$ is the series resistance.

We extract the reverse saturation current density ($J_S$) from the forward bias current between 0.1 V to 0.4 V over the measured temperature range. Assuming the $J_S$ is due to TE, we calculate the corresponding barrier height. The TE saturation current is related to the barrier height through equation 3.

$$I_{TEO} = SA^{**}T^2 \exp\left(-\frac{\phi_{B0}}{kT}\right) \qquad (3)$$

Where S is the area, $A^{**}$ is the effective Richardson constant, T is temperature and $\phi_{B0}$ is the barrier height.

By taking natural log on both sides of equation 3, and finding the slope with respect to temperature, we calculated the barrier height is 0.6 V. This is higher than ideal barrier height of 0.5 V between Cr and intrinsic nc-Si:H. The work function of Cr is 4.5 V and the electron affinity of nc-Si:H is assumed to be 4.01 V [6]. The higher than expected barrier height indicates that this nc-Si:H diode is not a Schottky diode, but a p-n junction diode.

Furthermore, the extracted non-ideality factor "n" is 1.85. The value is calculated using the measurement results between 0.1 V to 0.4 V where the diode exhibits exponential behavior. The value of "n" is consistent over a range of temperature from 243 K to 398 K with slight variations. The values of "n" over different temperature range are shown in Figure 3.

The non-ideality factor being higher than 1 indicates that the contribution of the TE current to the total current is not the dominant factor. The actual TE component is only a portion of the reverse saturation current but we assumed in our barrier height calculation that TE is the only source of current. Thus we can draw the conclusion that the actual barrier height is even higher than our calculated 0.6 V.

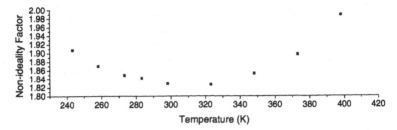

**Figure 3.** Non-ideality factor of diode with respect to measurement temperature.

Another method to calculate the barrier height is through the activation energy measurement method as outline in [7]. By plotting the forward current normalized versus temperature $I_F/T^2$ vs $1/T$ and then finding the slope, the barrier height can also be calculated. Through this method, the calculated barrier height of the diode is approximately 0.7 V at 0.4 V bias. This is similar to the aforementioned barrier height and it is also higher than the expected Schottky diode barrier. Thus we conclude that the nature of this diode is a p-n junction diode between the undoped and n+ nc-Si:H layers.

**Rectification using nc-Si:H diodes**

Simple rectification measurements are carried out using the two circuits seen in Figure 4. The two styles of rectifiers are connected to a 0.22 μF capacitor and a 1 MΩ resistor acting as the load. A sine wave AC signal with amplitude of 2 $V_{RMS}$ of varying frequency is generated using a function generator and the DC output voltage is measured using a Fluke 179 multimeter.

The measurement results of different rectifier configurations, single or full-wave bridge rectifier and different diode sizes, can be seen in Figure 5. There is a general trend where the output DC voltage drops as frequency increases due to decrease in performance of the diodes. As expected, larger diodes are able to output higher voltage than smaller diode. Full-wave bridge rectifiers are also able to produce higher DC voltage output at the same frequency and same diode size. In single diode configuration, the largest 250*250 μm² diode is able to output 2 V DC at 14 MHz. In full-wave bridge rectifier configuration, the largest diodes used had area of 200*200 μm² and this full-wave bridge rectifier can output 2.6 V DC at 14 MHz.

**DISCUSSION**

The nc-Si:H diode presented here is able to achieve higher current density than a-Si diodes due to higher conductivity of the material. At 2 V forward bias, this nc-Si:H p-n junction diode shows a current density of 10 A/cm² versus approximately $7*10^{-5}$ A/cm² for an a-Si:H Schottky diode reported in [5]. When compared to organic full-wave rectifiers, this nc-Si:H full-wave rectifier is able to outperform its organic-based counter-part. The nc-Si:H diode is able to output 2 V DC voltage, at 14 MHz frequency with a input signal at 2 $V_{RMS}$ while the organic rectifier had a 10 $V_{RMS}$ input signal to achieve the same output DC voltage [4].

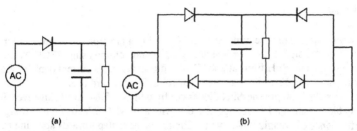

(a)                                           (b)

**Figure 4.** Setup for rectification measurement (a) with a single diode and (b) with 4 diodes in full-wave bridge rectifier.

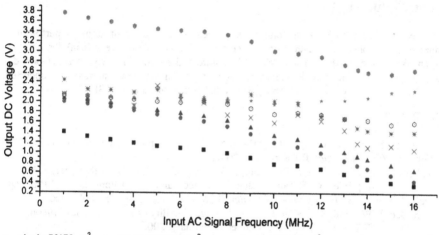

- single 50*50μm$^2$   • single 100*100μm$^2$   ▲ single 150*150μm$^2$   ● single 200*200μm$^2$

★ single 250*250μm$^2$   × full-wave 100*100μm$^2$   ✳ full-wave 150*150μm$^2$   • full-wave 200*200μm$^2$

**Figure 5.** Output DC voltage of single diode rectifiers and full-wave bridge rectifiers using different sized diodes with input AC signal of varying frequency and constant 2 $V_{RMS}$ amplitude.

Other advantages of using nc-Si:H for active device manufacturing include the possibility of having true CMOS circuits in the RFID tags. Our research group has demonstrated using nc-Si:H to create NMOS and PMOS thin-film transistors [8,9]. The use of organic semiconductors limits the transistors to PMOS only. The use of integrated CMOS transistors can lead to lower power draw versus PMOS only implementations of equivalent digital circuits.

Furthermore, our of nc-Si:H diode fabrication process uses existing PECVD technology which is already being used to manufacture large-area electronics such as liquid crystal displays and photovoltaic cells. PECVD is also compatible with roll-to-roll manufacturing techniques, hence the possibility of low cost manufacturing is also possible with nc-Si:H.

## CONCLUSION

We demonstrated the use of nc-Si:H in fabricating a junction diode suitable for use for rectification of an AC signal. The vertical diode was manufactured using PECVD deposited undoped and n-doped nc-Si:H films at 260 °C. The diode showed a current density of greater than 10 A/cm$^2$ and ON/OFF ratio of greater than 10$^6$ at 2 V forward bias.

With four 200*200 $\mu$m$^2$ nc-Si:H diodes configured as a full-wave bridge rectifier, the rectifier was able to output a DC voltage of 2.6 V with a 14 MHz 2 V$_{RMS}$ input AC signal which beat the performance of organic full-wave rectifiers. The next step towards applying nc-Si:H for manufacturing low cost RFID tags will be to combine the diode with CMOS circuits also fabricated with nc-Si:H.

## ACKNOWLEDGMENTS

We would like to thank Dr. John Hamel for lending us equipment for use in performing temperature-controlled IV measurements of the diode. We would also like to thank Dr. Yuriy Vygranenko for the insightful discussion on the nature of the diode. This work was performed using the Giga-to-Nanoelectronics Centre facilities at the University of Waterloo and supported by the National Sciences and Engineering Research Council of Canada (NSERC).

## REFERENCES

1. N. Huber, K. Michael, L. McCathie, "Barriers to RFID Adoption in the Supply Chain," *RFID Eurasia, 2007 1st Annual*, IEEE, pp. 1-6, 2007.
2. V. Subramanian, P. C. Chang, D. Huang, J. B. Lee, S. E. Molesa, D. R. Redinger, S. K. Volkman, "All-printed RFID Tags: Materials, Devices, and Circuit Implications," *Proc. Of the 19$^{th}$ International Conf. on VLSI Design (VLSID'06)*, IEEE, pp. 709-714, 2006.
3. B. S. Bae, J.-W. Choi, S-H. Kim, J-H. Oh, J. Jang, "Stability of an Amorphous Silicon Oscillator," *ETRI Journal*, Vol. 28, Number 1, February, 2006.
4. R. Rotzoll, et al., "Radio frequency rectifiers based on organic thin-film transistors," *Applied Physics Letters*, American Institute of Physics, vol. 88, 2006.
5. I. Ay, H. Tolunay, "The influence of ohmic back contacts on the properties of a-Si:H Schottky diodes," *Solid-State Electronics 51*, issue 3, pp. 381-386, 2007.
6. D. A. Neamen, *Semiconductor Physics and Devices: Basic Principles, Third Edition*, McGraw-Hill Higher Education: Boston, 2003, pp. 328, 713.
7. S. M. Sze, *Physics of Semiconductor Devices, Second Edition*, John Wiley & Sons: New York, 1981, ch. 5, pp 284-285.
8. H. J. Lee, A. Sazonov, A. Nathan, "Evolution of Structural and Electronic Properties in Boron-Doped Nanocrystalline Silicon Thin Films," MRS Symp. Proc., vol. 989, A21, 2007.
9. C. H. Lee, A. Sazonov, A. Nathan, "High-performance n-channel 13.56 MHz plasma-enhanced chemical vapor deposition nanocrystalline silicon thin-film transistors," J. Vac. Sci. Technol. A, American Vacuum Society, vol. 24, issue 3, pp. 618-623.

Mater. Res. Soc. Symp. Proc. Vol. 1066 © 2008 Materials Research Society 1066-A06-10

# Low Temperature Synthesis of Nanocrystalline Silicon and Silicon Oxide Films by Plasma Chemical Vapor Deposition

Atsushi Tomyo[1], Hirokazu Kaki[1], Eiji Takahashi[1], Tsukasa Hayashi[1], Kiyoshi Ogata[1], and Yukiharu Uraoka[2]

[1]Process Research Center, R & D Laboratories, Nissin Electric Co., Ltd., 47 Umezu, Takase, Ukyo-ku, Kyoto, 615-8686, Japan
[2]Graduate School of Materials Science, Nara Institute of Science and Technology, 8916-5 Takayama, Ikoma, Nara, 630-0192, Japan

## ABSTRACT

We have investigated the control method of nanocrystalline silicon (nc-Si) structures prepared by inductively coupled plasma chemical vapor deposition (ICP-CVD). In order to determine the diameter and surface density of nc-Si dots, plan-view transmission electron microscopy (TEM) was carried out. It was showed that spatially isolated nc-Si dots were synthesized and that the diameter and the standard deviation could be well controlled with substrate temperature, gas pressure and synthesis time. In particular, the mean diameter of 3.3 ± 0.6 nm and surface density of $1 \times 10^{12}$ cm$^{-2}$ were achieved with the optimum condition. Using the same ICP-CVD system, we have also obtained a $SiO_2$ film which has good electrical characteristics.

## INTRODUCTION

Nanocrystalline silicon (nc-Si) dot has attracted much attention because it exhibits unique features, such as Coulomb blockade, light emission and absorption [1,2]. For example, quantum-dot floating gate memory devices have advantage of the high reliability with respect to the breakdown of tunnel oxide [3]. In the practical applications of nc-Si dots in quantum effect device, it is important to control the size and surface density. In case of floating gate memory devices with nc-Si dots embedded in $SiO_2$, the diameter is theoretically expected to be less than 10 nm for room-temperature operation and the surface density should be as high as $10^{12}$ cm$^{-2}$ to gain a sufficient voltage shift for memory operation [4]. In addition, low temperature synthesis of nc-Si dots and $SiO_2$ films is desirable for high throughput in mass production and use of substrates which consist of low melting point materials.

The authors have already reported that nc-Si dots could be synthesized at the initial stage of microcrystalline silicon (μc-Si) films, with which a bottom-gate thin-film transistor exhibited a field-effect mobility of 3 cm$^2$/(V·sec) [5]. The deposition of μc-Si films was performed by ICP-CVD with low inductance antenna (LIA) units which were installed inside of a vacuum container so that the induced electric field from an antenna could be used effectively. By reducing the antenna inductance and fully covering antenna conductors with insulator, the plasma potential would decrease and thus the plasma damage to the underlayer could be suppressed [6]. It is worthy to mention that this plasma source could be applicable for the large-area deposition by increasing the number of antenna units.

In this study, nc-Si synthesis and SiO$_2$ deposition were also performed by ICP-CVD with LIA units, expecting that high density plasma is quite useful for synthesizing high density nc-Si dots and for lowering process temperatures with contrast to other methods [7,8]. We mainly focused on developing the control method of the size of nc-Si dots.

## EXPERIMENTAL PROCEDURE

An n-type 6-inch CZ-silicon substrate of (100) orientation with a resistivity of 5 Ωcm was first etched in a buffered 1%-HF (BHF) solution, and then transferred into the SiO$_2$ deposition chamber. After removing the natural oxide on the top surface of A schematic diagram of the experimental setup is shown in elsewhere [9]. While the substrate was heated up to the certain deposition temperature, which was fixed at 300°C in the present work, the pressure in the chamber was reduced to approximately 10$^{-4}$ Pa. A mixture of SiH$_4$ and O$_2$ gases was introduced into the chamber and then a silicon oxide layer was deposited on the top surface of the substrate by ICP-CVD. Radio frequency (13.56 MHz) power was supplied through LIA units. To maintain the plasma stability, optical emission spectrometer was used. Conditions of SiO$_2$ deposition are summarized in Table 1. Subsequently the substrate was transferred to the nc-Si synthesis chamber which had the same structure as the SiO$_2$ deposition chamber, keeping a vacuum state. Nc-Si dots were also synthesized by ICP-CVD under the conditions shown in Table 2. Substrate temperature, gas pressure and synthesis time were varied in order to investigate the optimum condition for nc-Si synthesis.

Structures of nc-Si dots were observed by a high-resolution transmission electron microscopy (HR-TEM). The diameter and surface density were determined by the plan-view TEM image. In addition, electronic properties of a single SiO$_2$ film were examined with a metal oxide semiconductor (MOS) capacitor formed by vacuum evaporation of aluminum electrodes. Post-metallization annealing was performed in N$_2$ and H$_2$ atmosphere at 450°C for 30 minutes.

Table 1: Conditions of SiO$_2$ deposition.

| | |
|---|---|
| SiH$_4$/O$_2$ gas flow rate ratio | 4.2 |
| RF power (mW/cm$^3$) | 24 |
| Deposition temperature (°C) | 300 |
| Total gas pressure (Pa) | 0.17 |
| Deposition rate (nm/sec) | 0.88 |

Table 2: Conditions of nc-Si synthesis.

| | |
|---|---|
| SiH$_4$/H$_2$ gas flow rate ratio | 15 |
| RF power (mW/cm$^3$) | 63 |
| Substrate temperature (°C) | 100 - 400 |
| Total gas pressure (Pa) | 0.67 - 4.0 |
| Synthesis time (sec) | 5 - 20 |

## RESULTS AND DISCUSSION

Typical cross-sectional and plan-view TEM images of nc-Si dots synthesized at the temperature of 100°C are shown in Fig. 1. Nc-Si dots were clearly observed with a diameter of sub-10 nm.

Figure 2(a) shows the size distribution of nc-Si dots synthesized with pressure of 4.0 Pa and synthesis time of 5 sec at various substrate temperatures. It was found that the diameter of nc-Si dots decreased with lowering the substrate temperature. It has been reported that the diffusion of radical precursors on the $SiO_2$ surface contribute the nc-Si synthesis [10]. Our result could be also interpreted that the diffusion of radical precursors was suppressed at low temperature, and thus the growth of nc-Si dots would not proceed. According to this trend, it is worthy to attempt the nc-Si synthesis at the room temperature for smaller nc-Si dots.

Figure 2(b) shows the size distribution of nc-Si dots synthesized with substrate temperature of 200°C and synthesis time of 5 sec at various pressures. It was found that small and uniform nc-Si dots could be synthesized at higher pressure. In this experiment, it was also found that the peak ratio of $H_\alpha$ (emission at 656 nm) to $Si^*$ (emission at 288 nm) in optical emission spectra decreased with increasing pressure as shown in Fig. 3. This result suggests that hydrogen radicals are reduced by collisions with source gases in transit to the $SiO_2$ surface. It has been reported that hydrogen radicals produce Si-OH bonds on the $SiO_2$ surface which play a role for nucleation sites of nc-Si dots [11,12]. Therefore, we considered that the decrease of nucleation sites led to suppress the coalescence of contiguous nc-Si dots. On the basis of this mechanism, we expect that the formation pattern of nc-Si dots is controllable with pre-surface treatments. Here, it should be mentioned that the polymerization reaction with radical precursors in plasma becomes outstanding with increase of gas pressure [13]. The succession of this reaction would produce large nc-Si dots, whose diameter is more than 10 nm. This suggests that optimum pressure would exist for synthesis of small and uniform nc-Si dots.

Figure 2(c) shows the size distribution of nc-Si dots synthesized with substrate temperature of 200°C and pressure of 4.0 Pa at various synthesis times. It was found that the standard deviation of the nc-Si diameter distribution increased with synthesis times of 10 and 20 sec. This result indicates that the growth of nc-Si dots on the $SiO_2$ surface proceeds as synthesis time passed, and the combination of contiguous nc-Si dots would occur.

Figure 4(a) shows current density to electric field (J-E) curve of a MOS capacitor using an ICP-CVD $SiO_2$ film deposited at 300°C. Breakdown voltage of 7.5 MV/cm at $1.0 \times 10^{-6}$ A/cm$^2$ and leakage current of $1.0 \times 10^{-9}$ A/cm$^2$ at 2.0 MV/cm were obtained for a $SiO_2$ film with a thickness of 12 nm. Capacitance-voltage (C-V) curves of the MOS capacitor were obtained as shown in Fig. 4(b) when the voltage was applied at 100 kHz from accumulation to inversion and retrace. Voltage shift in the hysteresis of C-V curves attributed to the oxide trapped charge was very small. In the previous work, we have found that an $O_2$ plasma density of $3 \times 10^{10}$ cm$^{-3}$ was obtained by ICP-CVD method with LIA units [14]. Such high density plasma would enable the oxidation of silicon effectively and deposition of a high quality $SiO_2$ film.

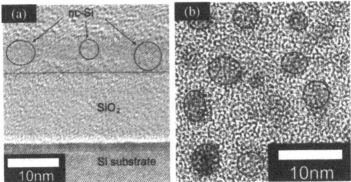

Fig. 1. (a) Cross-sectional and (b) plan-view TEM images of nc-Si dots synthesized at the substrate temperature of 100°C.

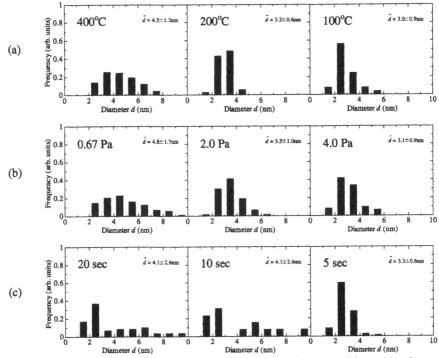

Fig. 2. Diameter distributions of nc-Si dots prepared at the different conditions. (a) substrate temperatures of 400, 200 and 100°C, (b) pressures of 0.67, 2.0 and 4.0 Pa, (c) synthesis times of 20, 10 and 5 sec.

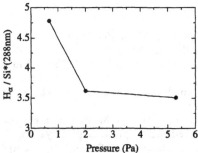

Fig. 3. H$_\alpha$(656nm)/Si$^*$(288 nm) peak ratio in optical emission spectra at the pressures corresponding to Fig. 2 (b).

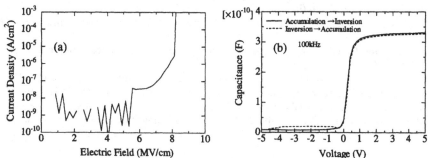

Fig. 4. (a) J-E and (b) C-V curves of a MOS capacitor using a ICP-CVD SiO$_2$ film with a thickness of 12 nm.

## CONCLUSIONS

This study showed that spatially isolated nc-Si dots were synthesized by ICP-CVD with LIA units and that the diameter and the standard deviation could be controlled with substrate temperature, gas pressure and process time. The diameter of almost all nc-Si dots under the present conditions was less than 10 nm. The experimental results suggest that deposited radical precursors and reactive nucleation sites on the SiO$_2$ surface have an important role in the synthesis of nc-Si dots. It is also shown that the electrical properties of a SiO$_2$ film deposited with the same ICP-CVD apparatus are capable of the thin dielectric layer of nc-Si devices. We confirmed that ICP-CVD method should be applicable for the preparation of many quantum devices.

## ACKNOWLEDGMENTS

The authors would like to thank Y. Setsuhara of the Osaka University with many useful discussions about the ICP-CVD method, and A.Ebe of EMD Corporation for technical assistance.

## REFERENCES

1. L. T. Canham, Appl. Phys. Lett. **57**, 1046 (1990).
2. S. Oda, Mater. Sci. Eng. **B00**, 1 (2003).
3. S. Tiwari, F. Rana, H. Hanafi, A. Hartstein, EF Crabbe and K. Chan, Appl. Phys. Lett. **68**, 1377(1996).
4. S. Naito, M. Satake, H. Kondo, M. Sakashita, A. Sakai, S. Zaima and Y. Yasuda, Jpn. J. Appl. Phys. **43**, 3779 (2004).
5. E. Takahashi, Y. Nishigami, A. Tomyo, M. Fujiwara, H. Kaki, K. Kubota, T. Hayashi, K. Ogata, A. Ebe and Y. Setsuhara, Jpn. J. Appl. Phys. **46**, 1280 (2007).
6. Y. Setsuhara, S. Miyake, Y. Sakawa and T. Shoji, Jpn. J. Appl. Phys. **38**, 4263 (1999).
7. N. Sugiyama, T. Tezuka and A. Kurobe, J. Crystal Growth **192**, 395 (1998).
8. Y. Numasawa and N. Koshida, U.S. Patent No. 7 091 138 (15 Aug. 2006).
9. A. Tomyo, H. Kaki, E. Takahashi, T. Hayashi and K. Ogata, Proc. of 4th Thin Film Materials & Devices Meeting, Kyoto (in press).
10. S. Miyazaki, Y. hamamoto, E.Yoshida, M.Ikeda and M.Hirose, Thin Solid Films **369**, 55 (2000).
11. Y. Inoue and O. Takai, J. Jpn. Soc. of Plasma Science and Nuclear Fusion Research 76, 1068 (2000).
12. Y. Morimoto, T. Igarashi and T. Okanuma, J. Non-Crys. Solids. **179**, 260 (1994).
13. M. Shiratani, T. Fukuzawa and Y. Watanabe, Jpn. J. Appl. Phys. **38**, 4542 (1999).
14. A. Tomyo, H. Kaki, E. Takahashi, T. Hayashi, K. Ogata, K. Ichikawa, Y. Uraoka and Yuichi Setsuhara, Proc. of the 24th Symp. of Plasma Processing, Osaka, Japan, 2007, p. 35.

Mater. Res. Soc. Symp. Proc. Vol. 1066 © 2008 Materials Research Society     1066-A06-11

# Cyclically Varying Hydrogen Dilution for the Growth of Very Thin and Doped Nanocrystalline Silicon Films by Hot-Wire CVD

Fernando Villar, Aldrin Antony, Delfina Muñoz, Fredy Rojas, Jordi Escarré, Marco Stella, José Miguel Asensi, Joan Bertomeu, and Jordi Andreu

XaRMAE, Física Aplicada y Óptica, University of Barcelona, Martí i Franqués 1-11, Barcelona, 08028, Spain

## ABSTRACT

Hot-Wire Chemical Vapour Deposition (HW-CVD) technique has demonstrated to be a good alternative to deposit quality thin films at low temperature. In this paper we focus our study on very thin (50 nm) n- and p-doped nc-Si:H films deposited at low substrate temperature around 100°C. We have observed that, in this low temperature deposition conditions, the promotion of an a-Si:H incubation layer leads to a poor doping efficiency and poor electrical properties of the films. Hence, in addition to the optimization of the deposition conditions, we deposited doped layers by cyclically varying the hydrogen dilution (CVH) during deposition process. This CVH method promotes a layer-by-layer growth and inhibits the formation of the incubation layer. Several doped nc-Si:H layers have been deposited with and without this CVH method. The structural, electrical and optical properties of these films and advantage of CVH in improving the device quality of the thin doped layers are reported.

## INTRODUCTION

The rapid growth of the world energy consumption and the increasing $CO_2$ has led the scientific community to look for new strategies and materials for the power production. In this context, the photovoltaic sector is one of the promising alternatives for a clean energy production. The main limitation in the growth of the photovoltaic industry is the high production cost of the solar cells. Recently, the scientific community has shown great interest in the low cost thin film solar cells based on amorphous (a-Si:H) and nanocrystalline silicon (nc-Si:H). These thin film devices need only a lower amount of material than those with c-Si; thus reducing the total cost of the solar cell production, In addition, these silicon thin film solar cells permit the possibility of roll-to-roll deposition onto flexible substrates, thereby reducing again the production cost [1].

Nevertheless the use of flexible plastic substrates imposes some restrictions on the deposition parameters, mainly the low substrate temperature ($T_s < 120°C$) are required in all the processes of the thin film solar cells fabrication. In this sense, the Hot-Wire Chemical Vapour Deposition (HW-CVD) is presented as a suitable technique to deposit at low temperature. This technique allows higher hydrogen dilution leading to the growth of nc-Si:H material without any additional substrate heating [2]. In previous works, we demonstrated the possibility to deposit ZnO:Al and intrinsic nc-Si:H on polyethylene naphthalene (PEN) at very low substrate temperature with good electrical and optical properties [2,3]. The substrate temperature limitation of plastic substrates needs a special attention when we consider the deposition of very thin nc-Si:H doped layers by HW-CVD, because the low temperature deposition of the doped layers may result in an undesired growth of an incubation layer with poor electronic properties [4]. The high versatility of the HW-CVD process permits a lot of flexibility in varying different deposition conditions at a low substrate temperature. Here we report the use of the cyclically varying hydrogen dilution (CVH) method for the growth of very thin and highly doped nc-Si:H films.

EXPERIMENTAL

Very thin doped nc-Si:H films were deposited onto Corning 1737F glass substrates in our HW-CVD reactor which is described elsewhere [5]. The deposition of thin films was done by the hot-wire catalytic dissociation of silane and hydrogen gas mixture at different pressures and for different hydrogen dilutions ($D_H = \phi_{H2} \times 100/(\phi_{SiH4} + \phi_{H2})$). Small amounts of phosphine or diborane gas were added to dope (n or p type) the films. The dissociation of the gases was achieved by means of two straight tantalum filaments with a diameter of 0.5 mm, placed under the substrate at a distance of 4 cm, and resistively heated at a temperature of 1750°C. The effective substrate temperature ($T_s$) during deposition was calibrated by measuring the temperature of the glass substrate with a thermocouple attached to it under the typical deposition conditions. In this work we have deposited samples at 100°C as the set point temperature ($T_c$) and without any intentional heating (WIH). Due to the radiation of the filaments the $T_s$ rises to higher values during the first steps of the deposition. The deposition time of the very thin layers is usually less than five minutes and this variation in substrate temperature ($T_s$) during the deposition is very important since it may affect the properties of the film. Figure 1 shows the variations in substrate temperature during the first 10 minutes of the deposition process, for an initial set temperature of 100 °C and that for without any intentional heating.

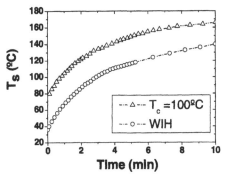

**Figure 1.** Evolution of the substrate temperature ($T_s$) during the first 10 minutes of the deposition process for substrates preheated at 100°C and for substrate without intentional heating.

Table 1 gives the different process conditions used for the deposition 6 different n- and p-doped layers. We developed a new cyclically varying hydrogen dilution (CVH) method in order to deposit very thin doped nc-Si:H thin films at low substrate temperature. In the table, the samples $N_2$ and $P_3$ were deposited by CVH method. The method consists in varying cyclically the hydrogen dilution during the deposition process between a maximum ($D_H$=98%) and a minimum ($D_H$=93.3%) with a cyclic period around 1 min. In this way, a layer-by-layer growth as shown in Figure 2 is achieved in order to avoid the undesired incubation layer. The instantaneous change of flow set point was achieved by the automation of the process using a PC, while the physical process took a few seconds delay.

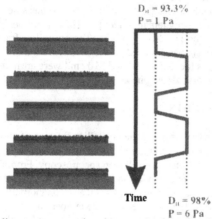

**Figure 2.** A schematic wave diagram representing the growth of the film by cyclically varying hydrogen dilution method (CVH).

**Table 1.** Deposition conditions for the doped layers. The wide range of depositions conditions used for various n- and p-doped layers are shown in first two columns. To calculate the gas phase doping level we used the formula $\phi_{doping} \times 100 / \phi_{SiH4}$. The samples N2 and P3 are deposited by CVH method.

| Parameters | | n-layers | | | p-layers | | |
|---|---|---|---|---|---|---|---|
| | | N1 | N2 | N3 | P1 | P2 | P3 |
| **Pressure (Pa)** | 1 – 8 | 1 | 1 and 6 | 1 | 1 | 1 | 1 and 6 |
| $\phi_{SiH4}$ (sccm) | 1 – 4.5 | 2 | 2 | 1.5 | 1 | 1.5 | 2 |
| $\phi_{PH3}$ (sccm) | 0.01 – 0.1 | 0.04 | 0.04 | 0.075 | - | - | - |
| $\phi_{B2H6}$ (sccm) | 0.04 – 0.2 | - | - | - | 0.04 | 0.06 | 0.08 |
| Gas phase dopig level (%) | 2 – 5 | 2 | 2 | 5 | 4 | 4 | 4 |
| $D_H$ (%) | 0 – 98 % | 98 | 93.3 and 98 | 93.3 | 0 | 0 | 93.3 and 98 |
| $T_c$ (°C) | WIH-200 | WIH | WIH | 100 | 100 | 100 | 100 |
| CVH system | | No | Yes | No | No | No | Yes |

175

The electrical conductivity of the samples was measured by four probe method with evaporated aluminium contacts. The thicknesses of the samples were measured with a profilometer. The microstructure of the six samples given in Table 1, were analysed by fitting the imaginary part of the pseudo-dielectric function ($\varepsilon = \varepsilon_1 + i\varepsilon_2$) measured by spectroscopic ellipsometry (SE) using the Bruggeman model [5]. The crystalline fraction of the samples were estimated from Raman spectra of these samples. The crystalline fraction defined as $X_c = (I_{520} + I_{510})/(I_{520} + I_{510} + I_{480})$ was deduced by the deconvolution of the spectra into three Gaussian bands. The peaks located at 520 and 510 cm$^{-1}$ were attributed to crystalline phases and the one at 480 cm$^{-1}$ to the disordered phase [6].

## RESULTS AND DISCUSSION

The dark conductivity ($\sigma_D$) and the crystalline fraction of all the deposited samples were calculated. Figure 2a shows the variation of electrical conductivity ($\sigma_D$) as a function of the crystalline fraction ($X_c$) estimated from Raman spectroscopy. From the figure, it is clear that nc-Si:H films show acceptable conductivity. In addition, irrespective of the deposition temperature, the films with $X_c > 0.1$ did not show much difference in conductivity. The results suggests that the doping incorporation is effective at these substrate temperatures. A special attention is required for the a-Si:H samples which show high conductivity. As we had already explained, it can be attributed to the sensitivity limit of Raman Spectroscopy measurements [2]. In this case, the very small crystallite size and thickness of the samples lead to $X_c \sim 0$ values, but measurements with

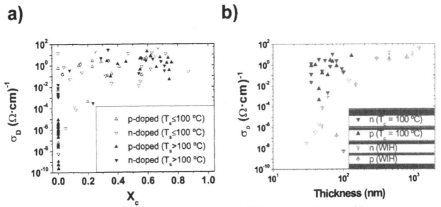

**Figure 2.** a) Distribution of dark conductivity ($\sigma_D$) of different n- and p-doped layers as a function of crystalline fraction ($X_c$). Open symbols are used for samples deposited at Ts<100°C and closed symbols for $T_s > 100$°C. b) The distribution of $\sigma_D$ of both n- and p-doped layers as a function of the thickness for the samples deposited at $T_s = 100$ °C and without intentional heating.

other techniques (eg. Spectroscopic ellipsometry and transmission electron microscopy) revealed small crystallites in these samples. Anyway, as we have expounded earlier, the importance here falls in the initial growing steps. Figure 2b shows the dark conductivity ($\sigma_D$) of highly p- and n-doped thin film samples as a function of the thickness. High variation in the electrical properties of these samples for thickness lower than 100 nm suggests an undesired incubation layer in the films deposited without intentional heating. Nevertheless a small preheating of the substrate at 100°C (see the calibration curve as shown in Fig.1 to know the real substrate temperature during deposition), considerably improves the value of $\sigma_D$ for all the thickness and for both n- and p-doping. This substrate temperature regime is compatible with most of the plastic substrates, and we have deposited these doped layers over PEN substrates and achieved similar values as shown in Fig2b [2]. These results are in agreement with the reported values [7,8].

In order to deposit at lower substrate temperature, we developed the CVH method as explained in the experimental part. Figure 3 shows the dark conductivity ($\sigma_D$) of the samples deposited by CVH method as function of the thickness. As shown in the plot, the n-doped layers deposited without intentional heating show good electrical properties. On the contrary, the p-doped layers deposited without intentional heating exhibit low conductivity. However, a preheating of the substrate at 100°C improves the electrical properties of these films significantly. One of the interesting and important facts is that, during the thin n-layer deposition by CVH, the real substrate temperature does not exceed 100 °C.

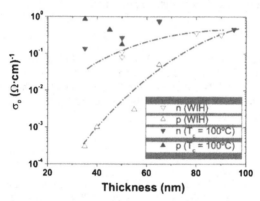

**Figure 3.** Variation of dark conductivity ($\sigma_D$) of n- and p-doped samples deposited by CVH at 100°C as well as by WIH as a function of the thickness.

SE measurements were performed on the six samples shown in table 1. From the fitting of the imaginary part of the pseudo-dielectric function we have observed that in the case of n-doped layers deposited with preheating the substrate (N3) and CVH (N2), the incubation layer growth was suppressed and more compact films were obtained. In the case of the first sample (N1) deposited without intentional heating, an undesired incubation layer of 7 nm has been detected. In addition, the high porosity observed in the layer (void fraction ~ 35%) might be the reason for the degradation in the electrical properties of this sample. In the case of p-doped layers (P1 and P2), an incubation layer of around 20 nm thickness was detected, though these samples were deposited with the pre-heating of the substrate. These two samples exhibit degradation in the conductivity by two orders on magnitude which can be attributed to the Staebler-Wronski effect or an oxidization process of the sample. The SE measurements of these samples show an amorphous fraction of ~ 80% in these film. The deposition of the p-layer by CVH method using preheated substrates resulted in decreasing the formation of the incubation layer to 8nm and more microcrystalline layers were obtained. These films show more compactness with acceptable values of void fraction (~15%) and exhibit higher conductivity and lower activation energy.

## CONCLUSIONS

Very thin n- and p-doped nc-Si:H films were deposited by Hot-Wire Chemical Vapor deposition. The samples were deposited at substrate temperature of 100 °C and also without intentional heating. The films deposited without intentional heating showed undesirable incubation layers of 7 nm for the n-layer and 20nm in the case of p-layer. These films showed a strong decrease of the conductivity. By implementing our cyclically varying hydrogen dilution (CVH) method, the formation of the incubation layer was suppressed in the case of n-layer. On the other hand, the CVH with preheating of the substrate for the p-layer deposition resulted in decreasing the incubation layer thickness to 8 nm. These films showed significant increase in electrical conductivity and crystallinity. The CVH method has been proved to be an efficient deposition technique for the deposition of very thin n- and p-doped layers at low substrate temperature with good electrical and structural properties.

## ACKNOWLEDGMENTS

This work has been supported by the EU through the FLEXCELLENCE project (contract 019948) and the Spanish National R&D Plan (projects ENE2005-25268-E, ENE2006-27250-E, ENE2007-67742-C04-03, and PSE-120000-2007-7). We would like to acknowledge LPICM of the École Polytechnique for the SE measurments, and also to P. Roca for his helpful discussions in interpreting the SE measurements.

## REFERENCES

1. J. Bailat, V. Terrazzoni-Daudrix, J. Guillet, F. Freitas, X. Niquille, A. Shah, C. Ballif, T. Scharf, R.Morf, A. Hansen, D. Fischer, Y. Ziegler, A. Closset. Proceedings of the 20th EU Photovoltaic Solar Energy Conference, ISBN 3-936338-19-1, WIP Renewable Energies, pp. 1529-1532, 2005.
2. F. Villar, J. Escarré, A. Antony, M. Stella, F. Rojas, J.M. Asensi, J. Bertomeu and J. Andreu. Thin Solid Films, Volume 516, Issue 5, 15 January 2008, Pages 584-587
3. M. Fonrodona, J. Escarré, F. Villar, D. Soler, J.M. Asensi, J. Bertomeu and J. Andreu. Sol. Energy Mater. Sol. Cells 89 (2005) p.37
4. Chisato Niikura, Romain Brenot, Joelle Guillet and Jean-Eric Bourée. Thin Solid Films, Volume 516, Issue 5, 15 January 2008, Pages 568-571.
5. A. Fontcuberta i Morral, P. Roca i Cabarrocas, C. Clerc, Phys. Rev., B 69 (2004) 125307.
6. L. Houben, M. Luysberg, P. Hapke, R. Carius, F. Finger, H. Wagner, Philos. Mag A 77 (6) (1998) 1447.
7. S.A. Filonovich, M. Ribeiro, A.G. Rolo and P. Alpuim. Thin Solid Films, Volume 516, Issue 5, 15 January 2008, Pages 576-579
8. P. Kumar and B. Schroeder. Thin Solid Films, Volume 516, Issue 5, 15 January 2008, Pages 580-583.

Mater. Res. Soc. Symp. Proc. Vol. 1066 © 2008 Materials Research Society

# Microstructure Effects in Hot-wire Deposited Undoped Microcrystalline Silicon Films

Wolfhard Beyer, Reinhard Carius, Dorothea Lennartz, Lars Niessen, and Frank Pennartz
IEF5 Photovoltaik, Forschungszentrum Jülich GmbH, Leo Brandt Strasse, Jülich, 52425, Germany

## ABSTRACT

The microstructure of hot-wire microcrystalline silicon films prepared at a wide range of deposition conditions was characterized by both the microstructure parameter from infrared absorption data (analyzing the Si-H stretching modes) and the effusion spectra of (low dose) implanted He and Ne. Parameter ranges leading to the growth of a dense material are identified. A (relatively) high silane flow at rather high filament temperature is found to result in a dense material at high deposition rate. The microstructure data obtained by the two microstructure characterization methods are found to be largely correlated.

## INTRODUCTION

With the success of plasma-grown thin film silicon for application in solar cells and other large area devices, the interest in alternative deposition methods for the base material is rising. Of particular interest is the hot-wire (HW) deposition method with high solar cell efficiencies achieved on small area [1], good potential for upscaling to large areas [2] and with no ion bombardment involved. On the other hand, it has been found that HW a-Si:H films, compared to plasma grown material, have a different and more pronounced void-related microstructure [3]. In this work, we focus on the (void-related) microstructure of undoped microcrystalline silicon using the microstructure parameter from IR absorption measurements [4] and the effusion spectra of (low dose) implanted helium and neon [5,6] for characterization. Aim is, to identify deposition conditions leading to the growth of a dense material since a void related microstructure may be detrimental for the long term material stability required in solar cells. The second aim of this work is a comparison of microstructure data of the two microstructure characterization methods for a large set of different Si:H samples.

## EXPERIMENTAL DETAILS

The hot wire microcrystalline silicon films (HW μc-Si:H) were grown in a stainless steel reactor, using two tantalum filaments (length 38 mm, diameter 1 mm) at a distance of about 3 cm from the temperature-stabilized substrate holder. The quartz /crystalline silicon substrates were fixed to the substrate holder by silver paste ensuring a good thermal contact. The filament temperature was measured by a pyrometer. The gases silane ($SiH_4$) and hydrogen were fed into the reactor by mass flow controllers and pumped (via a butterfly valve) by a turbomolecular/ roughing pump system. Various series of films were deposited with one deposition parameter (filament temperature $T_{fil}$, substrate temperature $T_S$, pressure p, silane and hydrogen flows) varying while other parameters were fixed. Film thickness determined by a step profiling instrument was typically 1 μm. Infrared absorption was measured using a Fourier transform spectrometer. The Si-H stretching absorptions centered near 2000 and 2100 $cm^{-1}$ were analyzed to obtain both the absolute concentration of silicon-bound hydrogen (using an absorption

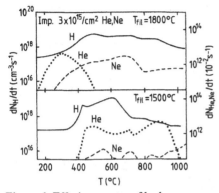

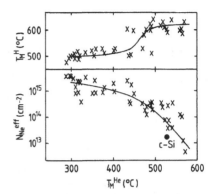

**Figure 1.** Effusion spectra of hydrogen and of implanted He and Ne for HW Si:H deposited at $T_{fil}$ = 1500 and 1800°C.

**Figure 2.** Temperature of maximum hydrogen effusion rate $T_M^H$ and of effused Ne, $N_{Ne}^{eff}$, versus $T_M^{He}$.

strength of A = $10^{20}$ cm$^{-2}$ [7]) and the microstructure parameter R. The latter is defined by R = I(2100)/ (I(2000) + I(2100)) with I (2000) and I (2100)) the integrated absorptions near 2000 and 2100 cm$^{-1}$ [4]. For microstructure analysis by He, Ne (and hydrogen) effusion, the films were implanted by 40 keV He$^+$ and 100 keV Ne$^+$ ions at a dose of $3 \times 10^{15}$cm$^{-2}$. According to the TRIM (transport of ions in matter) routine, the maxima of He and Ne distributions are at a depth of 3500 Å and 2100 Å, respectively. Gas effusion measurements were performed as reported elsewhere [8]. In a quartz tube evacuated by a turbomolecular pump the samples were heated at a rate of 20°C/min to 1050°C and the effusing gases hydrogen, helium and neon were detected using a quadrupole mass spectrometer.

## RESULTS

Typical effusion spectra of two materials which we term "dense" and "rich in interconnected voids" are shown in the lower and upper part, respectively, of Fig. 1. These samples were deposited at the filament temperatures $T_{fil}$ = 1500°C and 1800°C, respectively, while the other deposition parameters ($T_S$ =200°C, hydrogen flow 50 sccm, silane flow 3 sccm, p= 0.1 mbar) were the same. Plotted is the effusion rate of hydrogen, helium and neon as a function of temperature. From previous work [5,6] it is known that besides hydrogen effusion, the effusion of (implanted) rare gases provide a sensitive tool for microstructure in amorphous and microcrystalline silicon materials. Since rare gas atoms do not form bonds to silicon, their out-diffusion behaviour is strongly dependent on network openings in the microcrystalline or amorphous silicon material. Due to its small size, implanted He diffuses even in (dense) crystalline Si in the investigated temperature range and shows an effusion peak near 500°C. Ne, on the other hand, which has a larger size than He and is approximately equal in size to molecular hydrogen [5] does not show any significant effusion from c-Si wafers and from compact plasma-grown a-Si:H films [5] up to 1050°C. Thus He effusion provides a sensitive tool to study microscopic material density, while the effusion of neon provides information on the presence of rather large-size void channels. In the material with interconnected voids (Fig. 1,

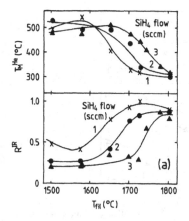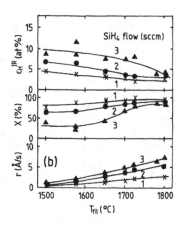

**Figure 3.** (a) $T_M^{He}$ and $R^{IR}$ and (b) hydrogen concentration $c_H^{IR}$, crystallinity X and deposition rate r versus filament temperature $T_{fil}$ for three $SiH_4$ flow rates ($H_2$ flow = 50 sccm, p = 0.1 mbar; $T_S$ =200°C fixed).

upper part), the hydrogen effusion peak lies near 500°C and the He effusion peak near 300°C. The implanted neon starts to evolve near 300°C. In the compact material, on the other hand (Fig. 1, lower part), the hydrogen effusion peak is near 600°C in agreement with other compact Si:H materials [6]. The He effusion shows a first maximum near 500°C followed by a second peak near 800-900°C which has been assigned to He trapped in isolated voids [5,6]. The structure near 700°C can be attributed to crystallization effects [5]. For this dense material there is almost no Ne effusion, in agreement with the results for single crystalline Si and compact a-Si:H [5].

An interconnection between the temperature $T_M^{He}$ of maximum He effusion rate and the Ne and hydrogen effusion characteristics is found for all samples. This is demonstrated in Fig. 2 where the total amount of effused Ne, $N_{Ne}^{eff}$, as well as the temperature of maximum hydrogen effusion rate, $T_M^H$, are plotted as a function of $T_M^{He}$. For low $T_M^{He}$, the material structure is quite open so that the effused amount of neon approaches the implanted dose of $3 \times 10^{15}/cm^2$. At high $T_M^{He}$, on the other hand, the material gets as dense as crystalline silicon (see data point for c-Si) and most implanted Ne remains stable up to 1050°C. The hydrogen effusion peak near 600°C (observed for high $T_M^{He}$) is explained by the presence of compact material with H effusion limited by H diffusion on a length scale of film thickness [9]. The lower $T_M^H$ for void-rich material (observed for lower $T_M^{He}$) is attributed to a reduction of the average H diffusion length which determines the H effusion temperature. A smaller H diffusion length is expected when hydrogen diffuses to the next (interconnected) void and not to the actual film surface.

The influence of various deposition parameters on the microstructure of HW Si:H is demonstrated in the Figs 3, 4 and 5. $T_M^{He}$ and the microstructure parameter $R^{IR}$ as well as concentration $c_H^{IR}$ of bound hydrogen (from IR data), crystallinity X (from Raman measurements) and deposition rate r are plotted as a function of filament temperature $T_{fil}$. For all samples, the $H_2$ flow was kept fixed at 50 sccm. Fig.3 refers to samples of three different $SiH_4$ flows of 1, 2 and 3 sccm with fixed p= 0.1 mbar and $T_S$ = 200°C. Fig. 4 shows for a $SiH_4$ flow of 2 sccm (and $T_S$ = 200°C) the influence of different pressure and in Fig. 5 the influence of

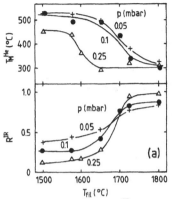
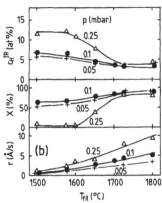

**Figure 4.** (a) $T_M{}^{He}$ and $R^{IR}$ and (b) hydrogen concentration $c_H{}^{IR}$, crystallinity X and deposition rate r versus filament temperature $T_{fil}$ for three different pressures p ($H_2$ flow = 50 sccm, SiH$_4$ flow 2 sccm and $T_S$= 200°C fixed).

different substrate temperatures is demonstrated (for a SiH$_4$ flow of 2 sccm and p= 0.1 mbar). Thus, all three figures have data of a reference series deposited with a flow of 2 sccm, a pressure of 0.1 mbar and $T_S$= 200°C in common (closed circles). For all series, $T_M{}^{He}$ and $R^{IR}$ show a mirror-like behavior, i.e. a (anti-) correlation is suggested. Also common to all series is the presence of relatively dense material at low filament temperature and of void-rich material at higher $T_{fil}$ as well as an increasing deposition rate with increasing $T_{fil}$. The crystallinity X is found to increase with rising $T_{fil}$. The concentration of bound hydrogen, $c_H{}^{IR}$, appears to be mainly related to crystallinity, i.e. a high crystallinity leads to low $c_H{}^{IR}$. We note, however, that by H effusion in almost all cases higher hydrogen concentration values are measured suggesting that either a large fraction of molecular hydrogen is present or that the absorption strength A is

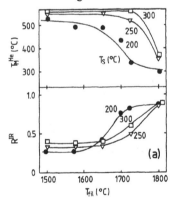
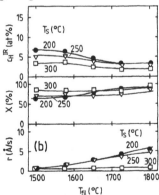

**Figure 5.** (a) $T_M{}^{He}$ and $R^{IR}$ and (b) $c_H{}^{IR}$, X and r versus $T_{fil}$ for three different substrate temperatures $T_S$ ($H_2$ flow (50 sccm), SiH$_4$ flow (2 sccm) and p = 0.1 mbar fixed).

182

different for amorphous and microcrystalline silicon materials. An important result of Fig 3 is that a higher $SiH_4$ flow leads to compact material up to the high $T_{fil} = 1700°C$, i.e. up to a high deposition rate. Thus, the growth of relatively dense Si:H at fairly high crystallinity of 70% and a deposition rate of 6 Å/s is possible. The results of Fig. 4 suggest that the structurally denser films are grown at lower pressure (but the deposition rate decreases) while a higher pressure leads to higher deposition rates but leads also to a transition to amorphous material at low $T_{fil}$. Finally, the results of Fig. 5 show that higher substrate temperatures lead to the growth of dense material up to rather high $T_{fil}$. However, a higher substrate temperature results generally in a decrease of deposition rate. Furthermore it is seen in Figs. 4 and 5, that changes in $T_M^{He}$ are not always reflected in $R^{IR}$.

## DISCUSSION

The results suggest that the growth of dense HW material is favored by a (relatively) low filament temperature, a low pressure, a rather high $SiH_4$ flow and a high substrate temperature. However, if also a high deposition rate is required, a rather high $T_{fil}$ along with a high $SiH_4$ flow are appropriate. At a fixed $H_2$ flow, the $SiH_4$ flow rate is then limited by the transition from microcrystalline to amorphous growth. Note that this deposition regime has previously been identified by Klein et al. [10] for achieving good quality material and solar cells at rather high deposition rate. As is well known, flows of atomic Si and H are generated at the hot wire at $T_{fil} = 1500 - 1800°C$ [11,12]. These flows are expected to increase with rising $T_{fil}$, as the decomposition of $SiH_4$ and $H_2$ requires energy. Thus, as long as the $SiH_4 -H_2$ gas mixture is not depleted of $SiH_4$ and as long as no H-related etching effects occur, a higher filament temperature will result in a higher deposition rate, in agreement with the experimental results. For a higher crystallinity, on the other hand, a high hydrogen coverage is required on the growth surface [13]. Since the H-H binding energy exceeds that of Si-H, the relative flow ratio of (atomic) H / (atomic) Si will increase with rising $T_{fil}$. This effect may explain why higher $T_{fil}$ favor crystallinity. Aiming for the growth of a dense material, on the other hand, high filament temperatures are not favorable, because the grain boundaries are no longer passivated by a-Si:H. Another possible effect is that at high $T_{fil}$, i.e. high Si flow rate, the growth surface may be reached by silicon atoms [11,12] which reduce the surface mobility. It seems that both effects can be avoided by a high flow of $SiH_4$, since a higher $SiH_4/H_2$ ratio in the gas phase generally favors a-Si growth and, furthermore, gas phase reactions of Si atoms with $SiH_4$ become more likely.

Concerning the relation between IR microstructure parameter and $T_M^{He}$ we note that a complete correspondence cannot be expected. While the IR microstructure parameter is based on the different vibrational frequencies of the Si-H stretching modes of surface and volume bound hydrogen [14] and thus detects voids via the presence of surface bound hydrogen, the helium effusion measures directly a

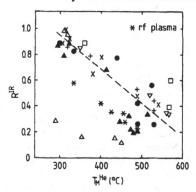

**Figure 6.** $R^{IR}$ versus $T_M^{He}$.

material density. Thus, it is conceivable that voids not covered by hydrogen could exist under certain deposition conditions, with the result of a low IR microstructure parameter for a void-rich material. In Fig. 6, the IR microstructure parameter is plotted as a function of $T_M^{He}$ for all series of this study. The same symbols as in Figs 3, 4 and 5 were used. A clear trend for a higher microstructure parameter $R^{IR}$ at a lower $T_M^{He}$ is observed. Plasma-grown μc-Si:H films (asterisks) show a similar trend. However the data points scatter appreciably and, in particular, samples of the high pressure series of Fig 4 barely show the trend at all. This result suggests some caution concerning the use of a low IR microstructure parameter as an evidence for the presence of a dense Si:H material.

## CONCLUSIONS

Microstructure characterization of various series of HW μc-Si:H samples shows the growth of rather dense material (with a low concentration of interconnected voids) at rather low filament temperatures, low pressure, high silane flow and relatively high ($T_S$ = 250-300°C) substrate temperatures. Since higher substrate temperatures, lower pressure and lower filament temperature tend to reduce the growth rate, our results suggest for the growth of dense material at rather high deposition rate the use of a relatively high filament temperature along with a relatively high silane flow. The microstructure parameter $R^{IR}$ and the helium effusion temperature $T_M^{He}$ are found to be largely correlated, i.e. a reduction of $T_M^{He}$ results in an increase of $R^{IR}$. Deviations were observed for a small number of samples only.

## ACKNOWLEDGMENTS

The authors wish to thank A. Dahmen for the ion implantations and M. Hülsbeck for the Raman measurements.

## REFERENCES

1. B. Schroeder, *Thin Solid Films* **430** (2003) 1.
2. K. Ishibashi, M. Karasawa, G. Xu, N.Yokokawa, M. Ikemoto, A. Masuda, H. Matsumura, *Thin Solid Films* **430** (2003) 58.
3. D.L. Williamson, D.W.M. Marr, E. Iwaniczko, B.P. Nelson, *Thin Solid Films* **430** (2003) 192.
4. A.H. Mahan, Raboisson, D.L. Willamson, R. Tsu, *Solar Cells* **21** (1987) 117.
5. W. Beyer, *MRS Symp. Proc.* **664** (2001) A9.2.1.
6. W. Beyer, *Phys. Stat. Solidi C* **1** (2004) 1144.
7. W. Beyer, M.S. Abo Ghazala, *MRS Symp. Proc.* **507** (1998) 601.
8. W. Beyer, J. Herion, H. Wagner, U. Zastrow, *Philos. Mag.* **B63** (1991) 269.
9. W. Beyer, *Solar Energy Materials and Solar Cells* **78** (2003) 23.
10. S. Klein, F. Finger, R. Carius, M. Stutzmann, *J. Appl. Phys.* **98** (2005) 024905.
11. C. Horbach, W. Beyer, H. Wagner, *J. Non-Crystalline Solids* **137-138** (1991) 661.
12. A.H. Mahan, Y. Xu, D.L. Williamson, W. Beyer, J.D. Perkins, M. Vanecek, L.M. Gedvillas, B.P. Nelson , *J. Appl. Phys.* **90** (2001) 5038.
13. M. Kondo, H. Fujiwara, A. Matsuda, *Thin Solid Films* **430** (2003) 130.
14. H. Wagner, W. Beyer, *Solid State Communications* **48** (1983) 585.

Mater. Res. Soc. Symp. Proc. Vol. 1066 © 2008 Materials Research Society      1066-A06-14

# Seeding Solid Phase Crystallization of Amorphous Silicon Films with Embedded Nanocrystals

Curtis Anderson, and Uwe Kortshagen
Department of Mechanical Engineering, University of Minnesota, 111 Church St. S.E., Minneapolis, MN, 55455

## ABSTRACT

Silicon nanocrystals with diameters up to 30 nm are used as nucleation seeds for fast solid phase crystallization of amorphous silicon films. Purely amorphous films required an incubation time of up to 12 hours at 600°C prior to the onset of nucleation, while films with nanocrystals embedded between layers of amorphous silicon grew immediately upon annealing in a quartz tube furnace. Structural characterization was performed by heated-stage transmission electron microscopy and Raman spectroscopy.

## INTRODUCTION

In recent years, thin film Si photovoltaic (PV) cells have gained in popularity compared to wafer-based modules. The last several years have seen a significant increase in the number of thin film PV modules sold [1]. While partially attributed to the recent shortage of crystalline Si wafers, the fast production of high quality Si thin films is critical for the advancement of cost-effective PV devices. Further, the conversion efficiency of amorphous Si (a-Si) solar cells will need to improve dramatically if they will ever compete with mono-crystalline wafer devices [2]. Large grain polycrystalline Si films (poly-Si) are viewed as a possibility to fill this role. The techniques used to produce poly-Si range from the classic direct deposition via thermal chemical vapor deposition [3], to the more recent "recrystallization" methods [4,5]. Furnace recrystallization of amorphous Si films (a-Si) has been studied for many years [6], and is understood as a classical nucleation and grain growth mechanism. The limitation of the process comes from the steady-state nucleation rate; one needs to limit or control the nucleation process in order to maximize the final annealed grain size. Other approaches have been successful in producing poly-Si with grains larger than several μm, such as excimer laser annealing [7], metal-induced crystallization [8], or even epitaxial growth over poly-Si seed layers [9]. These approaches find their respective limitations due to either fabrication scaling issues or additional processing steps to remove metal impurities. We introduce here a radically new yet simple approach for fast recrystallization of a-Si, wherein the nucleation of grains is controlled. This was accomplished by embedding freestanding Si nanocrystals in between separately deposited layers of a-Si film. Upon annealing in a typical tube furnace at 600°C, the embedded nanocrystals act as nucleation "seeds" for the surrounding amorphous film. The structure of the films is reported, along with initial measurements of the crystallization kinetics compared to films without nanocrystals embedded.

## EXPERIMENT

A-Si:H films with embedded Si nanocrystals were produced using a multi-layer deposition process. Initially, a 100 nm thick a-Si:H film was deposited at 250°C in a parallel-plate plasma-enhanced chemical vapor deposition (PECVD) reactor using diluted silane (5% in helium) at 100 milliTorr. These conditions were found to produce high quality a-Si:H, evidenced by the dominance of isolated Si-H bonds measured by FT-IR abosprtion; the ratio of $SiH_2/SiH$ was measured as 0.14, consistent with high quality films. Silicon nanocrystals with average diameter of around 30 nm were then produced in a separate non-thermal plasma chamber [10] that was coupled to the PECVD chamber via a 1 mm orifice. In this reactor, a mixture of the 5/95% silane/helium was further diluted in argon carrier gas at a pressure of 1.5 Torr, with RF power of 140 W applied to excite the plasma. This reactor can be modified to control the nanoparticle diameter as a function of the gas residence time. For this study, it was desirable to use large nanocrystals as they were more easily observed directly in the transmission electron microscope (TEM). Using a shutter mechanism to control the particle deposition time, the particles were injected through the 1 mm orifice for 10 seconds, impacting on the film surface. The deposition time can be used to control the number density of nanocrystals in the film, and was found to be around 4 particles/$\mu m^2$ for a 10 second deposition time. Finally a second 100 nm thick a-Si layer was deposited over the particles under the same conditions as the first. Figure 1 shows a schematic of this system. Samples were deposited for TEM measurements on molybdenum grids with a 10 nm carbon support film. For Raman spectroscopy, films were deposited under identical conditions on Corning 1737 glass substrates.

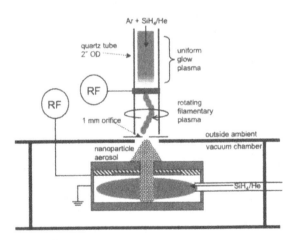

**Figure 1.** Schematic of apparatus used to deposit a-Si:H films with embedded nanocrystals

A Tecnai T12 TEM (120 kV acceleration, LaB$_6$ filament) equipped with a heating stage allowed us to record the morphology evolution of the embedded nanocrystals as the film was annealed. The bright-field TEM micrograph in Figure 2 shows the as-deposited structure of a 200 nm thick film with embedded particles. The dark contrast of the nanoparticles seen in the image suggests that the particles embedded within the film are single-crystalline, consistent with the synthesis conditions reported in Ref. [10]. The region surrounding the nanocrystals comes from the deposition of the amorphous film on top of the particles, resulting in hemispherical features on the film surface whose height and number density, measured by atomic force microscopy (AFM), corresponded to the observed diameter of the embedded nanocrystals. In Figure 2, it is seen that some of the grey contrast regions clearly contain a nanocrystal, while others do not. Detailed examination of the film by tilting the TEM stage revealed that each of these contrast regions indeed contains a single nanocrystal, making these surface features a useful indicator for the location and area density of the seed particles. Figures 3(a) and (b) show bright-field TEM micrographs of a seeded film prior to annealing and after 0.5 hours of annealing at 650°C on the heated TEM stage. The images show clearly that the growth of crystal grains in the film were initiated at the locations of the embedded nanocrystals, evidenced by the light and dark contrast regions in the bright-field image. Further, it is evident that every embedded crystal has formed a small grain after annealing, suggesting that the nature of the interface between the particles and surrounding amorphous film is quite uniform.

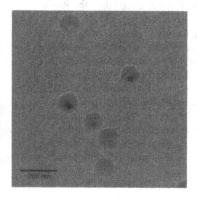

**Figure 2**. Bright-field TEM micrograph of 200 nm thick film with embedded Si nanocrystals prior to annealing.

**Figure 3.** Bright-field TEM micrographs of seeded film (a) as deposited and (b) after 0.5 hours of annealing at 650°C on the heated TEM stage.

To quantify the crystallization mechanism, a Witec Alpha 300 R Raman spectrometer was used with an Ar ion laser excitation source (514.5 nm). The film samples deposited on glass were annealed for 2 hr time intervals at 600°C in a quartz tube furnace, recording the Raman spectra at each interval. In Figure 4, a time series of Raman spectra is shown from the as-deposited film through the first 6 hours. Prior to annealing, the presence of the nanocrystals is not detectable in the Raman spectra; the amorphous signal at 480 cm$^{-1}$ dominated the spectra. This is to be expected considering the low seed density of around 4 nanocrystals/$\mu$m$^2$ observed in both TEM and AFM measurements. After one hour at 600°C, a measurable contribution from the crystalline component was evident in the spectra, observed near the bulk crystalline Si signal at 516 cm$^{-1}$. As the growth of the crystal grains continued, another component from grain boundaries and/or crystal defects was observed between 505-510 cm$^{-1}$. The growth of the crystalline grains continued until it dominated the spectrum and no further changes were observed, around 14 hours for this film structure. By resolving these spectral peaks and comparing their integrated intensities according to Equation 1, the Raman crystalline fraction was obtained.

$$X_C = (A_{516} + A_{505}) / (A_{516} + A_{505} + A_{480}) \tag{1}$$

Where $X_C$ is the Raman crystalline fraction, and $A_{516}$, $A_{505}$, and $A_{480}$ are the integrated intensities of the crystalline, grain boundary/crystal defect, and amorphous phases of the film respectively. It is noted that because the scattering cross-sections for these components have not been rigorously measured in this study, $X_C$ should be taken as a relative measure of the crystalline fraction rather than the true atomic percentage.

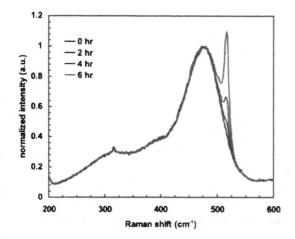

**Figure 4.** Normalized Raman spectra of seeded film annealed in a tube furnace at 600°C.

In Figure 5 the time dependence of $X_C$ for both seeded and non-seeded films is plotted on a semi-logarithmic scale. It is immediately noticed that the overall crystallization of the seeded film is faster than for purely amorphous films. A small but immediate increase in $X_C$ is seen in the seeded films, and is not observed in the a-Si:H sample. Non-seeded films required around 11 hours of incubation time prior to the onset of nucleation. A basic fit of the crystallization curves was performed according to Ref. [6], using Equation 2.

$$X(t) = 1 - \exp [ (t - \tau_0)^3 / \tau_c^3] \qquad (2)$$

Where t, $\tau_0$, and $\tau_c$ are the annealing, transient, and characteristic crystallization times respectively. The result of this fitting showed that for the seeded film the crystallization time $\tau_c$ was around 600 min, and around 900 min for the non-seeded film. Based on TEM observations, crystallization of the seeded films was dominated by the growth of the embedded nanocrystals; formation of additional nuclei was not directly observed. However in the a-Si:H sample, theory holds that both nucleation and grain growth are steady-state processes following the incubation period, and would both contribute to the total crystalline fraction during the annealing process. One could expect that after incubation a purely amorphous film should crystallize faster owing to the contribution of both mechanisms. That behavior was not observed, however, indicating a possible reduction in the energy required to grow the embedded crystals compared to those nucleated in pure a-Si:H. A more rigorous investigation of this hypothesis is currently underway.

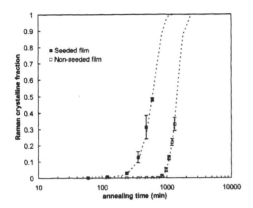

**Figure 5**. Plot of crystalline fraction vs. annealing time for film samples deposited on glass substrates, annealed at 600°C in a tube furnace.

ACKNOWLEDGMENTS

Funding for this work was provided by the NSF under the Integrative Graduate Education Research Traineeship (IGERT) grant DGE-0114372, grant DMI-0403887, grant DMR-0705675 and MRSEC award DMR-0212302. Parts of this work were carried out in the University of Minnesota I.T. Characterization Facility, which receives partial funding from the NSF through the NNIN program. Additional funding was provided by the University of Minnesota Center for Nanostructured Applications.

REFERENCES

1. www.eia.doe.gov,
2. M.A. Green, K. Emery, Y. Hisikawa, W. Warta, Prog. Photovolt: Res. Appl., **15** (2007) 425-430.
3. S. A. Campbell (2001). The Science and Engineering of Microelectronics Fabrication. New York, Oxford University Press.
4. Y.-L. Jiang, Y.-C. Chang, Thin Solid Films, **500** (2006) 316-321.
5. A. H. Mahan, B. Roy, Jr. R. C. Reedy, D. W. Readey, D. S. Ginley, J. Appl. Phys., **99** (2006) 023507.
6. R. B. Iverson, R. Reif, J. Appl. Phys., **62** (1987) 1675-1681.
7. C.-C. Kuo, W.-C. Yeh, J.-F. Lee, J.-W. Jeng, Thin Solid Films, **515** (2007) 8094-8100.
8. Y. Ishikawa, A. Nakamura, U. Uraoka, T. Fuyuki, Jpn. J. Appl. Phys., **43** (2004) 877-881.
9. S.R. Lee, K.M. Ahn, B.T. Ahn, J. Electrochem. Soc., **154** (2007) H778-H781.
10. A. Bapat, C. Anderson, C.R. Perrey, C.B. Carter, S.A. Campbell, U. Kortshagen, Plasma Phys. Control. Fusion, **46** (2004) B97-B109.

# Poster Session:
# Alloys, Structural Properties
# and Solar Cells

Mater. Res. Soc. Symp. Proc. Vol. 1066 © 2008 Materials Research Society 1066-A07-02

# Nanostructures with Group IV Nanocrystals Obtained by LPCVD and Thermal Annealing of SiGeO Layers

Bruno Morana[1], Andrés Rodríguez[1], Jesús Sangrador[1], Tomás Rodríguez[1], Óscar Martínez[2], Juan Jiménez[2], and Andreas Kling[3,4]

[1]Tecnología Electrónica, Universidad Politécnica de Madrid, E.T.S.I.T., Madrid, 28040, Spain
[2]Física de la Materia Condensada, U. de Valladolid, E.T.S.I.I., Valladolid, 47011, Spain
[3]Centro de Física Nuclear, Universidade de Lisboa, Lisbon, 1649-003, Portugal
[4]Instituto Tecnológico e Nuclear, Sacavém, 2686-953, Portugal

## ABSTRACT

Nanocrystals embedded in an oxide matrix have been fabricated by annealing SiGeO films deposited by LPCVD. The composition of the oxide layers and its evolution after annealing as well as the presence and nature of nanocrystals in the films have been studied by several experimental techniques. The results are analyzed and discussed in terms of the main deposition parameters and the annealing temperature.

## INTRODUCTION

Group IV semiconductor nanocrystals embedded in a dielectric matrix are of interest due to their potential application in Si based optoelectronics and non-volatile memories [1, 2]. The fabrication of these nanostructures has been undertaken, among other methods, by oxidation of polycrystalline SiGe layers [3] and crystallization of amorphous SiGe nanoparticles embedded in $SiO_2$ deposited by Low Pressure Chemical Vapour Deposition (LPCVD) [4]. In this work, the possibility of fabricating structures of this kind by LPCVD of SiGeO films and subsequent annealing to segregate the possible excess of Si and/or Ge in excess in the form of nanocrystals embedded in an oxide matrix has been studied. The nature and evolution of the oxide matrix and the presence of nanocrystals have been analyzed by several experimental techniques as a function of the deposition and annealing conditions.

## EXPERIMENTAL

The deposition of the SiGeO films was carried out on Si wafers using $Si_2H_6$, $GeH_4$ and $O_2$ as reactant gases. The pressure was set at 240 mTorr, the deposition temperature was 450 °C and several $GeH_4$:$Si_2H_6$:$O_2$ gas flow ratios were selected in the 0:10:2 to 20:10:2, which correspond to $GeH_4/Si_2H_6$ flow ratios (F) ranging from F=0 to F=2. The as-deposited samples were annealed at temperatures from 600 to 1000 °C for 1 hour in a $N_2$ atmosphere. Special care has been taken to avoid any oxidation of the films during these processes by using a high flow (12 l/min) of high purity $N_2$ (N55) and introducing and extracting the gas from the furnace through small diameter tubes.

The overall structure and composition of the films was studied by Rutherford backscattering spectrometry (RBS). The RBS spectra were taken in Cornell geometry at an incidence angle of 78 ° using a 2 MeV He$^+$ beam and interpreted by means of the RUMP code [5]. The number of atoms of Si, Ge and O per unit area present in the different films, labelled [Si], [Ge] and [O] respectively, was determined from the spectra fittings. Using these data, the R (Si) = [Si]/(2×[O]) and R (Ge) = [Ge]/(2×[O]) ratios were calculated. These ratios are useful to determine if Si and Ge oxides are able to be formed and if it is possible to find Si and/or Ge atoms in excess in the films. The composition of the as-deposited and annealed oxide matrices was analyzed by Fourier transform infrared spectroscopy (FTIR). The FTIR spectra were acquired using a Perkin Elmer Spectrum 100 spectrometer in the 200 to 7800 cm$^{-1}$ wavenumber range. The bare Si substrates were used as references to acquire the background in each case. The presence of amorphous and/or crystalline Si and Ge incorporated to the films as well as the existence of nanocrystals in the as-deposited and annealed samples was studied by Raman spectroscopy using visible (514.5 nm) and UV (325 nm line from a HeCd laser) excitations and a HR Labram Jobin-Yvon Raman spectrometer. Photoluminescence (PL) spectra, also excited using the 325 nm line of the same laser, were acquired at room temperature (RT).

## RESULTS AND DISCUSSION

### As-deposited films

Figure 1 shows the RBS spectra of some of the as-deposited samples. Growth rates from 20 to 40 nm / hour are obtained, which are appropriate for the controlled deposition of a few nanometers thick layers. The R (Si) and R (Ge) atomic fractions defined above are represented in figure 2 as a function of the GeH$_4$/Si$_2$H$_6$ gas flow ratio, ranging from F=0.2 to F=2. R (Ge) shows a linear dependence on F in the whole considered interval, indicating that an increasing amount of Ge atoms is being incorporated to the film as F increases. The value of R (Si), however, is almost constant and equal to 1 for F<1, pointing to the existence of a SiO$_2$ matrix with almost no Si in excess, and increases with F only for F>1.

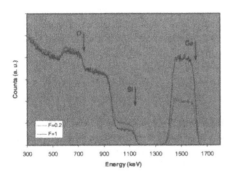

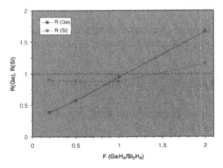

**Figure 1.** Selected RBS spectra of samples deposited using different values of F.

**Figure 2.** Atomic ratios (see text) determined from the RBS spectra as a function of F.

The FTIR and Raman spectra, acquired using visible excitation, of the samples deposited with different values of F are displayed in figures 3 and 4 respectively. In the sample with F=0, absorption bands were found in the FTIR spectra at around 463 (not included), 814 and 1070 cm$^{-1}$, all of them related to Si–O–Si bonds. Another band appears around 875 cm$^{-1}$, which is attributed in the literature to the bending mode of Si-H bonds and also to the SiH$_2$ scissors vibration. As F increases, the Si-O-Si related bands are also present in the spectra with similar intensities, while the intensity of the Si-H related band decreases significantly.

For F≥1 the Si-H band disappears and another band (clearly visible in the spectrum of the sample deposited with F=2) starts to appear at 995 cm$^{-1}$, as a shoulder of the 1070 cm$^{-1}$ one, which is related to Si-O-Ge bonds. The absorption bands around 1200 cm$^{-1}$ are related to the existence of distortions in the Si-O-Si bonds and are found to increase in intensity with increasing F in the whole interval of values of F. The existence of Ge-O-Ge bonds, which should introduce a band at around 885 cm$^{-1}$ due to their stretching mode, is discarded since in the samples with a greater amount of Ge atoms no absorption band at all is observed in the vicinity of this wavenumber.

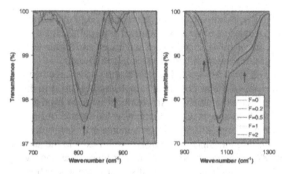 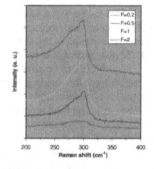

**Figure 3.** FTIR transmittance spectra of samples deposited with different values of F.

**Figure 4.** Ge-Ge band of the Raman spectra (see figure 3).

With regard to the Ge-Ge signal of the Raman spectra, the only one arising from the film that is visible, its presence in the spectra is a clear indication of the incorporation of Ge to the films. Moreover, the band is formed by a broad halo and a sharp peak located at 300 cm$^{-1}$, which account for the simultaneous presence of amorphous and crystalline Ge respectively, with a relative weight which depends on the value of F. The band corresponding to crystalline Ge presents phonon confinement, which suggests that crystalline Ge is arranged in clusters of a few nanometers in diameter [6]. The analysis of the Raman parameters concerning the intensities of the amorphous and crystalline bands and the full width at half maximun (FWHM) of the crystalline band shows that a) the total amount of Ge-Ge bonds increases as F does up to F=1, while for higher values of F it decreases, b) the relative amount of crystalline Ge respect to amorphous Ge also increases with increasing F up to F=1 and decreases for higher F, and c) The Ge cluster size increases with F, at least up to F=1, as the FWHM is found to decrease from 17 to around 8 cm$^{-1}$.

These results are interpreted as follows. For low values of F (F<1), the deposited material consists of almost stoichiometric Si oxide with an amount of Ge atoms incorporated to the film, in the form of amorphous and crystalline Ge, which depends on F. No evidences of the formation of Ge oxides are found. A small excess of Si probably exists, as the Si-H bands of the FTIR spectra indicate. Note that the possible excess of Si in the samples deposited with F>0, related to the intensity of the Si-H band, is much smaller than in the sample with F=0 although the flow of $Si_2H_6$ is the same, 10 sccm, in all of them. The presence of Ge therefore reduces or almost inhibits the incorporation of Si in excess into the Si oxide films. For high values of F (F>1) the deposited material consists of a matrix of mixed nature, formed by a mixture of Si and SiGe oxides of unknown proportions, also incorporating amorphous and crystalline Ge. Since part of the incoming Ge atoms are now incorporated to form the oxide matrix, the amount of Ge-Ge bonds in the samples decreases with respect to the previous cases. In all the interval of values of F, the incorporation of Ge and/or the formation of mixed oxides lead to the observed distortion of the structure of the Si oxide. The case F=1, which corresponds to the situation where at least one atom of each kind, Si and Ge, is present in the solid for each couple of oxygen atoms (see figure 2), represents the transition between the two different types of deposited materials.

### Annealed films

Figures 5 and 6 display selected FTIR and Raman spectra, in this case acquired using UV excitation to enhance the sensitivity of the technique to small amounts of amorphous and crystalline material, of the samples deposited with F=0.2 and F=0.5 and annealed at 600, 800 and 1000 °C (the spectra of the corresponding as-deposited samples are displayed in figure 3).

The FTIR spectrum of the sample with F=0.2 annealed at 600 °C shows the same bands and with almost the same intensities as the one of the as-deposited sample. Annealing at higher temperatures causes a slight but noticeable increase in the intensity of the absorption band at 875-885 cm$^{-1}$. The band at 995 cm$^{-1}$ also starts to appear. Considering now the sample of F=0.5, it is found that the intensity of the 875-885 cm$^{-1}$ and at 995 cm$^{-1}$ bands increase substantially after annealing at a temperature of 600 °C or higher. Since no hydrogen could exist in the sample after annealing at these temperatures, the 875-885 cm$^{-1}$ band should be attributed to an increasing amount of Ge-O-Ge bonds in the film. On the other hand, the band at 995 cm$^{-1}$ evidences the presence of Si-O-Ge bonds and therefore the formation of a mixed oxide.

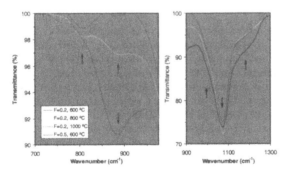
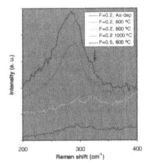

**Figure 5.** FTIR transmittance spectra of the samples deposited with F=0.2 and F=0.5 and annealed at different temperatures.

**Figure 6.** Ge-Ge signal of the Raman spectra (see figure 5).

The Raman spectra of the sample deposited with F=0.2 show that after the 600 °C annealing process the intensity of the Ge-Ge band has only slightly decreased, indicating that the annealed film contains a slightly smaller amount of Ge-Ge bonds than the as-deposited one. Annealing at 800 °C causes the Ge-Ge band to almost disappear (the small peak around 325 cm$^{-1}$, indicated by an arrow in figure 6, is an instrumental artifact). Increasing the temperature up to 1000 °C causes the Ge-Ge band to be fully absent in the Raman spectrum. With regard to the sample with F=0.5, annealing at 600 °C is enough to cause the Ge-Ge band to almost vanish (note from figure 4 that the Ge-Ge band in the spectrum of the as-deposited sample is much more intense than in the one of the sample with F=0.2).

The interpretation of these results is as follows. The sample deposited with F=0.2 holds its structure, consisting of a matrix of Si oxide incorporating amorphous and crystalline Ge, after annealing at 600 °C. The slight reduction in the amount of Ge atoms is also accounted by RBS (not shown) to correspond to Ge loss by outdiffusion, and no evidences of the formation of Ge oxides are found. However, the matrix is transformed into a mixture of Si, Ge and SiGe oxides for higher annealing temperatures. In the sample with F=0.5, this transformation is observed, to a much greater extent, even at 600 °C. A similar description holds for the samples with higher values of F. In the cases where Ge related oxides appear upon annealing, a clear correlation with the reduction of the intensity of the Ge-Ge band in the Raman spectra is found, since some Ge is now bonded to open oxygen bridge bonds.

Figure 7 shows the PL spectra of the samples deposited with different values of F from F=0 to F=2 and annealed at 600 °C. A spectrum of a bare Si wafer is also included as a reference of the background. The spectra present two bands, a blue-violet band around 400 nm (VL) and a broad band between 500 and 600 nm (GYL). The VL band was described in a previous paper [7], and its origin is attributed to oxygen deficient centers at the interface between the Ge nanocrystals and the oxide shells. The presence of Ge oxide in the films after annealing does not enhance the VL band, which confirms that it does not arise from the bulk of the Ge oxide. On the other hand, the GYL band is related to defects in the Si oxide. Figure 8 shows the evolution of the intensity of the VL band with the annealing temperature for samples deposited with different values of F.

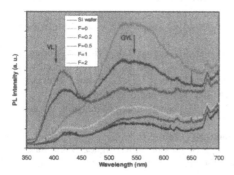

**Figure 7.** PL spectra of samples deposited with different values of F and annealed at 600 °C.

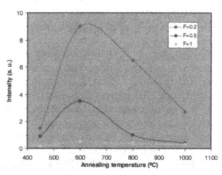

**Figure 8.** Intensity of the VL (400 nm) PL band as a function of the annealing temperature.

The VL band does not exist in the sample with F=0, which does not incorporate Ge, and its intensity is, at all the annealing temperatures, very small in the sample with F=1 and negligible in the sample with F=2. This band is observed in all the samples where the Raman spectra reveal the presence of Ge nanocrystals. Its intensity reaches a maximum value after annealing a 600 °C and then starts to decrease for higher annealing temperatures. The maximum intensity of the VL band is observed for the sample deposited with F=0.2 and annealed at 600 °C, in which Ge nanocrystals are observed and the matrix does not evidence the formation of mixed oxides.

## CONCLUSIONS

The fabrication of nanostructures with nanocrystals embedded in an oxide matrix by LPCVD of SiGeO films and annealing has been demonstrated. The influence flow ratio of the reactant gases and the annealing temperature on the composition of the oxide matrix and the nature of the nanocrystals have been studied. The nanocrystals are predominantly of Ge for all compositions of the gas, while the matrix is Si oxide only if the flow of the precursor of Ge kept low. The maximum intensity of the photoluminescence band (400 nm) related to the presence of nanocrystals has been obtained in the sample deposited with a low flow of the precursor of Ge and annealed at 600 °C.

## ACKNOWLEDGMENTS

This work was funded by the Spanish CICYT Project MAT2004-04580.

## REFERENCES

1. U. V. Desnica *et al*. Superlattices and Microstructures (2008), in press (DOI: 10.1016/j.spmi.2008.01.021).
2. E. W. H. Kan, W. K. Choi, W. K. Kim, E. A. Fitzgerald, D. A. Antoniadis. J. Appl. Phys. **95** (2004) 3148.
3. A. Rodríguez, M. I. Ortiz, J. Sangrador, T. Rodríguez, M. Avella, Á. C. Prieto, A. Torres, J. Jiménez, A. Kling, C. Ballesteros. Nanotechnology **18** (2007) 065702 and references threin.
4. A. Rodríguez, M. I. Ortiz, J. Sangrador, T. Rodríguez, M. Avella, Á. C. Prieto, J. Jiménez, A. Kling, C. Ballesteros. Phys. Stat. Sol. (a) **204** (2007) 1639 and references therein.
5. L. R. Doolittle, Nucl. Instr. and Meth. **9** (1985) 344, The latest version of this program can be found at http://www.genplot.com
6. H. Campbell, P. M. Fauchet. Solid State Commun. **58** (1986) 739.
7. M. Avella, Á.C. Prieto, J. Jiménez, A. Rodríguez, J. Sangrador, T. Rodríguez. Solid State Communications **136** (2005) 224 and references therein.

Mater. Res. Soc. Symp. Proc. Vol. 1066 © 2008 Materials Research Society 1066-A07-03

# On determination of properties of ultrathin and very thin silicon oxide layers by FTIR and X - ray reflectivity

Martin Kopani[1], Matej Jergel[2], Hikaru Kobayashi[3], Masao Takahashi[3], Robert Brunner[2], Milan Mikula[4], Kentarou Imamura[3], Stanislav Jurecka[5], and Emil Pincik[2]

[1]School of Medicine, Comenius University, Sasinkova 4, Bratislava, 811 08, Slovakia
[2]Institute of Physics, Slovak Academy of Science, Dubravska cesta 9, Bratislava, 845 11, Slovakia
[3]Japan Science and Technology Organization, Institute of Scientific and Industrial Research, Osaka University and CREST, 8-1, Mihogaoka, Ibaraki, Osaka, 567-0047, Japan
[4]Department of Graphic Art Technology and Applied Photochemistry, Slovak University of Technology, Faculty of Chemical and Food Technology, Radlinskeho 9, Bratislava, 812 37, Slovakia
[5]Department of Engineering Fundamentals, University of Zilina, Faculty of Electrical Engineering, kpt. J. Nalepku 1390, Liptovsky Mikulas, 031 01, Slovakia

## ABSTRACT

We analyze properties of ultra-thin $SiO_2$ + very thin $SiO_x$ double layer structure formed on high-doped n-type Si (100) wafers using FTIR, and X-ray reflectivity. The observed absorption band around 1230 cm$^{-1}$ is attributed to the longitudinal optical mode of $SiO_x$ precipitates incorporated in silicon matrix. In particular, the corresponding peak positions indicate that there are precipitates of $SiO_x$ with x >1.8. The absorption band around 1070 cm$^{-1}$ is attributed to the Si–O–Si stretching bond. This position is characteristic for stoichiometric $SiO_2$. From the results it can be concluded that differently shaped particles co-exist in the samples. This assumption is supported by the oxide density measurements performed by FTIR and X-ray reflectivity. We determined density of oxide layers, roughness of corresponding interfaces, and surface roughness by the X-ray reflectivity. Additionally, we present the results of multifractal analysis on a complete set of six samples.

## INTRODUCTION

Silicon very thin dioxide layers (thickness below 10 nm) belong among the most studied materials because their dielectric and interface properties are suitable for LSI technology and for production of advanced types of LCDs. The IR absorption spectra contain more absorption bands, e.g. at 800 and 1080 cm$^{-1}$ [1]. Infrared spectroscopy can give structural information, information on different types of bonding, quantitative information and information about kinetics of sample.

For the determination of parameters such as surface and interface roughness and density of investigated layers, X-ray reflectivity measurements can be used. Statistical characteristics of the surface roughness and fractal properties of the sample surfaces are reported and discussed in connection with the determination of the optical parameters of the studied structure.

We discuss the results obtained by FTIR, X-ray reflectivity on double oxide layer structure prepared on Si (100) crystalline substrate. Total oxide thickness does not exceed 12 nm.

The first high-dense $SiO_2$ layer of approx. 1.6 nm thickness is localized at the interface with c-Si. All samples were annealed at 250 °C for 60 minutes.

In addition, results of multifractal analysis of AFM records of surfaces of the full set of samples are shown.

## EXPERIMENT

Double layer ultra-thin $SiO_2$ + very thin $SiO_x$ structure was prepared on phosporus doped n-type Si (100) wafers with a ~10 Ω cm resistivity. The Si wafers were cleaned by standard RCA method before formation of oxide layers followed by etching in dilute hydrofluoric acid and post-passivation (post-oxidation) annealing (PPA). Such samples are marked in this paper as reference surfaces. The first ultra-thin $SiO_2$ layer was prepared by nitric acid oxidation of silicon (NAOS) according to [2]. In addition, Si dangling bonds in the formed oxide/Si interface and also some defects of oxide layer were passivated in HCN aqueous solution in some samples. After formation of oxide layers and passivation in the HCN solution, all samples were annealed in forming gas. The second layer of $SiO_x$ of approx. 10 nm thickness was deposited on the top of the first layer by e-beam evaporation of $SiO_2$ target.

Double layer $SiO_2$ + $SiO_x$ structure was analyzed using the following three methods:
a) FT-IR measurement in reflectance mode with Nicolet Magna 750 in the range 400 – 4000 cm$^{-1}$ for obtaining information on different types of bonding in the structure with the main aim to determine the density of the oxide layers, b) X-ray reflectivity to determine the density of oxide layers, roughness of corresponding interfaces, and surface roughness. The result fitting was based on Fresnel optical approach using recursive formalism worked out by [3]. Interface roughness was included by an attenuation factor [4]. Program package Leptos (Bruker AXS) and genetic algorithm were used, c) AFM (Keyence production) results of which were used in fractal analysis.

## DISCUSSION

FT-IR spectra of double layer of $SiO_2$ + $SiO_x$ structure exhibit peaks at approx. 1064 cm$^{-1}$ (transverse optical phonons, TO) and 1232 cm$^{-1}$ (longitudinal optical phonons, LO). In Fig. 1 representative FT-IR spectrum in Si-O-Si asymmetric stretching vibrational region for the $SiO_2$ + $SiO_x$ structure is shown.

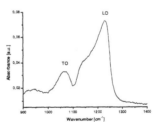

Fig.1: FT-IR spectrum of Si-O-Si asymmetric stretching vibrational region of double layer of $SiO_2$ + $SiO_x$ structure exhibits transverse optical phonons - TO at 1064 cm$^{-1}$ and longitudinal optical phonons - LO at 1232 cm$^{-1}$.

Table 1 shows calculated atomic density of $SiO_2$ + $SiO_x$ double layer obtained from IR measurements according to Imai at al [2]. The FT-IR spectra of samples had differences concerning absolute values of intensities of absorption lines because NAOS, $SiO_x$ and native oxide have different properties but ratio of intensities of absorption maxima for different samples were approximately the same therefore it is not possible directly formulate an idea on dominance disc-shaped or spherical precipitates. For such purposes there is needed additional detail analysis of corresponding part of spectra and also additional measurements. Another dominant absorption peaks were not observed in the given set of samples.

**Tab. 1.** The density calculated of $SiO_2$ + $SiO_x$ double layer from FT-IR measurements in region 900 – 1300 cm$^{-1}$.

| No. | Sample structure | Calculated density $\rho$ (g/cm$^3$) |
|---|---|---|
| 1 | NAOS + HCN + SiO$_x$ | 2,21 |
| 2 | NAOS + SiO$_x$+ HCN | 2,26 |
| 3 | NAOS + SiO$_x$ | 2,23 |
| 4 | SiO$_x$ | 2,28 |
| 5 | NAOS | 2.29 |

FT-IR of reference sample has been used at evaluation of densities of samples No.1-5. Density, thickness and roughness of the reference sample were possible to determine more precisely by XRR – see Table 2. Surface of the reference sample was treated before measurements by standard RCA method therefore its thickness is only 1.33 nm.

Table 2 shows calculated density and roughness of $SiO_2$/$SiO_x$ double layer obtained from X-ray reflectivity. The differences in density found using FT-IR are not significant with respect to uncertainty in the values because using $SiO_x$ e-beam evaporation slightly inhomogeneous oxide layers are formed. In addition, evaporation of $SiO_x$ on NAOS oxide layer forms between both types of oxides porous interlayer which properties depends on roughness of the NAOS surface and, consequently, density of sample No.3 can have lower density in comparison with dominant two components (No. 4 and No. 5). Deposition of $SiO_x$ was not performed in situ, i.e. immediately after formation of NAOS. The sample was moved to other vacuum equipment.

**Table 2.** Parameters of $SiO_2$ + $SiO_x$ double layer obtained from X-ray reflectivity.

| No. | Sample structure | NAOS thickness (nm) | NAOS density (g/cm$^3$) | SiO$_x$ thickness (nm) | SiO$_x$ density (g/cm$^3$) | Si/NAOS roughness (nm) | NAOS/SiO$_x$ (or Si/SiO$_x$) roughness (nm) | SiO$_x$ (or NAOS) roughness (nm) |
|---|---|---|---|---|---|---|---|---|
| 1 | NAOS+ HCN + SiO$_x$ | 1.20 | 1.83 | 10.77 | 2.51 | 0.02 | 0.49 | 0.69 |
| 2 | NAOS+SiO$_x$+ HCN | 1.57 | 2.14 | 11.33 | 2.11 | 0.03 | 0.09 | 1.46 |
| 3 | NAOS + SiO$_x$ | 1.22 | 1.95 | 11.85 | 2.37 | 0.07 | 0.38 | 0.65 |
| 4 | SiO$_x$ | - | - | 18.78 | 1.73 | - | 0.45 | 0.42 |
| 5 | NAOS | 1.62 | 2.71 | - | - | 0.02 | - | 0.66 |
| 6 | reference surf. | - | - | 1.33 | 1.95 | | 0.11 | 0.41 |

Table 2 contains more numerical information, because they document results obtained of fitting procedure using program package Leptos (Bruker AXS) and genetic algorithm. Program package uses electron structure of atoms in fitting procedure. Therefore we suppose that all set of data is needed to publish even if they are not used in discussion. Differences in thicknesses of $SiO_x$ document inhomogeneous evaporation of $SiO_x$ layer by e-beam source (compare e.g. samples No. 4 and No. 2). Generally roughness of Si/NAOS interface is very low (oxide grows homogeneously) in comparison with $NAOS/SiO_x$ due to the same reason – quality of evaporated $SiO_x$ layer is questionable. During evaporation group of $SiO_x$ molecules can be deposited on the sample surface. Densities published in Tab. 2 were obtained by fitting procedure of data recorded in grazing incidence mode of measurement of XRR. The tested area of the sample was large in comparison with FTIR based method. In addition, also density of double layer system was determined by FTIR. These reasons lead to difference in the obtained results. Of course, main reason is in used measured probing beam – IR light and X-ray irradiation – they have different depth of penetration to the testing sample.

The first absorption band around 1230 cm$^{-1}$ is attributed to longitudinal optical phonons of $SiO_x$ precipitates incorporated in silicon matrix where x is close to 2 [5]. Position of absorption band around 1230 cm$^{-1}$ gives information regarding stoichiometry of such precipitates and their density [6]. From the peak position in our samples, the precipitates are formed from $SiO_x$ with x > 1.8. The second absorption band around 1070 cm$^{-1}$ is attributed to the Si–O–Si stretching bond [7]. After annealing, this absorption band shifts to a higher wavenumber (1080 cm$^{-1}$). This position is characteristic for stoichiometric $SiO_2$ [8].

Stoudek and Humlicek [9] observed the disc-shaped particles at about 1250 cm$^{-1}$, spherical precipitates at about 1100 cm$^{-1}$ and interstitial oxygen band at 1107 cm$^{-1}$. Interstitial oxygen band at 1107 cm$^{-1}$ was not observed in this study. From the results it can be drawn that both disc-shaped and spherical-shaped particles coexist in this sample. On the basis of shift of the 1070 cm$^{-1}$ absorption one may suggest that the break of the Si–O–Si bonds and separation into Si and $SiO_2$ of the resulting phase occur [10]. When the material is annealed, reaction $2SiO \longrightarrow SiO_2 + Si$ can appear which leads to the formation of silicon nanoclusters embedded in a silicon oxide matrix [11].

We suppose that difference between the values of densities obtained for the same sample by X-ray and FTIR methods is determined by surface inhomogeneities formed during the chemical cleaning of surfaces and/or during the evaporation (oxidation) process and by the size of investigated areas of FTIR and X-ray reflectivity. The X-ray method uses grazing incidence angles. Unfortunately, we did not measure different size of sample to confirm this argument. The X-ray reflectivity results of surface roughness of several samples were compared with AFM results and consequently fractal analysis was made.

The basic concept for the description of rough surface with symmetric scaling is fractal dimension $D_q$ [12-13]. Fractal is a geometric shape that can be divided into subparts, each of which is a copy of the whole. This property of fractals is called self-similarity. This means that self-similarity of the chosen element at a variable magnification scale can be observed. A magnified view of one part of the rough surface does not precisely reproduce its whole structure but has the same qualitative appearance. A complex character of the semiconductor rough surface is often described by more than one fractal dimension. Because the rough surface does not exhibit a form of purely self-similar fractal, the self-similarity is local only. The surface irregularity distribution changes in dependence on the studied region. Concentration of large surface irregularities often occurs in a few regions and concentration of small irregularities in many regions. Therefore the most suitable method for the surface properties description is a multifractal analysis. In this work, we use the method of computing the $f(\alpha)$ multifractal spectrum developed by Chabra and Jensen [14]. The $f(\alpha)$ spectrum is the dimension of the theoretical support of a particular measure. In this method the probability of finding an $i$th fragment of the analyzed surface region is expressed by the formula $P_i(\delta)=A_i(\delta)/A_r(\delta)$, where $A_i(\delta)$ is the area in an $i$th box with $\delta$ scale and $A_r(\delta)$ is the total area measured in $\delta$ scale. Fractal dimension is computed by formula

$$D_{Fq} = \frac{1}{q-1}\lim_{q\to 0}\frac{\log\sum_{i=1}^{N}[P_i(\delta)]^q}{\log\delta} \tag{1}$$

where $q\in(-\infty,\infty)$, $N$ is total number of boxes necessary for covering $A_r$. In terms of probability $P_i$, a one-parameter family of normalized measure $\mu$ is constructed

$$\mu(q,\delta) = \frac{P_i(\delta)^q}{\sum_{i=1}^{N}P_i(\delta)^q}. \tag{2}$$

The fractal dimension of the subset $f(q)$ is determined by equation

$$f(q) = \lim_{\delta\to 0}\frac{\sum_{i=1}^{N}\mu_i(q,\delta)\log\mu_i(q,\delta)}{\log\delta} \tag{3}$$

indexed with the exponent $\alpha(q)$

$$\alpha(q) = \lim_{\delta\to 0}\frac{\sum_{i=1}^{N}\mu_i(q,\delta)\log P_i(q,\delta)}{\log\delta} \tag{4}$$

The multifractal analysis describes the statistical properties of the measure in terms of its distribution of the singularity spectrum $f(\alpha)$ corresponding to its singularity strength $\alpha$. If the shape of $f(\alpha)$ curve is humped, the scaling of the surface is considered multifractal. If the spectrum $f(\alpha)$ converges, the surface is considered mono- or non-fractal. Curve $f(\alpha)$ is convex with a single inflexion point at the maximum with $q = 0$.

One of the calculated results of the multifractal analysis is illustrated in Fig. 2.

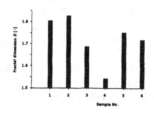

Fig. 2: Calculated fractal dimensions of six samples of Table 2.

Fractal dimension of $SiO_x$ layers evaporated on clean (100) c-Si substrate is the lowest from the set of samples. Compare, please, last column of the Tab. 2 of samples No. 3 and No. 4.

We suppose that ultrathin 1.6 nm thick chemical $SiO_2$ layer induces formation of fractal structures of larger dimension.

## CONCLUSIONS

We analyzed properties of $SiO_2$ + $SiO_x$ double layer structure formed on high-doped n-type Si (100) wafers. The FT-IR absorption spectra contain absorption bands around 1230 cm$^{-1}$ and 1070 cm$^{-1}$. We calculated the densities of $SiO_2$ + $SiO_x$ double layers from the $SiO_x$ stretching vibration absorption band. The values obtained from the X-ray reflectivity were slightly higher in more cases. We assume it is due mainly to different size of investigated areas. We used a multifractal analysis for the surface properties description. From the results we suppose that ultrathin 1.6 nm thick chemical $SiO_2$ layer induces formation of fractal structures of larger dimension

## ACKNOWLEDGMENTS

The support of this work of Japan Society for Promotion of Science (JSPS) and Slovak Scientific Grant Agency (VEGA) grants no. 2/7120/27 and 2/0047/08 is acknowledged.

## REFERENCES

1. S.E. Babayan, J.Y. Jeong, V.J. Tu, J. Park, G.S. Selwyn, and R.F. Hicks, Plasma Source Sci. Technol. **7**, 286-288 (1998).
2. S. Imai, M. Takahashi, K. Matsuba, Asuha, Y. Ishikawa, and H. Kobayashi, Acta Phys. Slovaca **55**, 305-313 (2005).
3. J.H. Underwood, and T.W. Barbee, AIP Conf. Proc. **75**, 170 (1981).
4. L. Nevot, and P. Croce, Rev. Phys. Apl. **15**, 761 (1980)
5. A. Borghesi, A. Piaggi, A. Sassella, A. Stella, and B. Pivac, Phys. Rev. B **46**, 4123-4127 (1992).
6. B. Pivac, A. Borghesi, M. Geddo, A. Sassella, and A. Stella, Appl. Sur. Sci. **63**, 245-248 (1993).
7. R.S. Yu, K. Ito, K. Hirata, K. Sato, W. Zheng, and Y. Kobayashi, Chem. Phys. Lett. **379**, 359-363 (2003).
8. J.T. Fitch, E. Kobeda, G. Lucovsky, and E.A. Irene, J. Vac. Sci. Technol. B **7**, 153-162 (1989)
9. R. Stoudek, and J. Humlicek, Phys. B. **376-377**, 150-153 (2006).
10. A. Banerjee, and G. Lucovsky, MRS Symp. Proc. **420,** 405 (1995).
11. F. Rochet, G. Dufour, H. Roulet, B. Pelloie, J. Perrière, E. Fogarassy, A. Slaoui and M. Froment, Phys. Rev. B **37**, 6468 - 6477 (1988).
12. B.B. Mandelbrot, The Fractal of Nature, Freeman, New York 1982.
13. J.F. Gouyet, Physics and Fractal Structures, Springer-Verlag, New York 1996.
14. A. Chabra, R.V. Jensen, Phys. Rev. Lett. 62 1327-1330 (1989).

Mater. Res. Soc. Symp. Proc. Vol. 1066 © 2008 Materials Research Society     1066-A07-04

## Structural and Opto-Electronic Properties of a-Si:H/a-SiN$_x$:H Superlattices

Stefan L. Luxembourg[1], Frans D. Tichelaar[2], Peter Kúš[3], and Miro Zeman[1]
[1]DIMES, Delft University of Technology, Feldmannweg 17, Delft, 2628 CT, Netherlands
[2]Kavli Institute of NanoScience, Faculty of Applied Sciences, Delft University of Technology, Lorentzweg 1, Delft, 2628 CJ, Netherlands
[3]Department of Experimental Physics, Comenius University, Bratislava, SK-842 15, Slovakia

### ABSTRACT

A series of multilayer structures consisting of alternating layers of hydrogenated amorphous silicon (a-Si:H) and amorphous silicon nitride (a-SiN$_x$:H) was fabricated using plasma enhanced chemical vapor deposition. The overall thickness and a-Si:H-to-a-SiN$_x$:H ratio was kept constant for the different multilayer samples. A blue shift of the optical bandgap was observed with decreasing a-Si:H layer thickness. High-Resolution Transmission Electron Microscopy was used to estimate the abruptness of the layer-to-layer transitions. The thickness of the interface mixing layer for transitions from a-Si:H to a-SiN$_x$:H was estimated to be 0.5 – 1 nm, while for the reverse transition a thickness of 2-2.5 nm was found. These findings are supported by results from Fourier Transform Infrared Spectroscopy.

### INTRODUCTION

Thin-film hydrogenated amorphous silicon (a-Si:H) based solar cells are promising candidates for low-cost photovoltaic energy generation. Currently, the achieved efficiencies are for an important part limited by loss processes associated with the non-optimal utilization of the solar energy spectrum; electrons excited above the absorber materials' conduction band edge loose part of their energy through thermalization, photons with energy smaller than the optical bandgap are not absorbed. A more efficient use of the solar energy spectrum is achieved in tandem or multi-junction solar cells, which combine several single-junction solar cells with absorber materials having different optical bandgaps. The number of semiconductor materials, which are both suited for harvesting the sun's energy in different parts of the spectrum and compatible with the low-temperature thin film deposition process, are limited. Recently, the concept of bandgap engineering using amorphous silicon based superlattices and quantum dots has received considerable attention as a generic approach to further extend the range of materials available to realize multi-junction solar cells with improved spectrum utilization [1].

Amorphous silicon based superlattices consist of alternating layers of a-Si:H and one of its alloys, which can be deposited using plasma enhanced chemical vapor deposition (PECVD). The high band gap material is referred to as the barrier layer, the low bandgap material as the well layer. Amorphous silicon based superlattices exhibit a blue shift of the optical absorption with decreasing well layer thickness, which offers a means of optical bandgap engineering [2, 3]. Controversy exists if in amorphous superlattices the observed shift of absorption can be attributed to effects of quantum confinement or is merely caused by the occurrence of interface mixing regions and an artifact of using the Tauc law to derive the optical bandgap at low layer thicknesses [4, 5]. A necessary condition to achieve quantum confinement is the formation of abrupt interfaces between subsequent layers in the superlattice structure [6].

We have studied the shift of the optical bandgap of superlattice / multilayer structures composed of a-Si:H and amorphous silicon nitride (a-SiN$_x$:H) with decreasing a-Si:H layer thickness. The presence and thickness of interface mixing layers was investigated using High-Resolution Transmission Electron Microscopy (HR-TEM) and Fourier Transform Infrared Spectroscopy (FTIR) spectroscopy.

## EXPERIMENT

A-SiN$_x$:H films were prepared from a gas mixture of SiH$_4$ and NH$_3$ in a capacitively coupled rf PECVD deposition system equipped with a showerhead electrode. The nitrogen content in the alloys was controlled by varying the fraction of NH$_3$ in the total gas mixture, % NH$_3$, from 5 to 96 % (%NH$_3$ = ([NH$_3$]/ ([NH$_3$] + [SiH$_4$]))×100). The films were deposited at an rf power density of 34 mW/cm$^2$, an inter-electrode distance of 14 mm, a deposition pressure of 80 Pa and a substrate temperature of 235 °C. Multilayer (ML) / superlattice structures were fabricated by alternating deposition of a-Si:H well layers and a-SiN$_x$:H barrier layers using the deposition conditions described above. A-SiN$_x$:H layers were deposited using 80% NH$_3$, resulting in barrier layer materials with an optical bandgap of approximately 3 eV. The ML structures were grown in a continuous manner: the plasma was not switched off in between the deposition of subsequent layers. Individual layer thicknesses were varied between 1 and 28 nm, such that the well-to-barrier layer thickness ratio was kept constant at 1:1.67. In this way the average composition of the different ML samples remains unchanged; only the number of interfaces present in the structure is varied. All samples were deposited on Corning Eagle 2000$^{TM}$ glass and silicon wafers in the same run.

Reflection and transmission (RT) spectra were recorded in the 250–1200 nm spectral range using a Perkin–Elmer 950 spectrophotometer. Subsequently, the Tauc and E$_{04}$ optical bandgaps were determined.

High Resolution Transmission Electron Microscopy (HR-TEM) images were recorded using a Tecnai F20/STEM FEI 200 kV transmission electron microscope without objective aperture.

The periodicity of the ML structures was determined from x-ray reflectivity (XRR) measurements. The deposition rates for the a-Si:H and a-SiN$_x$:H materials as part of ML structures were derived by varying the a-SiN:H layer thickness for constant a-Si:H thickness.

Infrared (IR) absorption spectra (400 – 4000 cm$^{-1}$) for both bulk materials and ML structures were recorded on a Thermo Nicolet 5700 Fourier Transform Infrared (FTIR) Spectrometer. Baseline subtraction is performed taking into account both coherent and incoherent interference.

## DISCUSSION

### HR-TEM analysis

In Figure 1.A a HR-TEM recording of an amorphous ML structure consisting of 47 layers is shown. The sample was grown at 180 °C, but otherwise equal conditions as described in the experimental section. The high reproducibility of the PECVD deposition process results in a regular stack of layers without significant distortions. In previous studies significant undulation, increasing with the number of layers in the stack, has been reported [7]. In our present study no

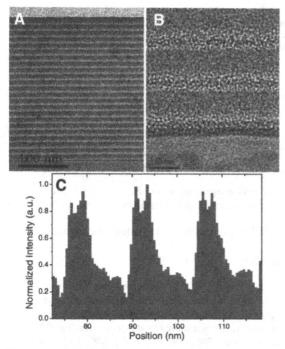

**Figure 1.** (A) Contains a TEM image displaying the full range of 47 layers present in the structure (a-SiN$_x$:H: high intensity, a-Si:H – low intensity). In (B) a high magnification image of the first few layers deposited is shown, (C) Displays an example of an intensity line profile used to determine the stacks' periodicity (for details see text).

significant undulations were observed. The contrast between a-Si:H and a-SiN$_x$:H results mainly from differences in the densities of the two materials. The higher hydrogen content of the a-SiN$_x$:H layers, as confirmed from Elastic Recoil Detection measurements, make them more susceptible to electron beam damage, which probably causes the coarser structure of these layers, as is visible in Figue 1.B. The periodicity of the structure can be determined from (electron) intensity line profiles, which were calculated perpendicular to the interfaces (Figure 1.C.). Each intensity data point is averaged over at least 600 pixels in horizontal direction. For the structure in Figure 1 the periodicity was found to be 13.9 ± 0.3 nm, with an a-SiN$_x$:H thickness of 5.8 ± 0.5 nm. Clearly, intensities at the transitions from layer-to-layer do not change abruptly but display a gradual change from one material to the other, which indicates the occurrence of an interface mixing layer. Moreover, the intensity gradient is dependent on the nature of the transition. At the end of the a-Si:H layers repeatedly an area of low intensity is observed. This might indicate an increase in density of the material. The dip around the middle of the a-SiN$_x$:H intensity peaks most likely results from electron beam damage. From the intensity gradients the thickness of the interface mixing layer were estimated. For a transition from a-Si:H to a-SiN$_x$:H it was estimated to be 0.5 – 1.0 nm, for the transition from a-SiN$_x$:H to a-Si:H 2-2.5 nm. The difference in the transition layer thicknesses might arise from the difference in deposition rates of

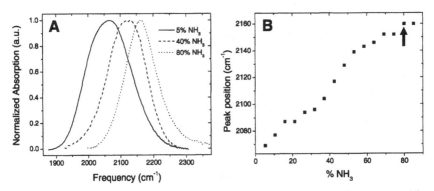

**Figure 2.** (A) Si-H stretching vibration absorption region for a-SiN$_x$:H films prepared for different % NH$_3$. (B) Si-H stretching vibration absorption peak position plotted versus % NH$_3$.

the two materials (0.8 Å/s for a-Si:H versus 2.3 Å/s for a-SiN$_x$:H) as determined from XRR.

## FTIR spectroscopy of bulk and ML samples

The Si-H stretching vibration region of the a-SiN$_x$:H IR absorption spectrum, roughly between 2000 and 2300 cm$^{-1}$, is sensitive to both the hydrogen concentration and nitrogen incorporation. With increasing nitrogen content the center position of the absorption in this region of the spectrum shifts to higher frequency. Six vibrational modes have been found to contribute to the absorption [8]: 2005 cm$^{-1}$ (Si$_3$Si-H), 2065 cm$^{-1}$ (Si$_2$Si-H$_2$), 2082 cm$^{-1}$ (NSi$_2$Si-H), 2140 cm$^{-1}$ (NSiSi-H$_2$, N$_2$SiSi-H), 2175 cm$^{-1}$ (N$_2$Si-H$_2$), 2220 cm$^{-1}$ (N$_3$Si-H). Roxlo *et al.* [9] showed that the Si-H stretching modes can be used to distinguish between the a-Si:H, a-SiN$_x$:H and interface mixing layers in ML structures. For this purpose they deconvoluted the ML absorption in the Si-H stretching region using modes centered at 2000 and 2175 cm$^{-1}$, characteristic of the a-Si:H and a-SiN:H absorption, and at 2080 and 2155 cm$^{-1}$ to describe bonds occurring at the interfaces.

In Figure 2 the shift of the center of the Si-H stretching absorption is shown for bulk a-SiN$_x$:H layers. As detailed above the peak shift is indicative of the relative increase of the contribution of Si-H stretching modes involving Si atoms bound to (multiple) N atoms. The obtained results are in close agreement with the work of Demichelis *et al.* [10] in which similar deposition conditions were used. The arrows in the figure indicate the peak position of the a-SiN$_x$:H material used in the ML structures.

In Figure 3.A the 2000 – 2300 cm$^{-1}$ region of the absorption spectrum of a ML structure with a period of 41 nm is compared to the spectra of its bulk material constituents. The a-Si:H and a-SiN$_x$:H contributions to the ML double peak structure are readily distinguished. The a-Si:H bulk absorption accurately matches the first peak position and rising edge of the ML double peak structure. In comparison with the 80% NH$_3$ bulk material absorption the a-SiN$_x$:H related peak in the ML is shifted to lower frequency. In addition an extended high wavenumber tail exists in the bulk a-SiN$_x$:H absorption. Clearly, the composition of the a-SiN$_x$:H material incorporated in the multilayer stack differs from the bulk material obtained at similar deposition conditions. A-SiN$_x$:H material deposited at 70% NH$_3$ displayed an accurate match of the second peak position and falling edge of the ML absorption.

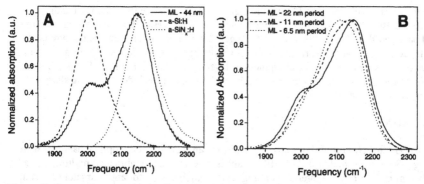

**Figure 3.** Comparison of (A) the Si-H stretching vibration absorption of a 44 nm period ML and its bulk material constituents, (B) ML stacks with different periodicity.

In Figure 3.B the Si-H stretching region is compared for ML structures with different periods. It is apparent that at a period of 6.5 nm the double peak structure of the ML absorption is lost. For small individual layer thicknesses the interface mixing layer dominates the ML spectrum. From these results an interface layer thickness on the order of 2-3 nm is expected, which supports the findings from HR-TEM.

## ML optical bandgap

In Figure 4 the change in optical bandgap is plotted as a function of the a-Si:H layer thickness. The blue shift with decreasing layer thickness is apparent in both the $E_{04}$ and Tauc bandgap. The observed shift, solely, is not considered as evidence for the occurrence of quantum confinement of charge carriers. It has been debated that the observed shift can be attributed to an artifact arising from the use of Tauc's law to determine the optical bandgap, which results in an increase of the derived optical gap for decreasing film thicknesses [4]. In the present work, the ML structures studied have constant average composition and thickness, thereby thickness

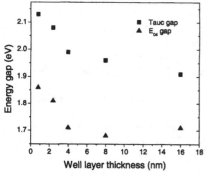

**Figure 4.** The $E_{04}$ and Tauc optical band gap plotted versus the well layer thickness; below 4 nm thickness the bandgap is blueshifted.

related artifacts are excluded. In addition, the bandgap shift is visible from the $E_{04}$ bandgap as well. Nevertheless, the interface mixing regions observed in the HR-TEM analysis can add to the observed bandgap shift. From this respect, hydrogen dilution of the reaction gas mixture is of interest. Dilution of the gas flow results in lower residence times and deposition rates, which should have a positive effect on the abruptness of layer-to-layer transitions. Preliminary results from FTIR analysis of ML structures prepared by hydrogen dilution indeed indicate sharper transitions (results not shown).

## CONCLUSIONS

A-Si:H / a-SiN$_x$:H multilayer structures display an increase of the optical bandgap with decreasing a-Si:H layer thickness. HR-TEM analysis reveals that the fabricated multilayer stacks are highly regular, but contain interface mixing layers which extend for approximately 1-2 nm. Absorption by these interface layers cannot be excluded as contribution to the observed blue shift of the optical bandgap. Preliminary results indicate that the interface thickness can be reduced using hydrogen dilution of the reaction gas mixture. A systematic study of the optical bandgap shift as function of the interface mixing layer thickness, controlled by varying deposition conditions, should result in more clarity on the occurrence of quantum confinement in amorphous multilayer / superlattice structures.

## ACKNOWLEDGMENTS

This work was carried out in the framework of the SELECT project financed by SenterNovem. The authors would like to acknowledge Tomas Roch from the Comenius University in Bratislava for the XRR analysis of the ML structures.

## REFERENCES

1. G. Conibeer, M. Green, R. Corkish, Y. Cho, E.-C. Cho, C.-W. Jiang, T. Fangsuwannarak, E. Pink, Y. Huang, T. Puzzer, T. Trupke, B. Richards, A. Shalav and K.-l. Lin, Thin Solid Films, **511-512**, 654 (2006)
2. B. Abeles and T. Tiedje, Phys. Rev. Lett., **51**, 2003 (1983)
3. N. Ibaraki and H. Fritzsche, Phys. Rev. B, **30**, 5791 (1984)
4. M. Beaudoin, M. Meunier and C. J. Arsenault, Phys. Rev. B, **47**, 2196 (1993)
5. N. Bernhard and G. H. Bauer, Phys. Rev. B, **52**, 8829 (1995)
6. J. P. Conde, V. Chu, D. S. Shen and S. Wagner, J. Appl. Phys, **75**, 1638 (1994)
7. H. Itoh, S. Matsubara, S.-i. Muramatsu, N. Nakamaru, T. Shimida and T. Shimotsu, Jap. J. Appl. Phys., **27**, L24 (1988)
8. E. Bustarret, M. Bensouda, M. C. Habrard, J. C. Bruyère, S. Poulin and S. C. Gujrathi, Phys. Rev. B, **38**, 8171 (1988)
9. C. B. Roxlo, B. Abeles and P. D. Persans, J. Vac. Sci. Technol. B, 4, 1430 (1986)
10. F. Demichelis, F. Giorgis and C. F. Pirri, Phil. Mag. B, **74**, 155 (1996)

Mater. Res. Soc. Symp. Proc. Vol. 1066 © 2008 Materials Research Society       1066-A07-06

# Transient Photoconductivity Study of the Distribution of Gap States in 100°C VHF-deposited Hydrogenated Silicon Layers

Monica Brinza[1], Guy J. Adriaenssens[2], Jatindra K. Rath[1], and Ruud E.I. Schropp[1]

[1]Debye Institute of Nanomaterials Science, Utrecht University, Department of Physics and Astronomy, SID - Physics of Devices, P.O.Box 80000, Utrecht, 3508 TA, Netherlands
[2]Halfgeleiderfysica, University of Leuven, Celestijnenlaan 200D, Leuven, B-3001, Belgium

## ABSTRACT

The energy distribution of gap states has been examined by means of transient photocurrent measurements in a series of 100°C VHF-deposited Si:H samples that spans the amorphous to microcrystalline transition. The 'amorphous' distribution, consisting of a continuous background and a prominent dangling-bond-induced peak, remains largely intact across the transition. The transport path located at the conduction band edge in a-Si:H, some 0.63 eV above the dangling bond D⁻ energy, moves down to ~0.55 eV above the corresponding D⁻ level in the microcrystalline samples.

## INTRODUCTION

By varying the silane to hydrogen flow ratio in a very-high-frequency (VHF) plasma reactor, a series of Si:H layers has been grown around the amorphous-to-microcrystalline transition, with Raman crystalline ratios varying from 0% to 67%. To obtain some detailed information about the modification of the silicon network structure across this transition, we investigated the distribution of gap states in the upper half of the band gap by means of transient photoconductivity (TPC). It has been generally accepted that in disordered silicon, and at moderate applied fields, electron transport is of the trap-limited band transport type, also known as the multiple-trapping process. For TPC this means that the carriers that are created at t=0 by a laser flash will thermalize through the density of gap states (DOS) and that the transient photocurrent is, therefore, sampling this DOS.

## TRANSIENT PHOTORESPONSE OF AN a-Si:H MODEL SYSTEM

The distribution of localized electronic states in the bandgap of disordered silicon can be represented by the sum of a more or less exponentially decreasing background density, caused by the lattice disorder, and a superimposed discrete defect band due to the silicon dangling bonds. The response of such system to pulsed across-gap excitation is well known from experiment [1], computer modelling [2,3] and analytical formulation of the problem [4].

Figure 1 shows an example of the photoresponse to be expected as a function of the relative strength of the defect peak with respect to the background state density. The parameter $\gamma$ indicates the density ratio between the two components at the defect peak energy $E_r$. For the case of a purely exponential DOS ($\gamma = 0$), the current transient follows a power law $I(t) \sim t^{-(1-\alpha)}$, where

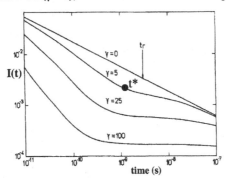

**Figure 1.** Simulated TPC signals generated for an exponentially decreasing DOS to which a discrete peak at $E_r$ is added at various strengths. $\gamma$ indicates the ratio of peak to background state density, $t_r$ the release time from the peak and t* the characteristic time of the peak-induced depression in the current transient. (after Seynhaeve *et al.* [3])

$\alpha$ is temperature dependent according to $\alpha = T/T_0$ and $T_0$ describes the exponential slope in $g(E) = g(0) \exp(-E/kT_0)$. It is seen in figure 1 that the presence of an additional discrete feature in the DOS causes a depression of the current transient below the $\gamma = 0$ power law, with the strength of the depression matching the value of $\gamma$. The point t*, as shown on the $\gamma = 5$ curve, marks the kink in the curves. It is a reference point that can be identified on experimental current decays, either directly on the I(t) trace, or more readily by plotting $I(t) \times t^{1-\alpha}$ versus time. For values of $\gamma$ near 1, i.e. when the depression is weak and the density in the defect peak is comparable to the background density, the value of t* can be used as a good approximation for $t_r = v^{-1}\exp(E_r/kT)$, the characteristic carrier emission time from the defect located at $E_r$. Measurements of t* versus temperature T then allow the deduction of the energy $E_r$ and the attempt-to-escape frequency $v$ of the defect center.

## Si:H SAMPLES

A series of Si:H layers has been grown at 100°C around the amorphous-to-microcrystalline transition by varying the hydrogen dilution ratio, defined as the hydrogen to silane flow ratio, from 5 to 25 in the 'ASTER' very-high-frequency (VHF) plasma reactor. Table 1 summarizes some of the sample properties [5]. The indicated crystalline ratios are based on Raman spectroscopy and are calculated as $R_c = (I_{510}+I_{520})/(I_{480}+I_{510}+I_{520})$, where $I_{480}$, $I_{510}$, $I_{520}$ are the intensities of the transverse optic mode in a-Si, grain boundaries and/or small grains, and c-Si, respectively. The samples used for the TPC measurements were deposited on Corning 2000 glass, with non-blocking Ag contacts separated by 0.5 mm in a gap cell configuration.

**Table 1.** Sample parameters showing the deposition rate D, the hydrogen dilution ratio, the Raman crystalline ratio $R_c$ and the dark current activation energy $E_a$.

| Sample | D (nm/s) | Dilution ratio | $R_c$ (%) | $E_a$ (eV) |
|--------|----------|----------------|-----------|------------|
| A3076  | 0.158    | 5              | 0         | 0.90       |
| A3093  | 0.106    | 15             | 0         | 0.91       |
| A3095  | 0.092    | 20             | 0.31      | 0.86       |
| A3097  | 0.080    | 22.5           | 0.60      | 0.53       |
| A3094  | 0.075    | 25             | 0.67      | 0.45       |

## RESULTS

Transient photocurrents were measured for all samples under an applied voltage of 400 V after optical excitation with a pulsed nitrogen laser coupled to a 540 nm dye. Figure 2(a) shows, for the microcrystalline ($\mu$c-Si:H) sample A3097, an example of the currents measured at temperatures between 30°C and the deposition temperature of 100°C. A modest dip in the current can be discerned around temperature-dependent times ranging from about $10^{-4}$ to $10^{-2}$ s. The same data are plotted in a $I(t).t^{0.7}$ versus t format in figure 2(b) to facilitate the determination of

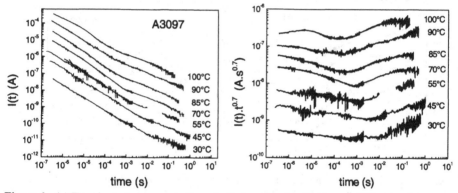

**Figure 2.** (a) Transient photocurrents through the sample with 60 % crystallinity for the indicated temperatures; (b) The observed currents multiplied by $t^{0.7}$. Individual curves were offset vertically for clarity.

the dip times t*. The experimental current dips are sufficiently weak in figure 2(a), e.g. in comparison with the $\gamma$ =5 trace in figure 1, to make $t^* \approx t_r$ a good approximation. A very similar set of TPC traces was obtained from the sample A3094 that was found to be 67% crystalline. From the regression line $\log t^* = \log \nu^{-1} + E_r/2.30kT$ through the t* data, as shown in figure 3 for both samples A3097 and A3094, estimates for the defect parameters $E_r$ and $\nu$ can be obtained. For A3097, $E_r = (0.53 \pm 0.02)$ eV and $\nu = 7 \times 10^{10}$ Hz (with an uncertainty range from 3 to 17 $\times$ $10^{10}$ Hz) are found, while the values for the A3094 sample are $E_r = (0.57 \pm 0.05)$ eV and $\nu = 4 \times 10^{11}$ Hz (range 0.6 to 26 $\times 10^{11}$ Hz).

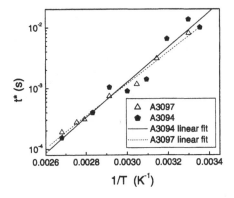

**Figure 3.** Temperature dependence of the characteristic dip times t* observed in the samples with 60% (A3097) and 67% Raman crystalline ratio (A3094).

At the other end of our deposition series are the fully amorphous samples A3076 and A3093. For these cases, no current depression is observed in the experimental TPC traces while the overall current decay remains similar to the $t^{-0.7}$ power law that is observed as background with the microcrystalline samples. The TPC data in figure 4 illustrate this behavior.

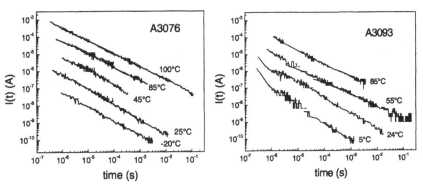

**Figure 4.** Transient photocurrents at various temperatures in the amorphous samples A3076 and A3093. Individual curves have been offset for clarity.

The sample A3095, which has a Raman crystalline ratio of 31%, proved more difficult to measure as may be judged from the somewhat irregular shape of individual current decays displayed in figure 5. Ignoring the somewhat different 50°C trace, the results are not inconsistent with a defect level some 0.5 eV from the transport path and with an attempt frequency of the order $10^{11}$ Hz, those being the values obtained above for the more crystalline samples.

## DISCUSSION

The energy position of the defect structure that is resolved for the largely crystalline samples, some 0.55 eV below the transport path, differs somewhat from the position deduced for

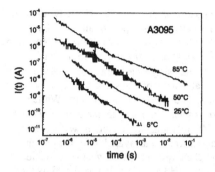

**Figure 5.** Transient photocurrents at indicated temperatures as measured in partly crystalline sample A3095. Individual traces have been offset for clarity.

the occupied dangling bond (D⁻) in amorphous hydrogenated silicon (a-Si:H) as extracted from post-transit time-of-flight (TOF) analysis. In pulsed experiments such as TPC and TOF, photo-excited carriers are quickly trapped and only contribute to the current upon being released from a trap. Any resolved defect structure consequently refers to the occupied state of the trap. The a-Si:H D⁻ band is centered some 0.63 eV below the conduction band mobility edge [6,7]. Since the depression on the experimental TPC curves is sufficiently weak to justify using $t^* \approx t_r$, the ~0.55 eV to ~0.63 eV difference cannot be ascribed to our use of the approximate $t^*$ values. Indeed, even the $t^*$ to $t_r$ difference at the $\gamma = 5$ curve of figure 1 would only result in a $kTln3$ difference for the calculated energy, i.e. only ~0.028 eV at room temperature. One has to conclude that, with the appearance of the crystallites, the transport level moves closer to the energy of the dangling bond. Since for the samples with ~60% crystalline ratio the dark current activation energy turns out to be around 0.5 eV [5], one may even propose that the Fermi level has shifted there into the D⁻ band.

The fact that, for the amorphous samples A3076 and A3093, no dip is seen in the TPC traces does agree with a dangling bond position at 0.63 eV in a-Si:H. Retaining the $\nu \sim 10^{11}$ Hz obtained for the crystalline samples as attempt-to-escape frequency, $t_r \approx 0.4$ s would follow, which is at the very edge of our measurable time range. There is an additional reason why no dip is observed in our TPC data for the amorphous samples: The ratio of dangling bond to background state density is higher at the 0.63 eV D⁻ position in the amorphous samples than is the case for the comparable values in the crystalline ones. Confirmation of this fact is provided by the small differences in overall slope of the TPC decays between crystalline and amorphous samples. These slopes are on the average somewhat shallower than $t^{-0.75}$ in the crystalline and somewhat steeper in the amorphous ones. Specifically, samples A3097 and A3094 have averages of $t^{-0.67}$, respectively $t^{-0.69}$, while those values are $t^{-0.81}$ for A3076 and $t^{-0.88}$ for A3093. As may be seen from the simulations in figure 1, the initial slope of the current decay becomes somewhat steeper with increasing values of $\gamma$. Moreover, the current drops farther below the $\gamma = 0$ line for higher $\gamma$ values, a phenomenon that explains why TPC reaches the lower detection limit of our measuring system sooner for the amorphous than for the crystalline samples. Support for the significance of the modest changes in the TPC slopes is further provided by the fact that the steepest overall slopes are observed with the A3093 sample. This is the amorphous sample for which the deposition conditions are closest to the amorphous to crystalline transition, a situation that is known to produce good material properties. The A3093 therefore exhibits the lowest Urbach parameter $E_0$ of the series [5] and consequently has the more steeply declining tail state density. It thus will have the largest ratio of peak to background density at $E_r$: While $E_0$ and the defect density are linearly correlated, the background density varies exponentially with $E_0$.

According to the expressions cited in the introductory discussion of transient photocurrent signals, a power-law decay such as the $\sim t^{-0.75}$ seen here could be interpreted as originating with an exponential background DOS where $\alpha = T/T_0 \approx 0.25$ then suggest a characteristic temperature $T_0$ of well over 1000 K. Such slowly varying exponential density of gap states has been used in the past for modelling purposes [8], but our results cannot be interpreted in this way. While a small decrease in the $t^{-(1-\alpha)}$ slope, hence increase in $\alpha$, is seen with rising temperature in the TPC traces, that change is largely insufficient to fit a $\alpha = T/T_0$ relationship. In fact, as was shown by Grabtchak *et al.* [9], experimental data from much shorter times than available for the present study are required to accurately model a DOS from TPC measurements. All we may conclude from our results is that there is a gap state background in Si:H that is largely unaffected by the degree of crystallinity of the sample.

Examining the combined TPC results from the five Si:H samples, a remarkably consistent pattern emerges with a silicon dangling bond protruding from a continuous distribution of gap states. The most obvious change across the series is the reduction of the energy distance between the $D^-$ level and the transport path in the microcrystalline samples with respect to that distance in a-Si:H. The results suggest that the dangling bond density, due to either the amorphous fraction or the grain boundaries, decreases with rising sample crystallinity. Those dangling bond centers nevertheless remain the dominant charge traps irrespective of the crystalline fraction.

## CONCLUSIONS

The density and distribution of gap states, as seen by transient photoconductivity, does not change drastically across a series of VHF-deposited Si:H samples that span the amorphous to crystalline transition region. Si dangling bonds, either in the amorphous network or at the grain boundaries, retain their role as prominent charge trap, but the smaller band gap in the crystalline regions does lead to a decrease of the energy distance between the electron transport path and the appropriate dangling bond level in samples on the crystalline side of the transition. In as much as that transport path is generally accepted to coincide with the conduction band mobility edge in a-Si:H, the above observation suggests a lowering of that mobility edge in $\mu$c-Si:H.

## REFERENCES

[1] C. Main, Mat. Res. Soc. Symp. Proc. **467**, 167 (1997).
[2] J.M. Marshall and R.A. Street, Solid State Commun. **50**, 91 (1984).
[3] G. Seynhaeve, G.J. Adriaenssens and H. Michiel, Solid State Commun. **56**, 323 (1985).
[4] D. Monroe, Solid State Commun. **60**, 435 (1986).
[5] M. Brinza, J.K. Rath and R.E.I. Schropp, Technical Digest of the International PVSEC-17, Fukuoka, Japan, 2007, p1298; to be published in Solar Energy Materials and Solar Cells.
[6] B. Yan, D. Han and G.J. Adriaenssens, J. Appl. Phys. **79**, 3597 (1996).
[7] I. Sakata, T. Kamei and M. Yamanaka, Phys. Rev. B **76**, 075206 (2007).
[8] M. Hack and M. Shur, J. Appl. Phys. **58**, 997 (1985).
[9] S. Grabtchak, C. Main and S. Reynolds, J. Non-Cryst. Solids **266**, 362 (2000).

Mater. Res. Soc. Symp. Proc. Vol. 1066 © 2008 Materials Research Society     1066-A07-11

# Characterization and light emission properties of osmium silicides synthesized by low energy ion implantation

P. R. Poudel[1], K. Hossain[1], J. Li[1], B. Gorman[2], A. Neogi[1], B. Rout[1], J. L. Duggan[1], and F. D. McDaniel[1]

[1]Physics, University of North Texas, 211 Avenue A, Denton, TX, 76203
[2]Material Science and Engineering, University of North Texas, 3940 North Elm St., Denton, TX, 76207

## ABSTRACT:

Low energy (55 KeV) Osmium ( $Os^-$ ) negative ion beam was used to implant ($5 \times 10^{16}$ atoms/cm$^2$) into p-type-Si (100). The implantation was performed with the ion source of a National Electrostatic Corp. 3 MV Tandem accelerator. The implanted sample was subsequently annealed at 650 °C in a gas mixture that was 4% $H_2$ + 96% Ar. Rutherford Backscattering spectrometry (RBS) analysis with 1.5 MeV Alpha particles was used to monitor the precipitate formation. Photoluminescence (PL) measurements were also performed to study possible applications of silicides in light emission. Cross-sectional Scanning Electron Microscopy (X-SEM) was performed for topographic image of the implanted region. RBS along with PL measurements indicate that the presence of osmium silicide ($Os_2Si_3$) phase for light emission in the implanted region of the sample.

## INTRODUCTION:

The study of metal silicides had taken rapid initiation in the late 1970s and early 1980s with the expectation of their device applications [1]. Since then, semi-conducting transition metal silicides have been drawing considerable interest as potential candidates for optoelectronics and microelectronic devices [2-4]. The iron disilicide β-FeSi$_2$ has been reported as one of the most promising silicides for its potential use as an optoelectronic material [5, 6]. The silicides have also been implemented as contacts for many novel devices such as nanowires/nanotubes of various materials [7]. The ultimate application of the silicides depends on how accurately the silicide phase can be controlled and how controllably the band gap is tailored. Many approaches have been employed in the formation of efficient light emitting silicon based devices [8, 9]. There has been a considerable progress in the growth of epitaxial layers and the single crystals of semi-conducting silicides for light emission and photovoltaic applications [10]. Ion implantation is one of many techniques to grow optically active crystals of metal silicides. This technique applies an accurate and predictable control of doping impurities within the surface layers [11]. In this work, we have tried to synthesize and investigate the light emission properties of osmium silicide (Os$_x$Si$_y$). Currently there are only a few literature reporting the synthesis of Os$_x$Si$_y$ with the severe difficulties in the synthesis process due to low vapor pressure of osmium[12,13]. Measuring the effect of low energy Osmium negative ion implantation, in the formation of silicide phases by different compositional, structural and optical characterization techniques such as RBS, Cross-sectional Scanning Electron Microscopy (X-SEM) and PL spectroscopy is the main purpose of this paper.

## EXPERIMENTAL:

Single crystal, 5-10 Ωcm, p-type (100) Si wafer was used in the present experiment. The sample was cleaned with acetone and rinsed in ultrasonic bath for about 5 minutes. This process was repeated several times in order to remove the surface contaminants on the sample. Prior to loading into the ion implantation chamber the sample was etched in a dilute HF solution in order to remove the native oxide layers on Si substrate. The sample was then implanted at a fluence of $5 \times 10^{16}$ atoms/cm$^2$ by 55 KeV negatively charged $^{76}$Os$^-$ ion beam. The ion implantation was performed in the Ion Beam Modification and Analysis Laboratory (IBMAL) at the University of North Texas (UNT). A National Electrostatics Corporation cesium negative ion sputter source was used to perform the implantation [14]. The Low current densities were maintained in order to avoid the sample heating during ion implantation. The implantation profile of the Osmium ions were simulated using a Monte Carlo simulation method (The Stopping and Range of Ions in Matter, SRIM-2008) [15]. For the 55 KeV Os$^-$ implantation, the projected range (R$_p$) and straggling of the ions was simulated to be 31.5 nm and 7 nm respectively, confirming the confinement of Osmium in the near surface layer. The implantation profile of Os$^-$ ion beam in Si substrate by using SRIM simulation is shown in Fig. (1).

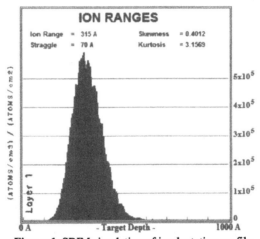

**Figure 1.** SRIM simulation of implantation profile

The RBS measurement was carried out to monitor the precipitates formation of implanted osmium in silicon substrate after subsequent annealing. The annealing was performed at 650 °C in a gas mixture (4% H$_2$ + 96% Ar). The Rutherford backscattering experiment was performed by using a Van de Graaff accelerator in IBMAL with 1.5 MeV He$^+$ beam incident at an angle of 5° with respect to the sample normal. A solid state detector was set at an angle of 150° with respect to the incident beam direction ro detect the backscattered He beam. The RBS spectra were simulated by using a code "SIMNRA" [16, 17]. The High Resolution SEM technique was used for cross sectional topography of the implanted sample. The cross-sectional sample was prepared by focused ion beam (FIB) milling technique [18]. The FEI Co. Nova 200, Dual Beam

FIB/ Field Emission SEM was used for ion beam milling. The various steps for cross-sectional sample preparation are illustrated in Figure 2. The 30 KeV Ga$^+$ beam, a current of 5nA-100pA was used for milling. Initially a thin layer of platinum as shown in Fig. 2(a) was deposited on the sample surface in order to protect the sample from the Ga$^+$ beam. The 5 keV e$^-$ beam was used to scan the sample cross section and the "Through the Lens Detector" was used to collect the backscattered signal from sample cross-section. The Fig. 2(b) and 2(c) show the SEM images for the beginning and the end of cross section milling process respectively.

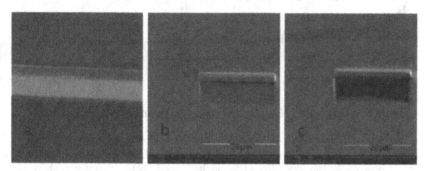

**Figure 2.** Various steps for X-SEM sample preparation (a) Platinum deposition before milling, (b) SEM image for the beginning of milling and (c) SEM image at the end of milling.

The PL measurement was conducted at low temperature (37 K) by using 325 nm HeCd laser as an excitation source. The emission of the scattered light from the sample was detected with a charged-coupled detector (CCD) camera.

**RESULTS AND DISCUSSIONS:**

RBS spectrum obtained from the sample after the 55 KeV Os$^-$ ion beam implantation with a fluence of 5 X10$^{16}$ atoms/cm$^2$ followed by annealing at 650 $^0$C for 3 hours is shown in Figure 3. The SIMNRA simulated spectrum is also presented along with the experimental RBS spectrum in Figure 3. A very close matching between the simulated and the experimental spectra was found with a Os: Si stoichiometry ratio of 2:3, indicating the formation of osmium silicide

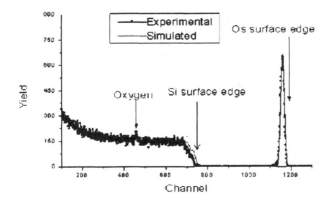

**Figure 3.** RBS spectrum of implanted sample along with SIMNRA simulation.

layer. In the spectrum, the relatively sharp signal of osmium indicates the formation of a thin layer of silicide (~50 nm). No apparent tail at the low energy edge of the osmium peak indicates that the interface between the substrate silicon site and the formed silicide layer is very sharp as observed earlier in ref [19]. The step formation near the surface edge of Si and strong oxygen peak indicate, the presence of $SiO_2$ (~50 nm) on top of the osmium-silicide layer. The surface silicon dioxide layers are formed due to presence of oxygen contaminations in the annealing furnace. The High Resolution X-SEM image of the implanted sample is shown in Fig.4. As we see in the X-SEM image, the osmium rich layer (silicide phase) of 50 nm phase is formed in between the surface silicon dioxide layer and substrate silicon layer. Figure 5. shows the PL spectrum from the osmium silicide layer. The surface silicon-dioxide layer was etched out using dilute HF solution before the PL measurements were performed.

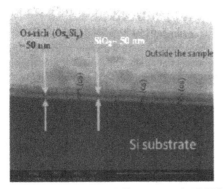

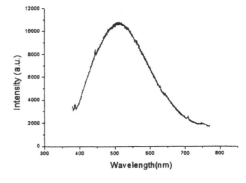

**Figure 4.** Backscattered Electron-mode X-SEM image of the sample

**Figure 5.** Photoluminescence Spectrum

The peak in the PL spectrum is centered at wavelength 525 that corresponds to a band energy of 2.36 eV. This peak is well within the range of experimentally determined band energy of 2.3 eV [20] and was as reported in ref [21] to be the result of direct transition for $Os_2Si_3$ phase of the silicide layer.

## CONCLUSIONS:

We have synthesized the osmium silicide precipitates in silicon substrate by low energy negative osmium ion implantation into Si(100) followed by high temperature annealing. The RBS measurements indicates formation of stoichiometric layers of $Os_2Si_3$. The X-SEM image of the implanted regions of the sample showed well defined layers of silicide. The PL measurement showed the band gap of 2.36 eV which is the result from direct transition osmium silicide $(Os_2Si_3)$ phase. In future, we plan to perform Transmission Electron Microscopy (TEM) Diffraction and Time Resolve photoluminescence to confirm the osmium phase and the nature of electronic transition.

## ACKNOWLEDGMENTS:

This work at university of North Texas is partially supported by 'The Robert A. Welch Foundation' under grant # 70634.

## REFERENCES:

1. L.J. Chen, Metal Silicides: An integral Part of Microelectronics, JOM, 57 N. 9(2005) 24
2. H. Lange, Silicide Thin Films-Fabrication, Properties, and Application, Materials Research Society, Boston (1996), 307
3. N.E. Christensen, Physicsl Review B, 42 (1990), 7148
4. H.Lange, Solid-State and Integrated Circuit Technology (1998) 247, Proceedings of 1998 IEEE
5. D. Leong, M. Harry, K. J. Reeson, and K. P. Homewood, Nature (London), 387, 686(1997)
6. S. Chu, T. Hirohada, and H. Kan, Jpn. Journal of Applied physics, Part 2 41, L299 (2002).
7. L.J. Chen, Silicide Technology for Integrated Circuits, The institution of Engineering and Technology publishing, London, UK (2004)
8. M.A. Lourenco, M. Milosavljevic, G, Shao, R.M. Gwilliam, K.P. Homewood, Thin Solid Films, 504(2006) 36
9. H.T. Lu, L.J. Chen, Y.L. Chueh, L.J. Chou, Journal of Applied physics, 93(2003) 1468
10. P.D. Townsend, P.J. Chandler, L. Zhang, Optical effects of Ion Implantation, Cambridge University Press (1994)
11. Yoshihito Maeda, Kenji Umezawa, Yoshikazu Hayashi, Kiyoshi Miyake, Kenya Ohashi, Thin Solid Films 381(2001) 256
12. F.Z. Amir, R.J. Cottier, T.D. Golding, W. Donner, N. Anibou, D.W. Stokes, Journal of Crystal Growth 294 (2006) 174-178

13. L.J. Mitchell, O.W. Holland, K Hossain, E.B. Smith, Jerome L. Duggan, and F.D. McDaniel, Nucl. Inst. and Meth., **241**, (2005) 548.

14. R. Middleton, A Negative Ion Cookbook, HTML Version: M. Wiplich, < http://tvdg10.phy.bnl.gov/COOKBook>, October 1989.

15. J. P. Biersack and L. G. Haggmark, Nucl. Inst. and Meth., **174**, (1980) 257. Recently updated, the package and its documentation are available at http://www.srim.org.

16. M. Mayer, SIMNRA user's guide, < www.rzg.mpg.de/~man>.

17. M. Mayer, SIMNRA, in: Proceeding of the 15th International Conference on the Application of Accelerators in Research and Industry, Vol. 475, 1999, 541.

18. D.P. Adams, M.J. Vasile, Journal of Vacuum Science and Technology, B 24(2006) 836

19. X Q Cheng, H N Zhu, R S Wang, J. Phys.: Condens. Matter 12(2000) 9195

20. L. Schellenberg, H.F. Braun, J. Muller, Journal of less common metals, Vol144, 2(1988)341

21. D.B. Migas, L. Miglio, V.L. Shaposhnikov and V.E. Borisenko, Phys. Status Solidi B 231 (2002) (1), 171

# Novel Applications

Mater. Res. Soc. Symp. Proc. Vol. 1066 © 2008 Materials Research Society          1066-A08-01

# Multilayered a-SiC:H Device for Wavelength-Division (de)Multiplexing Applications in the Visible Spectrum

Manuela Vieira[1,2], Miguel Fernandes[1], Paula Louro[1,2], Manuel Augusto Vieira[1,3], Manuel Barata[1,2], and Alessandro Fantoni[1]

[1]Electronics Telecommunication and Computer Dept., ISEL, Rua Conselheiro Emídio Navarro, Lisbon, 1959-007, Portugal
[2]CTS, UNINOVA, Monte da Caparica, Caparica, 2829-516, Portugal
[3]Traffic Dept., CML, Lisbon, 1049-001, Portugal

## ABSTRACT

A multiplexer is a device that combines two or more signals onto a single output without losing their specificity. In this paper we present results on the use of multilayered a-SiC:H heterostructures either as wavelength-division multiplexing or demultiplexing device (WDM). The WDM is a glass/ITO/a-SiC:H (p-i-n)/ a-SiC:H(-p) /Si:H(-i)/SiC:H (-n)/ITO heterostructure which faces the modulated light incoming together from different beams, each one with a specific wavelength and transmission rate. By reading out, at different applied bias, the photocurrent generated by all the incoming optical carriers, the information is multiplexed or demultiplexed and can be transmitted and recovered again. The devices were characterized through spectral response measurements, under different electrical bias and frequencies. Results show that in the multiplexing mode the output signal is balanced by the wavelength of each incoming optical carrier and modulated by their frequencies. In the demultiplexing mode the photocurrent is controlled by the applied voltage allowing to regain the transmitted information. An electrical model is presented to explain the device operation.

## INTRODUCTION

Recent advances on the production of polymer optical fibers (POF) [1] have opened possibilities for short distance communication systems at low cost. POF (Polymer Optical Fibers) present acceptable attenuation in light transmission within the visible spectrum and are presently used as a single channel transmission line [2]. To improve the data transmission rate, Wavelength Division Multiplexing (WDM) [3] can be employed for signal multiplexing and demultiplexing in the visible spectrum, which demands the conception of new devices. For WDM two key elements are crucial, a multiplexer and a demultiplexer. Although these two components are well known for infrared telecommunication systems, they must be completely renewed for the different transmission windows in the POF fibers, that work within the visible spectrum. Digital home appliance interfaces, home and car network and traffic control applications are foreseen due to the low cost associated to the a-SiC:H and POF technologies.

In this paper we present results on the optimization of multilayered a-SiC:H heterostructures either for wavelength-division multiplexing or demultiplexing applications in the visible spectrum.

## DEVICE OPERATION

The element sensor is a glass/ITO/a-SiC:H (p-i-n)/ a-SiC:H(-p) /Si:H(-i)/SiC:H (-n)/ITO double heterostructure produced by PECVD. Deposition conditions are described elsewhere [4]. The thickness and the absorption coefficient of the front photodiode are optimized for blue collection and red transmittance, and the thickness of the back one adjusted to achieve full absorption in the greenish region and high collection in the red spectral one. As a result, both front and back diodes act as optical filters confining, respectively, the blue and the red optical carriers, while the green ones are absorbed across both [5]. In Fig.1 the device configuration is depicted in both multiplexing and demultiplexing modes. Here, multiple monochromatic (Fig 1a) or a single polychromatic (Fig. 1b) beams are directed to the device where they are absorbed, accordingly to each wavelength, giving rise to a time and wavelength dependent electrical field modulation across it [6]. By reading out, under different applied bias, the total photocurrent generated by all the incoming optical carriers the information (wavelength, modulation frequency) is multiplexed or demultiplexed and can be transmitted or recovered again.

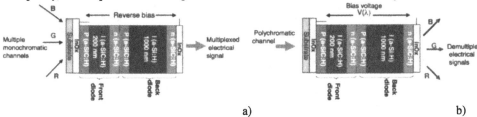

a)                                                                          b)

**Figure 1** - WDM device configuration: a) multiplexing mode; b) demultiplexing mode.

In the multiplexing mode the device faces the modulated light incoming together from different monochromatic channels. The combined effect is converted to an electrical signal via the device. In the demultiplexing mode a single modulated polychromatic light beam (mixture of different wavelength) impinges the device and the spectral sensitivity, which is voltage controlled, allows the recognition of the different color channels. In both modes a programmed logic circuit drive LEDs can be used to send out light into appropriated optical fibers (POF) for transmission to a destination where they can be split again using the demultiplexing mode.

## OPTOELECTRONIC CHARACTERIZATION

The devices were characterized through spectral response measurements (400-800 nm), under different modulated light frequencies (15 Hz to 2 KHz) and electrical bias (-5V to +2V). In Fig. 2a it is displayed, under reverse bias, the spectral photocurrent at different frequencies and, in Fig. 2b, the trend with the applied voltages is shown at 2 KHz.

Results show that, under reverse bias and low frequencies (f<400Hz), the spectral response increases with the frequency of the light that impinges the device, suggesting different capacitive effects during the device operation. In the blue/green spectral regions the photocurrent remains constant and increases in the reddish region (Fig. 2a). For higher frequencies (f>400Hz) the spectral response does not depend on the modulated light frequency and, as the applied voltage changes from forward to reverse (Fig. 2b), the blue/green spectral collection is enlarged while the red one remains constant.

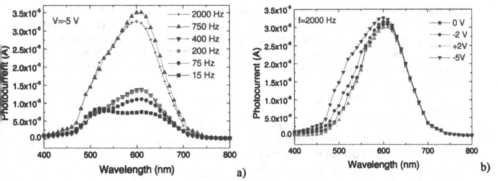

**Figure 2** – Spectral photocurrent under: a) reverse bias (-5V) and different frequencies; b) different applied voltages and at a modulated frequency of 2000 Hz.

## WAVELENGTH DIVISION MULTIPLEXING

In Fig. 3a it is displayed the multiplexed signals (solid lines). The transient signals were acquired at different applied voltages (-5V <V<+2V). The blue ($\lambda_L$=450 nm; dotted line) and the red ($\lambda_L$=650 nm; dash line) input channels are superimposed to guide the eyes across the monochromatic R and B input channels (Fig. 1a). The red signal frequency was 1.5 KHz and the blue one was half of this value. In Fig. 3b the same output signals of Fig. 3a are shown, but using for the input frequencies two orders of magnitude lower.

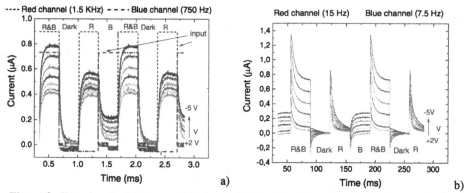

**Figure 3** - Wavelength division multiplexing (solid lines) at different applied voltages, obtained using the WDM device: a) High frequency regime. The blue (dotted blue line) and the red (dash-dot red line) guide the eyes into the input channels; b) Low frequency regime.

Both figures show that the multiplexed signal depends on the applied voltage and on the frequency regime of the input channels. There are always four levels. The higher occurs when both red and blue channels are ON (R&B) and the lower when both are OFF (dark). The red level (R) appears if the blue channel is OFF and is higher then the blue level (B) that occurs when the red channel is OFF. The step among them is higher under reverse bias and increases as the negative bias increases. Results show also that in the high frequency regime (Fig. 3a) the multiplexer acts as a charge integrator device while in low frequency regime it works as a differentiator. In both regimes, the output signals (multiplexed signals) show the potentiality of using the device for WDM applications since it integrates every wavelength to a single one retaining the input information.

In Fig. 4 the *ac* current-voltage characteristics under different wavelengths: 650 nm (R); 450 nm (B); 650 nm &450 nm (P&B), and frequencies regime are displayed. Results show that in both regimes, under red modulated light, the collection efficiency is independent on the applied voltage being higher at high frequencies, as expected from Fig. 2. Under blue irradiation the collection slowly increases as the applied voltage changes from forward to reverse, while, under combined red and blue irradiation, a quick increase is observed with the increase of the reverse bias. Under forward bias the R and the R &B signals are almost the same. This behavior is indicative of the blindness of the device to the blue component of the multiplexed signal under forward bias.

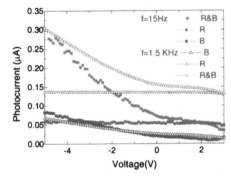

**Figure 4** – *ac* IV characteristics under R($\lambda_L$=650 nm); B($\lambda_L$=450 nm), R($\lambda_L$=650 nm) & B($\lambda_L$=450 nm) modulated light and different light frequencies (15Hz;1.5KHz).

## WAVELENGTH DIVISION DEMULTIPLEXING

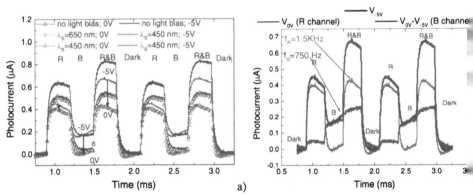

**Figure 5** - a) Transient photocurrent under different optical ($\lambda_S$=650nm, $\lambda_S$=450nm) and electrical bias (V=0V;V=-5V). b) Blue and red wavelength division demultiplexing output channels for the input signal (black line) obtained using the WDM device.

Different wavelengths which are jointly transmitted must be separated to retain all the information. A multiple time dependent wavelength combination (Fig. 3), **R** (650 nm), **B** (450 nm), **R&B** (650 nm & 450 nm), and **Dark** was projected on the active surface of the device. Through the back and front sides, a steady state red ($\lambda_L$=650 nm) or blue ($\lambda_L$=450 nm) optical biases were in that order superimposed (Fig. 1b) to increase, respectively, the blue collection of the front diode or the red collection of the back one [3]. The generated photocurrent was measured to readout the combined spectra. The result is plotted in Fig. 5a, in short circuit (open symbol) and under reverse bias (straight line). Results show that the blue optical bias has no effect on the readout signal, while the presence of the red bias shifts the **Blue** (B) component of the combined spectra to the dark level (see arrows), tuning the **Red** (R) component. Thus, by switching, under red light bias between short circuit and reverse bias the **Red** and **Blue** channels were regained. The multiplexed and the recovered signals are displayed in Fig. 5b.

## ELECTRICAL MODEL

An electrical model was developed (Fig. 6a) and supported by a SPICE simulation (Fig. 6b). The WDM device is considered to be as two photo transistors connected back to back, modeling respectively the a-SiC:H p-i-n-p and a-Si:H n-p-i-n sequences. One transistor, Q1, is *pnp* type and the second, Q2, *npn*. Capacitors, C1 and C2, are used to simulate the capacitance transient due to the minority carrier trapped in both p-i-n junctions. A voltage source was applied through a resistor, giving rise to a current I (R3). Two *ac* current sources, I1 and I2, are used to simulate the input blue and red channels. The frequencies are the same as the ones used in the experimental work (Fig. 3a). In Fig. 6b it is shown the input channels, I(I1), I(I2), the multiplexed signal, I(R3) and the current across the capacitors, I(IC1), I(IC2).

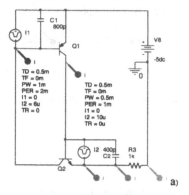
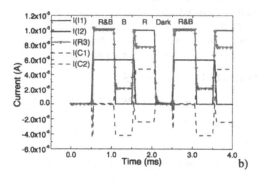

**Figure 6** - a) Equivalent electrical circuit of the pinpin photodiode. b) Signals obtained using SPICE simulation when the red (I2) and the blue (I1) modulated lights are impinging the device.

Good agreement with the experimental data (Fig.3a) is achieved. Results show that the internal junction (n-p) controls the current across the device. If the device is biased negatively (-5V), the p-n internal junction is forward-biased and the external voltage appears mainly across the front and back reverse-biased junctions. So, the current, I, depends not only on the balance between both blue and red photocurrents (I1, I2) but also on the end of each half-cycle of the frequency of the modulated current. Thus, the movement of charge carriers with an increase/decrease in the irradiation, results in a charging or a displacement current similar to the current (i=CdV/dt) that charges the capacitors.

Under negative bias and in the beginning of the cycle (R&B), the carriers generated by the blue photons flow across Q1 collector toward the base of Q2 and together with the ones generated by the red photons recombine or are collected (R&B level). Under blue irradiation, $I(I2)=0$, only the carriers generated by the blue photons flow across the device (B level). C1 charges positively and C2 negatively as a reaction to the decrease in the red irradiation. The opposite occurs under red irradiation. When both red and blue lights are simultaneously off, the current is limited by the leakage current (dark level). So, once triggered, the device continues to conduct until the current through it drops below a certain threshold value, such as at the end of a half-cycle, keeping the information of the wavelength (R&B, R, B, Dark) and frequency of the impinging light. When a positive voltage is applied, the junction capacitance across the internal n-p junction is charged. The charging current flows through the emitter of the two transistors. The device behaves essentially as a *npn* phototransistor with the *pnp* transistor acting like a emitter-follower with a very small gain. So, under lower positive voltages the only carriers collected come from the red channel enabling the demultiplexing of the previous multiplexed signal (Fig.5).

## CONCLUSIONS

We present a new device based in an a-SiC:H p-i-n-p-i-n heterostructure for multiplexing/demultiplexing applications. Two modulated input channels were transmitted together, each one located at different wavelengths and frequencies. The combined optical signal (multiplexed signal) was analyzed by reading out the photocurrent generated across the device. Results show that the multiplexed signal keeps the memory (wavelength and transmission rate) of the input channels. By switching between positive and negative voltages the input information can be recovered.

More work has to be done in the optimization of the device configuration in other to enlarge the input channels to the green wavelength range.

## ACKNOWLEDGEMENTS

This work supported by POCTI/FIS/58746/2004 and by Fundação Calouste Gulbenkian.

## REFERENCES

1. Mark G. Kuzyk, Polimer Fiber Optics, Materials Physics and Applications, Taylor and Francis Group, LLC; 2007.
2. S. Randel, A.M.J. Koonen, S.C.J. Lee, F. Breyer, M. Garcia Larrode, J. Yang, A. Ng'Oma, G.J Rijckenberg, H.P.A. Boom. "Advanced modulation techniques for polymer optical fiber transmission". proc. ECOC 07 (Th 4.1.4). (pp. 1-4). Berlin, Germany, 2007.
3. Michael Bas, Fiber Optics Handbook, Fiber, Devices and Systems for Optical Communication, Chap, 13, Mc Graw-Hill, Inc. 2002.
4. M. Vieira, A. Fantoni, M. Fernandes, P. Louro, G. Lavareda and C.N. Carvalho, Thin Solid Films, 515, Issue 19, 2007, 7566-7570.
5. P. Louro, M. Vieira, Yu. Vygranenko, A. Fantoni, M. Fernandes, G. Lavareda, N. Carvalho Mat. Res. Soc. Symp. Proc., 989 (2007) A12.04.
6. M. Vieira, M. Fernandes, J. Martins, P. Louro, R. Schwarz, and M. Schubert, IEEE Sensor Journal, 1, 2001, 158-167.

Mater. Res. Soc. Symp. Proc. Vol. 1066 © 2008 Materials Research Society 1066-A08-02

## Floating-Gate a-Si:H TFT Nonvolatile Memories

Yue Kuo, and Helinda Nominanda
Thin Film Nano & Microelectronics Research Laboratory, Texas A&M University,
College Station, TX, 77843-3122

## Abstract

Charge and discharge phenomena of the floating-gate amorphous silicon thin film transistor have been studied under dynamic operation conditions. The charge storage capacity decreases with the increase of the drain voltage because it is easier for electrons to be transported to the drain electrode than to be injected into the gate dielectric layer. The discharge efficiency with respect to the drain voltage has been investigated using three different discharge methods: negative gate bias voltage, light exposure, and thermal annealing, separately. The channel length affected both the charge capacity and the discharge efficiency due to the charge storage mechanism and the channel resistance. Majority of the stored charges were removed with the above method through various mechanisms. The low temperature favors the charge storage but the high temperature favors the discharge. This study revealed key parameters for the optimum operation of the low temperature fabricated nonvolatile memory device.

## Introduction

A charge storing device based on embedding a thin layer of hydrogenated amorphous in the gate dielectric of a hydrogenated amorphous silicon thin film transistor (a-Si:H TFT) had been demonstrated recently [1,2]. The TFT's modified gate dielectric consisted of a sandwich of bottom silicon nitride ($SiN_x$) as the control dielectric, a-Si:H as the floating gate, and top $SiN_x$ as the tunnel dielectric. The magnitude and polarity of the gate voltage ($V_g$) control the behavior of charges injected into and released from the a-Si:H floating gate. Memory characteristics could be estimated from the hysteresis of transfer characteristic curves [1,2]. This kind of memory device is attractive for applications on flexible substrates, displays, sensors, etc., due to its low temperature fabrication process.

In this study, authors report the charging and discharging efficiency of the TFT memory under gate and drain bias using different methods such as opposite $V_g$ stress, light exposure, and thermal annealing. Recently, the charge and discharge efficiencies of the TFT under the source and drain grounded condition were reported [3]. However, for many practical applications, the drain electrode is not grounded. Therefore, it is desirable to know the charge and discharge characteristics with the drain electrode biased. In this paper, authors investigated the charge storage capability and the discharge efficiency under the above condition. Mechanisms of both processes were also investigated.

## Experimental

Figure 1 shows the cross-sectional view of a floating-gate a-Si:H TFT fabricated on a Corning 1737 glass substrate. The complete self-aligned TFTs were prepared with 2 photomasks [4]. The plasma deposition and etching processes are the same as those used in preparing a conventional a-Si:H TFTs with field effect mobility > 0.3 cm²/V-s and $I_{on}/I_{off}$ current ratio > $10^6$. Detail of the fabrication process can be found in refs. 1 and 2. The control TFTs samples, i.e., without the embedded a-Si:H layer in the gate dielectric structure, were also fabricated for comparison.

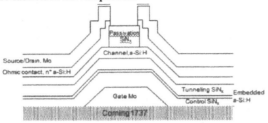

Figure 1. Cross-sectional view of an a-Si:H embedded floating-gate a-Si:H TFT

The measurement steps performed for the charge storage experiment were (1) measurement of transfer characteristic from $V_g$ = -5 V to 35 V at drain voltage ($V_d$) =10 V, (2) stress bias at $V_g$ = 35 V, for 10 s at $V_d$ = 0, 0.1, 1, or 10 V with the source electrode grounded, (3) measurement of transfer characteristic from $V_g$ = 35 V to -5 V at $V_d$ =10 V. Compared with the transfer characteristic of the pre-stress TFT, the transfer characteristic of step 3 shifted toward the negative direction due to the delay of the inversion layer formation from process of releasing negative charges [1]. The amount of charge stored at the TFT, Q, was estimated from the threshold voltage difference, $\Delta V_t$, between the forward (in step 1) and the backward (in step 3) transfer characteristic using the following equation [5]

$$Q = \Delta V_t \, \varepsilon \, / \, d \qquad (1)$$

where $\varepsilon$ and d are dielectric constant and the thickness of the tunnel $SiN_x$, respectively.

For discharging experiments, the TFT was first charged and then discharged. The charge step was the same as that described in the previous section: (1) measurement of the transfer characteristic from $V_g$ = -5 V to 35 V at $V_d$ =10 V and (2) stress bias at $V_g$ = 35 V for 10 s at $V_d$ = 0, 0.1, 1, or 10 V, separately with the source electrode grounded. For discharge with the negative gate bias method, the charged TFT was biased at $V_g$ = -35 V for 10 s with $V_d$ at 0, 0.1, 1, or 10 V independently. For discharge with the light exposure method, the charged TFT was exposed to a halogen lamp (GE Quartzline EKE 35200) at 585 μW/cm² for 5 s with $V_d$ at 0, 0.1, 1, or 10 V and $V_g$ = 0 V. For discharge with the thermal annealing method, the charged TFT was annealed at 210°C for 10 min. During the annealing, $V_g$ = 0 V and pulsed $V_d$ at 0, 1, or 10 V was applied repeatedly for 10 1-minute cycles with each cycle including 10 s on and 50 s off. After discharge, the TFT's transfer characteristic was measured from $V_g$ = -5 V to 35 V at $V_d$ = 10 V. The charge remained in the TFT was estimated using equation 1 where $\Delta V_t$ is defined as the difference of the $V_t$ before charge and that after the above discharge step. The percent discharge was calculated as follows

$$\text{Percent discharge} = \left( \frac{\text{Amount of charge stored}}{\text{Amount of charge remained}} \right)_{V_d} \times 100\% \qquad (2)$$

For the temperature effect on the charge capacity of the TFT, the transistor was heated to a temperature of 298, 308, 323, or 373 K. At this temperature, first, the TFT's transfer characteristic was measured. Then, the TFT was charged at $V_g = 35$ V and $V_d = 0$ V for 10s, which was followed by transfer characteristic measurement. The TFT's charge storage capacity at each temperature was calculated.

For the annealing discharging, the TFT was first charged at room temperature and the transfer characteristic was measured. Then, it was annealed at 423, 443, 463, or 483 K with $V_g = V_d = 0$ V for 10 minutes. The TFT was subsequently cooled down to room temperature. The transfer characteristic was measured again. The amount of charges remained in the TFT was subsequently calculated. All electrical measurements were performed in a black-box using an Agilent 4155C Semiconductor Parameter Analyzer.

## Results and Discussion

Figure 2 shows the transfer characteristics of a 9-nm embedded TFT with W/L of 99/58 µm at various stage of operations with $V_d = 0.1$ V. A large hysteresis between the transfer characteristics of the TFT before and after charging was observed. After discharging by the $-V_g$ or by light exposure, the transfer characteristics shift toward the before charging transfer characteristic, which results in the smaller hysteresis between the before charging and the after discharging transfer characteristics. With the application of $V_d$ bias during the $-V_g$ bias or light exposure discharging, the electron tunneled from the floating gate or electrons from the electron hole pairs are attracted to the drain electrode. On the forward direction sweep, the existence of such electron accumulation layer corresponds to a shift of the post discharging transfer characteristic to the right, i.e., towards the original transfer characteristic before charging. A more thorough discussion on the effect of $V_d$ on the charge and discharge process will be discussed in the following sections.

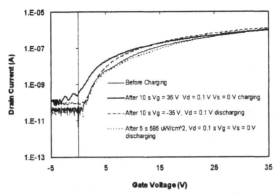

Figure 2. Transfer characteristics of a 9-nm embedded a-Si:H TFT (W/L=99 μm /58 μm) before charging, after charging and after discharging with various methods.

*Charging and Discharging under Various $V_g$ and $V_d$ Bias Conditions*

Previously, the charge storage capacility of the floating-gate a-Si:H was measured with the source and drain electrodes grounded. The result showed that the charge storage capacity was dependent on the stress time following the power law. It also increased with the increase of the bias $V_g$, which affected the electron accumulation layer formation and energy available for electrons to overcome the energy barrier between the channel a-Si:H layer and the control $SiN_x$ layer [3]. The charge efficiency is also related to the amount of available sites for charge trapping and retention in the gate dielectric structure [3]. Figure 3 shows that amount of charges stored in the TFT decreases with the increase of the $V_d$ value. When $V_d = 0$ V, electrons in the accumulation layer have to overcome the barrier height between the control $SiN_x$ and channel a-Si:H layer before being injected to the gate dielectric structure. However, when a positive $V_d$ is applied, electrons in the accumulation layer are constantly transferred to the drain electrode, which requires a lower energy than that of the energy barrier in the former case. Since in the linear region, the magnitude of $I_d$ increases with the magnitude of $V_d$, the amount of electrons trapped to the gate dielectric layer decreased with the increase of $V_d$. The amount of charges that can be stored in the TFT may reach a saturation value if the pinch-off phenomenon occurs [6]. The pinch-off point is dependent on the magnitude of $V_d$.

Fig. 3 also shows that the long channel TFT has a smaller amount of stored charges than the shorter channel TFT. This may be related to the structure of the TFT. The source contact-to-gate overlap region has a MIS capacitor structure, which contributes a large portion of electrons in the accumulation layer. The long channel TFT has a higher channel resistance than the short channel TFT has. Therefore, the former has less supply of electrons for injection to the gate dielectric layer.

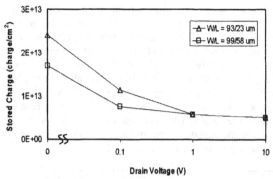

Figure 3. Stored charges of a 9-nm a-Si:H embedded TFT, biased at $V_g = 35$ V for 10 s, as a function of $V_d$.

For discharging by opposite gate polarity, the TFT was first stressed with at positive $V_g$ for a period of time and then stressed at a $-V_g$ at various $V_d$'s. Figure 4 shows the percent stored discharges, i.e., charged for 10 s at $V_g = 35$ V and $V_d = 0$, 0.1, 1, and 10 V, separately, being discharged for 10 s at $V_g = 35$ V and corresponding $V_d$'s. The high $-V_g$ value induced the formation of an inversion layer, which neutralized the stored electrons through injection of holes or expel of electrons from the gate dielectric layer. Fig. 4 shows that more than 80% of the stored charges can be removed by this method. In addition, the discharge efficiency decreases with the increase of $V_d$. The increase of $V_d$ corresponds to the increase of removal of holes, which is opposite of the electron flow direction, in the inversion layer to the source electrode. The smaller the $V_d$ is, the less effective these charge carriers are removed from the inversion layer and, therefore, the higher the discharge efficiency is. The discharge efficiency may also reach a saturation value. This is also probably due to the pinch-off phenomenon that limits the transport of the charges to the drain electrode.

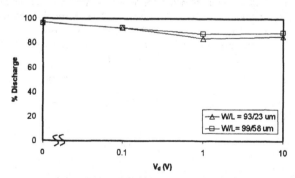

Figure 4. Discharge efficiency as a function of $V_d$, biased at $V_g = -35$ V for 10 s.

Figure 5 shows the discharge efficiency of the light exposure process, i.e., halogen at 585 $\mu$W/cm$^2$ for 5 s. During light exposure, the applied $V_d$ was the same as that applied during charge storage. The $V_d$ bias effect is the same as that in Fig. 4, i.e., the discharge efficiency decrease with the increase of the $V_d$ value. However, based on the same $V_d$, the light discharge is slightly more effective than that the negative gate bias discharge. During light exposure, both a-Si:H and SiN$_x$ films absorb light because the incident light has energy larger than the films' optical gaps, i.e., 1.75-1.85 eV for a-Si:H film and 1.95 eV-4.01 eV for SiN$_x$ film [7,8]. Electron-hole pairs were generated in the a-Si:H film [9,10]. In the embedded a-Si:H film, the newly generated holes combine with the trapped electrons and leave the unbounded electrons in the film and eventually leaked through the tunnel SiN$_x$ toward the channel a-Si:H layer [3].

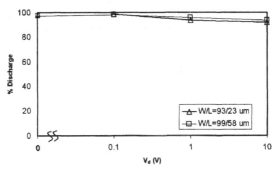

Figure 5. Percent stored discharges (after bias at $V_g$ = 35 V for 10 s) being discharged at different $V_d$'s at $V_g$ = 0 V under light exposure at 585 $\mu$W/cm$^2$ for 5 s.

Figure 6 shows the amount of discharged storage as a function of $V_d$ for 10-min thermal annealing at 210°C. Thermal annealing resulted in the energetic movement of stored charges for self-neutralization. The elevated temperature anneals the dangling bonds at the a-Si:H layers which facilitates the formation of accumulation layer. The increase in $V_d$ then corresponds to a delay of the inversion layer formation during post discharging transfer characteristic measurement. The hysteresis between the initial transfer characteristics and that after thermal annealing is larger. The large hysteresis corresponds to lower discharge efficiency, as observed in Fig. 6.

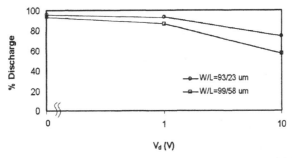

Figure 6. Percent discharge as a function of $V_d$, thermal annealing at 210°C for 10 min.

*Temperature Effects on Charge Capacity and Discharge Efficiency*

Charge and discharge processes of the floating-gate TFT are controlled by different factors. For instance, when the TFT is charged at $V_g = 35$ V, electrons in the accumulation layer have to obtain energy higher than the barrier height between the a-Si:H channel layer and the tunnel $SiN_x$ layer to be injected to the gate dielectric structure [3]. They are then retained by two mechanisms, i.e., the band well of the embedded a-Si:H layer or defects in the gate dielectric layers. These electrons are trapped in the deep or shallow states depending on the defect state. For example, during the injection of the high energy electrons, dangling bonds can be generated from the breakage of Si-Si or Si-H bond [11]. Equation 3 shows an example of dangling bond formation [12].

$$2 \, Si_4^0 + e \rightarrow Si_3^- + Si_3^0 \tag{3}$$

The dangling bond formation process is sensitive to temperature. Therefore, the charge storage capacity of the TFT is related to the temperature. Figure 7 shows the charge storage density decreases with the increase of temperature. This is consistent with the thermal annealing discharge effect. The high temperature favors the passivation of dangling bonds, the release of trapped charges, the migration of charges, and the transportation of charges through the $SiN_x$ dielectric layer. At the low temperature, charges have low mobilities and can be stored at the shallow state.

The discharge process is more complicated than the charge process because it is process dependent. For example, when discharge is carried out by the negative $V_g$ bias, the process involves expelling of trapped electrons or injection of holes from the inversion layer. When discharge is done by light exposure, the process involves electron-hole pair generation and free electron transportation through the tunnel $SiN_x$ layer. When discharge is accomplished by the thermal annealing step, the process involves hydrogen passivation and neutralization of defects and electron migration [13].

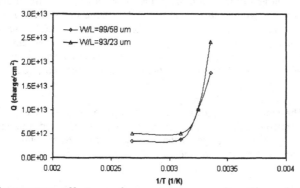

Figure 7. Temperature effect on charge storage capacity. Charged at $V_g$=35 V, $V_d$=$V_s$= 0 V for 10 s.

New defects or dangling bonds can be generated during discharge. For example, the Si-Si or Si-H bonds can be broken by the injected high-energy holes or the absorption of short wavelength light. Hydrogen in the PECV film can be released during a high annealing temperature step or a long annealing period. Therefore, it is impossible or very difficult to delineate the exact mechanism of each discharge process. However, the temperature is an important parameter for the discharge efficiency.

Figure 8 shows the discharge efficiency of thermal annealing as a function of the temperature. The TFTs were first charged at room temperature for 10 s with $V_g$ = 35 V, $V_d$ = 0 V before annealing at 423, 443, 463, and 483 K. The transfer characteristic after annealing then was measured at room temperature. The discharge efficiency increases with the increase of annealing temperature. This is consistent with dangling bonds passivation and charge neutralization at high temperature.

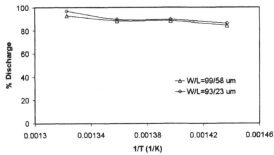

Figure 8. Temperature effect on discharge efficiency by thermal annealing. Charged at $V_g$ = 35 V, $V_d = V_s = 0$ V 10 s. Discharged at each temperature 10 min, $V_g = V_d = V_s = 0$ V.

## Conclusion

Charge and discharge of the floating-gate amorphous silicon thin film transistor has been studied under dynamic operation conditions. The charge storage decreases with the increase of the $V_d$ because it is easier for electrons to be transported to the drain electrode than to overcome the energy barrier between the channel amorphous silicon/tunnel silicon nitride gate dielectric. The discharge efficiency also decreases with the increase of the drain voltage in spite of the different discharge methods, i.e., negative gate bias voltage, light exposure, and high temperature thermal annealing, separately. The channel length affected the discharge efficiency, too, which was dependent on the discharge method. The temperature affects both the charge capacity and the discharge efficiency. The low temperature favors the charge storage but the high temperature favors the discharge. This study revealed key parameters for the optimum operation of the low temperature fabricated nonvolatile memory device.

### Acknowledgment

Authors would like to thank Chen-Han Lin and Jiong Yan for metal sputtering and Guojun Liu for plasma etching of TFT samples.

# References

1. Y. Kuo and H. Nominanda, *Appl. Phys. Lett.* **89**, 1 (2006).
2. Y. Kuo and H. Nominanda, Mat. Res. Soc. Symp. Proc. Vol. 989, Warrendale, PA, A10-03 (2007).
3. H. Nominanda and Y. Kuo, *J. Korean Phys. Soc.* (submitted)
4. Y. Kuo, *J. Electrochem. Soc.*, **139**, 1199-1204 (1992).
5. D. Kahng and S. M. Sze, *Bell Syst. Tech. J.* **46** 1283 (1967).
6. Y. Kuo, *Thin Film Transistors, Materials and Processes, Volume 1: Amorphous Silicon Thin Film Transistors*, New York, Kluwer, pp. 105, 32, 42 (2004).
7. W. L. Warren, J. Kanicki, F.C. Rong and E.H. Poindexter, *J. Electrochem. Soc.* **139**, 880 (1992).
8. R. Vernhes, O. Zabeida, J.E. Klemberg-Sapieha, and L. Martinu, *J. Appl. Phys.* **100**, 063308 (2006).
9. D. L. Staebler and C. R. Wronski, *Appl. Phys. Lett.* **31**, 292 (1977).
10. Y. Nakayama, P. Stradins, and H. Fritzsche, *J. Non-Cryst. Solids* **164**, 1061 (1993).
11. R. B. Wehrspohn, S. C. Deane, I. D. French, I. Gale, J. Hewitt, M. J. Powell, and J. Robertson, *J. Appl. Phys.*, **87**, 144 (2000).
12. R. A. Street, *Phys. Rev. Lett.* **49**, 1187 (1982).
13. W. B. Jackson, C. C. Tsai, and R. Thompson, *Phys. Rev. Lett.* **64**, 56 (1990).

# Thin-Film Transistors I

Mater. Res. Soc. Symp. Proc. Vol. 1066 © 2008 Materials Research Society       1066-A09-02

# Micro Crystalline Silicon TFT by the Metal Capped Diode Laser Thermal Annealing Method

Toshiaki Arai, Narihiro Morosawa, Yoshio Inagaki, Koichi Tatsuki, and Tetsuo Urabe
Corporate R&D, Sony Corporation, 4-14-1 Asahi-cho, Atsugi-shi, 243-0014, Japan

## ABSTRACT

A novel crystallization method for silicon based thin film transistor (TFT) is proposed for the fabrication of high performance large size flat panel displays. In spite of using almost the same TFT fabrication process as that of hydrogenated amorphous silicon (a-Si:H) TFT, the proposed metal capped laser thermal annealing method realizes the formation of uniform and dense micro crystalline silicon ($\mu$c-Si), and provides mobility of 3.1 $cm^2$/V•s, threshold voltage (Vth) of 2.3 V, and sub threshold slope (S) of 0.93 V/decade. Moreover, proposed stacked n+ amorphous silicon structure realizes extremely low off-current maintaining high on-current. As the reliability of TFT, a threshold voltage shift ($\Delta$Vth) under the high current bias stress test (BTS) condition was investigated, and realized the extrapolated $\Delta$Vth of +1.77 V after 100,000 hours stress of 10 $\mu$A and 50°C. This value is 2 orders smaller than that of a-Si:H TFT and only three times larger than that of low temperature poly silicon (LTPS) TFT.

We believe that our $\mu$c-Si TFT technology is the suitable solution for the high quality, large size flat panel display manufacturing.

## INTRODUCTION

Today, an active matrix flat panel displays (AM-FPDs) are becoming increasingly commonplace in the commercial display devices such as TV, OA monitors, notebook PCs or cellular phones. For the backplane as a switching device of the active matrix array, the hydrogenated amorphous silicon thin film transistor (a-Si:H TFT) technology leads the growth of AM-FPD industry, because of its superior electrical switching properties and simple production process.

In recent years, higher electrical property is required for the high resolution, high frame rate liquid crystal displays (LCDs) or the organic light emitting diode (OLED) displays. For the LCDs, higher mobility is required to switch in shorter writing period. For the OLED displays, higher mobility is required to flow high current for the bright photo emission, and higher reliability is required not to change the current flow for the stable display.

Low temperature poly silicon (LTPS) TFT, which is mainly used for the backplane of cellular phones, is the primary candidate replacing a-Si:H TFT. LTPS TFT realizes over 100 times of current flow and current stability against a-Si:H TFT, and is active as the driving circuits on the surrounding area of the backplane. However, LTPS has fatal demerits in the production cost, the limitation of the mother glass size (: ≤ Generation 4), or the short-range non-uniformity of TFT performance due to the crystallization method using excimer laser annealing (ELA). To improve these negative properties, many crystallization techniques had been developed such as a sequential lateral solidification method [1], a metal induced crystallization

(MIC) method [2, 3], or a thermal crystallization method using field enhanced rapid thermal processor [4], etc. However, there were not any technologies, which satisfy all of the requirements in the electrical properties and manufacturing facility.

In this paper, we propose new crystallization technique using diode laser thermal annealing (dLTA) system. According to this technique, extremely uniform hydrogenated micro crystalline silicon ($\mu$c-Si:H) TFT is provided, with a high and uniform electrical properties, high reliability matching with that of LTPS TFT, and high productivity employing large mother glass and simple manufacturing process matching with that of a-Si:H TFT [5, 6].

## EXPERIMENT

Figure 1 shows the cross sectional view of proposed TFT, and it was fabricated by the following process. Molybdenum was deposited on a glass substrate by means of dc magnetron sputtering apparatus, and patterned to form the gate electrode using conventional photo lithography and dry etching method. Silicon nitride, silicon oxide, and amorphous silicon are deposited by means of plasma enhanced chemical vapor deposition (PE-CVD) method to form a gate insulator and a precursor of channel layer. Molybdenum was deposited again to form the photo-thermal conversion layer. And then, the amorphous silicon film was crystallized by dLTA system, which is composed of laser heads and a scanning stage (Figure 2). More than 1W diode laser with 800 nm wavelength is used for the laser head. The exposed laser beam width was controlled to the width of channel region of each pixel, and the stage was scanned at a speed of about 150 mm/sec. After this crystallization process, the capping molybdenum film was etched off. Then, silicon nitride was deposited by means of PE-CVD method and patterned by the wet etching method to form an etching stopper of the TFT. Phosphorous-doped amorphous silicon (n+ a-Si) was deposited by means of PE-CVD method. After the patterning of the silicon island by the dry etching method, titanium-aluminum-titanium tri-layer were deposited by means of dc magnetron sputtering apparatus, and patterned by dry etching method to form the source and drain electrode of the TFT. In this process, n+ a-Si film on the etching stopper was also etched off. At last, silicon nitride was deposited by means of PE-CVD method as a passivation of the TFT.

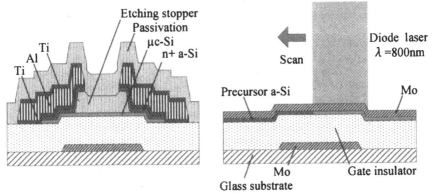

Figure 1. Cross sectional view of $\mu$c-Si TFT. Figure 2. Cross sectional view of laser thermal crystallization.

All the electrical properties were measured by the computer controlled Agilent 4156C precision semiconductor parameter analyzer. The crystallinity of the micro crystalline silicon was confirmed by Raman spectroscopy. A Jobin Yvon's LabRAM HR-800 system was used and the crystal fraction was calculated from the peak areas of decomposed peaks P1, P2, and P3 (P1: a-Si, P2: micro crystalline Si, P3: poly Si).

## DISCUSSION

### Crystallization method and film properties of obtained micro crystalline silicon

As a TFT structure, we employed an etching stopper (ES) type bottom gate TFT. In the LCD industry, a back channel cut (BCC) type TFT became to be widely used because of its shorter manufacturing steps and its narrower channel length than that of ES type TFT. But we chose ES type TFT because of its uniform TFT performance and less off-current. The channel length (L) is determined by the etching stopper width 8~10 micron, and the channel width (W) is determined by the source and drain width 20~100 micron, which are designed according as the panel specification, and channel thickness (d) is determined by deposited precursor thickness. In this design, TFT performance uniformity is mainly dominated by the dispersion of channel length L, promising high performance uniformity in large glass area.

To realize uniform electrical properties, we decided to use micro crystalline silicon as a channel. The poly crystalline silicon, which is formed by the crystallization method such as excimer laser annealing (ELA) method, has superior electrical properties and reliability, but its current uniformity is considerably worse because of the crystal size dispersion due to the pulse energy fluctuation of excimer laser. For the crystallization, we employed the metal capped diode laser thermal annealing (dLTA) method shown in Figure 2. This laser has very stable output (<±1%) and long life (over 10000h), resulting in low maintenance cost in production. As for the a-Si:H film does not have large absorbance at the laser wavelength of 800 nm, molybdenum was deposited as a capping metal of the precursor a-Si:H film. The capping metal works as the photo-thermal conversion layer, and also works as a thermal diffusion layer contributing to the uniform crystallization. After removing the capping metal, the crystallinity of micro crystalline silicon was confirmed by Raman spectroscopy. Figure 3 shows the spectrum and decomposed three peaks located at 481 cm$^{-1}$(P1), 509 cm$^{-1}$(P2), and 518 cm$^{-1}$(P3), corresponding to the amorphous silicon phase, the micro crystalline silicon phase and the poly crystalline silicon phase, respectively. The crystal fraction was calculated using following formula, and was 70%.

$$\text{Crystal fraction} = (P2+P3)/(P1+P2+P3) \qquad (1)$$

Figure 4 shows the Secco etched micro crystalline silicon film. Uniform 10~30 nm diameter crystalline was observed.

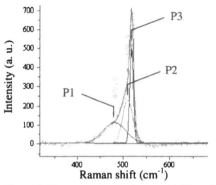

Figure 3. Raman peak of micro crystalline silicon and decomposed three peaks P1, P2 and P3.

Figure 4. SEM view of Secco etched micro crystalline silicon film.

## Electrical properties of micro crystalline silicon TFT

Figure 5 shows I-V characteristics of the typical TFT. Its field effect mobility, threshold voltage (Vth), and sub threshold slope (S) were 3.1 $cm^2$/V•s, 2.3 V, and 0.93 V/decade, respectively. The crystallinity of micro crystalline silicon is mainly affected by the thickness of precursor a-Si:H film or the laser power of dLTA process, and those effects to the on-current are shown in Figure 6 and 7. The thickness dispersion of the precursor a-Si:H film over the area of industrial manufacturing level is ±3%, meaning the ±2% variability in on-current. And the laser power dispersion of the dLTA system (, which includes the laser power dispersion and the scanning speed dispersion) is ±1%, meaning the ±3% variability in on-current. From these facts, we can mention that this dLTA process is enough robust as a manufacturing process.

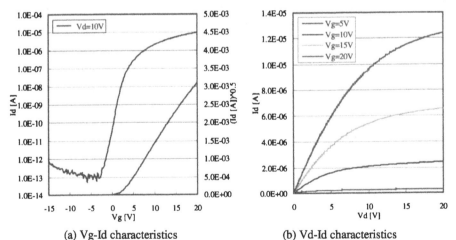

(a) Vg-Id characteristics

(b) Vd-Id characteristics

Figure 5. Electrical properties of micro crystalline silicon TFT.

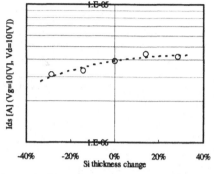

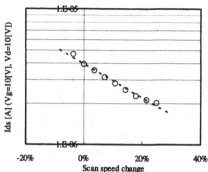

Figure 6. Effect of precursor a-Si thickness to the on-current of μc-Si:H TFT

Figure 7. Effect of laser scanning speed to the on-current of μc-Si:H TFT

This TFT has superior characteristics in the off-current even though using micro crystalline silicon as channel. The laser thermal crystallization using metal capped structure and dLTA system realizes low lattice defect density, and the stopper type TFT structure realizes low process damage to the channel region after the crystallization process, and the thin channel has an advantage in off-current from the structural point of view. Moreover, our proposed stacked n+ a-Si:H structure (Figure 8) contributes to the extra low off-current even in the state of high drain voltage (Figure 9). The off-current was easily reduced by reducing the phosphorous concentration of n+ a-Si:H layer with the reduction of on-current. This reduction of on-current was considered to be caused by the increase of the series resistance between the drain metal and the channel region. Therefore, we stacked high phosphorous concentration layer on the low phosphorous concentration layer, and realized extra low off-current maintaining high on-current.

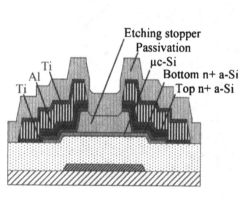

Figure 8. Stacked n+ a-Si:H structure TFT

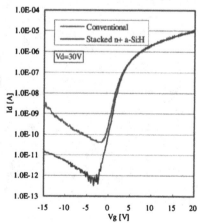

Figure 9. Vg-Id characteristics of conventional and stacked n+ a-Si:H structure TFT

Figure 10 shows the bias temperature stress (BTS) test results of three types of TFTs: amorphous Si, micro crystalline Si, and LTPS TFTs. For the OLED TV application as the most tough case, we assumed 100,000 hours reliability with an on-current of 10E-6 [A] at a temperature of 50 [°C] for each TFT. Assuming the drive scan TFT in the compensation circuit of the pixel, we applied same bias for gate (Vg) and drain (Vd) keeping 10E-6 [A]. A threshold voltage shift (ΔVth) was observed for 100,000 [sec] and extrapolated to 3.6E8 [sec]. The calculated ΔVth of micro crystalline silicon TFT after 3.6E8 [sec] stress was only 1.77 [V]. Its value was two orders

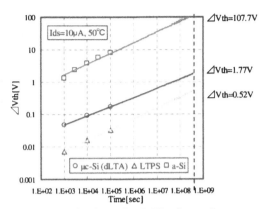

Figure 10. Vth shifts (after BTS) of amorphous silicon, micro crystalline silicon, and low-temperature poly crystalline silicon TFTs.

smaller than that of a-Si:H TFT (107.7 [V]), and only three times of that of LTPS TFT (0.52 [V]). Figure 11 shows the cross sectional TEM view of the micro crystalline silicon. About 10 nm diagonal of grains are observed, and those grains are formed from the interface of the gate insulator. This fact agrees with the superior reliability.

Figure 11. Cross sectional TEM view of μc-Si.

## Low resistivity bus line for the laser thermal annealing process

For the large size high resolution OLED display, both vertical and horizontal bus lines should have low resistivity for the high current flow and the complicated driving signal. In the laser thermal annealing process, the gate metal of the bottom gate TFT works as a heat sink. Therefore, a thermal capacity of gate metal is better to be small. Moreover, the conventional aluminum alloy gate metal which is widely used in liquid crystal display can not be used in laser thermal annealing process, because aluminum alloy makes hillocks, whiskers, or voids as a result of thermal migration in the laser thermal annealing process (Figure 12). We employed 90nm-thick molybdenum (Mo) as a gate metal and the partial Mo/aluminum-neodymium alloy (AlNd) clad structure for the bus line (Figure 13). Although this process requires additional photo-lithography process against the conventional a-Si:H TFT process, it realizes stable and high crystallinity of Si, and low resistivity bus line for large size, high resolution OLED display.

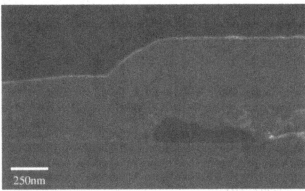

Figure 12. Cross sectional SEM view of laser thermal annealed Mo/AlNd line. A void was created by the thermal stress migration.

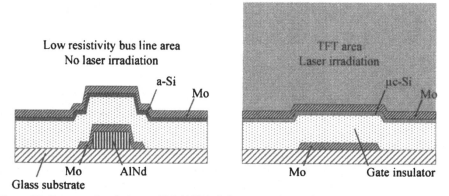

Figure 13. Cross sectional views of Mo/AlNd clad structure in non-irradiation area and Mo gate structure in irradiation area.

## CONCLUSIONS

We developed a novel crystallization technology, which realizes uniform and superior electrical properties: field effect mobility of 3.1 $cm^2/V \cdot s$, threshold voltage (Vth) of 2.3V, sub threshold slope (S) of 0.93 V/decade, and 8 orders of magnitude lower off-current than the on-current. From the BTS test, the calculated $\Delta$Vth of micro crystalline silicon TFT after 3.6E8 [sec] of 10E-6 [A] stress was only 1.77 [V]. This value is 2 orders of magnitude smaller than that of a-Si TFT and only three times of that of LTPS. Except for the novel dLTA system, all the process can be executed by the conventional manufacturing apparatus, which are commonly used in the manufacturing of a-Si TFT array. This technology is promised to contribute to the large size, high resolution, high frame rate LCDs or OLED displays.

## ACKNOWLEDGMENTS

We would like to thank Tatsuya Sasaoka for supporting our development and giving effective suggestions. We also thank all the project members in SONY and affiliated company for their devoted supports and valuable discussions.

## REFERENCES

1. Robert S. Sposili and James S. Im, Appl. Phys. Lett. 69, p.2864 (1996).
2. L. Hultman, A. Robertsson, H.T.G. Hentzell, I. Engstrom, and P.A Psaras, J. Appl. Phys., 62, p.3647 (1987).
3. H. K. Chung and K. Y. Lee, SID '05 Digest, p.956 (2005).
4. S.-J. Lee, D.-H. Shin, B.-S. Bae, and H.-J. Kim, AMLCD 2004, p.41 (2004).
5. T. Arai, N. Morosawa, Y. Hiromasu, K. Hidaka, T. Nakayama, A. Makita, M. Toyota, N. Hayashi, Y. Yoshimura, A. Sato, K. Namekawa, Y. Inagaki, N. Umezu and K. Tatsuki, SID '07 Digest, p.1370 (2007).
6. N. Morosawa, T. Nakayama, T. Arai, Y. Inagaki, K. Tatsuki, and T. Urabe, IDW '07, p.71 (2007).

# Alloys: Microcrystalline Silicon

Mater. Res. Soc. Symp. Proc. Vol. 1066 © 2008 Materials Research Society

# Analysis of Compositionally and Structurally Graded Si:H and Si$_{1-x}$Ge$_x$:H Thin Films by Real Time Spectroscopic Ellipsometry

Nikolas J. Podraza[1], Jing Li[1], Christopher R. Wronski[1], Mark W. Horn[1], Elizabeth C. Dickey[1], and Robert W. Collins[2]

[1]Materials Research Institute, The Pennsylvania State University, University Park, PA, 16802
[2]Department of Physics and Astronomy, University of Toledo, Toledo, OH, 43606

## ABSTRACT

Hydrogenated silicon (Si:H) and silicon-germanium alloy (Si$_{1-x}$Ge$_x$:H) thin films have been prepared by plasma enhanced chemical vapor deposition of SiH$_4$ and GeH$_4$ and measured during growth using real time spectroscopic ellipsometry (RTSE). A two-layer virtual interface analysis has been applied to study the structural evolution of Si:H films prepared in multistep processes utilizing alternating layers of high and low H$_2$-dilution materials, which have been designed to produce predominantly amorphous silicon (a-Si:H) films with a controlled distribution of microcrystallites. The compositional evolution of alloy-graded a-Si$_{1-x}$Ge$_x$:H has been studied as well using similar methods. In each study, the depth profile of the microcrystalline silicon (μc-Si:H) content, $f_{\mu c}$, or the Ge content, $x$, has been extracted. Additionally, RTSE has been used to monitor post-deposition exposure of a-Si:H, a-Si$_{1-x}$Ge$_x$:H, and a-Ge:H films to hydrogen plasmas in situ in order to study the sub-surface modification and etching that would be anticipated when a highly H$_2$-diluted layer is deposited on a layer prepared with lower dilution. These analyses provide guidance for enhancing the performance of Si:H based solar cells through controlled fractions of microcrystallites in bulk amorphous i-layer materials using modulated H$_2$ dilution, through controlled bandgap profiling using compositionally graded a-Si$_{1-x}$Ge$_x$:H, and through a better understanding of the modification of underlying layers during the deposition of subsequent layers in multilayer stacks.

## INTRODUCTION

A two-layer (roughness/outer-layer) virtual interface analysis (VIA) [1,2] has been applied to real time spectroscopic ellipsometry (RTSE) data collected during two plasma-enhanced chemical vapor deposition (PECVD) processes for Si-based thin film photovoltaics. From VIA, the evolution of (i) structure or composition, (ii) deposition rate, and (iii) surface roughness layer thickness can be extracted. In the first process, the structural evolution of hydrogenated silicon (Si:H) films prepared in a multistep sequence utilizing layers of alternating high and low H$_2$-dilution materials have been determined. In this case, a depth profile in the Si microcrystallite volume fraction $f_{\mu c}(d_b)$ with bulk layer thickness $d_b$ can be deduced [2]. By studying variations in the crystallite volume fraction evolution as a function of deposition parameters, it is possible to apply this method quite generally to tailor the microstructure and optical properties of such films using single and multistep deposition processes. In fact, purely amorphous and microcrystalline materials can be optimized, in addition to a wide variety of mixed-phase amorphous + microcrystalline materials. In the second process, graded a-Si$_{1-x}$Ge$_x$:H has been prepared by intentionally varying the flow ratio $G =$

[GeH$_4$]/{[SiH$_4$]+[GeH$_4$]} versus time, and in this case, a depth profile in Ge content $x(d_b)$ can be deduced [1]. As both of these approaches involve multilayer structures for which the gas flows are varied during the deposition, sub-surface modification and etching of underlying layers in the initial stages of high H$_2$-dilution deposition processes are important issues. For this purpose, RTSE has been used to study the effects of H$_2$-plasma post-deposition treatments of $a$-Si:H, $a$-Si$_{1-x}$Ge$_x$:H, and $a$-Ge:H films in the absence of complications due to overlying film growth.

## EXPERIMENT

The Si:H, Si$_{1-x}$Ge$_x$:H, and Ge:H films in this study were deposited onto native-oxide/$c$-Si substrates using single-chamber rf (13.56 MHz) PECVD and were measured in real time using a rotating compensator multichannel ellipsometer [3]. The fixed parameters were selected for the most part at values used in previous studies of pure Si:H PECVD [4], including the minimum rf power for a stable plasma ($P \sim 0.08$ W/cm$^2$), a low partial pressure of the source gases SiH$_4$+GeH$_4$ ($p_{par} \sim 0.06$ Torr), a low total pressure ($p_{tot} < 1.0$ Torr), and a substrate temperature $T_s$ = 200°C. The variable parameters in this study include the electrode configuration, either conventional anode (which is grounded) or cathode (which stabilizes at a dc self-bias of $\sim -25$ V), the H$_2$-dilution ratio $R$ = [H$_2$]/{[SiH$_4$]+[GeH$_4$]}, and the alloying flow ratio $G$ = [GeH$_4$]/{[SiH$_4$]+[GeH$_4$]}, which was varied from $G$ = 0 to 0.167, leading to $x$ values up to 0.4 as deduced by x-ray photoelectron spectroscopy (XPS) [5] (or fixed at $G$ = 0 or 1 to produce $a$-Si:H or $a$-Ge:H). For RTSE studies of the etching and near-surface modification of $a$-Si:H, $a$-Si$_{1-x}$Ge$_x$:H, and $a$-Ge:H films, the H$_2$-plasma conditions are identical to the film growth conditions, with the exception that only H$_2$ is used with a flow of 120 sccm.

The analysis of these thin films with modulated stucture and graded composition can be accomplished through the use of a four medium optical model consisting of (i) the ambient; (ii) a surface roughness layer; (iii) an outer-layer which contains the most recently deposited material; and a virtual interface to (iv) the pseudo-substrate which contains the past history of the deposition. The roughness layer is modeled using the Bruggeman effective medium theory as a 0.5/0.5 volume fraction mixture of the outer-layer material and void. The overall analysis approach is based on least-squares regression, with the free parameters being the outer-layer composition, $f_{\mu c}$ or $x$, the outer-layer growth rate $r$, and the surface roughness layer thickness $d_s$. A depth profile can be determined by integrating the instantaneous growth rate $r(t)$ determined from VIA with respect to time. For Si:H films with mixed amorphous and microcrystalline phases, microstructural changes are observed through the crystallite volume fraction $f_{\mu c}$ in the outer-layer of the film [2]. For $a$-Si$_{1-x}$Ge$_x$:H alloy films prepared with varying GeH$_4$ flow, compositional changes are observed during deposition through the Ge content $x$ in the outer-layer of the film, similar to earlier studies of compositionally graded $a$-Si$_{1-x}$C$_x$:H [1].

## RESULTS AND DISCUSSION

### Structurally modulated ($a+\mu c$)-Si:H

Films prepared with high H$_2$-dilution may initially nucleate and grow in the amorphous phase if the substrate is an $a$-Si:H film, but subsequently microcrystallites nucleate within the growing amorphous matrix and eventually coalesce into a single-phase microcrystalline material. This is the case when an $R$ = 40 Si:H film is prepared at the grounded anode, using as a substrate

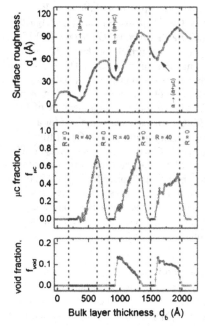

**Figures 1 and 2:** (left, top) Surface roughness, (left, center) microcrystalline fraction, and (left, bottom) void fraction evolution as determined by RTSE through a virtual interface analysis (VIA) technique; the results are depicted as functions of the bulk layer thickness for a multilayer film structure prepared by alternating high ($R = 40$) and low ($R = 0$) $H_2$-dilution layers; (above) dark-field cross sectional transmission electron microscopy image (TEM) for a multilayer Si:H film.

a native-oxide/$c$-Si wafer overdeposited with a $R = 0$ $a$-Si:H. The latter film is designed to erase the memory of the crystalline substrate [4]. Consequently, these films exhibit fully amorphous and fully microcrystalline regions within the film with distinctly different dielectric functions, as well as a structurally graded mixed-phase (amorphous+microcrystalline) region between the two where the microcrystallite fraction increases with distance from the substrate.

A multilayer structurally-modulated Si:H film was deposited and monitored by RTSE with the intention of fabricating a controlled mixed-phase film with a uniform distribution of microcrystallites by using three alternating low $R = 0$ and high $R = 40$ layers deposited onto a native oxide-covered $c$-Si substrate. Figure 1 shows the evolution of the surface roughness thickness, the microcrystalline fraction, and the void volume fraction as functions of bulk layer thickness, whereby the ($\varepsilon_1$, $\varepsilon_2$) spectra used in the analysis procedure have been determined from individual $R = 0$ and $R = 40$ depositions.

Figure 2 shows a dark-field cross sectional transmission electron microscopy (TEM) image of this multilayer structure. As shown in the image, the amorphous and mixed-phase layer thicknesses determined by VIA are consistent with those determined from the TEM. The VIA and TEM results indicate that it is possible to produce alternating amorphous and mixed-phase ($a$+$\mu c$) layers, implying that the low dilution $R = 0$ material is sufficient to suppress further microcrystallite growth on top of the $R = 40$ mixed-phase material.

### Compositionally graded $a$-Si$_{1-x}$Ge$_x$:H

In the second study to be reported here, a series of $a$-Si$_{1-x}$Ge$_x$:H films was prepared on $T_s$ = 200°C $c$-Si mounted on the rf powered cathode ($V_{dc} \sim -25$ V) using a fixed $H_2$-dilution ratio of

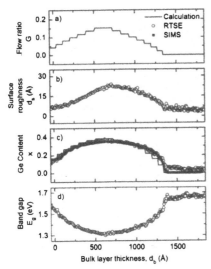

**Figure 3:** (a) Step-wise variations in flow ratio $G = [GeH_4]/\{[SiH_4]+[GeH_4]\}$, designed for a multilayer structure ($G = 0.042 \rightarrow 0.15 \rightarrow 0$) and (b) the surface roughness thickness, both as functions of accumulated thickness; also shown are depth profiles in (c) the Ge content in the outerlayer ($\sim$ 7 Å), and (d) band gap $E_g$ at the measurement temperature of $T = T_s = 200°C$. All results are obtained for a compositionally-graded $a$-Si$_{1-x}$Ge$_x$:H film prepared on a $G = 0.042$, $R = 10$ $a$-Si$_{1-x}$Ge$_x$:H substrate film. The open circles represent the experimental data obtained from VIA. In (c), the black solid line represents the predicted value based on individual depositions, and the solid squares represent the Ge content x as determined by SIMS.

$R = 10$. The GeH$_4$ ratio was varied from $G = 0$ to 0.167 for individual depositions in order to produce a database of $\varepsilon$ spectra for $a$-Si$_{1-x}$Ge$_x$:H in which the Ge content $x$ has been determined within the range $0.0 \leq x \leq 0.4$ by XPS [5]. These $\varepsilon$ spectra are then fit to an oscillator model based on the assumption of a constant dipole matrix element [6] in order to provide a database of dielectric functions dependent only on $x$.

After this database development, an $a$-Si$_{1-x}$Ge$_x$:H film was then prepared under the same conditions as the database series, but with varying ratio $G$ in order to grade the Ge content $x$ and the optical band gap $E_g$. Before grading was initiated, an opaque film with $G$=0.042 was deposited on the $c$-Si wafer as a substrate layer. The graded film was studied by RTSE and analyzed using the VIA technique with an average 7 Å outer-layer thickness in order to extract the time evolution of the Ge content in the outer-layer $x(t)$, the surface roughness thickness $d_s(t)$, and the growth rate $r(t)$. Figure 3 depicts the bulk layer thickness evolution of (a) the flow ratio $G(d_b) = [GeH_4]/\{[SiH_4]+[GeH_4]\}$, designed in a stepwise fashion for a multilayer structure ($G = 0.042 \rightarrow 0.15 \rightarrow 0$), and (b) the surface roughness thickness, as well as depth profiles in (c) the Ge content and (d) band gap $E_g$ at the measurement temperature $T_s = 200°C$. Here $d_b = 0$ Å corresponds to the instant when $G$ is increased above the substrate film value $G = 0.042$. Secondary ion mass spectroscopy (SIMS) depth profiling has been performed on this sample to obtain the Ge content $x$ as a function of bulk layer thickness, also shown in Figure 3(c) (solid squares), and the results are very close to the RTSE profile.

## Sub-surface modification of $a$-Si:H, $a$-Si$_{1-x}$Ge$_x$:H, and $a$-Ge:H by H$_2$-plasma

A series of $a$-Si:H, $G = 0.167$ $a$-Si$_{1-x}$Ge$_x$:H, and $a$-Ge:H films prepared without H$_2$-dilution ($R = 0$) have been studied by RTSE during post-deposition exposure to a H$_2$-plasma. Figure 4 depicts the variation in the film thickness, here represented by the effective thickness $d_{eff} = d_b + 0.5$ $d_s$, for an $a$-Si:H film exposed to the H$_2$-plasma. In this case, the film etches completely away at a rate of $-0.19$ Å/s. This type of experiment is of interest to simulate the

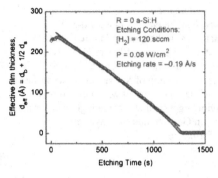
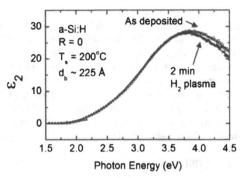

**Figures 4 and 5:** (left) Effective layer thickness versus time during the H$_2$-plasma etching of an $R = 0$ $a$-Si:H thin film with an initial thickness of ~230 Å. Etching conditions include a H$_2$-flow [H$_2$] = 120 sccm, plasma power $P$ = 0.08 W/cm$^2$, total pressure $p_{tot}$ < 1 Torr, and a substrate temperature $T_s$ = 200°C; (right) ($\varepsilon_1$, $\varepsilon_2$) obtained by numerical inversion and fit to an oscillator model -- for a $R = 0$ $a$-Si:H film at the end of deposition and then after 2 minutes of exposure to a H$_2$-plasma.

conditions under which a high $R$ Si:H layer is deposited onto a low $R$ Si:H layer, such as in the multilayer structure described previously (in which case $R = 40$ Si:H is deposited onto $R = 0$ $a$-Si:H). The low $R$ layer may be briefly exposed to a H$_2$-rich plasma, but the depositing film will soon cover the underlying material and prevent prolonged exposure to the H$_2$-plasma. Additionally, underlying material may be intentionally exposed post-deposition to a H$_2$-plasma in order to create a surface that promotes improved order or microcrystallinity in the subsequent layer [7]. In order to quantify changes in the optoelectronic quality of the underlying material, ($\varepsilon_1$, $\varepsilon_2$) were extracted for this material after a two-minute exposure to the H$_2$-plasma and compared to those obtained at the end of $a$-Si:H deposition. The imaginary parts $\varepsilon_2$ are highlighted in Figure 5. Both sets of dielectric functions were fit to an oscillator model for parameterization. The Lorentz broadening parameter, $\Gamma$, in this parameterization has been observed to decrease from $\Gamma$ = 2.41 eV after film deposition to $\Gamma$ = 2.27 eV after two minutes of exposure to the H$_2$-plasma. The decrease in $\Gamma$ has been correlated with an improvement in order in the amorphous film [1,5] and demonstrates that a short plasma exposure improves the electronic quality of a low $R$ underlying material in a multistep process.

$R = 0$ $a$-Ge:H and $G = 0.167$ $a$-Si$_{1-x}$Ge$_x$:H thin films were fabricated at the anode under standard deposition conditions and exposed to the

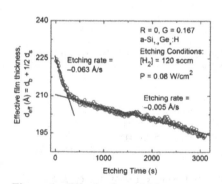

**Figure 6:** Effective layer thickness versus time during H$_2$-plasma etching of $R = 0$, $G = 0.167$ $a$-Si$_{1-x}$Ge$_x$:H thin film with an initial thickness of ~225 Å. Etching conditions include a H$_2$-flow [H$_2$] = 120 sccm, plasma power $P$ = 0.08 W/cm$^2$, total pressure $p_{tot}$ < 1 Torr, and substrate temperature $T_s$ = 200°C.

same $H_2$-plasma conditions as $R = 0$ $a$-Si:H. It is observed that $R = 0$ $a$-Ge:H etches at a rate of –0.002 Å /s, which is ~ two orders of magnitude slower than the etching rate for $a$-Si:H. The consequence of the different etching rates for these materials is observed in Figure 6, as the $G = 0.167$ $a$-Si$_{1-x}$Ge$_x$:H material initially etches at a faster rate in the first ~ 400 s of $H_2$-plasma exposure and at a slower rate subsequently. The change in etching rate during the course of the exposure to the plasma is most likely due to initial selective etching of Si over Ge at the surface of the film. As the Si at the surface is depleted, a Ge rich layer is formed which then inhibits further etching of the underlying material. This behavior suggests that when a high $R$ material is deposited onto low $R$ $a$-Si$_{1-x}$Ge$_x$:H in a device, a very thin Ge-rich interface layer may form which could be detrimental to the resulting device performance.

## CONCLUSIONS

Real time spectroscopic ellipsometry has been applied to study the growth of structurally modulated Si:H and compositionally graded $a$-Si$_{1-x}$Ge$_x$:H films. Virtual interface analysis (VIA) was used to track the microcrystallite volume fraction as a function of film thickness for a multilayer $(a+\mu c)$-Si:H film, and transmission electron microscopy was used to confirm the presence of a multilayer mixed-phase structure. Study of an $a$-Si$_{1-x}$Ge$_x$:H film prepared with graded Ge content $x$, has indicated the ability of RTSE to determine the depth profile in Ge content, which has been corroborated by secondary ion mass spectroscopy. Studies of post-deposition $H_2$-plasma exposure of $a$-Si:H has indicated that limited plasma exposure, similar to that occurring during multilayer deposition procedures, improves the order of underlying materials. In further studies of $a$-Ge:H and $a$-Si$_{1-x}$Ge$_x$:H, it has been observed that $a$-Ge:H etches at a much slower rate than $a$-Si:H. This difference in etching rate can result in selective removal of Si over Ge in $a$-Si$_{1-x}$Ge$_x$:H materials when exposed to a $H_2$-plasma, resulting in the formation of a Ge-rich surface layer.

## ACKNOWLEDGMENTS

This research was supported by the NREL Thin Film Photovoltaics Partnership Program Subcontract Nos. NDJ-1-30630-01 and ZXL-5-44205-06.

## REFERENCES
1. H. Fujiwara, J. Koh, and R. W. Collins, *Thin Solid Films* **313-314**, 474 (1998).
2. A. S. Ferlauto, G. M. Ferreira, R. J. Koval, J. M. Pearce, C. R. Wronski, R. W. Collins, M. M. Al-Jassim, and K. M. Jones, *Thin Solid Films* **455-456**, 665 (2004).
3. I. An, J. Zapien, C. Chen, A. Ferlauto, and R. W. Collins, *Thin Solid Films* **455-456**, 132 (2004).
4. R. W. Collins, A. S. Ferlauto, G. M. Ferreira, C. Chen, J. Koh, R. Koval, Y. Lee, J. M. Pearce, and C. R. Wronski, *Solar Energy Mater. Solar Cells* **78**, 143 (2003).
5. N. J. Podraza, C. R. Wronski, M. W. Horn, and R. W. Collins, *Mater. Res. Soc. Symp. Proc.* **910**, A.10.1.1 (2006).
6. A. S. Ferlauto, G. M. Ferreira, J. M. Pearce, C. R. Wronski, R. W. Collins, X. Deng, and G. Ganguly, *J. Appl. Phys.* **92**, 2424 (2002).
7. J. M. Pearce, N. Podraza, R. W. Collins, M. M. Al-Jassim, K. M. Jones, J. Deng, and C. R. Wronski, *J. Appl. Phys.* **101**, 114301 (2007).

Mater. Res. Soc. Symp. Proc. Vol. 1066 © 2008 Materials Research Society     1066-A10-03

# Two-Step Capacitance Transients From an Oxygen Impurity Defect

Shouvik Datta[1], J. David Cohen[1], Yueqin Xu[2], and Howard M. Branz[2]

[1]Department of Physics, University of Oregon, 1371 E 13th Avenue, Eugene, OR, 97403
[2]Silicon Materials Group, National Renewable Energy Laboratory, 1617 Cole Boulevard, Golden, CO, 80401

## ABSTRACT

This paper describes the study of an electron-trapping defect which underwent significant configurational relaxation in oxygen contaminated hydrogenated amorphous silicon-germanium (a-Si,Ge:H) alloys grown by hot-wire chemical vapor deposition. An unusual two-step electron emission from this relaxed defect is studied using junction-capacitance-based measurements. In this work, we monitor the recovery of the relaxed defect after filling it by photoexcited electrons and also by electrons injected with a voltage filling pulse. The dependence of the transient shape on filling pulse time is described. We have also performed experiments which clearly demonstrate that this is a bulk defect and exclude contributions from any additional blocking junctions.

## INTRODUCTION

We earlier reported [1,2] the discovery of an oxygen related defect with a large configurational relaxation energy [3,4] (roughly 0.8 eV) in hydrogenated amorphous silicon-germanium (a-Si,Ge:H) alloys grown by hot-wire chemical vapor deposition (HWCVD) with roughly 30at.% Ge and high levels (~5×10²⁰ cm⁻³) of intentional oxygen contamination. The existence of this defect was revealed by a negative feature in the transient photocapacitance (TPC) spectra [5-6] near 1.35 eV, which indicated that valence band electrons photoexcited into these oxygen defects remained deeply trapped even when the optical excitation threshold is very close to the conduction band. The detailed behavior of those defects in these a-Si,Ge:H alloys were further characterized [2] using a variety of techniques including transient photocurrent (TPI) spectroscopy, as well as by monitoring the capacitance recovery in the dark following the excitation with 1.2 eV light. These latter studies [2] of the time evolution of the capacitance transients due to release of electrons from these oxygen defects into the conduction band as a function of temperature clearly demonstrated that significant configurational relaxation must have occurred. This may be the *best documented & most clear cut* observation to date of such a substantial relaxation of defect energy following the change of charged state of a defect level in amorphous silicon or related materials.

Recently, we have been studying in greater detail the kinetics of this oxygen-related defect relaxation in samples with Ge fractions of both 30% and 15% with different levels of oxygen. For example, we have examined dark capacitance recovery due to electron emission transients obtained after selective light pulses or voltage filling pulses. Indeed, under certain conditions we can observe a distinct 'two-step' capacitance recovery transient in the dark. This suggests an additional configurational rearrangement of the amorphous network associated with the electron emission from these oxygen defects. Plausible origins of the observed kinetics of these electronic transitions from the oxygen related defect will be discussed.

## SAMPLES

A series of a-Si,Ge:H alloys with nominal Ge fractions of 15% and 30%, as determined by secondary ion mass spectrometry (SIMS), were deposited by the HWCVD technique on both specular stainless steel and $p^+$ crystalline Si substrates at the National Renewable Energy Laboratory (NREL). These HWCVD films were grown by decomposing silane and germane gases with a tantalum filament at 1800 °C during the growth. The intentional oxygen contamination was systematically introduced using a controlled air leak into the growth chamber during the hot-wire CVD process. Details of these samples were published elsewhere [1,2]. Our SIMS measurements indicated that the oxygen impurity levels for a series of four 30at.% Ge alloy samples progressively varies from $8\times10^{18}$ cm$^{-3}$ to $5\times10^{20}$ cm$^{-3}$. However, the nitrogen contents were found to be nearly identical and very low ($< 4\times10^{16}$ cm$^{-3}$) in all these samples. Fourier transform infrared spectroscopy (FTIR) signatures of Si-H (2000 cm$^{-1}$) and Ge-H (1876 cm$^{-1}$) stretching vibration modes were also unaffected by the sequential increase in oxygen concentration [7]. We thereby rule out significant changes in the degree of hydrogen incorporation in these series of samples. More importantly, the 980 cm$^{-1}$ FTIR line characteristic of Si-O-Si stretching vibration mode increased with oxygen and was very prominent in the sample with $\sim5\times10^{20}$ cm$^{-3}$ oxygen. Semi-transparent palladium dots ($\sim0.0088$ cm$^2$) were thermally evaporated on top of the intrinsic a-Si,Ge:H layer as front contact and In-Ga eutectic liquid was used as back ohmic contact for our junction capacitance based measurements.

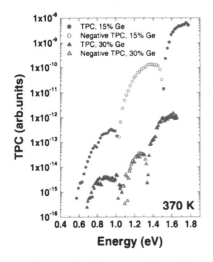

We utilized the dominant c-Si($p^+$)/a-Si,Ge:H depletion junction under high reverse bias ( typically -2.5 V to -4 V) for all the transient measurements unless otherwise specified. At the end of the following section, we will also discuss observations which exclude contributions from any additional blocking junctions.

## RESULTS AND DISCUSSION

We recently reported [1,2] the observation of negative photocapacitance spectra associated with the optical transition of an electron from valence band to a oxygen impurity defect present in these a-Si,Ge:H alloy films. Figure 1 shows the TPC spectra (i.e., the transient photocapacitance signal normalized to light flux *vs.* photoexcitation energy) of HWCVD grown a-Si,Ge:H alloy films with 30% and 15% Ge content respectively. Such optical transitions

Figure 1. Transient Photocapacitance Spectra of HWCVD a-Si$_{0.7}$Ge$_{0.3}$:H & a-Si$_{0.85}$Ge$_{0.15}$:H alloys grown on p$^+$ c-Si substrates. The absolute magnitude of the negative TPC signal (hollow symbols) represents the photo-induced capture of electrons by the oxygen impurity defect. The presence of the usual deep defect transition associate with Si and/or Ge dangling bonds is also evidenced by the positive signal 0.8 to 0.9 eV below the conduction band.

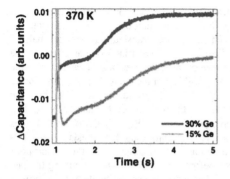

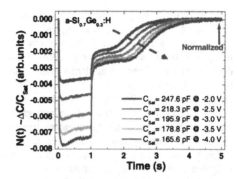

Figure 2. Dark capacitance recovery after light exposure for 1 sec showing a two stage capacitance recovery in both samples. The curves have been vertically shifted for clarity. The very fast initial rise in the dark for the 15at.% Ge sample is likely due to fast emission of some trapped bandtail holes.

Figure 3. Time dependence of the trapped charge N(t) during a 1s exposure to 1.2 eV light and the subsequent recovery under dark. Reverse bias is increased from -2 V to -4 V as indicated by the dashed arrow. These transients are plotted as $[C(t)-C_{sat}]/C_{sat}$, where $C_{sat}$ is the value of capacitance at 5 seconds as indicated in the legend.

contribute *negatively* to the TPC spectra (as shown in Fig 1) since the trapped electrons reduce the depletion capacitance of such marginally n-type samples having a positive depletion charge. We have *hypothesized* that this oxygen impurity defect state is associated with a positively charged oxygen level, most likely the three fold coordinated oxygen ($O_3^+$) center proposed by Shimuzu et al. in case of a-Si:H [8-10]. In our a-Si,Ge:H alloys [2], the $O_3^+$ state traps the optically excited valence band electron ($O_3^+ \Rightarrow O_3^0 + h$) such that the residual valence band hole leaves the depletion region within the time window of our TPC measurement. The observation that the electron trapped in the defect is not quickly released thermally when it lies so close to the conduction band clearly indicates that the $O_3^+$ defect must undergo significant configurational relaxation after the electron capture.

However, more information has been obtained concerning this photo-induced electron trapping; namely, we also observe an unusual dark capacitance recovery transient due to the emission of these trapped electrons from the configurationally relaxed state of the defect. Figure 2 shows the dark recovery transients of the junction capacitance for both a-Si,Ge:H alloys with 30% and 15% Ge contents as measured at 370 K. Up to the 1 s mark, these samples were exposed to either 1.2 eV light (for the a-Si$_{0.7}$Ge$_{0.3}$:H sample) or 1.3 eV (for the a-Si$_{0.85}$Ge$_{0.15}$:H sample). These photon energies were chosen to selectively excite electrons from the valence band into the oxygen defect in the negative photocapacitance signal regime while also not significantly causing excitation into bandtail states (see Figure 1). Afterwards, we clearly see two different stages during the emission of trapped electrons as evidenced by the two-step dark capacitance recovery in both a-Si,Ge:H alloys with 30at.% and 15at.% Ge. Between the first and the final change of the dark capacitance there is a time period during which the capacitance does not change very much. We shall call this stage as the "latency state". Before discussing the possible origin of these unusual dark transients we first describe experiments which prove that these kind of capacitance transients must truly be ascribed to bulk defects.

In Figure 3, we plot the 1.1 kHz capacitance transients of the 30% Ge sample at 370 K for a series of different reverse bias applied to back c-Si/a-Si,Ge:H junction during and after the application of a 1.2 eV light pulse of about 1 s. Each capacitance transient has been normalized to its own final value at t=5 s to allow a more direct visual comparison of the effect of the different values of reverse applied voltage. Initially, we observe the decrease in capacitance due to the photo-induced electron capture into the defects (resulting in the negative photocapacitance signal, as in Fig. 1). Following exposure to this 1.2 eV light for 1 second, we then clearly see the presence of the two-step capacitance recovery transients in the dark. At each value of the applied voltages, the decrease of the negative trapped charge density ( $N(t) \propto \Delta C(t)/C \ll 1$ ) goes through the 'latency state' before the final recovery. An intriguing hypothesis is that this 'latency state' indicates a substantial configurational rearrangement of the amorphous neighborhood of the oxygen impurity state required before it can release the remaining trapped electrons to the conduction band. The duration of the latency state and the total emission time is seen to decrease as we decrease the reverse bias. The fact that we see an *increase* in emission time for the higher reverse biases that have lower capacitance values indicates that this effect has nothing to do with the "RC time constant" issues of the blocking junction.

Next, we demonstrate that this effect is due to a bulk defect rather than from some spatially localized or interface associated defect which might cause such an apparent 'latency state' by pinning the band bending, and hence the depletion capacitance, for a period of time. In Figure 4 we plot the capacitance change $\Delta C = C(@1.5 \text{ s}) - C(@ 5 \text{ s})$ taken from Fig. 3 (i.e, the difference before and after the second transient) normalized to $C_{Sat} \equiv C(@ 5 \text{ s})$ for all the different biases. The linearity of the plot [11, 12] means that the spatial distribution of the trapped charges $N(t)$ follows the space charge density. This establishes that the oxygen related defect contributing the negative photocapacitance signal is a *bulk defect*.

We have also filled these oxygen-related states with electrons from the conduction band at 370 K by subjecting the a-Si$_{0.7}$Ge$_{0.3}$:H / p+ c-Si junction to zero bias pulses before restoring the -3V reverse bias. Just as for the case with the light pulses, we see ( Figure 5a ) a two step capacitance recovery transient after the -3 V reverse bias is re-established provided the 0V filling pulse durations is greater than 0.25 s. Moreover, we notice that the two step transient gradually evolves from a single stage capacitance recovery to a two stage capacitance recovery as we increase the zero bias voltage pulse duration above 0.25 s. This may imply a *'cooperative'*

Figure 4. The capacitance differences before and after the second transient scale linearly with the ambient depletion capacitance at long times. This indicates that the second transient originates from a bulk defect level.

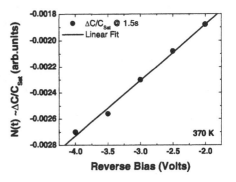

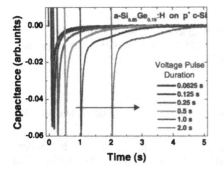 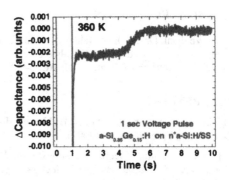

Figure 5. (a) Capacitance transients at -3V taken after 0V filling pulses applied for the indicated time. Capacitance recovery begins after the end of respective filling pulses indicated by the sharp vertical changes. The arrow indicates the direction of the increase of voltage pulse duration. (b) Capacitance transients at -4V reverse bias for a 0V filling pulse applied for 1 s to front Pd/a-SiGe:H junction grown with a $n^+$ a-Si:H back ohmic contact on stainless steel substrate.

nature of the defect relaxation and recovery. It also signals the onset of the two step capacitance transients only when the defect system is driven strongly by the external perturbation which, in this case, is the voltage filling pulse. Moreover, at the large reverse bias around -3 V, the front Pd contact is in substantial forward bias and thus effectively shorted so that the dominant junction is at the a-$Si_{0.7}Ge_{0.3}$:H/ c-Si($p^+$) interface. However, to confirm that the two-step capacitance recovery is in no way due to the presence of any possible second blocking contact in the Pd/a-$Si_{0.7}Ge_{0.3}$:H/c-Si($p^+$) structure we have repeated the same experiment on a similar a-$Si_{0.7}Ge_{0.3}$:H alloy grown with an ohmic back contact of 400 Å $n^+$ type a-Si:H deposited onto a stainless steel substrate. Figure 5b, displays the capacitance recovery transients of the Pd/a-$Si_{0.7}Ge_{0.3}$:H junction with this ohmic back contact. We again clearly see the presence of two stage capacitance recovery after pulsing the Pd /a-$Si_{0.7}Ge_{0.3}$:H junction to zero bias for 1 second.

## CONCLUSIONS

We have established that the unusual two step capacitance recovery transient originates from a bulk defect in HWCVD grown a-Si,Ge;H. This feature is also present in samples with a single blocking junction and proved to be unrelated with any RC time constant effect. Most likely, the 'latency state' between these two step electron emission processes is related to the configurational rearrangement of the amorphous network that enables the remnant electrons to be emitted. The microscopic detail of these unusual electron emission transients from the oxygen related defect is not yet resolved. It may indicate a complex configurational rearrangement of the defect's environment that facilitates the second stage of emission of trapped electrons. However, it might also indicate an electron transport pathway to escape from the depletion region (for example, hopping through neighboring defects) that takes some time to re-establish itself after the first emission process. Further measurements will be carried out to resolve these issues.

## ACKNOWLEDGEMENTS

We thank Dr. A. Harv Mahan of NREL for helpful discussions. Work at the University of Oregon was supported by NREL Subcontract ZXL-5-44205-11. Work at NREL was supported by the U.S. DOE under Contract DE-AC36-99GO10337.

## REFERENCES

[1] Shouvik Datta, Yueqin Xu, A. H. Mahan, Howard M. Branz, and J. David Cohen, Mat. Res. Soc. Symp. Proc **989**, A4.3 (2007).

[2] Shouvik Datta, Yueqin Xu, Howard M. Branz, and J. David Cohen, 22nd International Conference of Amorphous Semiconductors, Colorado, USA, August 2007. J. Non Cryst. Solids. **354**, 2126 (2008).

[3] P. W. Anderson, Phys. Rev. Lett. **34**, 953 (1975).

[4] R. A. Street and N. F. Mott, Phys. Rev. Lett. **35**, 1293 (1975).

[5] A.V. Gelatos, K.K. Mahavadi, J.D. Cohen, and J.P. Harbison, Appl. Phys. Lett. **53**, 403 (1988).

[6] J. David Cohen and Avgerinos V. Gelatos, in *Amorphous Silicon and Related Materials*, Vol. A, ed. by Hellmut Fritzsche (World Scientific, Singapore, 1989), p 475.

[7] G. Lucovsky, J. Yang, S. S. Chao, J. E. Tyler, and W. Czubatyj, Phys. Rev. B. **28**, 3225 (1983).

[8] Tatsuo Shimizu, Minoru Matsumoto, Masahiro Yoshita, Masahiko Iwami, Akiharu Morimoto, and Minoru Kumeda, J. Non-Cryst. Solids. **137&138**, 391 (1991).

[9] H. Fritzsche, J. Non-Cryst. Solids. **190**, 180 (1995).

[10] David Adler, Solar cells. **9**, 133 (1983).

[11] D. V. Lang, J. Applied Physics, **45**, 3014 (1974).

[12] G. L. Miller, D. V. Lang, and L. C. Kimmerling, Ann. Rev. Mater. Sci. **7**, 377 (1977).

Mater. Res. Soc. Symp. Proc. Vol. 1066 © 2008 Materials Research Society          1066-A10-04

# Micro Photovoltaic Modules for Micro Systems

Nicolas Wyrsch, Sylavain Dunand, and Christophe Ballif
Institut de Microtechnique, University of Neuchatel, Breguet 2, Neuchatel, 2000, Switzerland

## ABSTRACT

Amorphous silicon based solar modules are very attractive for the powering of various microsystems for both indoor and outdoor applications. This technology offers a lot of flexibility in terms of module design, output voltage, shape, size, choice of substrates and offers also the possibility to embed sensors such as photodiodes. This paper focuses on the development of micro-solar modules with area $\leq 0.15$ cm$^2$. Several micro-modules with output voltage of up to 180 V (for a total area of 0.1 cm$^2$) were designed and fabricated. The performance limitation introduced by the segment monolithic interconnection and the design of the latter is presented and discussed. An example of a micro-module with a total size of 3.9x3.9 mm$^2$ developed for a micro-robot with dual voltage outputs and embedded photodiodes is also presented.

## INTRODUCTION

Development of various types of Microsystems (e.g. autonomous micro-systems, sensor networks, micro-robots, etc) requires the parallel development of energy scavenging solutions. Given the energy consumption of most devices, the use of photovoltaic module is a valuable option for many applications [1]. In this context, amorphous silicon (a-Si:H) solar cell technology using micro-fabrication processes offers several advantages. Efficiency values higher than the ones of typical c-Si cells can be obtained due to the more optimal value of the band gap of a-Si:H for most cases of indoor illuminations; for example, an efficiency of 23% can be obtained for a-Si:H in the case of fluorescent light illumination as illustrated in Fig. 1. Voltage and current values can be tailored to the application needs with monolithic interconnections of the module segments, and additional sensors can also be integrated on the same substrate. Furthermore, the solar module can also be designed in any shape in order to maximize the use of the available area. Finally, fabrication on flexible substrate is also possible.

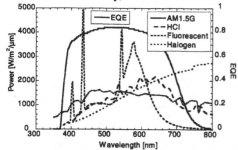

| Light source | Efficiency |
|---|---|
| AM1.5G | 10.4% |
| Metal halide HCI | 14.8% |
| Halogen | 6.1% |
| Fluorescent | 23.0% |

**Figure 1.** Light spectra for sun AM1.5G and various indoor illumination sources (metal halide HCI, fluorescent and halogen lamps) adjusted for the same short-circuit current $I_{sc}$ as for AM1.5G. The external quantum efficiency (EQE) of 0.25 cm$^2$ a-Si:H cell (cf. Fig. 2b) is also plotted. The corresponding cell performances for all illuminations sources are indicated in the table for the same $I_{sc}$ of 15.9 mA/cm$^2$, assuming that fill factor FF of 74.3% and open-circuit voltage $V_{oc}$ of 886 mV are the same for all sources (as plotted in Fig. 2a for AM1.5G).

In this paper several examples of a-Si:H based micro photovoltaic modules are presented. Modules with a total area ≤15 mm$^2$, developed on glass wafers using micro-fabrication procedures (including photolithography). The monolithic interconnections of the module segments add parasitic serial and parallel resistance and, as the segments size is reduced, may critically affect the module performance. Therefore various interconnection geometries have been tested and compared in order to devise design rules. For the same purpose, high voltage modules were also designed and fabricated. The analysis of the performance of these high voltage modules offers additional information for optimizing interconnection schemes.

Finally, the practical development of solar modules for an application in micro-robotics led to a further optimization of the design and processing and illustrates the potential of this technology for energy scavenging for micro-systems.

## EXPERIMENTAL

All micro photovoltaic modules have been fabricated on 4" borofloat glass wafers. For these modules, the same cell structure was used as for large photovoltaic cells, namely glass/ZnO:B/p-a-Si:H/i- a-Si:H/n- μc-Si:H/ ZnO:B configuration. ZnO:B transparent conductive oxide (TCO) layers were deposited by LP-CVD (low pressure chemical vapor deposition); details on the deposition technique can be found in Refs. 2, 3. P-i-n diodes were deposited by Very High Frequency plasma-enhanced chemical vapor deposition (VHF PE-CVD) at 40 MHz and 200°C in a KAI S or KAI M reactor [4]. Thicknesses of the front ZnO layer and of the i-layer were 2 μm and 270 nm, respectively. For large reference cells a back contact using a 2 μm thick ZnO:B layer and a white paste was used while for micro-modules, Al contact pads were further evaporated, covering also the back thin (50 nm thick) ZnO:B.

Finally for some of a the micro-modules, a 200 nm thick SiO$_x$ passivation layer was deposited by VHF PE-CVD and opened by photolithography on the Al contact pads. Cell patterning for monolithic interconnection of the segments or for additional implementation of photodiodes in the modules was performed by photolithography and etching steps. Mask design and alignment precision was better than 3 μm. Wet etching using diluted HCl acid was used for the front ZnO:B contact, dry etching using SF$_6$/O$_2$ mixture was used for the a-Si:H and SiO$_x$ layers while the back ZnO:B/Al back contact was patterned using Al etch solution.

Typical current-voltage I(V) under illumination of reference cells (co-deposited with the micro-modules) as well as external quantum efficiency (EQE) curves are given in Fig. 2. These reference cells were systematically deposited together with the micro-modules.

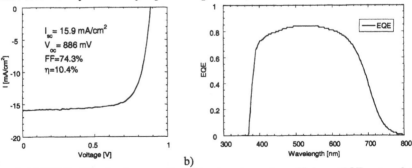

**Figure 2.** (a) I(V) for illumination as well as (b) external quantum efficiency EQE curves for 0.25 cm$^2$ reference cells deposited under similar conditions as for the micro-modules.

I(V) was measured either under (at 100 mW/cm$^2$) using a 2 light sources Wacom solar simulator (for relatively large solar cells) or on an illumination test bench using an halogen lamp with IR filter with an illumination adjusted to give an identical current for a-Si:H reference cells as under the solar simulator (the actual illumination was approximately 100 kLux).

## RESULTS AND DISCUSSION

For standard large area a-Si:H solar modules, every layer is deposited on the entire substrate surface and then pattern by laser scribing in order to isolate the individual cells and realize the monolithic serial interconnection. In the present case of micro-modules fabrication, we have more freedom in terms of the stacking of the various layers, as well as for the definition of the active layer. As illustrated in Fig. 3, the active area of each module segment can be defined either by the ZnO layer (Fig. 3b) or by the metallic back contact (Fig. 3a, c). An overlapping of the a-Si:H layer on the ZnO one is also possible. All stacking configurations shown in Fig. 3 have been tested and compared.

**Figure 3.** Schematic side view of the possible stacking of the layers to form module segments.

The monolithic interconnection implies that the metal back contact of each segment must at some point of the segment periphery cover the side of the p-i-n diode layer stack before reaching the ZnO front contact of the adjacent segment. The rather conductive back contact layer is therefore expected to create a shunt between the p- and n-layer of the diode, as illustrated in Fig. 4. In large area modules, this local shunt usually plays a negligible role on the module performance, due to the relative large distance between this local shunt and the active area. For micro-modules this shunt is expected to reduce considerably the performance as the active area of segment is decreased. Several interconnection schemes have therefore been designed and tested to study the effect of the gap between the position of the side of the a-Si:H layer stack (expected position of the shunt, as see in Fig. 4) and the active area and the effect of the width and length of the interconnection area (contact area between the ZnO and metal back contact).

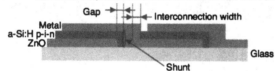

**Figure 4.** Schematic side view of the interconnection between two segments and position of the expected shunt.

### Test modules

In order to study various layer stacking configurations, as well as interconnection designs, test structures as shown in Fig.5a were fabricated. Each test structure comprised 6 modules of 4 segments, with 2 length of the interconnection area on a 3x3 mm$^2$ total area. The width (15 μm) of the interconnection was the same for all modules and the gap between the interconnection area and the active area was changed from 10 to 40 μm. One set of test structures was designed with an active layer define by the ZnO top contact (Fig. 3b) and a second with an active layer defined by the bottom metallic contact (Fig. 3c).

I(V) characteristics of the different structures show very little differences which are within the scattering of the results under the used light illumination used (see experimental details). No significant change in parallel or serial resistance was observed. However, a large scattering of the performance was observed from run to run (see Fig. 5.b) indicating that such small monolithic interconnections are quite sensitive to the quality of the processing (alignment errors, under-etching of the layers, etc). Nevertheless, slightly better yields were obtained on structures with a higher interconnection length and on some of them very high performance could be obtained with FF better than 72% (cf. Fig. 5b).

 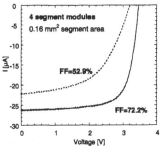

a)                                b)

**Figure 5.** (a) Picture of the back of a test structure ($3\times3$ mm$^2$ total area) with 6 modules of 4 segments with identical active area but various interconnection design, and (b) I(V) characteristics measured on two similar modules from 2 different deposition and processing batches. Dark areas of the picture correspond to flat aluminum area (mainly bonding pads) while white area correspond to Al deposited on rough surface (due to ZnO roughness).

High voltage modules

In order to further study the effect of segment size and interconnect on the performance, several modules were designed with number of segments ranging from 23 to 223 segments on the same $3\times3$ mm$^2$ total area. The geometry of the interconnection was kept constant for all modules. Two examples of modules are shown in Fig 6.

a)                                b)

**Figure 6.** (a) Front side (glass side) picture of a module with 34 segment and (b) back side (module side) picture of a module with 98 segment fabricated on $3\times3$ mm$^2$ total area.

$V_{oc}$ may be limited by the parallel resistance when the photo-generated current is reduced. In our case, the anticipated presence of a shunt in the interconnection (i.e. a fixed parallel resistance) should be revealed if one reduces the segment area. As seen in Fig. 7, no limitation in $V_{oc}$ is noticed as the segment size is decreased down to 800 μm$^2$ under 100 kLux illumination and a maximum voltage of 180 V is obtained for 223 segments. From this behavior, one can conclude that the interconnection does not introduce large shunts, in agreement with the observation done on the test structures. In Fig. 7, one can also observe that the optimization of the processing led to a significant improvement in the reproducibility in the fabrication of those modules. Nevertheless, measurement of the I(V) characteristics should be carried out to quantitatively evaluate the parallel resistance introduced by the monolithic interconnection.

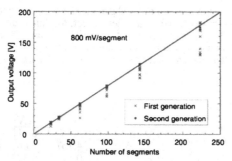

**Figure 7.** Open circuit voltage $V_{oc}$ as a function of the segment number (and a fixed total module size of $3\times3$ mm$^2$) for 2 different generations of modules. The bold solid line corresponds to an output voltage of 800 mV/segment under approx. 100 kLux illumination. First and second generations share the same module designs; the change illustrates the progress in the processing.

Micro-system modules

For a specific application in micro-robotics (European project I-Swarm [5], cf. Fig. 8c), a modules with two voltage sources (1.5 and 3.3 V) as well as two photodiodes had to be integrated on a $3.9\times3.9$ cm$^2$ glass substrate. The robot comprises a locomotion and sensing module, an IR communication module, a power and sensor module as well as an ASIC (application specific integrated circuit) for the robot control. The solar module was designed to deliver a power of ca. 1 mW for the 3.3 V source and 0.5 mW for 1.5 V source under 100 kLux illumination from a HCI metal halide light source. The same device structure as for the test modules was used and in this context several inter-connection designs were also compared. The segment areas for the "high" voltage (HV), "low" voltage (LV) and photodiodes are 5.2, 0.9 and 0.039 mm$^2$, respectively. A schematic electrical diagram of the two voltage sources, as well as pictures of the fabricated modules, are shown in Fig. 8.

**Figure 8.** (a) Electrical diagram of the I-Swarm solar modules dual voltage sources with 5 segments in series, (b) front side (glass side) picture of a module and (c) picture of an autonomous microrobot developed in the framework of the I-Swarm project on a Swiss 5 cent coin. The $3.9\times3.9$ mm$^2$ solar modules can be seen on the top with 2 photodiodes in the corners. The flexible connector is used for initial programming and testing of the robot and is then cut away.

As already observed for the test structures, very limited effect of the stacking and definition of the active area was observed. Nevertheless a slightly better yield was obtained with an active area defined by the bottom Al contact and an a-Si:H layer overlapping the ZnO top contact (see Fig. 3c). Most of the fabricated devices exhibited rather high serial resistance, mostly due to an insufficient interconnection area. Best results were obtained by increasing the interconnection width to 40 µm, and by choosing a length of 1200 µm for HV segments and 600 µm for LV ones. Nevertheless, the length was found to have less effect. Module characteristics of the best modules are plotted in Fig. 9. We can observe that the serial resistance is still limiting the performance (mostly with a reduction of the fill factor FF). A further increase of the interconnection area as well as the implementation of small bus bars are expected to further improve the behavior.

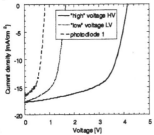

|  | HV | LV | Photo 1 | Photo 2 |
|---|---|---|---|---|
| $I_{sc}$ [mA] | 0.159 | 0.911 | 0.0065 | 0.0065 |
| $V_{oc}$ [V] | 4.157 | 1.667 | 0.822 | 0.828 |
| FF | 0.613 | 0.572 | 0.517 | 0.522 |
| Power[mW] | 0.41 | 0.87 | 0.0028 | 0.0028 |

**Figure 9.** Current density I(V) as a function of voltage of the HV, LV voltage sources as well as for one photodiode (the I(V) of second photodiode not shown here is almost identical) of a I-Swarm solar modules under 100 kLux illumination (halogen lamp). The corresponding short-circuit $I_{sc}$, open-circuit voltage $V_{oc}$, FF and output power are also indicated.

CONCLUSIONS

Several micro solar modules with total area of 3-15 mm$^2$ were designed and fabricated on glass substrates using micro-fabrication processes. Monolithic interconnection and flexibility of the cell design allow for a wide range of module output voltage values. Even though the interconnect design of the device is critical for the performance, the latter is more dependant on the fabrication process that the geometry itself. It is observed that in most case these small modules are limited due to serial resistance and small interconnections and small connection pads.

High voltage modules able to generate up to 20 V/mm$^2$ under 100 kLux illumination have been successfully fabricated. No limit in the voltage density was so far found. This type of modules may have applications for example in MEMS (micro electromechanical systems), electrostatic or piezoelectric actuators. The fact that this type of device can be deposited on various types of substrates is an asset.

Finally, micro-modules for a micro-robot application were designed and successfully fabricated. This example illustrates the possibility of this a-Si:H technology to tailor the output voltage to the needs of the application and to allow also the integration of sensors such as photo-diodes. It also highlights the potential of this technology for various micro-system powering.

ACKNOWLEDGMENTS

The authors acknowledge J.-M. Breguet of the Robotic System Laboratory of the Ecole Polytechnique Fédérale de Lausanne and the EU project I-Swarm for financial support.

REFERENCES

[1]    E. Cantatore; M. Ouwerkerk, Microelectronics Journal 37 (2006) 1584.
[2]    J. Meier et al., Proc. of the 3rd World Conf. on PV Energy Conversion, (2004) 2801
[3]    S. Faÿ et al., Solar Energy Materials & Solar Cells 86 (2005) 385.
[4]    S. Benagli et al., Proc. of the 22nd EU Photovoltaic Conference, Milan, 2007 (2007) 2177.
[5]    http://www.i-swarm.org

**Film Growth and Characterization I**

Mater. Res. Soc. Symp. Proc. Vol. 1066 © 2008 Materials Research Society        1066-A11-01

# Magnetic Resonance in Hydrogenated Nanocrystalline Silicon Thin Films

Tining Su[1], Tong Ju[2], Baojie Yan[3], Jeffrey Yang[3], Subhendu Guha[3], and P. Craig Taylor[1]
[1]Department of Physics, Colorado School of Mines, Golden, CO, 80401
[2]Department of Physics, University of Utah, Salt Lake City, UT, 84112
[3]United Solar Ovonic LLC, Troy, MI, 48326

## ABSTRACT

We have investigated the localized electronic states in mixed-phase hydrogenated nanocrystalline silicon thin films (nc-Si:H) with electron-spin-resonance (ESR). The dark ESR signal most likely arises from defects at the grain boundaries or within the crystallites. With illumination with photon energies ranging from 1.2 eV to 2.0 eV, there is no evidence of photo-induced carriers trapped in the bandtail states within the amorphous region. Dependence of the light-induced ESR (LESR) upon the exciting photon energy reveals that, at different excitation photon energies, different regions dominate the optical absorption. This behavior may have potential consequences for understanding the light-induced degradation in nc-Si:H.

## INTRODUCTION

Hydrogenated nanocrystalline silicon thin films (nc-Si:H) may be the most promising materials for the next generation solar cells. Due to the strong optical absorption at lower energies, these films are ideal for collecting light with energies below the optical gap of hydrogenated amorphous silicon (a-Si:H) [1-3]. The nc-Si:H films for the best device performance typically have a mixed phase, with a volume fraction of crystallinity of about 50% [4,5]. As a consequence, the material is highly inhomogeneous. The optical absorption and consequently the dynamics of the photo-induced carriers are complicated and not well understood, especially for energies below the optical gap of a-Si:H and crystalline silicon.

Previously, we reported that the mixed-phase nc-Si:H sample shows negligible light-induced degradation as measured by electron-spin-resonance (ESR) [6]. The optical absorption of low energy light near $E = 1$ eV is most likely due to localized states at the grain boundaries [6]. However, when the sample is illuminated with light of higher energies, the absorption can occur in all three regions, namely the crystalline region, the amorphous region, and the grain boundaries. As a result, the dynamics of the photo-induced carriers can be very complicated. The carriers can diffuse across the grain boundaries, they can also be trapped at the localized states at the grain boundaries, and probably also in the bandtail states in the amorphous region. Measurements of the optical coefficients alone do not provide information on the details of the absorption, nor do they reveal details of the carrier dynamics. On the other hand, ESR provides details of the localized states, such as those trapped in bandtails and grain boundaries. We report an ESR study of the localized states in nc-Si:H, with or without illumination, and the dependence of the production of paramagnetic states on the exciting photon energy.

## EXPERIMENT

All nc-Si:H samples were made at United Solar Ovonic LLC. These samples were deposited under the similar conditions as the samples we studied previously [6]. The nc-Si:H films were deposited on quartz ESR substrates for ESR and optical measurements. A large powdered sample was made from a large area film deposited on aluminum foil. The films on quartz substrates have thicknesses of about 1.5 μm. Raman measurements on samples deposited under similar conditions showed that the crystalline volume fraction is about 50% [5]. The crystallites have an average grain size of about 20 nm as indicated by AFM measurements on sample deposited under similar conditions [7].

ESR measurements were performed on a Bruker EMX ESR spectrometer, operating at X-Band. The dark ESR was measured at room temperature, using the first harmonic detection method. For low temperature measurements, the saturation of ESR signal makes it difficult to use the first harmonic detection. In order to increase the signal-to-noise ratio for the saturated ESR signals, we also used the second harmonic detection method. During the low temperature measurement, microwave power level was varied over two orders of magnitude in search of all possible ESR signals.

For LESR measurements, an Nd:YAG laser and high power LED's from Roithner were used. Typical light intensities were about 3 to 15 mW/cm$^2$ for the LED's, and about 80 mW/cm$^2$ for the Nd:YAG laser. There is no observable heating of the samples during the illumination.

## RESULTS AND DISCUSSION

### Dark ESR

Figure 1 shows the dark ESR in a powdered sample at room temperature. The dark ESR from a typical a-Si:H powder is also shown for comparison. For both signals, the zero crossing fields correspond to a g-value of $g \approx 2.005$.

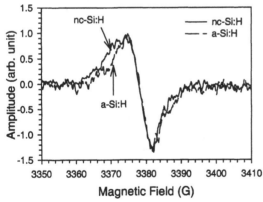

**Figure 1** Comparison of the dark ESR in nc-Si:H and a-Si:H at 300 K. Solid line represents data in nc-Si:H, dotted line represents that in a-Si:H.

The dark ESR signal in nc-Si:H is significantly broader than that in a-Si:H, and the asymmetric lineshape is probably the result of an anisotropic powder average (powder pattern) [8]. This signal is very similar to that in a sample with a crystallinity of > 90% [8]. Since the volume fraction of the amorphous phase differs by about a factor of five, the nearly identical lineshapes from these two samples suggest that this signal arises from localized states that are either on the grain boundaries or within the nano-crystallites, and the amorphous region is mostly defect free, at least as seen by ESR. This result is probably because that the average size of the amorphous region (~ 20 nm) is much smaller than the average distance between defects in typical a-Si:H ( ~ 60 nm for a defect density of $10^{16}$ cm$^{-3}$), and the disorder probably can be largely accommodated by the hydrogenation of the grain boundaries without having to create a dangling bond within the amorphous region. Since Optical absorption measurements in nc-Si:H do not delineate the absorption in the amorphous phase, it is not clear if there is any effect on the localized bandtail states within the amorphous region.

## Light-induced localized states in nc-Si:H

Figure 2 shows the light-induced LESR signals in nc-Si:H films on quartz substrates at different photon energies. The measurements were preformed at $T = 20$ K. The signals were detected by the second harmonic method, which produces a lineshape similar to the integral of the typical derivative lineshape. Nevertheless, subtle differences exist between the actual integrated lineshape and that from the second harmonic detection, particularly when multiple lines with different saturation values are present. The arrows in Fig. 2(a) and (b) indicate where the LESR signals occur in pure a-Si:H films [9,10].

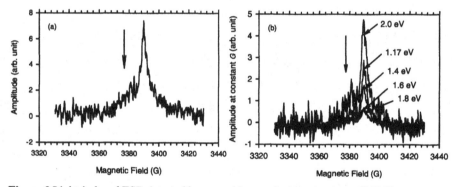

**Figure 2** Light-induced ESR detected by second harmonic detection in nc-Si:H films on quartz substrates at $T = 20$ K. (a) LESR under illumination with 780 nm ($E = 1.6$ eV) light, (b) comparison of lineshapes under illumination with different photon energies. The linewidths of the narrow and broad lines are about 6 and 20 gauss, respectively. The arrows in (a) and (b) indicate where the LESR signals occur in pure a-Si:H.

As shown in Fig. 2(a), only two signals with different widths were detected at g-values close to $g = 1.998$. The g-values may be slightly different for the two lines, but the asymmetric lineshape prohibited a very accurate estimate of the g-values, particularly for the broader line. The values of full-width-at-half-maximum (FWHM) of the two components are about 6 gauss for the narrow line, and about 20 gauss for the broader component. The g-values of these signals are similar to those excited by the Nd;:YAG laser at $T = 7$ K as reported previously [6]. However, the width of the broad component is larger than that excited by the Nd:YAG laser at 7

K [6]. This may be due to a partial saturation of the broad component using second harmonic detection. These signals are also similar to those observed in micro-crystalline silicon thin films (μc-Si:H) with high crystallinity [8]. They are most likely due to the electrons and holes trapped at the grain boundaries, as suggested previously [6].

In Fig. 2(b), LESR signals under illumination with different photon energies are plotted. The signals are normalized to a constant generation rate. The LESR lineshapes are rather similar for all photon energies. In all cases, within experimental error, we did not observe the LESR signals arising from trapped carriers in the amorphous silicon bandtails, as is observed in typical a-Si:H films [9,10].

In addition, Fig. 2(b) shows that the LESR at the same generation rate strongly depends on the photon energy of the excitation light. This observation contrasts the results from previous photoluminescence (PL) measurements in a-Si:H, in which the PL efficiency under constant generation rate is independent of the photon energy of the excitation light. Since LESR in nc-Si:H probably arises from the states at the grain boundaries, the photo-generated carriers in both the crystalline and amorphous regions will have to take extra steps to reach the grain boundaries. The difference in the efficiency of producing the trapped states at the grain boundaries, as shown in Fig. 2(b), probably reveals the different carrier dynamics following the carrier generation.

In Fig. 3, we plot the ESR amplitude at constant generation rate as a function of excitation energy, and compare the dependence on excitation photon energy to that of the optical absorption coefficients of a-Si:H, nc-Si:H, and bulk crystalline silicon (c-Si).

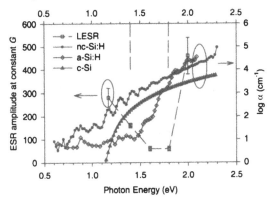

**Figure 3** Dependence of the LESR in nc-Si:H on the photon energy of the exciting light. Also plotted are the optical absorption in nc-Si:H, a-Si:H, and c-Si. Open squares represent LESR signals that are normalized to constant generation rate. Solid circles, open diamonds, and open triangles represent absorption coefficients in nc-Si:H, a-Si:H, and c-Si, respectively. All the lines are aids to the eye. See text for details.

As can be seen from Fig. 3, the optical absorption in nc-Si:H can be approximately divided into three regions. At higher energies ($E > \sim 1.8$ eV), absorption in amorphous region dominates. At energies between 1.4 eV and 1.8 eV, most of the absorption likely occurs within the crystalline region. At even lower energies ($E < \sim 1.4$ eV), the absorption is probably due to the absorption at the grain boundaries, as suggested previously [6]. The two dashed lines on the upper half of the figure are rough estimates of the demarcation energies of the different dominant absorption mechanisms. By considering the different carrier dynamics in different regions, the dependence of the LESR on excitation energy can be partially explained.

In a-Si:H, once the carriers are excited into the extended states, they will rapidly thermalize and cross the mobility edge. They are then trapped in the bandtail states. During the thermalization process, they can diffuse across a considerable distance. An estimate of the diffusion length is given by [11],

$$L_T = (D\Delta E / \hbar\omega_0^2)^{1/2},$$ (1)

where $\Delta E$ is the carrier energy above the mobility edge, $D$ is the diffusion coefficient, and $\omega_0$ is the phonon energy. For electrons, assuming $\Delta E = 0.2$ eV, $D = 0.5$ cm$^2$ s$^{-1}$, and a phonon energy $\hbar\omega_0 \sim 0.025$ eV, the diffusion length $L_T$ is estimated as $L_T \sim 5$ nm. This is within the same order of magnitude of the distance the carriers have to travel to reach the grain boundaries. In other words, when excited high above the mobility edge, most of the carriers will be able to reach the grain boundaries before they are trapped in the localized bandtail states. This can explain the higher efficiency of producing LESR with a photon energy of 2.0 eV. On the other hand, if the carriers are excited into states very close to the mobility edge, they probably will be trapped very quickly. These carriers will not be able to diffuse far enough to reach the grain boundaries, and will not contribute to producing the LESR signal. This is probably the case with light near the a-Si:H band gap. At low energies ($E < 1.4$ eV), most of the optical absorption is probably at grain boundaries. The LESR is produced "locally", and hence with a higher efficiency. For the particular sample we used, the crystallinity (~50%) is above the percolation threshold for the conductivity. The carrier dynamics in the crystalline region is more complicated. There is no obvious reason for the observed low efficiency in producing LESR at energies between 1.5 eV and 1.8 eV.

One difficulty of this explanation is the holes trapped in the bandtail states in the amorphous region. Due to the lower mobility of holes in a-Si:H, it may take significantly higher energy for the holes to reach the grain boundaries. This could result in trapped holes within the amorphous region. Within the signal-to-noise ratio of our measurements, we did not observe this signal ($g = 2.01$) [9].

One potential implication of this process is to explain the very low photo-induced degradation in these samples. If most of the carriers excited in the amorphous region reach the grain boundaries before they cross the mobility edge, then there will be significantly less recombination in the amorphous region, resulting in a lower production of metastable defects. Further study with PL is ongoing to test this assumption.

## CONCLUSIONS

We investigated the localized electronic states in mixed-phase nc-Si:H with ESR. Dark ESR at room temperature suggests that the amorphous region in these samples is nearly defect free. Under illumination with photon energies from 1.2 eV to 2 eV, light-induced ESR provides no evidence of trapped carriers in the amorphous region. Light-induced ESR at constant generation rate strongly depends on the exciting photon energy, suggesting that the production of the LESR depends on the dynamics of the carrier diffusion to the grain boundaries.

## ACKNOWLEDGMENTS

The authors thank P. Straidins for helpful discussions. The work done at the Colorado School of Mines is partially supported by NSF under grant number DMR-0702351, and by DOE under subcontract number DE-FC36-07G017053. The work done at United Solar Ovonic is also partially supported by DOE under subcontract number DE-FC36-07G017053.

## REFERENCES

1.  J. Meier, R. Flückiger, H. Keppner, A. Shah, Appl. Phys. Lett. **65**, 860 (1994).
2.  O. Vetterl, F. Finger, R. Carius, P. Hapke, L. Houben, O. Kluth, A. Lambertz, A. Mück, B. Bech, H. Wagner, Sol. Energy Mater. Sol. Cells **62**, 97 (2000).
3.  K. Yamamoto, IEEE Trans. Electron. Dev. **46**, 2041 (1990).
4.  T. Roschek, T.Repman, J. Müller, B. Tech, H. Wagner in *Proceedings of the 28th IEEE Photovoltaic Specialists Conference*, (IEEE, NY, 2000) pp150.
5.  G. Yue, B. Yan, G. Ganguly, J. Yang, and S. Guha, Appl. Phys. Lett. **88**, 263507 (2006).
6.  T. Su, T. Ju, B. Yan, J. Yang, S. Guha, and P. C. Taylor, J. Non-cryst. Solids, (in press).
7.  C. Inglefield, private communication.
8.  M. M. de Lima, P. C. Taylor, S. Morrison, A. Legeune, F. C. Marwues, Phys. Rev. B **65**, 235324 (2002), and references therein.
9.  G. Schumm, W. B. Jackson, and R. A. Street, Phys. Rev. B **48**, 14198 (1993).
10. N. A. Schultz and P. C. Taylor, Phys. Rev. B **65**, 235207 (2002), and references therein.
11. R. A. Street, *Hydrogenated Amorphous Silicon*, (Cambridge University Press, Cambridge, 1991), pp285.

Mater. Res. Soc. Symp. Proc. Vol. 1066 © 2008 Materials Research Society     1066-A11-02

# Voids in Hydrogenated Amorphous Silicon: A Comparison of ab initio Simulations and Proton NMR Studies

Sudeshna Chakraborty[1], David C. Bobela[2], P. C. Taylor[3], and D. A. Drabold[1]

[1]Department of Physics and Astronomy, Ohio University, Athens, OH, 45701
[2]Department of Physics, University of Utah, Salt Lake City, UT, 84112
[3]Department of Physics, Colorado School of Mines, Golden, CO, 80401

## ABSTRACT

Recently, a new hydrogen NMR signal has been observed in a number of PECVD prepared hydrogenated amorphous silicon (a-Si:H) films of varying quality. It is speculated that the signal is the consequence of a dipolar-coupled hydrogen pair separated, on average, by $1.8 \pm 0.1$ Å. To elucidate the possible bonding configurations responsible for the NMR data of ref. [1], we have used *ab initio* simulation methods to determine a set of relaxed structures of a-Si:H with varying void sizes and H-concentrations. Models containing two isolated hydrogen atoms indicate a preferred H-H distance of approximately 1.8 Å when the two atoms bond to nearest neighbor silicon atoms. This separation also occurs for models containing small, hydrogenated voids, but the configurations giving rise to this H-H distance do not appear to be unique. For larger voids, a proton separation of about 2.4Å is seen, as noted previously [2]. There appears to be consistency between the computed structures and the NMR data for configurations consisting of isolated hydrogen pairs or for clusters of an even number of hydrogen atoms with the constraint that the average H-H distance is 1.8 Å. In this paper, we will discuss the most probable bonding configurations of clustered hydrogen based upon the extent of the NMR data and simulated structures.

## INTRODUCTION

The microstructure of hydrogenated voids in a-Si:H, and their potential role in the Staebler-Wronski effect, are not well understood. Early nuclear magnetic resonance (NMR) studies on films prepared by plasma enhanced chemical vapor deposition (PECVD) revealed that hydrogen occurs in isolated and clustered environments with a small fraction of $H_2$ molecules existing in larger voids [3]. The intuitive idea of a hydrogenated divacancy gained support when multiple quantum NMR studies on PECVD films showed evidence of clusters of six hydrogen atoms [4]. Recently, a new hydrogen NMR signal has been observed by Bobela *et al.* in a number of PECVD prepared a-Si:H films [1]. It is speculated that the signal is the consequence of a dipolar-coupled hydrogen pair separated, on average, by $1.8 \pm 0.1$ Å. However, more complex hydrogen configurations, such as those resulting from hydrogenated voids or poly-silane like chains, $(SiH_2)_n$ can also explain the sometimes unusual dipolar-dipole powder patterns that have consistently emerged in a number of a-Si:H films. To elucidate the possible bonding configurations responsible for the NMR data of Ref. [1], we have used *ab initio* simulation methods to determine an ensemble of structures of a-Si:H with varying void sizes and H-concentrations. Studies are still underway as to whether or not the configuration responsible for this H-H separation is associated with the Staebler-Wronski effect (SWE).

This work also addresses the preliminary results for the kinetics of hydrogen molecules, $H_2$, when they form in voids. Recent studies by C. Longeaud have linked the interstitial $H_2$ trapped in the voids of a-Si:H with metastability [5]. Although the existence of $H_2$ in a-Si:H has been well-documented by NMR experiments, Ref [6], the exact location of these entities within the network and its relation to the SWE is still a mystery.

## MODELS AND METHODS

We studied different configurations of a-Si:H models with varying concentrations of hydrogen and void sizes in them. Starting with two defect free CRN models of amorphous silicon of 216 atoms and 64 atoms, silicon atoms are removed and spheres of radii of about 2-6 Å are cut and voids are created which results in dangling Si bonds in the interface of the void and the CRN. These bonds are then terminated with hydrogen atoms at a Si-H distance of 1.5 Å and then the system is annealed at a temperature of 700°K and 300°K. Finally, a conjugate gradient relaxation is applied to reduce the forces to almost zero. The model of a-Si:H with a divacancy is created by removing two silicon atoms which are nearest neighbors and terminating the resulting dangling Si bonds with hydrogen. We generated 4 models with a divacancy at different sites in a network with 62 Si atoms and 6 H atoms and 214 Si atoms and 6 H atoms. Models of a-Si:H with larger voids are created by cutting spheres of radii 4-6 Å and removing 8, 16, 32 and 64 Si atoms. The interface Si dangling bonds are then passivated with hydrogen.

Another set of models was generated where the H atoms are put at a distance of ~3 Å from each other and at a distance of at least ~2 Å from the silicon atoms so that they are not bonded to the silicon atoms. These hydrogen atoms would be referred to as mobile H atoms and are introduced to study how the network evolves with the introduction of the H atoms and to study how Si–H bonds are formed. All the models are equilibrated at temperatures of 300°K and 500°K for 3ps.

In our simulations, we used SIESTA (Spanish Initiative for Electronic Simulations with Thousands of Atoms) which is a DFT based first principles pseudopotential method based on Linear Combination of Atomic Orbitals (LCAO) [7]. SIESTA uses norm-conserving Troullier-Martins pseudopotentials [8] in its fully non-local (Kleinman-Bylander) [9] form. In this work, self consistent Kohn-Sham functional in the General Gradient Approxinmations (GGA) with the parametrization of Perdew, Burke and Ernzerhof [10] is used. Both double-$\varsigma$ polarized basis sets (DZP) for silicon and hydrogen are used with two s and three p orbitals for the H valence electrons and two s, six p and five d orbitals for Si valence electrons.

## RESULTS AND DISCUSSIONS

Representative proton NMR spectra of dipolar coupled hydrogen pairs are shown in Fig. 1 for "device" grade (Fig. 1a) and "poor" quality (Fig. 1b) a-Si:H. The spectra differ only by the intensity of the central region, which constitutes the best clue as to the identity of the H system contributing to the lineshape. From the form of the spin-spin interactions of the dipolar coupled hydrogen system, the spectrum in Fig. 1 is most likely due to a cluster of hydrogen atoms whereby the nearest neighbor H atoms are separated by 1.8 Å and are less strongly coupled to their neighbors. An intuitive candidate is the hydrogenated divacancy, since the geometry allows for H separations of up a few angstroms, which would constitute a "loosely" coupled H-pair when compared to the nearest neighbor spacings. The dotted line represents a simulation of such a geometry and does well to match the gross features of the NMR data. Other geometries cannot be ruled out. However evidence from multiple quantum NMR experiments do support the preferred cluster size of six atoms [4]. Figure 1b is reminiscent of the spectra presented in Ref. [1], where the interpretation of the data required the H-H distance to be 1.8 Å regardless of the broadening affects. For spectra with this characteristic shape, it is likely that the H-H pair is isolated from other clusters.

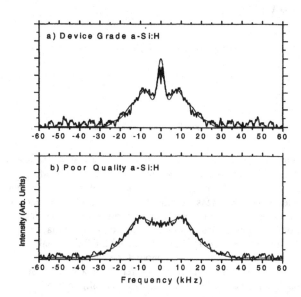

**Figure 1.** Proton NMR spectra of dipolar coupled hydrogen atoms in a) device grade and b) poor quality a-Si:H. Dotted lines denote simulations of the spectra, see text for details.

For the models with a divacancy with no dangling bonds present, the H-H distance is tracked for a Molecular Dynamics (MD) run for 3 ps. We observed a peak at a H-H distance of 1.8 ± 0.2 Å for all the models. Although dihydride, (SiH$_2$), structures giving a H-H separation of 2.4 Å as reported earlier [2] is present in some of the models, a mean H-H distance of ~1.8 Å is obtained from two distinct scenarios: hydrogen connected to silicon atoms that are next nearest neighbors, a monohydride (SiH HSi) configuration, and hydrogen clustering at the interface of the voids. The void surface comprises either Si-H$_2$ or Si-H bonding configurations. Not only do (Si-H H-Si) configurations contribute to a H-H distance of ~1.8 Å, but H atoms from these configurations separated from H atoms in the adjacent dihydride configurations are responsible. However, there is no unique configuration responsible for the H-H distance of ~1.8 Å.

We performed similar studies on the models with larger voids. In these scenarios, a H-H separation of 2.4 Å consistently emerges where both complexes are formed. This result supports the physically intuitive idea that space constraints restrict the H atoms from assuming their natural separation of 2.4 Å. The reduced H-H distance is simply a consequence of clustering in small divacancy-like voids. A representative result for the distribution of H-H separation is shown in the Fig.2. The H-H separation is distributed about 1.8 Å with an FWHM of 0.53 Å.

This agrees very well with the proton- proton separation extracted from the NMR data in Fig. 1b by Bobela *et al.* [1].

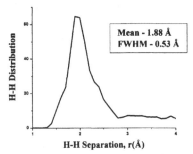

**Figure 2.** Plot of H-H distribution vs. H-H separation for a divacancy like void with 214 Si atoms and 6 H atoms. MD simulation temperature is 700°K.

Similarly, for the models with unbonded H, a MD simulation is performed at 300°K and 500°K for 3 ps to let the H atoms form bonds on their own. We observed that the unbonded hydrogen atoms tend to form monohydride configurations in pairs. Four out of the six hydrogen atoms form monohydrides and cluster together whereas the other two are isolated Si-H bonds. In small voids like divacancies, it is observed that there is a tendency to form more monohydride configurations than dihydrides. Another complex where hydrogen bonded to silicon that are nearest neighbors is also formed. In these models, the H-H distance of $1.8 \pm 0.2$ Å is confirmed. Snapshots of the MD simulation showing the formation of a monohydride configuration is shown in Fig. 3. The hydrogen atom comes into the bond center whereby it breaks the Si-Si bond and sticks to one silicon. A dangling bond remains after the monohydride configuration has formed.

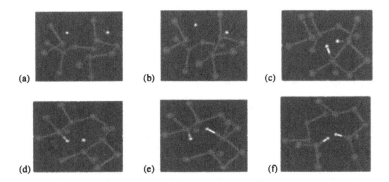

**Figure 3.** Creation of dangling Si bonds and formation of monohydride (SiH HSi) complexes at a particular site in a-Si:H with 62 Si atoms and 6 H atoms with a divacancy from *ab initio* MD simulation. H atoms are denoted by white color and Si atoms by dark color. (a) Mobile hydrogen atoms come towards a dangling bond, (b) one mobile H terminates the Si dangling Bond, (c) Si atoms become fivefold coordinated, (d) H comes between the bond center, (e) H atom breaks Si-Si bond and hops between Si, (f) H finally forms the (SiH HSi) complex.

In these models, the mobile hydrogen atoms play an important role in creating dangling bonds. In the a-Si models with a divacancy-like void consisting of 6 dangling bonds, the interface of the void reconstructs to leave 2 dangling bonds upon relaxation. Now, when unbonded H atoms are included in the network, the mobile H atoms tend to form Si-H bonds. When hydrogen comes within a 2 Å radius of a silicon atom which has 4 nearest neighbors, it bonds with the silicon making it 5 fold coordinated. Hence, the bond between two silicon atoms breaks and results in a dangling bond at a distance of ~4-5 Å from the H atom. A Si-H bond is formed and as a result one or more dangling bonds are created. In all the models, H atoms hop between two Si atoms till it chooses one.

Formation of $H_2$ molecule inside the voids is also seen in some cases. When two unbonded hydrogen atoms come in close proximity (~2 Å) to each other, they tend to form molecules instead of forming Si-H bonds. In certain cases, as shown in the Figure below, the diffusive mobile H atom come close to a bonded H atom and break Si-H bond. These atoms then switch bonds and finally form a $H_2$ molecule. These molecules tend to get trapped inside the voids. After the formation of the $H_2$ molecule, the bond breaking and bond switching mechanism slows down. Snapshots of the formation of a hydrogen molecule inside the void are shown in Fig. 4.

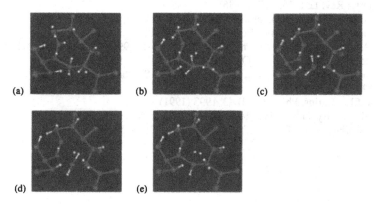

**Figure 4.** Formation of $H_2$ molecules inside a big void in a-Si:H with 152 Si atoms and 67 H atoms from *ab initio* MD simulations. H atoms are denoted by white color and Si atoms by dark color. (a) H atoms become mobile, (b) H comes between bond center, (c) H breaks Si-Si bond, (d) while in between the bond center, the H atom comes closer to another bonded H atom, (e) the H atom breaks Si-H bond and finally forms $H_2$ molecule.

## SUMMARY

We have studied the various bonding configurations and the dynamics of hydrogen in a-Si:H with various sizes of voids. Our results are indicative of a proton-proton distance of 1.8 ± 0.2 Å that consistently emerges for all models with a divacancy. For models with dangling Si bonds around the divacancy decorated with hydrogen, this H-H distance appears to result from monohydride configurations as well as from hydrogen clustering in the inner surfaces of the divacancy. For models with unbonded hydrogen, it is observed that hydrogen tend to form monohydride configurations in pairs which is an indication of hydrogen clustering. However, for large voids a proton-proton separation of ~2.4 Å is observed. In some cases, formation of $H_2$ molecules is also observed. Whether these structures are relevant to SWE is still unclear.

## ACKNOWLEDGMENTS

The research is supported by NSF under grant number # DMR 0600073, 0605890. We would like to thank F. Inam for helpful conversations and the specific suggestion to explore the dynamics of unbonded H in the network. The work at Colorado School of Mines is supported by NSF under grant number # DMR-0702351 and DOE (through United Solar Ovonics) under contract number DE-FC36-07G017053.

## REFERENCES

1. D. C. Bobela, T. Su, P. C. Taylor and G. Ganguly, J. Non. Cryst. Sol., **352** (2006).
2. Tesfaye A. Abtew, D.A. Drabold, P.C. Taylor, Appl. Phys. Lett. **86**, 241916 (2005).
3. W.E. Carlos and P.C.Taylor, Phys. Rev. B **26**, 3605 (1982)
4. J. Baum *et.al*, Phys. Rev. Lett. **56**, 1377 (1986)
5. C.Longeaud, Journal of Optoelectronics and Advanced Materials 3, 461 (2002)
6. T. Su, S. Chen, and P. C. Taylor, Phys. Rev. B **62**, 12849 (2000).
7. P. Ordejón, E. Artacho, and J.M. Soler, Phys. Rev. B **53**, 10441 (1996); D. Sánchez-Portal, P. Ordejón, E. Artacho, and J.M. Soler, Int. J. Quantum Chem. **65**, 453 (1997); J. M. Soler, E.Artacho, J.D. Gale, A. García, J. Junquera, P. Ordejón, and D. Sánchez-Portal, J. Phys, Condens. Matter **14**, 2745 (2002).
8. N. Troullier and J.L. Martins, Phys. Rev. B **43**, 1993 (1991).
9. L. Kleinman and D.M. Bylander, Phys. Rev. Lett. **48**, 1425 (1982).
10. J.P. Perdew, K. Burke, and M. Ernzerhof, Phys. Rev. Lett. **77**, 3865 (1996)

Mater. Res. Soc. Symp. Proc. Vol. 1066 © 2008 Materials Research Society          1066-A11-06

# Quality and Growth Rate of Hot-Wire Chemical Vapor Deposition Epitaxial Si Layers

Charles W. Teplin, Ina T. Martin, Kim M. Jones, David Young, Manuel J. Romero, Robert C. Reedy, Howard M. Branz, and Paul Stradins
NCPV, National Renewable Energy Laboratory, 1617 Cole Blvd., Golden, CO, 80401

## ABSTRACT

Fast epitaxial growth of several microns thick Si at glass-compatible temperatures by the hot-wire CVD technique is investigated, for film Si photovoltaic and other applications. Growth temperature determines the growth phase (epitaxial or disordered) and affects the growth rate, possibly due to the different hydrogen coverage. Stable epitaxy proceeds robustly in several different growth chemistry regimes at substrate temperatures above 600°C. The resulting films exhibit low defect concentrations and high carrier mobilities.

## INTRODUCTION

Recently, new film silicon solar cell technologies are being developed to reduce the amount of Si used in solar cells and their manufacturing costs. Even with indirect-gap crystalline Si, absorber layers less than 10 micron thick are sufficient to achieve reasonable external quantum efficiencies. A promising approach is to start with high-quality c-Si seed layer on a relatively inexpensive substrate such as borosilicate glass and epitaxially thicken it up to 10 microns [1]. To achieve good cell performance, the epi-layer material quality should allow for photocarriers to be collected before they recombine via bulk defects. Roughly, this means the areal density of defects (such as bulk dislocations originating from seed interface) should be below $10^6$ cm$^{-2}$ for 10 micron film. The quality of the epitaxial interface and surfaces must be high because recombination in thin layers can be dominated by surfaces. In addition, the impurity levels in the layers should be low to reduce recombination and permit controlled doping at low levels. In past, we have demonstrated epitaxial growth by hot-wire chemical vapor deposition (HWCVD) above 620°C with defect densities ~$10^8$ cm$^{-2}$ [2, 3], grew epitaxially on the difficult (111) Si surface [2] and successfully thickened a Si seed on borosilicate glass with this technique [3]. In this work, we report further systematic study and improvements of the technique leading to films that show promise for the future PV device work.

## EXPERIMENTAL

Epitaxial Si films were grown on (100) oriented, RCA-cleaned c-Si wafers in our hot-wire chemical vapor deposition (HWCVD) reactor from silane gas decomposing on a W filament heated to 2100 °C. Diborane and phosphine gases were introduced during deposition through flow-controlled orifices to achieve controlled dopant levels in the $10^{16} - 10^{18}$ cm$^{-3}$ range. Growth temperature, a critical epi-growth parameter (see below), was monitored by ellipsometry, pyrometry, and thermocouple response. Film structure was monitored in-situ by real-time spectral ellipsometry. Deposited films were analyzed by transmission electron microscopy (TEM), secondary ion mass spectrometry (SIMS), X-ray diffraction, cathodoluminescence and other techniques. Carrier concentration and mobility in doped films on intrinsic wafers was measured by Hall effect technique.

## RESULTS

Table 1 shows the growth phase and growth rate of Si films deposited on (100) c-Si as function of substrate temperature, $T_s$. Silane flow and pressure were kept at 20 sccm and 10 mTorr, respectively. At temperatures below approximately 450 °C, epitaxial growth takes place but is unstable: after ~100 nm of epitaxy, amorphous Si cones develop that eventually take over the whole film because of their faster growth rate [4]. In the intermediate 450 – 570°C range, no epitaxy was observed even during the very first stages of growth: the film was typically polycrystalline. Finally, above about 650°C, continuous, fast epitaxial growth reproducibly ensued with no signs of breakdown.

**Table 1.** Representative Si film growth rates, R, and the resulting phase at several substrate temperatures $T_s$. Silane flow and system pressure were 20 sccm and 10 mTorr, respectively.

| $T_s$ (°C) | 392 | 494 | 640 |
|---|---|---|---|
| **R (nm/min)** | 59 | 78 | 102 |
| **Phase** | Unstable epi (Ref. 4) | Small grain poly c-Si | Stable epi |

The thickest films grown at ~10 micron did show surface texturing [2], however, the growth orientation remained (100) as confirmed by XRD. Table I shows that film growth rate increases with substrate temperature, a trend observed over many growth conditions.

Fig. 2 shows cross-sectional TEM picture of such an epitaxial film grown at 670 °C. The electron microdiffraction patterns of film and substrate confirm epitaxial growth of the film. The defect density evaluated from this film's TEM data is ~ $10^7$ cm$^{-2}$. In our most recent films, cathodoluminescence data do not show dislocations and the limits of detection mean that defect densities are below $10^6$ cm$^{-2}$. Impurity levels of O, C, and H measured by SIMS are in $10^{18}$ cm$^{-3}$ range; W levels are below SIMS detection limit of $10^{15}$ cm$^{-3}$ .

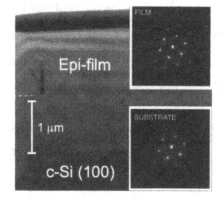

**Fig. 2.** Cross-sectional TEM picture of HWCVD epitaxial Si film grown on c-Si (100) wafer at 670°C. Inserts show electron microdiffraction patterns from film and the substrate. Growth conditions: silane flow 20 sccm, system pressure 10 mTorr. Dashed line denotes the original growth interface.

Fig. 3 shows an example of epitaxial growth rate $R$ as function of silane flow $f$ at 670 C. In this case, two W filaments were used. System pressure $p$ and flow $f$ were the same as in Table 1. At silane flow below 30 sccm, $R$ increases with $f$ indicating silane depletion regime where system pressure is due to both silane and hydrogen produced from the decomposed silane. Above 30 sccm, depletion is no longer observed suggesting that system pressure corresponds to silane pressure. Doubling the pressure in this regime resulted in higher deposition rate of 300 nm/min due to the increased collision rate of silane molecules with the W filament. This deposition rate means that a 6 micron film can be deposited in 20 mins.

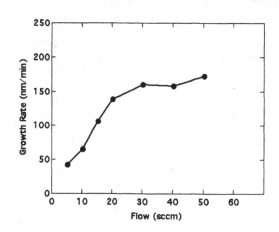

**Fig. 3.** Epitaxial growth rate as function of silane flow. System pressure was maintained at 10 mTorr during deposition.

Table 2 shows doping results from n- and p-type films. The measured Hall mobility values are 12-20% below the impurity/phonon scattering limit for our dopant contents, as calculated by Si transport property calculator [5]. We also observed good correlation of carrier concentrations with SIMS dopant concentrations.

**Table 2.** Two examples of Hall mobility in n- and p- type doped HWCVD epitaxial Si films.

| Film type | Carrier concentration $(cm^{-3})$ | Hall mobility $(cm^2/Vs)$ | Impurity/phonon scattering $(cm^2/Vs)$ (Ref. 5) |
|---|---|---|---|
| n-type | $10^{17}$ | 613 | 730 |
| p-type | $8 \times 10^{17}$ | 164 | 184 |

## DISCUSSION

The temperature dependence of the growth phase was already noted by us in [2]. Here, we note that the onset of the fast and stable epitaxy above about 620 - 650C is rather sharp and we have not yet achieved stable epitaxy below 600C. This suggests a strong connection to surface dehydrogenation process during the epitaxial growth. Lin et al. [6] directly measured c-Si (100) surface dehydrogenation rate as function of temperature. Using their activation energy and prefactor values, we have estimated that when $T_s$ rises above 600C, the residence time of monohydride hydrogen at the growing surface becomes less than the typical time to grow our one epitaxial monolayer. As suggested in [2], the surface is completely dehydrogenated in the stable epitaxial regime. This dehydrogenation and increased surface mobility also explain the increase of the growth rate with temperature seen in Table 1.

The films grown at the different flows in Fig. 3 correspond to epitaxial growth at 670°C. Therefore, epitaxy proceeds equally well both in silane depletion and non-depletion regimes (below and above $f = 30$ sccm). In both regimes, above 10 mTorr, the Si atoms from the filament experience enough collisions (>10) to react with $SiH_4$ and form $Si_2H_2$ [7] or other precursor. These are likely the main growth radicals in both regimes; however, in the depletion regime there is also significant accumulation of atomic and molecular hydrogen in the system due to their long residence time. As a result, the growth chemistries at the surface are expected to differ. Yet, epitaxy takes place in both regimes suggesting a universal growth mechanism. Moreover, at pressures below 5 mTorr, the mean free path of the Si atoms from the filament becomes comparable with the filament-substrate distance. Thus, growth in this regime proceeds directly from Si radicals. Yet, stable epitaxy still takes place. Our observation of epitaxy in all these regimes show that HWCVD epitaxial growth is a rather robust process as long as growth temperature is sufficiently high. Further optimization of growth conditions are expected to bring further increase in the growth rate above the 300 nm/min we have already achieved.

High carrier mobilities in doped films (Table 2), close to impurity/phonon scattering limit, suggest that the carrier scattering from lattice defects such as dislocations will not be a limiting factor for the future photovoltaic device work. This agrees with good structural quality of the films seen in TEM (Fig. 2). The next important issue -- the photocarrier lifetime -- is under study in our laboratory.

## SUMMARY

The phase of Si film growth on (100) wafers by the hot-wire CVD technique strongly depends on the growth temperature. The strong temperature effect on the resulting growth phase (epitaxial or disordered) and the growth rate is likely due to the strongly temperature-dependent hydrogen residence time at the growing surface. Pressure-flow dependences suggest that the epitaxy is relatively insensitive of growth radical chemistry. HWCVD epitaxial Si films have low defect concentrations and high carrier mobilities.

## ACKNOWLEDGEMENTS

The authors thank James Doyle for providing insights into HWCVD Si growth chemistry and suggestion for the low pressure experiment. This work is supported by the U.S. Department of Energy under Contract No. DE-AC36-99G010337.

## REFERENCES

1. A. G. Aberle, Thin Solid Films 511 (2006) 26.
2. Q. Wang, C. W. Teplin, P. Stradins, B. To, K. M. Jones and H. M. Branz, Journal of Applied Physics 100 (2006) 5.
3. C. W. Teplin, H. M. Branz, K. M. Jones, M. J. Romero, P. Stradins and S. Gall, Mat. Res. Soc. Symp. Proc. 989 (2007) 133.
4. C. W. Teplin, E. Iwaniczko, B. To, H. Moutinho, P. Stradins and H. M. Branz, Physical Review B 74 (2006) 5.
5. U. of Delaware - IGERT, http://www.udel.edu/igert/pvcdrom/APPEND/SILICON.HTM.
6. D. Lin and C. R., Physical Review B 60 (1999) R8461.
7. W. Zheng and A. Gallagher, Thin Solid Films 516 (2008) 929.

# Crystallization Techniques

Mater. Res. Soc. Symp. Proc. Vol. 1066 © 2008 Materials Research Society 1066-A12-02

# Influence of the Structural Properties of Microcrystalline Silicon on the Performance of High Mobility Thin-Film Transistors

Kah-Yoong Chan[1,2], Dietmar Knipp[1], Reinhard Carius[2], and Helmut Stiebig[2]

[1]School of Engineering and Science, Jacobs University Bremen, Bremen, 28759, Germany
[2]IEF5-Photovoltaics, Research Center Juelich, Juelich, 52425, Germany

## ABSTRACT

The influence of the crystalline volume fraction of hydrogenated microcrystalline silicon ($\mu$c-Si:H) on the performance of thin-film transistors (TFTs) processed at temperatures below 180 °C was investigated. TFTs employing $\mu$c-Si:H channel material prepared near the transition to amorphous growth exhibit the highest electron charge carrier mobilities exceeding 50 cm$^2$/Vs. The influence of the crystalline volume fraction of the intrinsic $\mu$c-Si:H material on the transistor parameters like the charge carrier mobility and the contact resistance will be discussed.

## INTRODUCTION

Thin-film transistors (TFTs) are key elements in large area electronics. To date, TFTs based on amorphous silicon (a-Si:H) are widely used as pixel switches for display backpanels [1]. However, the realization of more complex peripheral circuitry is not possible due to the low charge carrier mobility and device instability of a-Si:H [1-2]. So far external drivers are needed or the circuitry has to be realized by polycrystalline silicon (poly-Si) TFTs with high charge carrier mobilities and stable threshold voltages [1]. However, the fabrication cost of poly-Si TFTs is higher due to high fabrication temperatures or additional laser crystallization steps.

Hydrogenated microcrystalline silicon ($\mu$c-Si:H) is a promising alternative to existing technologies due to its high device charge carrier mobility [3]. The material consists of amorphous phases, crystallites and voids [4], and is usually deposited at low temperature by plasma-enhanced chemical vapor deposition (PECVD) using a high hydrogen dilution. The material properties are influenced by the silane concentration (SC) during the deposition (SC = SiH$_4$/(SiH$_4$+H$_2$), where SiH$_4$ is the silane flow rate and H$_2$ is the hydrogen flow rate), the applied plasma power [4] and the plasma excitation frequency [5]. The microstructure of the $\mu$c-Si:H can be varied from highly crystalline to material where amorphous growth prevails. In recent years $\mu$c-Si:H has been deployed as active material in TFTs and high device charge carrier mobilities have been demonstrated [6-9].

This paper reports on the characterization of top-gate staggered TFTs employing intrinsic (i) $\mu$c-Si:H as channel layer. The properties of the channel layer were varied from highly crystalline to material where amorphous growth prevails. We investigated the dependence of the transistor parameters like the electron charge carrier mobility and the contact resistance on the crystalline volume fraction ($X_C$) of the $\mu$c-Si:H channel material.

## EXPERIMENT

Fig. 1 shows the schematic cross-section of the investigated μc-Si:H TFT. The drain and source contacts were realized by chromium. Afterwards, a n-type μc-Si:H film with a dark conductivity of 10 S/cm (activation energy = 17 meV) was deposited by PECVD at 180 °C to form ohmic contacts between the drain and source electrodes and the i-μc-Si:H channel material. The i-μc-Si:H film with a thickness of 100 nm was prepared by PECVD at 160 °C, in the high pressure (1330 Pa) and high power (0.3 W/cm²) regime, which facilitates the deposition of material at high deposition rates of up to 25 nm/min [10]. The thickness of the channel layer was chosen to be 100 nm to ensure high quality μc-Si:H on the film surface, which forms the channel of the transistor [4,11]. On the other hand the channel layer should be thin to minimize the series resistance between the drain/source electrode and the accumulation region close to the gate dielectric [12]. The n- and i-layers were prepared at an excitation frequency of 13.56 MHz. Following the deposition of the i-layer, a gate dielectric of 300 nm was prepared by PECVD at 150 °C. Finally, the gate electrode was formed by an aluminum film. To allow for the fast evaluation of the materials and the device properties, a simple two-mask photolithographic process was developed. In order to improve the device behavior all transistors were annealed at an elevated temperature of 150 °C for 30 minutes under ambient conditions [13].

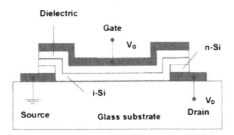

**Figure 1.** Schematic cross-section of a top-gate staggered microcrystalline silicon thin-film transistor.

The crystalline volume fraction of the channel layer of the TFTs was varied by changing the silane concentration from 0.5 % to 1.5 %. Raman spectroscopy (laser wavelength = 488 nm) was employed to estimate the crystalline volume fraction of the i-layers according to the formula:

$$X_C = \frac{I_{500} + I_{520}}{I_{480} + I_{500} + I_{520}}, \tag{1}$$

where $I_{480}$, $I_{500}$ and $I_{520}$ are the areas of the corresponding Gaussian peaks used to fit the measured Raman spectra near 480 cm⁻¹ (associated with the amorphous phase), 500 cm⁻¹ and 520 cm⁻¹ (associated with the crystalline phase), respectively. Upon increase of the SC from 0.5 % to 1 % and 1.2 %, the i-μc-Si:H layer exhibits a decrease in the crystalline volume fraction from 66 % to 54 % and 47 %. The crystalline volume fraction drops dramatically to below 6 % when the

SC is increased to 1.3 % and 1.5 %. A silane concentration in the range of 1 % to1.2 % is considered as the transition to the amorphous growth regime.

## RESULTS AND DISCUSSION

The transfer characteristics of a μc-Si:H TFT (i-layer, $X_C$ = 54 %) with a channel length of 200 μm and a channel width of 1000 μm are shown in Fig. 2, for drain voltages, $V_D$, of 0.1 V and 1 V. A linear device charge carrier mobility of 55 cm²/Vs and a threshold voltage of around 3 V were extracted from the transfer characteristics. The μc-Si:H TFT exhibits a subthreshold slope of 0.34 V/decade. The on/off ratio of the TFT for low drain voltages is larger than $10^5$.

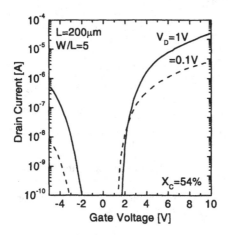

**Figure 2.** Transfer characteristics of a microcrystalline silicon thin-film transistor with a channel length of 200 μm and channel width to channel length ratio of 5. The microcrystalline silicon channel material was grown near the transition to amorphous growth.

The dependence of device charge carrier mobility on the channel length for different crystalline volume fractions is examined in the following. Fig. 3 shows the extracted device charge carrier mobility as a function of the channel length for crystalline volume fractions of 5.9 %, 54 % and 66 %. The device charge carrier mobilities were extracted from the transfer characteristics measured at a drain voltage of 1 V. The TFTs employing an i-μc-Si:H channel layer with a crystalline volume fraction of 5.9 % exhibit the lowest device charge carrier mobilities. The extracted device charge carrier mobilities are below 1 cm²/Vs, resembling the mobility values obtained for a-Si:H TFTs. The TFTs with a crystalline volume fraction of 54 % exhibit the highest device charge carrier mobilities. Apparently, the device charge carrier mobility is highest for i-μc-Si:H grown near the transition to amorphous growth regime. This behavior differs from the investigations of the Hall mobility for μc-Si:H bulk layers, which show that the Hall mobility continuously increases with increasing crystalline volume fraction [14,15]. For the investigated μc-Si:H TFTs, the drop of the device charge carrier mobility for high $X_C$ can be likely attributed to the high void fraction and the defects associated with cracks on the surface between the crystalline columns in the highly crystalline material [4,16,17]. This can affect the electron transport along the channel. No clear correlation is observed between the threshold voltage and the crystalline volume fractions, since the threshold voltage depends not only on the defect density of the i-µc-Si:H, but also on the work function difference between the gate metal and the semiconductor and trapped charges within the gate oxide and its interface with the channel material [18].

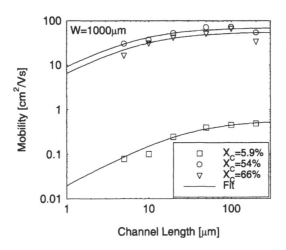

**Figure 3.** Extracted device charge carrier mobility as a function of channel length for TFTs with crystalline volume fraction of 5.9 %, 54 % and 66 %.

The data in Fig. 3 were fitted according to the following equation, which takes the influence of the contact resistance on the device charge carrier mobility into account [9]:

$$\mu = \mu_0 \cdot \frac{L}{L + \mu_0 \cdot C_G \cdot r_C \cdot \left( V_G - V_T - \frac{V_D}{2} \right)}. \tag{2}$$

Here µ is the experimentally extracted device charge carrier mobility, $\mu_0$ is the intrinsic charge carrier mobility which is free of contact effects, L is the channel length, $C_G$ is the gate capacitance per unit area, $r_C$ is the normalized drain and source contact resistance, $V_G$ is the gate voltage and $V_T$ is the threshold voltage. The extracted normalized contact resistance, $r_C$, and the intrinsic charge carrier mobility, $\mu_0$, of the TFTs are plotted as a function of the crystalline volume fraction of the intrinsic channel material in Fig. 4. The intrinsic charge carrier mobility obtained from the fit for the data in Fig. 3 is highest for the TFTs with crystalline volume fraction of 54 %. For the TFTs with $X_C$ higher than 47 %, the normalized contact resistances are below 1 kΩ·cm. The contact resistance increases dramatically to above 100 kΩ·cm for the TFTs with $X_C$ of 5.9 %.

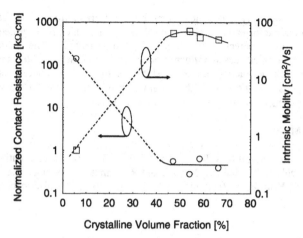

**Figure 4.** Extracted intrinsic charge carrier mobility and normalized contact resistance as a function of the crystalline volume fraction.

The contact resistance can be described by [19]:

$$r_C = \sqrt{\rho_C \cdot R_s} \cdot \frac{\cosh(d/L_T)}{\sinh(d/L_T)}, \tag{3}$$

where $\rho_C$ is the specific contact resistivity of the drain/source to channel interface and $R_S$ is the sheet resistance of the microcrystalline silicon channel layer. $L_T$ is the characteristic length over which the current spreads under the drain and source electrode, and d is the length of the overlap region between gate and drain/source. Since the length of the overlap of the gate and the drain/source contacts of the investigated TFTs is significantly larger than $L_T$ the normalized contact resistance, $r_C$, can be simplified by [19]:

$$r_C = \sqrt{\rho_C \cdot R_s}. \tag{4}$$

Based on the Eqn. (4) the distinctly high contact resistance of the TFTs with low crystalline volume fraction of 5.9 % can be explained by the increased sheet resistance of the microcrystalline silicon channel layer. The increase of the sheet resistance is caused by the drop of the charge carrier mobility. Therefore, a microcrystalline channel layer with low crystalline volume fraction does not only lead to a decrease of the charge charier mobility. The contact resistance, $r_C$, is also increased since the contact resistance is directly correlated with the charge carrier mobility.

## CONCLUSIONS

Top-gate μc-Si:H TFTs were realized with high electron charge carrier mobilities exceeding 50 cm²/Vs at temperatures below 180 °C. The experimental results reveal that the highest charge carrier mobility is obtained when the i-μc-Si:H is grown near the transition to amorphous growth regime instead of highly crystalline regime, which is suggested as due to the high void fraction and the defects in the highly crystalline material. The crystalline volume fraction of the channel material has an influence on the contact resistance of the TFTs. The TFTs with low crystalline volume fraction of 5.9 % exhibit a distinctly high contact resistance above 100 kΩ·cm in comparison to the TFTs with crystalline volume fractions of 47 % to 66 %, which exhibit significantly reduced contact resistances below 1 kΩ·cm.

## ACKNOWLEDGMENTS

The authors are thankful to S. Bunte and Y. Mohr (IBN-PT) for preparation of the $SiO_2$, M. Hülsbeck, J. Kirchhoff, T. Melle, S. Michel, and R. Schmitz for technical assistances and E. Bunte, A. Gordijn, D. Hrunski, S. Reynolds and V. Smirnov for helpful discussions.

## REFERENCES

1. T. Tsukada, *Technology and Applications of Amorphous Silicon, Springer Series in Material Science, 37,* edited by R. A. Street (Springer-Verlag, Berlin, Germany, 2000).
2. B. Stannowski, R. E. I. Schropp, R. B. Wehrspohn and M. J. Powell, *J. Non-Cryst. Solids* **299-302,** 1340 (2002).
3. J. I. Woo, H. J. Lim, and J. Jang, *Appl. Phys. Lett.* **65,** 1644 (1994).
4. O. Vetterl, F. Finger, R. Carius, P. Hapke, L. Houben, O. Kluth, A. Lambertz, A. Mück, B. Rech and H. Wagner, *Sol. Energy Mater. Sol. Cells* **62,** 97 (2000).
5. F. Finger, P. Hapke, M. Luysberg, R. Carius, H. Wagner, M. Scheib, *Appl. Phys. Lett.* **65,** 2588 (1994).
6. I.-C. Cheng and S. Wagner, *Appl. Phys. Lett.* **80,** 440 (2002).
7. C.-H. Lee, A. Sazonov and A. Nathan, *Appl. Phys. Lett.* **86,** 222106 (2005).
8. K. Kandoussi, A. Gaillard, C. Simon, N. Coulon, T. Pier and T. Mohammed-Brahim, *J. Non-Cryst. Solids* **352,** 1728 (2006).
9. K.-Y. Chan, E. Bunte, H. Stiebig and D. Knipp, *Appl. Phys. Lett.* **89,** 203509 (2006).
10. B. Rech, T. Roschek, T. Repmann, J. Müller, R. Schmitz and W. Appenzeller, *Thin Solid Films* **427,** 157 (2003).
11. H. Shirai, T. Arai and T. Nakamura, *Appl. Surf. Sci.* **113&114,** 111 (1997).
12. I. -C. Cheng, S. Allen and S. Wagner, *J. Non-Cryst. Solids* **338-340,** 720 (2004).
13. K.-Y. Chan, E. Bunte, H. Stiebig and D. Knipp, *J. Appl. Phys.* **101,** 074503 (2007).
14. T. Bronger and R. Carius, *Thin Solid Films* **515,** 7486 (2007).
15. K. Shimakawa, *J. Non-Cryst. Solids* **266-269,** 223 (2000).
16. M. Tzolov, F. Finger, R. Carius and P. Hapke, *J. Appl. Phys.* **81,** 7376 (1997).
17. R. W. Collins and B. Y. Yang, *J. Vac. Sci. Technol. B* **7,** 1155 (1989).
18. D. W. Greve, *Thin-Film Transistors, Field Effect Devices and Applications: Devices for Portable, Low-Power, and Imaging Systems,* 1st Ed. (Prentice Hall, New Jersey, 1998), Chap. 7, p. 286.
19. D. K. Schroder, *Semiconductor Material and Device Characterization,* 2nd Ed. (John Wiley & Sons, Inc., New York, Chichester, Weinheim, Brisbane, Singapore, 1998), Chap. 3, p. 152.

# Thin-Film Transistors II

Mater. Res. Soc. Symp. Proc. Vol. 1066 © 2008 Materials Research Society          1066-A13-03

# The Positive Gate Bias Annealing Method for the Suppression of a Leakage Current in the SPC-Si TFT on a Glass Substrate

Sang-Geun Park, Joong-Hyun Park, Seung-Hee Kuk, Dong-Won Kang, and Min-Koo Han
School of Electrical Engineering Science, Seoul National University, 130Dong-305Ho San 56-1
Shilim 9 Dong Gwanak-Gu, Seoul, 151-742, Korea, Republic of

## ABSTRACT

We fabricated PMOS SPC-Si TFTs which show better current uniformity than ELA poly-Si TFTs and superior stability compare to a-Si:H TFT on a glass substrate employing alternating magnetic field crystallization. However the leakage current of SPC-Si TFT was rather high for circuit element of AMOLED display due to many grain boundaries which could be electron hole generation centers. We applied off-state bias annealing of $V_{GS}$=5V, $V_{DS}$=-20V in order to suppress the leakage current of SPC-Si TFT. When the off-state bias annealing was applied on the SPC-Si TFT, the electron carriers were trapped in the gate insulator by high gate-drain voltage (25V). The trapped electron carriers could reduce the gate-drain field, so that the leakage current of SPC-Si TFT was reduced after off-state bias annealing. We applied AC-bias stress on the gate node of SPC-Si TFT for 20,000 seconds in order to verify that the leakage current of SPC-Si TFT could be remained low at actual AMOLED display circuit after off-state bias annealing. The suppressed leakage current was not altered after AC-bias stress. The off-state bias annealed SPC-Si TFT could be used as pixel element of high quality AMOLED display.

## INTRODUCTION

Hydrogenated amorphous silicon (a-Si:H) TFTs are considered as the pixel element of active matrix organic light emitting diode (AMOLED) due to excellent uniformity in large areas. However the threshold voltage of a-Si:H TFT is increased easily under electrical bias stress [1]. Poly-Si TFTs employing excimer laser annealing (ELA) show very good electric characteristics such as high mobility and good stability. But the non-uniformity of poly-Si TFT caused by inherent fluctuation of excimer laser should be improved. Recently solid phase crystallized silicon TFT (SPC-Si TFT) on the glass substrate has gained a considerable attention for pixel element of AMOLED due to a better current stability than a-Si:H TFTs, and an improved current uniformity compare to ELA poly-Si TFTs. The leakage current of SPC-Si TFT is rather high due to fairly many grain boundaries in the channel which behaves generation and recombination centers[2].

The purpose of our work is to propose a new method to suppress a leakage current of SPC-Si TFT employing the positive gate bias annealing. Recently the pixel circuit which can suppress the leakage current effect on the AMOLED display was reported [3]. However the method which can reduce the leakage current of SPC-Si TFT would be more efficient for high quality AMOLED display. When the positive gate-source bias(5V) on the gate node and negative source-drain bias(-20V) were applied in the SPC-Si TFT, the leakage current was decreased due to the drain field reduction caused by trapped electron charges in a gate insulator. The reduced leakage current was remained low even after the AC bias stress for 20,000 seconds.

## EXPERIMENT

### Fabrication of device

We fabricated PMOS SPC-Si TFTs of coplanar structures on the glass substrate. Amorphous Si was deposited by 50nm. It was crystallized at 700°C for 15 minutes employing alternating magnetic field crystallization (AMFC). The glass substrate was not damaged during the crystallization. The grain size of crystallized Si was around 300Å. The hydrogen plasma treatment was applied in order to improve the s-slope and threshold voltage of SPC-Si TFT [4]. $SiO_2$ was deposited by 59nm as a gate insulator after active layer pattering. Then the gate electrode was patterned and source, drain were defined by ion implantation (self-aligned structure). The channel width of fabricated SPC-Si TFT was 10μm and the channel length was 7 μm. Figure 1 shows the cross section view of the fabricated SPC-Si TFT.

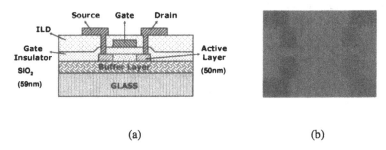

(a)                                                    (b)

Figure 1. (a) The schematic of the fabricated SPC-Si TFT. (b) The microscopic picture of the fabricated SPC-Si TFT

### Off-sate bias annealing

Figure 2 shows transfer characteristics of SPC-Si TFT. When the leakage current of SPC-Si TFT was measured at $V_{GS} = 5V$, $V_{DS} = -5V$, it was about 1.49 pA. It was rather high for pixel circuit of AMOLED display. We applied off-state bias annealing with $V_{GS} = 5V$, $V_{DS} = -20V$ for 1,000 seconds in order to suppress the leakage current of SPC-Si TFT. The leakage current of the SPC-Si TFT was reduced from 1.49 pA to 0.47 pA at $V_{GS} = 5V$, $V_{DS} = -5V$ due to gate filed reduction caused by trapped electron carriers in the gate insulator. The leakage current was not increased even at the high gate voltage while it was increased as the gate voltage increased at the initial SPC-Si TFT.

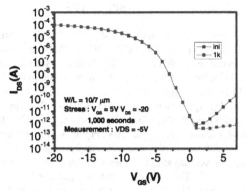

Figure 2. The transfer characteristics of SPC-Si TFT at initial and after off-state bias annealing. The leakage current was reduced significantly after off-state bias annealing.

## Thermal annealing of off-state bias annealed SPC-Si TFT

We applied thermal annealing at the off-state bias annealed SPC-Si TFT to verify that the trapped electron carriers could be de-trapped by thermal energy under practical AMOLED display application. We applied the off-state bias annealing of $V_{GS}=10V$, $V_{DS} = -20V$ for 1,000seconds in order to suppress the leakage current of SPC-Si TFT. The gate-drain voltage(30V) during off-bias annealing, was sufficient for reducing the leakage current of SPC-Si TFT. Then we annealed it at 60 °C, 90°C and 120°C for 1 hour respectively. As shown in figure 3, the leakage current of the SPC-Si TFT was not altered after thermal annealing. We can deduce that the trapped electron carriers could not be released by thermal energy under 120°C. Since the practical operation temperature of AMOLED display is under 60°C, the reduced leakage current could be remained low at the practical application.

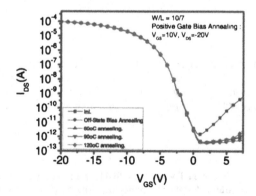

Figure 3. Thermal annealing of off-state bias annealed SPC-Si TFT. The leakage current of SPC-Si TFT was remained low after the thermal annealing.

## AC bias stress at the off-state bias annealed SPC-Si TFT

We applied AC gate bias stress in the SPC-Si TFT after the off-state bias annealing in order to show that the leakage current of a SPC-Si TFT would be remained low when it used as the practical pixel element of the AMOLED display. When it used practical pixel circuit of AMOLED display, the AC signal is applied on the gate node continuously. When it is used as driving pixel of AMOLED display, it has to drive up to 2μA current for sufficient OLED illumination. We applied AC signal on the gate node of SPC-Si TFT for 20,000 seconds, after off-state bias annealing of $V_{GS}$=5V, $V_{DS}$=-20V. The frequency of AC signal was 60Hz with 50% pulse width. The value of AC bias was -6V and 0V. Figure 4(b) shows the AC signal diagram. The $V_{DS}$ was fixed as -6V during AC bias stress, so that the source-drain current was about 2.5 μA for on period during the AC bias stress. Figure 4(a) shows the AC bias stress result on off-state bias annealed SPC-Si TFT. The leakage current was not altered under the AC bias stress. The off-state bias annealed SPC-Si TFT could be operated with low leakage current at the practical application of AMOLED display.

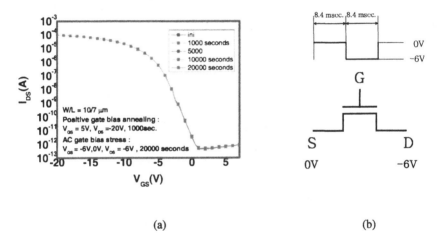

(a)                                        (b)

Figure 4. (a) The AC-bias stress result in SPC-Si TFT after off-state bias annealing. The leakage current of SPC-Si TFT was remained low even after AC-bias stress for 20,000 seconds. (b) The AC signal of stress condition. The gate voltage was alternated from 0V to -6V with 50% duty cycle and frequency of 60Hz. The drain-source current of SPC-Si TFT was about 2.5 μA during on period.

## DISCUSSION

The mobility of SPC-Si TFT was about 28.41 V/cm$^2$ sec. It was increased slightly to 29.38 V/cm$^2$ sec (3.4%) after off-state bias annealing. The on-current was not altered by off-sate bias annealing. It could be explained trapped electron carriers in the $SiO_2$ gate insulator. The electron carriers were accumulated at the active layer by positive gate bias during the positive bias annealing. These electron carriers were affected by the gate-drain field near the drain junction. The electron carriers were trapped at the $SiO_2$ insulator near the drain junction. This trapped electron carriers could reduce the gate field on the drain junction. As a result the leakage current was reduced. This trapped electron carriers are shown in figure 5. The leakage current at the high gate voltage could be explained by band to band tunneling mechanism [5]. The energy band of valence band and conduction band was bended by high gate voltage. The energy gap of valence band and conduction band became narrow so that many electron tunneling could occur. However after positive gate bias annealing the trapped electron charges in the gate insulator reduced gate-drain field, the energy band could not be banded so much. The leakage current induced by band to band tunneling was suppressed successfully employing positive gate bias annealing.

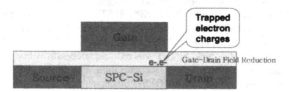

Figure 5. The trapped electron carriers at the $SiO_2$ gate insulator. It could reduce gate-drain field so the leakage current of SPC-Si TFT could be reduced.

In order to identify the trapped electron carriers in a gate insulator by the off-state bias annealing, we also measured C-V characteristics of SPC-Si TFT before and after off-state bias annealing. The result is shown in Figure 6. The normalized value of gate-drain capacitance was shifted positively after the off-state bias annealing due to the trapped electron charges in a gate insulator near the drain junction [6].

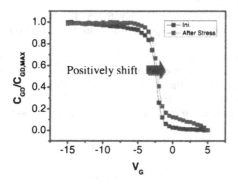

Figure 6. The normalized value of gate-drain capacitance in SPC-Si TFT before and after the off-sate bias annealing. The normalized value of capacitance was shifted positively after the off-state bias annealing due to the trapped charges in a gate insulator

## CONCLUSIONS

We fabricated PMOS SPC-Si TFT on a glass substrate employing alternating magnetic field crystallization. The SPC-Si TFT showed better current uniformity than excimer laser annealing poly-Si TFT and better stability than a-Si:H TFT. However the leakage current of SPC-Si TFT was rather high for circuit element of AMOLED display. In this work we applied off-state bias annealing of $V_{GS}$=5V, $V_{DS}$=-20V in order to suppress the leakage current of SPC-Si TFT. When the off-state bias annealing was applied in the SPC-Si TFT, the electron carriers were trapped in the gate insulator by high gate-drain field. The trapped electron carriers could reduce the gate-drain field, so that the leakage current of SPC-Si TFT was decreased even at the high gate voltage. We applied AC-bias stress on the gate node of SPC-Si TFT for 20,000 seconds in order to verify that the leakage current of SPC-Si TFT could be remained low at actual AMOLED display circuit after off-state bias annealing in the SPC-Si TFT. The suppressed leakage current was not altered after AC-bias stress.

The leakage current of SPC-Si TFT was reduced successfully and it was not altered even at the AC-bias stress employing off-state bias annealing. The annealed SPC-Si TFT could be used as pixel element of high quality AMOLED display.


## REFERENCES

1. M.J.Powell, "Charge trapping instabilities in amorphous silicon-silicon nitride thin-film transistors", *Appl. Phys. Lett. 43(6)*, pp.597-599, 1983.
2. F. V. Farmakis, J. Brini, G. Kamarinos, C. T. Angelis, C. A. Dimitriadis, and M. Miyasaka, "On-Current Modeling of Large-Grain Polycrystalline Silicon Thin-Film Transistors", *IEEE TRANSACTIONS ON ELECTRON DEVICES*, VOL. 48, No. 4, Apr. 2001
3. J-H Lee, H-S Park, J-H Jeon, and M-K Han, "Suppression of TFT leakage current effect on active matrix displays by employing a new circular switch", *Solid-State Electronics*, vol.52, Issue 3. pp.467-472, Mar., 2008.
4. F. Kail, A. Hadjadj, P. Roca i Cabarrocas, "Hydrogen diffusion and induced-crystallization in intrinsic and doped hydrogenated amorphous silicon films", *Thin Solid Films*, vol. 487, Issues 1-2, 1 pp. 126-131, Sep, 2005
5. M. Yazakis, S. Takenaka1 and H. Ohshima, "Conduction Mechanism of Leakage Current Observed in Metal-Oxide-Semiconductor Transistors and Poly-Si Thin-Film Transistors", *Jpn. J. Appl. Phys.* Vol. 31 pp.206-209 Part 1, No. 2A, 15 Feb., 1992
6. K-C Moon, J-H Lee, and M-K Han, "The Study of Hot-Carrier Stress on Poly-Si TFT Employing C-V Measurement", IEEE Transaction on Electron Device, vol.52, No.4, pp.512-517, April, 2005

Mater. Res. Soc. Symp. Proc. Vol. 1066 © 2008 Materials Research Society       1066-A13-04

# High Performance Bottom Gate μc-Si TFT Fabricated by Microwave Plasma CVD

Akihiko Hiroe, Akinobu Teramoto, and Tadahiro Ohmi
New Industry Creation Hatchery Center, Tohoku University, Aza-Aoba 6-6-10, Aramaki,
Aobaku, Sendai, Japan

## ABSTRACT

Deposition trend of μc-Si was investigated using microwave (2.45GHz) plasma enhanced CVD. μc-Si films with the preferential orientation of (111) and (220) were deposited and compared. Raman scattering results show that the (111) preferentially oriented film has higher crystallinity while ESR measurements result in the fact that the (220) preferentially oriented film has smaller dangling bond density. Bottom gate thin film transistors (TFT's) were fabricated using these μc-Si films as channel layer and evaluated. $H_2$ plasma post-treatment has been found to be effective to improve the TFT characteristics. Mobility of about $1.4cm^2/Vsec$ and on/off ratio of more than $10^5$ have been achieved.

## INTRODUCTION

Microwave (2.45GHz) plasma enhanced CVD has been studied for the deposition systems of large area devices [1-4] since plasma density is high ($>10^{11}cm^{-3}$), electron temperature is low ($<2eV$), and with proper antenna configurations it is easy to realize large area plasma [5]. Regarding the electrical properties of devices fabricated by microwave plasma CVD, however, only preliminary result for solar cell has been presented [3]. In this paper, we report the investigation results of the deposition trend of μc-Si using microwave plasma enhanced CVD together with the TFT fabrication results.

## EXPERIMENT

Figure1 (a) shows the cross sectional view of the microwave plasma CVD chamber used in this experiment. Microwave is introduced into the chamber through quartz window which sits at the top of the chamber. Two stage shower plate is situated below this quartz window, where plasma and/or radical generation gas is supplied through upper shower plate while process gas is fed through the lower shower plate. In our experiment, Ar and/or $H_2$ are used as plasma/radical generation gas and $SiH_4$ as process gas. Alumina coated susceptor is located 50mm below the lower shower plate. $H_2$ plasma treatment was also done by microwave plasma chamber. Basic concept of this chamber is the same as the one used for CVD with the difference being the capability to apply substrate bias and the substrate size of 200mm instead of 33mm for CVD chamber.

Figure 1 (b) shows the schematic illustration of the cross sectional view of TFT we fabricated. Low resistive $p^+$ Si wafer (~8mΩcm) was used as substrate which also worked as bottom gate of the transistor. On top of Si substrate, $SiO_2$ layer, which worked as gate dielectric of transistor, was deposited by atmospheric pressure CVD at 400°C, and then annealed at 800°C in $O_2$ ambient for 30min in quartz tube furnace. Non-doped μc-Si layer and P-doped $n^+$ layer were deposited on the gate dielectric after Ar/$H_2$ plasma pre-treatment. Resistivity of $n^+$ layer was 0.1~0.3Ωcm. After island patterning, aluminum layer was deposited by vacuum evaporation.

Aluminum etching and $n^+$ etching to form channel region were carried out before the removal of native oxide on the back side of Si substrate, and then aluminum layer is deposited by vacuum evaporation to form backside contact. After as-deposited measurements, samples were treated with $H_2$ plasma, passivated with silicon nitride, and then annealed at 450°C/40Pa in Ar for two hours. Prior to the TFT fabrication experiments, μc-Si deposition trend was investigated using Si wafer with thermal oxide of 100nm thick. We investigated deposition temperature of 250 to 350°C, pressure of 4 to 13.3Pa, and microwave power of 1700 to 2300W.

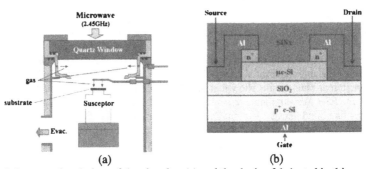

(a)                                              (b)

**Figure 1** Cross sectional view of the chamber (a) and the device fabricated in this experiment (b)

## RESULTS AND DISCUSSION

Deposited films usually show preferential orientation of (111) (figure 2). Flow rate ratio of this sample is $SiH_4/H_2$=3/35sccm. Figure 3 shows $H_2$ flow rate dependence of $I_{(220)}/I_{(111)}$ ratio together with $Ar/H_2$ pre-treatment sample. In this figure, $SiH_4$ flow rate is fixed at 0.48sccm, and two flow rate cases of 10 and 35sccm for Ar+$H_2$ are plotted. With Ar+$H_2$ flow rate of 35sccm, film stays at (111) oriented even as $H_2$ flow rate increases while film becomes more to (220) oriented when Ar+$H_2$ flow rate is set to 10sccm, and peak intensity ratio ($I_{(220)}/I_{(111)}$) markedly increases with $Ar/H_2$ plasma pre-treatment. Figure 5 shows the XRD spectrum of the film with $Ar/H_2$ plasma pre-treatment. It is clearly seen that the preferential orientation of this film is (220). The μc-Si film deposited by the flow rate of $SiH_4/H_2$=3/35sccm will be referred to as "high flow rate sample" and the μc-Si film deposited by the flow rate of $SiH_4/H_2$=0.48/10sccm with $Ar/H_2$ plasma pre-treatment will be referred to as "low flow rate sample" hereinafter. Fabricated TFT's using these films will be called likewise (high flow rate sample, low flow rate sample).

Figure 4 shows Raman (a) and electron spin resonance (ESR) spectra of the high flow rate sample and low flow rate sample. In figure 4(a), crystalline fraction ($X_c$) was estimated by the following equation (1).

$$X_c = \frac{I_c}{(I_c + I_a)} \tag{1}$$

Where $I_c$ is the integrated peak intensity from crystalline structure (~520cm$^{-1}$) and $I_a$ is that from amorphous structure (~480cm$^{-1}$). This is a semi-quantitative estimation of the crystalline fraction because it is known that several intermediate peaks exist between these two peaks, and also Raman cross section is not considered. Still, the difference of crystallinity between these two samples is clear, i.e. crystallinity of the high flow rate sample is high compared with the low flow rate sample. On the other hand, dangling bond densities estimated by ESR measurements show that low flow rate sample has less dangling bonds density compared with high flow rate sample by more than one order of magnitude.

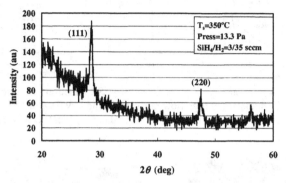

**Figure 2** Typical XRD spectrum

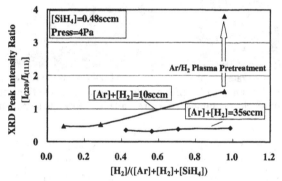

**Figure 3** $H_2$ flow rate ratio dependence of XRD peak intensity ratio ($I_{(220)}/I_{(111)}$) together with the sample with Ar/H₂ plasma pretreatment.

Figure 5 shows the cross sectional SEM images of high flow rate sample (a) and low flow rate sample (b). High flow rate sample shows clear columnar structure while low flow rate sample has smooth surface morphology, which is considered to be consistent with the fact that high flow rate sample has more dangling bonds within the film.

These experimental results show that when the crystallinity of each grain within the film is high, grain boundaries tend to produce more dangling bonds. This is the case with high flow rate sample while in the case of low flow rate sample, crystallinity of each grain is low or in other words net work is relaxed thereby causing less dangling bonds at the grain boundaries (figure 6).

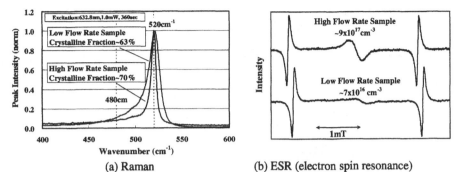

(a) Raman                    (b) ESR (electron spin resonance)

**Figure 4** Raman and ESR spectra of high flow rate sample and low flow rate sample. High flow rate sample shows higher crystallinity but also has more dangling bonds compared with low flow rate sample.

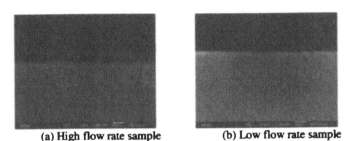

(a) High flow rate sample        (b) Low flow rate sample

**Figure 5** Cross sectional SEM images of high flow rate sample (a) and low flow rate sample. High flow rate sample shows clear columnar structure while low flow rate sample has smooth morphology.

High flow rate sample              Low flow rate sample

**Figure 6** Schematic illustrations of film structures of high flow rate sample (left) and low flow rate sample. Grains of high flow rate sample have better crystallinity compared with low flow rate sample. Because of the high crystallinity, network of each grain of high flow rate sample tends to be more rigid causing more dangling bonds between the grains compared with low flow rate sample.

Firstly, bottom gate μc-Si TFT using low flow rate condition results is presented since dangling bond density is thought to be more important for electrical properties rather than the crystallinity. Results for the high flow rate sample will be mentioned later.

Some TFT's show mobility of about 0.6~7cm$^2$/Vsec while others show equal to or less than 0.1cm$^2$/Vsec. This may be due to the instability of the surface condition. However, it is possible to improve the mobility of the sample with very low initial mobility by H$_2$ plasma treatment. Figure 7 shows the substrate bias dependence of the mobility during the H$_2$ plasma treatment. Sample was treated with H$_2$ plasma with different substrate bias power incrementally from 5W to 45W at the susceptor temperature of 300°C. Mobility improves to about 1cm$^2$/Vsec when treated at bias power of 30W.

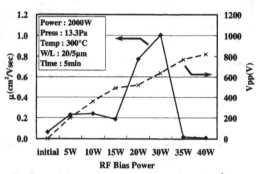

**Figure 7** H$_2$ plasma treatment trend. Mobility improves to about 1cm$^2$/Vsec when treated at 30W.

Figure 8 shows the mobility before and after the H$_2$ plasma treatment for three different samples i.e. low flow rate sample with high initial mobility, low flow rate sample with low initial mobility, and high flow rate sample. As mentioned above, low flow rate sample with low initial mobility shows large improvement after the H$_2$ plasma treatment whereas other two samples show only small or little improvements.

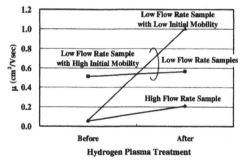

**Figure 8** Mobility of before and after the H$_2$ treatment for three different samples i.e. low flow rate sample with low initial mobility, low flow rate sample with high initial mobility, and high flow rate sample. Low flow rate with low initial mobility sample shows large improvement while other two samples show only small or little improvements.

Figure 9 shows the transfer and output characteristics of the low flow rate sample with low initial mobility after H$_2$ plasma treatment, SiNx passivation, and thermal annealing at 450°C/40Pa in Ar ambient. Mobility of about 1.4cm$^2$/Vsec and on/off ratio of more than 10$^5$ have been achieved.

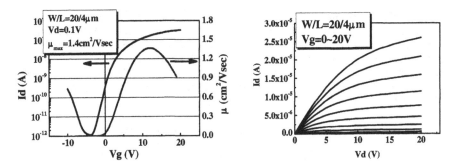

**Figure 9** Transfer and output characteristics of low flow rate sample with low initial mobility after hydrogen plasma treatment, SiNx passivation, and thermal annealing.

## REFERENCES

1. H. Shirai, T.Arai, and H.Ueyama, *Jpn. J. Appl. Phys.* **37,** 1078 (1998)
2. S.Somiya, H.Toyoda, Y.Hotta, and H.Sugai, *Jpn. J. Appl. Phys.* **43,** 7696 (2004)
3. H.Jia , J.K.Saha, N.Ohse,and H.Shirai, J. Non-Cryst. Solids **352,** 896 (2006)
4. T.Takeda, K.Tanaka, H.Inoue, M.Hirayama, T.Tsumori, H.Aharoni, and T.Ohmi, *Jpn. J. Appl. Phys.* **46,** 2542 (2007)
5. T.Ohmi, M.Hirayama, and A.Teramoto, *J. Phys. D, 39* **R1** (2006)

# Solar Cells II

Mater. Res. Soc. Symp. Proc. Vol. 1066 © 2008 Materials Research Society    1066-A14-01

Production Technology of Large-Area, Light-Weight, Flexible Solar Cell and Module

Makoto Shimosawa, Shinichi KAWANO, Takamasa ISHIKAWA, Tetsuro NAKAMURA, Yasushi SAKAKIBARA, Shinji KIYOFUJI, Hirofumi ENOMOTO, Hironori NISHIHARA, Tomoyoshi KAMOSHITA, Masahide MIYAGI, Junichiro SAITO, and Akihiro TAKANO
Photovoltaic Power Div., Fuji Electric Systems, 4003-1, Koei, Nankan-town, Tamana-county, Kumamoto-Pref., 861-0814, Japan

## ABSTRACT

Production technologies of amorphous silicon / amorphous silicon germanium (a-Si/a-SiGe) flexible film solar cells have been established. The solar cells having a unique monolithic device structure using through-hole contacts are continuously fabricated on flexible plastic films. We have developed various production technologies, such as high speed high quality a-Si deposition method, very high speed laser patterning method and so on. We constructed a new solar cell factory in Kumamoto, Japan, and started commercial production in 2006. We have production capacity of 12MW/Y using the new production line that can process 1m-wide and 2000m-long film substrates. We are planning to install two more production lines and have 40MW/Y production capacity in FY 2008.

## INTRODUCTION

We developed a unique monolithic device structure named "SCAF" (Series-Connection through Apertures formed on Film substrate) (Fig.1) [1]. The most prominent feature of "SCAF" structure is through hole contacts made by mechanical punching hole formation and thin film metal film deposition on both sides of the film. There are two kinds of connection holes formed on the plastic film. One is a series connection hole and the other is a current collection hole. Accumulated layers on both sides of the film are divided into unit solar cells by laser patterning lined. The series connected unit cells through the contact holes generate high voltage.

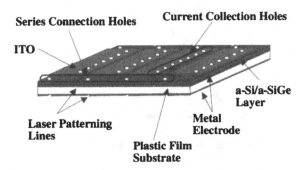

Fig.1 Schematic diagram of SCAF solar cell [1].

In order to fabricate such unique series connection structure on plastic film substrate, we have developed a production line using roll-to-roll method including an improved roll-to-roll film deposition process named "stepping-roll process" as shown in Fig.2. Outline of the process sequence is as follows:

(1) Series connection holes are made on the plastic film substrate by the punching apparatus.
(2) Metal electrodes are deposited on both sides of the film substrate by a conventional roll-to-roll sputtering apparatus.
(3) Current collection holes are made on the plastic film substrate by the punching process.
(4) a-Si, a-SiGe and related layers, ITO and backside metal electrodes are deposited in succession in the stepping-roll apparatus.
(5) Layers deposited on both sides of the substrate are divided by the laser-patterning technique to form unit cells.

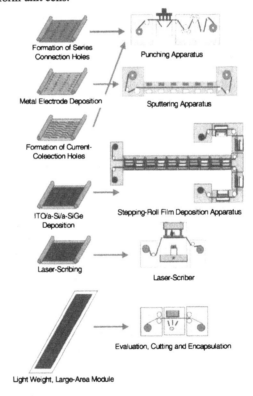

Fig.2 Schematic diagram of production line of flexible film solar cells.

The stepping-roll apparatus consists of plasma enhanced chemical vapor deposition (CVD) chambers for depositing of a-Si, a-SiGe and related layers, and sputtering chambers for depositing transparent metal electrode (ITO) and backside electrodes. Two film rolls, which have the length of 2000m, can be processed at the same time in the stepping-roll apparatus without breaking vacuum.

We have developed various production technologies. Some of the technologies are listed below:

(1) Low temperature and high speed textured electrode deposition method [2]
(2) High speed amorphous silicon deposition method [3-5]
(3) Modified roll-to-roll deposition apparatus technology (Stepping-roll apparatus) [1]
(4) Large-area high quality transparent electrode deposition method [1]
(5) Very high speed laser patterning method

These developed technologies were applied to the production line in our new Kumamoto PV factory (Fig. 3). Our new factory started commercial production in 2006. We have production capacity of 12MW/Y now, and are introducing two more production lines to have 40MW/Y production capacity in FY 2008.

Fig.3 Kumamoto PV factory.

We have already developed and shipped two types of modules. One is flexible type module and the other is steel plate based building-integrated module (Fig. 4 and 5). Both sides of the film solar cells are covered by ethylene tetrafluoroethylene (ETFE) and encapsulating layers in the flexible type module to show flexibility. In the steel plate based building-integrated module, top surface is covered by ETFE and another side is attached to the steel plate to make curved rooftop structure. Typical module size is 4m long, 0.5m wide and 1-2mm thick. The module consists of 8 unit modules (4 series, 2 parallel connection). Each unit module generates 11.5W at the maximum power point. Nominal power, Voc and Isc of the 4m long module are 92W, 430V, and 0.39A, respectively.

In this paper, we focus on one production technology of "High speed amorphous silicon deposition method" and propose a practical parameter to investigate plasma conditions. The parameter, which can be monitored easily even in production apparatus, represents plasma parameters, such as plasma potential and self-bias.

Fig.4 Flexible type module.

Fig.5 Steel plate based building-integrated module mounted on gymnasium.

## EXPERIMENTAL DETAILS

Single-junction a-Si and a-SiGe solar cells with 300nm thick intrinsic layers were fabricated on metal electrode coated polyimide films using plasma enhanced CVD. Many small area (0.12cm$^2$) a-Si and a-SiGe single-junction solar cells were made on large area (0.4m by 0.8m) patterned metal electrodes deposited on the polyimide films in the stepping-roll apparatus [1]. Each CVD chamber in the stepping-roll apparatus has a capacitively coupled plasma reactor. Amplitude of voltage oscillation (Peak-to-peak voltage) at the cathode is called "Vpp" which can be measured by an oscilloscope easily.

Various deposition conditions, such as excitation frequency of the plasma and working pressure, were applied to deposit intrinsic layers in the single-junction a-Si and a-SiGe solar cells. Substrate temperature was set at 200 °C for intrinsic layer deposition. Two excitation frequencies of 13.56MHz and 27.12MHz were used for plasma genen. Working pressures was changed from 65Pa to 450Pa to deposit intrinsic layer. Applied rf power density to sustain plasma was ranged from 8 to 130 mW/cm$^2$. Optical band gap of a-SiGe intrinsic layer in the single-junction solar cells was adjusted to 1.5eV by controlling mixing of SiH$_4$ and GeH$_4$ gases. All solar cells have the same p- and n-layers.

Photovoltaic properties of single-junction solar cells were characterized by a conventional current-voltage (I-V) measurement under $1000W/m^2$ illumination (AM1.5) before and after the light soaking test. I-V curves of a-SiGe single-junction solar cells were also collected using a red color filter in order to simulate bottom layer of tandem solar cells. Bond densities of Si-H and Si-H$_2$ were measured by Fourier transform infrared spectroscopy (FT-IR) to evaluate film quality.

## RESULTS AND DISCUSSION

Figure 6 shows initial and stabilized efficiencies of a-Si single-junction solar cells as a function of deposition rate of intrinsic layer. Figure 6 includes conversion efficiency data of solar cells with intrinsic layers deposited various conditions, such as plasma excitation frequency and working pressure. The best conversion efficiency is obtained at low deposition rate region. In general, conversion efficiency deteriorates with increase of deposition rate. However, several solar cells show relatively high conversion efficiency even though the intrinsic layers are deposited at high deposition speeds around 30nm/min as shown in Fig. 6. Increasing plasma excitation frequency and optimizing working pressure realize such deposition conditions.

Relationships between bonding density (SiH$_2$/SiH) determined by FT-IR and initial and stabilized efficiencies of a-Si solar cells are shown in Fig.7. All intrinsic layers examined in this study have SiH$_2$/SiH below 0.5, which indicates all the intrinsic layer show good photoelectric properties. Si-H$_2$ rich bonding configuration is formed by inappropriate chemical reactions such as insertion of higher silane reactive species (represented as Si$_x$H$_y$) at the growing top surface of the film [6]. If SiH$_2$/SiH value is above 0.5, SiH$_2$/SiH value has a strong correlation with conversion efficiency. Whereas, SiH$_2$/SiH value is below 0.5 like in this study, the value does not have a strong correlation with conversion efficiency as shown in Fig. 7. Figure

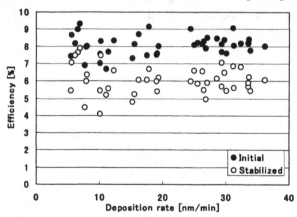

Fig.6 Initial and stabilized conversion efficiencies of a-Si solar cells as a function of deposition rate of intrinsic layer.

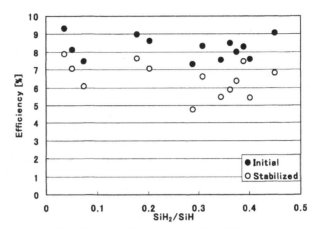

Fig.7 Relationships between SiH₂/SiH bonding and initial and stabilized efficiencies of a-Si solar cells.

6 and 7 suggest that the tendency of conversion efficiency cannot be expressed by deposition rate and SiH₂/SiH value.

We considered ion bombardment effect on the growing top surface of a-Si and a-SiGe films. Figure 8 shows schematic diagram of potential distribution in a rf discharge. Peak-to-peak voltage (Vpp) and self-bias voltage (Vdc) at the cathode (rf) electrode can be measured by oscilloscope. Vs is defined as time mean average of plasma potential. Vs in the large-area capacitively coupled plasma can be expressed using Vpp and Vdc as follows [5]:

$$Vs = \frac{1}{2}\left(\frac{Vpp}{2} - Vdc\right) = \frac{Vpp}{4} - \frac{Vdc}{2}$$

where Vdc is negligible small value in the case of large-area electrodes. In the case of our stepping-roll apparatus, Vdc value is less than 1/30 of Vpp value because the area of the electrode is so large. Consequently, Vs is expressed approximately as follows:

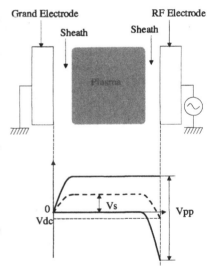

Fig.8 Schematic diagram of potential distribution in rf discharge.

$$Vs = \frac{Vpp}{4}$$

The mean free path of ions in plasma is estimated about 50-300 micrometers when working pressure is set between 60 and 300 Pa. On the other hand, the length of sheath is about 5mm and does not depend on deposition conditions significantly. Ions may collide with molecules over ten times while they cross the sheath. The collision frequency increases with increasing working pressure. Ions lose energy at each collision. Average energy (E) of ion reaching growing film surface is inversely related to working pressure. E is expressed as follows:

$$E = \frac{kVpp}{P}$$

where k is a proportional constant. Vpp/P is a measurable practical value that shows an index of ion bombardment. Increasing Vpp/P value corresponds to increasing energy of ion bombardment to film growing surface, resulting in deterion film properties.

Figure 9 shows relationship between Vpp/P value and initial and stabilized efficiency of a-Si solar cells. Figure 10 shows relationship between light induced degradation rate of a-Si solar cells and Vpp/P value.

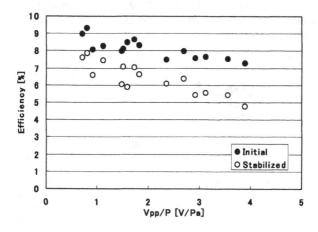

Fig.9 Relationship between Vpp/P value and initial and stabilized efficiency of a-Si solar cells.

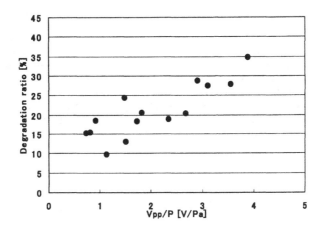

Fig.10 Relationship between light induced degradation rate of a-Si solar cells and Vpp/P value.

Figure.11 shows initial and stabilized efficiencies of a-SiGe solar cells as a function of Vpp/P. Conversion efficiency of a-SiGe solar cells was measured using a short wavelength light cut color filter. Light induced degradation rate of a-SiGe solar cells as a function of Vpp/P is shown in Fig.12. Vpp/P has strong correlations with solar cell performances in both cases of a-Si and a-SiGe solar cells. The ion bombardment model affecting on the growing film top surface and cell performances was validated by these results.

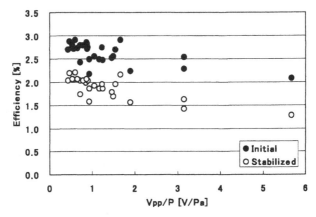

Fig.11 Initial and stabilized efficiencies of a-SiGe solar cells as a function of Vpp/P.

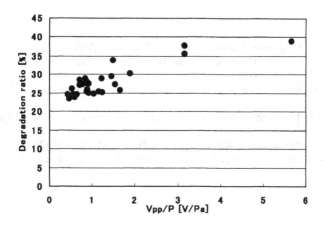

Fig.12 Light induced degradation rate of a-SiGe solar cells as a function of Vpp/P.

## CONCLUSIONS

Roll-to-roll based solar cell fabrication processes using plastic film substrates have been established. The first production line having a production capacity of 12MW/Y is in operation. The production capacity is going to be increased in FY 2008.

As one of the developed production technologies, control of plasma conditions monitored by Vpp/P in a-Si and a-SiGe CVD is introduced. Lowering Vpp/P by changing deposition conditions, such as plasma excitation frequency and working pressure, is effective to have higher efficiency solar cells even if the intrinsic layer is deposited at relatively high deposition rate.

## ACKNOWLEDGMENTS

This work described here was partly supported by the New Energy and Industrial Technology Development Organization (NEDO) under the New Sunshine Program of the Ministry of Economy, Trade and Industry.

## REFERENCES

1. A. Takano and T. Kamoshita, *Jpn. J. Appl. Phys.*, **43**, 7976 (2004).
2. A. Takano, M. Uno, M. Tanda, S. Iwasaki, H. Tanaka, J. Yasuda and T. Kamoshita, *Jpn. J. Appl. Phys.*, **43**, L277 (2004).
3. A. Takano, T. Wada, T. Yoshida, Y. Ichikawa and K. Harashima, *Jpn. J. Appl. Phys.*, **41**, L323 (2002).
4. A. Takano, M. Tanda, M. Shimosawa, T. Wada and T. Kamoshita, *Jpn. J. Appl. Phys.*, **42**, L1312 (2003).
5. S. Fujikake, A. Takano and T. Kamoshita, IEEF Trans. FM, **124**, 837 (2004).
6. A. Matsuda, M. Takai, T. Nishimoto and M. Kondo, Sol. Energy Mater. & Sol. Cells, **78**, 3 (2003).

Mater. Res. Soc. Symp. Proc. Vol. 1066 © 2008 Materials Research Society

# Study of Large Area a-Si:H and nc-Si:H Based Multijunction Solar Cells and Materials

Xixiang Xu[1], Baojie Yan[1], Dave Beglau[1], Yang Li[1], Greg DeMaggio[1], Guozhen Yue[1], Arindam Banerjee[1], Jeff Yang[1], Subhendu Guha[1], Peter G. Hugger[2], and J. David Cohen[2]

[1]United Solar Ovonic LLC, 1100 West Maple Road, Troy, MI, 48084
[2]Department of Physics, University of Oregon, Eugene, OR, 97403

## ABSTRACT

Solar cells based on hydrogenated nanocrystalline silicon (nc-Si:H) have demonstrated significant improvement in the last few years. From the standpoint of commercial viability, good quality nc-Si:H films must be deposited at a high rate. In this paper, we present the results of our investigations on obtaining high quality nc-Si:H and a-Si:H films and solar cells over large areas using high deposition rate. We have employed the modified very high frequency (MVHF) glow discharge technique to realize high-rate deposition. Modeling studies were conducted to attain good spatial uniformity of electric field over a large area (15"x15") MVHF cathode for nc-Si:H deposition. A comparative study has been carried out between the RF and MVHF plasma deposited a-Si:H and nc-Si:H single-junction and a-Si:H/nc-Si:H double-junction solar cells. By optimizing the nc-Si:H cell and the tunnel/recombination junctions, we have obtained an initial aperture-area (460 cm$^2$) efficiency of 11.9% for a-Si:H/nc-Si:H double-junction cells using conventional RF (13.56 MHz) plasma deposition. The deposition rate was 3 Å/s. Results on solar cells made with MVHF will also be presented.

## INTRODUCTION

Research efforts in recent years have outlined improvements for solar cells incorporating nc-Si:H, and their usefulness in attaining high stabilized conversion efficiency [1,2]. However, one characteristic of nc-Si:H is that relatively thick layers, 1-3 μm, are necessary for efficient light absorption due to the indirect nature of its optical bandgap. In order to cost-effectively manufacture the thick nc-Si:H cells on large-area substrates, an MVHF glow discharge technique is employed in the current work to achieve the high deposition rates required, since traditional RF (13.56 MHz) excitation typically limits useful deposition rates to the range of ≤ 3 Å/s. The MVHF technique is then extended to a-Si:H cells, which also benefit by increased rate and reduced deposition times. Developments in large-area MVHF were preceded by small-area experimental studies, which are also reviewed.

Several important considerations exist for the application of MVHF glow discharge. First, the initial performance characteristics of these cells should not be compromised due to the high deposition rate. Second, the use of thick nc-Si:H intrinsic layer films grown by the MVHF technique present potential issues with uniformity of both film thickness and cell performance over large area substrates. A third consideration is whether MVHF-deposited films can produce cells with lower light-induced degradation than conventional RF-deposited films. Finally, microstructural properties and film defect analyses are desired to understand differences in the two techniques. In this study we address these issues to demonstrate scalability to production.

## EXPERIMENTAL

Using two large-area multi-chamber batch deposition machines, comparative experiments were conducted to fabricate three types of *n-i-p* solar cell structures in each machine. One machine uses a conventional RF glow discharge for deposition of intrinsic absorbing layers while the other uses MVHF glow discharge. Both machines use RF excitation for the doped layers. The solar cell structures included single-junction a-Si:H cells on stainless steel (SS) substrates, single-junction nc-Si:H cells on Ag/ZnO back reflector-coated stainless steel (BR), and double-junction a-Si:H/nc-Si:H cells also on BR. Deposition rates and film thickness uniformity were optimized by process parameters such as cathode substrate spacing, input power density, gas mixture dilution ratio and total gas pressure.

Subsequent to Si depositions, top contacts of indium-tin-oxide (ITO) were deposited defining a cell active area of $0.25 \text{ cm}^2$. Metal grids were evaporated on top of the ITO. Current density-voltage (J-V) measurements and quantum efficiency (QE) measurements were performed for solar cell characterization. Open-circuit voltage ($V_{oc}$) and Fill Factor (FF) were determined from the J-V measurements, while short-circuit current density ($J_{sc}$) was taken from integral of the QE curves with the AM1.5 solar spectrum. The J-V measurements were conducted under an AM1.5 solar simulator at 25 °C. The QE measurements were performed under a short-circuit condition for single-junction cells and proper electrical and optical biases for multi-junction cells at room temperature in the wavelength range from 300 to 1000 nm. QE data were collected in the initial and stabilized states. The stabilized efficiency of the solar cells is expected to be reached by light-soaking under an open-circuit condition with $100 \text{ mW/cm}^2$ white light at 50 °C for 1000 hours. Quantification of the light-induced degradation is given at 300 hours of light-soak duration.

The a-Si:H single-junction cells were deposited with intrinsic layer thickness of ~300 nm. These cells were deposited directly on SS substrates, and doped layers were all of similar character. Normal deposition rates for the intrinsic layers were ~2 Å/s for the RF-produced films and 10~15 Å/s for the VHF-produced films. Separately, cells were fabricated for higher rate RF (~4 Å/s) for comparison of stability.

The nc-Si:H single-junction cells were deposited on Ag/ZnO back reflectors with a 1.0-1.5 μm intrinsic layer thickness. The a-Si:H/nc-Si:H double-junction cells were composite structures of the two types of component cells, with minor adjustments for overall performance.

## RESULTS AND DISCUSSION

### Small-area MVHF-deposited multi-junction cells

With small-area VHF plasma batch deposition machines, we have demonstrated that nc-Si:H is a potential replacement for a-SiGe:H layer(s) in a-Si:H/a-SiGe:H/a-SiGe:H triple-junction or a-Si:H/a-SiGe:H double-junction structures [1-3]. As shown in Table I, both the initial and stable cell performance of the triple-junction cells incorporating one or two nc-Si:H component cells is comparable to our best a-Si:H/a-SiGe:H/a-SiGe:H triple-junction cell. With the improved light-soaking stability, one can expect the multi-junction cell incorporating nc-Si:H may achieve higher stable efficiency than their all amorphous counterparts.

**Table I.** Comparison of small-area (0.25 cm$^2$) multi-junction cells deposited on Ag/ZnO substrate produced by RF and VHF plasma excitation. Both initial and stabilized efficiencies (after 1000 hours of light soaking) are listed for comparison.

| Plasma Excitation | Cell Structure | Initial Eff. (%) | Stable Eff. (%) |
|---|---|---|---|
| All RF | a-Si/a-SiGe/a-SiGe triple junction | 14.6 | 13.0 |
| RF a-Si/a-SiGe VHF nc-Si | a-Si/a-SiGe/nc-Si triple junction | 15.4 | 13.0 |
| All VHF | a-Si/nc-Si/nc-Si triple junction | 14.1 | 13.3 |
| All VHF | a-Si/nc-Si double junction | 13.4 | 11.6 |

## Large-area VHF chamber design and uniformity

VHF reactor design was facilitated in the current work by modeling studies of MVHF plasma electric field distributions over a large-area (15"x15") cathode. We input different cathode structures and MVHF application techniques. With the help of the model, we designed a new cathode that results in electric field uniformity better than ±5% over the entire area. We also designed a gas distribution assembly to attain desirable gas flow pattern in the high flow regime, and heating assembly to obtain uniform heating over the deposition area. In addition, process conditions such as cathode-substrate spacing, gas flow and pressure, hydrogen dilution ratio need to be optimized to obtain high rate uniform film deposition. The film thickness uniformities of both a-Si:H and nc-Si:H films deposited in a retrofitted MVHF batch plasma deposition machine show good agreement with that predicted from the model.

Nine smaller samples (~2"x2") were taken from a large-area-deposited single-junction a-Si:H cell run (substrate size 14"x15") to determine uniformity of film thickness and of cell efficiency over an area of >450 cm$^2$ using the MVHF technology. Figure 1 shows (a) the film thickness uniformity and (b) the cell performance uniformity. The results showed that under the current deposition conditions, the thickness uniformity is within ± 7% and efficiency uniformity is ± 2%, demonstrating the scalability of the process to large-area batch substrates.

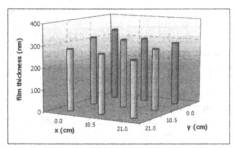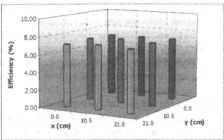

**Figure 1.** Uniformities of (a)VHF-deposited a-Si:H film thickness and (b) a-Si:H cell efficiency over large-area substrate.

## Single-junction a-Si:H cells

With satisfactory uniformity reached, we made a-Si:H cells using VHF plasma to compare with RF deposited cells. Two RF samples, one at the lower deposition rate and another at higher deposition rate, as well as one VHF sample at high deposition rate were selected for evaluation of initial performance and light-induced degradation. All the three samples are ~3000 Å thick. Averaged cell data for these runs are given below in Table II, both initial and as a function of light-soaking time up to 300 hours. (Stabilized cell data will be taken after light soak duration of 1000 hours).

**Table II.** Comparison of a-Si:H cells produced by RF (2.0 Å/s), RF (3.5 Å/s) and VHF (13.5 Å/s) in the current study. Cells are 0.25 cm² active area, taken from large-area depositions of ~1600 cm².

| Excitation | Sample | Soak Time (hours) | $V_{oc}$ (V) | Fill Factor | $J_{sc}$ (mA/cm²) | Efficiency (%) | % Change in Efficiency |
|---|---|---|---|---|---|---|---|
| RF | 12789 | 0 | 0.986 | 0.720 | 10.50 | 7.46 | |
| | (2.0 Å/s) | 300 | 0.942 | 0.628 | 10.17 | 6.01 | -19.4 |
| RF | 12791 | 0 | 0.972 | 0.706 | 10.08 | 6.91 | |
| | (3.5 Å/s) | 300 | 0.919 | 0.573 | 9.23 | 4.86 | -29.7 |
| VHF | 3652 | 0 | 0.991 | 0.737 | 9.69 | 7.08 | |
| | (13.5 Å/s) | 300 | 0.952 | 0.654 | 9.18 | 5.72 | -19.2 |

As illustrated in Table II, the high rate (13.5 Å/s) MVHF deposited a-Si:H cells show similar performance to the low rate (2 Å/s) RF deposited cells, but better light soaked stability than higher (3.5 Å /s) rate RF cells. Additionally, it is seen that when the RF deposition rate is increased by less than a factor of two, the rate of degradation in light soak increases, making this approach less desirable.

## Raman measurements on nc-Si:H

In order to obtain multi-junction cells incorporating nc-Si:H, the micro-structure of nc-Si such as crystalline volume fraction and grain size evolution needs to be investigated. Raman spectroscopy measurements were made on three nc-Si:H samples by using the green laser (532 nm), and the specifications of the three samples are listed in Table III. Figure 2 (a) and (b) show the Raman spectra for a RF nc-Si:H sample (2B12705) and a VHF sample (3D3782). Though both RF and VHF nc-Si:H samples show similar Raman spectra, VHF nc-Si:H clearly shows a higher crystalline volume fraction. As shown in Figure 2 (a) and (b), the Raman spectra are deconvoluted into three components of amorphous TO (~ 480 cm⁻¹), intermediate (~ 510 cm⁻¹), and crystalline (~520 cm⁻¹) modes using three Gaussian functions. The results of each peak position and percentage of each component are summarized in Table III. The crystalline volume fractions are 77~80% for the VHF samples and 56% for the RF sample. In this case, there are no significant differences between 0.9 μm and 1.5 μm thick VHF nc-Si:H samples, indicating that nanocrystalline evolution was under control. Smit et al. developed a new fitting procedure of Raman spectra to extract the crystalline fractions in nc-Si:H films using two components, an amorphous and a crystalline feature [4]. However, the new fitting procedure would not change the conclusion of this work.

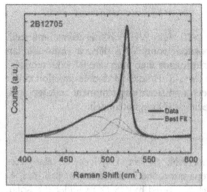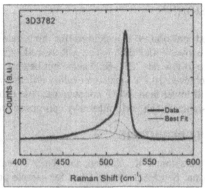

**Figure 2.** Raman spectra for (a) RF nc-Si:H sample, 2B12705, and (b) VHF nc-Si:H sample, 3D3782.

**Table III.** Raman sample specifications and deconvolution results of the Raman spectra, where p and f denote the Raman shift peak position (cm$^{-1}$) and percentage of each component, respectively. R represents the ratio of the areas of the c-Si and intermediate peaks to the total areas of the three peaks.

| Sample No. | Deposition method | Materials | Thickness (μm) | amorphous | | intermediate | | crystalline | | R (%) |
|---|---|---|---|---|---|---|---|---|---|---|
| | | | | $p_a$ (cm$^{-1}$) | $f_a$(%) | $p_i$ (cm$^{-1}$) | $f_i$(%) | $p_c$ (cm$^{-1}$) | $f_c$(%) | |
| 2B12705 | RF | nc-Si:H | 1.2 | 483 | 44 | 511 | 19 | 523 | 37 | 56 |
| 3D3781 | VHF | nc-Si:H | 0.9 | 488 | 20 | 512 | 18 | 522 | 62 | 80 |
| 3D3782 | VHF | nc-Si:H | 1.5 | 489 | 23 | 513 | 25 | 522 | 52 | 77 |

## a-Si:H/nc-Si:H double-junction cell performance

After we attained good uniformity and cell performance of a-Si:H and promising Raman results of nc-Si:H deposited using MVHF plasma, we started making a-Si:H/nc-Si:H double-junction cells, aiming to achieve better cell performance than the RF deposited cells.

By optimizing the RF deposited nc-Si:H cell and the tunnel/recombination junctions, an initial efficiency of 13.6% for small area (0.25 cm$^2$ active area) [5] and 11.9% for large area (460 cm$^2$ aperture area) have been attained on a-Si:H/nc-Si:H double-junction cells using conventional RF (13.56 MHz) plasma deposition. In contrast, the best efficiency for the double-junction cell deposited by MVHF plasma is only 10.7% for a 0.25 cm$^2$ active area. The 77-80% crystalline volume fractions for the MVHF samples may also indicate the MVHF nc-Si:H is further from amorphous to crystalline transition edge. Work is in process to further optimize nc-Si:H cells made by MVHF.

## CONCLUSION

Good uniformity is achieved for high rate (>10 Å/s) VHF a-Si:H films and cell performance over >450 cm². The a-Si:H cells deposited using VHF plasma show similar performance to low rate RF a-Si:H cells, but better performance than high rate RF counterparts. VHF has achieved 10.7% small-area initial efficiency for a-Si:H/nc-Si:H double-junction cells, which is still lower than its RF counterpart. We expect significant improvement in large area VHF multi-junction cell performance by optimizing the nc-Si:H component cell.

## ACKNOWLEDGEMENTS

The authors thank K. Younan, D. Wolf, T. Palmer, N. Jackett, L. Sivec, B. Hang, R. Capangpangan, S. Ehlert, and E. Chen for sample preparation and measurements, S. Jones, J. Doehler, and A. Kumar for discussions. The work was supported by US DOE under the Solar America Initiative Program Contract No. DE-FC36-07 GO 17053.

## REFERENCES

[1]  B. Yan, G. Yue, and S. Guha, Mater. Res. Soc. Symp. Proc. Vol. 989, 2007, p. 335.
[2]  G. Yue, B. Yan, G. Ganguly, J. Yang, S. Guha, and C. Teplin, Appl. Phys. Lett. 88, 263507 (2006)
[3]  J. Yang, A. Banerjee, and S. Guha, Appl. Phys. Lett. 70, 2975(1997).
[4]  C. Smit, R. A. C. M. M. van Swaaij, H. Donker, A. M. H. N. Petit, W. M. M. Kessels, and M. C. M. van de Sanden, J. Appl. Phys. 94, 3582(2003).
[5]  G. Ganguly, G. Yue, B. Yan, J. Yang, S. Guha, Conf. Record of the 2006 IEEE 4th World Conf. on Photovoltaic Energy Conversion, Hawaii, USA, May 7-12, 2006, p.1712.

Mater. Res. Soc. Symp. Proc. Vol. 1066 © 2008 Materials Research Society 1066-A14-04

# Improved Photon Absorption in a-Si:H Solar Cells using Photonic Crystal Architectures

Rana Biswas[1,2], and Dayu Zhou[2]

[1]Department of Physics & Astronomy, Ames Laboratory, Iowa State University, Ames, IA, 50011

[2]Microelectronics Research Center and Department of Electrical and Computer Engineering, Iowa State University, Ames, IA, 50011

## ABSTRACT

Improved light-trapping is a major route to improving solar cell efficiencies. We design a combination of a 2-dimensional photonic crystal and a one-dimensional distributed Bragg reflector as the back reflector for a-Si:H solar cells. This configuration avoids inherent losses associated with textured back-reflectors. The photonic crystals are composed of ITO and can easily serve as a conducting back contact. We have optimized the geometry of the photonic crystal to maximize absorption using rigorous scattering matrix simulations. The photonic crystal provides strong diffraction of red and near-IR wavelengths within the absorber layer and can enhance the absorption by more than a factor of 10 relative to the case without the photonic crystal. The optical path length with the photonic crystal can improve over the limit for a random roughened scattering surface.

## INTRODUCTION

A critical need for all solar cells is to maximize the absorption of the solar spectrum. Optical enhancements and light trapping is a cross-cutting challenge applicable to all types of solar cells.

Traditionally optical enhancements have involved use of anti reflecting coatings coupled with a metallic back reflector. Solar cell efficiencies are improved by textured metallic back reflectors which scatter incident light through oblique angles, thereby increasing the path length of photons within the absorber layer [1]. A completely random *loss-less* scatterer is predicted [2] to achieve an enhancement of $4n^2$ (n is the refractive index of the absorber layer), which has the value near 50 in a-Si:H. However, the idealized limit of loss-less scattering is not possible to achieve in solar cells, and it is estimated that optical path length enhancements of ~10 are achieved in practice [3]. Textured back reflectors of Ag coated with ZnO have intrinsic losses resulting from surface plasmon modes of the granular interface. Optical measurements by Springer et al [4] have estimated losses of 3-8% with every reflection. Such losses accumulate rapidly. With a loss of 4% with each pass, 34% of the light is lost in 10 passes. It is thus necessary to examine alternative schemes for light trapping.

Although the analysis in this paper can be applied to any semiconductor absorber, we focus on a-Si:H, where the optical constants have been well-determined [5]. For a-Si:H with an energy gap of 1.6 eV typical of mid-gap cells, photons with wavelengths below the band edge of 775 nm are absorbed. Short wavelength solar photons in the blue and green regions of the spectrum have absorption lengths less than 0.5 μm and are effectively absorbed within the thin absorber layer. However, the absorption length of photons grows rapidly for red light ($\lambda$ >600 nm) and even exceeds 2-3 μm for photons near the band edge. These red and near-IR photons are very difficult to absorb in thin a-Si:H layers and light-trapping schemes are critical to harvest these long-wavelength photons. Similar physical considerations apply to the band-edge photons in c-Si absorber layers [6] which are also difficult to harvest.

## APPROACH

We develop a scheme for photon-harvesting based on the use of photonic crystals as back-reflector which diffract the near-band edge photons. Such photonic crystals are loss-less, thereby providing an immediate advantage over textured metals. The simulations are performed with a rigorous scattering matrix algorithm where Maxwell's equations are solved in Fourier space[7,8,9]. The entire structure is divided into slices in the z-direction. Within each layer, the dielectric function can be periodic in x and y. Both polarizations are included. The continuity of the parallel components of E and H at each interface leads to the scattering matrices of each layer from which we obtain the scattering matrix S for the entire structure. Using the S-matrix, we simulate the reflection, transmission and absorption for incident light. The advantage this approach is that any number of layers of differing width can be easily simulated since a real-space grid is not necessary. Since the solutions of Maxwell's equations are independent for each frequency, the computational algorithm has been parallelized where each frequency (or group of frequencies) is simulated on a separate processor. The individual layers utilize realistic frequency dependent dielectric functions that include absorption and dispersion. This theoretical approach has been very successful in both photonic crystals and solar cell structures [7,8,10].

We have simulated a photonic crystal (PC) based solar cell in which the metallic back-reflector is replaced by a photonic crystal. This configuration consists of a two dimensional photonic crystal (analogous to a diffraction grid) placed on a distributed Bragg reflector (DBR) or 1-dimensional photonic crystal, with sufficient dielectric contrast between the two components. In the solar cell configuration (Fig. 1), we have 1) A traditional antireflective coating as a top contact (thickness $d_0$), 2) An absorber layer (thickness $d_2$), 3) A two-dimensional photonic crystal (thickness $d_3$) of ITO a base layer of ITO (thickness $d_6$), and 4) the DBR.

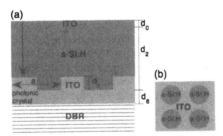
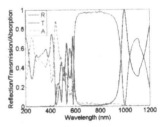

Fig. 1 a) Schematic solar cell configuration with 2-d photonic crystal and distributed Bragg reflector (DBR) in side view. b) Top view of 2-d photonic crystal layer.

Fig. 2 Reflection (R), transmission (T) and absorption A of the ITO-Si DBR.

In this configuration the diffraction of light occurs at the back of the cell where light absorption is very low. The 2D PC diffracts light within the absorber layer and the DBR specularly reflects light without loss. The combination of the two PCs leads to multiple total internal reflections within the absorber layer where light is tightly trapped inside [6,11].

In previous work we modeled an insulating DBR consisting of silicon and silicon dioxide layers [10]. Such insulating photonic crystals have already been demonstrated to enhance the efficiency of c-Si solar cells[6]. However they can not serve as a back contact and complex side contacts are necessary. Hence we use a conducting photonic crystal as a back contact. This consists of a two dimensional indium tin oxide (ITO) photonic crystal on a DBR of ITO and c-Si. To achieve a reflective band (Fig. 2) over red and near-IR wavelengths (600-780 nm) the thicknesses of the ITO/Si layers are 90 nm and 50 nm respectively, close to one quarter of a wavelength. Such a DBR is conductive and is an n-type contact.

The PC is a square lattice (with lattice spacing a) of a-Si:H cylinders (of radius r) in the ITO background. Since the dielectric contrast of the two materials is large (~4) considerable diffraction can occur from the 2D photonic crystal. In modeling pin solar cells a significant portion of the short wavelengths are absorbed within the thin p-layer located at the ITO interface. To account for p-player losses we chose a p layer thickness of $d_p=20$ nm. The absorption in the i-layer is then $A(\lambda) = A_t(\lambda)\exp(-d_p/L_d(\lambda))$, where $A_t(\lambda)$ is the total absorption and $L_d$ the wavelength dependent absorption length in a-Si:H.

As in most solar cells we utilize a thin ($d_2$=500nm) absorber layer. For a-Si:H we utilize the wavelength-dependent dielectric functions determined by spectroscopic ellipsometry by Ferlauto et al [5]. We also use the experimental dielectric functions for c-Si and ITO.

## RESULTS

We calculated the total absorption within the i-layer <A> weighted by the AM 1.5 solar spectrum as a convenient measure of the solar cell performance.

$$< A >= \int_{\lambda_{min}}^{\lambda_g} A(\lambda) \frac{dI}{d\lambda} d\lambda \qquad (1)$$

Here $dI/d\lambda$ is the incident solar radiation intensity per unit wavelength and the integration is from 280 nm ($\lambda_{min}$) to 775 nm ($\lambda_g$).

The physics underlying the photonic crystal diffraction is that the diffraction is controlled by the Fourier components of the dielectric function of the 2D-PC. For an array of cylinders the Fourier components are

$$\varepsilon(G) = f\left(\varepsilon_{Si} - \varepsilon_{ITO}\right)\frac{2J_1(GR)}{GR} \qquad (2)$$

Here f is the filling ratio $f=\pi(R/a)^2$, and $J_1$ is the first Bessel function. The principal Fourier component (with $G_1=2\pi/a$) is maximized for a radius R=0.38a.

We systematically performed a large number of simulations to optimize each of the layer thicknesses ($d_0$, $d_3$, $d_6$) along with the lattice spacing a, and radius R [12]. These variables are somewhat coupled to each other. We optimized one variable keeping others constant. For the optimized value of the first variable we then re-optimize the remaining variables one at a time. This gives us a solution which we checked for robustness for changing variables independently.

The optimal solar cell configuration we found had $d_0$~65 nm, $d_3$~100 nm, radius

R/a~0.39-0.4 and lattice spacing a~0.7 μm. The thickness ($d_6$) of the base ITO layer is very small for best absorption but a thin layer with thickness ~10 nm does not change results and offers a better back contact. The antireflection coating thickness $d_0$ is close to a quarter wavelength in the ITO near the middle of the optical range (500 nm). The optimal radius R is close to that expected from the strength of the Fourier component.

The absorption of the optimal photonic crystal enhanced solar cell is compared with the absorption from the configuration without the 2D photonic crystal, but with a DBR back-reflector (Fig. 3a). Short wavelengths (below 600 nm) are effectively absorbed in both configurations since the photon absorption length is less than the absorber thickness (0.5 μm). At wavelengths near the band edge (600-775 nm) the PC-enhanced solar cell has considerably higher absorption than without the PC. At these near IR wavelengths photons with long absorption lengths are not effectively absorbed within the absorber layer by the flat reflector. The PC-enhanced solar cell is able to effectively diffract these photons to increase their path length and dwell time within the absorber layer. The absorption of the PC-solar cell is more than a factor of 5 larger (and even a factor of 10 larger) than without the PC (Fig. 3b).

a

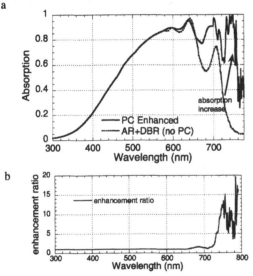

b

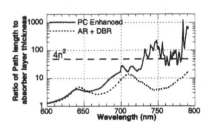

Fig. 4 The ratio of the optical path length to the absorber layer thickness, for the optimized PC enhanced solar cell (solid) compared to the reference cell (dotted). The horizontal dashed line is the theoretical limit of path length enhancement from a randomly roughened back reflector (~50).

Fig. 3. a)The absorption of the photonic crystal based solar cell as a function of wavelength, compared with a reference cell without the photonic crystal but with an antireflective coating (AR) and DBR reflector.b) Ratio of enhancement between the

The PC-solar cell has resonant wavelengths where the absorption reaches ~100% near the band edge. These resonances occur at wavelengths where the round trip optical phase change in the absorber layer $\Delta\phi$ is $2m\pi$ which occurs when the z-component of the wave-vector $k_\perp = \pi m/d_2$, where m is an integer. Accounting for diffraction, the wavelengths at which this occurs are

$$\lambda = 2\pi n(\lambda) / \sqrt{G_x^2 + G_y^2 + \left(m\pi / d_2\right)^2} \qquad (3)$$

n is the refractive index of a-Si:H and $G_x$, $G_y$ are the components of reciprocal lattice vectors ($G_x = i(2\pi/a)$; $G_y = j(2\pi/a)$). The diffraction resonances occur for integer values of i, j and m and exhibit peaks in the absorption for wavelengths near the band-edge for values of a larger than ~ 0.25 µm. The peaks overlap and form the overall absorption enhancement. It is necessary to have several diffraction resonances within the wavelength window (0.6 – 0.775 µm) below the band edge, where the absorption length of photons is longer than the absorber layer thickness.

## DISCUSSION

We have used the calculated absorption and absorption length of a-Si:H to extract an optical path length of photons in the absorber layer (Fig. 4). The ratio of the path length with the PC to without the PC shows considerable enhancement at near IR wavelengths. The optical path length can be enhanced by more than a factor of 50 and can exceed the classical limit of $4n^2$ for near band edge wavelengths, indicating that diffraction can be more effective than scattering from randomized metallic components.

One critical parameter is the lattice spacing a. The absorption is optimized for a lattice spacing of 0.7 µm (Fig. 5). For diffraction to be effective it is necessary for the lattice spacing to be larger than the wavelength of light in the absorber layer ($a > \lambda/n(\lambda)$ ). This condition is satisfied for $a > 0.25$ µ. However for larger lattice spacings (a>1 µm), there is still improvement of the absorption over a flat reflector, and may be easier for fabrication. Fabrication may require holographic lithography.

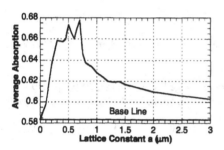

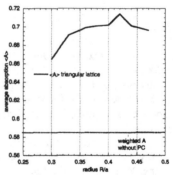

Fig. 5 Variation of absorption with lattice spacing a Compared to the cell without photonic crystal (base line).

Fig. 6 Variation of absorption with radius of cylinders in photonic crystal for the triangular lattice. A metallic back reflector is used.

We have also investigated replacing the DBR at the back with a flat metallic back reflector. We find comparable results for the PC on an idealized *loss-less* flat metal. Both square and triangular lattice photonic crystals are equally effective for absorption enhancement. The triangular lattice photonic crystal (Fig. 6) also provides enhancement of the absorption for a radius R/a~0.4-0.42 and similar lattice spacing as in the square case. For comparison the metallic back reflector calculation is shown (Fig. 6).

We investigated using a 2-d photonic crystal in the front of the cell[13] between the absorber layer and the anti-reflection coating. Although there is considerable diffraction, this can suffer from a large absorption from the p-layer at the front of the cell. Light absorbed in a thick p-layer forming a photonic crystal can not be utilized for collection of carriers. It is much more beneficial to utilize a thicker n-layer at the back of the cell where the absorption is low.

## CONCLUSIONS

We develop a novel light trapping scheme in a-Si:H solar cells with conducting photonic crystals. By combining a 2D photonic crystal diffraction grating and DBR, we simulate efficient harvesting of optical and near IR photons, without losses associated with metallic reflectors. The absorption at longer wavelength ($\lambda = 0.65$-$0.78$ $\mu$m) up to the bandgap is increased by more than a factor of 10 at certain wavelengths.

## ACKNOWLEDGEMENTS

We acknowledge support from the Catron Solar Foundation. The Ames Laboratory is operated for the Department of Energy by Iowa State University under contract No. W-7405-Eng-82. It is a pleasure to thank V. Dalal for many stimulating discussions. We also acknowledge support from the NSF under grant ECS-06013177. We thank R. Collins and N. Podraza for kindly supplying optical data for amorphous silicon films and most helpful discussions.

## REFERENCES

[1] B. Yan, J. M. Owens, C. Jiang, S. Guha, Materials Res. Soc. Symp. Proc. **862**, A23.3 (2005).

[2] E. Yablonovitch, J. Opt. Soc. Am. **72**, 899 (1982).

[3] J. Nelson, The Physics of Solar Cells, (Imperial College Press, London, 2003), p. 279.

[4] J. Springer, A. Poruba, L. Mullerova, M. Vanecek, O. Kluth and B. Rech, J. Appl. Phys. **95**, 1427 (2004).

[5] A.S. Ferlauto, G. M. Ferreira, J. M. Pearce, C. R. Wronski, R. W. Collins, X. Deng, G. Ganguly, J. Appl. Phys. **92**, 2424 (2002).

[6] L. Zeng, Y. Yi, C. Hong, J. Liu, N. Feng, X. Duan, L.C. Kimmerling, B.A. Alamariu, Appl. Phys. Lett. **89**, 111111 (2006); Materials Res. Soc. Symp. **862**, A12.3 (2005).

[7] R. Biswas, C.G. Ding, I. Puscasu, M. Pralle, M. McNeal, J. Daly, A. Greenwald, E. Johnson, Phys. Rev. B. **74**, 045107 (2006).

[8] R. Biswas, S. Neginhal, C. G. Ding, I. Puscasu, E. Johnson, J. Opt. Soc. of Am. B **24**, 2489 (2007).

[9] Z. Y. Li and L. L. Lin, Phys. Rev. E **67**, 046607 (2003).

[10] R. Biswas and D. Zhou, Mater. Res. Soc. Symp. Proc. **989**, A03.02 (2007).

[11] P. Bermel, et al, Opt. Express **15**, 16161 (2007).

[12] D. Zhou, and R. Biswas, to appear in J. Appl. Phys. **103**, (2008).

[13] K.R. Catchpole and M. A. Green, J. Appl. Phys. **101**, 063105 (2007).

# Film Growth and Characterization II

Mater. Res. Soc. Symp. Proc. Vol. 1066 © 2008 Materials Research Society 1066-A15-04

# Mechanical Properties and Reliability of Amorphous vs. Polycrystalline Silicon Thin Films

Joao Gaspar[1], Oliver Paul[1], Virginia Chu[2], and Joao Pedro Conde[2,3]

[1]Dept. Microsystems Eng. (IMTEK), University of Freiburg, Freiburg, 79110, Germany
[2]INESC Microsistemas e Nanotecnologias, Lisbon, 1000-029, Portugal
[3]Dept. Chemical and Biological Eng., Instituto Superior Tecnico, Lisbon, 1049-001, Portugal

## ABSTRACT

This paper presents the mechanical characterization of both elastic and fracture properties of thin silicon films from the load-deflection response of membranes, also known as the bulge test. Properties extracted include the plane-strain modulus, prestress, fracture strength and Weibull modulus. Diaphragms made of low-temperature, hydrogenated amorphous and nanocrystalline silicon films (a-Si:H and nc-Si:H, respectively) deposited by plasma enhanced chemical vapor deposition (PECVD) and, for comparison, membranes composed of high-temperature polycrystalline silicon (poly-Si) deposited by low pressure chemical vapor deposition (LPCVD) have been fabricated and characterized. The structures are bulged until failure occurs. From the stress profiles in the diaphragms at fracture, the brittle material strength is analyzed using Weibull statistics. The bulge setup is fully automated for the sequential measurement of several membranes on a substrate realizing the high-throughput acquisition of data under well controlled conditions. A comprehensive study of the mechanical properties of low-temperature silicon films as a function of deposition parameters, namely substrate temperature, RF power, hydrogen dilution and doping, is presented.

## INTRODUCTION

Microelectromechanical systems (MEMS) use planar microelectronics fabrication techniques to produce 3D structures with electronic and mechanical functionality. MEMS sensors and actuators can be based on a variety of different physical, chemical and biological principles [1]. Most MEMS devices are fabricated using bulk micromachining of crystalline silicon (c-Si) substrates or surface micromachining of poly-Si films, requiring processing temperatures in the range of 550-1100°C [2]. Thin-film MEMS use thin-film technologies. Advantage has been taken of the low temperatures at which some of thin films can be deposited. Low temperature processing makes it possible to use a wide variety of substrates such as glass, plastic or stainless steel sheet. Another important feature of thin-film technologies is that the film properties (optical, mechanical, optoelectronic and chemical) can be tuned by varying the deposition conditions. Low-temperature thin-film MEMS may even be CMOS-compatible, allowing their integration with control electronics as part of a backend process. Thin-film MEMS devices based on a-Si:H have been reported [3-6].

The characterization of the mechanical properties of thin films used in MEMS is of great importance since these determine device functionality and performance. Parameters such as the plane-strain modulus $E_{ps}$ and prestress $\sigma_0$ can be extracted from the measurement of the pressure-deflection response of membranes composed of those films [7-9], as well as their fracture characteristics [10,11]. The mechanical properties of c-Si and poly-Si materials have been extensively characterized [9-11]. However, there exist few published results about the mechanical properties of low-temperature silicon films. In this paper, the wafer-scale bulge technique [11] is applied to extract both elastic and fracture constants of a-Si:H and nc-Si:H films, deposited at temperatures as low as 100°C. The mechanical properties obtained are studied as a function of the deposition conditions and compared to those of standard high-$T$ poly-Si films.

## MECHANICAL MODEL

The deflection profile $w(x,y)$ of a membrane of width $a$ and length $b$, be it a single layer or a multilayer, bulged with a pressure $P$, is schematically shown in Fig. 1. For aspect ratios $b$:$a$ larger than 4:1, an extended middle section of the structure responds with a plane strain deformation $w(x)$ that does not depend on $y$ [8,10,11]. From the model in [10,11], which takes into account the prestress, structural stretching, and bending stiffness of the films, and the compliance of the supporting edges, the relation between the center deflection $w_0$ and $P$ is

$$w_0 = \frac{Pa^2}{8S} + \left[\frac{P}{S}\left(D_2 + \frac{Ka}{2}\right) + S_1\right] \frac{1 - \cosh\left(\sqrt{S/D_2}\, a/2\right)}{S\cosh\left(\sqrt{S/D_2}\, a/2\right) + K\sqrt{S/D_2}\,\sinh\left(\sqrt{S/D_2}\, a/2\right)}, \quad (1)$$

where $K$, $S$, $S_1$ and $D_2$ denote the rigidity of the membrane supporting edges, effective line force, initial bending moment per unit length and bending stiffness, respectively, which depend on the thickness $h$ and elastic constants of each individual layer in the stack composite [10,11].

The distribution of the stress $\sigma_{xx,n}$ in layer $n$ of a general stack is given by [10,11],

$$\sigma_{xx,n}(x,z) = \sigma_{0,n} + \frac{E_{ps,n}}{2a}\int_{-a/2}^{a/2}\left(\frac{dw}{dx}\right)^2 dx - E_{ps,n}(z - z_0)\frac{d^2w}{dx^2}, \quad (2)$$

where $E_{ps,n}$ and $\sigma_{0,n}$ denote the plane-strain modulus and prestress of layer $n$, respectively, and $z_0$ denotes the $z$-coordinate of the reference plane.

## EXPERIMENT

The fabrication process of thin silicon membranes is schematically shown in Fig. 2. It

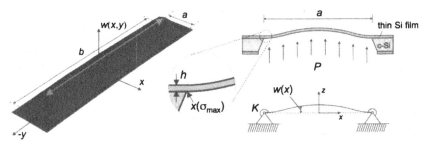

**Figure 1.** Schematics of the plane-strain deformation of a long membrane (and its extended middle section) under the influence of a uniform differential pressure $P$.

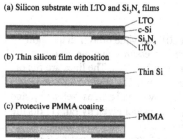

(a) Silicon substrate with LTO and Si$_3$N$_4$ films
- LTO
- c-Si
- Si$_3$N$_4$
- LTO

(b) Thin silicon film deposition
- Thin Si

(c) Protective PMMA coating
- PMMA

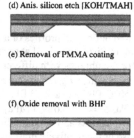

(d) Anis. silicon etch [KOH/TMAH]

(e) Removal of PMMA coating

(f) Oxide removal with BHF

**Figure 2.** Schematics of the fabrication of thin-film silicon membranes by bulk micromachining of silicon substrates. In case of strongly compressive films, these are stacked on top of LPCVD Si$_3$N$_4$ films in order to obtain overall tensile membranes.

begins with the patterning of 770°C-LPCVD 100-nm-Si$_3$N$_4$ / 450°C-LPCVD 500-nm-low-temperature oxide (LTO) layers on the rear side of a c-Si wafer, Fig. 2 (a). These layers serve as a mask for the later bulk-micromachining. The thin silicon film is then deposited on the front side 500-nm-LTO film, Fig. 2 (b), followed by a protective PMMA coating, Fig. 2 (c). The c-Si substrate is etched with KOH/TMAH solutions, Fig. 2 (d), and the PMMA resist is then removed, Fig. 2 (e). The fabrication is concluded with the selective removal of the oxide layer using a BHF-solution, Fig. 2 (f). In case of strongly compressive silicon films, these are stacked with highly tensile LPCVD Si$_3$N$_4$ layers ($\sigma_0 > 1$ GPa [8-11]) yielding plane-strain membranes without post-buckling deflection at $P = 0$. Diaphragms with an aspect ratio of 10:1 and widths $a$ ranging from 400 to 800 μm were processed. The characterized low-temperature, PECVD silicon films include layers with thicknesses in the 200-300 nm range, grown at substrate temperatures $T$ between 100 and 250°C, with RF powers $P$ from 5 to 30 W, pressures $p$ between 0.1 and 0.9 Torr, hydrogen (H$_2$) dilution series of silane (SiH$_4$) from 0 to 98%, and inclusion of dopant phosphine (PH$_3$, 1 sccm) and Argon (Ar, 500 sccm) gases. For comparison, high-temperature 625°C-LPCVD poly-Si films with thicknesses up to 2 μm, annealed at 1050°C for 1 hour, were processed as well. Per film, a maximum of 80 diaphragms is obtained on 4-inch silicon wafers, except for the layers grown with 5 W, where 25 membranes are released on 5 cm × 2.5 cm substrates.

The measurement of the deflection of the membranes under uniform differential pressure is made using the recently developed wafer-scale bulge test setup shown in Fig. 3 [10,11]. The sensing part of the setup consists of an optical auto-focus sensor and motorized x-y-θ table. Point, line and area scans are possible. The actuation part consists of a pressure controller that applies pressures between –80 and 800 kPa to solenoid valves controlled by relays that redirected the pressure to the membranes via appropriate lines, feedthroughs, and customized wafer fixture. The entire setup is automated for the sequential characterization of all diaphragms on a wafer.

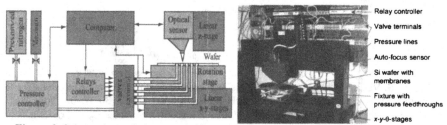

**Figure 3.** Schematics and photograph of automated wafer-scale bulge test setup.

341

## RESULTS AND DISCUSSION

Figure 4 shows the measured deflection profiles of a 801-μm-wide bilayer diaphragm (270 nm of a-Si:H on top of 98 nm of $Si_3N_4$) obtained at several pressures, and the center deflection $w_0$ as a function of $P$ for membranes with different width values. The lines are fits of Eq. (1) to the experimental data, from which $E_{ps}$ and $\sigma_0$ are extracted. The wafer distribution of thus obtained values of $E_{ps}$ of a 250°C-PECVD a-Si:H film is shown in Fig. 5, as well as $E_{ps}$ and $\sigma_0$ histograms [$E_{ps}$ = 114.9±9.4 GPa, $\sigma_0$ = −305.6±9.5 MPa]. The model given by Eq. (1) fits the data well, regardless of membrane thickness and width, indicating a correct geometrical scaling. It is worth mentioning that, because of the relatively high levels of compressive prestress of the a-Si:H films, the data shown in Fig. 5 have been obtained from the measurement of stacks of

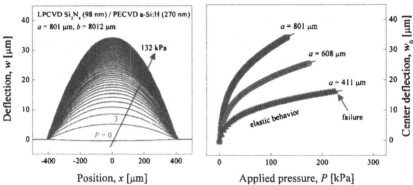

**Figure 4.** On the left, deflection profiles of a a-Si:H/$Si_3N_4$ membrane obtained at different pressures. On the right, center deflection as a function of $P$ for membranes with different widths, bulged up to fracture. The lines are fits to Eq. (1).

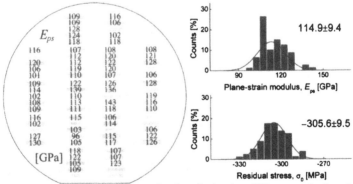

**Figure 5.** Values of plane-strain modulus $E_{ps}$ obtained at different wafer locations for membranes with different sizes for a-Si:H films [PECVD 250°C, 30 W, 0.9 Torr, 20 sccm $SiH_4$, 20 sccm $H_2$, 500 sccm Ar] and histograms of the extracted $E_{ps}$ and $\sigma_0$ data.

these films with a highly tensile reference LPCVD Si₃N₄ layer, as described in detail elsewhere [10,11].

The membranes are bulged until fracture and the stress distribution within the diaphragms at the moment of failure is computed using Eq. (2). This is then used to analyze the brittle material strength via Weibull distributions [11]. When dealing with composites of silicon on top of nitride layers, it is verified that the stress levels in the $Si_3N_4$ are lower than the characteristic strengths of this layer (8-10 GPa [10,11]), implying that fracture originates in the thin silicon film. The fracture strength $\sigma_f$ of the thin silicon layers is thus obtained by evaluating the maximum of Eq. (2) within these films at the rupture pressure. Pooled Weibull data and fits are shown in Fig. 6, from which the mean fracture strength $\mu$ and Weibull modulus $m$ are obtained. From Fig. 6, the poly-Si films show higher values of $\mu$ and are more predictable from the fracture point of view, with $m$ values reaching 20.1.

The detailed dependence of $E_{ps}$, $\sigma_0$ and $\mu$ of the low-temperature PECVD silicon films on $H_2$ dilution of $SiH_4$ gas [flow ratio: $100 \times F_{H_2} / (F_{H_2} + F_{SiH_4})$] used during deposition is shown in Fig. 7. Poly-Si data are represented as well. Regardless of $H_2$ dilution, substrate temperature, power, doping with phosphorous (inclusion of $PH_3$) and addition of Ar, and within the experimental uncertainty, the PECVD films have a plane-strain modulus $E_{ps}$ of ~120, lower than that of the poly-Si layers. The prestress of most of the low-temperature films is compressive, except for the layers grown at 100°C with 0 and 98% $H_2$ dilution, a-Si:H and nc-Si:H films, respectively [12]. As the temperature increases from 100 to 250°C, the residual stress of films deposited without $H_2$ dilution shifts from tensile to compressive. Similar trends of $\sigma_0$ with $H_2$ dilution are observed for both 100°C and 250°C series, including in the amorphous-to-nanocrystalline transition [12], with more compressive values for the 100°C layers. From Fig. 7, the single doped sample did not show any measurable effect of doping on $\sigma_0$. Films grown with higher reactor powers (30 W) with 500 sccm of Ar also show compressive stress. The strength of the PECVD silicon films, with values between ~0.3 and ~1.1 GPa, is lower than that of poly-Si and shows similar trends as $\sigma_0$ with $H_2$ dilution. Films grown at 100°C are stronger than 250°C layers, which have better optoelectronic properties [12]. Moreover, it can be concluded from Fig. 7 that increasing the RF power, inclusion of Ar and dopants, while keeping the other parameters constant, enhances $\mu$.

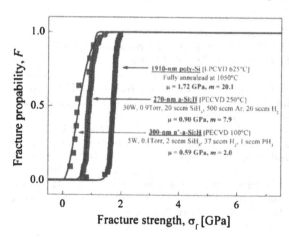

**Figure 6.** Probability of fracture $F$ as a function of the fracture stress $\sigma_f$ for high-temperature poly-Si and several low-temperature PECVD thin silicon films. The solid curves are fits of the Weibull distribution to the experimental data from which both mean fracture strength $\mu$ and Weibull modulus $m$ are extracted.

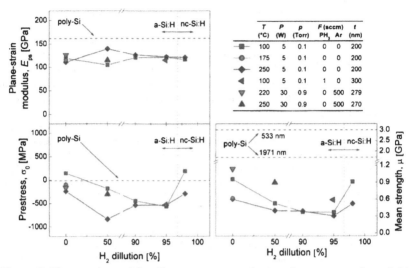

**Figure 7.** Plane strain modulus $E_{ps}$, prestress $\sigma_0$ and mean fracture strength $\mu$ of thin silicon films deposited by PECVD as a function of $H_2$ dilution. For comparison, the horizontal dashed lines represent the mechanical data from poly-Si films.

## CONCLUSIONS

The detailed dependence of both elastic and fracture properties on deposition conditions, extracted from bulge tests, of low-temperature PECVD silicon films is presented. These films have lower elastic and strength constants and are less predictable from the fracture point of view, in comparison to high-temperature poly-Si layers. Nevertheless, their properties still allow their integration in reliable MEMS devices with the benefits of low-temperature processing.

## REFERENCES

1. Jack W. Judy, *Smart Mater. Struct.* **10**, 1115 (2001).
2. N. Maluf, *An Introduction to MEMS Engineering*, Artech House, Boston, 2000.
3. J. Gaspar, V. Chu, J. P. Conde, *J. Appl. Phys.* **93**, 10018 (2003).
4. J. Gaspar, V. Chu, N. Louro, R. Cabeça, J.P. Conde, *J. Non-Cryst. Sol.*, **299-302**, 1224 (2002).
5. J. Gaspar, V. Chu, J. P. Conde, *17th IEEE MEMS 2004 Conf. Tech. Dig.*, 633 (2004).
6. A. J. Syllaios, T. R. Schimert, R. W. Gooch, W. L. McCarde, B. A. Ritchey, J. H. Tregilgas, *Mat. Res. Soc. Symp. Proc.* **609**, A14.4.1 (2000).
7. E. I. Bromley, J. Randall, D. Flanders, R. Mountain, *J. Vac. Sci. Technol. B* **1**, 1364 (1983).
8. V. Ziebart, O. Paul, U. Münch, J. Schwizer, H. Baltes, *J. Microelectrom. Syst.* **7**, 320 (1998).
9. O. Paul, and P. Ruther, *Material Characterization*, Ch. 2 in *CMOS – MEMS*, Advanced Micro & Nanosystems Series, Eds. O. Brand, G. K. Fedder, Wiley-VCH, Weinheim, 2005.
10. J. Gaspar, M. Schmidt, J. Held, O. Paul, *Mat. Res. Soc. Symp. Proc.* **1052**, DD1.2 (2008).
11. O. Paul, and J. Gaspar; *Thin-Film Characterization Using the Bulge Test*, Ch. 3 in *Reliability of MEMS*, Eds. O. Tabata, T. Tsuchiya, Wiley-VCH, Weinheim, 2007.
12. P. Alpuim, V. Chu, J. P. Conde, *J. Appl. Phys.* **86**, 3812 (1999).

Mater. Res. Soc. Symp. Proc. Vol. 1066 © 2008 Materials Research Society        1066-A15-05

## In-Situ Transmission Electron Microscopy Investigation of Aluminum Induced Crystallization of Amorphous Silicon

Ram Kishore[1], Renu Sharma[2], Satoshi Hata[3], Noriyuki Kuwano[3], Yoshitsuga Tomokiyo[3], Hameed Naseem[4], and W. D. Brown[4]

[1]Electron Microscope Section, National Physical Laboratory, Dr KS Krishnan Road, New Delhi, 110012, India
[2]Center for Solid State Science, Arizona State University, Tempe, AZ, 85287
[3]ASTEC, Kyushu University, Kasuga, 816-8580, Japan
[4]Electrical Engineering Department, University of Arkansas, Fayetteville, AR, 72701

## ABSTRACT

The interaction of amorphous silicon (a-Si) and aluminum (Al) has been examined using in-situ transmission electron microscopy. Carbon coated nickel grids were used for depositing thin (~50nm) amorphous silicon films using ultra high vacuum cluster tool and a thin film of Aluminum (~50nm) was deposited subsequently on a-Si film by sputtering. The grid containing a-Si and Al films was mounted on a heating holder of FEI 200kV TEM and loaded in the TEM for viewing the microstructural and phase transformations during the in-situ heating process. The microstructural features and electron diffraction patterns in the plain view mode were observed with increase in temperature starting from 30 °C to 275 °C. The temperatures used in this experiment were 30, 100, 150, 200, 225, 275°C. A sequential change in microstructural features and electron diffraction pattern due to interfacial diffusion of boundary between Al and amorphous Si was investigated. Evolution of polycrystalline silicon with randomly oriented grains as a result of a-Si and Al interaction was revealed. After the in-situ heating experiment the specimen was taken out and etched to remove excess of Al and the subjected to high resolution imaging under TEM and EDS analysis. The EDS analysis of the crystallized specimen was performed to locate the Al distribution in the crystallized silicon. It has been shown that Al induced crystallization can be used to convert sputtered a-Si into polycrystalline silicon as well as nanocrystalline silicon at a temperature near 275 °C by controlling the in-situ annealing parameters. The mechanism of AIC has been discussed from the experimental results and the phase diagram of Al-Si system.

## INTRODUCTION

Interaction of amorphous silicon with aluminum results in the formation of nano-, micro- and polycrystalline silicon at low temperatures. These materials are important due to their application in thin film transistors (TFTs), active matrix liquid crystal displays (AMLCD), solar cells and various other applications. Aluminum-induced crystallization (AIC) and doping of a-Si:H has been of great interest [1-6] because of its ability to produce polycrystalline silicon with good crystallographic and electrical properties at low temperatures. It is believed that dissolution of Al atoms in a-Si may weaken the Si bonds and enhance nucleation and grain growth. We had shown in our earlier work [3,7] that interaction of Al with α-Si:H takes place at temperatures as low as 150 °C and also reported for the first time the direct evidence of AIC at 150 °C using TEM and XRD [7]. Epitaxial silicon films from amorphous silicon can also be obtained using AIC [12]. We had reported microstructural changes resulted by in-situ heating of amorphous silicon aluminum layers using scanning electron microscopy and also proposed a model for the low temperature crystallization by AIC [11]. Recently Wang et al.[13] have also given an interpretation of the kinetics of the AIC. Very little work has been done on the microstructural changes ovserved by transmission electron microscopy using in-situ annealing of the amorphous silicon aluminum layers [11,12]. In the present work we have investigated the microstructural and crystallization changes due to interaction of amorphous silicon and aluminum at elevated temprartures using plan view transmission electron microscopy (TEM).Carbon coated nickel grids were used for TEM studies. An ultra high vacuum cluster tool was used for the deposition of a 50nm a-Si films and a 50 nm thin film of Al was sputter coated on a-Si film inside the same system using Al sputtering chamber. The microstructural features and electron diffraction in the plain view mode were observed with increase in temperature starting from room temperature to 275 °C and the crystallization behaviour was investigated using microstruture and electron diffraction pattern acquired at different temperatures during the in-situ annealing process.

EXPERIMENTAL

Deposition of Amorphous Silicon and Aluminum

Carbon coated Nickel grids were loaded inside a high vacuum PECVD system. A thin layer (~50nm) of amorphous silicon was sputter coated by using silicon target. The sputtering chamber was evacuated to $1 \times 10^{-7}$ torr and the specimen were transferred to the sputtering chamber. A flow rate of 20 sccm of high purity Argon gas was maintained in the deposition chamber and 130 Watt RF power was used to sputter amorphous silicon. After amorphous silicon deposition the grids were transferred to Al sputtering chamber. The Al sputtering chamber was evacuated to a base pressure if $5 \times 10^{-8}$ torr. The chamber pressure was maintained at 5 mtorr during the sputtering and Ar flow rate was kept at 20sccm during the deposition. A 150 W RF power was used to sputter the a 50 nm thick Al layer on top of a-Si layer.

## In-situ TEM Analysis

The specimen was loaded inside TEM heating holder. The temperature was increased and kept constant for 5-10 minutes at a particular temperature and then the microstructure at fixed magnification of x63K was acquired, the electron diffraction pattern of the same area was also recorded. The temperature was then increase and fixed at desired value and microstructure and EDP were again recorded. The temperatures used in this experiment were 30, 100, 150, 200, 225, 275 °C.

## RESULTS AND DISCUSSIONS

A sequential change in the microstruture has been shown in Fig. 1 (a-g). The microstructure of the film at 30 °C (Fig. 1a ) shows distinct aluminum grains with amorphous silicon in the background. No significant changes in the microstructure and EDP were observed till 250 °C (Fig. a-f). The EDP showed characteristic hallos of amorphous silicon with sharp rings belonging to aluminum. No significant changes were observed in EDP from 30 to 250 °C and therefore the EDP has been shown as inset in Fig. 1a and Fig. 1f only. At 275 °C it was observed that the halos in the electron diffraction pattern started disappearing and high contrast dark spots in the microstructure starts appearing. The specimen was kept at this temperature for 20 minutes and then microstructure and EDP were recorded after every 5 minutes. Fig. 1g shows one such micrograph which indicates that the a-Si film has crystallized. The EDP as inset shows that the silicon halos have been converted into sharp rings with spots in them showing crystallization of amorphous silicon.

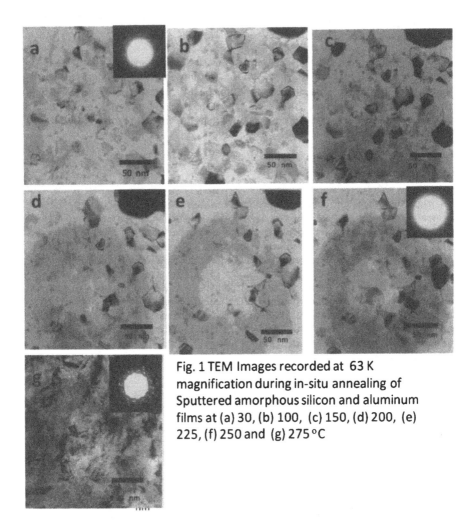

Fig. 1 TEM Images recorded at 63 K magnification during in-situ annealing of Sputtered amorphous silicon and aluminum films at (a) 30, (b) 100, (c) 150, (d) 200, (e) 225, (f) 250 and (g) 275 °C

The specimen was taken out from TEM and unloaded from the specimen holder. The grid was etched in dilute KOH solution to remove aluminum from top of crystallized silicon layer. The same grid was then again loaded into the microscope for examination. Fig. 2a shows microstructure and EDP of the specimen after etching. It may be noticed that the film is polycrystalline with randomly oriented grains and subgrains of size varying between 20-500 nm. The diffraction pattern shows various rings belonging to polycrystalline silicon. The EDP (Figure 2 b) also shows number of sharp rings with dots in them which belong to aluminum indicating that Al could not be removed completerly from the top of the grid. Figure 2c shows one of the high resolution image of the specimen. This image show that the silicon is crystallized with randomly oriented grains. The fringes in the HR images reveal the single crystalline structure of the individual grains with different orientations. The EDS spectra of the specimen in Fig. 3 shows the presence of silicon and aluminum. The spectrs shows the presence of Cu, Fe and Ni peaks coming mainly from TEM grid material.

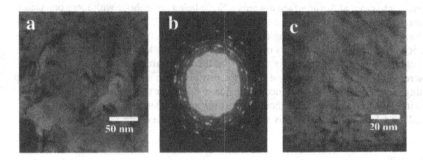

Fig. 2 (a) TEM Image of the specimen annealed in-situ at 275 °C and recorded after Al Etching (b) EDP of the same area (c) High Resolution Image of Etched specimen

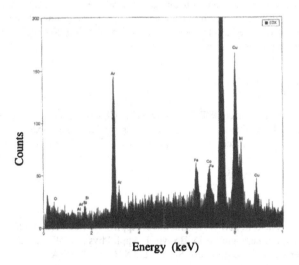

Fig. 3 EDS Spectra of the specimen after Al etching

## Mechanism of Crystallization

As discussed in our earlier publication [11] the mechanism of AIC may be explained on the basis of Al diffusion in a-Si. Al atoms, being lighter than silicon, try to diffuse in an open structure of a-Si that has weaker link. The phenomenon starts only when the flux of Al atoms has sufficient energy to cross at the interface towards a-Si. Si being amorphous with more open sites engulfs these Al atoms by process similar to interstitial diffusion. The engulfed Al atoms act as heterogeneous nuclei for the surrounding cluster of Si atoms to reorganize themselves in well defined lattice form. These small clusters of Si atoms arranged in a lattice are the precursors for the growth of crystallized Si in the matrix of a-Si. The thermal entropy of the system increases with the increase of annealing temperature. Thus the process keeps simulating and also speeds up as the heating is raised further. A continuous diffusion of Al across the interface into a-Si also spreads the individual Al entities throughout. A situation arises when the Al layer is seen depleted and the a-Si is rich in Al. At this stage one can see that the entire a-Si is transformed into polycrystalline Si.

## CONCLUSIONS

These studies show that the interaction of sputtered amorphous silicon and Al starts near 275 °C temperature and amorphous silicon is converted into polycrystalline silicon with randomly oriented grain structure. The mechanism of crystallization is diffusion based. These results also reveals that by controlling the in-situ annealing parameters the aluminum induced crystallization can be used to convert a-Si into nano as well as polycrystalline silicon. The results are very useful because these days nanocrystalline silicon is being investigated for its use in designing high efficiency silicon solar cells.

## ACKNOWLEDGEMENTS

The authors would like to thanks DST, India and NSF, USA for financial support for this work. One of the authors (RK) is thankful to Kyushu University, Fukuoka, Japan for providing financial assistance as well as microscopy facilities to carry out in-situ TEM experiments.

## REFERENCES

1. Ishihara and T. Hirao, Thin Solid Films 155, (1987) 325
2. M. S. Haque, H. A. Naseem and W. D. Brown, J. Appl. Phys., 76 (1994) 3928
3. M. S. Haque, H. A. Naseem and W. D. Brown, J. Appl. Phys.,79 (1996) 7529
4. O.Nast, T. Puzzer, L.M. Koschier, A.B. Sproul and S.R. Wenham, Appl. Phys. Lett., 73 (1998) 3214
5. O. Nast, S. Brehme, D. H. Neuhaus, and S. R. Wenham, IEEE Trans. on Electron Devices, 46 (1999) 2062
6. J. H. Wei and S. C. Lee , J. Vac Sci Technol A 16, (1998) 587
7. Ram Kishore, Chris Hotz, , H. A. Naseem and W. D. Brown, Electrochemical and Solid State Letters 4(2), (2001) G14

8. W.S. Liao and S.C. Lee J. Appl. Phys 81 (1997) 7793
9. Ram Kishore, K.N.Sood and H.A. Naseem, Jr. of Mater Sci Lett 21 ( 2002) 647
10. Ram Kishore, Chris Hotz, H. A. Naseem and W. D. Brown, Jr. Applied Crystallography 36 (2003), 1236
11. A.K.Srivastava, K.N.Sood, R.Kishore and H.A.Naseem, Electrochemical and Solid State Letters, 9 (2006) G219-G221
12. Abu-Safe, Husam H., Khalil Sharif, Hameed A. Naseem, William D. Brown, Ram Kishore, and Mowafak Al-Jassim, Proc.. of Symp A: Amorphous and Polycrystalline Thin-Film Silicon Science and Technology-2006, MRS Proceedings Volume 910, 0910-A21-04
13. T. J. Kono and R. Sinchair, *Philos. Mag.*, **66(1992)** 749
14. J.Y. Wang, Z.M. Wang and E.J. Mittemeijer, J. Appl. Phys. ,102 (2007) 113523

Mater. Res. Soc. Symp. Proc. Vol. 1066 © 2008 Materials Research Society       1066-A15-06

# Blue and Yellow Electroluminescence of MOSLED Made on Si-rich SiO$_x$ Film with Detuning Buried Si Nanoclusters Size

Chung-Hsiang Chang[1], Chi-Wee Liu[2], Chin-Hua Hsieh[3], Li-Jen Chou[3], and Gong-Ru Lin[1]

[1]Graduate Institute of Photonics and Optoelectronics and Department of Electrical Engineering, National Taiwan University, No. 1, Roosevelt Rd. Sec. 4, Taipei, 10617, Taiwan

[2]Graduate Institute of Electronics Engineering, National Taiwan University, No. 1, Roosevelt Rd. Sec. 4, Taipei, 10617, Taiwan

[3]Department of Materials Science and Engineering, National Tsing Hua University, No. 101, Section 2, Kuang Fu Rd., Hsinchu, 300, Taiwan

## ABSTRACT

We have demonstrated the blue and yellow electroluminescence of MOSLEDs made on Si-rich SiOx film with buried Si nanoclusters of different sizes. The situation of dehydrogenation of Si nanocrystals within the SiOx film becomes more pronounced then the re-growth of SiO$_2$ matrix along with the prolongation of annealing time period. A linear variation on the O/Si composition ratio of the Si-rich SiOx film related to the deposition recipe is reported, giving rise to the precipitation of Si nanocrystals with different size. With such synthesis conditions, the SiOx films result in relatively strong photoluminescence at blue and yellow colors. From the comparison of the I–V curves we can conclude that there is a linear decrease on the threshold voltage of the SiOx based MOSLEDs by decreasing the thickness of the SiOx layer. According to EL pattern, we could demonstrate that the yellow- and blue-light pattern can be observed at 5.5 and 7.25 MV/cm, respectively.

## INTRODUCTION

Light-emitting silicon nanostructures have been studied intensively due to possible application for Si-based optoelectronic devices. One of these structures produced by high temperature annealing of silicon-rich silicon oxide (SRSO) is Si nanoclusters (nc-Si) embedded within the SiO$_2$ matrix. This method precipitates high-density oxide passivated nc-Si in robust matrix of SiO$_2$, which has advantage of compatibility with standard Si processing technology. Different preparation methods and annealing conditions have been studied intensively for improving of light emission from nc-Si. As an indirect band gap material, bulk Si is known to be inefficient in light emission. Besides, the low band gap of bulk Si at about 1.1 eV allows only for infrared emission, instead of visible light [1-5]. Therefore, optoelectronic devices are currently built on compound semiconductors due to their high efficiency in light emission. Compound semiconductors, however, are hard to integrate into cheap and versatile silicon circuits due to lattice mismatch problems. Once Si devices can be made exhibiting high electroluminescence (EL) efficiency, the low cost and powerful optoelectronic integrated circuit Si devices will become feasible. Besides, due to mature processing technologies, the Si-based light emitting diode (LED) may become a candidate for the next generation flat-panel display. In this study, we report for the first time the blue and yellow electroluminescence of MOSLEDs made on Si-rich SiOx film with buried Si nanoclusters of different sizes.

## EXPERIMENTAL DETAILS

The silicon nitride films studied in this experiment were deposited by plasma-enhanced chemical vapor deposition (PECVD) technique on (100) p-type silicon substrate. The growth temperature and the RF plasma power were 350 °C and 130 W, respectively. The background pressure was 67 Pa. After the standard RCA clean procedures, the substrate was loaded into the reaction chamber. $SiH_4$ and $N_2O$ were used as the reactant gas resources to deposit the silicon dioxide films. We hereafter separate the samples into B and C groups by their different recipes on the $N_2O$ fluence. B and C correspond to $N_2O$ fluence of 100 and 50 sccm. The other indexes of B or C samples denote their growth durations. For sample B3, the flow rate of $SiH_4$ and $N_2O$ gases was 220 and 100 sccm, respectively, moreover, for sample C4, the flow rate of $SiH_4$ and $N_2O$ gases was 62 and 50 sccm, respectively. After deposition the films were annealed at high temperature (1100 °C) for optimum condition in $N_2$ atmosphere to induce the separation of the Si and $SiO_2$ phases with the formation of Si nanocrystals embedded in the insulating matrix. A 2000-Å ITO film layer was deposited on the top of Si-rich SiOx with contact diameter of 0.8 mm to form the diode. A 5000-Å Al contact electrode was coated on the bottom of the Si substrate. The MOS diode was driven by either a programmable electrometer (Keithley, model 6517) with resolution as low as 100 fA or a voltage source meter (Keithley, 236), which uses microprobes (Karr Suss, 253).

## DISCUSSION

Fig. 1 shows the thickness of the Si-rich SiOx film grown by PECVD at different time, which are 2, 4 and 8 minutes along the x-axis respectively. In comparison with B and C, it is important to note that the variation of thickness was in proportion to the deposited time by PECVD. From these experimental results, it means that a linear variation of thickness was in control of fabrication. That is, the deposited time could control exactly the thickness of the SRSO film, this result is advantageous to fabricate the optimum thickness for application. Besides, the variation of thickness is mainly attributed to dehydrogenation, which occurred by annealing. The situation of dehydrogenation becomes more obvious along with the prolongation of annealing time period which leads to a thinner layer. Therefore, because the annealing time of sample B was longer than C, the thickness of sample B was smaller. Fig. 2 shows the room-temperature PL spectra of sample C4 grown at different time by annealing for optimum condition, the deposited time are 2, 4 and 8 min. The normalized PL intensity of Sample C4 increases when the deposited time is increased from 2 to 8 min. It was found that there is a PL spectrum between 350 and 410 nm, which is attributed to the small size of nc-Si in the Si-rich SiOx sample. It is interesting to note that the normalized PL at different deposited time is not equal to each other. It means that the structure of SRSO film grown at different condition is not uniform. In comparison with 2 and 8 min, the maximum normalized PL intensity decreases from 65 to 35, therefore, the result indicates that the excess amount of nc-Si decreases with deposited time reducing. In the right-hand side of Fig. 2 shows the room-temperature PL spectra of sample B3 grown at different time by annealing for optimum condition, the deposited time are 2, 4 and 8 min. At the same time, the appearance of the peak is attributed to the improvement of the structure.

Therefore, it is important to note that variation of structure in SRSO film relates to the deposited time.

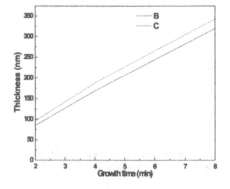 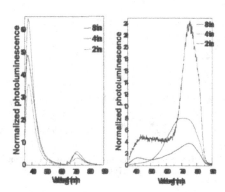

Fig. 1 The thickness of the Si-rich SiOx film grown by PECVD at different time, which are 2, 4 and 8 minutes along the x-axis respectively.

Fig. 2 The normalized PL of sample C4 grown at different time, and the normalized PL of sample B3 grown at different time (the right-hand side).

Fig. 3 shows the current-voltage (I-V) characteristics of sample B3 with different deposited time, which is 2, 4 and 8 min, respectively. The threshold currents were determined to be 150, 50 and 10 μA for 8, 4 and 2 min, respectively. The threshold voltages were determined to be 5, 2.25 and 1.25 MV/cm for 8, 4 and 2 min, respectively. Higher reverse-bias condition is required to form an inverse n-channel at SiOx/p-Si interface. The tunneling-based carrier transport mechanism is dominated due to the exponential like I-V behavior. It should be noted that the threshold voltage of sample B3 was linearly decreased from 8 to 2 min, it meant that the thickness in the SRSO film was proportion to the threshold voltage. In the right-hand side of Fig. 3 shows the current-voltage (I-V) characteristics of sample C4 with different deposited time, which is 2, 4 and 8 min, respectively. The threshold currents were determined to be 75, 50 and 25 μA for 8, 4 and 2 min, respectively. The threshold voltages were determined to be 6.75, 3.125 and 1.25 MV/cm for 8, 4 and 2 min, respectively. Compared with the left side with right side of Fig. 3, it is interesting to note that the threshold voltage was almost linearly decreased from 8 to 2 min. From the comparison of the I-V curves we can conclude that by decreasing the thickness in the SiOx layer there is a linear decrease in the threshold voltage. It is obvious that with the increasing of deposition time, the position of the Si-O peak moves towards the higher binding energy, it means that the silicon-rich SiOx structure transfers gradually to the $SiO_2$ structure, and the area percent ratio of Si-O increase to 100%. That is, the longer deposited time was useful to get high ratio of O/Si in the SRSO film. At the same time, according to the Fig. 4, the lower ratio of O/Si did not necessarily enhance the excess amount of nc-Si in the SRSO film. It should be noted that the lower ratio of O/Si of C4 sample still have a blue emission at 390 nm. Compared with B and C, there is a discrepancy in the $N_2O$ fluence, therefore, we can suppose that the variation of $N_2O$ fluence is possible to effect on the nc-Si. From these results, the structure in SRSO film has a discrepancy with the variation of deposited time, in addition, there is a linear decay in the threshold voltage while providing the deposited time decreasing linearly.

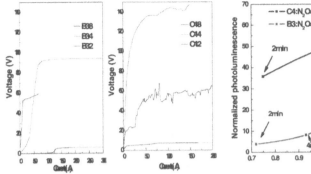 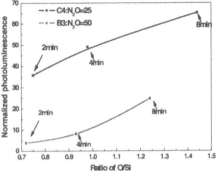

Fig. 3 I–V of sample B3 and C4 grown at different time responses of the ITO/ Si-rich SiOx / p-Si/Al MOS diode.

Fig. 4 Normalized Photoluminescence of B3 and C4 sample grown at different time vs. ratio of O/Si.

The EL patterns of the B3 MOS diode at forward bias from 4.5 to 6.5 MV/cm are shown in Fig. 5. The EL patterns become bright as voltage increase. The dark pattern was observed at 4.5 V, 4.75 V and 5 MV/cm, the voltage increasing to 5.25 MV/cm, the red-light far-field pattern can be observed. When the applied voltage increases to 5.5 MV/cm, the obviously red light emerges from ITO pattern. The EL pattern from red light becomes origin-red light as applied voltage increases from 5.5 to 6.5 MV/cm. However, under the high voltage of 6.5 MV/cm, the device reveals obviously soft-breakdown phenomenon and breakdown soon in few minute (1~3 min). In Fig. 6 shows that the EL patterns of the C4 MOS diode at forward bias from 6.75 to 7.755 MV/cm. Compared with sample B3 and C4, it is important to note that the turn-on voltage was higher than B3. That is, it means that the structure of C4 was too strong to enhance the carrier tunneling. The dark pattern at 6.75 MV/cm and 7 MV/cm has less blue emission, the voltage increasing to 7.25 MV/cm, the blue-light far-field pattern can be observed. The EL pattern from blue light becomes blue-white light as applied voltage increases from 7.25 to 7.75 MV/cm. However, under the high voltage of 8 MV/cm, the device reveals obviously soft-breakdown phenomenon and breakdown soon in few second.

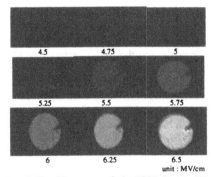 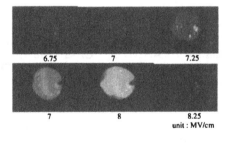

Fig. 5 The EL pattern of the ITO/ Si-rich SiOx (sample B3) / p-Si/Al MOS diode biased at 4.5 -6.5 MV/cm.

Fig. 6 The EL pattern of the ITO/ Si-rich SiOx (sample C4) / p-Si/Al MOS diode biased at 6.75-8.25 MV/cm.

## CONCLUSIONS

In conclusion, the situation of dehydrogenation of Si nanocrystals within the SiOx film becomes more pronounced then the re-growth of $SiO_2$ matrix along with the prolongation of annealing time period. A linear variation on the O/Si composition ratio of the Si-rich SiOx film related to the deposition recipe is reported, giving rise to the precipitation of Si nanocrystals with different size. With such synthesis conditions, the SiOx films result in relatively strong photoluminescence at blue and yellow colors. This phenomenon has been confirmed by analysis of XPS and RBS, meanwhile, Compared with B and C, there is a discrepancy in the $N_2O$ fluence, therefore, we can conclude that the variation of $N_2O$ fluence is possible to effect on the nc-Si. It is interesting to note that the threshold voltage was almost linearly decreased from 8 to 2 min. From the comparison of the I–V curves we can conclude that there is a linear decrease on the threshold voltage of the SiOx based MOSLEDs by decreasing the thickness of the SiOx layer. According to EL pattern, we could demonstrate that the yellow- and blue-light pattern can be observed at 5.5 and 7.25 MV/cm, respectively.

## ACKNOWLEDGMENTS

This work was supported in part by the National Science Council (NSC) of the Republic of China under grants NSC96-2221-E-002-099 and NSC97-ET-7-002-007-ET.

### References
1. N.-M. Park, T.-S. Kim, and S.-J. Park, "Band gap engineering of amorphous silicon quantum dots for light-emitting diodes," *Appl. Phys. Lett.*, vol. **78**, pp. 2575-2577, (2001).
2. S. Furukawa and T. Miyasato, "3-Dimensional quantum well effects in ultrafine Silicon particles," *Jpn. J. Appl. Phys.*, vol. **27**, L2207-2209, (1988).
3. L.-S. Liao, X.-M. Bao, X.-Q. Zhang, N.-S. Li, and N.-B. Min, "Blue luminescence from $Si^+$-implanted $SiO_2$ films thermally grown on crystalline silicon," *Appl. Phys. Lett.*, vol. **68**, pp. 850-852, (1996);
4. M. Ehbrecht, B. Kohn, F. Huisken, M. A. Laguna, and V. Paillard, "Photoluminescence and resonant Raman spectra of silicon films produced by size-selected cluster beam deposition," *Phys. Rev. B*, vol. **56**, pp. 6958-6964, (1997).
5. L. Patrone, D. Nelson, V. I. Safaov, M. Sentis, W. Marine, and S. Giorgio," Photoluminescence of silicon nanoclusters with reduced size dispersion produced by laser ablation," *J. Appl. Phys.*, vol. **87**, pp. 3829-3837, (2000).

# Poster Session:
# Thin-Film Transistors

Mater. Res. Soc. Symp. Proc. Vol. 1066 © 2008 Materials Research Society                    1066-A16-02

# The Effect of Electrical Stress on the New Top Gate N-type Depletion Mode Polycrystalline Thin Film Transistors Fabricated by Alternating Magnetic Field Enhanced Rapid Thermal Annealing

Won-Kyu Lee[1,2], Sang-Myeon Han[1], Sang-Geun Park[1], Sung-Hwan Choi[1], Joonhoo Choi[2], and Min-Koo Han[1]

[1]School of Electrical Engineering, Seoul National University, Gwanak-ro 599, Sillim9-dong, Gwanak-gu, Seoul, 151-742, Korea, Republic of

[2]LCD Business, Samsung Electronics Co. Ltd., Yongin, 449-711, Korea, Republic of

## ABSTRACT

We have fabricated the new top gate depletion mode n-type alternating magnetic field enhanced rapid thermal annealing (AMFERTA) polycrystalline silicon (poly-Si) thin film transistors (TFTs), which show the excellent electrical characteristics and superior stability compared with hydrogenated amorphous silicon (a-Si:H) TFTs and excimer laser crystallized (ELC) low temperature polycrystalline silicon (LTPS) TFTs. The fabricated AMFERTA poly-Si TFTs were not degraded under hot-carrier stress, and highly biased vertical field stress. The considerably large threshold voltage shift ($\Delta V_{TH}$) and trap state density reducing were occurred when the gate bias and drain bias were both large enough. The dominant mechanism of instability in the fabricated depletion mode AMFERTA poly-Si TFTs may be due to carrier induced donor-like defects reduction within the channel layer, especially near the drain junction.

## INTRODUCTION

LTPS TFTs based on an ELC method have high field-effect carrier mobility for pixel elements of active matrix organic light-emitting diode (AMOLED) displays. But the non-uniformity of the TFT characteristics caused by the laser shot characteristics needs to be improved. And the degradation of the ELC TFTs under the strong lateral field was severe. In order to obtain uniform TFT characteristics and reliable devices without using a laser, several crystallization methods have been proposed [1,2]. The solid phase crystallization (SPC) of amorphous silicon (a-Si) is a useful method of obtaining poly-Si film and it has many advantages over ELC such as simplicity, low cost, higher uniformity, good reliability, and large-area applicability. However, the heat treatment of the SPC requires a high temperature above 600 °C and a long annealing time (typically longer than 10 hours), which prevent its use on a thermally susceptible glass substrate [3,4].

Recently, the results of AMFERTA on a glass substrate have been reported [5-7]. AMFERTA employs field enhanced rapid thermal annealing in which the rapid thermal annealing induced by halogen lamps is combined with alternating magnetic fields. Induction of alternating magnetic field inside the silicon film leads to generation of electromagnetic force, which is the driving force for the kinetic enhancement. The electromagnetic force generates a selective joule heating of the silicon films and the movement of the silicon atoms through the effect of the applied field on the charged defects (such as vacancies, interstitial atoms and impurities) [8-10]. This methodology successfully decreased the crystallization temperature and time, so that a uniform poly-Si film is easily obtained on the glass substrate without deformation [8-10].

The purpose of this paper is to report the effect of electrical stress and the degradation mechanism on the new top gate depletion mode n-type AMFERTA poly-Si TFTs.

## EXPERIMENT

We have fabricated top gate n-type depletion mode poly-Si TFTs employing AMFERTA method on the glass substrate. Figure 1 shows the vertical layer structure of a fabricated AMFERTA poly-Si TFT. Silicon oxide was deposited with a thickness of 500 nm on the glass substrate for the buffer layer. Then, 100 nm-thick a-Si and 5 nm-thick n+ a-Si layers were deposited successively by plasma enhanced chemical vapor deposition (PECVD). It should be noted that this process is the same as the conventional a-Si:H TFT process [11]. The n+ a-Si layer is expected to improve the contact resistance between the active Si layer and S/D layer. The a-Si and n+ a-Si layers were crystallized by AMFERTA at 750 °C within 30 minutes. After poly-Si islands were defined, S/D electrodes were patterned. The crystallized a-Si and n+ Si layers were dry-etched masking the S/D electrodes to define the channel region. 100 nm-thick silicon oxide ($SiO_x$) gate insulator was deposited and a gate electrode was patterned. In order to prevent contamination, 600 nm-thick $SiN_x$ passivation layer was deposited. Finally, contact holes were opened and 70 nm-thick indium-tin-oxide (ITO) electrode was patterned. The width and length of AMFERTA TFT were 200 μm and 4 μm, respectively.

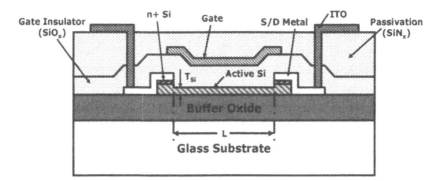

**Figure 1.** The vertical layer structure of a fabricated top gate n-type AMFERTA poly-Si TFT.

## DISCUSSION

### Electric characteristics of top gate n-type depletion mode AMFERTA poly-Si TFT

We have measured the transfer characteristics of a fabricated top gate n-type AMFERTA poly-Si TFT. Figure 2 shows the drain current ($I_{DS}$) – gate voltage ($V_{GS}$) characteristics of the AMFERTA poly-Si TFT. The drain current flowed from the drain to the source even at zero gate voltage, which suggests that the n-type AMFERTA poly-Si TFT operates in depletion-mode.

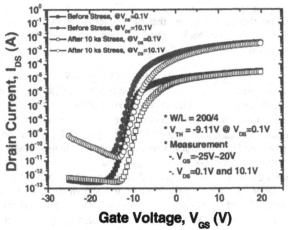

**Figure 2.** The transfer characteristics of a fabricated top gate n-type depletion mode AMFERTA poly-Si TFT at $V_{DS}$ = 0.1 V and $V_{DS}$ = 10.1 V.

The threshold voltage ($V_{TH}$) was -9.11 V, the field-effective mobility ($\mu_{FE}$) was 6.16 cm²/V•s, the subthreshold slope (S) is 0.2697 V/dec and the on/off current ratio was 1.12 × 10⁸. The $V_{TH}$ at $V_{DS}$ = 0.1 V is defined as the gate voltage for a drive current of (W/L) × 10⁻⁸ A. The S value is the measured reciprocal slope of the transfer characteristics in the switching region. Figure 3 shows the secondary ion mass spectrometry (SIMS) data of the phosphorous doping density profile in the a-Si and n+ a-Si layers in the vertical direction after AMFERTA

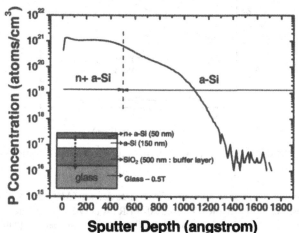

**Figure 3.** The P+ ion concentration profile in the AMFERTA poly-Si thin film measured by SIMS. The inset is a schematic diagram of measured AMFERTA poly-Si thin film.

crystallization. During the crystallization process, the P+ ions in the n+ a-Si layer were diffused into the a-Si layer. From this data, we knew that P+ ions remained in the channel surface of AMFERTA poly-Si TFT in spite of back channel Si (crystallized a-Si and n+ a-Si layers) over-etching. That is caused the negative shift of $V_{TH}$ [7].

## Stability of top gate n-type depletion mode AMFERTA poly-Si TFT

In order to investigate the stability of depletion mode AMFERTA poly-Si TFTs, we have measured the transfer characteristics with various stress conditions and also calculated $V_{TH}$ shifts ($\Delta V_{TH}$). The each stress time was 10,000 seconds, after that we measured I-V characteristics at $V_{DS} = 0.1$ V. In the previous studies on the stability of poly-Si TFTs, several degradation mechanisms were proposed [12,13], which were hot-carrier degradation, trapped charges by gate bias, and defects creation by either hot-carrier or gate bias.

Figure 4 shows the $\Delta V_{TH}$'s versus stressed time for each stress conditions in the fabricated depletion mode AMFERTA poly-Si TFTs. It is shown that lateral field only ($V_{GS} = 0$ V, $V_{DS} = 15$ V) or vertical field only ($V_{GS} = 14$ V, $V_{DS} = 0$ V) stress condition did not cause the degradation of the AMFERTA poly-Si TFT. Under those conditions, $\Delta V_{TH}$'s were negligible during 10,000 seconds. The proposed AMFERTA poly-Si TFT was degraded when the both vertical bias and lateral bias were sufficiently high enough ($V_{GS} = 10$ V, $V_{DS} = 15$ V). At that condition, the $\Delta V_{TH}$ is 3.76 V after 10,000 seconds stressed. Therefore, we were able to recognize that the degradation of fabricated depletion mode AMFERTA poly-Si TFT was due to the donor-like defects reduction by sufficiently strong gate bias and drain bias.

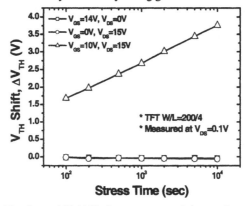

**Figure 4.** The threshold voltage shift ($\Delta V_{TH}$) versus stressed time under various stress bias conditions.

Figure 5 shows the rates of normalized trap state density ($\Delta N_t$) variation versus stressed time for each stress conditions in the proposed AMFERTA poly-Si TFTs. The trap state density ($N_t$) can be calculated from $\ln (I_D/V_G)$ versus $1/V_G$ characteristics by using equations 1 [14].

$$I_D = \mu_0 \left( \frac{W}{L} \right) C_{ox} V_G V_D \exp \left[ \left( -q^3 N_t^2 t \right) / \left( 8 \varepsilon_{si} kTC_{ox} V_G \right) \right] \tag{1}$$

The trap state density was considerably changed when both gate bias and drain bias were sufficiently high enough (about 16% reduced after 10,000 stressed). In the n-type accumulation mode poly-Si TFTs, the $V_{TH}$ was increased with increasing $N_t$ [15]. But the $V_{TH}$ of measured AMFERTA poly-Si TFTs was increased with decreasing $N_t$, which is due to depletion mode operation of the proposed AMFETA poly-Si TFT. In the channel of proposed AMFERTA poly-Si TFT, there are sufficiently many P+ ions, which act donor-like trap states. Therefore, after sufficiently high vertical and lateral bias stressed, donor-like trap states are neutralized. So measured trap state density was reduced.

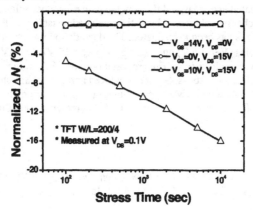

**Figure 5.** The normalized trap state density versus stressed time under various stress bias conditions.

Figure 6 shows the normalized capacitances between gate and source electrode ($C_{GS}$) and the normalized capacitances between gate and source electrode ($C_{GD}$), when the gate bias and drain bias were sufficiently high ($V_{GS}$ = 5 V, $V_{DS}$ = 15 V). The $C_{GS}$ was not changed after 10,000 seconds stress. But $C_{GD}$ was parallel shifted after stress. This means that the degradation was occurred near the drain junction of AMFETA poly-Si TFT.

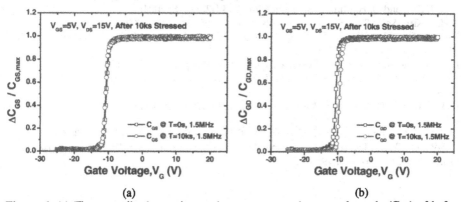

(a)                                     (b)

**Figure 6.** (a) The normalized capacitances between gate and source electrode ($C_{GS}$) of before and after 10,000 seconds stress, (b) The normalized capacitance between gate and drain electrode ($C_{GD}$) of before and after 10,000 seconds stress.

From these results, we could believe that the dominant instability mechanism in the top gated n-type depletion mode AMFERTA poly-Si TFT may be due to carrier induced donor-like defects reduction within the channel layer, especially near the drain junction.

## CONCLUSIONS

We have fabricated and measured the new top gate n-type depletion mode AMFERTA poly-Si TFTs. The mechanism of electric bias stress induced instability in the fabricated AMFERTA poly-Si TFTs has been investigated. The degradation was observed when the sufficiently high gate bias ($V_{GS} > 5$ V) and drain bias ($V_{DS} > 15$ V). The $\Delta V_{TH}$ was 3.76 V and the trap state density was reduced 16%. The proposed poly-Si TFTs were not degraded in the gate bias only or drain bias only stress. Stressing induced parallel voltage shift for AMFERTA poly-Si TFTs in high frequency (1.5 MHz) capacitance ($C_{GD}$) measurement. Therefore, the lateral field or vertical field reducing structure could be increased the stability of AMFERTA poly-Si TFTs.

## REFERENCES

1. S.-W. Lee and S.-K. Joo, *IEEE Electron Device Lett.*, vol. 17, pp. 160-162, 1996.
2. S. Y. Yoon, K. H. Kim, C. O. Kim, J. Y. Oh, and J. Jang, *J. Appl. Phys.*, vol. 82, pp. 5865-5867, 1997.
3. T. Sameshima, *J. Non-Cryst. Solids*, 227-230, pp. 1196-1201, 1998.
4. S. Y. Yoon, S. J. Park, K. H. Kim, and J. Jang, *Thin Solid Films*, vol. 383, pp. 34-38, 2001.
5. B. S. So, Y. H. You, H. J. Kim, Y. H. Kim, J. H. Hwang, D. H. Shin, S. R. Ryu, K. Choi, and Y. C. Kim, *Mater. Res. Soc. Symp. Proc.*, Vol. 862, pp. 275-280, 2005.
6. W.-K. Lee, S.-M. Han, S.-G. Park, Y.-J. Chang, K.-C. Park, C.-W. Kim, and M.-K. Han, *Mater. Res. Soc. Symp. Proc.*, Vol. 989, pp. 405-410, 2007.
7. W.-K. Lee, J.-H. Park, J. Choi, and M.-K. Han, *IEEE Electron Device Lett.*, vol. 29, pp. 174-176, 2008.
8. H. J. Kim and D. H. Shin, U.S. Patent 6 747 254 B2, Jun. 8, 2004.
9. H. J. Kim, *Solid State Phenomena*, Vols. 124-126, pp. 447-450, 2007.
10. S. H. Park, H. J. Kim, K. H. Kang, J. S. Lee, Y. K. Choi, and O. M. Kwon, *J. Phys. D: Appl. Phys.* 38, pp. 1511-1517, 2005.
11. M. J. Powell, *IEEE Trans. Electron Devices*, vol. 36, pp. 2753-2763, 1989.
12. M. Hack, A. G. Lewis, and I-W. Wu, *IEEE Trans. Electron Devices*, vol. 40, pp. 890-897, 1993.
13. J. C. Kim, J. H. Choi, S. S. Kim, and J. Jang, *IEEE Electron Device Lett.*, vol. 25, pp. 182-184, 2004.
14. S. Seki, O. Kogure, and B. Tsujiyama, *IEEE Electron Device Lett.*, vol. 8, pp. 368-370, 1987.
15. I-W. Wu, W. B. Jackson, T.-Y. Huang, A. G. Lewis, and A. Chiang, *IEEE Electron Device Lett.*, vol. 11, pp. 167-170, 1990.

Mater. Res. Soc. Symp. Proc. Vol. 1066 © 2008 Materials Research Society                    1066-A16-03

# Negative Bias Temperature Instability for P-channel of LTPS Thin Film Transistors with Fluorine Implantation

Chyuan-Haur Kao, and W. H. Sung

Electronics Engineering, Chang Gung University, 259 Wen-Hwa 1st Road, Kwei-Shan, Tao-Yuan, 333, Taiwan

## ABSTRACT

This paper studies the impact of LTPS (low temperature polycrystalline silicon) TFTs with fluorine implantation under NBTI (Negative bias temperature instability) stress. The fluorinated TFTs' devices can obtain better characteristics with samller threshold voltage shift, lower trap states and lower subthreshold swing variation. Therefore, the fluorine implantation does not only improve initial electrical characteristics, but also suppresses the NBTI-induced degradation.

## INTRODUCTION

Low-Temperature Polycrystalline Silicon thin-film transistors (LTPS TFTs) have attracted much attention due to high possibility for the integration of peripheral circuits and active matrix [1-2]. Compared with amorphous TFTs, the LTPS TFTs have better performance, higher field-effect mobility, and better reliability stability [3].

Negative bias temperature instability (NBTI) has become one increasingly important issue especially for thin gate oxide in p-channel MOSFETs [4-5]. In general, the degradation mechanism about the NBTI stress in MOSFETs is mainly attributed to the generation of fixed oxide charges and interface trap states [6]. In polycrystalline silicon TFTs, some studies have also found the performance degradation after NBTI stress, which may generate interface states and grain boundary trap states in the polysilicon/polyoxide interface and polysilicon grain boundaries [7-9].

This paper studies the fluorine implantation into the LTPS P-channel TFTs after the NBTI stress can obtain better characteristics with small threshold voltage shift, lower interface states and lower subthreshold swing variation. Since the implanted fluorine into the TFT channel poly-Si can terminate the poly-Si defects or trap states to form more strong bonds for reliability improvements. Therefore, the fluorine implantation did not only improve the initial electrical characteristics, but also suppress the NBTI-induced degradation.

## EXPERIMENTAL DETAILS

The P-channel LTPS TFTs were fabricated on thermally oxidized silicon wafers. An amorphous-silicon film of about 130-nm was initially deposited at $550^0$C by low-pressure chemical vapor deposition (LPCVD) using pure $SiH_4$ gas. It was then furnace-annealed at $600^0$C for 24 hours in $N_2$ ambient gas to recrystallize the silicon films. Fluorine ions with a dosage of $2 \times 10^{15}$ cm$^{-2}$ was implanted at 11 keV, and then activated at $600^0$C in an $N_2$ ambient. After defining the active region, a 50 nm thickness of gate-oxide was deposited by a PECVD system with gas mixtures of $SiH_4$ and $N_2O$ at $300^0$C. Another 200-nm polysilicon film was deposited at $620^0$C and then patterned as the gate electrode. A self-aligned boron implantation at a dose of $5 \times 10^{15}$ cm$^{-2}$ was used to dope the drain, source, and gate areas. The dopants were then activated at $600^0$C for 8 hr in an $N_2$ ambient. After a plasma-enhanced CVD (PECVD) of $SiO_2$ was deposited with a thickness of 250 nm, the contact holes were opened and 500-nm Al films were deposited and defined.

## RESULTS AND DISCUSSION

Figure 1 shows the transfer Id-Vg characteristics of the p-channel TFTs devices without (control) and with fluorine implantation at a dosage of $2 \times 10^{15}$ cm$^{-2}$ under $25^0$C room temperature. It can be seen that the electrical characteristics of the on-current, subthreshold swing and threshold voltage were improved significantly for the device with fluorine implantation. This is believed to be due to the reduction of interface states and trap states effectively. Figure 2 shows the output Id-Vd characteristics of the control and fluorine implanted devices of LTPS TFTs. It is also found that the drain current of the fluorine-implanted device is larger than that of the control device.

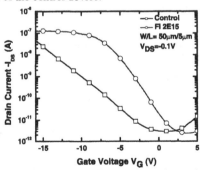

Figure 1. The transfer Id-Vg characteristics of the p-channel TFTs' devices without (control) and with fluorine implantation at a dosage of $2 \times 10^{15}$ cm$^{-2}$.

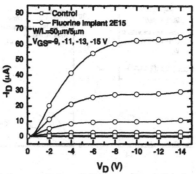

Figure 2. The output Id-Vd characteristics of the control and fluorine implanted devices.

During the Negative bias temperature instability stress, the substrate was heated to set the stress temperature at $100^0$C, and the stress voltage of - 15V was applied to the gate with source/drain grounded. The transfer Id-Vg characteristics of the p-channel TFTs' devices without (control) and with fluorine implantation after the NBTI stress for 1600 sec are shown in Fig. 3. It's seen that the threshold voltage of the control device shifts to the negative direction with larger leakage current (Ioff) after the NBTI stress. Furthermore, other device characteristics such as the subthreshold swing and field-effect mobility were also degraded after the NBTI stress. This is due to the generation of positive fixed-oxide charges and more interface states in the polysilicon/polyoxide interface after the NBTI stress.

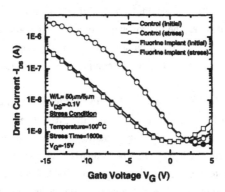

Figure 3. The transfer Id-Vg characteristics of the p-channel TFTs' devices without (control) and with fluorine implantation after the NBTI stress for 1600 sec

369

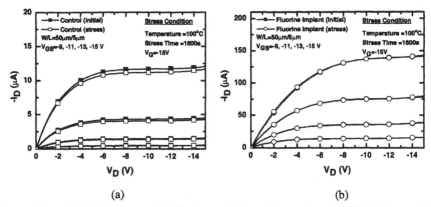

(a)                                                                (b)

Figure 4(a) and 4(b). The output Id-Vd characteristics of the p-channel TFTs' devices without (control) and with fluorine implantation after the NBTI stress for 1600sec.

The fluorine implantation we used can effectively improve the device performance and suppress the NBTI-induced degradation. Compared electrical characteristics with the control device, the fluorine-implanted device has better electrical characteristics such as higher on-current, smaller subthreshold swing and smaller threshold voltage. Since the fluorine creates negative charges in the gate oxide. These charges make it possible to control the Vth and lower potential, which results in the improvement of the ON-current, subthreshold swing and the leakage current for the TFT's. By the way, the fluorine atoms can passivate the defects in the channel poly-Si layer, which can increase the carrier mobility and also suppresses the leakage current. Therefore, after the NBTI stress, the variation of performance degradation is also small for the fluorine-implanted device. The improvements of electrical characteristics are believed to be due to that the implanted fluorine can passivate the dangling bonds and break the strained bonds to reduce trap states existed in the polysilicon/polyoxide interface and polysilicon grain boundaries.

Figure 4(a) and (b) show the output Id-Vd characteristics of the p-channel TFTs' devices without (control) and with fluorine implantation after the NBTI stress for 1600 sec. It is also seen that the drain current of the fluorine implanted device has a larger driving drain current with very small variation after the NBTI stress in comparison with the control sample. Since the driving current is related to the threshold voltage and field-effect mobility. Therefore, the reduction of the driving current after the NBTI stress is due to the threshold voltage shift and field-effect mobility decease.

Figure 5 shows the threshold voltage shift of both devices versus stressing time from 1 to 1600 sec under the NBTI stress at $V_G$= -10 and -15V. It is seen that the fluorine implanted device also has smaller threshold voltage shift compared with the control device after the NBTI stress.

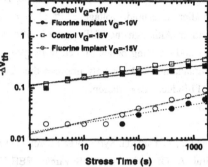

Figure 5. The threshold voltage shift of both devices versus stressing time from 1 to 1600 sec under the NBTI stress at $V_G$= -10 and -15V.

The schematic cross section of the control and fluorine-implanted devices after the NBTI stress is shown in Fig. 6. There is a lot of weak Si-H bonds existed in the poly-Si/polyoxide interface and polysilicon gain boundaries for conventional TFT's device. After the NBTI stress, the weak Si-H bonds are dissociated, and the dissociated hydrogens diffuse into the gate oxide and break Si-O bonds, and then create positive fixed charges in the gate oxide and induce worse characteristics and performances [10-11]. However, the implanted fluorine can form more strong Si-F bonds to replace the weak Si-H bonds in the polysilicon/polyoxide interface and grain boundaries, and then few charges are generated in the gate oxide and better characteristics are obtained. Therefore, the fluorine-implanted device does not only improve the initial electrical characteristics, but also suppresses effectively the NBTI-induced degradation.

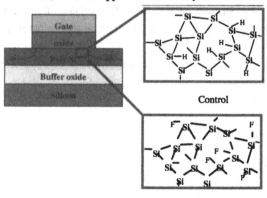

Control

Fluorine Implantation

Figure 6. The schematic cross section of the control and fluorine-implanted devices.

## CONCLUSION

The devices of LTPS TFTs with fluorine implantation under NBTI stress can obtain better characteristics with smaller threshold voltage shift, lower interface states and lower subthreshold swing variation. Although the TFTs' devices after the NBTI stress may create more interface trap-states and dangling bonds existed in the polysilicon/polyoxide interfaces; however, the fluorine implantation did not only improve the initial electrical characteristics, but also suppress the NBTI-induced degradation.

## REFERENCES

[1]  K. Yoneda, R. Yokoyama, T. Yamada: Symp. VLSI Circuits, Tech. Dig., 2001, p. 85.

[2]  T. Serikawa, S. Shirai, A. Okamoto, and S. Suyama: IEEE Trans. Electron Devices 36, 1929 (1989).

[3]  I. W. Wu, W. B. Jackson, T. Y. Huang, A. G. Lewis, and A. Chiang: IEEE Electron Device Lett. 11, 167 (1990).

[4]  C. E. Blat, E. H. Nicollian, and E. H. Poindexter: J. Appl. Phys. 63 (1991) 1712.

[5]  D. K. Schroder and J. A. Babcock: J. Appl. Phys. 94, 1 (2003).

[6]  A. T. Krishnan, V. Reddy, and S. Krishnan: IEDM Tech. Dig., 2001, p. 865.

[7]  S. Maeda, S. Maegawa, T. Ipposhi, H. Nishimura, T. Ichiki, J. Mitsuhashi,M. Ashida, T. Muragishi, and T. Nishimura: VLSI Symp. Tech. Dig., 1993, p. 29–30.

[8]  K. Okuyama, K. Kubota, T. Hashimoto, S. Ikeda, and A. Koike: IEDM Tech. Dig., 1993, p. 527.

[9]  Chih-Yang Chen, Jam-Wem Lee, Wei-Cheng Chen, Hsiao-Yi Lin, Kuan-Lin Yeh, Po-Hao Lee, Shen-De Wang, and Tan-Fu Lei: IEEE Electron Device Lett. 27, 893 (2006).

[10] Chih-Yang Chen, Jam-Wem Lee, Shen-De Wang, Ming-Shan Shieh, Po-Hao Lee, Wei-Cheng Chen, Hsiao-Yi Lin, Kuan-Lin Yeh, Po-Hao Lee, and Tan-Fu Lei: IEEE. Trans. Electron Devices 53, 2993 (2006).

[11] Chyuan Haur Kao, C. S. Lai and C. L. Lee: Journal of Electrochemical Society 154, 259 (2007).

Mater. Res. Soc. Symp. Proc. Vol. 1066 © 2008 Materials Research Society　　1066-A16-04

# Hysteresis Phenomenon in Sequential Lateral Solidification poly-Si Thin Film Transistor at Low Temperature (213K)

Sung-Hwan Choi, Sang-Geun Park, Won-Kyu Lee, Tae-Jun Ha, and Min-Koo Han
School of Electrical Engineering and Computer Sciences, Seoul National University, San 56-1, Shillim 9-Dong, Gwanak-gu, Seoul, 151-742, Korea, Republic of

## ABSTRACT

We have investigated temperature dependence on the hysteresis phenomenon of SLS poly-Si TFT on a glass substrate, extremely at low temperature (213K). The p-type sequential lataral solidification (SLS) polycrystalline Silicon (poly-Si) TFT was fabricated on glass substrate. As the temperature was reduced, it was observed that hysteresis phenomenon was increased, whereas the hysteresis was suppressed at high temperature. This could be explained by a difference of initially electron and hole trapped charges into gate insulator is much larger in low temperature than in high temperature. And we have verified that drain current was changed with a different previous gate starting voltage even at same bias condition by experimental results due to the hysteresis phenomenon of SLS poly-Si TFT. Hysteresis of SLS poly-Si TFT should be improved for a pixel element of high quality AMOLED display.

## INTRODUCTION

Active matrix organic light emitting diode (AMOLED) employing thin film transistor (TFT) pixel circuits has been considered as future display due to its high brightness, compactness, fast response time and wide viewing angle [1]. Therefore a quality of TFTs as well as a life time of electroluminescence should be improved. Recently, hydrogenated amorphous silicon (a-Si:H) TFTs and low temperature silicon (LTPS) TFTs are considered as a pixel element of driving circuit in the AMOLED. From the practical point of view, low temperature poly-Si TFT which employing laser crystallization method such as Sequential Lateral Solidification (SLS) has gained a lot of attention due to its excellent current stability, high field effect mobility and large driving current.

However, it also suffers from hysteresis phenomenon similar to other devices. Residual image sticking has still served as a critical problem in the AMOLED display. Residual image sticking might be observed for a few minutes due to the hysteresis phenomenon of the driving transistor in the AMOLED display [2]. Therefore, the hysteresis is needed for driving a high quality AMOLED display based on SLS poly-Si TFT. And because an operating temperature of AMOLED panel is varied with its application (from very low temperature to high temperature), the investigation of the temperature dependence on the hysteresis of SLS poly-Si TFT is also required. Hysteresis phenomenon including very low temperature has not been reported yet in poly-Si TFT.

The purpose of our work is to investigate the hysteresis characteristics of p-type poly-Si TFTs employing sequential lateral solidification (SLS) crystallization on the glass substrate with various temperatures (213K-400K). Our experimental results showed that the drain current of the SLS poly-Si TFT was altered even at same bias condition according to previous gate voltage due to the hysteresis of SLS poly-Si TFT. At reduced temperature (213K), the hysteresis phenomenon of SLS poly-Si TFT was increased compared with room temperature (300K), whereas that of SLS poly-Si TFT was decreased at elevated temperature (400K).

## FABRICATION OF SLS POLY-SI TFT

We have fabricated low temperature SLS poly-Si TFT following the conventional low temperature (450°C) poly-Si TFT process. The device has a conventional coplanar top-gate structure. And we deposited a 50-nm-thick amorphous silicon (a-Si) active layer by plasma enhanced chemical vapor deposition (PECVD), and dehydrogenation procedure was carried out to prevent the explosion of residual hydrogen during laser crystallization. Then, the a-Si active layer was crystallized by the SLS method with 308 nm XeCl excimer laser, and the silicon grain grew up in a fixed lateral direction. After the irradiation, a typical LTPS process sequences were followed. The structure was fabricated to function as the drain and source electrodes by the self-aligned doping process without using an additional mask. With active and gate patterning step, gate patterning were adjusted to the lateral grain growth direction as shown in Fig. 1.

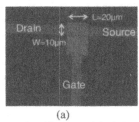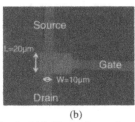

(a)                                    (b)

**Figure 1.** Photograph of fabricated SLS Processed poly-Si TFT on glass substrate.
(a) Single gate TFT with perpendicular channel to lateral grain growth direction
(b) Single gate TFT with parallel channel to lateral grain growth direction

## RESULTS AND DISCUSSION

In order to investigate the mechanism of hysteresis phenomenon with various temperatures, we have measured $I_{DS}$-$V_{GS}$ transfer characteristics by a double sweep mode at different temperature. The image sticking caused by hysteresis phenomenon could be evaluated by the threshold voltage variation ($\Delta V_{TH}$) with gate voltage sweep direction. Figure. 2 shows that hysteresis in the transfer characteristics (measured at $V_{DS}$ = -0.1V) was observed between forward ($V_{GS}$ = 10V to -15V, gate voltage sweep step = 0.2V) and reverse gate voltage sweeps ($V_{GS}$ = -15V to 10V).

Figure. 2 (a) and (b) shows that transfer characteristics of SLS poly-Si TFT (for perpendicular channel and parallel channel) at 213K, 300K and 400K respectively (measured at $V_{DS}$ = -0.1V). From now on, we are intending to explain mainly the hysteresis characteristics of SLS poly-Si TFT with perpendicular channel to lateral grain growth direction. When applied drain bias and temperature was -0.1V and 300K respectively, a threshold voltage of forward voltage sweep direction (from +10V to -15V) was -3.135V while that of reverse voltage sweep direction (from -15V to +10V) was -3.857V due to the hysteresis phenomenon of the SLS poly-Si TFT. And at elevated temperature (400K), the threshold voltage of forward and reverse gate voltage sweep direction was -2.552V and -2.779V. At reduced temperature (213K), it was measured that the threshold voltage with forward and reverse gate bias direction was -4.357V, -5.072V each other.

The threshold voltage ($V_{TH}$) of SLS poly-Si TFT was increased because a fermi energy level of the poly-Si was decreased as the temperature was reduced. Figure. 3 shows measured threshold voltages of poly-Si TFT with various temperatures. The threshold voltages of SLS poly-Si TFT according to the gate voltage sweep direction were quite different due to the hysteresis phenomenon of SLS poly-Si TFT as shown in Fig. 3. The difference of threshold voltage ($\Delta V_{TH}$) in perpendicular channel was increased from 0.227V (400K) to 0.715V (213K). And $\Delta V_{TH}$ in parallel channel was also increased from 0.144V (400K) to 0.594V (213K).

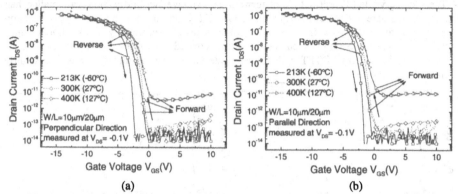

(a)                                                                                     (b)

**Figure 2.** Transfer Characteristics of the hysteresis phenomenon of SLS poly-Si TFT with various temperature ranges (From 213K to 400K). (a) Perpendicular channel to lateral grain growth direction, (b) Parallel channel to lateral grain growth direction.

The variation of $V_{TH}$ with gate voltage sweep direction was increased considerably at low temperature. As the temperature was increased, a difference of the threshold voltage between forward and reverse gate sweep direction was decreased. And the threshold voltage of SLS poly-Si TFT with forward and reverse gate voltage sweep direction was increased as temperature was reduced. However as shown in Fig. 3, the difference of threshold voltage was not increased simply, because at some points (such as 225K, 213K in perpendicular channel, and 213K in parallel channel) the difference was decreased slightly compared with that above 225K. For these reasons, an increase of the hysteresis phenomenon at reduced temperature might not be explained mainly by the increase of threshold voltage.

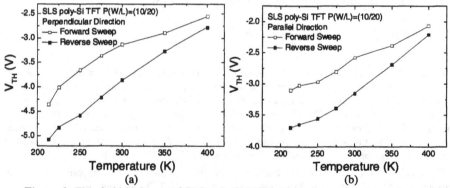

(a)                                                                                     (b)

**Figure 3.** Threshold Voltages of SLS poly-Si TFT with various temperature ranges (From 213K to 400K). (a) Perpendicular channel, (b) Parallel channel.

Therefore in order to study the temperature dependence of the hysteresis phenomenon in SLS poly-Si TFT fully, sub-threshold swing (s-swing) of SLS poly-Si TFT was extracted from a measured transfer characteristic with various temperatures from 213K to 400K as shown in Fig. 4. The s-swing value ($dV_G/d(\log I_{DS})$) of forward gate voltage sweep direction was smaller than that of reverse gate voltage sweep direction above 275K, because the initially trapped electron charges with forward sweep direction are de-trapped faster than initially trapped hole charges with reverse sweep direction. As temperature was decreased at temperatures below about 300K, the s-swing of SLS poly-Si TFT with forward gate voltage sweep direction was also slightly decreased.

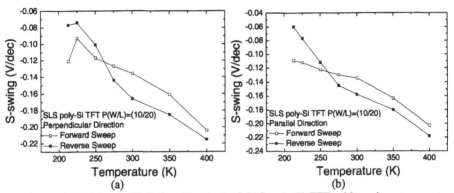

**Figure 4.** Subthreshold Swing (S-swing) of SLS poly-Si TFT with various temperature ranges (From 213K to 400K). (a) Perpendicular channel, (b) Parallel channel.

But the s-swing of reverse voltage sweep direction was decreased sharply, resulting in an increase of hysteresis phenomenon. The s-swing of forward bias sweep direction was decreased from -0.136V/dec (300K) to -0.121V/dec (213K) while that of reverse bias sweep direction was steeply decreased from -0.166V/dec (300K) to -0.077V/dec (213K) as a temperature was decreased. If there is a significant interface-trap density $D_{it}$ between silicon and silicon dioxide, s-swing could be approximately formulated as below equation [3].

$$S \cong (\ln 10)\frac{dV_G}{d(\ln I_D)} = (\ln 10)\frac{dV_G}{d(\beta\psi_s)} = (\ln 10)(\frac{kT}{q})(1+\frac{(C_D+C_{it})}{C_{ox}}), \text{ where } C_{it} \text{ is } qD_{it.} \qquad (1)$$

The s-swing of the TFT could be decreased as initially trapped charges were reduced in a gate insulator by an above equation. Because ($kT/q$) term in the above equation was altered as the temperature (T) was changed, we could not exactly analyze the factor of interface-trap density from s-swing value directly. Therefore in order to investigate the temperature dependence of SLS poly-Si TFT by analyzing the s-swing value, we have revised it to verify a quantity of interface-trapped charges by multiplying the s-swing value by '(300K) / (each temperature)'. The modified s-swing value (for perpendicular and parallel channel) of SLS poly-Si TFT is shown in Fig. 5.

The mechanism of the s-swing variation could be explained by the trapping of carriers between poly-Si and gate oxide. In the case of p-type poly-Si TFTs, trapping and de-trapping of each carriers occurs with applied voltages. When a gate voltage was shifted from positive bias

(+10V) to negative bias (-15V) (forward voltage sweep), the electron charges were initially trapped in a silicon dioxide. If the temperature was increased or decreased from 300K, a quantity of electron trapped charges was more gained in gate insulator than that at 300K. This is because electron charges were excited by thermal energy at high temperature, so that more trapping of electron charges was occurred in a gate insulator than at 300K. And electron trapping probability was also increased slightly by defect scattering at low temperature (Thermal velocity of electron carriers was reduced). The electric characteristics of TFTs are thoroughly influenced by the traps and lattice structures in the grains at low temperature [4]. Because the interface-trapped charge capacitance $C_{it}$ in above equation was affected by interface-trapped charges, s-swing of SLS poly-Si TFT was increased as the temperature was increased or decreased from 300K with a forward gate voltage direction.

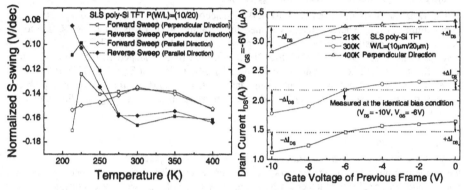

**Figure 5.** Normalized s-swing values of SLS poly-Si TFT (for perpendicular and parallel channel with lateral grain growth direction) with various temperatures (213K-400K).

**Figure 6.** Drain current of SLS poly-Si TFT at an identical bias condition ($V_{GS}$ and $V_{DS}$) according to previous gate starting voltages with 213K, 300K and 400K.

Then when the gate voltage was shifted from negative bias (-15V) to positive bias (10V) (reverse voltage sweep), hole charges were initially trapped in gate insulator. At high temperature, hole charges could not be gained sufficient energy from the electric field compared with electron charges due to its heavier effective mass [3]. It was known that activation energy of electron (to inject into the Si/SiO$_2$ barrier) is about 3.13eV and that of hole is about 4.25eV [5]. It means that the probability of trapping for electrons is rather larger than that of trapping for holes.

At low temperature, a quantity of trapped hole charges was reduced in a reverse gate voltage sweep direction. The decrease of hole trapping in the Si-SiO$_2$ interface at low temperature leads to reduction of interface trap state density, so that s-swing value in reverse voltage sweep direction was decreased as shown in equation (1). The s-swing value of device was decreased rapidly at temperatures below about 300K in reverse gate voltage sweep direction due to the decrement of initially trapped hole charges in a SiO$_2$ gate insulator. If a gate voltage for high current (for example, -15V for p-type device) is applied, initial trapped hole charges induced in channel region. So threshold voltage of SLS poly-Si TFT would be increased due to electron would be induced in the channel, s-swing value and on current would be reduced as a result.

In order to investigate the relation between hysteresis and OLED current error, we have measured a drain current from different starting gate-voltages in SLS poly-Si TFT with various temperatures. Figure. 6 shows that measurement results of the drain current at identical bias condition (same $V_{GS}$ and $V_{DS}$) with different previous gate starting voltages from -10V to 0V. As shown in Fig. 6, even at same gate-source voltage ($V_{GS}$ = -6V) and drain-source voltages ($V_{DS}$ = -10V) the drain current was altered with a previous gate-source voltage due to the hysteresis of SLS poly-Si TFT. The drain current of 1.78μA was measured at $V_{GS}$ = -10V, while that of 2.34μA was measured at $V_{GS}$ = 0V at 300K. The variation of drain current (($I_{max}$-$I_{min}$)/$I_{min}$) was about 31.4% even at same bias condition according to the previous gate voltage. When it was measured at 213K, the drain current was 1.12μA with a previous gate voltage (-10V). It was 1.65μA with a previous gate-source voltage (0V).

The variation of drain current was about 47.2% according to previous gate voltage at reduced temperature (213K), whereas that at elevated temperature (400K) was about 18.7% due to the decrease of hysteresis phenomenon in SLS poly-Si TFT. If SLS poly-Si TFT is employed as a pixel element of AMOLED display, the OLED current will be changed with the previous scan signal due to the variation of drain current caused by hysteresis of SLS poly-Si TFT. Because the difference of drain current with previous gate voltage results in image sticking problem, hysteresis of p-type SLS poly-Si TFT needs to be improved to realize the high quality AMOLED display.

CONCLUSIONS

We have investigated hysteresis phenomenon of low temperature SLS poly-Si TFT with various temperature including very low temperature. A quantity of initially trapped charge in a poly-Si and gate oxide interface could be changed by gate voltage sweep directions, so that it is attributed to cause the hysteresis phenomenon. In addition, we have discussed the drain current variation ($\Delta I_{DS}$) according to a previous gate voltage. Larger drain current error in the conventional 2-TFT pixel circuit results in serious image sticking. It should be reduced an interface trap density by novel process engineering [6] as well as design a new pixel circuit suppressing the hysteresis phenomenon for high quality AMOLED panel composed of SLS poly-Si TFTs.

REFERENCES

1. M. Stewart, "Polysilicon TFT Technology for Active Matrix OLED Display," IEEE Trans. on Electron Devices, Vol. 48, No. 5, pp. 845–851 (2001).
2. B.K. Kim, O.H. Kim, H.J. Chung, J.W. Chang, and Y.M. Ha, "Recoverable Residual Image Induced by Hysteresis of Thin Film Transistors in Active Matrix Organic Light Emitting Diode Display," Japanese Journal of Applied Physics, vol.43, no.4A, pp. L482-L485 (2004).
3. S.M. Sze and Kwok. K. NG, "Physics of Semiconductor Devices (3rd Edition)," (Wiley-Interscience, 2007).
4. L. Michalas, G. J. Papaioannou, D. N. Kouvatsos, and A. T. Voutsas, "Role of Bandgap States on the Electrical Behavior of Sequential Lateral Solidified Polycrystalline Silicon TFTs," J. Electrochem. Soc., Volume 155, Issue 1, pp. H1-H5 (2008).
5. K.M. Han and C.-T. Sah, "Positive oxide charge from hot hole injection during channel-hot-electron stress," IEEE Trans. on Electron Devices, Vol. 45, No. 7, pp. 1624-1627 (1998).
6. S. Luan and G.W. Neudeck, "Effect of NH₃ plasma treatment of gate nitride on the performance of amorphous silicon thin-film transistors," J.Appl.Phys.68 (7), 1 Oct. (1990).

Mater. Res. Soc. Symp. Proc. Vol. 1066 © 2008 Materials Research Society 1066-A16-07

# Hydrogenated Nanocrystalline Silicon Thin Film Transistor Array for X-ray Detector Application

Kyung-Wook Shin[1], Mohammad R. Esmaeili-Rad[1], Andrei Sazonov[1], and Arokia Nathan[2]
[1]Department of Electrical and Computer Engineering, University of Waterloo, 200 University Avenue West, Waterloo, Ontario, N2L 3G1, Canada
[2]London Centre for Nanotechnology, University College London, 17-19 Gordon Street, London, WC1H 0AH, United Kingdom

## ABSTRACT

Hydrogenated nanocrystalline silicon ($nc$-Si:H) has strong potential to replace the hydrogenated amorphous silicon ($a$-Si:H) in thin film transistors (TFTs) due to its compatibility with the current industrial $a$-Si:H processes, and its better threshold voltage stability [1]. In this paper, we present an experimental TFT array backplane for direct conversion X-ray detector, using inverted staggered bottom gate $nc$-Si:H TFT as switching element. The TFTs employed a $nc$-Si:H/$a$-Si:H bilayer as the channel layer and hydrogenated amorphous silicon nitride ($a$-SiN$_x$) as the gate dielectric; both layers were deposited by plasma enhanced chemical vapor deposition (PECVD) at 280°C. Each pixel consists of a switching TFT, a charge storage capacitor ($C_{px}$), and a mushroom electrode which serves as the bottom contact for X-ray detector such as amorphous selenium photoconductor. The chemical composition of the $a$-SiN$_x$ was studied by Fourier transform infrared spectroscopy. Current-voltage measurements of the $a$-SiN$_x$ film demonstrate a breakdown field of 4.3 MV/cm. TFTs in the array exhibit a field effect mobility ($\mu_{EF}$) of 0.15 cm$^2$/V·s, a threshold voltage ($V_{Th}$) of 5.71 V, and a subthreshold leakage current ($I_{sub}$) of $10^{-10}$ A. The fabrication sequence and TFT characteristics will be discussed in details.

## INTRODUCTION

The commercialization of the direct conversion flat panel X-ray detector is in progress with the $a$-Si:H TFT technology [2, 3]. The direct conversion flat panel X-ray detector has advantages over the indirect conversion counterpart in the image quality, due to lack of lateral optical scattering in photodiode, and simple fabrication of photoconductor. On the other hand, the inverted staggered bottom gate $nc$-Si:H TFT with stable $V_{Th}$ over 3 to 5 hours of stress has been reported by our group [1]. Compatibility with $a$-Si:H production technology, along with the stability in $V_{Th}$ allow simpler readout scheme compared to that with $a$-Si:H counterpart. Therefore, we have fabricated a direct conversion X-ray detector backplane to evaluate the suitability of the $nc$-Si:H TFT for this application.

## EXPERIMENT

The TFT cross section and an optical micrograph of a single pixel are shown in Figs. 1 (a) and (b), respectively. We used 100 nm molybdenum (Mo) as the gate, and 300 nm $a$-SiN$_x$ as the gate dielectric. The bilayer channel comprised of 15 nm nc-Si:H capped with 35 nm a-Si:H. Subsequently, n+ doped layer and aluminum source/drain contacts are formed. As shown in Figure 1 (b), each pixel consists of a TFT (In-pixel $nc$-Si:H TFT) and the $C_{px}$. Here, the aspect ratio (WL) of the In-pixel TFT is 300 μm/50 μm. The square shaped mushroom electrode serves as the bottom contact of the photoconductor, and also protects the TFT from incident X-ray photons. Total pixel area is $500 \times 500$ μm$^2$.

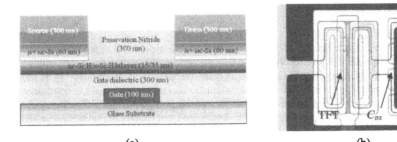

(a)                                                   (b)

**Figure 1.** (a) Cross section of the bottom gate $nc$-Si:H TFT and (b) optical micrograph of a pixel in the array.

The fabrication steps are illustrated in Figs. 2 (a) to (e), which is based on a fully wet etched process developed in our group [4]. The entire process is composed by seven mask steps which require a photolithography process for each stage. The photolithography process was performed by Karl Suss MJB3 contact mask aligner using emulsion masks. The first step, Fig. 2 (a), is the formation of the Mo gate and $C_{px}$ electrodes on a Corning 1737 glass, by sputtering. Subsequently, the tri-layer was deposited using a multi chamber 13.56 MHz PECVD system, manufactured by MVSystems Inc. Prior to the deposition of the $nc$-Si:H, the gate dielectric surface was treated by hydrogen plasma for 5 minutes to improve the $a$-SiN$_x$/$nc$-Si:H interface [5]. The passivation $a$-SiN$_x$ was then patterned to define active area and to open windows for the doped layer (Fig. 2 (b)). Next, $n+$ $nc$-Si doped layer and another protection nitride were deposited and patterned as illustrated in Fig. 2 (c). Contact holes for source/drain electrodes were opened in the protection nitride. Then, 300 nm aluminum was sputtered and patterned to form source/drain electrodes and $C_{px}$ top electrode (Fig. 2 (d)). Another 300 nm $a$-SiN$_x$ was deposited to passivate the In-pixel TFT and $C_{px}$. Finally, the TFT passivation $a$-SiN$_x$ was patterned and 300 nm aluminum was sputtered and patterned for the mushroom electrode.

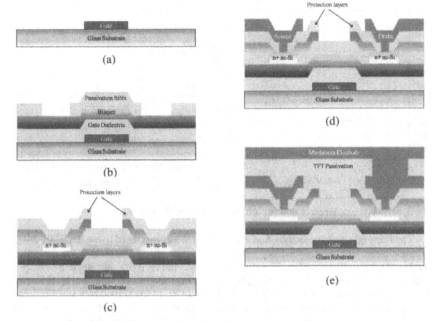

**Figure 2.** Fabrication sequence of the direct conversion X-ray detector array: (a) gate electrode formation, (b) tri-layer deposition followed by source/drain contact hole formation, (c) $n+$ $nc$-Si:H contact formation, (d) source/drain electrode formation, and (e) TFT passivation followed by mushroom electrode formation.

## RESULTS AND DISCUSSIONS

Current-Voltage ($I$-$V$) characteristics of the $a$-SiN$_x$ layer is illustrated in Fig. 3 (a). The measurement was performed using the Keithley 4200 semiconductor characterization system. Measurements were performed on MIS structures with the 100 nm aluminum as the metal, the 300 nm nitride as the insulator, and highly p+ doped silicon wafer as the semiconductor. The MIS area is 490 cm$^2$. From the $I$-$V$ characteristics, we see that the nitride layer breaks down at 4.3 MV/cm electric field and the leakage current is around $10^{-11}$ A for the fields below 2 MV/cm that TFTs usually operate. The breakdown field of 4.3 MV/cm seems to be acceptable for the passivation nitride, which also acts as the interlayer dielectric (see Fig. 2 (e)) and could be exposed to high voltages in direct x-ray detectors. For example, if the voltage between the mushroom electrode and TFT source reaches the breakdown field, 4.3 MV/cm, or 135 V, it can lead to the loss of signal and generated charges by the photoconductor. However, we think that such a condition is rare in normal operations.

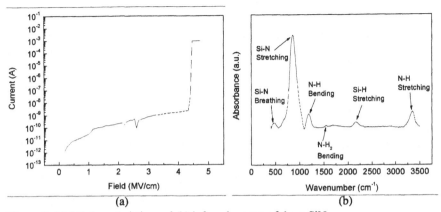

**Figure 3.** (a) *I-V* characteristics and (b) infrared spectra of the *a*-SiN$_x$.

The Infrared spectrum is shown in Fig. 3 (b), and was measured using the FTIR-8400S spectrophotometer from Shimadzu Corporation. In this case, the MIS structure was formed on lightly doped p-type silicon wafers. The absorption peaks in Fig. 3 (b), are due to Si–N Breathing, Si–N Stretching, N–H Bending, N–H$_2$ Bending, Si–H Stretching, and N–H Stretching modes, which are located at 430–490, 880–900, 1180–1190, 1550, 2150–2180, and 3340 cm$^{-1}$, respectively [6,7]. The relatively stronger peak due to N–H bonds than that due to Si–H bonds indicates that the nitride layer is nitrogen rich and can be used as the gate dielectric with minimum charge trapping [8].

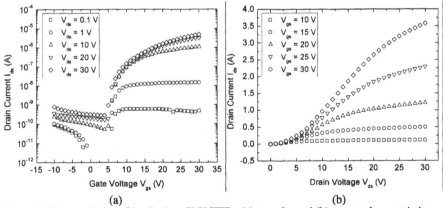

**Figure 4.** Characteristics of in-pixel *nc*-Si:H TFTs: (a) transfer and (b) output characteristics.

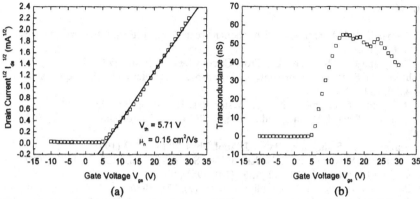

**Figure 5.** (a) $\mu_{EF}$ and $V_{Th}$ extraction from the TFT saturation characteristics and (b) transconductance of the pixel TFT.

Transfer and output characteristics of the In-pixel $nc$-Si:H TFT are illustrated in Figs. 4. (a) and (b), respectively, while the $V_{Th}$ and $\mu_{EF}$ extraction scheme and transconductance are shown in Figs. 5 (a) and (b). The characteristics of In-pixel $nc$-Si:H TFT were obtained at the process step depicted in Fig. 2 (d). The TFT exhibits a $V_{Th}$ of 5.71 V, a $\mu_{EF}$ of 0.15 cm$^2$/V·s, and $I_{sub}$ of $10^{-10}$ A, at 10 V drain bias ($V_{ds}$). From the output characteristic, severe current crowding at drain biases of less than 10 V was observed [9]. The transconductance of the TFT, at $V_{ds}$ = 10 V, exhibited a degradation at gate biases ($V_{gs}$) higher than 15 V. This degradation mainly originates from parasitic resistances, and surface roughness scattering at the gate dielectric/channel layer interface [10]. The current crowding can also be attributed to parasitic resistances.

## CONCLUSIONS

In this work, we evaluated the performance of the inverted staggered bottom gate $nc$-Si:H TFT for the direct X-ray detector application. The evaluation included the $I$-$V$ characteristics and the Infrared spectrum of $a$-SiN$_x$ dielectric, as well as the In-pixel $nc$-Si:H TFT transfer and output characteristics. The $a$-SiN$_x$ was suitable for the gate dielectric and TFT passivation purposes, and for the structure investigated in this work. However, the TFT characteristics showed current crowding and transconductance degradation. To remedy these two effects, it requires optimizations of the contact layer to eliminate the current crowding, and further analysis of transconductance degradation to estimate the parasitic resistance effect.

## ACKNOWLEDGMENTS

This work was performed using the Giga-to-Nanoelectronics Centre facilities at the University of Waterloo and funded by the Natural Sciences and Engineering Research Council of Canada (NSERC).

**REFERENCES**

1. M. R. Esmaeili-Rad, A. Sazonov and A. Nathan, Tech. Dig. - IEEE Int. Electron Devices Meet. **2006**, 303.
2. S. O. Kasap and J. A. Rowlands, IEE Proc.-Circuits Devices Syst. **149** (2), 85 (2002).
3. B. Polischuk, S. Savard, V. Loustauneau, M. Hansroul, S. Cadieux, and A. Vaque, Proc. SPIE **4320**, 582 (2001).
4. A. M. Miri and S. G. Chamberlain, Proc. Mater. Res. Soc. Symp. **377**, 737 (1995).
5. B. C. Lim, Y. J. Choi, J. H. Choi, and J. Jang, IEEE Trans. Electron Devices **47** (2), 367 (2000).
6. G. N. Parsons, J. H. Souk, and J. Batey, J. Appl. Phys. **70** (3), 1553 (1991).
7. D. Stryahilev, A. Sazonov, and A. Nathan, J. Vac. Sci. & Technol. A **20** (3), 1087 (2002).
8. N. Lustig and J. Kanicki, J. Appl. Phys. **65** (10), 3951 (1989).
9. M. J. Powell and J. W. Orton, Appl. Phys. Lett. **45** (2), 171 (1984).
10. A. Valletta, L. Mariucci, G. Fortunato, and S. D. Brotherton, Appl. Phys. Lett. **82** (18), 3119 (2003).

Mater. Res. Soc. Symp. Proc. Vol. 1066 © 2008 Materials Research Society          1066-A16-08

# Temperature and humidity effects on the stability of on-plastic a-Si:H thin film transistors with various conduction channel layer thicknesses

Jian Z Chen[1], and I-Chun Cheng[2]

[1]Institute of Applied Mechanics, National Taiwan University, No.1 Sec.4 Roosevelt Rd., Taipei, 10617, Taiwan

[2]Department of Electrical Engineering and Graduate Institute of Photonics and Optoelectronics, National Taiwan University, No.1 Sec.4 Roosevelt Rd., Taipei, 10617, Taiwan

## ABSTRACT

Stability is an important issue for the application of TFTs. In this paper, we present the effects of humidity and temperature on the stability of inverted-staggered back-channel-cut a-Si:H TFTs with various conduction channel layer thicknesses. We evaluated the stability of on-plastic TFTs of different conduction layer thicknesses made at a process temperature of 150°C on 51-$\mu$m thick Kapton polyimide foil substrates. With conduction channel layer thickness of 50nm, humidity reversibly varies the characteristics of TFTs, but TFTs of conduction layer thickness greater than 100 nm are pretty immune to the humidity change. The temperature dependent stability and characteristics of TFTs were analyzed from 20°C to 60°C. Rising temperature from 20°C to 56°C, the threshold voltage ($V_t$) drops about 2 volts; on-off current ratio decreases by one order of magnitude mainly due to thermally excited carriers in the off-region.

## INTRODUCTION

In spite of the rapid development of organic electronics and metal-oxide transistors, hydrogenated amorphous silicon thin film transistors (a-Si:H TFTs) are still the mainstream technology for active matrix backplanes of displays. This is because its process is robust, low-cost, and has been scaled up for large area deposition with high throughput. Therefore, a lot of researches are focused on developing high quality low temperature processed a-Si:H TFTs fabricated on plastic substrate to achieve nonbreakable, comfortable, and elastic flexible electronics [1-6]. Thin channel layer TFTs are attracting increasing interest since it can further reduce the production time. In this paper, we present the effects of temperature and humidity on stability of a-Si:H TFTs with various conduction channel layer thicknesses fabricated on plastic foils at 150°C.

## EXPERIMENT

The silicon thin-film transistors (a-Si:H TFTs) investigated are in inverted-staggered back-channel-cut structure. The TFTs were all fabricated on flexible substrates of 51$\mu$m thick Kapton at 150°C. The cross-section is shown in Fig. 1. The detailed processing procedure is stated in other paper [7-8]. With the same structure, the conduction channel layer thickness varies as 50nm, 100nm, and 200nm. We characterized TFTs with a HP 4155B semiconductor parameter analyzer. Two drain-source voltages $V_{ds}$, 0.1 V and 10 V, were applied to TFTs while gate voltage $V_{gs}$ was scanned from -10 V to 25 V with medium integration time setting under dark environment. For measurement at elevated temperature, a thermoelectric device is used as the heating stage with temperature controlled by varying the input current. A thermal couple is used to measure the temperature during TFT characterization.

The room temperature characteristics of as-fabricated TFTs with 50nm, 100nm and 200nm conduction layer thicknesses are shown in Fig. 2. Off-current is about two orders of magnitude larger in TFT with 50 nm thick channel layer than in TFTs with thicker channel layers. This may be due to discrete tiny amount of n+-Si left on the surface. Nevertheless, further plasma etching was not performed since it damages the conduction channel layer. The other possibility could be back channel conduction influenced by the environment. Quantitative comparison of threshold voltage, on-current, off-current, and mobility is shown in Fig. 3. We extracted threshold voltage $V_t$ at $I_{ds} = 10^{-8}$ A, on-current $I_{on}$ at $V_{gs} = 25$ V, and off-current at $V_{gs} = -5$ V from transfer curves of $V_{ds} = 10$ V. Field-effect mobility $\mu$ was calculated from the least-squares slope fitted to $I_{ds} > 10^{-8}$ A with $V_{ds} = 0.1$ V.

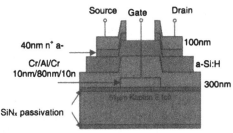

Figure 1 Inverted-staggered a-Si:H TFT structure on plastic substrate.

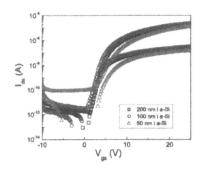

Figure 2 Transfer curves of as-fabricated a-Si:H TFTs (W/L = 400μm/40μm).

From Fig. 3, TFT with 50 nm a-Si:H channel layer has higher threshold voltage, lower field-effect mobility, and higher off-current (lower on-off ratio). Its performance is the worst among the three. On the other hand, the TFTs with 100 nm conduction channel layer thickness perform best. The further increase of conduction channel layer thickness to 200 nm causes the reduction of field-effect mobility and increase of threshold voltage. The reasons are not clear yet.

## Discussion

### Effects of humidity on TFTs with thin conduction channel layer thickness

We then analyze TFT along with time. The TFT performance varies noticeably with time for the TFT thickness of 50 nm conduction channel layer, while it is fairly stable for TFTs with the thicker layers, as shown in Fig. 4. From these transfer curves, the major change is located at the off transition region due to its low drain-source current level. The effect is mainly caused by the change of humidity and the mechanism is explained by Fig.5. In humid environment, water molecules cause fixed positive charges at surface. These positive charges induce compensating negative charges which form conducting back channel in the a-Si:H layer and provide additional conduction path. As a result, conductance and off-current both increase. This effect shows up in the off region because of conductivity is much smaller in dark environment. This effect is most significant in the case of high quality film [9]. The other possibility is that water molecules cause defects inside a-Si:H; however, this could be excluded because the observed phenomena is reversible.

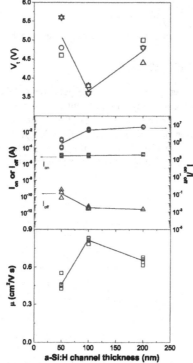

Figure 3 Characteristics of on-plastic a-Si:H TFTs.

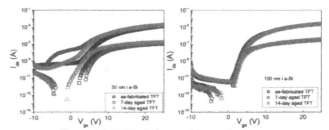

Figure 4 Stability influenced by environment.

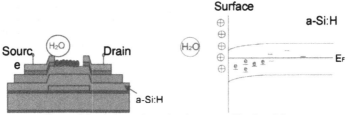

Figure 5 Backchannel conduction caused by humidity.

### Effects of operation temperature variation

Figure 6 shows the operational temperature dependent stability. Rising the temperature from 24 °C to 56 °C, the threshold voltage drops about 2 V. Meanwhile, the on-current increases ~ 30 μA from 10 μA while off-current increases about one order of magnitude, which is mainly caused by the thermal excitation of carriers added in the off-regime. This leads to the decrease of on-off ratio at higher temperature [10]. The rise of off-current for more negative $V_{gs}$ indicates conduction of hole-current [11], with holes supplied from thermal excitation.

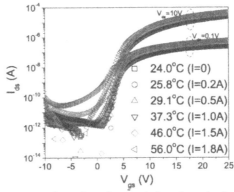

Figure 6 Temperature dependent characteristics of on-plastic a-Si:H TFTs.

## CONCLUSIONS

We studied the electrical characteristics of inverted-staggered back-channel-cut amorphous silicon thin-film transistors (a-Si:H TFTs) on flexible substrates of 51μm thick Kapton. The TFTs had conduction channel layer thicknesses of 50nm, 100nm, and 200nm. The electrical characteristics of TFTs with 100nm and 200nm channel layers are similar. However, in the TFT with active layer thickness of 50nm the off-current is about two orders of magnitude higher than with the thicker layer, and the mobility is lower. In addition, the TFT performance varies noticeably with time for the 50nm active layer, while it is fairly stable for TFTs with the thicker layers. We also evaluated TFTs in the temperature range between 20°C to 60°C. The threshold voltage ($V_t$) drops about 2 volts as the temperature rises from 20°C to 56°C; meanwhile, on-off current ratio decreases by one order of magnitude. This is mainly due to thermally excited carriers in the off-region.

## ACKNOWLEDGMENTS

The thin film transistors were fabricated and characterized in Macroelectronics Laboratory and Princeton Institute for the Science and Technology of Materials (PRISM), both at Princeton University. The authors, J.-Z.C. and I-C.C., thank National Science Council, Taiwan (ROC) for partial financial support, under Grants NSC 97-2218-E-002 -024 and NSC 96-2218-E-002-032.

## REFERENCES

1. J. P. Conde, P. Alpuim, V. Chu, *Thin Solid Films* **430**, 240 (2003).
2. J.-H. Ahn, H-Sik Kim, K. J. Lee, Z. Zhu, E. Menard, R. G. Nuzzo, J. A. Rogers, *IEEE Electron Dev. Lett.* **27**, 460 (2006).
3. Y.-H. Kim, C-H. Chung, J. Moon, G. H. Kim, D-J. Park, D-W. Kim, J. W. Lim, S. J. Yun, Y-H. Song, J. H. Lee, *IEEE Electron Dev. Lett.* **27**, 579 (2006).
4. S. H. Won, J. K. Chung, C. B. Lee, H. C. Nam, J. H. Hur, and J. Jang, *J. Electrochem. Soc.* **151**, G167 (2004).
5. A. Nathan, A. Kumar, K. Sakariya, P. Servati, S. Sambandan, D. Striakhilev, *IEEE J. Solid State Circuit* **39**, 1477 (2004).
6. A. Sazonov, C. McArthur, J. Vac. Sci. Tech. A **22**, 2052 (2004).
7. H. Gleskova, S. Wagner, W. Soboyejo, and Z. Suo, *J. Appl. Phys.* **92**, 6224 (2002).
8. J. Z. Chen, K. Cherenack, C. Tsay, I.-C. Cheng, S. Wagner, *Electrochem. Solid-State Lett.* **11**, H26 (2008).
9. R. A. Street, "Hydrogenated amorphous silicon," Cambridge University Press, 1991.
10. J. Z. Chen, I.-C. Cheng, submitted to *Journal of Applied Physics*.
11. F. Lemmi, R. A. Street, *IEEE Trans. Electron Dev.* **47**, 2404 (2000).

# Poster Session:
# Crystallization Techniques

Mater. Res. Soc. Symp. Proc. Vol. 1066 © 2008 Materials Research Society　　　1066-A17-04

# Laser Fabrication of Sharp Conical Microstructures on Si Thin Films by Nd:YAG Laser Single Pulse Irradiation

Joe Moening, and Daniel Georgiev
University of Toledo, Department of Electrical Engineering and Computer Science, Mail Stop 308, Toledo, OH, 43606

## ABSTRACT

Conical microstructures with nanoscale sharpness form on silicon films as a result of single-pulse, localized UV irradiation using a solid-state, Q-switched Nd:YAG laser. Projection imaging of pinhole apertures was employed to obtain micron-sized irradiation spots on the surface of silicon-on-insulator samples. The formation of these structures requires melting of the silicon film and was followed at different laser fluence levels and irradiation spot sizes. Atomic force microscopy (AFM) was used to characterize the structures. After fabricating small arrays of such micro-cones, the silicon top layer was selectively etched away in order to understand the role of the underlying silicon oxide. AFM images of such etched samples revealed that the topography of the oxide material below the cones had been significantly modified: bumps with heights that represent a significant fraction of the original Si cone height have formed. This suggests that substrate melting plays an important role in the mechanism of formation of the silicon cones.

## INTRODUCTION

Reliable, simple and low-cost techniques for fabrication of sharp micro- and nanotips of silicon and other semiconductor and metal materials, as well as large, high-density arrays of such tips, are desirable in technological applications. Such applications include emitters for field-emission-based devices such as high-definition displays [1,2] and other vacuum microelectronics devices and systems [3], probes for scanning probe microscopy techniques, and novel designs of high-density nonvolatile phase-change memory devices . There is also a considerable amount of ongoing research on surface patterning of materials for biomedical applications [4-6], which could benefit from new developments in the area of materials surface micro- and nano-structuring.

High-intensity pulsed laser radiation can be used to modify surfaces of materials and fabricate technologically desirable structures on a micrometer and sub-micrometer level. The processes involved in such radiation/matter interactions are complex and usually non-equilibrium due to high heating and cooling rates, large temperature gradients and a variety of chemical and photochemical transformations. These processes and their interplay are not well understood, and therefore systematic studies of laser irradiation of materials as a function of a certain set of parameters are essential.

Laser fabrication techniques that utilize high-energy pusled lasers with large-area homogenized beam cross-sections can provide the advantages of high-resolution, high throughput, uniformity, highly localized heating, simplicity and reproducibility. In addition, combinations of laser techniques with other established technological steps could evolve into new, more flexible technologies. Remarkable Si columns with heights of about 20 μm and widths of 2-3 μm have been produced by multiple-pulse, large-spot, nano-second excimer laser irradiation (KrF laser, $\lambda$ = 248nm) of Si wafers in oxygen and oxygen-containing ambient [7]. Conditions for nano-structuring of Si surfaces have been identified in a study of the irradiation of bulk Si as a function of the number and the fluence of the applied KrF excimer laser pulses in different gas environments [8]. In all of these cases, single-crystal Si has been subjected to multiple-pulse, large-laser-spot-area irradiation and there has been little or no control over resulting surface topography. KrF excimer laser irradiation by large-area scanning of amorphous Si thin films on Mo-coated glass substrates has been reported [9] to result in a sharp surface morphology that can be viewed as consisting of randomly situated, densely packed, nano-sized crystallites.

Direct laser fabrication of sharp conical structures (referred to as nanotips) on silicon thin films [10], or bulk silicon [11], was recently reported. This work is related to the technique described in Ref. 10 .

## EXPERIMENTAL

The laser source used in the irradiation experiments was an oscillator-amplifier Q-switched Nd:YAG system, model Powerlite 8010 by Continuum Lasers, Inc. emitting pulses with duration of 8ns at a wavelength of 266nm (4th harmonic). Individual pinholes, or small arrays of such pinholes, were imaged to form uniformly illuminated circular spots onto the sample surface. The projection system was operated at a demagnification factor of 10 and had a resolution limit of about 2.0 μm. A schematic of the irradiation set up is shown in Fig. 1. The laser pulse energy was measured to be reproducible within 10% . Most of the work was done on commercially acquired silicon-on-insulator (SOI) wafers that consisted of a 250nm-thick (100) single-crystal Si on a 1.0 μm-thick silicon oxide layer. The laser irradiation was performed in ambient conditions, and the sample surface topography was then studied by contact-mode atomic force microscopy (AFM) on a Veeco MultiMode AFM system using AppNano tips, model SICONA. Some samples were immersed in a $HNO_3$/HF (Transene, RSE, 100:1 ratio, with high Si-to-$SiO_2$ selectivity) etching solution in order to etch away the top silicon layer and then examine the topography of the underlying (oxide) material.

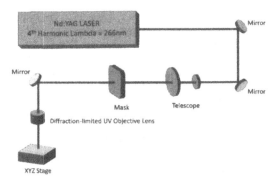

**Fig. 1.** Laser set-up used in the radiation experiments.

## RESULTS AND DISCUSSION

The AFM images in Fig. 2a and 2b show tips that formed upon using fluence levels of 0.46 J/cm² (a) and 0.32 J/cm² (b) and a spot diameter of 7.5μm. The z-scale of the images has been expanded to better show the size and shape and of the tips and the changes in the surrounding irradiated area. We observe round and relatively flat circular depressions, the diameters of which roughly correspond to the size of the laser spot. This depression is, typically, several tens of nanometers below the original surface, and almost-conic tips appears in the center of the depressions (see Fig 2c). As evident from the figure, the fluence can be used to control the size of the tip. The range within which this is possible for (for this particular spot size and SOI sample) is from approximately 0.26 J/cm², below which only minor surface changes are observed, to about 0.50 J/cm², above which the cones become much less sharp or small craters develop at their apices. At even higher fluences the central crater develops into a wider ablation hole with raised edges. All these observations are similar to earlier work that employed excimer laser pulses [10]. The apparent radius of curvature of the tip apex is 40-50nm as estimated using the AFM scan, which, again, is in agreement with Ref. 10.

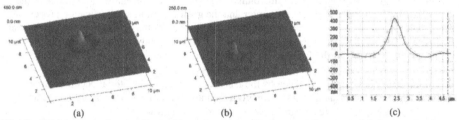

(a)                                    (b)                                    (c)

**Fig. 2.** AFM images of cones formed by using an irradiation spot diameter of 7.5μm and pulse fluence of 0.46 J/cm² (a) or 0.32 J/cm² (b). In (c) is shown a cross-sectional scan taken near the apex of the cone shown in (a).

Fig. 3 shows a series AFM images of tips obtained by using three different spot sizes: with diameters of 0.25μm, 0.50μm, and 0.75μm. The fluence thresholds for observable surface changes, and the entire ranges within which nanotips form, shift to higher values upon decreasing the diameter of the irradiation spot. This is also evident from the fluence needed to form the nanotips in Fig. 3: the tips are close in size (i.e. height) but require very different fluence values ranging from 1.65J/cm² for a spot diameter of 2.5μm to 0.46J/cm² for a spot diameter of 7.5μm.

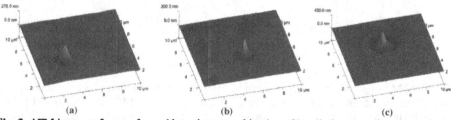

(a)                                    (b)                                    (c)

**Fig. 3.** AFM images of cones formed by using a combination of irradiation spot diameter and fluence of: 2.5μm and 1.65 J/cm² (a); 5.0μm and 0.55 J/cm² (b); 7.5μm and 0.46J/cm² (c).

To understand this dependence, one needs to first understand the mechanism of formation of these structures. A complete analysis of the involved physical processes would be overwhelmingly difficult and would need to account for [12]: temperature dependence of all material parameters; heat dissipation through radiation; evaporation and plasma formation just above the surface together with the corresponding laser radiation shielding; surface tension; overheating and overcooling; chemical transformations such as oxidation. It has been argued [10,13] that the mechanism of formation relies on two main factors: first, the predominantly lateral heat flow upon resolidification of the laser- melted silicon, which sits on a thermally insulating substrate, causing directed solidification from the periphery of the molten circular pool towards its center; and second, the anomalous higher volume density of molten silicon with respect to solid-state silicon (whether crystalline or amorphous). The laser radiation is absorbed entirely in a thin surface layer of thickness that is less than 10nm (the absorption coefficient of Si at $\lambda=266$nm is in the order of $10^6$cm$^{-1}$). This surface layer becomes a source of heat that causes melting of the entire thickness of the film: under the conditions of irradiation, the depth of melting of bulk silicon would be more than the thickness of the film [14]. Because the underlying silica has much lower thermal conductivity than Si, heat is dissipated predominantly laterally through the surrounding Si film volume (especially once the entire film volume in the laser-irradiated spot is melted). The dynamics and the geometry of this solidification process, is responsible, at least as a first approximation, for the formation of the tips. Due to the lateral heat dissipation, the periphery of the spot is more rapidly cooled and have lower temperature than the spot's central region. Thus the freezing front moves from the edges to the center and pushes the remaining liquid silicon toward the center. This process of fast displacement of liquid silicon toward the center is strongly enhanced by the fact that the solidified material occupies larger volume. An important issue that is not directly accounted for in the above simplified model is the role of any substrate melting. Surface tension is also not considered.

From basic heat transfer considerations, the reason for the strong dependence of the required nanotip-fromation fluence on the irradiation spot size, as illustrated in Fig. 3, is in the surface to volume ratio increase upon decreasing the spot size. The smaller the spot, the smaller the diameter of the disk-like molten pool of silicon and the more efficient is the heat transfer in the lateral (i.e. within the Si film) direction, which then translates into laser pulse higher energy needed for the formation of the conical structure

Figures 4a and 4b show AFM images of arrays of nanotips that formed upon imaging arrays of pinholes with a single laser pulse. Separately, in Fig. 4c is shown an array of nanotips

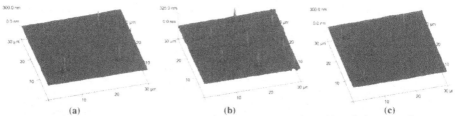

(a)                    (b)                    (c)

**Fig. 4.** AFM images of arrays cones formed by using a combination of irradiation spot diameter and fluence of: 7.5μm and 0.41 J/cm$^2$, square array (a); 5.0μm and 0.55 J/cm$^2$, hexagonal array (b). In (c) is shown an array of cones formed wit sequential pulses, diameter 5.0μm and fluence 0.46 J/cm$^2$.

that were obtained by using sequential single pulses and a single pinhole together with sample stage movement. In both cases it is evident that the nanotips formation is reproducible as long as the beam used is spatially homogeneous and the pulse-to-pulse energy stability is good. Such results demonstrate the potential of this laser-based fabrication method to serve as the basis of a simple, cost-effective technology for the fabrication of large, dense arrays of such conical structures. In fact, the pulse-to-pulse stability of the system that we used (10%) is substantially inferior to what other, more specialized but still commercially available, laser systems can offer. More involved beam homogenization could be used as well.

Fig. 5 shows AFM images of SOI samples on which first nanotips were formed and then the top Si layer was etched away by using a Si enchant solution which has significantly slower etch rate for $SiO_2$. This allows for the complete etching of the Si layer while stopping the etching process at the interface with the underlying oxide. Thus the underlying oxide material morphology is revealed. As can be seen, the topography of the oxide material below the cones was significantly modified. Bumps with heights of up to one third of the original Si cone have formed, indicating that substrate melting plays an important role in the mechanism of formation of the silicon cones.

These results show the need for a more complete and precise model for the mechanism of formation of these structures, which would accounts for the substrate melting and would explain the apparent transport of substrate material (or at least the topography changes) in the underlying oxide substrate material.

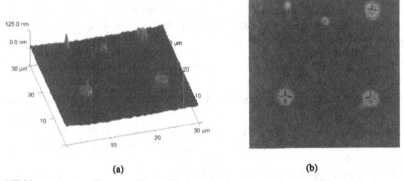

(a)          (b)

Fig.5. AFM images a small array of laser-fabricated micro-cones the top Si layer of which was etched entirely to reveal the morphology of the underlying oxide material: a 3D view (a); a top view (b). Some of the observed small spikes are artifacts and the high-roughness texture is due to the etching procedure. The "cross" morphology, seen in (b), is due to cracks or channels that may be related to any specifics in the crystalline structure of the Si cones (intergrain boundaries, crystalline facets etc.)

## CONCLUSIONS

We observed the formation of sharp conical microstructures on Si films as a result of *single-pulse, localized UV irradiation* using a solid-state, Nd:YAG laser. We examined the topography of the oxide material below these microstructures after complete etching of the Si film, and found that substrate melting plays an important role in the mechanism of formation of the silicon cones. Further work on understanding the mechanism of formation of these structures is therefore needed. Our results show that the formation process is controllable. This direct laser micro-/nano-structuring method can be used as the basis of a simple, cost-effective technology for the fabrication of large, dense arrays of such structures.

## ACKNOWLEDGEMENTS

Our work has benefited from AFM expertise and technical assistance provided by Joseph Lawrence and Prof. Arun Nadarajah from the University of Toledo's Department of Bioengineering.

## REFERENCES

1. C.A. Spindt, I. Brodie, L. Humphrey, E.R. Westerberg, J. Appl. Phys. 47, 5248 (1976)
2. L. Dvorson, I. Kymissis, A.I. Akinwande, J.Vac.Sci.Technol. B, 21, 486 (2003)
3. H. H. Busta, J. Micromech. Microeng. 2 , 43 (1992)
4. S. Turner, L. Kam, M. Isaacson, H.G. Craighead, W. Shain, J. Turner, J.Vac.Sci.Technol. B 15, 2848 (1997)
5. A.M.P. Turner, N. Dowel, S.W.P. Turner, L. Kam, M. Isaacson, J.N. Turner, H.G. Craighead, W. Shain, J. Biomedical Res. 51, 430 (2000)
6. M.P. Maher, J. Pine, J. Wright, Y.C. Tai, J. Neuroscience Methods 87, 45 (1999)
7. A.J. Pedraza, J.D. Fowlkes, D.H. Lowndes, Appl.Phys.Lett. 74, 2322 (1999)
8. A.J. Pedraza, J.D. Fowlkes, Y.F. Guan, Appl.Phys. A 77, 277 (2003)
9. Y.F. Tang, S.R.P. Silva, B.O. Boskovic, J.M. Shannon, Appl. Phys. Lett. 80, 4154 (2002)
10. D.G. Georgiev , R.J. Baird, I. Avrutsky, G. Auner , G. Newaz, Appl. Phys. Lett., 84 (2004) 4881
11. G. Wysocky, R. Denk, K. Piglmayer, N. Arnold, D. Bauerle, Appl. Phys. Lett, 82 (2003) 692
12. D. Bäuerle, *Laser Processing and Chemistry* (Springer-Verlag, Berlin, 2000).
13. J. Eizenkop, I. Avrutsky , G. Auner , D.G. Georgiev , V. Chaudhary, J. Appl. Physics, 101 (2007) 94301
14. S. De Unamino, E. Fogarassy, Appl. Surf. Sci., 36, 1 (1989)

Mater. Res. Soc. Symp. Proc. Vol. 1066 © 2008 Materials Research Society       1066-A17-05

## Impurities and Grain Size Modeling in Recrystallized Silicon

Valeri V. Kalinin[1], Alexandre M. Myasnikov[1], and Vladislav E. Zyryanov[2]

[1]Department of Single Crystals and Silicon Structures, Institute of Semiconductors Physics, 13 Lavrent'ev Avenue, Novosibirsk, AK, 630090, Russian Federation

[2]Department of Radiotechnic, Electronics and Physics, Novosibirsk State Technical University, 20 Marx Avenue, Novosibirsk, 630049, Russian Federation

### ABSTRACT

In previous publications [1, 2 and 3], spreading resistance probe (SRP) measurements for quality control of metal induced lateral crystallization (MILC) of amorphous silicon (a-Si) were studied, and the mechanism of nickel diffusion was simulated using technology computer-aided design (TCAD) modeling.

Now, we continue to present the explanation of experimental results by modeling with the Synopsys TCAD package, whereby models for resistivity vs. grain size in implanted recrystallized silicon layers are implemented and compared with experiments.

Findings show that the SRP method can be used for the characterization of the MILC process of amorphous silicon and that a comparison of experimental and calculated data allows both a turn from qualitative to quantitative analysis of recrystallized silicon film and an estimate of grain size. It has been found that grain size depends on location in the MILC region and on the time and temperature of recrystallization.

### INTRODUCTION

Amorphous silicon has attracted a great deal of interest as material for large area applications such as thin film transistors [4-9]. However, the field-effect mobility in amorphous silicon is very low. Solid phase crystallization (SPC) is a typical method to improve its structure, but the temperature of SPC is too high for large-area glass substrates to be used.

Several metals, like Pd [10], Al [11] and Ni [12], crystallize amorphous silicon, upon contact, much faster than SPC does. Unfortunately, the addition of a great quantity of seed metal can induce the degradation of devices formed in amorphous silicon or reduce to zero almost all of the advantages of amorphous silicon recrystallization. A small quantity of seed metal, in turn, can cause non-optimal conditions of recrystallization.

At the moment, MILC films consist of separated recrystallized regions, which can be studied by SRP measurements [2]. As known, the given method is used to provide data about spreading resistance in the layers and to receive, with great accuracy, the distribution of resistance in ultra-shallow junction (USJ) layers [13].

Using experimental data of SRP measurements received and described in [2], it is shown that carrier mobility in the boron ion implanted layer formed in the MILC region is up 65 % in comparison with mobility in the boron ion implanted layer formed in single crystalline silicon. It is also shown that, under different conditions, mobility can change from 24 $cm^2$/Vs to 34 $cm^2$/Vs, depending upon the time of recrystallization and the distance from the seeds.

The program DIOS from package Synopsis TCAD allows modeling of the process flow for semiconductor technology [14]. DIOS is used for the calculation of properties of single crystalline and polycrystalline silicon layers ion implanted with subsequent annealing and for defining the resistivity of layers. In this paper, DIOS is used for the simulation of silicon film properties in the same conditions as prepared and measured in [2].

## RESULTS AND DISCUSSION

Simulation was carried out for a layer of 0.3 µm as single crystalline and recrystallized (polycrystalline) silicon deposed on silicon substrate oxidized up to 0.7 µm. Silicon films were calculated as ions implanted with boron, phosphorus and arsenic at doses of $10^{14}$ - $10^{16}$ cm$^{-2}$ at an energy of 40 keV in order to introduce p- and n-type carriers. Then, annealing of samples was simulated at a temperature of 1100 °C for 10 seconds. Impurities distributions for polycrystalline silicon layers were calculated for different grain sizes, from 0.001 µm to 0.7 µm. The resistance for these layers doped with boron, phosphorus and arsenic was also calculated.

The calculated impurities distributions in single crystalline and polycrystalline silicon for a dose of $10^{15}$ cm$^{-2}$ are shown in Figures 1 and 2.

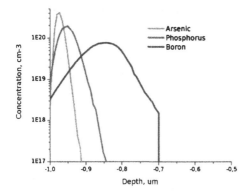

Figure 1. Impurities Distributions in Single Crystalline and Recrystallized Silicon as Implanted with a Dose of $10^{15}$ cm$^{-2}$ and an Energy of 40 keV. (There are no differences in distributions for recrystallized and single crystalline silicon)

There are no differences in impurities distributions in implated layers for single crystalline or polycrystalline silicon (See Figure 1). However, in Figure 2, one can see the difference for phosphorus and arsenic distributions after annealing. This difference in distribution is related to penetrating phosphorus and arsenic along grain boundaries at annealing and formatting almost flat shelves above the concentration of $10^{18}$ cm$^{-3}$. These differences in phosphorus and arsenic distributions do not allow direct comparison using the experimental and calculated data for resistance. For boron-doped layers, the situation is the opposite; for implanted and annealed boron distributions in single crystalline and polycrystalline silicon, we have the same, or almost the same, curves, allowing comparison of the calculated results of resistance.

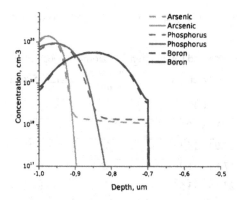

Figure 2. Impurities Distributions in Single Crystalline (solid lines) and Recrystallized (dotted lines) Silicon Implanted with a Dose of $10^{15}$ cm $^{-2}$ and an Energy of 40 keV and Annealed for 10 s at 1000° °C.

Referencing the Synopsys TCAD package, simulated data are calibrated for ion implantation and post-implantation annealing; therefore, calculations predict the data with very high accuracy and reliability for layers of single crystalline and polycrystalline silicon. We may reliably use these calculated data for comparison with experimental data of SRP measurements [14]. However, the results of SPR measurements depend upon the size of probe diameter and other correction factors, which have been specified in [2] and which interfere with the use of absolute values.

In carrying out SRP measurements, the main mistake in the value of spreading resistance is due to uncertainty of the contact area of probes with a measured layer. However, in carrying out one set of measurements, it is possible to consider the contact area as a constant or as slightly changing, and the values of resistance in the given set are independent from the contact area, depending only on the properties of a measured layer.

In this case, the measurements on recrystallized and single crystalline silicon are spent in one set, and the ratio of spreading resistance values for a recrystallized layer and a single crystalline layer is created entirely by the layer properties, without any influence of the contact area.

In Table 1, partially taken from [2], the spreading resistance ratios of experimental values are shown in the fourth column. Based on these values, grain sizes of polycrystalline silicon at equal resistivity ratios have been defined by calculated curve from Figure 3 and included in the fifth column of Table 1. These results show that even MILC regions with sizes of about 10 - 20 μm have grain sizes which do not exceed several tenths of a micrometer - 0.3 - 0.5 μm. However, this size is more than the thickness of deposed amorphous silicon. The given calculated grain sizes are average values in causing such resistivity; however, sizes correlate with grain sizes in MILC regions shown in article [3]. Given values of the grain sizes are from 0.15 μm to 0.48 μm, which seem small compared to the sizes of MILC regions; however, they make up 50% to 160 % from 0.3 μm thickness of a silicon film. These grain sizes, defined by the layer resistivity, may be considered "electrical" grain sizes, correlating with real grain sizes in MILC regions and defined by a calculated curve.

Table 1. Data Summary of SRP Measurements and Calculations for Boron-Doped Layer.

| Sample/Region On Surface | R, Ohm | Mobility, cm$^2$/Vs | Resistance Ratio | Grain Size, μm |
|---|---|---|---|---|
| Single Crystalline Silicon | 194 - 196 | 51.3 | 1 | - |
| A-Si/MILC | 400 - 424 | 23.58 - 25 | 2.11 | 0.3 |
| A-Si/Poly-Si | 661 - 670 | 14.93 - 15.13 | 3.41 | 0.15 |
| A-Si/"Nickel" | 459 - 565 | 17.64 - 21.79 | 2.62 | 0.21 |
| A-Si/MILC | 298 - 363 | 27.55 - 33.56 | 1.7 | 0.48 |
| A-Si/Poly-Si | 469 - 593 | 16.86 - 21.32 | 2.72 | 0.2 |
| A-Si/"Nickel" | 347 - 515 | 19.42 - 28.82 | 2.21 | 0.28 |

This calculated resistance ratio vs. grain size for boron-doped layers at doses from $10^{14}$ to $10^{16}$ cm$^{-2}$ is shown in Figure 3.

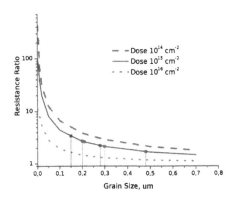

Figure 3. Resistance Ratio for Recrystallized and Single Crystalline Silicon Regions for Different Boron Implanted Doses vs. Grain Size.

## CONCLUSIONS

Thus, it was demonstrated the application of spreading resistance probe measurements for the characterization of the recrystallization process of amorphous silicon in comparison with calculated resistivity data. Grain sizes can be estimated by comparing resistance data for boron-doped layers. Resistance ratio vs. grain size is calculated for the layers boron-implanted with doses from $10^{14}$ to $10^{16}$ cm$^{-2}$ and for different grain sizes from 0.001 to 0.7 μm.

Results show that grain sizes vary from 0.15 to 0.5 μm at different conditions of MILC for the experimental SRP data from [2] and depend on location in the MILC region and on the time and temperature of recrystallization.

## ACKNOWLEDGMENTS

Authors are grateful to Dr.M.C.Poon for the data of SRP measurements, and also wish to express especially thanks Lara H. for preparation and editing the text of this paper.

## REFERENCES

1. A.M.Myasnikov, M.C.Poon, P.C.Chan, K.L.Ng, M.S.Chan, W.Y.Chan, S.Singla, C.Y.Yuen, Edited by J.R.Abelson, J.B.Boyce, J.D.Cohen, H. Matsumura and J. Robertson, (Mater. Res. Soc. Proc. 715, Pittsburgh, PA, 2002) A22.11.
2. Alexandre M. Myasnikov, Vincent M.C. Poon, Vincent T.C. Leung, Mansun Chan, and Lawrence C.F. Cheng, Edited by J.R.Abelson, G.Ganguly, H. Matsumura, J.Robertson and E.A.Schiff (Mater. Res. Soc. Proc. 762, Pittsburgh, PA, 2003) A17.2.
3. Aleksey M.Agapov, Valeri V.Kalinin, Alexandre M.Myasnikov, Vincent M.C.Poon, Bert Vermeire, Edited by R.W.Collins, P.C.Taylor, M.Kondo, R.Carius and R.Biswas (Mater. Res. Soc. Proc. 862, Pittsburgh, PA, 2005) A6.6.
4. S.-I.Jun, Y.-H.Yang, J.-B.Lee, D.-K.Choi, Appl. Phys. Lett., 75, 1999, 2235.
5. H.M.Wang, M.Chan, S.Jagar, M.C.Poon, M.Qin, Y.Y.Wang, P.K.Ko, IEEE Trans. Electron Dev., 47, 2000, 1580.
6. J.N.Lee, Y.W.Choi, B.J.Lee, B.T.Ahn, Journ. Appl. Phys., 82, 1997, 2918.
7. K.H.Lee, Y.K.Fang, S.H.Fan, Electron. Lett., 35, 1999, 1108.
8. Z.Meng, M.Wang, M.Wong, IEEE Trans. Electron Dev., 47, 2000, 404.
9. S.Y.Yoon, S.J.Park, K.H.Kim, J.Jang, C.O.Kim, Journ. Appl. Phys. 87, 2000, 609.
10. G.Liu and S.J.Fonash, Appl. Phys. Lett., 55, 1989, 660.
11. G. Radnoczi, A. Robertson, H. T. G. Hentzell, S.F.Gong, and M.A.Hasan, J. Appl. Phys., 69, 1991, 6394.
12. C.Hayzelden, J.L.Batstone, and R.C.Cammarata, Appl. Phys. Lett., 60, 1992, 225.
13. T.Clarysse, P.De Wolf, H.Bender, W.Vandervorst, J. Vacuum Sci. Techn., B14, 1996.
14. Dios User's Guide, version X-2005.10, Synopsys TCAD Package

# Poster Session:
# Imagers, Sensors and
# Novel Applications

Mater. Res. Soc. Symp. Proc. Vol. 1066 © 2008 Materials Research Society        1066-A18-02

# Improvement in pinpin Device Architectures for Imaging Applications

P. Louro[1,2], A. Fantoni[1], M. Fernandes[1], G. Lavareda[3,4], N. Carvalho[3,4], and M. Vieira[1,2]

[1]DEETC, ISEL, R. Conselheiro Emídio Navarro, 1, Lisbon, 1959-007, Portugal
[2]CTS, Uninova, Monte da Caparica, Caparica, 2829-516, Portugal
[3]DCM, Caparica, 2829-516, Portugal
[4]IST, C1, Lisboa, 1049-001, Portugal

## ABSTRACT

In this paper we present results on the optimization of device architectures for color and imaging applications, using a device with a TCO/pinpi'n'/TCO configuration. A set of different devices with different intrinsic back layers (i') are analysed.

The effect of the applied voltage on the color selectivity is discussed. Results show that the spectral response curves demonstrate rather good separation between the red, green and blue basic colors. Combining the information obtained under positive and negative applied bias a color image is acquired without color filters or pixel architecture. A low level image processing algorithm is used for the color image reconstruction.

## INTRODUCTION

The use of multilayered structures based on a-SiC:H alloys as color sensors has been an important topic of research in the field of sensing applications [1, 2, 3]. In these multilayered devices the light filtering is achieved through the use of different band gap materials, namely a-$Si_{1-x}C_x$:H. In these devices the spectral sensitivity in the visible range is controlled by the external applied voltage. Thus, proper tuning of the device sensitivity along the visible spectrum allows the recognition of the absorbed light wavelength, and consequently the identification of the RGB components of a colored image [4].

In this paper color pinpi'n' sensitive detectors are tested using the laser scanned photodiode technique (LSP) [5]. This technique allows a complete color analysis to be performed with a single two terminal detector element and an optically addressed readout technique. The image to acquire is optically mapped onto the sensing photodiode and a low-power light spot scans the device by the opposite side. The photocurrent generated by the moving spot corresponds to the image signal as its magnitude depends on the light pattern localization and intensity. For image color acquisition the device is biased at different voltage values, which modulates the output image signal and allows the reconstruction of the color image [6, 7].

## EXPERIMENTAL DETAILS

### Device Architecture

A series of stacked devices are analyzed in this work. The front p-i-n structure is common to all devices while the thickness of the back intrinsic absorber layer was varied in order to optimize the device. The geometry of the stacked sensors is sketched in Fig. 1. The sensor is a double heterostructure device and consists of a glass/ITO/p-i-n a-SiC:H photodiode which faces

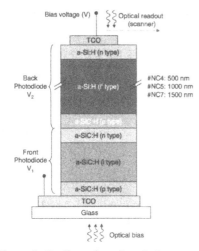

**Figure 1.** Configuration of stacked structures.

the incident illumination, followed by a-SiC:H(-p) /a-Si:H(-i')/a-Si:H (-n')/ITO heterostructure, that allows the optical readout. The doped layers are based on a-SiC:H to increase resolution and to prevent image blurring. To profit from the light filtering properties of the active absorbers, the intrinsic layer of the front diode is thinner and based on a-SiC:H and the back one is thicker and based on a-Si:H.

In the polychromatic operation mode different sensitivity ranges are selected by using a voltage scan waveform, and the photocurrent generated by the optical scanner is measured to sample the image signals. This approach leads to different collection regions resulting in voltage controlled multispectral photodiodes, coding for red (R), blue (B), and green (G) components.

## RESULTS AND DISCUSSION

### Device operation

In Fig. 2 it is displayed the simulated generation/recombination profile within device #NC7 for different wavelengths of the incident light at short circuit conditions. Typical values of band tail and gap state parameters for amorphous material were used [6]. The doping level was adjusted in order to obtain approximately the same conductivity of the individual layers. In the a-SiC:H film the optical band gap of 2.1 eV was chosen in compliance with the obtained experimental values.

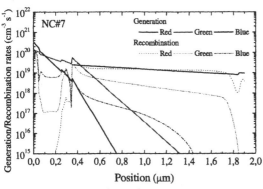

**Figure 2.** Numerical simulation of the generation and recombination rates profiles under different light wavelengths within device #NC7 (short circuit).

Results show that the light penetration depth controls the generation of carriers accordingly to the nature of the incident light. The thin a-SiC:H front absorber prevents the red light to generate carriers along the front photodiode. The photo generated carriers in this absorber arise from higher energy light photons (in the green/blue spectral range). The high generation across the structure due to blue photons is due to the reduced penetration depth of the blue light caused by the higher value of the absorption coefficient in this spectral range.

## Optoelectronic characterization

In Fig. 3 it is displayed the collection efficiency of device #NC5, in the voltage range of -6V up to +6V, under different steady state illumination conditions ($\lambda_L$ = 650 nm, 550 nm and 450 nm) for a red probe pulsed light ($\lambda_S$ = 650 nm). Results show that the collection efficiency, at 650 nm depends on the optical bias conditions. The trend under illumination is different from the one observed in dark. When the applied bias changes from reverse to forward bias the carrier collection decreases, with different slopes in dark and under illumination, leading to voltage controlled color discrimination.

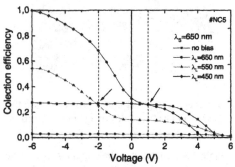

**Figure 3.** Collection efficiency in dark and under red, green and blue steady state illumination using a red scanner ($\lambda$s=650 nm).

The elimination of the green/blue signals at different voltage values (arrows in Fig. 3) is the key to enable color sensing properties to these devices. By tuning the voltage to the value where the green signal is suppressed (-2 V), the measured signal results only from contributions of the red and blue components, whereas that at +1V the blue signal is eliminated and the measured signal results only from contributions of the red and green components. At any reverse voltage value, the measured signal results from the contribution of the three components. Combining the information obtained at three applied voltages a color image can be acquired. Thus, the recognition of the fundamental colors (red, green and blue) is possible through the correct selection of the applied voltage values. A good color separation is achieved with device #NC5 because when the green signal is suppressed the blue signal exhibits low amplitude. A similar situation occurs when the blue signal is suppressed, as at +1V the green signal is still low in amplitude. Thus, the configuration of sensor #NC5 with an intermediate thickness of the absorber layer allows better separation of the fundamental colors.

## Image acquisition and reconstruction

Fig. 4a) shows the optical image of a graded wavelength mask (rainbow) that simulates the visible spectrum in the range between 400 and 700 nm. This image was created using a Matlab program that computes the digital image. The picture was then printed in a transparent paper using the subtractive color system of the printers. This optical image was projected onto the a-SiC:H front diode and acquired through the a-Si:H back one with a moving red scanner. For image acquisition two applied voltages were used to sample the image signal: + 1V and -6 V. In Fig. 4 b) it is displayed the 2-D color image reconstruction using the acquired image signals. The algorithm used for image color reconstruction took into account that at -6 V the positive signals correspond to the blue/green contribution and the negative ones to the red inputs (Fig. 3). The green information was extracted from the image signal sampled at +2 V, where the blue signal is almost suppressed (Fig. 3) and the green and red signals are negative.

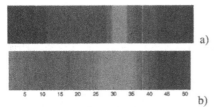

**Figure 4.** a) Optical image projected onto the device and b) 2-D color image reconstruction.

The combined integration of this information allows the fulfillment of the R, G and B channels good agreement between the optical image and the electrical reconstructed color image was obtained although further optimization of the sampling threshold voltage values seems to be needed in order to improve the R, G and B channels distributions. The device was then further analysed using as optical bias the light from a monochromator (in the 400 nm to 800 nm range) which corresponds to single wavelengths and the optical image of the printed rainbow mask where the colors result from multiple wavelength combinations. Under these conditions, the image signal was then measured. Results are plotted in Fig. 5 and show that under these different optical biasing conditions the photocurrent of the device exhibits different trends. With monochromatic light (Fig. 5a) the photocurrent exhibits a maximum around 465 nm, then decreases for longer wavelengths, and reverses in sign around 550 nm.

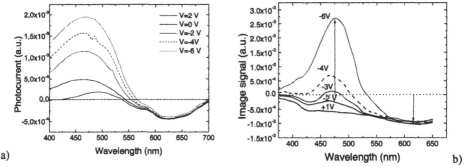

**Figure 5.** a) Photocurrent obtained with the light from the monochromator and b) Image signal obtained with the rainbow mask.

Here, in the blue range the photocurrent is always positive, increasing as the reverse voltage increases. In the reddish part of the spectrum, the photocurrent reverses sign and its magnitude is independent on the applied voltage bias. When compared with the signal without light impinging the front diode (dark level at 750 nm) at short circuit conditions, the red and blue signals have opposite signs and the green signal is almost suppressed, allowing blue and red recognition. The green information is obtained for the voltage bias that suppresses the blue signal, which occurs at slightly positive bias.

The analysis of the image signal obtained with the same device using as optical bias multiple wavelength combinations (Fig. 5b) is similar to the one obtained with single wavelengths only in the reddish region of the visible spectrum, where in both cases the signal is reversed and independent on the applied bias. Main differences occur in the remaining regions of the spectrum, namely on the blue region where the signal exhibits sign reversal due to the presence of a red component in this part of the mask. This effect shows up also in the greenish region, as in this case the signal is similar to the one observed in the reddish part of the spectrum. Thus, we can conclude that with this mask all the colors result from a significant red component which influences the device photocurrent characteristics.

## Low level image processing

Using these assumptions the RGB channels were fulfilled (Fig. 6, solid lines) and the corresponding digital image representation reconstructed (Fig. 7a). Results show that the distribution is adequate, although not yet correctly balanced, as the region of intersection between the red and the green curves, as well as the overlap between the blue and the green channels occurs at moderate values of intensity.

This makes difficult the representation of light cyan and light yellow/orange colors (Fig. 7a). As already stated before the blue, green and red colors are adequately distributed along the spectrum, but the cyan, yellow and orange colors are missing (arrows of the picture), due to the fact that in the spectrum region where they should appear the RGB intensities are not high enough. Thus some further correction must be done in order to change the relative intensities of the channels. This can be done using the gamma correction procedure, which is a nonlinear operation used to code and decode luminance or tristimulus values in video or still images. In the simplest cases it is defined by the following power-law expression: $V_{out} = V_{in}^{\gamma}$, where the input ($V_{in}$) and output values ($V_{out}$)

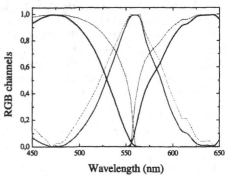

**Figure 6.** Distribution of the Red, Green and blue channels (solid lines: without gamma correction; dashed lines: using different values of the coefficient for the gamma correction, $\gamma_R=0.5$, $\gamma_G=0.7$, $\gamma_B=0.3$).

are non-negative real values, typically in a determined range such as 0 to 1. The case $\gamma < 1$ is often called gamma compression or gamma encoding and $\gamma > 1$ is called gamma expansion or gamma decoding. Gamma encoding helps to map data into a more perceptually uniform domain, so as to optimize perceptual performance of a limited signal range, such as a limited number of bits in each RGB component.

The use of the gamma correction using power coefficients less than one results in an enhanced increase of the image signal intensity in the region of low intensity. This effect grows with the decrease of the gamma power coefficient. Thus low values of $\gamma$ are responsible for an enhanced mapping of the original low intensity values. The gamma correction was then implemented, and the new distribution of the RGB channels is displayed in Fig. 6 (dashed lines). The corresponding digital color image representation obtained by using the gamma correction is displayed in Fig. 7b.

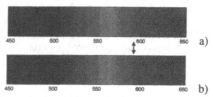

**Figure 7.** Digital image representation of the rainbow picture acquired by the device: a) without gamma correction; b) with gamma correction ($\gamma_R=0.5$, $\gamma_G=0.7$, $\gamma_B=0.3$).

The use of the gamma correction results in a brighter image that allowed the representation of some colors that were missing in the original, uncorrected image. This was mainly due to the use of different mapping of the channels intensities that allowed for the case of the cyan and yellow colors an overlap, respectively, between the blue and the green and the red and the green channels, at higher values of intensities (0.8 with gamma correction, and 0.5 without).

## CONCLUSIONS

The various design parameters and image reconstruction algorithms are discussed. A physical model supported by electrical and numerical simulations gives insight into the methodology used for image representation and color discrimination.

A low level image processing algorithm was used for the reconstruction of the color image. Obtained results show that further optimization on the color image reconstruction is necessary, as there is some mismatch in the reconstructed color images. In order to advance more in this domain, some low level image processing algorithms were used to enhance the image intensity of each channel, which resulted in a better balanced colored reconstructed image. These preliminary results based on the implementation of the gamma correction technique, show that with further image processing techniques better results can be achieved.

## ACKNOWLEDGEMENTS

This work has been financially supported by POCI/CTM/56078/2004 and by Fundação Calouste Gulbenkian.

## REFERENCES

1. G. de Cesare, F. Irrera, F. Lemmi, F. Palma, IEEE Trans. on Electron Devices, Vol. 42, No. 5, May 1995, pp. 835-840.
2. A. Zhu, S. Coors, B. Schneider, P. Rieve, M. Bohm, IEEE Trans. on Electron Devices, Vol. 45, No. 7, July 1998, pp. 1393-1398.
3. M. Mulato, F. Lemmi, J. Ho, R. Lau, J. P. Lu, R. A. Street, J. of Appl. Phys., Vol. 90, No. 3 (2001), pp. 1589-1599.
4. H. Stiebig, J. Gield, D. Knipp, P. Rieve, M. Bohm, Mat. Res. Soc. Symp. Proc 337 (1995) 815-821.
5. M. Vieira, M. Fernandes, J. Martins, P. Louro, A. Maçarico, R. Schwarz, M. Schubert, IEEE Sensors Journal, 1 (2) (2001) 158-167.
6. P. Louro, M. Vieira, A. Fantoni, M. Fernandes, C. Nunes Carvalho, G. Lavareda, Sensors and Actuators A 123-124 (2005) 326-330.
7. P. Louro, M. Vieira, Y. Vygranenko, M. Fernandes, A. Garção, Mat. Res. Soc. Symp. Proc., S. Francisco, 989 (2007) A12.04.

Mater. Res. Soc. Symp. Proc. Vol. 1066 © 2008 Materials Research Society          1066-A18-04

# Noise Analysis of Image Sensor Arrays for Large-Area Biomedical Imaging

Jackson Lai[1], Denis Striakhilev[2], Yuri Vygranenko[2], Gregory Heiler[1], Arokia Nathan[3], and Timothy Tredwell[1]

[1]Carestream Health Inc., Rochester, NY, 14615

[2]Department of Electrical and Computer Engineering, University of Waterloo, Waterloo, N2L 3G1, Canada

[3]London Centre for Nanotechnology, London, United Kingdom

## ABSTRACT

Large-area digital imaging made possible by amorphous silicon thin-film transistor (a-Si TFT) technology, coupled with a-Si photosensors, provides an excellent readout platform to form an integrated medical image capture system. Major development challenges evolve around the optimization of pixel architecture for detector fill factor and manufacturability, all the while suppressing that noise stems from pixel array and external electronics. This work discusses the behavior and modeling of system noise that arises from imaging array operations. An active pixel sensor (APS) design with on-pixel amplification was studied. Our evaluation demonstrates that a $17 \times 17$ inch array with $150 \times 150$ μm pixels can achieve system noise as low as 1000 electrons through proper design and optimization.

## INTRODUCTION

Large-area, digital-image sensors are revolutionizing medical imaging by enabling electronic storage capability, immediate feedback, and possibilities to support previously unachievable applications related to computer-aided image processing. Hydrogenated amorphous silicon thin-film transistor (a-Si:H TFT) technology, frequently used in liquid crystal display applications, is extended to perform backplane readout for large-area detector systems. Technological attributes of a-Si:H, such as uniformity over a large area, compatibility with various substrate materials, and research maturity, have provided an excellent development platform for high performance, low noise, and fully integrated digital detector systems.

The signal-to-noise ratio (SNR) of the readout backplane is one the key performance evaluation attributes. Biomedical imaging applications place an even more stringent requirement on SNR compared to other imaging systems because of safety standards in the medical arena. A typical imaging array implemented with a-Si:H technology has one TFT in the pixel level acting as a row access switch. However, the design has no inherited signal amplification, making it prone to electronic noise and unsuitable for low-dosage applications such as fluoroscopy [1-2]. Active pixel sensors (APSs) have been investigated to alleviate some design concerns. In a conventional APS design, each pixel contains a read, a reset, and an amplifier TFT [2], and the composite source-follower amplifier setup provides the signal gain to overcome external and pixel noise. The design involves higher circuit complexity that potentially leads to higher noise, hence it is critical to investigate the noise performance of the design.

This work presents a systematic noise analysis of an APS pixel array using a-Si:H technology. First, the most significant noise sources are discussed individually, and their respective contributions to APS pixel operation are investigated, followed by insights into design optimization in an attempt to maximize SNR.

## TFT and PIXEL NOISE

### Flicker noise

Field effect transistors, such as MOSFETs and TFTs, exhibit flicker noise similar to other conductors, and many authors have studied their effects [3-5]. Since then, two main theories for the physical origin of 1/f noise in transistors have been developed, namely, the carrier number fluctuations theory and the mobility fluctuation theory. The number fluctuation theory suggests that the 1/f noise is attributed to the random trapping and de-trapping of carriers in surface states, while the latter theory considers the 1/f noise as a result of the fluctuation in the carrier mobility. Unfortunately, a-Si:H TFTs suffer from the disordered lattice orientation in the active layer and do not benefit from any unified 1/f noise model studied for their crystalline silicon counterpart [5]; hence, the two theories are both considered and experimentally verified here.

Table 1 lists the power spectral density (PDF) models for both 1/f noise models, where $k^*$ and $\alpha_H$ are constants from the number and mobility fluctuation model, respectively [6-7], $n = \alpha-1$, and $\alpha$ is the power parameter for a-Si:H TFT. Through extracting the slopes from the log-log plot of $S_I(f)$ vs. $(V_{GS} - V_T)$ of TFT in both regimes of the operation, one can determine which model generates a better fit to experimental data. Figure 1 illustrates the plot used to extract the characteristic slopes, and it demonstrates an exponent of 2.81 for saturation and 1.23 for the linear regime. The power parameter is separately extracted to be 2.3 for in-house fabricated a-Si:H TFTs. Our TFT samples appear to match the mobility fluctuation theory better, and this is also in agreement with other authors [8-10]. Figure 2 shows the power spectra for a-Si:H TFT in the saturation regime at different gate biases. Here, flicker noise dominates most of the spectrum at low frequencies. However, thermal noise drowns out flicker noise near 7 kHz and become indistinguishable thereafter. Flicker noise increases as the gate bias increases, which agrees with predictions that it is directly proportional to direct current. Measured data show a 21% maximum error with the theoretical (mobility fluctuations) prediction.

Table 1: Flicker noise, power spectral densities for number and mobility fluctuation models [9-11].

| | Number fluctuation | Mobility fluctuation |
|---|---|---|
| $S_{I\_linear}$ | $\dfrac{k^*}{f}\dfrac{\mu_{EFF}}{C_G L^2}\left[ K \cdot (V_{GS} - V_T)^{n-1} V_{DS}^2 \right]$ | $\dfrac{\alpha_H}{f}\dfrac{q\mu_{EFF}}{L^2}\left[ K(V_{GS} - V_T)^n V_{DS}^2 \right]$ |
| $S_{I\_saturation}$ | $\dfrac{k^*}{f}\dfrac{\mu_{EFF}}{C_G L^2}\left[ K \cdot (V_{GS} - V_T)^{n+1} \right]$ | $\dfrac{\alpha_H}{f}\dfrac{q\mu_{EFF}}{L^2}\left[ K(V_{GS} - V_T)^{n+2} \right]$ |

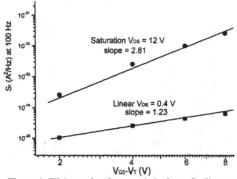

**Figure 1: Flicker noise characteristic slopes for linear and saturation regimes.**

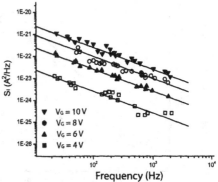

**Figure 2: Flicker noise spectra for a-Si:H TFT in the saturation regime.**

## Thermal noise

Active devices, such as TFTs, conduct current through an induced electronic channel, thus the drain-source current also suffers from thermal noise [11]. The power spectra of thermal noise in a-Si:H exhibits similar trends as other field-effect transistors, such as MOSFETS and such findings, have been confirmed by various authors [10][12]. The thermal noise PDF for TFTs is:

$$S_I(f) = A \cdot (4kT) \cdot K \cdot (V_{GS} - V_T)^n \text{ where } K = \mu_{EFF} C_G \frac{W}{L} \text{ and } A = \begin{cases} 1 \text{ in linear} \\ \frac{2}{3} \text{ in saturation} \end{cases}. \quad (1)$$

The measurement of thermal noise for a-Si:H TFT is not a trivial task for a few reasons. First, it is difficult to isolate the contribution of flicker noise from thermal noise in a-Si:H, largely due to the trap states in the channel, which cause a significant 1/f component. Therefore the drain-source current has to be minimized in order to suppress flicker noise for successful measurement. Measurement results, described in Figure 3, were taken at frequencies well beyond the 1/f corner frequency. Second, the intrinsically low free carrier count in a-Si:H materials cause the thermal fluctuation to be small, in comparison to MOSET devices, making it difficult to measure without any significant amplification. It is for these reasons, thermal noise measurement for TFTs in the saturation regime is not recommended.

Using the above method and reasoning, thermal noise is measured, and the result is shown in Figure 3 with the y-axis in the log-2 scale. Gate voltage is varied from 4 to 10 V in 2 V increments. The theoretical current noise power exhibits almost a power-law relation with gate bias, and the shape draws similarity to TFT $I_{DS}$ current. Here, theoretical prediction and measurement results show reasonable agreement for low gate biases. As the gate bias increases, however, the discrepancies grow, and the measured noise power is typically larger than the predictions. This is mainly due to the larger TFT-series resistance contribution to total drain-source resistance, because the channel resistance is reduced as gate bias increases. The channel resistance dominates the total drain-source resistance at small $V_{GS}$, hence, the lower discrepancy.

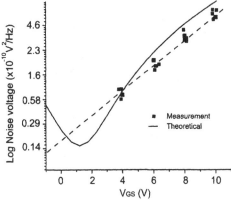

**Figure 3: Thermal noise measurement results.**        **Figure 4: Reset noise measurement results.**

## Reset noise

Charge detection in imaging circuits requires the photosensitive element to be reset to a known reference value after readout, allowing the signal to be integrated for the next image frame. However, according to thermodynamic principles, the capacitive node cannot be reset to the same voltage. Even for the scenarios when there is no accumulated charge, reference voltage still suffers from slight fluctuations, which is related to the thermal conductance variations of the reset device. This fluctuation is non-deterministic and is termed "reset noise" or "$kTC$ noise" [13]. The power spectral density of reset noise is:

$$S_c(\omega) = S_o |H(\omega)|^2 \left[ 1 + \zeta + \frac{1}{2}(\omega_f \tau)\zeta^2 F_1 \cdot \mathrm{sinc}^2\left(\frac{\omega\delta}{2}\right) + \zeta F_2 \mathrm{sinc}^2\left(\frac{\omega\delta}{2}\right) \right], \tag{2}$$

where the various quantities are defined as:

$$S_o = \frac{1}{\pi}kTR, \ \zeta = \frac{\delta}{\tau}, \ \omega_f = \frac{1}{RC}, \ |H(\omega)|^2 = \frac{\omega_f^2}{\omega_f^2 + \omega^2}, \tag{3}$$

$$F_1 = \frac{\omega_f^2}{\omega_f^2 + \omega^2} \cdot \frac{\sinh\left[\omega_f(\tau-\delta)\right]}{\cosh\left[\omega_f(\tau-\delta)\right] - \cos(\omega\tau)}, \ \text{and} \tag{4}$$

$$F_2 = 2|H(\omega)|^2 \frac{\cos\left(\frac{\omega\tau}{2}\right)\cosh\left[\omega_f(\tau-\delta)\right] - \cos\left[\omega_f\left(\tau-\frac{\delta}{2}\right)\right]}{\cosh\left[\omega_f(\tau-\delta)\right] - \cos(\omega\tau)}. \tag{5}$$

The derivation of the above formulas and expressions are fully documented and published by the author in [14]. The measurement results for reset noise are shown in Figure 4. The PDF exhibits a sinc function behavior with respect to frequency. In particular, the term with $\mathrm{sinc}^2\left(\frac{\omega\delta}{2}\right)$ dominates the expression as reset duration (i.e., $\delta$) becomes smaller. The power spectrum in Figure 4 is measured for a duty cycle of 10% and a reset pulse width of 10 ms. Here, the experimental data is plotted against the model and shows reasonable agreement. Larger discrepancy between the model and theory is observed near the harmonics of the reset pulse (~100 Hz) and is attributed to clock feedthrough of the gate-pulsed TFT.

## Data line thermal noise

In imaging array systems, the data line is commonly shared between all of the pixels in the same column. The data line parasitic resistance ($R_{DL}$) and the capacitance ($C_{DL}$) not only provide a low pass-filtering action to signal readout, they also contribute to thermal noise. The noise variance stored in the data-line capacitance is given by:

$$\sigma^2_{DL\_th} = \frac{\pi f_o}{2} 4kTR_{DL}.$$ (6)

where $f_o$ is the amplifier noise bandwidth. The validity of the amplifier noise bandwidth limitation is reasonable because the corner frequency imposed by the $R_{DL} C_{DL}$ product is typically much smaller than that of the amplifier. This assumption is contrary to the TFT thermal noise analysis where the noise bandwidth is limited by the in-pixel amplifier or the pixel time constant.

## Total noise

The various noises discussed so far are assumed to be uncorrelated, so they can be added in quadrature. The noise appearing at the intermediate stages of the imaging array readout chain are first converted to the output, and then referred back to the signal-integrating node using the total system gain. The input-referred noise variance is then converted to units of electrons for a meaningful comparison to the input signal charge.

Figure 5 shows the PSD of the pixel readout chain (reset TFT not pulsed). The noise is clearly 1/f dominated due to uninterrupted direct current. This agrees with predictions (see Figure 5) that flicker noise is the major noise source at the pixel level. The noise dependency on amplifier gate voltage also confirms the dominance of the 1/f component because flicker noise is proportional to direct current. In practical imaging array scenarios, the readout chain will be pulsed, and this leads to a reduction in the 1/f noise component.

The total noise figures are tabulated in Figure 6 for a 17 × 17 inch pixel array with $150^2 \mu m^2$ pixel size. Our evaluation shows that an array using an 18 μm channel length process can achieve 1800 electrons of noise performance. The noise model is also extended to predict the performance of the same design scaled to a 6 μm channel length process. It is evident that the increased pixel amplification from critical dimension shrinkage enables the total noise to be kept close to 1000 electrons, which approaches the X-ray quantum noise limit. It is important to point out that as the pixel design is optimized, the total noise is dominated by reset noise stemming from the photosensor. The use of additional TFTs (compared to a 1T design) for signal amplification is justified because it suppresses data line thermal and external amplifier noise without introducing excessive noise components.

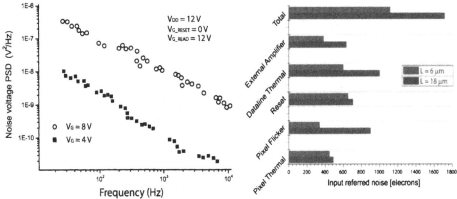

**Figure 5: APS noise PSD for different AMP TFT gate biases.**

**Figure 6: Total input-referred noise in electrons for channel lengths of 6 and 18 μm.**

## CONCLUSIONS

This work analyzes the various noise contributors in an APS pixelated array using a-Si:H technology. Analytical models for flicker, thermal, reset, and miscellaneous external noise are presented and compared to measurement results with reasonable accuracy. Reset noise is explained to be the major noise contributor to pixel operation, while an APS pixel design can assist in the suppression of other spurious noise components. It is also pointed out that with an industry-standard, 6 μm channel length process, the total input-referred noise of a pixilated array can be as low as 1000 electrons, making it amenable for a low-dosage, large-area imaging detector for biomedical applications.

## REFERENCES

1. G. Weckler, *IEEE L. Solid-State Circuits,* SC-2, 65, 1967.
2. K.S. Karim, A. Nathan, J.A. Rowlands, *JVST A,* 20(3), pp. 1095-1099, 2002.
3. F. Reif, *Fundamentals of Statistical and Thermal Physics,* McGraw-Hill, 1st edition, 1965.
4. J.M. Boundry, L.E. Antonuk, *J. Appl. Phys.,* 76, pp. 2529-2534, 1994.
5. J.B. Johnson, *Phys. Rev.,* 26, pp. 71-85, 1925.
6. A.L. McWhorter, "1/f noise and germanium surface properties," *Semiconductor Surface Physics,* University of Pennsylvania Press, Philadelphia, pp. 207-228, 1956.
7. F.N. Hooge, L.K.J. Vandamme, *Phys. Lett.,* 66A, pp. 316-326, 1978.
8. H. Mikoshiba, *IEEE Trans. Elec. Dev.,* 29(6), pp. 965-970, 1982.
9. KK. Hung, P.K. Ko, C.Hu, Y.C. Cheng, *IEEE Trans. Elec. Dev.,* 37(3), pp. 654-665, 1990.
10. K.S. Karim, A. Nathan, J.A. Rowlands, in *Opto-Canada: SPIE Regional Meeting on Optoelectronics, Photonics and Imaging,* SPIE TD01, pp. 358-360, 2002.
11. T. Hui, E.G. Abbas, *Proc. SPIE,* 3965, pp. 168-176, 2000.
12. J.M. Boundry, L.E. Antonuk, *J. Appl. Phys.,* 76, pp. 2529-2534, 1994.
13. J. Hynecek, *IEEE Trans. Elec. Dev.,* 37 (3), pp. 640-647, 1990.
14. J. Lai, A. Nathan, *IEEE Trans. Elec. Dev.,* 52(10), pp. 2329-2332, 2005.

Mater. Res. Soc. Symp. Proc. Vol. 1066 © 2008 Materials Research Society        1066-A18-05

# Noise in Different Micro-bolometer Configurations with Silicon-Germanium Thermo-sensing Layer

Mario M Moreno, Andrey Kosarev, Alfonso J Torres, and Ismael Cosme
Electronics, National Institute for Astrophysics, Optics and Electronics, L.E.Erro No. 1, col.Tonatzintla, Puebla, 72840, Mexico

## ABSTRACT

We have studied noise in four configurations of un-cooled micro-bolometers: three of them were built in a planar structure with a) a-$Si_x$ Ge $_y$, y = 0.88, b) a-$Si_x Ge_y B_z$:H, y = 0.67 and z = 0.26; c) y = 0.71, z = 0.23 and the fourth d) sandwich structure with y = 0.88. These samples were characterized by SIMS (composition), FTIR (H-bonding and H content), conductivity measurements ($\sigma(T)$, activation energy, TCR), current-voltage characteristics in dark and under illumination enabling in this way the determination of responsivity. The power noise spectral density (PNSD) versus frequency S(f) was studied in the range of frequency f=1 to f=$10^3$ Hz under IR illumination and constant bias. The measurements were performed in a vacuum chamber with pressure P=10 mTorr. In general the S(f) measured on our devices, demonstrated three regions separated by two corner frequencies: $f_{c1}$ and $f_{c2}$. The regions are: 1) f $\leq$ $f_{c1}$ $S_1 \sim f^{\beta}$ and $\beta$ = 0.1 to 0.3, 2) $f_{c1} \leq$ f $\leq f_{c2}$, $S_2 \sim f^{\gamma}$ and $\gamma$ = 0.75 to 1.5 and 3) f $\geq f_{c2}$ and S3 – const (f). The different samples here studied showed different values of $f_{c1}$, $f_{c2}$, $\beta$, $\gamma$ and $S_3$ level. The noise characteristics experimentally observed experimentally are used for determining the detectivity $D^*$ of the devices and these data are analyzed and compared with the data reported in literature.

## INTRODUCTION

A renewed recent interest in thermal detectors has resulted from to plasma deposited materials with large temperature coefficient of resistance (TCR), which in conjunction with micro-machining technology for thermal isolation, have provided a fabrication process compatible with the dominating Si CMOS technology that makes it possible the development of system -on-chip configurations. In our previous works we have reported on the study of fabrication and characterization of single cell micro-bolometers based on silicon-germanium thermo-sensing films deposited by low frequency plasma [1-3]. Noise measurements are very important part of device characterization, which have been poorly reported in literature with only a few publications [4-8]. In ref. [4] the spectral dependence of noise has been reported in $Ge_x Si_{1-x} O_y$ films deposited by sputtering. The currently proposed models for noise description in non-crystalline samples (either in films or in device structures) are still debated and no one can be considered as the only accepted [9].

The goal of this work is to study experimentally noise spectral density in several configurations of micro-bolometers built with silicon-germanium (a-$Si_x Ge_y$:H) as thermo-sensing films. The power noise spectral density (PNSD) has been measured in four configurations of micro-bolometers: a) planar structure with a silicon-germanium intrinsic (a-$Si_x Ge_y$:H, y =0.88) thermo-sensing film, b) planar structure with a silicon-germanium-boron alloy (a-$Si_x Ge_y B_z$:H, y =0.67, z =0.26) thermo-sensing film, c) planar structure with a silicon-germanium-boron alloy (a-$Si_x Ge_y B_z$:H, y =0.71, z =0.23) thermo-sensing film , d) sandwich structure with an intrinsic (a-$Si_x Ge_y$:H, y =0.5) thermo-sensing film.

## EXPERIMENT

The micro-bolometers structures studied have been fabricated on Si-wafer by using the surface micro-machining technique. The fabrication process flow has been previously described in ref.[2, 3]. The IR absorber, support and protective layers ($SiN_x$), and the thermo-sensing layer (a-$Si_xGe_yB_z$:H) were deposited by low frequency plasma (LF PECVD). The metal electrodes are deposited by electron beam evaporation.

Patterning is performed with photolithography and a combination of both RIE and wet chemical etching. The cross-sections of the structures are illustrated in Figure 1.

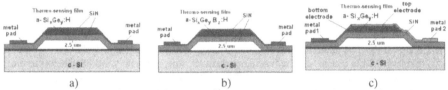

a)                                   b)                                   c)

Figure 1. Cross-section diagram of micro-bolometers: a) planar with the intrinsic (a-$Si_xGe_y$:H, y = 0.88), b) planar with Si-Ge-B alloys a-$Si_xGe_yB_z$:H (y = 0.67, z = 0.26 and y = 0.71, z = 0.23) thermo-sensing film, c) sandwich with the intrinsic (a-$Si_xGe_y$:H, y = 0.88) thermo-sensing film.

The deposition conditions for the thermo-sensing a-$Si_xGe_yB_z$:H films, deposited by LF PECVD were as follows: substrate temperature $T_s$= 300 °C, power W = 450 W, the discharge frequency f=110 kHz, pressure P= 0.6 Torr in a gas mixture consisting of silane ($SiH_4$), germane ($GeH_4$) and diborane ($B_2H_6$) and hydrogen. Composition of the films was changed through the variation of flows of the aforementioned gases (details see e.g. in ref. [10]). The characteristics of the thermo-sensing films in the experimental structures are listed in Table I.

Table I. Characteristics of the thermo-sensing films.

|  |  | Thermo-sensing films | | |
|---|---|---|---|---|
|  |  | Process 479 | Process 480 | Process 443 |
| Film Thickness ($\mu$m) | | 0.42 | 0.51 | 0.5 |
| Deposition rate ($\overset{o}{A}$ /s) | | 7 | 9.5 | 2.8 |
| Solid content obtained from SIMS | x (Si) | 0.05 | 0.04 | 0.11 |
|  | y(Ge) | 0.67 | 0.71 | 0.88 |
|  | Z(B) | 0.26 | 0.23 | $2.0x10^{-5}$ |
| Film properties | $E_a$ (eV) | 0.21 | 0.18 | 0.345 |
|  | TCR ($K^{-1}$) | -0.027 | -0.023 | -0.044 |
|  | $\sigma_{RT}$ ($\Omega$cm)$^{-1}$ | $1x10^{-2}$ | $2.5x10^{-2}$ | $6x10^{-5}$ |
|  | $\sigma_0$ ($\Omega$cm)$^{-1}$ | 36.46 | 24.55 | 34.85 |

The active area of the thermo-sensing layer is $A_b = 70 \times 66 \ \mu m^2$. The current-voltage I(U) characteristics of the structures were measured in vacuum (P = 10 mTorr) thermostat with electrometer ("Keithley" – 6517-A) at room temperature in dark and under IR illumination. From this measurement we were able to to determine current responsivity at different bias voltage. IR source provided an intensity $I_0 = 5.3 \times 10^{-2} \ W/cm^2$ on the surface of the sample. The noise spectral density (NSD) measurements were performed with lock-in amplifier ("Stanford Research Systems" – SR 530) used as a quadratic detector with frequency bandwidth $\Delta f = 1$ Hz. The detectivity $D^*$ was calculated from I(U) characteristics and noise level either in the region of "white" noise, when it is possible to identify this region or at the frequency corresponding to noise minimum (f= 100 , 200 Hz).

## DISCUSSION

The power noise spectral density (PNSD), is defined as $S^I_{cell}(f) = [I_{noise}/(\sqrt{\Delta f})]^2$, $I_{noise}$ is r.m.s. current noise was measured with $\Delta f = 1$ Hz and it is shown in the following figures. For the planar configuration with the a-$Si_xGe_y$:H (y=0.88) thermo-sensing film it is shown in Figure 3 a). The PNSD $S^I_{cell}(f)$ of the planar configuration with the a-$Si_xGe_yB_z$:H (y=0.67, z=0.26) thermo-sensing film is shown in Figure 3 b). Figure 3 c) shows the PNSD $S^I_{cell}(f)$ of the planar configuration with the a-$Si_xGe_yB_z$:H ( y =0.71, z =0.23) thermo-sensing film. Finally the PNSD $S^I_{cell}(f)$ of sandwich configuration with the a-$Si_xGe_y$:H ( y =0.88) thermo-sensing film is shown in Figure 3 d). The PNSD in the cell is obtained from $S^I_{cell}(f) = S^I_{cell+system}(f) - S^I_{system}$ , where $S^I_{cell+system}(f)$ is the total measured PNSD for the micro-bolometer with the measuring system and the $S^I_{system}$ is the PNSD measured in the system without the micro-bolometer.

As it is shown in Figure 3 $S^I_{cell}(f)$ generally presents 3 regions conventionally separated by corner frequencies $f_{c1}$ and $f_{c2}$. In two regions a spectral dependence of noise can be described by function $S^I_{cell}(f) \sim f^\gamma$. In the region f< $f_{c1}$ in both planar structure in Figure 3 a) and in sandwich structure in Figure 3 d) values of $\gamma$ =0.29 and 0.1, respectively, were observed. For the region $f_{c1} < \gamma < f_{c2}$ $\gamma$ was measured close to 1 (from 1.04 to 1.34) for three samples, but one sample (planar with a-$Si_xGe_yB_z$:H with y= 0,71, z=0.23) showed $\gamma \approx 0.67$. In the region f> $f_{c2}$ $S^I_{cell}(f) \approx$ const (f) i.e. it is "white" noise.

It is interesting that both noise level and $S^I_{cell}(f)$ depend on configuration of the micro-bolometer for example: compare Figures 3 a) and d) for the samples with the same thermo-sensing material, but with different configuration: planar and sandwich, respectively. Also they depend on the material for thermo-sensing film (compare PNSD of the planar configurations in Figure a), b) and c)).

The PNSD characteristics of the studied samples are summarized in Table II. It should be noted that the lowest noise was observed in planar structure with "intrinsic" a-$Si_xGe_y$:H having $\gamma \approx 1.34$ i.e. slightly higher than exponent of "classic" 1/f noise. The boron incorporation increased the PNSD by 1-2 orders of value compared with that measured in "intrinsic" material. The highest noise value for the studied planar structures was found for the structure with thermo-sensing material with z= 0.07 (Figure 3 c)). Even higher value of noise was observed in the sandwich configuration (Figure 3 d)). In low frequency region f<$f_{c1}$ the slopes with $\beta$= 0. 29 and 0.1 (i.e. less than 1) were observed in planar and in sandwich configurations, respectively. Thus the data obtained have demonstrated variation of noise (PNSD, its characteristics and values) with both the thermo-sensing film material and the device configuration. If the data presented in Figure 3 are compared with those published in literature, it will be seen that in the structures here studied $f_{c2} \approx 70$ to 350 Hz, a smaller value than the $f_{c2} > 1000$ Hz reported in [4]. Our results suggest the discussion/development of some noise models, but it is clearly that further experimental study and analysis is an important future task.

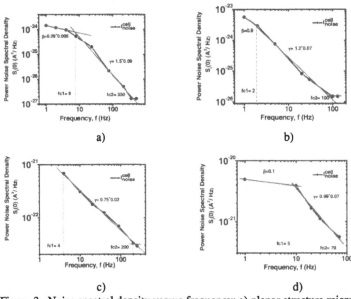

Figure 3. Noise spectral density versus frequency: a) planar structure micro-bolometer with an a-Si$_x$Ge$_y$:H (y =0.88) thermo-sensing film, b) planar structure micro-bolometer with an a-Si$_x$Ge$_y$B$_z$:H (y =0.67 z =0.26) thermo-sensing film, c) a planar structure micro-bolometer with an a-Si$_x$Ge$_y$B$_z$:H (y =0.71, z =0.23) thermo-sensing film, and d) sandwich structure micro-bolometer with an a-Si$_x$Ge$_y$:H (y=0.88) thermo-sensing film.

Table II. Noise characteristics in three regions of the micro-bolometers.

| Sample | Noise Regions | | | | | |
|---|---|---|---|---|---|---|
| | Region no. 1 | | Region no. 2 | | Region no. 3 | PNSD |
| | $\beta$ | $f_{c1}$ (Hz.) | $\gamma$ | $f_{c2}$ (Hz.) | I Noise level (AHz$^{-1/2}$) | $S_I(0)$ (A$^2$Hz$^{-1}$) |
| Planar structure a-Si$_x$Ge$_y$:H (y =0.88) | 0.29 | 8 | 1.5 ± 0.09 | 350 | $10^{-14}$ | $10^{-27}$ |
| Planar structure a-Si$_x$Ge$_y$B$_z$:H (y=0.67, z =0.26) | --- | 2 | 1.2 ± 0.07 | 100 | $10^{-13}$ | $10^{-26}$ |
| Planar structure a-Si$_x$Ge$_y$B$_z$:H (y=0.71, z=0.23) | ---- | 4 | 0.75 ± 0.02 | 200 | $10^{-12}$ | $10^{-23}$ |
| Sandwich structure a-Si$_x$Ge$_y$:H (y=0.88) | 0.1 | 9 | 0.99 ± 0.07 | 70 | $10^{-11}$ | $10^{-22}$ |

The detectivity is calculated with equation:

$$D^* = \frac{\mathfrak{R}_I \cdot \sqrt{A_{cell}}}{I_{noise} / \sqrt{\Delta f}} = \frac{R_I \cdot \sqrt{A_{cell}}}{\sqrt{S_{cell}^I}} \qquad \dots\dots\dots\dots\dots \quad (1)$$

Where: $\mathfrak{R}_I$ is the current responisivity, $A_{cell}$ is the detector area, $S_{cell}^I$ is the cell PNSD and $\Delta f = 1$ Hz is the bandwidth of the measurement system.

We have determined the detectivity values $D^*$ for the four structures. The planar structure with the a-Si$_x$Ge$_y$:H (y =0.88) film resulted in $D^* = 3.3 \times 10^8$ cmHz$^{1/2}$W$^{-1}$; for the planar structure with the a-Si$_x$Ge$_y$B$_z$:H (y =0.67, z =0.26) film $D^* = 1.7 \times 10^9$ cmHz$^{1/2}$W$^{-1}$; for the planar structure with the a-Si$_x$Ge$_y$B$_z$:H ( y =0.71, z =0.23) film $D^* = 8.9 \times 10^7$ cmHz$^{1/2}$W$^{-1}$ and for the sandwich structure micro-bolometer with the a-Si$_x$Ge$_y$:H (y =0.88) film $D^* = 4 \times 10^9$ cmHz$^{1/2}$W$^{-1}$.

Table III shows the current NSD, I$_{noise}$, and detectivity, D*, values for the four micro-bolometer configurations.

Table III. Current NSD (I$_{noise}$), detectivity (D*), current responsivity (R$_I$) and voltage responsivity (R$_U$) values for the four micro-bolometer configurations.

| | Planar structure with a-Si$_x$Ge$_y$:H film | Planar structure with a-Si$_x$Ge$_y$B$_z$:H film | Planar structure with a-Si$_x$Ge$_y$B$_z$:H film | Sandwich structure with a-Si$_x$Ge$_y$:H film |
|---|---|---|---|---|
| Film process No. | 443 | 479 | 480 | 443 |
| Voltage Responsivity R$_U$ (VW$^{-1}$) | $7.2 \times 10^5$ | $1.2 \times 10^5$ | $1.8 \times 10^5$ | $2.2 \times 10^5$ |
| Current Responsivity R$_I$ (AW$^{-1}$) | $2 \times 10^{-3}$ | $3 \times 10^{-2}$ | $7 \times 10^{-2}$ | 14 |
| $\sqrt{S_{cell}^I}$ (A Hz$^{-1/2}$) | $4 \times 10^{-14}$ | $1.2 \times 10^{-13}$ | $5.3 \times 10^{-12}$ | $2.3 \times 10^{-11}$ |
| Detectivity, D$^*$ (cmHz$^{1/2}$W$^{-1}$) | $3.3 \times 10^8$ | $1.7 \times 10^9$ | $8.9 \times 10^7$ | $4 \times 10^9$ |

From Table III it can be seen that material of thermo-sensing layer influence significantly on the noise characteristics. For the same planar configuration changes of "white" noise level of about two orders of value depending on thermo-sensing material are observed. Also such characteristics as corner frequencies $f_{c1}$ and $f_{c2}$ and PNSD showed remarkable changes. For instance, the sandwich configuration of micro-bolometer demonstrated different PNSD characteristics in comparison with planar configuration using the same thermo-sensing material. The "white" noise level, $f_{c2}$, $\beta$ and $\gamma$ values were also remarkably different.

## CONCLUSIONS

The measurements and study of noise in four configurations of the micro-bolometers with a-Si$_x$ Ge$_y$B$_z$:H thermo-sensing film have demonstrated that: a) both material of thermo-sensing film and configuration influence significantly on noise characteristics such as "white" noise level, $\beta$ and $\gamma$ values and corner frequencies $f_{c1}$ and $f_{c2}$; b) the sandwich configuration showed both the highest "white" noise level and the current responsivity and provided the highest value of detectivity, making this configuration promising for further study and optimization for 2D arrays in IR imaging.

## ACKNOWLEDGMENTS

Authors acknowledge CONACyT project No.48454F for financial support, Dr. Yu. Kudriavtsev (CINVESTAV, Mexico) for SIMS analysis, Dr. S.Rumiantsev (Rensselaer Polytecnic Institute, USA) for useful discussions of the results. I.Cosme acknowledges CONACyT scholarship No. 207900.

## REFERENCES:

1. A.Torres, A.Kosarev, M.L.Garcia Cruz, R.Ambrosio. *J.Non-Cryst. Solids*,**329**, 179 (2003).
2. M.Garcia, R.Ambrosio, A.Torres, A.Kosarev. *J.Non-Cryst. Solids,* **338-340**, 744 (2004)
3. A.Kosarev, M.Moreno, A.Torres, R.Ambrosio. Mater., Res., Symp., Proc., **910**, 0910-A17-05 (2006)
4. A.Ahmed and R.N.Tait. *Infrared Physics and Technology*, **46**, 468 (2005).
5. J.L.Tissot and F.Rothan, *Proc. SPIE*, **3379**, 139 (1998).
6. T.Schimert and D.Ratcliff, *Proc. SPIE*, **3577**, 96 (1999).
7. S.Sedky and P.Fiorini, *IEEE Trans. Electron Devices,* **46**, 675 (1999).
8. T.A.Enukova and N.L.Iranova, *Tech. Phys. Lett.* 2, 504 (1997).
9. R.E.Johanson, M.Guenes, S.O.Kasap. *IEE Proc.-Circuits Devices Syst.,* **149** (1),68 (2002).
10. A.Kosarev, L.Sanchez, A.Torres, T.Felter, A.Ilinskii, Y.Kudriavtsev, R.Asomoza. *Mat.Res. Symp.Proc.*, **910**, A07-02 (2006).

Mater. Res. Soc. Symp. Proc. Vol. 1066 © 2008 Materials Research Society            1066-A18-06

## Transient Current in a-Si:H-Based MIS Photosensors

Miguel Fernandes[1], Yuriy Vygranenko[1], Manuela Vieira[1], Gregory Heiler[2], Timothy Tredwell[2], and Arokia Nathan[3]

[1]DEETC, ISEL, Rua Conselheiro Emidio Navarro, Lisbon, 1959-007, Portugal
[2]Carestream Health Inc., 1049 Ridge Road West, Rochester, NY, 14615
[3]London Centre for Nanotechnology, UCL, London, WC1H 0AH, United Kingdom

### ABSTRACT

In this work we analyze the transient current in the metal/a-SiN$_x$/a-Si:H/n+/ITO structures under different biasing conditions and temperatures. The dark current decay was measured within an interval of 1 second in the temperature range from 294 to 353K. It was found that when the bias pulse amplitude is kept constant, the transient current strongly depends upon the offset voltage of the bias pulse. This result is in good agreement with device modeling performed using ATLAS. The detailed analysis shows that the transient dark current originates from traps in the i-layer bulk and traps at the semiconductor-insulator interface. Under optimized biasing conditions and elevated temperatures the bulk current component becomes dominant

### INTRODUCTION

The hydrogenated amorphous silicon (a-Si:H) PIN photodiode and the MIS photoelectric converter are two alternative sensing elements used in indirect-conversion flat panel X-ray detectors [1], [2]. The major advantage of the MIS structure over PIN is fact that this device has the same layer sequence as the a-Si:H TFT switch and therefore, they can be fabricated simultaneously resulting in an effective reduction in the lithography mask count [3]. Since the blocking p-layer is replaced by the insulator in the MIS structure, problems related with cross contamination are avoided leading to an increase on the device yield. The main disadvantage of the MIS structure is that the noise level may be high because of the transient dark current. The transient dark current originates from traps at the semiconductor-insulator interface and a-Si:H bulk trap states [4], [5]. The results reported in our previous work show that the noise component associated with the transient dark current can be largely eliminated by adjusting the biasing conditions [6]. In this work we analyze the transient current in a-Si:H-based MIS structures under different biasing conditions and temperatures.

### EXPERIMENTAL PROCEDURE

The samples used in this work are metal(Mo)/a-SiN$_x$ (100 nm)/a-Si:H (1 μm)/n+ (20 nm)/ITO (65 nm) structures on the glass substrate. Semiconductor and dielectric layers were deposited by plasma enhanced chemical vapor deposition (PECVD) at 300°C. Magnetron sputtering was used for the metal- and ITO films. The photosensitive area of the segmented structures is 1×1 mm$^2$. Further details on the device design, fabrication, and characterization under illumination including spectral response measurements can be found elsewhere [6].

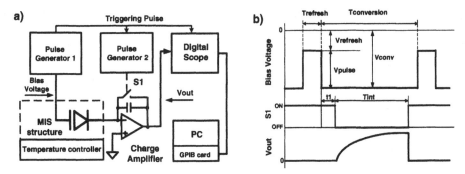

**Figure 1.** (a) Block diagram of the setup and (b) signal waveforms.

To measure the transient currents at elevated temperatures, an experimental setup was built. A block diagram of the measurement system and corresponding test signal waveforms are shown in Fig. 1a and 1b, respectively. The measurement setup includes a controlled temperature chamber, a charge amplifier, two pulse generators, a digital storage oscilloscope, and a personal computer (PC) with GPIB card and dedicated software. The first pulse generator provides bias pulses for the MIS structure and triggering pulses for the digital storage oscilloscope and the second pulse generator. The second pulse generator controls the charge amplifier reset switch. The digital oscilloscope captures the output signal of the charge amplifier during a preset time, and saves acquired waveforms in the computer.

Depending on the applied electrical bias, the MIS-type sensor operates in one of two modes: (1) refresh mode and (2) photoelectric conversion mode [2]. As it is shown in Fig. 1b, $V_{refresh}$ and $V_{conv}$ are the voltages applied to the gate of MIS structure during refresh and conversion periods, respectively. The cathode of the device remains at ground potential as forced by the charge amplifier. The amplitude of the refresh pulse ($V_{pulse}$) is defined as the difference between voltages in the two modes ($V_{pulse}= V_{refresh} - V_{conv}$). In the conversion mode, the charge amplifier integrates the input current from the MIS structure during the time period $T_{int}$. A delay $t_1$ is introduced to avoid detection of the charge induced by the geometrical capacitance of the MIS structure. To reduce the noise level, the output waveforms were captured by a digital oscilloscope in average mode with 64 readings. The time dependence of the input current was obtained by transforming the output waveform $V_{out}(t)$ according to the equation

$$I_{dark} = C_f \cdot \frac{dV_{out}}{dt},$$ (1)

where $C_f$ is the feedback capacitor of the charge amplifier. A feedback capacitor of $C_f=1$nF was used for testing 1x1 mm$^2$ MIS structures.

**RESULTS AND DISCUSSION**

The current transients were measured in the temperature range from 294 to 353K. The period of the biasing pulse was 1s, $T_{referesh}=2$ ms and $t_1=200$ μs. For measurements when $V_{refresh}$ was varied from 0 to -8 V, a 4 V biasing pulse was used. For measurements when $V_{refresh}$ was varied from 0 V to +4 V, the bias pulse was adjusted such that $V_{conv}$ was -4 V.

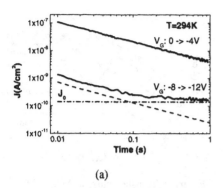 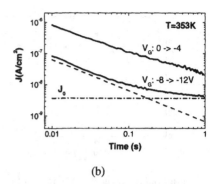

(a)                                                    (b)

**Figure 2.** Transient dark current at the temperatures (a) 293 and (b) 353K. The transients were measured after switching the bias voltage applied to the gate as that indicated in the figures.

Fig. 2a and 2b shows a *log-log* plot of the dark current $J_{dark}$ as a function of time measured at temperatures of 294 and 353K, respectively. The dark current can be defined as the sum of a transient component, which follows a power law time dependency, and a steady-state component

$$J_{dark}(t) = \Delta J(t_0)(t_0/t)^{\beta} + J_0, \tag{2}$$

where $\Delta J(t_0)$ is the transient current at a given time $t_0$, $\beta$ the dimensionless constant, and $J_0$ the steady-state current. The constant $\beta$ ranges from 0.64 to 0.92 depending on the $V_{refresh}$ and temperature as this is shown in Fig. 3.

Fig. 2 also shows the transient (dash line) and steady-state (dash-dot line) current components obtained for the $J_{dark}(t)$ at $V_{refresh} = -8$ V. The values of $J_0$ are 140 pA/cm² and 3.8 nA/cm² at 294 and 353 K, respectively. The steady-state current may be attributed to the thermal generation of the carriers in the i-layer bulk or/and leakage through the dielectric. The observed level of the steady-state current at T=294 K is factor of 2 to 3 higher than that in the state-of-the-art a-Si:H p-i-n diodes. The reason is likely a leakage through dielectric layer. Our silicon nitride has a resistivity of ~10¹⁶ Ohm-cm in electric fields below 2 MV/cm. Assuming that the voltage drop across the dielectric layer is close to the $V_{refresh}=-8V$, the leakage current through the SiNx layer with thickness of 100 nm is estimated to be 80 pA/cm2, i.e. consistent with the measured value. At elevated temperatures the level of the steady-state current is in agreement with expected values for the thermally generated bulk component.

The occupation of the trap levels at the semiconductor-insulator interface and in the i-layer bulk is determined by the bias voltage during the refresh period. When the $V_{refresh}$ changes from 0 to −8 V, the magnitude of the transient current decreases by 2 and 1 orders of magnitude at T=294 and 353K, respectively. Fig. 4 shows the $\Delta J(t_0)$ at $t_0 = 0.1$ s as a function of the $V_{refresh}$

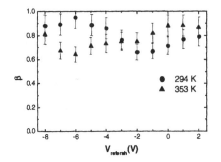

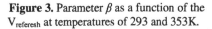

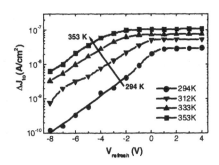

**Figure 3.** Parameter $\beta$ as a function of the $V_{referesh}$ at temperatures of 293 and 353K.

**Figure 4.** Transient current $\Delta J(t_0)$ at $t_0= 0.1$s as a function of the $V_{referesh}$ measured at different temperatures.

for different temperatures. The value of $\Delta J(t_0)$ increases with decreasing voltage applied to the gate and then saturates. The possible reason of such strong dependence of the transient current magnitude on the $V_{refresh}$ is that the transient dark current originates from traps in the i-layer bulk and traps at the semiconductor-insulator interface. The trapped charge at the interface is determined by the $V_{refresh}$.

The device modeling was performed using the Silvaco Atlas Tool, Version 5.12.1.R. The geometric mesh for the MIS structure was created using the command line features in the tool. The device dimensions were selected to be the same as in the tested samples, where the thicknesses of the SiN, i-Si:H and $n^+$ layers are 100, 1000 and 30 nm, respectively. The a-Si:H bulk trap densities were set using a combination of tail and gaussian distributions.

The Silvaco Atlas Tool allows DOS modeling using four bands: two tail bands (a donor-like valence band and an acceptor-like conduction band), and two deep level bands (one acceptor-like and the other donor-like) which are modeled using Gaussian distributions. The tail density parameters are nta, wta, ntd and wtd, while the gaussian parameters are ega, wga, nga, egd, wgd and ngd. These parameters and the simulation equations are well defined in the Silvaco Atlas User Guide. Specific settings follow: mun=15.0 cm$^2$/(V•s), mup=1.0 cm$^2$/(V•s), nc300= 5.0× $10^{20}$ cm$^{-3}$, nv300=$10^{21}$ cm-3, eg300=1.85 eV, nta=5.0× $10^{18}$ cm-3/eV, ntd=5.0 × $10^{18}$ cm-3/eV, wta=0.08 eV, wtd=0.10 eV, nga=1.1× $10^{15}$ cm-3, ngd=1.0 × $10^{15}$ cm-3, ega=0.75 eV, egd=0.85 eV, wga=0.1 eV, and wgd=0.1 eV. Due to simulation noise issues, the band tail distributions at the band edges (nta and ntd) were set to fairly low values, while their characteristic energy levels (wta and wtd) were set fairly high. These settings may affect the absolute accuracy of the simulation but should not significantly impact the behavior trend which is being studied. The density of states distribution is shown in Fig. 5

Fig. 6 shows simulated energy band diagrams of the MIS structure in the refresh mode at $V_{refresh}$ =0V, and in the photoelectric conversion mode at $t_0$ =10 ms and $V_{conv}$=–4V. Since the capacitance of the SiN layer is larger than the capacitance of the a-Si:H i-layer, the applied potential drops mainly across the a-Si:H i-layer.

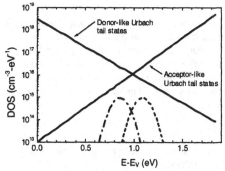

**Figure 5.** Density of states in the a-Si:H i-layer.

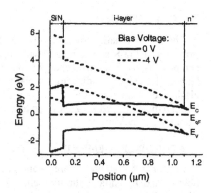

**Figure 6.** Energy band diagrams of the MIS structure at zero bias (solid lines) and in 10 ms after applying -4V to the gate (dashed lines).

The simulation plan first focused on the transient behavior of the bulk traps. Fig. 7 shows simulated bulk transient currents at 294 and 353K. The difference in the magnitudes of the currents at the same amplitude of the bias pulse in the range of milliseconds is lower than that in Fig. 2. Since in a real device there are SiN/a-Si:H interface defects, the next stage of this effort will be to include these in the model and assess their impact on transient behavior.

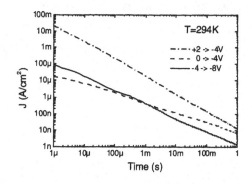
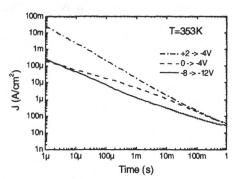

**Figure 7.** Simulated bulk transient current components.

## CONCLUSIONS

The transient dark current of a-Si:H MIS photosensor was measured in the temperature range from 294 to 353 K and compared with simulated data. It is found that the dark current component associated with charge trapping at the insulator-semiconductor interface can be largely eliminated by adjusting the bias voltage during the refresh period. Under optimized biasing conditions, at room temperature, the steady-state current is associated with a leakage through dielectric, while at elevated temperatures the charge carrier generation in i-layer bulk is a dominant mechanism.

### ACKNOWLEDGEMENTS

Authors are grateful to Carestream Health Inc., Rochester, and the Portuguese Foundation of Science and Technology through fellowship BPD20264/2004 and project POCI/CTM/56078/2004 for financial support of this research. Authors would like to thank Silvaco Application Engineers Steve Broadbent (Northeast, USA) and Sung-won Kong (Seoul, S. Korea), for their assistance with modeling and simulation.

### REFERENCES

[1]R. A. Street (Ed.), in Technology and Applications of Amorphous Silicon (Springer, Berlin, 2000) pp.147-221.

[2]M. Watanabe et al. *Proc. SPIE*, **4320**, 103 (2001).

[3]S. Taked et al. Patent US 5591963, Jan. 7, 1997.

[4]J. S. Choi, and G. W. Neudeck, *IEEE Trans. Electron Devices* **39**, 2515 (1992)

[5]G. Kawachi, *J. Appl. Phys.* **80**, 5786 (1996)

[6]N. Safavian, Y. Vygranenko, J. Chang, Kyung Ho Kim, J. Lai, D. Striakhilev, A. Nathan, G. Heiler, T. Tredwell, and M. Fernandes, *Mater. Res. Soc. Symp. Proc.* **989,** A12-06 (2007)

Mater. Res. Soc. Symp. Proc. Vol. 1066 © 2008 Materials Research Society          1066-A18-08

## Physically Based Compact Model for Segmented a-Si:H *n-i-p* Photodiodes

Jeff Hsin Chang[1], Timothy Tredwell[1], Gregory Heiler[1], Yuri Vygranenko[2], Denis Striakhilev[2], Kyung Ho Kim[2], and Arokia Nathan[3]
[1]Carestream Health Inc., 1049 Ridge Road West, Rochester, NY, 14615
[2]University of Waterloo, 200 University Avenue West, Waterloo, M1S 3H3, Canada
[3]London Centre for Nanotechnology, University College, London, WC1H OAH, United Kingdom

## ABSTRACT

Hydrogenated amorphous silicon (a–Si:H) *n–i–p* photodiodes are used as pixel sensor elements in large-area flat-panel detectors for medical imaging diagnostics. An accurate model of the sensor plays an imperative role in determining the performance of the detector systems. This work presents a compact model for segmented a–Si:H *n–i–p* photodiodes that is suitable for circuit-level simulation. The underlining equations of the model are based on device physics where the parameters are extracted from pertinent measurement results of previously fabricated a–Si:H *n–i–p* photodiodes. Furthermore, the implemented model allows photoresponse simulation with the addition of an external current source. Results of the simulation demonstrated excellent matching with measurement data for different photodiode sizes at various temperatures. The model was implemented in Verilog-A and simulated under a Cadence Virtuoso design environment using device geometry and extracted parameters as inputs. The model formulation and parameter extraction process are presented. Results of the simulation are in good agreements with measurement data.

## INTRODUCTION

Segmented a–Si:H *n–i–p* photodiodes serve as excellent sensor elements for large-area flat-panel detectors [1]. These photodiodes are typically arranged into two-dimensional arrays and coupled to a scintillation layer to form the suitable imager for medical imaging applications. While designing such detector systems, a fast and accurate photodiode model can assist engineers in assessing and addressing design issues prior to production. Studies of a–Si:H *p–i–n* and *n–i–p* structures have prompted the formulation of numerical and analytical models [2]–[6]. The models generally require the simultaneous solution of electron and hole continuity equations, and Poisson's equation. These models provide insight for understanding the photodiode device physics with their solutions based on various assumptions and simplifications of material and electrical parameters. To describe the carrier transport, however, the exact details of the localized state distributions must be considered. This information is often cumbersome to obtain. Furthermore, numerical models are computer-intensive; the duration of the simulation for a single photodiode can last more than a few hours with a multi-core computing cluster. For pixel- and array-level simulations, a compact model that is sufficiently accurate and consumes less computing resources is required.

This work describes the formulation of a compact model suitable for pixel- and array-level simulations. The underlining equations of the model are based on the physics of the device while parameters are extracted from simple measurements of fabricated photodiodes. This allows rapid calibration of the model. The a–Si:H *n–i–p* photodiodes used for this work are fabricated using a standard plasma-enhanced chemical vapor deposition (PECVD) technique. Details of the sample preparation and photodiode structure are presented elsewhere [7].

## MODEL IMPLEMENTATION

A simplified equivalent circuit for the photodiode model is shown in Figure 1. $C_D$ is the junction capacitance of the photodiode; $R_S$ is the lumped contact resistance; $I_D$ provides the quasi-static behavior of the photodiode, while $I_D(t)$ provides the dynamic behavior of the photodiode; and $I_{ph}$ is the current component generated due to photodiode illumination. Two additional terminals are introduced in the model to provide a means of simulating photogenerated current. An external current source with magnitude $q\varphi_{ph}$ can be connected to the terminals where $q$ and $\varphi_{ph}$ are the elementary charge and impinging photon flux, respectively. Given that the properties of the current source can be programmed easily in the simulator, this method provides additional flexibility into describing the behavior of the light source illuminating the photodiode.

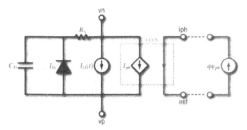

**Figure 1:** A simplified schematic of the equivalent circuit for the a–Si:H $n$–$i$–$p$ photodiode model.

## MODEL FORMULATION AND PARAMETERS EXTRACTION

To address each of the model parameter quantities given in Figure 1(a), four sets of measurements must be performed on fabricated photodiode samples. First, the quasi-static dark current-voltage ($I_D$–$V$) characteristics are measured for different photodiode sizes at different operating temperatures; the measured dark current values ($I_D$) are normalized to current density ($J_D$) because $I_D$ is proportional to the photodiode area ($A_D$). This assumption is validated for high-performance a–Si:H $n$–$i$–$p$ photodiodes where the central current component dominates [7]. At low forward bias ($0.05 < V < 0.60$ V), the photodiode current is dominated by the diffusion and subsequent recombination of excess carriers injected into the bulk $i$–layer from the $n^+$ and $p^+$ contacts. The photodiode current density at low forward bias increases exponentially with voltage and can be described by the equation

$$J_D(V,T) = J_{ol}(T)\left[\exp\left(q\frac{V - J_D R_S}{n_l(T)kT}\right) - 1\right] \qquad (1)$$

where $J_{ol}(T)$ and $n_l(T)$ are the temperature-dependent photodiode saturation current density and ideality factor, respectively; subscript $l$ indicates low bias regime. $J_{ol}$ and $n_l$ can be extracted from semi-logarithmic $J_D$–$V$ plots of the photodiodes via linear regression. The temperature dependence of $J_{ol}(T)$ can be extracted from the Arrhenius plots of the measurement data.

$$J_{ol}(T) = J_{ool} \exp\left( -\frac{E_{afl}}{kT} \right) \tag{2}$$

$J_{ool}$ is the pre-exponential factor and $E_{afl}$ is the forward bias activation energy. The details of their extraction are presented elsewhere [8]. $n_l(T)$ remains rather constant between 20–80°C.

The reduction of the bias voltage across the junction due to $R_S$ cannot be ignored at higher bias regime where the voltage drop $R_S$ becomes significant. At bias voltages larger than 0.8V, the photodiode current follows a power-law relationship.

$$J_D(V,T) = J_{oh}(T) V^{n_h(T)} \tag{3}$$

Parameters $J_{oh}$ and $n_h$ can be extracted via regression from a double logarithmic plot. The temperature dependence of these two parameters can be obtained using a similar method as those used for the low forward bias region.

$$J_{oh}(T) = J_{ooh} \exp\left( -\frac{E_{afh}}{kT} \right) \tag{4}$$

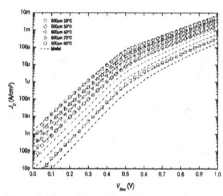

**Figure 2:** Forward $J_D$–$V$ characteristics at various temperatures. Measurement results are shown as hollow symbols and modeling results are shown as lines.

**Figure 3:** Photodiode capacitance with various substrate materials. Measurement results are shown as hollow symbols and modeling results are shown as a line.

Even though Verilog-A implementation allows abrupt concatenation of different operation regimes, it is often a good practice to ensure that no singularity occurs at the boundaries of these regions. The two forward-bias operation regions are combined using a weighting function and the total forward current is described by

$$J_D = \frac{J_{ool} \exp\left( \dfrac{qV}{n_o kT} - \dfrac{qE_{afl}}{kT} \right)}{1 + \exp\left[ w_{pl}\left( V - w_{ql} \right) \right]} + J_{ooh} \exp\left( -\frac{qE_{afh}}{kT} \right) V^{n_{oh}} \frac{\exp\left[ w_{ph}\left( V - w_{qh} \right) \right]}{1 + \exp\left[ w_{ph}\left( V - w_{qh} \right) \right]} \tag{5}$$

where $w_p$ and $w_q$ are two weighting function parameters influenced by the transition voltage between the two bias regions. Following the convention mentioned previously, subscript $l$ and $h$ denote low and high forward bias regions, respectively. A comparison between the measurement data and model result is shown in Figure 2.

The fundamental lower limit of the reverse dark current in a–Si:H $n$–$i$–$p$ photodiodes is determined by the thermally induced carrier emission from deep levels in the bulk a–Si:H layer [9]. Additional sources that contribute to the increase in leakage current include emissions of carriers from defect states at the $i$–$p$ and $n$–$i$ interfaces, contact injection, edge leakage, and macro-structural shunt paths. Dominating mechanisms can be categorized into central components that scale with $A_D$ and a peripheral current, which scales with the photodiode perimeter ($P_D$). For a square photodiode with side length $L$, $I_D$ can be described by [8]

$$I_D = J_B L^2 + J_P L + I_o \qquad (6)$$

where $J_B$ is the central current density, $J_P$ is the peripheral current density, and $I_o$ accounts for the current contributions from the photodiode corners and instrumentation leakage.

To extract $J_B$ and $J_P$, a full set of $I_D$–$V$ measurements must be made for a range of photodiode sizes (the range should include large-area photodiodes where a central component dominates and a small area where peripheral contribution is significant). The $I_D/L$ vs. $L$ plot for these devices yields $J_B$ as the slope of the linear regression and $J_P$ as the intercept. The $J_B$ component increases with larger reverse bias voltage due to field-assisted excitation either through thermal assisted tunneling (TAT) or Poole-Frenkel (PF) emission. Consequently, $J_B$ can be modeled as

$$J_B(V,T) = J_{Bo}(T)\exp\left(-\beta_t \frac{V}{t_i}\right)\exp\left(-\beta_{pf}\sqrt{\frac{V}{t_i}}\right) \qquad (7)$$

where $t_i$ is the $i$–layer thickness of the a–Si:H $n$–$i$–$p$ photodiode. $\beta_t$ and $\beta_{pf}$ are the exponential factors for TAT and PF, respectively. Parameters $J_{Bo}(T)$, $\beta_t$, and $\beta_{pf}$ can be extracted from the $\ln|J_B|$ vs. $(V/t_i)^{1/2}$ plot, via regression. The temperature dependence of $J_{Bo}(T)$ can be extracted from the Arrhenius plots of the measurement data.

$$J_{Bo}(T) = J_{Boo}\exp\left(-\frac{E_{ab}}{kT}\right) \qquad (8)$$

$J_{Boo}$ is the pre-exponential constant, and $E_{ab}$ is the reverse bias activation energy of the central current component. The details of their extraction are presented elsewhere [8]. Given that the quasi-Fermi level is located near the midgap, the value of $E_{ab}$ is approximately half the mobility gap of a–Si:H.

The peripheral current component is more challenging to model due to its strong dependence on processing conditions. For state-of-the-art photodiodes, $J_P$ can be modeled as a shunt resistance $R_P$. The contribution of $J_P$ toward $I_D$ diminishes with an increase in temperature and can be modeled by

$$J_P(V,T) = \frac{V}{R_o\left[1+\alpha(T-T_o)\right]} \qquad (9)$$

where $R_o$ is $R_P$ at temperature $T_o$ and $\alpha$ is the temperature coefficient of $R_P$. The reverse-leakage current measurement results and corresponding modeling results are presented elsewhere [8].

To model the junction capacitance of a–Si:H $n$–$i$–$p$ photodiodes, a set of $C$–$V$ measurements is performed. In reverse bias, the photodiode capacitance stays relatively constant beyond –1 V, which indicates full depletion across the $i$–layer. The depletion width increases with higher reverse-bias voltage; therefore, there is minor bias voltage dependence on the capacitance of the photodiode. The normalized photodiode capacitance ($C_D$) can be approximated by

$$C_D(V) = \frac{\varepsilon_o \varepsilon_r}{W_d} + \gamma_c V \qquad (10)$$

where $\varepsilon_o$ and $\varepsilon_r$ are the permittivity of free space and intrinsic a–Si:H, respectively. $W_d$ is the total depletion width including $t_i$ and the depletion width extending into the $p^+$ and $n^+$ contacts. $\gamma_C$ is the voltage dependence parameter for $C_D$, and its value can be extracted via linear regression from the C–V plot. Figure 3 shows C–V measurement results of a–Si:H n–i–p photodiodes deposited on various substrate materials. $C_D$ on a glass substrate matches well with the estimated value from device geometry.

A current-controlled current source (CCCS) is used to model the photogenerated current for the photodiode. The magnitude of the current source is

$$I_{ph}(\lambda) = q\eta_{qe}(\lambda)\varphi_{ph}A_D \qquad (11)$$

where $\eta_{qe}$ is the external quantum efficiency of the photodiode, $\varphi_{ph}$ is the impinging photon flux, and the magnitude of the connected current source is $q\varphi_{ph}$.

The current transient phenomenon due to trapping and releasing of carriers occurs with a shift in Fermi level ($E_F$) due to perturbations of the operating environment such as a change in the bias voltage. In the short time range (<100 msec), $I_D(t)$ is most sensitive to the position of $E_F$ while in the longer time range, $I_D(t)$ is most sensitive to the density of state (DOS) distribution. A model of the transient current for a reverse-biased photodiode is important because the phenomenon can impose ghost images for consecutive readout frames. For a–Si:H n–i–p photodiodes biased as low reverse voltages ($0 < V < -3$), the current decay can be approximated by

$$I_D(t, I_{ill}, \tau_{ill}) = I_{ill}\left(\frac{t}{\tau_{ill}} + 1\right)^{\beta} \qquad t > 0 \qquad (12)$$

where $I_{ill}$ is the highest current level achieved during illumination, $\tau_{ill}$ is the illumination duration, and $\beta$ is the power-law current decay parameter, and its value can be extracted from a double logarithmic plot of measured current values for $t \gg \tau_{ill}$. Note that for large $t$, $I_D(t) \approx 0$ and the total photodiode current approaches its steady-state dark current level. Figure 4 shows the transient current decay after various illumination intensities. Figure 4(a) shows the dependence of $I_{ill}$ and $\tau_{ill}$ for different illumination intensities; both parameters follow a stretched exponential relationship. An example of the simulation result is also shown in Figure 4(b). The photodiode is subjected to a periodic light pulse via a CCCS. The current of the photodiode decays after each pulse.

## CONCLUSIONS

The formulation of a compact model for segmented a–Si:H n–i–p photodiodes is presented. The underlining model equations are based on device physics. The presented model does not require computer-intensive solutions and is suitable for pixel- and array-level simulation. Features of the model include photodiode quasi-static $I_D$–V characteristics and its temperature dependence, photodiode capacitance, photoresponse, and current transient. The simple parameter-extraction method allows the model to be calibrated easily.

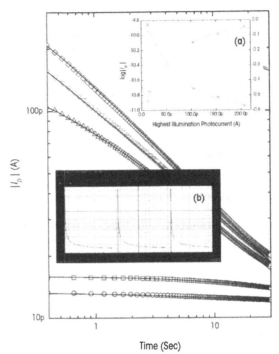

**Figure 4:** Transient current decay with various illumination intensities. Measurement results are shown as hollow symbols while modeling results are shown as lines.

## REFERENCES

1. R. A. Street, in *Technology and Application of Amorphous Silicon*, edited by R. A. Street, (Springer-Verlag, Heidelberg, 2000), p. 204.
2. I. Chen and S. Lee, *J. Appl. Phys.* **53**, 1045 (1982).
3. G. A. Swartz, *J. Appl. Phys.* **53**, 712 (1982).
4. M. Hack and M. Shur, *J. Appl. Phys.* **54**, 5858 (1983).
5. K. Misiakos and F. A. Lindholm, *J. Appl. Phys.* **64**, 383 (1988).
6. S. R. Dhariwal and S. Rajvanshi, *Sol. Energy Mater. Sol. Cells* **79**, 199–233 (2003).
7. Y. Vygranenko, J. H. Chang, A. Nathan, *Mater. Res. Soc. Symp. Proc.* **862**, A9.4 (2005).
8. J. H. Chang, T. C. Chuang, Y. Vygranenko, K. H. Kim, D. Striakhilev, A. Nathan, G. Heiler and T. Tredwell, *Mater. Res. Soc. Symp. Proc.* **989**, A19.4 (2007).
9. R. A. Street, *Appl. Phys. Lett.* **57**, 1334 (1990)

Mater. Res. Soc. Symp. Proc. Vol. 1066 © 2008 Materials Research Society    1066-A18-10

## Luminescent Colloidal Silicon Nanocrystals Prepared by Nanoseconds Laser Fragmentation and Laser Ablation in Water

Vladimir Svrcek[1], Davide Mariotti[2], Richard Hailstone[3], Hiroyuki Fujiwara[1], and Michio Kondo[1]

[1]Research Center for Photovoltaics, National Institute of Advanced Industrial Science and Technology (AIST), Central 2, Umezono 1-1-1, Tsukuba, 305-8568, Japan

[2]Department of Microelectronic Engineering, Kate Gleason College of Engineering Rochester Institute of Technology, 82 Lomb Memorial Drive, Rochester, NY 14623

[3]Department of Microelectronic Engineering, Chester F. Carlson Center for Imaging Science, Rochester Institute of Technology, 54 Lomb Memorial Drive, Rochester, NY 14623

## ABSTRACT

The surface states of silicon nanocrystals (Si-ncs) considerably affect quantum confinement effects and may determinate final nanocrystals properties. Colloidally dispersed Si-ncs offer larger freedom for surface modification compared to common plasma enhanced chemical vapor deposition or epitaxial synthesis in a solid matrix. The Si-ncs fabrication and elaboration in water by pulsed laser processing is an attractive alternative for controlling and engineering of nanocrystal surface by environmentally compatible way. We report on the possibility of direct silicon surface ablation and Si-ncs fabrication by nanosecond pulsed laser fragmentation of electrochemically etched Si micrograins and by laser ablation of crystalline silicon target immersed in de-ionized water. Two nanosecond pulsed lasers (Nd:YAG, and excimer KrF) are successfully employed to assure fragmentation and ablation in order to produce silicon nanoparticles . Contrary to the fragmentation process, which is more efficient under Nd:YAG irradiation, the laser ablation by both lasers led to the fabrication of fine and room temperature photoluminescent Si-ncs. The processing that has natural compatibility with the environment and advanced state of fabrication technologies may imply new possibilities and applications.

## INTRODUCTION

Silicon nanocrystals (Si-ncs) with grain size of less than 5 nm exhibit quantum confinement effects and visible photoluminescence (PL) at room temperature and are recognized as one of the potential materials in optoelectronic, bio-imaging, and photovoltaic applications [1, 2]. It is generally accepted that visible PL in Si-ncs originates from surface related recombination that occurs with quantum confinement effect [3]. Fabrication and elaboration of the Si-ncs in liquids may bring some new possibilities for modifying both surface states and quantum confinement [4]. In addition, the host liquid could modify the electron-hole interaction and can be used for optimization of carrier density population under excitation conditions. It has been shown recently that nanosecond pulsed laser ablation in aqueous media is suitable for fabrication of luminescent water soluble Si-ncs [5, 6]. Water offers advantageous fabrication conditions that are both environmentally friendly and with unique surface chemistry. The oxide shell that is formed in the fabrication of Si-ncs in water, provides a natural and stable form of surface passivation [5]. The hydrophilic oxide surface may serve as a dielectric shell for further surface functionalization by specific molecular species at the nanoscale level.

Here, we report on the formation of Si-ncs and surface modification directly in de-ionized water by lasers processing. For the synthesis of Si-ncs two laser processes at room temperature in ambient atmosphere are applied and investigated. The first process involves the fragmentation by nanosecond (ns) laser pulses in water with Si micrograins prepared by electrochemical etching. In the second process, Si-ncs were produced by ns laser ablation of a Si target immersed in de-ionized water. Two nanosecond pulsed Nd:YAG lasers and an excimer KrF laser are employed to ensure successful fragmentation and ablation and to produce silicon nanoparticles that are dispersed in water. We compare how the preparation with the two nanosecond lasers in aqueous solution at ambient pressure and temperature affects Si-ncs structure, morphology and most importantly luminescence properties.

## EXPERIMENT

The micrograins used for the first process, were prepared by electrochemical etching. For this purpose, silicon wafers (p-type boron doped, <100>) were etched 2 hours at constant current 1.6 mA/cm$^2$ in HF:ethanol electrolyte (1:4) and subsequently mechanically pulverized [4, 6]. The micrograins were harvested by sedimentations in ethanol. The 0.01 wt% Si micrograins in aqueous solution was then prepared. In order to obtain homogenous dispersion of the hydrophobic Si micrograins [6] in de-ionized water and to ensure micrograins fragmentation, few drops of ethanol have been used to wet the micrograins surface prior to the introduction of water. The colloidal solutions were sonicated for 10 min and 5 ml of the solution was used for the fragmentation process. Two different nanosecond pulsed lasers (third harmonic of Nd:YAG, Spectra Physics LAB-150-30, 355 nm, 30 Hz, 8 ns and excimer KrF, 245 nm 20 Hz, 10 ns) were employed and compared. The laser irradiation was set at ~ 6 mJ/pulse fluence for 2 hours at room temperature and ambient pressure. The laser beam was focused into 1 mm diameter spot on the liquid surface by a lens. For all cases, the glass containers were closed and rotated during the irradiation.

For the second process, a crystalline silicon wafer with the same parameters as used for micrograins fabrication, was glued at the bottom of the glass container and used as a target for ablation. Nd:YAG and excimer KrF lasers were applied to irradiate onto the target immersed in 10 ml of de-ionized water at room temperature and ambient atmosphere for 2 hours. The laser fluence was set at 5.3 mJ/pulse. To reveal the PL of the freshly prepared Si-ncs, the solution was kept for several weeks in ambient conditions. A small droplet of the obtained colloidal solutions (prepared by fragmentation or laser ablation) was then deposited onto a copper grid for high-resolution transmission electron microscopy (HR-TEM) and scanning electron microscope (SEM) observations. HR-TEM was performed using a microscope with a 200 kV acceleration voltage. Transmission electron diffraction and Raman spectroscopy were employed to perform more localized analyses of Si nanoparticle structure. The PL measurements were performed at room temperature and ambient atmosphere using a fluorophotometer (Shimadzu, RF–5300PC) with excitation by monochromatic light at 300 nm from a Xe lamp.

## LASER FRAGMENTATION

Silicon surface itself as well as the silicon micrigrains are hydrophobic. Si micrograins can be easily dispersed in almost any organic-based solution. On the other hand, direct dissolution in water is inefficient and micrograins accumulate at the water surface [6].

Fragmentation by pulsed laser requires a homogenous dispersion of the Si micrograin in the liquid media. In order to overcome the insolubility of the Si micrograins in water, a small amount of ethanol was used to wet the micrograin surface prior to the introduction of water. Image 1(a) shows a photo of 0.01 % Si grains wetted with ethanol and homogenously dispersed in de-ionized water. After applying nanosecond pulsed laser irradiation, fragmentation of the micrograins is induced. For both lasers used, it is observed that at low laser fluence (<1.1 mJ/pulse), the solution color changed to dark brown. A prolonged laser irradiation did not induce any additional color change. On the contrary, at higher laser fluence, the irradiation time affected the color of the solutions. In this case, after prolonged irradiation, the solutions became more transparent. Figure 1(b) and Figure 1(c) show solutions after laser ablation for 2 h at fluence of 5.9 mJ/pulse by Nd:YAG laser and excimer KrF laser, respectively. The colloid color changed from yellow to mostly transparent after pulsed laser processing. The results show that the fragmentation by Nd:YAG is more efficient and by naked eye the solution became more transparent (Figure 1(b)). This was further confirmed by absorbance measurements. Figure 1(d) shows corresponding absorbance of the solutions before and after laser irradiations. It can be seen that the absorbance of the solution fragmented by excimer KrF laser is higher compared to Nd:YAG laser. There could be several reasons that justify the different behaviors for the different lasers. One important observation is that the micrograins have quite significant absorption cross-section around 400 nm, which is closer to the 355 nm wavelength of the Nd:YAG if compared to the excimer KrF laser (249 nm). Therefore the Nd:YAG laser is expected to show better fragmentation. TEM and SEM images also show that the fragments are smaller after Nd:YAG irradiation with an average diameter of less than 60 nm. In the case of excimer KrF laser, fragments diameters exceed 100 nm. In both cases the silicon particles maintain the crystalline diamond-like structure. Unfortunately, the particle sizes are quite large and exceed the quantum confinement limit for silicon (<10 nm). Therefore the visible PL at room temperature in such particles is not observed.

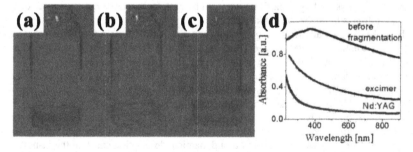

**Figure 1.** Photos of 0.01 % ethanol wetted Si micro-grains in water (a) before and (b, c) after laser irradiation for 2 hours at a fluence of ~ 6 mJ/pulse. Figure (b) and (c) display colloidal suspension after pulsed nanosecond Nd:Yag and excimer KrF laser irradiations, respectively. (d) Corresponding absorbance of aqueous colloidal solutions before and after laser fragmentation with Nd:Yag (red line) and excimer KrF laser (blue line) are shown.

Several simultaneous processes are responsible for the fragmentation of Si micrograins in both laser irradiation conditions. Similarities can be found with reported results of the fragmentation in transparent polymer [6, 7, 8]. For both laser irradiations, a bright spot near the surface level is observed. Optical breakdown of water and explosion sounds can also be heard. These local explosions are due to the collapse of vaporized cavitation bubbles, and of the micrograins fragments that have accidentally diffused under the laser beam. When the Si micrograin absorbs photons from the laser beam a strong electric field is generated causing an electron avalanche and consequent water breakdown. [6, 7]. In the same process, shock waves are also generated. Shockwaves can be accelerated at large distances with the explosion on the silicon micrograins. Despite the large distances traveled by shock waves, they still have sufficient energy to fragment particles [6]. These processes occur only for Si grains homogeneously dispersed in water. It has been verified for both laser conditions that fragmentation did not take place when the grains were not wetted by ethanol and accumulated on the top surface of the aqueous solution[6].

## LASER ABLATION

It has been reported elsewhere that Si-ncs surface oxidation in air improves and stabilizes the luminescent properties of nanocrystals [3]. Water is a natural oxidizing agent and surface modification of silicon directly in water might be an effective way to prepare stable luminescent Si-ncs. It has been reported that by nanosecond pulsed Nd:YAG laser ablation of Si target immersed in water, spherical particles containing Si-ncs can be formed during aging process in deionized water [5, 9]. Similar morphological changes by aging in de-ionized water of Si-ncs are observed for Si-ncs fabricated by excimer KrF laser as well. Irregular fragments progressively agglomerate and stabilize into separate spheres after few-weeks of aging. Figure 2(a) shows typical TEM image spherical silicon particles after aging in suspension for two weeks. The Si-ncs were prepared by laser ablation of silicon wafer with laser fluence at 5.3 mJ/pulse for 2 hours. We obtained spherical particles with the average size of 40 nm. Further prolonged aging leads to the formation of spheres with the larger diameters [9]. Detailed Raman spectra and HR-TEM structural analyses of aged particles prepared by both lasers confirm the presence of Si-ncs with the crystalline silicon diamond-like structure.

Laser heating, adiabatic cooling and final expansion are the main formation processes of Si-ncs by laser ablation in water. A theoretical description of Si-ncs formation was proposed to describe formation of blue luminescent crystalline Si-ncs stabilized in oxidizing liquid by nanosecond pulsed laser ablation [5, 10]. Similar description can be used to interpret the results obtained in this work. Irradiation of nanosecond laser pulses with photon energy (245 nm: ~ 5.06 eV, 355 nm: ~ 3.5 eV) over the band gap of silicon ($E_{gSi}$=1.1 eV) leads to a rapid heat generation. Immediately after the absorption of the laser light, a dense cloud of Si atoms spread from the laser forming a typical plasma plume. The confinement of the laser-generated plasma in water significantly enhances Gibbs energy as a dynamical description entity of the Si-ncs formation. In fact, the liberated Gibbs energy of the Si-ncs is kinetic energy subsequently dissipated during the Si-ncs ejection trajectory within the confined laser-induced plume.[8]. In first approximation, the Gibbs energy of Si-ncs formation inside of the water-confined plasma plume as a function of the pressure can be written as follow [10, 11]

$$G(P) = \frac{kT}{V}\ln\left(\frac{P(I)}{P_0}\right) \qquad (1)$$

where $k$ is the Boltzman constant, and $T$ is the plume temperature. The plasma pressure $P(I)$ in our case is the maximum pressure generated by the lasers in the water. Embryotic Si particles are formed together with ejected Si-nc particles. Si atoms rapidly aggregate onto the embryotic particles within the plume until the Si atom density in the surrounding of the forming Si-ncs is drastically reduced. [11]. When the Si-ncs escape from the plasma plume into the water, the growth is suppressed by stacking of silicon-based complexes on the surface. These complexes not only suppress the Si-nc growth but also passivate and stabilize the surface.

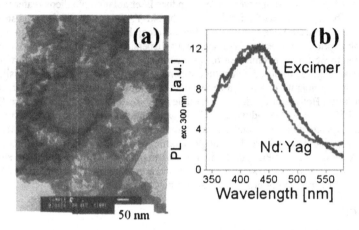

**Figure 2.** (a) Transmission electron microscopy (TEM) image of silicon nanocrystals (Si-ncs) prepared by nanosecond excimer KrF laser ablation of silicon wafer in de-ionized water with laser fluence 5.3 mJ/pulse for 2 hours. (b) Corresponding photoluminescence (PL) spectra from Si-ncs dispersed in water. Blue line represents the PL data of Si-ncs prepared by excimer KrF laser and aged in water for two weeks. The PL spectra of Si-ncs prepared by Nd:YAG laser aged for 4 months are shown for comparisons.

In both cases of laser irradiation, it is observed that as prepared Si-ncs by nanosecond laser ablation dispersed in water showed no visible PL at room temperature. Aging treatment in water for several weeks is necessary to reveal the PL [5]. Corresponding photoluminescence (PL) spectra from Si-ncs dispersed in water prepared by excimer KrF laser and Nd:YAG laser are compared in Figure 2 (b). Rather similar PL spectra with maximum located at ~ 400 nm are recorded after aging. The PL intensity of Si-ncs prepared by excimer KrF laser and aged for two weeks (blue line) is similar to the PL curve obtained for Si-ncs aged for about 4 months and prepared by Nd:YAG laser irradiation (red line). In addition, the KrF excimer laser ablation with shallower absorption produces Si-ncs with red shifted PL peak with maximum centered at 420 nm. The monoxide surface states of Si-ncs together with quantum confinement size effects are responsible for observed photoluminescence at room temperature [3, 5]. For Nd:YAG laser at least two weeks of aging in de-ionized water is necessary to observe room temperature visible blue PL . Our results indicate that modification and oxide passivation of Si-ncs surface states and defects in oxide layer can effectively shorten by excimer KrF laser ablation. Smaller aggregates of silicon nanocrystals are formed with excimer KrF laser ablation. Then, more Si-ncs are exposed to the water and oxidation may take place during aging more efficiently. This leads to an improvement of the exciton localization and shorter time required for producing visible PL.

## CONCLUSIONS

Nanosecond laser processing by excimer KrF and Nd-YAG laser was investigated as a cost-effective alternative to modify silicon nanoparticle surface and to synthesize silicon nanocrystals in de-ionized water. Two different lasers have been applied to induce homogeneous fragmentation of micrograins in water and/or to induce laser ablation of silicon wafer immersed in water. It has been shown that the fragmentation by Nd-YAG is more efficient and smaller particles could be produced due to the irradiation wavelength being a closer match to the micrograin absorption cross section. Homogeneous dispersion of Si-ncs is achieved after laser processing due to wetting phenomena of the new-formed particles. However, the particles size exceeds the size required to observe quantum confinement effects and therefore no visible PL is observed at room temperature. This can be overcomed by direct ablation of silicon target immersed in water. Both laser conditions were able to fabricate blue-luminescent ultra-fine Si-ncs. The aged Si-ncs in de-ionized water exhibit room temperature PL peaked at 400 nm that can be attributed to quantum confinement effects and oxide-based surface states. During the aging process rather similar morphological and agglomeration phenomena of the Si-ncs into spherical particles were observed. The PL maximum of Si-ncs prepared by excimer KrF is red shifted by 20 nm. It is believed that environmentally friendly and non-toxic silicon nanoparticle fabrication and surface modification might provide a basis for new applications.

## REFERENCES

1. L. T. Canham, Appl. Phys. Lett **57**, 1046 (1990).
2. V. Lehmann and U. Gosele, Appl. Phys. Lett. **58**, 856 (1991).
3. M. V. Wolkin, J. Jorne, P. M. Fauchet, G. Allan, and C. Delerue, Phys. Rev. Lett. **82**, 197 (1999).
4. V. Švrcek, A. Slaoui, and J.-C. Muller, J. Appl. Phys. **95**, 3158 (2004).
5. V. Švrcek, T. Sasaki, Y. Shimizu, and N. Koshizaki, Appl. Phys. Lett **89**, 213113 (2006).
6. V. Švrcek, T. Sasaki, Y. Shimizu, and N. Koshizaki, Chem. Phys. Lett. **429**, 483 (2006).
7. F. Mafune and T. Kondow, Chem. Phys. Lett. **343**, 383 (2004).
8. Y. B. Zeldovich and Y. P. Raizer, *Physics of Shock Waves and High-Temperature Hydrodynamic Phenomena* (Dover Publication, Inc,, New York, 2001).
9. V. Švrcek, T. Sasaki, Y. Shimizu, and N. Koshizaki, *Journal of Laser Micro/Nanoengineering* **2**, 15 (2007).
10. A. A. Oraevsky, V. S. Letoshkov, and R. O. Esenafiev, *Proceedings of the Workshop"laser ablation: Mechanism and Applications", Pulsed laser ablation of biotissue. review of ablation mechanisms* (Springer, Berlin, 1991).
11. V. Švrcek, T. Sasaki, Y. Shimizu and N. Koshizaki, J. Appl. Phys., **103**, 023101 (2008).

Mater. Res. Soc. Symp. Proc. Vol. 1066 © 2008 Materials Research Society          1066-A18-11

# Optimized O/Si Composition Ratio for Enhancing Si Nanocrystal Based Luminescence in Si-rich SiO$_x$ Grown by PECVD with Argon Diluted SiH$_4$

Chung-Hsiang Chang[1], Chin-Hua Hsieh[2], Li-Jen Chou[2], and Gong-Ru Lin[1]

[1]Graduate Institute of Photonics and Optoelectronics and Department of Electrical Engineering, National Taiwan University, No. 1, Roosevelt Rd. Sec. 4, Taipei, 10617, Taiwan
[2]Department of Materials Science and Engineering, National Tsing Hua University, No. 101, Section 2, Kuang Fu Rd., Hsinchu, 300, Taiwan

## ABSTRACT

Effect of O/Si composition ratio on near-infrared photoluminescence (PL) of PECVD grown Si-rich SiO$_X$ after 1100°C annealing are analyzed by Rutherford backscattering (RBS) and Fourier-transformed infrared spectroscopy (FTIR) to show nonlinear relationship with strongest PL at 760 nm at optimized O/Si = 1.24, total Si concentration of 44.6 atom.%, and N$_2$O/SiH$_4$ fluence ratio of 4.5. A nearly Gaussian function of the normalized PL intensity vs. O/Si composition ratio has been observed due to the significant variation on the Si nanocrystals size with the density of the excessive Si atoms.

## INTRODUCTION

Versatile technologies have been proposed for synthesizing Si-rich oxide (SiO$_X$) or nitride (Si$_x$N$_y$) with buried Si nanocrystals (nc-Si), such as electron-beam evaporation, rf-magnetron sputtering, Si-ion implantation, and plasma-enhanced chemical-vapor deposition (PECVD) [1-4]. The nc-Si related room-temperature PL obtained from electrochemically etched porous Si, PECVD-grown Si-rich SiO$_2$, and Si-ion-implanted SiO$_2$ (SiO$_2$:Si$^+$) has stimulated comprehensive investigations on versatile nc-Si based light-emitting devices. The most intriguing synthesis is the PECVD deposition associated with subsequent heat treatment, since it enables the easy deposition of a Si-rich SiOx film with a sufficiently high density of excess Si atoms by controlling the fluence of reactant gases. As SiO is thermodynamically less stable than the Si and SiO$_2$ phases, thermal annealing treatment leads to the formation of the silicon particles, as shown in a previous study [5]. Besides, Surface passivation techniques, such as oxidation and hydrogenation [5-7], have been applied to decrease defects and increase the PL efficiency. Cheylan and Elliman [7] studied the effect of hydrogen on the photoluminescence of Si nanocrystals embedded after the sample was exposed in forming gas (95% N$_2$ + 5% H$_2$, 500°C, 1 h) because of the hydrogen passivation of non-radiative defects in nanocrystals. Many researchers studied the reaction between Si–SiO$_2$ interface with atomic hydrogen or molecular hydrogen [8-9]. However, there are few detailed studies on the correlation between the N$_2$O/SiH$_4$ fluence ratio and the O/Si composition ratio for optimizing the nc-Si precipitation. In this work, the Rutherford backscattering (RBS), high-resolution transmission electron microscopy (HRTEM), Fourier transform infrared (FTIR), X-ray Photoelectron spectroscopy (XPS) and near-infrared PL are employed to study the effects of the N$_2$O/SiH$_4$ fluence ratio on the composition ratio of the PECVD-grown and thermally annealed Si-rich SiO$_X$ film with buried nc-Si.

## EXPERIMENT

Si-rich $SiO_X$ films were deposited on (100)-oriented p-type Si substrates with resistivities of 1-5 $\Omega \cdot$ cm using a conventional high-density PECVD system at pressure and forward RF power of 67 Pa and 30 W, respectively. The samples were prepared at constant substrate temperature of 350°C and different gas mixtures, in which several recipes of gas mixtures with two constant $N_2O$ fluences of 50 and 100 sccm, and with various ratios on the $N_2O/SiH_4$. The Si-rich $SiO_X$ samples were encapsulated annealing in a quartz furnace with $N_2$ atmosphere at 1100°C from 15 to 120 min to induce precipitation of nc-Si. Afterwards, the optimized growth condition was employed to grow the Si-rich $SiO_X$ sample at different deposition time. The analysis of the HRTEM image showed clearly that the amorphous $SiO_X$ contained a high density of small clusters uniformly distributed in the matrix which was identified as Si nanocrystals. On the other hand, the statistical analysis of the crystal size distribution obtained from the bright-field cross-sectional HRTEM pictures showed a distribution of crystal size ranging from 3 to 6 nm with a mean radius of 4.5 nm. As shown in Figs. 1 and 2, the observed nc-Si size correlated well the theoretical value of $\lambda=1.24/(1.12+3.73/d^{1.39})$ as reported by Delerue et al [10]. For HRTEM images, the higher contrast means that the material has better crystalline and larger atomic number (Z) [11]. In our samples, we have only Si and $SiO_2$ in the film, therefore the dark image is the crystalline Si.

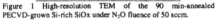

Figure 1 High-resolution TEM of the 90 min-annealed PECVD-grown Si-rich SiOx under $N_2O$ fluence of 50 sccm.

Figure 2 High-resolution TEM of the 120 min-annealed PECVD-grown Si-rich $SiO_X$ under $N_2O$ fluence of 100 sccm.

## RESULTS AND DISCUSSION

After a Si-rich $SiO_X$ film PECVD-grown at a constant $N_2O$ fluences of 100 and 50 (defined as conditions a and b) sccm, and a $N_2O/SiH_4$ ratio increasing from 4 to 5.5 was annealed at 1100°C for increasing from 15 to 120 min, the normalized PL at a wavelength of 730-760 nm was observed (see Fig. 3 and 4) [12]. The meaning behind these four graphs indicated that a variety of intensities at different annealing times merely occurred while initiating the precipitation processes of nc-Si. As a result of the excess amount of Si atoms in the films providing sufficient energy for recombining to nc-Si, a well accelerated expansion of nc-Si was presented. For $SiO_X$ samples with constant $N_2O$ fluence of 100 sccm and varying $N_2O/SiH_4$ fluence ratio, the near-infrared PL was enhancing after the 15-min annealing with a increasing trend of PL power vs. annealing time, indicating that the amount of nc-Si component in the sample was increasing. The comparison on wavelength and peak intensity of SiOx samples grown with higher $N_2O$ fluence reveals that there is a slight variation in wavelength as well as size of nc-Si at different annealing duration. The longer the annealing time, the more amount of the large size of nc-Si could be fabricated. As the $N_2O/SiH_4$ ratio decays more during growth, the PL becomes weaker. A blue-shifted wavelength from 760 to 720 nm corresponding to the nc-Si size shrinkage is observed for annealed samples grown at lower fluence ratio, which is attributed to the reoxidation occurred at outer surface of nc-Si. The nc-Si is unable to be presented at a $N_2O/SiH_4$ ratio of smaller than 4.

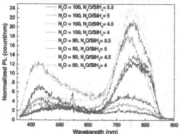

Figure 3 Thickness normalized PL of SiO$_X$ with different N$_2$O/SiH4 ratios by annealing at optimum condition.

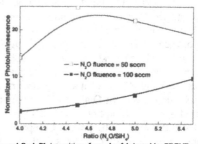

Figure 4 Peak PL intensities of samples fabricated by PECVD with different N$_2$O/SiH$_4$ ratios obtained at optimized annealing time.

By reducing $N_2O$ fluence, numerous defects were produced during as-grown to reveal the unstable condition of SiO$_X$ structures. After annealing for 15-30 min, a broadband orange PL was performed with the precipitation of mid-size nc-Si. As the annealing time lengthened to 90 min, the strongest PL intensity is red-shifted to 760 nm with the linewidth remaining consyant. Lengthening the annealing to 45 min results in larger nc-Si size but an opposite trend happens at annealing for 120 min, while the re-oxidation is initiated to shrink the nc-Si size and attenuate the blue-shifted PL intensity. Figure 4 illustrates the thickness-normalized PL intensity is the highest at the $N_2O/SiH_4$ ratio of 5.5 at larger $N_2O$ fluencies but the optimized ratio decreases when reducing $N_2O$ fluence. If we reduce the $N_2O/SiH_4$ ratio, a more intensive PL is observed after 90-min annealing. The optimized N$_2$O/SiH4 ratio of 4.5 for maximum PL response is found at $N_2O$ fluence of 50 sccm, as shown in Fig. 4. Under same $N_2O$ fluence, the variation of thickness (nm) is not significantly changed with a fluence ratio ranging from 4.0 to 5.5, indicating that the higher fluence in PECVD chamber could not affect thickness of Si-rich SiO$_X$ film. However, a thickness increasing trend with increasing $N_2O$ fluence ratio is also observed. The PL normalized to unit thickness reveal a steady increase with raising fluency ratio.

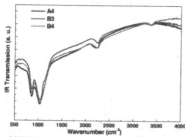

Figure 5 FTIR Spectra of the as-deposited SiO$_X$.

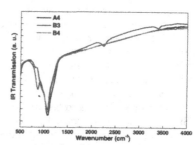

Figure 6 FTIR Spectra of the SiO$_X$ by annealing.

In comparison with lower $N_2O$ fluence grown SiO$_X$ samples, less excess Si atoms as well as nc-Si density can be observed in SiO$_X$ grown with $N_2O$ fluence of 100 sccm. Moreover, the FTIR spectra of the as-grown and annealed SiOx samples show distinct absorption peaks at 870

445

and 1020 cm$^{-1}$ as contributed by the bending of Si-H bonds [13] and the symmetric stretching vibration of Si-O bond [14], while another peak observed at 2250 cm$^{-1}$ supports the existence of vibration mode for the passivated Si-H$_X$ bonds normally observed in amorphous Si:H materials [15]. The characteristic of the Si-O-Si stretching vibration mode at 1070 cm$^{-1}$ was of particular interest, which reveals a minimum transmission for SiO$_X$ grown at larger N$_2$O fluencies. The 1070cm$^{-1}$ absorption can be attenuated with decreasing N$_2$O and N$_2$O/SiH$_4$ fluence ratio due to the oxygen deficiency in the grown SiO$_X$ film, which will be greatly reduced under a Si-rich growth condition and reveal a stoichiomrtric composition of the grown oxide film. Concurrently, the normalized PL is decreased as the 1070 cm$^{-1}$ intensity increases, whereas the FTIR peaks at 810 cm$^{-1}$ and 870 cm$^{-1}$ rises and fall each other, respectively, indicating the vicissitude between Si-O and Si-H bending bonds with in the Si-rich SiO$_X$ matrices. The Si-H related 870cm$^{-1}$ absorption peak becomes significantly by decreasing N$_2$O/SiH$_4$ fluence ratio, so as the 2250 cm$^{-1}$ FTIR peak. Compared with Fig. 5 and 6, it is easy to see that the Si-O vibration mode of all annealed samples increase towards 1070 cm$^{-1}$, with higher initial oxygen content samples going to a higher Si-O vibration frequency [16]. In Sample B3, all the absorption bands associated with hydrogen bonding show higher peak intensities, indicating that Sample B3 has a higher hydrogen concentration [17]. In addition, our experimental results also indicates that the PECVD grown oxide structure become more like SiO with O/Si ratio decreasing form 1.24 to 0.88.

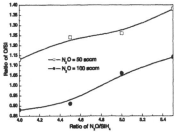 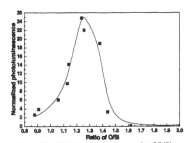

Figure 7 Comparing with various ratio of N O/SiH vs. ratio of O/Si.    Figure 8 Normalized Photoluminescence vs. ratio of O/Si.

In Fig. 7, the ratio of O/Si in the SiO$_X$ is increased when increasing ratio of N$_2$O/SiH$_4$ under same N$_2$O fluence. The Si Atoms is easier to be decomposed from SiH$_4$ that the oxygen from N$_2$O as the desorptional energies of N$_2$O and SiH$_4$ are 101.5 kcal/mol and 75.6 kcal/mol. Furthermore, raisng the total fluence of reacting gases could inevitable lead to the insufficient desorption phenomenon in PECVD chamber as the plasma energy is average by more reacting molecules. As the higher fluence of N$_2$O assists the desorption of more oxygen atoms with decreasing SiH$_4$ fluence, which leads to the increasing O/Si ratio of the SiO$_X$ sample. Therefore, more oxygen atoms can be desorbed from N$_2$O and react with Si atoms in the chamber, the ratio of O/Si at the sample is gradually increased with reducing N$_2$O or total fluences. The normalized PL at 760 nm increased when the O/Si ratio increasing from 0.88 to 1.24, and then decreased to 1.38 in Fig. 8, therefore, the maximum PL at 760 nm was observed in a Si-rich layer by annealing at 1100°C when the N$_2$O/SiH$_4$ ratio was 4.5 and N$_2$O fluence was 50, however, the calculated RBS ratio of O/Si in the SRSO layer was 1.24 [18]. That is, the ratio of O/Si in the Si-rich SiO$_X$ samples was not direct proportion to the normalized PL, indicating that a great deal of Si concentration could not enhance the precipitation of nc-Si effectively. From the analysis of RBS, we conclude that there is a maximum PL intensity with O/Si ratio decreasing from 2 to 1.24, whereas the PL intensity shows an opposite trend with O/Si ratio further decreasing from 1.24 to 0.88. The changing PL reponse with decreasing O/Si ratio has been elucidated by variation on the SiO and $SiO_2$ related bending and symmetric stretching vibration modes of Si–O bonds, which has been confirmed from the RBS and FTIR analyses of the $SiO_X$ grown at different $N_2O$ fluencies and $N_2O/SiH_4$ fluence ratios.

## CONCLUSIONS

The optimum annealing time at a temperature of 1100°C and the $N_2O/SiH_4$ ratio are 90 min and 4.5 at a constant $N_2O$ fluence (50), and the normalized photoluminescence has maximum intensity at the $N_2O/SiH_4$ ratio is 4.5. However, the calculated Rutherford backscattering (RBS) ratio of O/Si in the SRSO layer was 1.24, corresponding to a total Si concentration of about 44.64 atom %. In particular, the ratio of O/Si in the Si-rich $SiO_X$ samples was not direct proportion to the normalized PL, it meant that a great deal of Si concentration could not enhance the existence of nc-Si effectively. On the other hand, the strongest PL at 760nm at optimized O/Si=1.24. This phenomenon has been confirmed by RBS and FTIR absorption spectroscopic analysss. A nearly Gaussian function of the normalized PL intensity vs. O/Si composition ratio has been observed due to the significant variation on the Si nanocrystals size with the density of the excessive Si atoms. Increasing the $N_2O$ fluence oppositely reduces the O/Si composition ratio and results in a SiO-like matrix with few Si nanocrystals precipitation. In contrast, the increasing O/Si ratio with decreasing $N_2O$ fluence could lead to a dilute Si excessive condition, thus also attenuating the PL intensity.

## ACKNOWLEDGMENTS

This work was supported in part by the National Science Council (NSC) of the Republic of China under grants NSC96-2221-E-002-099 and NSC97-ET-7-002-007-ET.

## REFERENCES

1. Q. Ye, R. Tsu, and E. H. Nicollian, "Resonant tunneling via microcrystalline-silicon quantum confinement," *Phys. Rev. B*, vol. 44, pp. 1806 - 1811 (1991).
2. A. Pèrez-Rodrìguez, O. González-Varona, B. Garrido, P. Pellegrino, J. R. Morante, C. Bonafos, M. Carrada, and A. Claverie, "White luminescence from Si$^+$ and C$^+$ ion-implanted $SiO_2$ films," *J. Appl. Phys.*, vol. 94, pp. 254 - 262 (2003).
3. D. Pacifici, E. C. Moreira, G. Franzo, V. Martorino, and F. Priolo, "Defect production and annealing in ion-irradiated Si nanocrystals," *Phys. Rev. B*, vol. 65, 144109 (2002).
4. L. T. Canham, "Silicon quantum wire array fabrication by electrochemical and chemical dissolution of wafers," *Appl. Phys. Lett.*, vol. 57, pp. 1046-1048 (1990).
5. E. Neufeld, S. Wang, R. Apetz, C. Buchal, R. Carius, C.W. White and D.K. Thomas, "Effect of annealing and $H_2$ passivation on the photoluminescence of Si nanocrystals in $SiO_2$," *Thin Solid Films*, vol. 294, pp.238-241 (1997).
6. S.P. Withrow, C.W. White, A. Meldrum, J.D. Budai, D.M. Hembree and J.C. Barbour, "Effects of hydrogen in the annealing environment on photoluminescence from Si nanoparticles in $SiO_2$," *J. Appl. Phys.*, vol. 86, pp.396-401(1999).

7. S. Cheylan and R.G. Elliman, "Effect of hydrogen on the photoluminescence of Si nanocrystals embedded in a $SiO_2$ matrix," *Appl. Phys. Lett.*, vol. **78** (2001).

8. K.L. Brower, "Kinetics of $H_2$ passivation of $P_b$ centers at the (111) Si-$SiO_2$ interface," *Phys. Rev. B*, vol. **38**, pp.9657-9666 (1988).

9. J.H. Stathis and E. Cartier, "Atomic hydrogen reactions with $P_b$ centers at the (100) Si/$SiO_2$ interface," *Phys. Rev. Lett.*, vol. **72**, pp.2745-2748 (1994).

10. C. Delerue, G. Allan, and M. Lannoo, "Theoretical aspects of the luminescence of porous silicon," *Phys. Rev. B*, vol. **48**, pp. 11024-11036 (1993).

11. David B. Williams, and C. Barry Carter, *Transmission electron microscopy: a textbook for materials science*, Plenum Press, New York, 1996.

12. Y. Q. Wang, G. L. Kong, W. D. Chen, H. W. Diao, C. Y. Chen, S. B. Zhang and X. B. Liao, "Getting high-efficiency photoluminescence from Si nanocrystals in $SiO_2$ matrix," *Appl. Phys. Lett.*, vol. **81**, pp. 4174-4176 (2002).

13. P. Gonzalez, D. Fernandez, J. Pou, E. Garcia, J. Serra, B. Leon, M. Perez-Amor ,T. Szorenyi, "Study of the Gas-Phase Parameters Affecting the Silicon-Oxide Film Deposition Induced by an ArF Laser," *Appl. Phys. A*, vol. **57**, pp. 181-185 (1993).

14. O. Jambois, H. Rinnert, X.Devaux, and M.Vergnat, "Influence of the annealing treatments on the luminescence properties of SiO/$SiO_2$ multilayers," *J. Appl. Phys.*, vol. **100**, Art. No. 123504 (2006).

15. F. Ay, A. Aydinli, "Comparative investigation of hydrogen bonding in silicon based PECVD grown dielectrics for optical waveguides," *Optical Materials*, vol. **26**, pp.33-46 (2004).

16. F. Yuna, B.J. Hindsa, S. Hatatania, S. Odaa,U, Q.X. Zhaob, M. Willander, "Study of structural and optical properties of nanocrystalline silicon embedded in $SiO_2$," *Thin Solid Films*, vol. **375**, pp.137-141 (2000).

17. Chang-Hee Cho, Baek-Hyun Kim, Tae-Wook Kim, and Seong-Ju Park, Nae-Man Park and Gun-Yong Sung, "Effect of hydrogen passivation on charge storage in silicon quantum dots embedded in silicon nitride film," *Appl. Phys. Lett.*, vol. **86**, pp. 143107 (2005).

18. S. Guhaa, S. B. Qadri, R. G. Musket, M. A. Wall and Tsutomu Shimizu-Iwayama, "Characterization of Si nanocrystals grown by annealing $SiO_2$ films with uniform concentrations of implanted Si," *J. Appl. Phys.*, vol. **88**, pp. 3954 - 3961 (2000).

# Sensors, Transistors and
# Active Matrix Arrays I

Mater. Res. Soc. Symp. Proc. Vol. 1066 © 2008 Materials Research Society          1066-A19-02

# Fluorescence Detection of DNA Hybridization Using an Integrated Thin-Film Amorphous Silicon n-i-p Photodiode

A. C. Pimentel[1], R. Cabeça[1,2], M. Rodrigues[1,2], D.M.F. Prazeres[2,3], V. Chu[1], and J. P. Conde[1,3]

[1]INESC Microsistemas e Nanotecnologias, Rua Alves Redol, 9, Lisbon, 1000-029, Portugal
[2]IBB-Institute for Biotechnology and Bioengineering, Centre for Biological and Chemical Engineering, Instituto Superior Técnico, Av. Rovisco Pais, Lisbon, 1049-001, Portugal
[3]Dept. of Chemical and Biological Engineering, Instituto Superior Técnico, Av. Rovisco Pais, Lisbon, 1049-001, Portugal

## ABSTRACT

This paper presents the fluorescence detection of DNA hybridization with a surface immobilized probe using a hydrogenated amorphous silicon (a-Si:H) photosensor. This sensor integrates a $SiO_2$ layer for DNA probe immobilization, a *p-i-n* a-Si:H photodiode for fluorescence detection and a fluorescence filter of hydrogenated amorphous silicon carbide (a-SiC:H) to cut the excitation light. With this integrated photosensor system, a five order of magnitude difference was obtained in the signal measured at the emission wavelength and that measured at the excitation wavelength for the same incident photon flux. The fluorophore Alexa Fluor 430 was used to label the DNA target molecules and a laser at 405 nm and a photon flux of $5.7 \times 10^{16}$ cm$^{-2}$.s$^{-1}$ was used as the excitation light source. The detection limit achieved for fluorophores in solution in contact with the device and for fluorophores immobilized on the device surface is $5 \times 10^{-9}$ M and 0.4 pmol/cm$^2$, respectively. The hybridization of the tagged DNA target with a covalently or electrostatically immobilized probe was successfully detected at a surface density of ~3 pmol/cm$^2$.

## INTRODUCTION

Fluorescence is one of the most commonly used methods in biology and biomedical analysis for detection of nucleic acids, proteins and cells. This method uses an external light source to excite fluorophores attached to the biomolecules of interest and measures the resultant fluorescence.

Hydrogenated amorphous silicon (a-Si:H) photosensors are a promising candidate for integrated optoelectronic detection of fluorescently labeled biomolecules in microarray and lab-on-a-chip applications due to their high quantum efficiency in the visible light spectrum, low dark current, and low-temperature processing technology (below 250 °C) which allows the use of substrates such as glass and polymers [1].

In previous work, fluorescence detection of immobilized DNA was demonstrated using an amorphous silicon a-Si:H sensor in a parallel contact configuration with an integrated optical filter. The detection limit of this sensor was of the order of $1 \times 10^{12}$ molecules/cm$^2$ [2]. Kamei *et al.* used an a-Si:H *p-i-n* photodiode with a ZnS/YF$_3$ [3] and a $SiO_2$/Ta$_2$O$_5$ [4] fluorescence filter for DNA detection in solution. The detection limits reported were 17 nM [3] and 7 nM [4]. Caputo *et al* [5] reported the detection of DNA using a single and a multicolor amorphous silicon *p-i-n* photodiode for fluorescence detection in solution, using the glass substrate as the excitation light filter; the detection limit achieved was 3 nM. Schöler [6] integrated an amorphous silicon *p-i-n* photodiode in a Lab-on-Microchips (ALMs) platform for capillary electrophoresis detection of DNA labeled with rhodamine, achieving a detection limit of 10 μM in a 26 nL volume.

This paper presents a thin-film a-Si:H *p-i-n* photodiode with an integrated a-SiC:H filter for hybridization detection of DNA labeled with a fluorophore (Alexa Fluor 430). The detection of the fluorophore both in solution and immobilized on the device surface is described. The detection of the hybridization of a complementary DNA oligonucleotide target with a surface-immobilized DNA probe is demonstrated.

## EXPERIMENTAL PROCEDURES

### Fabrication of the a-Si:H *p-i-n* photodiode

A 200 μm x 200 μm a-Si:H *n-i-p* photodiode is microfabricated. The Al back contact is sputtered over a glass substrate (Schott AF45) and made by photolithography and wet-etching. The amorphous silicon (a-Si:H) *p-i-n* junction is deposited by plasma enhanced chemical vapor deposition (PECVD) at 250°C and 0.1 Torr. First, a 200 Å layer of n$^+$-a-Si:H (P doped) is deposited, followed by a 5000 Å of i-a-Si:H, and, finally, a 200 Å layer of p$^+$-a-Si:H (B doped). A 200 μm x 200 μm mesa junction is made by photolithography and etched by reactive ion etching (RIE). An insulating layer of SiN$_x$ (1000 Å) is deposited by PECVD at 100°C and 0.1 Torr as a sidewall passivation layer. Vias are opened in the passivation layer to allow electrical contact between the p$^+$-a-Si:H and the indium tin oxide (ITO) top electrode. The top electrode is a 1000 Å ITO layer deposited by sputtering and fabricated by lift-off. Al lines are defined to provide electrical contact between the sensor and the contact pads. The integrated fluorescence filter consists of a 1.96 μm layer of amorphous silicon carbon alloy (a-SiC:H) deposited by PECVD at 100°C and 0.1 Torr. The bandgap of the a-SiC:H filter is 2.25 eV and the value of E$_{04}$ is 2.46 eV. Finally, 750 Å of SiO$_2$ are deposited to allow the immobilization of the biomolecules. The inset in Figure 1(a) represents a cross sectional view of the *p-i-n* photodiode.

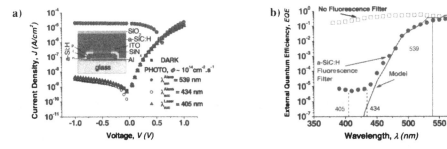

**Figure 1.** a) Dark and photo *J-V* characteristics of the a-Si:H *p-i-n* photodiode with the integrated filter. The inset in the figure represents a cross sectional view of the a-Si:H *p-i-n* photodiode with an integrated fluorescence filter used in fluorescence detection. b) External Quantum Efficiency (*EQE*) of the a-Si:H *p-i-n* photodiode with the integrated a-SiC:H fluorescence filter (●) and without the fluorescence filter (□). The solid line represents the *EQE* estimated by an optical model.

## Protocol for DNA target hybridization with an electrostatically immobilized DNA probe

The electrostatic immobilization of a single strand DNA (ssDNA) oligonucleotide (17 bp) probe was carried out by first cleaning the chip surface in cholic acid (2% w/v) for 12 h at room temperature (*RT*). Then, the chip surface is silanized with amino propyl triethoxy silane (2% w/v in de-ionized water) for 2 h at *RT*. After the silanization step the chip surface is covered with amino groups which are positively charged in aqueous solution at neutral pH. DNA probes can thus be immobilized electrostatically via the negatively charged phosphate groups. The immobilization was carried out by contacting the chip with a 0.5 M DNA probe aqueous solution during 1 h at *RT*. Then, a stringent wash is performed to remove the probes that were not immobilized. A pre-hybridization step is performed with bovine serum albumin (2% w/v, 2 h at *RT*) to avoid non-specific adsorption of the DNA target on the chip's surface during the hybridization procedure. An aqueous solution of ssDNA target labeled with the fluorophore Alexa Fluor 430 is prepared to a concentration of 0.5 M and a 10 μL drop is placed on top of the chip surface for hybridization reaction (1 h at *RT*). The chip surface is then submitted to a stringent wash in order to remove the DNA targets that did not hybridize, rinsed with de-ionized water and dried. The hybridized DNA pair is indicated schematically in Figure 3 (b). The photoresponse of the device was measured under $\lambda_{exc}^{Laser} = 405$ nm excitation light to detect the fluorophore-labeled hybridized DNA.

## Protocol for DNA target hybridization with a covalently immobilized DNA probe

The device surface was silanized as described in the previous section. Next, a cross-linking step was carried out using a 1 mM solution of N-(ε-maleimido caproyloxy) sulfosuccinimide ester (sulfo-EMCS) in phosphate buffer, pH 7.4 for 2 h at room temperature. The bi-functional cross-linker sulfo-EMCS will attach covalently onto the amino groups present on the chip surface. The chip is then washed twice for 5 minutes in phosphate buffer solution, rinsed with de-ionized water and dried. A 10 μL drop of 0.5 μM aqueous solution of DNA probe is placed on top of the cross-linked surface of the chip. The covalent DNA immobilization reaction will occur for 2 h at *RT*. The DNA probe is functionalized in the 5' end with a thiol group (-SH) that will covalently attach to the maleimide group present in the bi-functional cross-linker sulfo-EMCS. The washing step after immobilization, the pre-hybridization step, and its respective washing follow the same protocol as described in the previous section. The hybridization step follows the same protocol as for the electrostatic hybridization with the difference that it takes 2 h. The hybridized DNA pair is indicated schematically in Figure 3 (c). The photoresponse of the device was measured under $\lambda_{exc}^{Laser}$ excitation light to detect the Alexa-labeled hybridized DNA.

## RESULTS AND DISCUSSION

### Characterization of the a-Si:H *p-i-n* photodiode with an integrated fluorescence filter

An *I-V* measurement in the dark is used to obtain the shunt resistance, $R_{shunt}$, the saturation current density, $J_0$ and the ideality factor, $n_0$, of the microfabricated diodes. The values obtained are 0.88 TΩ, 460 pA/cm$^2$ and 1.84, respectively (Figure 1(a)). The fluorescence filter was designed to maximize the transmission at the emission wavelength while minimizing the transmission at the excitation wavelength. Assuming that each photon absorbed in the intrinsic a-Si:H layer of the device generates an electron-hole pair which is collected at the contacts, the theoretical External Quantum Efficiency (EQE is the number of electron-hole pairs detected per incident photon) can be estimated. The EQE of the device as a function of the wavelength corresponds to the light absorbed by the a-Si:H intrinsic layer $(A_{i\ a\text{-}Si:H})$ and is given by equation (1).

$$EQE(\lambda) = A_{i-a-Si:H}(\lambda) = \left(T_{p^+-a-Si:H}(\lambda) - R_{i-a-Si:H}(\lambda)\right) \times \left(1 - e^{(\alpha_{i-a-Si:H}(\lambda)d_{i-a-Si:H})}\right) \qquad (1)$$

where $T_{p^+-a-Si:H}$ corresponds to the light transmitted by the $p^+$-a-Si:H layer and $R_{i-a-Si:H}$ corresponds to the light reflected by the i-a-Si:H layer. $\alpha_{i-a-Si:H}$ and $d_{i-a-Si:H}$ correspond to the absorption coefficient and the thickness of the i-a-Si:H layer, respectively.
EQE measurements were performed using a lock-in technique at 13 Hz and no bias was applied to the photodiode. Figure 1(b) shows the experimental and calculated EQE, which show good agreement, except below 430 nm, where the loss of efficiency of the filter is tentatively attributed to the presence of pinholes and/or compositional inhomogeneities in the a-SiC:H film. Despite this, the fluorescence filter is able to cut by 5 orders of magnitude the excitation light at 405 nm. Figure 1 (a) also shows the photo J-V characteristics of the photodiode under a photon-flux $\Phi \sim 10^{14}$ cm$^{-2}$.s$^{-1}$ for the emission wavelength of Alexa Fluor 430 ($\lambda_{em}^{Alexa}$ = 539 nm), for the excitation wavelength of Alexa Fluor 430 ($\lambda_{exc}^{Alexa}$=434 nm) and also at the wavelength of the laser used as excitation light source ($\lambda_{exc}^{Laser}$=405 nm).

## Calibration of the a-Si:H photosensor system as a function of the quantity of Alexa Fluor 430 in solution and adsorbed on a solid surface

A stock solution of Alexa Fluor® 430 (carboxylic acid, succinimidyl ester) was prepared in dimethyl sulfoxide (DMSO) to a concentration of 0.01 M. The stock solution was then diluted in de-ionized water to concentrations ranging from 1 nM to 1 μM. The photodiode response was calibrated as a function of the number of moles of fluorophore adsorbed on the device surface and also in solution. The device response was acquired using a lock-in amplifier with the excitation light source illumination (laser at 405 nm wavelength, $5.7 \times 10^{16}$ cm$^{-2}$.s$^{-1}$ photon flux and a spot size of approximately 4 mm diameter) chopped at 13 Hz. The measurements were made with the diode unbiased.

The calibration curves of the photodiode response as a function of the number of fluorophores surface adsorbed (circles) and in solution (squares) are shown in Figure 2. For the measurement of the surface-adsorbed fluorophore, a 10 μL drop of fluorophore solution with a known concentration was placed on top of the chip surface and allowed to evaporate. Once the solvent had evaporated, the photocurrent was measured. The limit of detection (LOD) of the current device system to surface adsorbed fluorophores was 0.4 pmol/cm$^2$. Measurements of the fluorophore in solution were performed by acquiring the photocurrent signal after placing a 10 μL drop of fluorophore solution with a given concentration on top of the chip surface. The detection limit of the concentration of fluorophores in solution is 5 nM. A model that calculates the photon flux reaching the device at the fluorophore emission wavelength, $\Phi_{model}$ is used to describe the device response.

Fluorescence detection of ssDNA target hybridization with electrostatically (Figure 3(b)) and covalently (Figure 3(c)) immobilized probes was successfully achieved. The photodiode response was converted in DNA target surface concentration using Figure 2. For the electrostatically immobilized probe, it was possible to distinguish between the complementary DNA target and the non-complementary DNA target because the signal acquired using the complementary target, which gives a surface density of 3.37 pmol/cm$^2$ for the calibration curve, is approximately six times higher than the signal acquired with the non-complementary target, which gives a surface density of 0.54 pmol/cm$^2$.

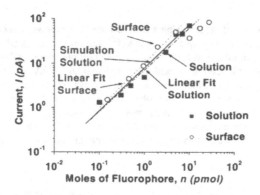

**Figure 2.** Response of the a-Si:H *p-i-n* photodiode with the integrated filter plotted as a function of the number of moles of fluorophore immobilized on the 12.57 mm$^2$ chip surface (○) or in a 10 μL solution (■). The dotted line and the dash-dotted line represent the linear fit to the experimental results for the surface immobilized fluorophores and for the fluorophores in solution, respectively. The solid line represents the simulation results which essentially coincide for both situations.

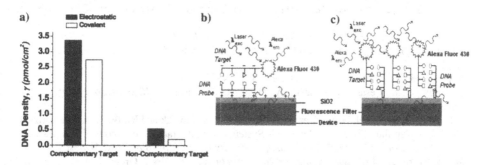

**Figure 3. a).** Surface concentration of DNA target after hybridization with the electrostatically immobilized and with the covalently immobilized DNA probe. Schematic for DNA hybridization with: b) electrostatically immobilized probes, and c) covalently immobilized probes.

For the covalently immobilized probe it was also possible to distinguish between the complementary and the non-complementary DNA target because the DNA target surface density detected was 2.74 pmol/cm$^2$ and 0.19 pmol/cm$^2$, respectively. Figure 3(a) summarizes the results obtained for DNA target hybridization. The values obtained for the hybridization of non-complementary targets for both electrostatically and covalently immobilized probes are below the limit of quantification, so the concentrations mentioned above give an upper limit to the concentration of non-complementary DNA targets present.

## CONCLUSIONS

DNA hybridization was detected using a fluorescence detection system based on an amorphous silicon $p$-$i$-$n$ photodiode with an integrated fluorescence filter. A DNA surface density of approximately 3 pmol/cm$^2$ was detected for both complementary DNA hybridization with an electrostatically immobilized probe and for complementary DNA hybridization with a covalently immobilized probe. It was also possible to distinguish between the complementary and the non-complementary target using both probe immobilization techniques. Further improvement in device characteristics such as decreasing the leakage current and filter design to reduce light transmission at the excitation wavelength can increase the sensitivity of the detector and increase the signal to noise ratio at lower fluorophore concentrations and surface densities.

## ACKNOWLEDGMENTS

The authors gratefully acknowledge V. Soares, J. Bernardo and F. Silva for clean-room device processing. This work was supported by Fundação para a Ciência e Tecnologia (FCT) through research projects and the Ph.D. grant SFRH / BD / 17379 / 2004. INESC MN acknowledges funding from FCT through the Associated Lab–IN. J.P. Conde thanks the Gulbenkian Foundation for a travel grant.

## REFERENCES

1. R. A. Street, *Hydrogenated amorphous silicon*: Cambridge University Press, Cambridge, UK, 1991.
2. F. Fixe, V. Chu, D.M. F. Prazeres and J.P. Conde, *Nucleic Acids Res.*, **32**, 70 (2004).
3. T. Kamei, B. Paegel, J. Sherer, A. Skelley, R. Street and R. Mathies, *Anal. Chem.*, **75**, 5300 (2003).
4. T. Kamei and T. Wada, *Appl. Phys. Lett.*, **89**, 114101 (2006).
5. Caputo, D., de Cesare, G., Nascetti, A., Negri, R., Scipinotti, R., *IEEE Sensors Journal*, **7**, Issue 9, 1274 (2007).
6. L. Schöler, K. Seibel, H. Schäfer, R. J. Püschl, B. Wenclawiak and M. Böhm: Characterization of a Micro Capillary Zone Electrophoresis System With Integrated Amorphous Silicon Based Optical Detectors in *Materials and Strategies for Lab-on-a-Chip—Biological Analysis, Microfactories, and Fluidic Assembly of Nanostructures*, edited by S. Grego, J.M. Ramsey, O. Velev, and S. Verpoorte (Mater. Res. Soc. Symp. Proc. **Volume 1004E**, Warrendale, PA, 2007), 1004-P03-20.

Mater. Res. Soc. Symp. Proc. Vol. 1066 © 2008 Materials Research Society        1066-A19-03

# Noise Characterization of Polycrystalline Silicon Thin Film Transistors for X-ray Imagers Based on Active Pixel Architectures

L. E. Antonuk, M. Koniczek, J. McDonald, Y. El-Mohri, Q. Zhao, and M. Behravan
Department of Radiation Oncology, University of Michigan, Ann Arbor, MI, 48109

## ABSTRACT
An examination of the noise of polycrystalline silicon thin film transistors, in the context of flat panel x-ray imager development, is reported. The study was conducted in the spirit of exploring how the 1/f, shot and thermal noise components of poly-Si TFTs, determined from current noise power spectral density measurements, as well as through calculation, can be used to assist in the development of imagers incorporating pixel amplification circuits based on such transistors.

## INTRODUCTION

Active matrix, flat panel imagers (AMFPIs) are based on large area arrays (presently up to ~43×43 cm$^2$) whose pixels typically consist of a single hydrogenated amorphous silicon (a-Si:H) thin film transistor (TFT) coupled to a pixel storage capacitor. While such imagers are used in many medical applications, the relatively modest size of the signal generated per detected x-ray by the imager, relative to the electronic (i.e., dark) noise, results in significant loss of imaging performance under conditions of low exposure or very small pixel sizes. To overcome this limitation, a significant increase in signal relative to noise is required [1]. One approach, which preserves the advantages of conventional AMFPIs (e.g., large area, compact), involves the use of polycrystalline silicon (poly-Si) TFTs to create an amplification circuit in every pixel – analogous to the pixel architecture of CMOS active pixel sensors. The considerably higher electron and hole mobilities of poly-Si TFTs, compared to those of a-Si:H TFTs, enable faster switching times and considerably more complex pixel circuits. Initial prototype arrays with 1- and 2-stage poly-Si pixel amplifiers have been developed and show encouraging results [2,3].

For such arrays, a critical issue is the degree to which noise properties of the individual poly-Si TFTs in the pixel circuit limit the performance of the entire imager. In order to fully explore this question, a detailed knowledge of the noise behavior of the individual transistors is required. While many studies of poly-Si TFT noise have been reported [4,5], it is highly desirable to acquire noise data from individual TFTs having the same design, and made in the same manner as those used in actual prototype arrays, since the results should be representative of their performance in pixel circuits. This paper describes a methodology for acquiring and utilizing noise data obtained from individual poly-Si TFTs for purposes of imager development. The methodology is illustrated through measurements of current noise power spectral density. This research was conducted in the spirit of a pilot study toward understanding the relationship between the range of conditions over which good quality data can be obtained, and how such information can aid the development of poly-Si imagers based on active pixel architectures.

## MEASUREMENT METHODOLOGY

The poly-Si imager array development underway in our group is being performed in collaboration with scientists at the Palo Alto Research Center, where the arrays are fabricated. The present noise measurements were performed on individual poly-Si TFTs located in regions

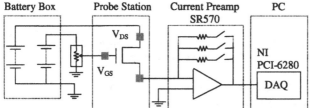

**Figure 1:** Schematic illustration of the apparatus used to measure source current from individual poly-Si TFTs for purposes of current power density determination. DAQ refers to the data acquisition card.

of array substrates reserved for "test devices". In this study, one transistor design was examined: an n-channel transistor with an 8 μm width (*W*), and two 5 μm long (*L*) gates. This design corresponds to the addressing transistor and the reset transistor used both in 3-TFT, as well as in 5-TFT prototype pixel amplifier arrays ("PSI-2" and "PSI-3", respectively) [2,3]. The transistors were fabricated through pulsed excimer-laser crystallization [2], with a 50 nm thick poly-Si film, and a 100 nm thick gate oxide with a gate capacitance of $3.45 \times 10^{-8}$ F/cm$^2$.

Current measurements on the TFTs were performed using a methodology described below, involving the apparatus shown in Figure 1. Micro-positioning probes made electrical contact to the TFT in a light-tight aluminum box constructed to reduce the effect of external electromagnetic noise. To further minimize noise, the voltages applied to the TFT gate, $V_{GS}$, and drain, $V_{DS}$, were supplied by NiMH batteries (housed in a shielded box), with fine adjustment of $V_{GS}$ performed using a high precision potentiometer. The source current, $I_{DS}$, was amplified by a high-sensitivity current preamplifier (SRS, model SR570), with selectable gain settings, and digitized by a standard data acquisition card (National Instruments, NI PCI-6280). The reported results correspond to measurements obtained from a single, high-quality TFT whose behavior is believed to be representative of properly functioning transistors in the prototype array designs.

## DETERMINATION OF CURRENT NOISE POWER SPECTRAL DENSITY, $S_I$

For this study, current data was acquired at a sampling frequency of ~606 kHz over an interval of ~443 s, resulting in a total of ~$2.68 \times 10^8$ samples. This allowed empirical examination of the poly-Si TFT noise properties over as large a frequency range as possible, limited only by practical considerations imposed by the preamplifier bandwidth and noise, as well as by limitations associated with our ability to maintain mechanically and electrically stable conditions over the course of a measurement. For a given set of current data acquired for a particular combination of $V_{DS}$ and $V_{GS}$ voltages, the data was subjected to a linear detrending correction, to account for small drifts in the measured current (e.g., arising from battery discharge).

The results presented in Figure 2 and Table I correspond to a $V_{DS}$ value of ~1.3 V, supplied by a single battery cell, and for $V_{GS}$ values ranging from ~0.9 to 10.5 V. The $V_{GS}$ values are listed in column 1 of Table I, where the lowest voltage corresponds to the lowest TFT current that allowed extraction of valid results, given the aforementioned preamplifier and measurement interval limitations. The magnitude of the source current, $I_{DS}$, and the corresponding choice of a preamplifier gain setting, are given for each $V_{GS}$ value in columns 2 and 3 of Table I. Figure 2(a) shows current power density results corresponding to several values of $V_{GS}$, obtained from the detrended current data via a fast Fourier transform (FFT). Data is presented up to the frequency at which the preamplifier bandwidth and/or noise becomes a limitation. The limiting frequency up to which valid data could be obtained at each $V_{GS}$ value is given in column 4 of Table I.

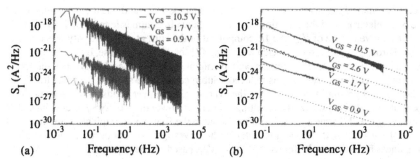

(a) Frequency (Hz)    (b) Frequency (Hz)

**Figure 2:** Plots of empirically determined current power density, $S_I$, as a function of frequency, for several values of $V_{GS}$. (a) The total power density of the TFT, prior to averaging and preamplifier noise correction. (b) The 1/f component of the power density, after averaging, and after correcting for other known noise contributions. See text for further details.

**Table I:** Summary of information and results related to determination of noise components for the poly-Si TFT at a $V_{DS}$ of 1.3 V. For each value of $V_{GS}$ (column 1): the corresponding source current (column 2); the preamplifier gain setting at which data was acquired (column 3); the limiting frequency of the preamplifier (column 4); the measured preamplifier noise level (column 5); calculational estimates of TFT shot and thermal noise (columns 6 and 7); and the parameters from the fit of the 1/f noise, $K$ and $\alpha$ (columns 8 and 9, respectively), are reported.

| $V_{GS}$ (V) | Source Current, $I_{DS}$ (A) | Preamp Gain Setting (A/V) | Preamp Limiting Freq. (Hz) | Preamp Noise (A²/Hz) | Shot Noise, $S_{SHOT}$ (A²/Hz) | Thermal Noise, $S_{TH}$ (A²/Hz) | 1/f Noise Parameter, $K$ (A²) | 1/f Noise Parameter, $\alpha$ |
|---|---|---|---|---|---|---|---|---|
| 0.9 | $3.9\times10^{-11}$ | $2\times10^{-11}$ | 0.5 | $4.1\times10^{-29}$ | $1.3\times10^{-29}$ | $4.8\times10^{-31}$ | $4.1\times10^{-27}$ | 0.93 |
| 1.3 | $5.9\times10^{-10}$ | $5\times10^{-10}$ | 3 | $5.0\times10^{-29}$ | $1.9\times10^{-28}$ | $7.2\times10^{-30}$ | $3.8\times10^{-25}$ | 0.91 |
| 1.5 | $9.1\times10^{-10}$ | $1\times10^{-9}$ | 5 | $4.7\times10^{-29}$ | $2.9\times10^{-28}$ | $1.1\times10^{-29}$ | $8.1\times10^{-25}$ | 0.97 |
| 1.7 | $2.5\times10^{-9}$ | $2\times10^{-9}$ | 15 | $1.9\times10^{-27}$ | $8.1\times10^{-28}$ | $3.2\times10^{-28}$ | $4.1\times10^{-24}$ | 0.86 |
| 2.6 | $5.7\times10^{-8}$ | $5\times10^{-8}$ | 100 | $2.2\times10^{-27}$ | $1.8\times10^{-26}$ | $7.0\times10^{-27}$ | $3.2\times10^{-22}$ | 0.93 |
| 3.9 | $4.0\times10^{-7}$ | $2\times10^{-7}$ | 1,000 | $3.6\times10^{-25}$ | $1.3\times10^{-25}$ | $5.0\times10^{-26}$ | $4.8\times10^{-21}$ | 0.95 |
| 5.2 | $1.2\times10^{-6}$ | $1\times10^{-6}$ | 1,000 | $3.9\times10^{-25}$ | $3.9\times10^{-25}$ | $1.5\times10^{-26}$ | $1.2\times10^{-20}$ | 0.92 |
| 6.6 | $2.5\times10^{-6}$ | $2\times10^{-6}$ | 10,000 | $5.1\times10^{-24}$ | $7.9\times10^{-25}$ | $3.1\times10^{-26}$ | $3.3\times10^{-20}$ | 0.97 |
| 7.8 | $3.9\times10^{-6}$ | $2\times10^{-6}$ | 10,000 | $5.1\times10^{-24}$ | $1.2\times10^{-24}$ | $4.8\times10^{-26}$ | $5.2\times10^{-20}$ | 1.00 |
| 9.1 | $5.7\times10^{-6}$ | $5\times10^{-6}$ | 10,000 | $5.2\times10^{-24}$ | $1.8\times10^{-24}$ | $7.1\times10^{-26}$ | $8.7\times10^{-20}$ | 1.01 |
| 10.5 | $7.9\times10^{-6}$ | $5\times10^{-6}$ | 10,000 | $5.2\times10^{-24}$ | $2.5\times10^{-24}$ | $9.7\times10^{-26}$ | $1.3\times10^{-19}$ | 1.03 |

The results shown in Figure 2(a) correspond to contributions from various noise sources: 1/f, shot and thermal noise from the TFT, as well as preamplifier noise. Towards obtaining the 1/f component, expected to be a major contributor to pixel circuit noise, the spectral density of preamplifier noise was determined through application of the same methodology described above, replacing the TFT with various resistors. The results are shown in column 5 of Table I. The spectral densities of the shot and thermal TFT noise, $S_{SHOT}$ and $S_{TH}$, were estimated using the expressions [6]:

$$S_{SHOT} = 2qI_{DS} \quad \text{(A}^2\text{/Hz)} \tag{1}$$

and

$$S_{TH} = 4k_B T \cdot I_{DS}/V_{DS} \quad \text{(A}^2\text{/Hz)} \tag{2}$$

respectively, where $k_B$ is Boltzmann's constant, $T$ is temperature and $q$ is the magnitude of the charge of an electron, and the resulting values are shown in columns 6 and 7 of Table I.

For a given value of $V_{GS}$, the 1/f component of the current power density was obtained as follows. Through application of Welch averaging [7] (in the present case, by dividing the detrended current data into 50 segments, and averaging the FFTs of each segment), the spread of the current power density results was reduced. Subtraction of the spectral densities for TFT shot, TFT thermal and preamplifier noise from the average noise power yields an estimate of the 1/f component. Figure 2(b) shows the resulting 1/f current power density for several values of $V_{GS}$. The relatively small degree of non-linearity visible in the figure may originate from the intrinsic behavior of the TFT itself [5] – and is under further investigation. For purposes of this pilot study, a linear fit to the results for each value of $V_{GS}$, based on the expression:

$$S_{1/f} = K/f^{\alpha} \qquad (A^2/Hz) , \qquad (3)$$

where $K$ and $\alpha$ are the fit parameters, was performed and is indicated by dashed lines in the figure. The values of $K$ and $\alpha$ obtained from these fits are given in the last two columns of Table I. The magnitude of $K$ scales closely with $I_{DS}$ over most of the investigated range of $V_{GS}$, as anticipated [8]. The average value of $\alpha$ is 0.95. Note that, while the variation of $\alpha$ with $V_{GS}$ at values above ~5 V is interesting, the data are insufficient to warrant any definite conclusion on a possible correlation between $\alpha$ and $V_{GS}$.


## USE OF Poly-Si NOISE RESULTS IN IMAGER ARRAY DEVELOPMENT

Poly-Si TFT noise characteristics, such as the results reported above, can assist in the development of x-ray imaging arrays, based on such transistors, in a variety of ways.
i) It is important to examine noise for different TFT widths, lengths and channel types, and to monitor how changes to the TFT fabrication process (e.g., increasing the laser power to increase TFT mobility) affect noise. Since the noise depends upon the quality (e.g., the trap density and grain size) of the poly-Si material [4,5], it is also useful to examine how noise varies across a given substrate – as this could introduce non-uniform performance across an array.
ii) Investigating changes in TFT noise as a function of radiation dose provides information about the suitability of poly-Si for applications with different lifetime doses to imaging arrays [9].
iii) TFT noise measurements can be compared against predictions from analytical models [10].
iv) Empirically determined TFT noise information can be used as input to circuit simulations that allow more extensive and detailed modeling of the noise behavior of complex circuits than is possible through analytical means alone. An example of such modeling is described below.

The modeling was conducted for a circuit corresponding to the output stage common to prototype poly-Si amplifier arrays [2,3]. The circuit in Figure 3(a) consists of a source-follower, $TFT_{SF}$, and an addressing transistor, $TFT_{ADDR}$, both in parallel with their respective current noise sources, $I_{NOISE}$ (i.e., $I_{SFNOISE}$ and $I_{ADDRNOISE}$). The input to the circuit simulation for each noise source is based on the measured 1/f noise power density results reported in Table I. The simulation was constructed to examine how 1/f noise affects charge transfer to the capacitance, $C_{DATA}$, of an array data line as a function of readout time, $t_{RO}$ (the time after $TFT_{ADDR}$ is switched on). The parameters for this circuit and simulation details are described in ref. [11].

The modeling involved three steps. In the first step, the circuit simulation is run without the current noise sources. For each 1 ns simulation timestep (the smallest timestep that was computationally practical for this pilot study), $I_{DS}$ currents for both TFTs, as well as the dataline voltage $V_{OUT}$, are recorded, the latter being shown in the top graph of Figure 3(b). Step 2 uses

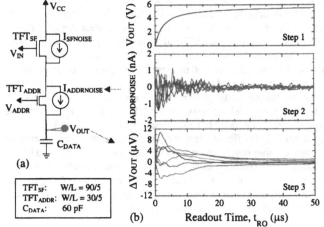

**Figure 3:** (a) Schematic diagram of the simulated circuit. (b) Results, plotted as a function of readout time, $t_{RO}$, at 0.5 µs intervals for clarity, for: (top) the output voltage, $V_{OUT}$ from Step 1 of the modeling; (middle) the input noise source, $I_{ADDRNOISE}$, from Step 2; and (bottom) the difference between the output voltage with, and without current noise sources, $\Delta V_{OUT}$, from Step 3.

these currents, along with the measured 1/f noise results, to generate random noise datasets that are input as $I_{SFNOISE}$ and $I_{ADDRNOISE}$ to another simulation run in Step 3. Six examples of such noise datasets for $I_{ADDRNOISE}$ are illustrated in the middle graph of Figure 3(b), and details of this noise generation step are described below. In Step 3, $V_{OUT}$ is recorded and subtracted from the $V_{OUT}$ values from Step 1, resulting in a waveform for the voltage difference, $\Delta V_{OUT}$. Steps 2 and 3 are repeated many times to build up a statistically meaningful number of $\Delta V_{OUT}$ waveforms – six of which are illustrated in the bottom graph of Figure 3(b).

For each simulated TFT, the generation of $I_{NOISE}$ in Step 2 involved creating a normalized spectral density, $S_N$, which is derived from the results presented in Table I. These results were scaled to match the dimensions and operating points of the TFT. It is assumed that the two 5 µm long gates of the measured device behave like a single 10 µm long gate. The relationship used to derive $S_N$ from Table I results is a first order approximation based on relations shown in references [8] and [12]:

$$S_N = S_{1/f} W^2 L^2 / I_{DS} .\qquad(4)$$

A temporal 1/f noise data set, $Z_N$, with one datapoint for each timestep over the simulated period, is generated by taking the inverse FFT of $\sqrt{S_N}$, with a random phase for each frequency. The current noise, $I_{NOISE}$, for each simulated TFT is created by scaling $Z_N$ by the $I_{DS}$ that corresponds to the TFT current for each timestep, and by the device dimensions according to the equation:

$$I_{NOISE}(t_{RO}) = Z_N(t_{RO})\sqrt{I_{DS}(t_{RO})}/(WL)\quad(A).\qquad(5)$$

The resulting $I_{NOISE}$ provides a reasonable approximation for 1/f noise current behavior of the simulated TFTs. Further improvements to this approximation would include obtaining noise data from a variety of TFTs, making measurements with finer adjustments of $V_{GS}$ and $V_{DS}$, using a more sensitive current preamplifier for low current measurements, and using a non-linear fit to the current power density to allow a more precise determination of the 1/f noise scaling factors.

In the present example, the parameters chosen in the circuit simulation correspond to a 3-TFT, one-stage pixel amplifier array [11] with 2000×2000 pixels, a pitch of 150 µm, and a pixel capacitance of 1.5 pF. In fluoroscopic operation, the inverse of $t_{RO}$ determines the maximum possible frame rate. In the study, Steps 2 and 3 were repeated a total of 100 times. The standard deviation of the resulting $\Delta V_{OUT}$ waveforms at a particular $t_{RO}$ corresponds to the rms noise, due to TFT 1/f noise contributions, on the data line at that readout time. From these runs, the rms noise is found to be 2.9 and 0.56 μV at readout times of 5 and 50 μs, corresponding to maximum frame rates of 100 and 10 frames per second (fps), respectively. For these two $t_{RO}$ values, ~89% and 97% of the signal is transferred to the data line, and the resulting magnitude of the noise from the modeled part of the pixel circuit, referred to the pixel storage capacitor, is ~31 and 5 e [rms], respectively. By comparison, the electronic pixel noise of a conventional AMFPI would be on the order of $10^3$ e [rms]. Thus, the model suggests that the 1/f noise contribution from the output stage of the pixel amplifier circuit, operated in charge readout mode, will not be a limiting factor in noise performance. Of course, significantly larger noise contributions from other circuit elements (e.g., reset TFT), as well as other noise components (i.e., shot and thermal) also need to be considered – and can be modeled using techniques similar to those described above. It is anticipated that such modeling will be valuable in poly-Si array development by helping to optimize noise, readout speed, linearity and other properties [11] through detailed examination of promising pixel circuits, and variation of device and processing parameters.

## DISCUSSION

For flat panel imagers based on poly-Si arrays with active pixel architectures, the degree to which large improvements in signal to noise performance can be realized will strongly depend upon the noise properties of the pixel circuit and its constituent TFTs. It is therefore anticipated that the development of such imagers will benefit from noise modeling involving circuit simulations parameterized by detailed information obtained from poly-Si TFT measurements of the type reported in this paper. Further refinement of this methodology is under investigation.

## ACKNOWLEDGEMENTS

We wish to thank Mike Yeakey, Chuck Martelli and Taeyjuana Curry for assistance with instrumentation and data acquisition, and Robert Street and JengPing Lu of PARC for useful discussions. This study was partially supported by NIH grant R01 EB000558.

## REFERENCES

1.  LE Antonuk et al., Med. Phys. 27(2), 289 (2000).
2.  JP Lu et al., Appl. Phys. Lett. 80(24), 4656 (2002).
3.  LE Antonuk et al., SPIE 5745, 18 (2005).
4.  M Rahal et al., IEEE Trans. Elect. Dev. 49(2), 319 (2002).
5.  CA Dimitriadis et al., J. Appl. Phys. 83(3), 1469 (1998).
6.  P Horowitz, "The Art of Electronics," (Cambridge University Press, 1989).
7.  SJ Orfanidis, "Introduction to Signal Processing," (Prentice Hall, 1995).
8.  C Jakobson et al., Solid-State Electron 42(10), 1807 (1998).
9.  Y Li et al., J. Appl. Phys. 99, 064501-1 (2006).
10. N Matsuura et al., Med. Phys. 26(5), 672 (1999).
11. LE Antonuk et al., SPIE 6913, 69130I-1 (2008).
12. CA Dimitriadis et al., J. Appl. Phys. 91(12), 9919 (2002).

Mater. Res. Soc. Symp. Proc. Vol. 1066 © 2008 Materials Research Society    1066-A19-04

# High Fill Factor a-Si:H Sensor Arrays with Reduced Pixel Crosstalk

Y. Vygranenko[1], A. Sazonov[2], D. Striakhilev[2], J. H. Chang[3], G. Heiler[3], J. Lai[3], T. Tredwell[3], and A. Nathan[4]

[1]Electronics Telecommunications and Computer Engineering, ISEL, Lisbon, 1950-062, Portugal
[2]Electrical and Computer Engineering, University of Waterloo, Waterloo, N2L 3G1, Canada
[3]Carestream Health, Rochester, NY, 14652-3487
[4]London Centre for Nanotechnology, UCL, London, WC1H OAH, United Kingdom

## ABSTRACT

In this paper, we report on low noise, high fill factor amorphous silicon (a-Si:H) image sensor structures for indirect radiography. Two types of the sensor arrays comprising n-i-p photodiodes and m-i-s photosensors have been fabricated. The device prototypes contain 100 x 100 pixels, with a pixel pitch of 139 μm. The active-matrix addressing is provided by low off-current TFTs. The sensors are vertically integrated onto the TFT-backplane, by implementing a 3-μm-thick low-k interlayer dielectric. This dielectric layer serves to reduce the data line capacitance and to planarize underlying topography. The detector was designed for reduced data-line resistance and parasitic coupling. Details of the device design and fabrication, along with sensor performance characteristics, are presented and discussed.

## INTRODUCTION

Active-matrix, flat-panel X-ray detectors based on amorphous silicon (a-Si:H) technology are widely used for medical imaging [1]. Here, an X-ray phosphor screen is coupled to an array of a-Si:H sensors and an a-Si:H thin-film transistor (TFT) backplane, which, in turn, are connected to external electronics. The design of the array needs to targets the most important performance features, such as the sensitivity, dynamic range and spatial resolution. In planar architecture imaging arrays, where the TFTs and photo-sensors are juxtaposed on the same plane, the fill factor is inherently low due to the limited area available for the sensor. Alternatively, in high fill-factor array designs, the sensor array is vertically integrated on top of the TFT backplane.

In this paper, we report on active matrix X-ray detectors, with high fill-factor a-Si:H sensor arrays. The novelty of our array design is that the data lines are formed with the top metallization layer in order to reduce the data line resistance and parasitic coupling. Furthermore, we used a segmented sensor design to reduce pixel crosstalk for improved spatial resolution. Two types of device prototypes, one with n-i-p photodiodes and one with m-i-s photosensors, were designed, and characterized for their electrical and optical performance.

## EXPERIMENT

The detector prototype has 100 x 100 pixels, with a pixel pitch of 139 μm. The active-matrix addressing is provided by low off-current TFTs. Figure 1(a) shows a cross-section view of the pixel. Here, a photosensor (either n-i-p or m-i-s structure), with metal bottom electrode and ITO top electrode, is located on top of a TFT array and separated by an insulation-planarization layer. The gate line serves to bias the TFT to switch on the pixel, and the data line is then used to transfer the charge to external readout electronics. The bias line is used for photosensor biasing. Figure 1(b) shows a top view optical micrograph of a single pixel. Here, unlike the planar arrays, the photosensitive area extends over the TFT region increasing the fill factor. In this particular design, the photosensor area is 126 μm by 112 μm, and the fill factor is 73%. The gate line and data line widths are 22 μm and 15 μm, respectively.

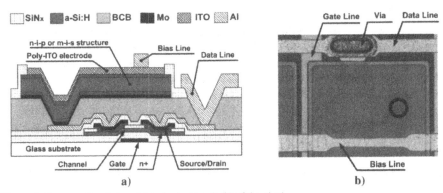

**Figure 1.** Cross-sectional view (a) and micrograph (b) of the pixel.

The detector fabrication starts with the TFTs backplane. The fabrication sequence of the back-channel-passivated TFTs includes 5 lithography steps [2]. A 100-nm-thick Mo layer is sputtered on the Corning glass substrates followed by lithography step to form gate lines (*Mask #1*). Then, the a-SiN$_x$ (250 nm)/a-Si:H (50 nm)/a-SiN$_x$ (200 nm) stack is deposited by an one-pump-down process in the multi-chamber PECVD system, and the top nitride layer is patterned using wet-etch to define the transistor islands (*Mask #2*). Next, a 50-nm-thick a-Si:H n$^+$-layer is deposited, followed by the sputtering of a 150-nm-thick Mo layer. The bi-layer is patterned to form source-drain contacts (*Mask #3*). Then, a 300 nm-thick a-SiN$_x$ passivation layer is deposited, and the vias are formed under the source-drain contacts and gate-line contact pads (*Mask #4*). A 500-nm-thick Al layer is then sputter deposited and patterned with *Mask #5* to form the top metallization. The TFT-backplane is then covered with a ~3-μm-thick Benzocyclobutene (BCB) insulating-planarization interlayer. Photosensitive BCB (Cyclotene 4022-35, Dow Chemical Co.) was used due to dielectric properties (dielectric constant of 2.65, volume resistivity of ~10$^{19}$ Ohm-cm), excellent gap-fill and planarization, low moisture absorption, and low-temperature (~200°C) cure. The process includes a lithography step to form vias in the BCB-layer (Mask #6). Figure 2 shows an optical micrograph with circular drain-to-sensor via (on top of the figure) and oval source-to-data line via (on the left). The SEM image of the via profile reveals that a slope of the via wall is close to 45 degrees, which allows the use of a thin (~100 nm) metal layer for the photosensor bottom electrode (see Figure 3).

The photosensor integration starts with the sputtering of a metal layer followed by the a-Si:H n-i-p or a-SiN$_x$/a-Si:H/n$^+$ photosensor deposition. The deposition conditions used have been described elsewhere [3]. The thickness of the i-layer was in the range from 0.5 to 1 μm, while the thicknesses of the a-SiN$_x$ and doped layers were typically ~120 and ~25 nm, respectively. All layers are patterned in one lithography step by dry etching (Mask #7). Then, a 500-nm-thick SiN$_x$ passivation layer is deposited at 150°C and patterned with Mask #8 by wet etching. A 70-nm-thick polycrystalline indium tin oxide (poly-ITO) layer is then deposited using RF sputtering at a substrate temperature of 150°C and patterned to form sensor electrodes (Mask #9). Finally, a 1-μm-thick Al layer is deposited by sputtering and patterned with *Mask #10* to form the bias and data lines, and contact pads.

For the array prototype under discussion, the capacitance and resistance of the data line are 17.4 fF per pixel and 26 Ω/cm, respectively. The detector design significantly reduces the noise level as a result of low data-line resistance and reduced parasitic coupling [4].

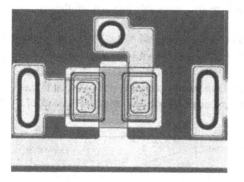 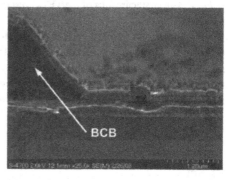

**Figure 2.** Micrograph of the switching TFT with the planarization layer.

**Figure 3.** SEM image of the via profile.

## RESULTS AND DISCUSSION

Figure 4 shows typical transfer characteristics of the TFTs and the gate leakage ($I_G$) dependence on the gate voltage ($V_G$) in logarithmic (a) and linear (b) scales. The channel length (L) and width (W) are 18 and 33 μm, respectively. To enable high-accuracy measurements, an assembly of 10 TFTs connected in parallel was tested. The off-current and the gate leakage were

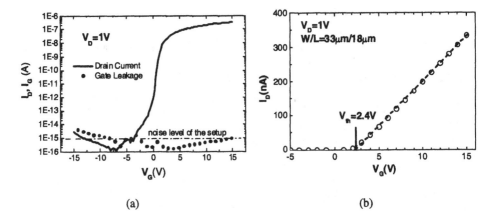

(a)                                                (b)

**Figure 4.** TFT transfer characteristics in the logarithmic (a) and linear (b) scales.

still below the noise level of our setup, i.e. less than 1 fA in the operating range of the gate bias between -7 V to + 15 V. The ON/OFF current ratio was estimated to be ~$10^9$. The ON-characteristic in Figure 4 (b) is linear suggesting a low contact resistance. The threshold voltage is 2.4 V. The characteristics of the TFTs presented here are comparable with state-of-the-art a-Si:H TFTs.

Figure 5 shows a typical current-voltage (J-V) characteristic of the $112 \times 126 \ \mu m^2$ n-i-p photodiode with an i-layer thickness of 1 $\mu m$. The diode shows exponential current growth (over seven orders of magnitude) under forward bias up to 0.8 V. Under reverse bias of 5 V, the dark current density is 130 pA/cm$^2$. The achieved level of the leakage current is similar to that in the state-of-the-art sensors with continuous i-layer and segmented diodes of comparable size [5, 6].

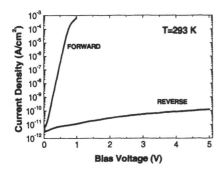

**Figure 5.** Current-voltage characteristics of the n-i-p photodiode.

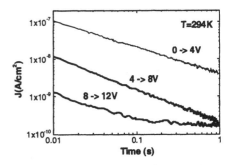

**Figure 6.** Transient dark current in the m-i-s sensor at different offset voltages of the bias pulse applied to the ITO electrode. The bias pulse amplitude is 4 V.

On the other hand, the m-i-s device requires a refresh pulse to be applied periodically, which causes a dark transient current. Hence, the noise level can be significantly higher than that in n-i-p photodiodes. The transient dark current originates from traps at the semiconductor-insulator interface and from a-Si:H bulk trap states. The associated components of the transient dark current can be largely eliminated by adjusting the biasing conditions. Figure 6 shows the transient dark current in the m-i-s sensor at different offset voltages of the bias pulse applied to the ITO electrode, when the bias pulse amplitude is kept constant at 4 V. The magnitude of the transient current exponentially decreases with increasing offset voltage reaching the level of the steady-state current, which is factor of 2 to 3 higher than that in the n-i-p diodes. The leakage through the silicon nitride is likely to be a limiting factor for the tested devices.

Figure 7 shows spectral response characteristics of the a-Si:H n-i-p photodiode and m-i-s photosensor with poly-ITO top electrodes. The samples have the same i-layer thickness of 1μm. The measurements were performed under similar biasing conditions: the reverse bias of 5 V was applied to the n-i-p photodiode, while the biasing pulse of +6 V with offset of +2 V was applied to the cathode of the a-SiN$_x$/a-Si:H/n$^+$/ITO structure.

The m-i-s structure shows strong interference fringes in the long-wavelength part of the spectra. They originate from an effective light reflection from the a-Si:H/a-SiN$_x$/Mo interface. The external quantum efficiency reaches peak value of 86% at 590 nm for the n-i-p photodiode, while for m-i-s structure it reaches 66% at 620 nm. Then, the quantum efficiency decreases monotonically at shorter wavelengths because of the absorption loss in the p$^+$- or n$^+$-layers. Since the thicknesses of the doped layers are about the same (~25 nm) in both p-i-n and m-i-s structures, the noticeable difference in the magnitude of the spectra cannot be ascribed to this mechanism. The disadvantage of the m-i-s sensor is that the charge generated in i-layer cannot be fully readout through the blocking dielectric layer [7]. The ratio of transferred ($Q_{out}$)-to-generated

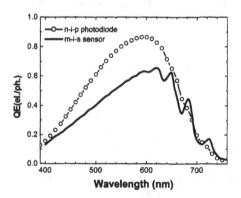

**Figure 7.** Spectral response characteristics of the n-i-p photodiode (circles) and m-i-s photosensor (solid line).

($Q_{sig}$) charges, $\eta_Q$, calculated using the equivalent circuit model of the m-i-s structure is:

$$\eta_Q = \frac{Q_{out}}{Q_{sig}} = \frac{C_{SiN}}{C_{SiN} + C_{a-Si}} = \frac{1}{1 + \dfrac{\varepsilon_{Si} d_{SiN}}{\varepsilon_{SiN} d_{Si}}},$$ (1)

where $\varepsilon_{Si} = 11.8$ and $\varepsilon_{SiN} = 6.4$ are the dielectric constants of the a-Si:H and a-SiN$_x$, respectively. For the actual layer thicknesses $d_{SiN} = 120$ nm and $d_{Si} = 1$ μm, the $\eta_Q$ is 0.82. This value is consistent with the observed differences in the quantum efficiency spectra between n-i-p and m-i-s sensors.

## CONCLUSIONS

We have developed novel pixel architectures with vertically integrated photosensors and TFTs architectures for low noise, high fill factor arrays for indirect radiography. Vertical integration of the TFT-backplane with light sensors was performed by implementing a thick low-k interlayer dielectric for the reduction of data line capacitance and to planarize underlying topography. The fabrication process of the imagers is compatible with existing industrial processes and equipment. The performance of the a-Si:H TFTs and the vertically integrated p-i-n and m-i-s photosensors reported here is state-of-the-art and suitable for digital radiography applications.

## ACKNOWLEDGEMENTS

The authors are grateful to the Natural Science and Engineering Research Council of Canada (NSERC), and the Portuguese Foundation of Science and Technology through fellowship BPD20264/2004 for financial support of this research. This work was performed using the Giga-to-Nanoelectronics Centre facilities at the University of Waterloo.

## REFERENCES

1. R.A. Street (Ed.), in *Technology and Applications of Amorphous Silicon*, Springer, Berlin, 2000 p. 147
2. A. Nathan, B. Park, A. Sazonov, S. Tao, I. Chan, P. Servati, K. Karim, T. Charania, D. Striakhilev, Q. Ma and R. V. R. Murthy, *Microelectronics Journal* **31**, 883 (2000).
3. J. H. Chang, Y. Vygranenko, and A. Nathan, *J. Vac. Sci. Technol. A*, **22**, 971 (2004).
4. J. Lai, Y. Vygranenko, G. Heiler, N. Safavian, D. Striakhilev, A. Nathan, T. Tredwell, , *Mater. Res. Soc. Symp. Proc.* **989**, A14.05 (2007).
5. J. A. Theil, *Mater. Res. Soc. Symp. Proc.* **762**, A21.4.1 (2003).
6. R. L. Weisfield et al. *Proc. SPIE*, **5368**, 338 (2001).
7. N. Safavian, Y. Vygranenko, J. Chang, Kyung Ho Kim, J. Lai, D. Striakhilev, A. Nathan, G. Heiler, T. Tredwell, and M. Fernandes, *Mater. Res. Soc. Symp. Proc.* **989**, A12.06 (2007).

# Sensors, Transistors and
# Active Matrix Arrays II

Mater. Res. Soc. Symp. Proc. Vol. 1066 © 2008 Materials Research Society       1066-A20-03

# Self-Aligned Amorphous Silicon Thin Film Transistors With Mobility Above 1 cm$^2$V$^{-1}$s$^{-1}$ Fabricated at 300°C on Clear Plastic Substrates

Kunigunde H. Cherenack, Alex Z. Kattamis, Bahman Hekmatshoar, James C. Sturm, and Sigurd Wagner
Princeton University, Princeton, NJ, 08540

## ABSTRACT

We have developed a fabrication process for amorphous-silicon thin-film transistors (a-Si:H TFTs) on free-standing clear plastic substrates at temperatures up to 300°C. The 300°C fabrication process is made possible by using a unique clear plastic substrate that has a very low coefficient of thermal expansion (CTE < 10ppm/°C) and a glass transition temperature higher than 300°C. Our TFTs have a conventional inverted-staggered gate back-channel passivated geometry, which we designed to achieve two goals: accurate overlay alignment and a high effective mobility. A requirement that becomes particularly difficult to meet in the making of TFT backplanes on plastic foil at 300°C is minimizing overlay misalignment. Even though we use a substrate that has a relatively low CTE, accurately aligning the TFTs on the free-standing, 70-micrometer thick substrate is challenging. To deal with this immediate challenge, and to continue developing processes for free-standing web substrates, we are introducing techniques for self-alignment to our TFT fabrication process. We have self-aligned the channel to the gate by exposing through the clear plastic substrate. To raise the effective mobility of our TFTs we reduced the series resistance by decreasing the thickness of the amorphous silicon layer between the source-drain contacts and the accumulation layer in the channel. The back-channel passivated structure allows us to decrease the thickness of the a-Si:H active layer down to around 20nm. These changes have enabled us to raise the effective field effect mobility on clear plastic to values above 1 cm$^2$V$^{-1}$s$^{-1}$.

## INTRODUCTION

Thin-film transistor backplanes made on optically clear plastic substrate foils could find universal use in flexible displays, because they may be employed with any kind of display frontplane, be it transmissive, emissive or reflective. Transistors [1] and displays [2,3,4] on clear plastic substrates have been demonstrated in the past. However, in order to accommodate the low process temperatures of commercial clear polymers [5], the deposition of the a-Si:H TFT stack has been reduced from ~ 300°C on glass [6,7] to as low as 75°C [8]. While the initial electrical performance of a-Si:H TFTs fabricated at such ultra-low temperatures is satisfactory, recent experiments have shown poor stability under gate-bias stress [9-12]. In response we have been raising the a-Si:H TFT process temperature on clear plastic [13, 14, 15, 16] to develop a "glass-like" process at 300°C. This enabled us to achieve "glass-like" TFT stability on plastic.

Our long-term goal is to enable roll-to-roll fabrication – therefore we are working with free-standing substrates. To obtain functional transistors on free-standing plastic foil substrates, the mechanical stress needs to be designed carefully, especially at high processing temperatures [17, 18]. We have used stress control to develop a crack-free TFT fabrication process at 300°C on a clear plastic substrate. However, even if the device layers are crack-free, the stress in the TFT stack causes the substrate to expand or contract (depending on the nature of the combined strain of the total structure), leading to misalignment between consecutive mask layers. After deposition at high process temperatures the subsequent misalignment between mask layers can be very large, causing the TFTs at the edges of the substrate to malfunction. If it is not possible to reduce the total strain in the substrate by engineering the strain, as mentioned above, it becomes necessary to investigate alternative methods to reduce strain-induced misalignment between mask layers. The misalignment can be reduced by laminating or electrostatically

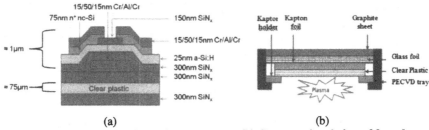

(a)                                     (b)

**Figure 1  (a) Schematic of transistor geometry (b) Cross-sectional view of face-down substrate mount for plasma-enhanced chemical vapor deposition.**

bonding the substrate to a stiff carrier plate [19], by clamping the substrate into a rigid frame [20], or by digitally compensating the masks for substrate distortion [21].

In our work we focused on developing a self-alignment method which would serve to eliminate overlay misalignment completely. One requirement necessary to implement self-alignment is the ability to expose the photoresist through the back of the substrate. Since the amorphous silicon layer in the TFT stack is very absorptive in the UV wavelength used to develop the photoresist during photolithography, we needed to reduce the thickness of our amorphous silicon channel region as much as possible. Therefore we chose the back-channel passivated TFT geometry shown in Figure 1(a) that allowed us to reduce the a-Si:H layer thickness from our conventional thickness of ~300nm down to ~25nm while still maintaining a rugged TFT fabrication process. In this self-aligned process, the self-alignment is therefore achieved between the gate (mask 1) and the channel passivation (mask 2). An additional benefit of this TFT geometry is that a thinner amorphous silicon layer also results in a lower contact resistance at the source/drain terminals, and therefore a higher measured TFT mobility. We also reduced the source/drain contact resistance by replacing the standard $n^+$ amorphous silicon (a-Si) in the source/drain contacts with a $n^+$ nano-crystalline silicon (nc-Si) layer using a layer-by-layer deposition method [22].

In this paper we discuss the fabrication of self-aligned a-Si:H TFTs at 300°C on a clear plastic substrate, the performance of these self-aligned TFTs and the improved alignment that was achieved between the gate and the channel passivation using our self-aligned process.

# EXPERIMENT

## Substrate preparation

The 7.5x7.5 cm$^2$ and 75-$\mu$m thick optically clear plastic (CP) foil substrates that we use have a working temperature of $\geq$ 300°C. Their in-plane coefficient of thermal expansion $\alpha_{substrate}$ is $\leq$ 10 ppm/°C, which is sufficiently low to obtain intact device layers in a 300°C process [23]. A rule of thumb for crack prevention is ($\alpha_{substrate}$ - $\alpha_{TFT}$) x ($T_{process}$ – $T_{room}$) $\leq$ 0.3%. During PE-CVD deposition the substrate is placed in a frame facing downward, and is backed first with Kapton E polyimide foil, then with a glass slide and finally a graphite sheet, as shown in Figure 1(b). The graphite serves as black body absorber for radiative heating in the nominally isothermal PE-CVD pre-heat and deposition zones. This mount lets the substrate expand and contract to some extent during PE-CVD. Following an outgassing anneal at 200°C in the load lock, the substrate is transferred to the SiN$_x$ deposition-chamber for deposition at 280°C of a 300-nm thick SiN$_x$ passivation layer on the future device side (front) of the substrate, at an RF (13.56 MHz) power density of 20 mW/cm$^{-2}$, which puts the SiN$_x$ under tensile stress. The substrate is transferred back to the load lock and flipped to expose its back side. It is then returned to the SiN$_x$chamber and a 300-nm thick SiN$_x$ passivation layer is deposited at 280°C on the back side of the substrate at a high plasma power density (90 mW/cm$^{-2}$), producing compressive stress in the SiN$_x$.

## Transistor Fabrication

Throughout the process the substrate is kept free-standing except that is precisely flattened for photolithography by temporary bonding to a glass plate with water. After our usual substrate preparation, a thermally evaporated tri-layer of 15 nm Cr, 50 nm A$\ell$, and 15 nm Cr. is deposited. This bottom metal layer is patterned using conventional photolithography (first mask level). Then the sample is loaded into the PE-CVD system and the following depositions are carried out: (i) a 300 nm thick SiN$_x$ gate dielectric at 300°C at a power density of 90 mW/cm$^{-2}$, (ii) a 25nm a-Si:H channel layer deposited at 17 mW/cm$^{-2}$ and (iii) a 150nm thick SiN$_x$ layer as the channel passivation. Now the sample is removed from the PE-CVD system and we carry out the self-alignment step:

First a layer of photoresist is spin-coated onto the sample. Then we then expose the substrate through the back in our mask aligner for 15 minutes at a power density of 3.5mW/cm$^2$. In this step the bottom gate electrode acts as the mask to self-align the channel passivation to the gate. The SiN$_x$ layer is now wet etched in buffered oxide etch (HF:NH$_4$F: H$_2$O) for 50 seconds. By slightly over-etching the back-channel SiN$_x$ protection layer during the patterning, the required overlap over the gate is created. A piranha clean (H$_2$O$_2$:H$_2$SO$_4$) and followed by a short buffered oxide etch dip ensure clean interfaces between the exposed a-Si:H channel layer and the subsequently deposited S/D layer. This concludes the self-alignment step.

Next, a 75-nm n$^+$ nc-Si:H and a tri-layer of 15/50/15-nm Cr/Al/Cr film are deposited and patterned to form the source/drain contacts (second mask level). This is followed by etching the a-Si:H to isolate individual devices (third mask level). Finally, holes are opened to contact the bottom gate (fourth mask level) completing the TFT process. Details of this process are given in [23]

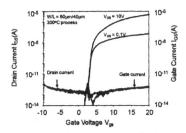

**Figure 3  Transfer characteristics of a self-aligned a self-aligned a-Si:H TFT made on clear plastic at 300°C.**

## DISCUSSION

### Transistor Evaluation

After fabricating the samples (as described in the previous section) the samples are annealed at 135°C for 30 minutes in air.  The TFTs are evaluated and gate-bias stressed using an HP4155A parameter analyzer.  For transfer characteristics the gate voltage is swept from 20V to -10V, at 10V drain-source voltage.  During gate bias stressing the source and drain are grounded and a positive voltage is applied to the gate for 600 seconds.  Then the transfer characteristic is measured again by sweeping the gate voltage from 20V to -10V.  This is done for gate bias voltages of 30V to 60V, corresponding to electric fields of (0.9 to 1.8) x $10^8$ V/m.  The shift in the threshold voltage was determined on the subthreshold slope of the transfer curves at the drain current value of 1 x $10^{-10}$ A.  We use TFTs with a W/L ratio of 80μm/40μm.

### Self-aligned TFT measurements

Typical transfer characteristics for back-channel passivated a-Si:H TFTs made using the self-aligned process are shown in Figure 3.  On clear plastic the linear mobility is 1.11cm²/Vs, the saturation mobility 1.08cm²/Vs, the threshold voltage ~4V, the on/off current ratio > 1x$10^7$, and the subthreshold slope 350mV/decade.  Using a standard photolithography process we achieved linear mobilities of ~ 0.95cm²/Vs, and saturation mobilities of ~0.96cm²/Vs mobilities.  Clearly the self-aligned process results in high-quality TFTs with higher mobilities compared to the standard photolithographic process.

Figure 4 (a) shows a close-up view of the channel-region is shown for a TFT made with the standard process at the edge of the plastic work-piece - and Figure 4(b) shows another TFT where the channel passivation is patterned using self-alignment.  In each picture the gate area is indicated by a green square and the channel passivation area is indicated by a light blue square.  In Figure 4(a) the channel passivation has shifted partially off the gate, and the TFT is not functional, while in Figure 4(b) the squares overlap perfectly - showing how the misalignment between the gate and the channel passivation layers has been eliminated. The effective over-etch of the channel passivation (the distance from the edge of the passivation layer to the edge of the gate) is 2μm.

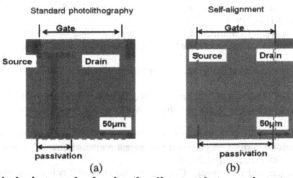

Figure 4 Optical micrographs showing the alignment between the gate and subsequent layers for (a) a TFT 3cm away from the center of the substrate fabricated using standard patterning, and (b) a TFT which the channel passivation patterned using self-alignment.

## CONCLUSIONS

Misalignment between the gate and channel passivation in a back-channel passivated TFT geometry is eliminated by using a self-alignment method. The back-channel passivated TFT geometry allowed us to reduce the amorphous silicon channel resistance down to 25nm, resulting in TFTs with an improved measured mobility. However, since flexible displays will ultimately be fabricated using roll-to-roll fabrication on free-standing web substrates we still need to develop new techniques for aligning the gate with the source/drain contact and the interconnects.

## ACKNOWLEDGMENTS

We gratefully acknowledge technical collaboration with the DuPont Company and the sponsorship of this research by the United States Display Consortium. K. H. C. thanks the Princeton Plasma Physics Laboratory for a PPST Fellowship.

## REFERENCES

1. Jang Yeon Kwon, Do Young Kim, Hans S. Cho, Kyung Bae Park, Ji Sim Jung, Jong Man Kim, Young Soo Park, and Takashi Noguchi, "Low Temperature Poly-Si Thin Film Transistor on Plastic Substrates", *IEICE Trans. Electron.*, vol. E88–C, no.4, pp. 667-671 (2005).
2. D.P. Gosain, T. Noguchi, and S. Usui, "High mobility thin film transistors fabricated on plastic substrates at a processing temperature of 110°C," *Jpn. J. Appl. Phys. 2, Lett.*, vol.39, no.3A/B, pp.L179–L181 ( March 2000).
3. K. Long, A. Z. Kattamis, I-C. Cheng, H. Gleskova, S. Wagner, J. C. Sturm, M. Stevenson, G. Yu, and M. O'Regan, "Active-Matrix Amorphous-Silicon TFTs Arrays at 180°C on Clear Plastic and Glass Substrates for Organic Light-Emitting Displays", *IEEE Trans. Elec. Dev.*, vol. 53, no. 8, pp. 1789-1796 (August 2006).
4. K. R. Sarma, "a-Si TFT OLED Fabricated on Low-Temperature Flexible Plastic Substrate", *Mat. Res. Soc. Symp. Proc.*, Vol. 814, pp I13.1.1-12 (2004).
5. William A. MacDonald, "Engineered Films for Display Technologies", *J. Mater. Chem.*, vol. 14, pp 4-10 ( 2004.)

6. C. Blaauw, "Preparation and Characterization of Plasma-Deposited Silicon Nitride", J. Electrochem. Soc., vol. 131, pp. 1114-1118 (1984).

7. S. Wagner, H. Gleskova, J. C. Sturm, and Z. Suo, "Novel processing technology for macroelectronics," in Technology and Application of Amorphous Silicon, R. A. Street, editor Springer, Berlin, pp.222-251 (2000).

8. C.R. McArthur, "Optimization of 75°C Amorphous Silicon Nitride for TFTs on Plastics", MASc thesis, University of Waterloo (2003).

9. R. B. Wehrspohn, S. C. Deane, and I. D. French et al., "Relative importance of the Si-Si bond and Si-H bond for the stability of amorphous silicon thin film transistors," J. Appl. Phys., vol. 87, issue 1, pp. 144–154 (January 2000).

10. C.-S. Yang, L. L. Smith, C. B. Arthur, and G. N. Parsons, "Stability of low-temperature amorphous silicon thin film transistors formed on glass and transparent plastic substrates," J. Vac. Sci. Technol. B, vol. 18, no. 2, pp. 683-689 (March/April 2000).

11. Y. Kaneko, A. Sasano, and T. Tsukada, "Characterization of instability in amorphous silicon thin-film transistors," J. Appl. Phys., vol. 69, pp. 7301–7305 (1991).

12. K. Long, A.Z. Kattamis, I-C. Cheng, H. Gleskova, S. Wagner, J.C. Sturm, "Stability of amorphous-silicon TFTs deposited on clear plastic substrates at 250°C to 280°C," IEEE Elec. Dev. Lett., vol.27, no.2, pp. 111- 113 (Feb. 2006).

13. Ke Long, "Towards Flexible Full-Color Active Matrix Organic Light-Emitting Displays: Dry Dye Printing For OLED Integration and 280°C Amorphous-Silicon Thin-Film Transistors on Clear Plastic Substrates", Ph.D thesis, Princeton University (2006).

14. A. Z. Kattamis, I-Chun Cheng, Ke Long, Bahman Hekmatshoar, Kunigunde Cherenack, Sigurd Wagner, James C. Sturm, Sameer Venugopal, Douglas E. Loy, Shawn M. O'Rourke, and David R. Allee. "Amorphous Silicon Thin Film Transistor Backplanes Deposited at 200°C on Clear Plastic", J. Display Technology, vol.3, no.3, pp. 304-308 (September 2007).

15. Kunigunde H. Cherenack, Alex Z. Kattamis, Bahman Hekmatshoar, James C. Sturm, and Sigurd Wagner, "Amorphous-Silicon Thin-Film Transistors Fabricated at 300°C on a Free-Standing Foil Substrate of Clear Plastic," IEEE Electron Device Lett., vol. 28, no. 11, pp.1004-1006 (November 2007).

16. Bahman Hekmatshoar, Alex Z. Kattamis, Kunigunde H. Cherenack, Ke Long, Jian-Zhang Chen, Sigurd Wagner, James C. Sturm, Kamala Rajan, and Michael Hack, "Reliability of Active-Matrix Organic Light-Emitting-Diode Arrays With Amorphous Silicon Thin-Film Transistor Backplanes on Clear Plastic", IEEE Electron Device Lett., vol. 29, pp. 63-66 (2008).

17. I-Chun Cheng, Alex Kattamis, Ke Long, Jim Sturm, Sigurd Wagner, "Stress control for overlay registration in a-Si:H TFTs on flexible organic-polymer-foil substrates", Journal of the SID, vol. 13, no. 7, pp. 563-568 (2005).

18. H. Gleskova, I. C. Cheng, S. Wagner, and Z. G. Suo, "Thermomechanical criteria for overlay alignment in flexible thin-film electronic circuits," Applied Physics Lett., vol. 88, pp. 011905-1-3 (2006).

19. F. Lemmi, W. Chung, S. Lin, P. M. Smith, T. Sasagawa, B. C. Drews, A. Hua, J. R. Stern, and J. Y. Chen, "High-performance TFTs fabricated on plastic substrates," IEEE Electron Device Lett., vol. 25, no. 5, pp. 486–48 ( 2004)

20. A. Kattamis, I.-C. Cheng, K. Long, J. C. Sturm, and S.Wagner, "Dimensionally stable processing of a-Si TFTs on polymer foils," in Proc. 47$^{th}$ Ann. TMS Electron. Mater. Conf., p. 73 (2005)

21. W. S. Wong, K. E. Paul, and R. A. Street, "Digital-lithographic processing for thin-film transistor array fabrication," J. Non-Cryst. Sol., vol. 338–340, pp. 710–714 (2004)

22. P. Roca i Cabarrocas, R. Brenot, P. Bulkin, R. Vanderhaghen, B. Drevillon, and I. French, "Stable microcrystalline silicon thin-film transistors produced by the layer-by-layer technique," J. Appl. Phys. 86, 7079-7082 (1999)

23. I-Chun Cheng, Alex Z. Kattamis, Ke Long, James C. Sturm, and Sigurd Wagner, "Self-Aligned Amorphous-Silicon TFTs on Clear Plastic Substrates", IEEE Trans. Elec. Dev, vol. 27, no. 3 (March 2006

Mater. Res. Soc. Symp. Proc. Vol. 1066 © 2008 Materials Research Society          1066-A20-04

## Aligned-Crystalline Si Films on Glass

Alp T. Findikoglu[1], Ozan Ugurlu[2], and Terry G. Holesinger[2]

[1]MPA-STC, Los Alamos National Laboratory, MS T004, Los Alamos, NM, 87544

[2]MPA-STC, Los Alamos National Laboratory, MS K763, Los Alamos, NM, 87544

## ABSTRACT

We report structural and electronic properties of Aligned-Crystalline Si (ACSi) films on glass substrates. These films show enhanced majority carrier mobilities and minority carrier lifetimes with increasing crystallinity, i.e., with improving alignment and connectivity of the grains. A 0.4-$\mu$m-thick ACSi film with a total grain mosaic spread of 4.2° showed Hall mobility of 47 cm$^2$/V.s for a p-type doping concentration of $1.9 \times 10^{18}$ cm$^{-3}$. A prototype n+/p/p+–type diode fabricated using a 4.2-$\mu$m-thick ACSi film showed minority carrier lifetime of ~3.5 $\mu$s and estimated diffusion length of ~30 $\mu$m in the p layer with a doping concentration of $5 \times 10^{16}$ cm$^{-3}$.

## INTRODUCTION

This study describes microstructural and electronic properties of highly-crystalline Si, Aligned-Crystalline Si (ACSi), films that are epitaxially deposited on glass substrates. Glass substrates, due to their transparency, durability, and chemical robustness, are of particular interest for use in applications such as sensors, photovoltaics, and displays. However, since glass is amorphous, it cannot be used directly as an epitaxial template in conventional growth processes. To achieve ACSi film growth on such an amorphous substrate, we first grow an ion-beam-assisted deposition (IBAD) textured buffer layer that is biaxially-oriented (i.e., with preferred out-of-plane and in-plane crystallographic orientations); this layer in turn serves as a crystalline template for the growth of subsequent epitaxial buffer layers and ACSi films. With improving crystalline quality, such ACSi films show enhanced majority carrier mobility and improved minority carrier lifetime.

## EXPERIMENTAL DETAILS

In the IBAD texturing process,[1,2] an off-normal ion beam establishes biaxially-oriented grains of a growing film of certain materials on a non-single-crystalline substrate, thus establishing a highly crystalline template for the epitaxial growth of subsequent layers. The details of the IBAD texturing process are still not well understood,[1,3,4] although some computational and analytical modeling work has been done in the area of ion-atom interactions and their effect on texture formation.[5,6]

We have previously reported our work on the preparation of ACSi films on metal tapes using an IBAD textured buffer layer.[7,8,9] In this letter, we extend our work to glass substrates, and expand our device fabrication and characterization studies to include thin-film diodes and their electrical characteristics. Various alternative approaches to obtain crystalline Si on glass have also been extensively explored by other research groups.[10,11,12,13]

In this study, all deposited layers were electron-beam (e-beam) evaporated in situ on 0.5-mm-thick glass substrates that were cut in 1 cm x 1 cm squares. The substrates were commercially-obtained, fusion-drawn, Boro-alumino-silicate glass, with the thermal expansion coefficient closely matched to that of Si, and with a softening point ($10^{7.6}$ poises) of 985°C. We first deposited a 5-nm-thick amorphous Si-O film at room temperature as a nucleation layer on the glass substrates. On this nucleation layer, we deposited a 3-to-5-nm-thick IBAD-textured MgO layer with a 750-eV $Ar^+$ assist beam (at ambient temperature, with ion/atom ratio of about 1, and a 45° angle between the ion beam and the substrate normal) to form a biaxially-oriented crystalline template layer. On the IBAD MgO layer, we deposited a 40-nm-thick homo-epitaxial MgO layer at about 500 °C,[3,4] followed by a hetero-epitaxially deposited 150-nm-thick γ-Al₂O₃ layer at about 750 °C to complete the buffer layer stack. This final layer of the buffer acted not only as an effective diffusion barrier, but also as a robust template layer for the hetero-epitaxial Si film growth at ~750 °C under an ambient chamber pressure of ~$10^{-6}$ Torr. The Si growth rates varied between 0.5 to 2 nm/s. The p-type doping in the Si films was achieved by adding appropriate amounts of B to the Si source material, whereas n-type doping was achieved by simultaneous evaporation of $P_2$ from a compound dissociation cell during Si growth.

The crystallographic orientation, epitaxial relationship, and microstructure of the multilayered samples were analyzed by X-ray diffraction (XRD), transmission electron microscopy (TEM), energy dispersive spectroscopy (EDS), scanning electron microscopy (SEM), and selected area diffraction (SAD). For these characterizations, we prepared two samples with different Si layer thicknesses (4 μm and 10 μm) under identical growth conditions.

## DISCUSSION

The XRD θ-2θ scan (Fig. 1a) shows both 4-μm and 10-μm-thick Si films grew epitaxially on the γ-Al₂O₃ buffer layer, with a (001) orientation perpendicular to the substrate surface. The rocking curve on the Si (004) peak yielded an average full width at half-maximum (FWHM) value, $\Delta\omega_{Si(004)}$, of about 1.5° (Fig. 1b), whereas the φ-scan on the Si (220) peak showed pure four-fold symmetry, with an average FWHM value, $\Delta\phi_{Si(220)}$, of about 3.2° (Fig. 2c), yielding out-of-plane ($\Delta\omega_{Si(004)}$) and in-plane ($\Delta\phi_{Si(400)}$) mosaic spreads of 1.5° and 2.8°, respectively. The average total mosaic spread $TMS$,[4,7] for these samples is thus ~3.5°.

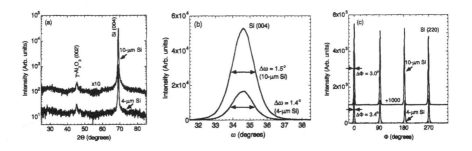

Figure 1. a) XRD θ-2θ scan of the multilayer structure, showing (001)-oriented ACSi and γ-Al₂O₃ layers, b) ω rocking curve on the Si (004) peak, and c) φ-scan of the Si (022) peak.

Figure 2a shows TEM/SEM/SAD composite micrograph of the 10-μm-thick ACSi sample. The SEM image (top left) shows smooth grain boundaries with a grain size of ~1 μm for the ACSi film. The TEM image (bottom) shows a columnar structure, in which misfit dislocations and stacking faults are visible within the Si film. Similar defects have been observed in silicon-on-sapphire (SOS) and Si/γ-Al₂O₃ double hetero-structures on Si single-crystals and are mostly due to lattice mismatch.[14] The SAD pattern (top right) confirms the grain alignment in the Si film that the X-ray rocking curves and phi-scans (see Figs 1b and 1c) indicate. Figure 2c shows an EDS line scan on the Si film, indicating that the buffer layer does not exhibit any discernible reaction/interdiffusion with either the glass substrate or the Si film.

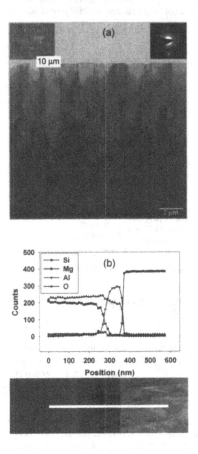

**Figure 2.** TEM/SEM/SAD composite micrograph of a) 10-μm-thick ACSi film on buffer layer on glass; b) an EDS line scan on the ACSi film on buffer layer on glass, indicating no discernible interdiffusion or chemical reaction.

To study the effects of inter-grain alignment on the majority carrier transport in ACSi films on glass, we prepared a series of samples with varying total mosaic spreads and patterned them for Hall mobility measurements. Figure 3 shows normalized Hall mobility vs total mosaic spread *TMS* for five ACSi film samples on glass. As reference, we have added to this plot our previous results on ACSi films on metal-alloy substrates. Also, to be able to compare the carrier mobility of samples with different doping concentrations, we have normalized the carrier mobility of the films with respect to bulk single-crystal Si, and used a Hall factor of 0.8 to convert drift mobilities to Hall mobilities.[15,16] These new results follow a common trend that we had observed previously for ACSi films on metal-alloy tapes;[7,8,9] namely, for a given film thickness and doping concentration, as the *TMS* is reduced (i.e., the inter-grain alignment is improved) the carrier mobility increases. For example, a 0.4-μm-thick ACSi film on glass with a *TMS* of 4.2° showed Hall mobility of 47 $cm^2$/V.s (or, normalized mobility of 0.60) for a p-type doping concentration of $1.9\times10^{18}$ $cm^{-3}$.

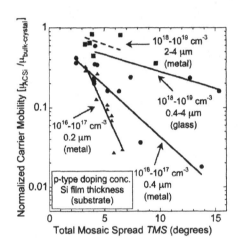

**Figure 3**. Normalized carrier mobility vs total mosaic spread *TMS* for ACSi films on glass and metal substrates (lines are guides to the eye).

To examine electronic properties further, we prepared a vertical diode of the type n+/p/p+ where n+, p, and p+ refer to a heavily-doped n-type (~$10^{19}$ $cm^{-3}$), a lightly-doped p-type (~$5\times10^{16}$ $cm^{-3}$), and a heavily-doped p-type (~$10^{18}$ $cm^{-3}$) ACSi layer, respectively. We used reactive ion etching to pattern ~0.5 mm x 0.5 mm mesas of n+ and p layers. We then evaporated ~200 nm of Ni and Al to form ohmic contacts on the n+ and p+ layers, respectively. The ACSi film for this diode had *TMS* ~ 5.9°.

Figure 4a shows the capacitance-voltage (C-V) characteristics of the diode prototype. A linear fit to the $C^{-2}$ vs V curve yields a uniform p-layer doping of ~$5\times10^{16}$ $cm^{-3}$ and a built-in voltage $V_{bi}$ of ~1.0 V. This built-in voltage is similar to what is expected in a good crystalline-film Si diode with the above doping levels.[16] Normalized capacitance (C/C₀) and conductance (G/G₀) vs frequency of diodes can be used to estimate minority carrier lifetime in the lightly doped layer of a diode, as shown in Figure 4b.[16] Combining this effective lifetime with an approximate estimate of electron mobility of ~150 $cm^2$/V.s in the p-layer (using the normalized hole mobility plot of Figure 3 and assuming that a similar dependence will hold for electrons) we estimate the effective diffusion length in the p layer of the diode prototype to be ~30 μm.

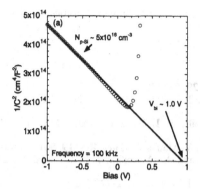 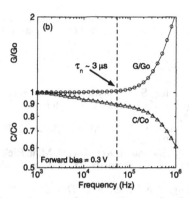

**Figure 4.** a) Capacitance$^{-2}$ vs bias, and b) normalized capacitance (C/Co) and conductance (G/Go) vs frequency for a 4.2-μm-thick ACSi film diode of the type n+/p/p+.

## CONCLUSIONS

We have adapted our IBAD texturing technique to grow biaxially-oriented buffer layers on glass substrates. This IBAD layer constitutes a transformation of the amorphous surface of the glass to a highly ordered surface amenable to epitaxial growth of subsequent layers. We have built an epitaxial buffer stack on this IBAD layer that allowed growth of ACSi films with varying concentrations of n- and p-type doping, and with thicknesses up to the explored range of 10 μm. Hall measurements confirmed that with improving crystallinity (i.e., with improving connectivity and alignment of the grains) in the ACSi film, the majority carrier mobility improves, asymptotically approaching values expected of single-crystalline Si films. We have also studied the minority carrier lifetime in ACSi films by fabricating a n+/p/p+ type diode. This diode prototype has provided evidence that our approach allows building of abrupt doping profiles of both p- and n-type crystalline Si films in situ on glass. Also, we have observed relatively long effective minority carrier lifetimes (~3 μs) and diffusion lengths (~30 μm) in the 3-μm-thick p layer of such a diode. We note that this diode was based on an ACSi film with a non-optimal *TMS* of ~5.9°.

Most photovoltaics and electronics applications of semiconductor films on glass substrates would benefit from improvements in performance and/or reductions in cost. Current technologies either use inexpensive but poor-performing amorphous semiconductor films, such as amorphous silicon,[17] or attempt to improve performance by transforming amorphous films to polycrystalline films by post-processing, such as solid phase crystallization,[18,19] or pulsed-laser crystallization.[20,21] These techniques lead to polycrystalline texture in the film with large-angle grain boundaries that could be detrimental to electrical and optical performance. Our approach uses an in situ process to grow grain-aligned Si films on glass substrates, thus retains the advantages of conventional epitaxial growth techniques, such as better doping profile control, lower processing temperatures, and ease of monolithic device integration.

## ACKNOWLEDGEMENTS

This work was funded by Los Alamos National Laboratory Directed Research and Development Project under the United States Department of Energy. The authors thank Woong

Choi for his prior work, Vadimir Matias and Ian H. Campbell for helpful discussions, and Jeffrey O. Willis for his useful comments on the manuscript.

## REFERENCES

1. P. N. Arendt, S. R. Foltyn, Mater. Res. Soc. Bull. **29**, 543 (2004).
2. Y. Iijima, K. Kakimoto, Y. Sutoh, S. Ajimura, T. Saitoh, Supercond. Sci. Tech. **17**, 264 (2004).
3. C. P. Wang, K. B. Do, M. R. Beasley, T. H. Geballe, R. H. Hammond, Appl. Phys. Lett. **71**, 2955 (1997).
4. A. T. Findikoglu, S. Kreiskott, P. M. te Riele, and V. Matias, J. Mater. Res. **19**, 501 (2004).
5. L. Dong, D. J. Srolovitz, G. S. Was, Q. Zhao, and A. D. Rollett, J. Mater. Res. **16**, 210 (2001).
6. L. Dong, L. A. Zepeta-Ruiz, D. J. Srolovitz, J. Appl. Phys. **89**, 4105 (2001).
7. A. T. Findikoglu, W. Choi, V. Matias, T. G. Holesinger, Q. X. Jia, D. E. Peterson, Adv. Mater. **17**, 1527 (2005).
8. W. Choi, J. K. Lee, A. T. Findikoglu, Appl. Phys. Lett. **89**, 262111 (2006).
9. A. T. Findikoglu, W. Choi, M. Hawley, M. J. Romero, K. M. Jones, M. M. Al-Jassim, in *Progress in Advanced Materials Research* (Ed: N. H. Voler, Nova Science Publishers, Hauppauge, New York, 2007), Ch. **6**.
10. K. Yamamoto, IEEE Trans. Electron Dev. **46**, 2041 (1999).
11. A. G. Aberle, P. I. Widenborg, D. Song, A. Straub, M. L. Terry, T. Walsh, A. Sproul, P. Campbell, D. Inns, B. Beilby, M. Griffin, J. Weber, Y. Huang, O. Kunz, R. Gebs, F. Martin-Brune, V. Barroux, S. H. Wenham, "Recent Advances in Polycrystalline Silicon Thin-Film Solar Cells on Glass at UNSW", presented at *Thirty-First IEEE Photovoltaic Specialists Conference* (Lake Buena Vista, FL, USA, 2005).
12. J. H. Werner, R. Dassow, T. J. Rinke, J. R. Kohler, R. B. Bergmann, Thin Sol. Films **383**, 95 (2001).
13. C. W. Teplin, D. S. Ginley, H. M. Branz, J. Non-Cryst. Solids **352**, 984 (2006).
14. T. Kimura, A. Sengoku, M. Ishida, Jpn. J. Appl. Phys. **35**, 1001 (1996).
15. J. F. Lin, S. S. Li, L. C. Linares, K. W. Teng, Solid State Electron. **24**, 827 (1981).
16. S. M. Sze, *Physics of Semiconductor Devices* (Wiley-Interscience, New York, 1981).
17. S. Toshiharu, J. Appl. Phys. **99**, 11 (2006).
18. L. Haji, P. Joubert, J. Stoemenos, N. A. Economou, J. Appl. Phys. **75**, 3944 (1994).
19. R. B. Bergmann, J. Kohler, R. Dassow, C. Zaczek, J. H. Werner, Physica Status Solidi A**166**, 587 (1998).
20. J. S. Im, R. S. Sposili, M. A. Crowder, Appl. Phys. Lett. **70**, 3434 (1997).
21. M. Tai, M. Hatano, S. Yamaguchi, T. Noda, S. K. Park, T. Shiba, M. Ohkura, IEEE Trans. Electron. Devices **51**, 934 (2004).

Mater. Res. Soc. Symp. Proc. Vol. 1066 © 2008 Materials Research Society      1066-A20-06

## Monolithic 3D Integration of Single-Grain Si TFTs

Mohammad Reza Tajari Mofrad, Ryoichi Ishihara, Jaber Derakhshandeh, Alessandro Baiano, Johan van der Cingel, and Cees Beenakker
Department of Electrical Engineering, DIMES/ECTM, Technical University of Delft, Delft, Netherlands

## ABSTRACT

Vertical stacking of transistors is a promising technology which can realize compact and high-speed integrated circuits (ICs) with a short interconnect delay and increased functionality. Two layers of low-temperature fabricated single-grain thin-film transistors (SG TFTs) have been monolithically integrated. NMOS mobilities are 565 and 393 cm$^2$/Vs and pMOS mobilities are 159 and 141 cm$^2$/Vs, for the top and bottom layers respectively. A three-dimensional (3D) inverter has also been fabricated, with one transistor on the bottom layer and the other on the top layer. The inverters showed an output voltage swing of 0 to 5 V with a switching voltage of around 2 V.

## INTRODUCTION

Down-scaling of transistors increases the density of the chip and the interconnects. Higher device density needs more space reserved for routing purposes, which decreases the advantage of down-scaling. Also the interconnect delay increases in such a way that becomes the limiting factor for circuit performance [1]. Three-dimensional integration is a solution for these issues as it decreases the interconnect length and power consumption. It also increases the functionality of a chip by enabling integration of a sensor layer on top of it. High-density ICs may be manufactured using the existing industrial infrastructure, eliminating the expenses required for developing new equipment or new sub-micron processes.

3D integration may be in package level, wafer level [2][3] or device level. The first two suffer from the low interconnect density between the stacked layers limited by the alignment accuracy of about 1 μm, between the packages or the wafers. Vertical stacking in device level, or monolithic integration, results in the largest decrease in interconnect length decrease, highest density of interconnects between the successive active silicon layers and offers new routing possibilities crucial to system-on-chip (SoC) design.

Device level integration of electronics started in the early eighties. The first attempts towards monolithic fabrication were not successful [4][5]. Regardless of the approach, the main issue was the high temperature (>600 °C) needed to form high quality silicon for the top layers, which caused doping redistribution in the bottom layers [6]. Another limitation was the usage of single-crystalline Si wafer as seeding the top layer Si growth in other successful results[7]. This limits the application to microelectronics only and is useless for large area electronics.

The SG TFT are aimed at obtaining characteristics comparable to CMOS devices by placing the channel of a TFT inside a single grain. Single-grain location-controlled silicon is obtained by the so-called μ-Czochrolski process which is a location-controlled crystallization method using excimer laser [8]. In this method the position of the grains is controlled and the channel of the TFTs are designed to fit inside a grain-boundary-less area. One of the great advantages of this technology is that no c-Si wafer is needed as the seeding layer. This advantage makes it attractive not only for microelectronics, but also for large area electronics like display applications. TFTs with high mobilities of the order of 600 $cm^2$/Vs are reported in the past using this process[9].

In this paper, monolithic stacking of two layers of SG TFTs is reported. Characteristics of p- and nMOS transistors fabricated on bottom or top silicon layer, using maximum temperature of 350 °C after the a-Si deposition, are presented. Two kinds of 3D CMOS inverters are constructed: n-channel SG TFT stacked on p-channel devices and vice versa. These inverters are capable of full voltage swing and have symmetric voltage characteristics. These results demonstrate that SG TFTs are promising for future 3D IC applications.

## FABRICATION

The process flow starts with μ-Czochrolski process: location-controlled crystallization of a-Si. Cavities of 700 nm depth and 100 nm diameter are made in 1.5 μm thick $SiO_2$ on a bulk-Si wafer, using dry etching and successive $SiO_2$ deposition grown from tetra-ethyl-ortho-silicate (TEOS) [8]. The cavities are filled with 250 nm thick low-pressure chemical vapor deposition (LPCVD) a-Si at 545 °C, implanted a suitable channel doping (~$10^{15}$ $cm^{-2}$) and crystallized by means of excimer laser pulses ($\lambda$=308, E=1200 mJ/$cm^2$). Islands are patterned in this single-grain silicon and serve as the SG TFT active region. 30 nm thick inductively coupled pressure-enhanced chemical vapor deposition (ICPECVD) oxide is used as gate dielectric, deposited at 250 °C. Next, a 250 nm a-Si is deposited and patterned as gate. Source, drain and gate are implanted (~$10^{18}$ $cm^{-2}$) and activated by the excimer laser with energy density of 300 mJ/$cm^2$. During this step, the a-Si gate is crystallized into 100 nm large grain poly silicon. Figure 2 shows the surface and the cross-section of a test structure, with an a-Si layer on top of an LPCVD p-silicon layer. After the dopant activation and the annealing step, the grain sizes of these two layers are of the same order of size. The bottom layer is then passivated by 2 μm thick PECVD oxide grown from a TEOS source at 350 °C and planarized using CMP. The CMP consist of two major planarizing and smoothening steps in which Rodel ILD1300 with PH of 10.7 is used as a slurry.

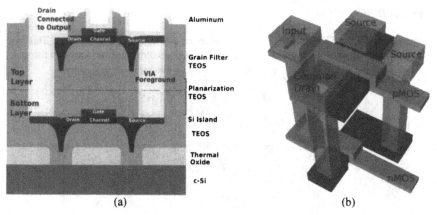

(a)                                                    (b)

Figure 1: (a) Shows a schematic overview of a 3D Inverter; (b) Shows a three-dimensional layout of an inverter used in a TCAD simulation. The underlying transistor which is an nMOS, has an extended source allowing the connection to the ground. The common connections are also shown.

Processing the second layer starts also with μ-Czochrolski process and follows the same process flow as the bottom layer until planarization. As the planarity of this surface is less crucial, a two step deposition and etch-back of SiO₂ is enough to smooth the sharp steps. The vias connect the bottom layer to the pads and to the top layer contacts. For the common contacts, a part of the active region of the top layer transistor is etched together with its underlying oxide, all the way to the active region of the bottom layer transistor (Figure 1). The interconnects are formed by a 675 nm thick Al, sputtered at 350 °C. The maximum process temperature is 350 °C after the a-Si LPCVD step.

(a)                                                    (b)

Figure 2: (a) A SEM image of the surface of an a-Si test layer on an LPCVD p-Si layer after the laser annealing. The grain sizes are of the same order; (b) A TEM cross-section image of the same structure. It is shown that the a-Si layer has been crystallized in the hole film depth.

485

## RESULTS and DISCUSSION

Figure 3 shows the TEM image of two transistors on two adjacent silicon layers. This cross section is made across the gate and shows the success of common contact between the two

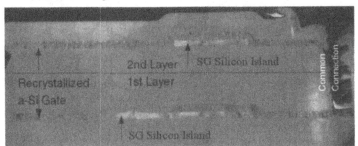

Figure 3: TEM cross-sectional image of a 3D inverter, with an nMOS device on the top layer and a pMOS on the bottom. The cross-section is made along the gate width. The channel area of the transistors is visible in the image

layers as it was discussed in the previous section. The depth of the bottom layer is 3 μm and the vias are 1.5 μm wide. The image shows that the thickness of Al at the bottom of the via (with aspect-ratio of 2) is very thin which increases the contact resistivity.

Several devices with different channel widths are fabricated. The channel length is 1.5 μm. NMOS and pMOS transistors on both layer show good matching. Table 1 reports the average characteristic values for different devices on each layer. Figure 4 shows the transfer characteristics of the top and bottom SG TFTs. The transistors have high mobilities, and threshold voltage variation of 0.1 V. A good matching between the characteristics of the top and bottom transistors is observed. The main reason for the high drain leakage ($I_{leakage}$) of the bottom layer TFT, is the poor gate patterning process which caused the over-etch of the islands at the source and drain regions. The maximum drain current ($I_{d(max)}$) is measured at $V_{ds}$ =0.1 V. $I_{leakage}$ is measured at $V_{gs}$= 2 V and -3 V for the p-and nMOS devices, respectively.

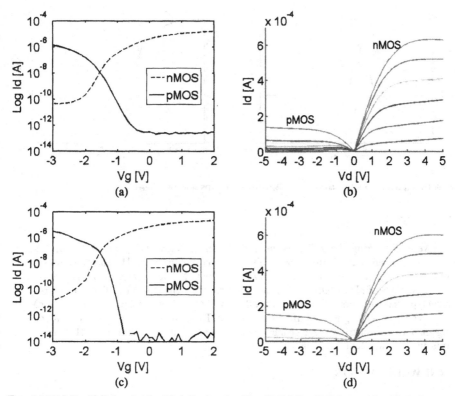

**Figure 4:** (a),(b) $I_d$-$V_g$ and $I_d$-$V_d$ characteristics of the bottom layer transistors; (c),(d) $I_d$-$V_g$ and $I_d$-$V_d$ characteristics of the top layer transistors

**Table 1: Characteristics of MOS devices**

| Type of device | Mobility [cm²/Vs] | S [mV/dec] | $V_{th}$ [V] | $I_{d(max)}$ [A] | $I_{leakage}$ [A] |
|---|---|---|---|---|---|
| nMOS Top | 565 | 245 | -0.8 | $2.19 \times 10^{-05}$ | $5.39 \times 10^{-12}$ |
| pMOS Top | 159 | 95 | -2.4 | $1.61 \times 10^{-05}$ | $2.80 \times 10^{-13}$ |
| nMOS Bottom | 393 | 280 | -0.6 | $1.81 \times 10^{-05}$ | $1.49 \times 10^{-10}$ |
| pMOS Bottom | 141 | 151 | -2 | $1.43 \times 10^{-05}$ | $9.2 \times 10^{-13}$ |

Two different 3D configurations have been fabricated: nMOS on pMOS and vice versa. The first type shows better device characteristics, which may be due to fabrication process variables. Switching voltage of the inverters lies around 2 V. The input signal swings from 0 to 5 V, and so does the output. The voltage and current characteristics of the inverters are give in Figure 5.

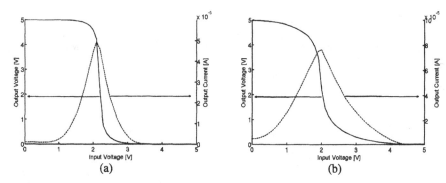

Figure 5: The output characteristics of (a) a pMOS on nMOS and (b) an nMOS on pMOS inverters

## CONCLUSIONS

Vertical stacking of SG TFTs is successfully demonstrated. The grain sizes of the bottom and top Si layers are 5 and 6 μm, respectively. Top layer mobilities are 159 and 565 cm$^2$/Vs and bottom layer mobilities are 141 and 393 cm$^2$/Vs, for the pMOS and nMOS devices respectively.

Two kinds of 3D CMOS inverters are constructed; n-channel SG TFT stacked on p-channel type and vice versa. The channel width of pMOS devices is 4.2 μm, twice that of nMOS devices. The channel length is 1.5 μm for all devices. These inverters are capable of full voltage swing and have symmetric voltage characteristics. These results suggest that SG TFTs are promising for future 3D IC applications.

## ACKNOWLEDGMENTS

The authors would like to thank the ICP members and staff of DIMES research center for their support during the fabrication of these devices and the Dutch Technology Foundation STW for their financial support. They also thank Dr. H. Shimada for useful discussions.

## REFERENCES

1- K.C. Saraswat, F. Mohammadi, "Effect of scaling of interconnections on the time delay of VLSI circuits", IEEE Transactions on Electron Devices 29 (1982)
2- V. W. C. Chen, P. C. H. C. and M. Chan, "Three Dimensional CMOS Integrated Circuits on Large Grain Polysilicon Films", International Electron Devices Meeting Technical Digest, 161-164 (2000)
3- T. Matsumoto, M. Satoh, K. Sakuma, H. Kurino, N. Miyakawa, H. Itani, and M. Koyanagi, "New three-dimensional wafer bonding technology using the adhesive injection method", Japanese Journal of Applied Physics Part 1-Regular Papers 37 (1998)
4- M. W. Geis, D. C. Flanders and H. I. Smith, "Crystallographic orientation of silicon on an amorphous substrate using an artificial surface-relief grating and laser crystallization", Applied Physics Letters 35, 71(1979)

5- R. Reif and J. E. Knott, J. E., "Low-temperature process to increase the grain size in polysilicon films", Electronics Letters 17, 586 (1981)

6- T.Nishimura, Y.Inoue, K.Sugahara, S.Kusunoki, T.Kumamoto, S.Nakagawa, M.Nakaya, Y.Horiba, and Y.Akasaka, "Three dimensional IC for high performance image signal processor", International Electron Devices Meeting 33, 111-114 (1987)

7- Yong-Hoon Son et al., "Laser-induced Epitaxial Growth (LEG) Technology for High Density 3-D Stacked Memory with High Productivity", IEEE Symposium on VLSI Technology, 80-81(2007)

8- Paul C. van der Wilt, B. D. van Dijk, G. J. Bertens, R. Ishihara, and C. I. M. Beenakker, "Formation of location-controlled crystalline islands using substrate-embedded seeds in excimer-laser crystallization of silicon films", Applied Physics Letters 79, 1819 (2001)

9- Vikas Rana et al., "High Performance P-Channel Single-Crystalline Si TFTs Fabricated Inside a Location-Controlled Grain by u-Czochralski Process", IEICE Transactions on Electronics, E87-C (11), 1943-1947 (2004)

# AUTHOR INDEX

Adjallah, Y., 29
Adriaenssens, Guy J., 211
Al-Jassim, M.M., 87
Anderson, Curtis, 29, 185
Andreu, Jordi, 173
Antonuk, L.E., 457
Antony, Aldrin, 173
Arai, Toshiaki, 243
Asensi, José Miguel, 173
Asomoza, Rene, 113

Baiano, Alessandro, 483
Ballif, Christophe, 265
Banerjee, Arindam, 325
Barata, Manuel, 225
Becker, Christiane, 139
Beenakker, Cees, 483
Beglau, Dave, 325
Behravan, M., 457
Behrends, Jan, 139
Bertomeu, Joan, 173
Beyer, Wolfhard, 179
Birkmire, R.W., 41
Biswas, Rana, 331
Blackwell, C., 29, 155
Blomberg, Tom, 131
Bobela, David C., 279
Bowden, S., 41
Branz, Howard M., 15, 259, 285
Brinza, Monica, 211
Brown, W.D., 345
Brunner, Robert, 199
Burrows, M.Z., 41

Cabeça, R., 451
Carius, Reinhard, 179, 293
Carvalho, N., 407
Chakraborty, Sudeshna, 279
Chan, Kah-Yoong, 293
Chang, Chung-Hsiang, 353, 443
Chang, Jeff Hsin, 431, 463
Checa, Nery D., 113
Chen, Jian Z., 385
Cheng, I-Chun, 385
Cherenack, Kunigunde H., 471

Choi, Joonhoo, 361
Choi, Sung-Hwan, 361, 373
Chou, Li-Jen, 353, 443
Chu, Virginia, 339, 451
Cohen, J. David, 149, 259, 325
Collins, Robert W., 253
Conde, Joao Pedro, 339, 451
Cosme, Ismael, 419

Das, U.K., 41
Datta, Shouvik, 259
DeMaggio, Greg, 325
Derakhshandeh, Jaber, 483
Devilee, Camile, 67
De Wolf, Stefaan, 49
Dickey, Elizabeth C., 253
Dippo, P., 87
Dogan, Pinar, 139
Dorenkamper, Maarten, 67
Drabold, D.A., 279
Duggan, J.L., 217
Dunand, Sylavain, 265

El-Mohri, Y., 457
Enomoto, Hirofumi, 315
Escarré, Jordi, 173
Esmaeili-Rad, Mohammad R., 379

Fantoni, Alessandro, 225, 407
Fenske, Frank, 139
Fernandes, Miguel, 225, 407, 425
Findikoglu, Alp T., 477
Fujiwara, Hiroyuki, 21, 49, 437

Gall, Stefan, 139
Gaspar, Joao, 339
Georgiev, Daniel, 393
Gorka, Benjamin, 139
Gorman, B., 217
Gudovskikh, A.S., 75
Guha, Subhendu, 61, 149, 273, 325

Ha, Tae-Jun, 373
Hailstone, Richard, 437
Han, Min-Koo, 301, 361, 373

Han, Sang-Myeon, 361
Hänel, Tobias, 139
Hata, Satoshi, 345
Haukka, Suvi, 131
Hayashi, Tsukasa, 167
Hegedus, S.S., 41
Heijna, Maurits C.R., 67
Heiler, Gregory, 413, 425, 431, 463
Hekmatshoar, Bahman, 471
Hernandez, Salvador G., 113
Hiroe, Akihiko, 307
Holesinger, Terry G., 477
Hollenstein, Christoph, 3
Horn, Mark W., 253
Hossain, K., 217
Howling, Alan A., 3
Hsieh, Chin-Hua, 353, 443
Hugger, Peter G., 149, 325

Imamura, Kentarou, 199
Inagaki, Yoshio, 243
Inayoshi, Yohei, 119
Ishihara, Ryoichi, 483
Ishikawa, Takamasa, 315

Jergel, Matej, 199
Jia, Haijun, 21
Jiang, Chun-Sheng, 61, 87
Jiao, Lihong (Heidi), 99
Jiménez, Juan, 193
Jones, Kim M., 285
Ju, Tong, 35, 273
Jurecka, Stanislav, 199

Kakalios, J., 29, 155
Kaki, Hirokazu, 167
Kalinin, Valeri V., 399
Kamoshita, Tomoyoshi, 315
Kang, Dong-Won, 301
Kao, Chyuan-Haur, 367
Kattamis, Alex Z., 471
Kawano, Shinichi, 315
Kim, Kyung Ho, 431
Kishore, Ram, 345
Kiyofuji, Shinji, 315
Kleider, J.P., 75

Kling, Andreas, 193
Knipp, Dietmar, 293
Kobayashi, Hikaru, 199
Kondo, Michio, 21, 49, 437
Koniczek, M., 457
Kopani, Martin, 199
Kortshagen, Uwe, 29, 155, 185
Kosarev, Andrey, 113, 419
Kudriavtsev, Yurii, 113
Kuk, Seung-Hee, 301
Kuo, Yue, 231
Kuraseko, Hiroshi, 21
Kúš, Peter, 205
Kuwano, Noriyuki, 345
Kwong, Ian Chi Yan, 161

Lai, Jackson, 413, 463
Lavareda, G., 407
Lee, Hyun Jung, 161
Lee, Won-Kyu, 361, 373
Leegwater, Hans, 67
Lennartz, Dorothea, 179
Li, J., 217
Li, Jing, 253
Li, Yang, 325
Lin, Gong-Ru, 353, 443
Lips, Klaus, 139
Liu, Chi-Wee, 353
Loffler, Jochen, 67
Louro, Paula, 225, 407
Luxembourg, Stefan L., 205

Mariotti, Davide, 437
Martin, Ina T., 285
Martínez, Óscar, 193
Matero, Raija, 131
Matsumoto, Mitsutaka, 119
McDaniel, F.D., 217
McDonald, J., 457
Mikula, Milan, 199
Miyagi, Masahide, 315
Moening, Joe, 393
Morana, Bruno, 193
Moreno, Mario M., 419
Morosawa, Narihiro, 243
Moutinho, H.R., 87

492

Muñoz, Delfina, 173
Myasnikov, Alexandre M., 399

Nakajima, Setsuo, 119
Nakamura, Tetsuro, 315
Naseem, Hameed, 345
Nathan, Arokia, 379, 413, 425,
    431, 463
Neogi, A., 217
Niessen, Lars, 179
Nishihara, Hironori, 315
Nominanda, Helinda, 231

Ogata, Kiyoshi, 167
Ohmi, Tadahiro, 307
Opila, R.L., 41

Park, Joong-Hyun, 301
Park, Sang-Geun, 301, 361, 373
Paul, Oliver, 339
Pennartz, Frank, 179
Pi, Xiaodong, 155
Pieters, Bart Elger, 93
Pimentel, A.C., 451
Pincik, Emil, 199
Podraza, Nikolas J., 253
Poudel, P.R., 217
Prazeres, D.M.F., 451

Rath, Jatindra K., 211
Rech, Bernd, 139
Reedy, Robert C., 285
Rodrigues, M., 451
Rodríguez, Andrés, 193
Rodríguez, Tomás, 193
Rojas, Fredy, 173
Romero, Manuel J., 87, 285
Root, Andrew, 131
Rout, B., 217
Ruske, Florian, 139

Saito, Junichiro, 315
Sakakibara, Yasushi, 315
Sameshima, Toshiyuki, 107
Sangrador, Jesús, 193
Sazonov, Andrei, 161, 379, 463

Schicho, Sandra, 93
Schropp, Ruud E.I., 211
Sharma, Renu, 345
Shimosawa, Makoto, 315
Shin, Kyung-Wook, 379
Soppe, Wim J., 67
Stella, Marco, 173
Stiebig, Helmut, 93, 293
Stradins, Paul, 15, 35, 285
Strahm, Benjamin, 3
Striakhilev, Denis, 413, 431, 463
Sturm, James C., 471
Su, Tining, 273
Suemitsu, Maki, 119
Sung, W.H., 367
Svrcek, Vladimir, 437

Tajari Mofrad, Mohammad Reza,
    483
Takahashi, Eiji, 167
Takahashi, Masao, 199
Takano, Akihiro, 315
Tatsuki, Koichi, 243
Taylor, P. Craig, 35, 273, 279
Teplin, Charles W., 15, 61, 285
Teramoto, Akinobu, 307
Tian, Longzhang, 125
Tichelaar, Frans D., 205
To, B., 87
Tomokiyo, Yoshitsuga, 345
Tomyo, Atsushi, 167
Torchynska, Tatyana V., 145
Torres, Alfonso J., 113, 419
Toyoshima, Yasutake, 119
Tredwell, Timothy, 413, 425, 431,
    463

Uehara, Tsuyoshi, 119
Ugurlu, Ozan, 477
Urabe, Tetsuo, 243
Urabe, Yuji, 107
Uraoka, Yukiharu, 167

Van Aken, Bas B., 67
van der Cingel, Johan, 483
Vieira, M., 407

Vieira, Manuel Augusto, 225
Vieira, Manuela, 225, 425
Villar, Fernando, 173
Vygranenko, Yuriy, 413, 425, 431,
    463

Wagner, Sigurd, 471
Wang, Xinwei, 125
Wronski, Christopher R., 99, 253
Wyrsch, Nicolas, 265

Xu, Xixiang, 149, 325
Xu, Yueqin, 259

Yan, Baojie, 61, 149, 273, 325
Yan, Yanfa, 61
Yang, Jeffrey, 61, 149, 273, 325
Young, David, 285
Yue, Guozhen, 61, 149, 325

Zeman, Miro, 205
Zhao, Q., 457
Zhou, Dayu, 331
Zyryanov, Vladislav E., 399

# SUBJECT INDEX

amorphous, 29, 49, 75, 99, 113,
    125, 145, 155, 259, 279,
    315, 325, 345, 367, 385,
    399, 413, 419, 451
annealing, 21, 41, 217
Ar, 217
atomic layer deposition, 131

barrier layer, 217
biomedical, 413
bonding, 279

chemical vapor deposition (CVD)
    (deposition), 173, 179, 193
crystal growth, 167, 185, 399, 483
crystalline, 3, 139, 243
crystallographic structure, 21

defects, 35, 99, 211, 259, 273
devices, 161, 293, 373, 463
display, 301

elastic properties, 339
electrical properties, 29, 75, 93,
    367, 379, 419
electron spin resonance, 35, 273
electronic
    material, 113
    structure, 211
energy generation, 265, 285
epitaxy, 285, 477

film, 353, 443

Ge, 113
grain boundaries, 87

infrared (IR) spectroscopy, 131,
    179, 205
insulator, 107, 425
ion-beam assisted deposition, 477

laser
    ablation, 437
    annealing, 125, 243, 393, 483

liquid, 107
lithography (removal), 471
luminescence, 145, 437

memory, 231
microelectro-mechanical
    (MEMS), 339
microelectronics, 161, 225, 265
microstructure, 179, 393

nanoscale, 437
nanostructure, 15, 61, 145, 155,
    193, 325
nuclear magnetic resonance
    (NMR), 279

optical properties, 199, 253
optoelectronic, 407, 425, 431, 451,
    463
oxide, 193

passivation, 41, 107
photoconductivity, 211, 225, 425
photovoltaic, 49, 61, 67, 93, 149,
    185, 205, 265, 325, 331
plasma-enhanced CVD (PECVD)
    (deposition), 3, 21, 61, 67, 119,
    167, 253, 307, 315, 353,
    367, 379, 385, 419, 443,
    471
polycrystal, 361, 457

scanning probe microscopy
    (SPM), 87
semiconducting, 407
sensor, 225, 407, 413, 431, 451
Si, 3, 15, 29, 35, 41, 49, 75, 87, 99,
    119, 125, 139, 149, 155,
    161, 167, 173, 185, 231,
    243, 253, 259, 285, 293,
    301, 307, 315, 331, 345,
    353, 361, 373, 385, 393,
    399, 443, 477
simulation, 331, 457
surface reaction, 131

thin film, 15, 67, 93, 119, 139, 149,
 173, 199, 231, 273, 293,
 301, 307, 339, 361, 373,
 379, 431, 457, 463, 471,
 483

transmission electron microscopy
 (TEM), 205, 345

x-ray reflectivity, 199

# The Economics and Uncertainties of Nuclear Power

Is nuclear power a thing of the past or a technology for the future? Has it become too expensive and dangerous, or is it still competitive and sufficiently safe? Should emerging countries invest in it? Can we trust calculations of the probability of a major nuclear accident? In the face of divergent claims and contradictory facts, this book provides an in-depth and balanced economic analysis of the main controversies surrounding nuclear power. Without taking sides, it helps readers gain a better understanding of the uncertainties surrounding the costs, hazards, regulation and politics of nuclear power. Written several years on from the Fukushima Daiichi nuclear disaster of 2011, this is an important resource for students, researchers, energy professionals and concerned citizens wanting to engage with the continuing debate on the future of nuclear power and its place in international energy policy.

FRANÇOIS LÉVÊQUE is Professor of Economics at Mines ParisTech and a part-time professor at the Robert Schuman Centre for Advanced Studies (European University Institute). He has advised many international bodies on energy policy and the economics of regulation, including the International Energy Agency, the OECD and the European Commission.

# The Economics and Uncertainties of Nuclear Power

FRANÇOIS LÉVÊQUE

CAMBRIDGE
UNIVERSITY PRESS

# CAMBRIDGE
## UNIVERSITY PRESS

University Printing House, Cambridge CB2 8BS, United Kingdom

Cambridge University Press is part of the University of Cambridge.

It furthers the University's mission by disseminating knowledge in the pursuit of education, learning and research at the highest international levels of excellence.

www.cambridge.org
Information on this title: www.cambridge.org/9781107087286

© François Lévêque 2015

First published 2015

*A catalogue record for this publication is available from the British Library*

ISBN 978-1-107-08728-6 Hardback

# Contents

*List of abbreviations*                                    *page* viii

Introduction                                                        1

**Part I  Estimating the costs of nuclear power: points of
          reference, sources of uncertainty**                       7

1  Adding up costs                                                  9
   The notion of cost                               10
   Social, external and private costs               12
   External effects relating to independence and security   13
   The price of carbon                              16
   Decommissioning and waste: setting the right discount rate   23
   Liability in the event of accident               34
   Technical and financial production costs         36
   Adding up the costs: the levelized cost method   39

2  The curse of rising costs                                       43
   The rising costs of nuclear power                44
   International comparisons                         49
   Is there no limit to escalating costs?           53

3  Nuclear power and its alternatives                              64
   The relative competitive advantage of nuclear power over
     gas or coal                           66
   The competitive advantages of nuclear power and
     renewable energies                    72

**Part II  The risk of a major nuclear accident: calculation
          and perception of probabilities**                        79

4 Calculating risk                                                 81
   Calculating the cost of major accidents                      81
   Calculating the frequency of major accidents                 86
   Divergence between real-world observation of accidents and
      their frequency as predicted by the models               91

5 Perceived probabilities and aversion to disaster                102
   Biases in our perception of probabilities                   103
   Perception biases working against nuclear power              111

6 The magic of Bayesian analysis                                  118
   The Bayes–Laplace rule                                      118
   Are we naturally good at statistics?                        121
   Choosing the right prior probability                        123
   Predicting the probability of the next event                125
   What is the global probability of a core melt tomorrow?     132

**Part III  Safety regulation: an analysis of the American,
           French and Japanese cases**                            139

7 Does nuclear safety need to be regulated?                       141
   Inadequate private incentives                              142
   Civil liability                                            147
   Civil liability for nuclear damage, in practice            149
   A subsidy in disguise?                                     151

8 The basic rules of regulation                                   157
   An engineer's view of safety regulation                    157
   Japanese regulation, an example to avoid                   160
   The Japanese regulator is a captive to industry            166

9 What goal should be set for safety and how is it
  to be attained?                                               172
   Technology versus performance-based standards in the
      United States                                             174

# Contents • vii

Choosing a goal for overall safety: words and figures 179
A French approach in complete contrast to the
   American model 186
Regulator and regulated: enemies or peers? 193
Pros and cons of American and French regulation 198

**Part IV National policies and international governance** 207

10 Adopting nuclear power 211
   Atoms for peace 212
   Pioneers and followers 216
   Aspiring nuclear powers 222

11 Nuclear exit 226
   German hesitation over a swift or gradual exit 226
   In France early plant closure and nuclear cutbacks 235

12 Supranational governance: learning from Europe 244
   Why is the national level not sufficient? 244
   European safety standards 247
   Europe's patchwork of liability 251
   Coexistence fraught by conflicting views on nuclear power 256

13 International governance to combat proliferation:
  politics and trade 265
   The IAEA and the NPT: strengths and weaknesses 266
   Nuclear trade 269
   Which model: the armament industry, or oil and gas
      supplies and services? 273
   Export controls 282

  Conclusion 296

*Notes* 302
*Index* 328

# Abbreviations

| | |
|---|---|
| ANRE | Agency for Natural Resources and Energy (Japan) |
| ASN | Autorité de Sûreté Nucléaire |
| CEA | Commissariat à l'Energie Atomique |
| EDF | Electricité de France |
| Ensreg | European Nuclear Safety Regulation Group |
| EPR | European Pressurized Reactor |
| GDP | Gross Domestic Product |
| IAEA | International Atomic Energy Agency |
| INES | International Nuclear Event Scale |
| INPO | Institute of Nuclear Power Operations |
| IRRS | International Regulatory Review Service |
| IRSN | Institut de Radioprotection et de Sûreté Nucléaire |
| JNES | Japan Nuclear Energy Safety Organization |
| MIT | Massachusetts Institute of Technology |
| NPT | Treaty on the Non-Proliferation of Nuclear Weapons |
| NRC | Nuclear Regulatory Commission |
| NSC | Nuclear Safety Commission (Japan) |
| NSG | Nuclear Suppliers Group |
| OECD | Organisation for Economic Cooperation and Development |
| UAE | United Arab Emirates |

# Introduction

Is nuclear power a thing of the past or a technology for the future? Has it become too expensive and dangerous, or is it still competitive and sufficiently safe? Should emerging countries develop nuclear power or look elsewhere? Can we trust calculations of the risk of a major nuclear accident, given that their results diverge? Is international cooperation on safety and non-proliferation bound to fail or is it in fact gathering strength? The views on all these subjects are contradictory. Often the only common ground between them is their uncompromising, categorical nature. A quick look at the facts certainly fails to yield any obvious answers. The construction projects for new nuclear plants in Europe are behind schedule and well over their original budgets; meanwhile similar projects in China are on target, both for their deadline and budget. Japan, a country renowned for its excellent technology, failed to prevent a major accident at Fukushima Daiichi. The United States, surfing on a shale gas boom, is turning its back on nuclear power. In Europe, the United Kingdom is planning to build several new reactors, whereas Germany is stepping up plans to retire existing plants. Depending on which source you accept, the disaster at Chernobyl caused several hundred fatalities, or several tens of thousands. The number of major accidents observed since the start of nuclear power is greater than the figure forecast by the experts' probabilistic studies. Similarly the perception of nuclear risk is very different from the value calculated by cool-headed scientists. After lengthy debate, the European Union has adopted a common framework for nuclear safety, but the safety authorities are still national agencies and civil liability regimes form a contrasting patchwork. Lastly, with regard to the non-proliferation of nuclear weapons, Iran has signed the international treaty designed to limit the use of civilian nuclear facilities for military ends, but at the same time it has launched a uranium-enrichment programme which bears no relation to its energy requirements.

Faced with so many divergent claims and apparently contradictory facts, we obviously need to take a closer look at what is going on. It is time for in-depth analysis of costs, hazards, risks, safety measures, decisions by specific countries to invest in nuclear power or pull out, and the rules for international governance of the atom. In short, it is time to study and understand the global economics of nuclear power. Such is the purpose of this book.

It is rooted in two convictions.

Firstly, that analysis which does not take sides, for or against nuclear power, can interest readers.

I hope to show that it is possible not to adopt a normative stance on this issue without handing out platitudes. The economic approach adopted here is deliberately positive, the aim being to understand particular situations, explain phenomena and foresee certain developments. In a word, to focus on consequences: the political consequences for a country which decides to invest in nuclear power or retire its existing plants; the effect of a carbon tax on the competitive position of nuclear power; the impact of observed accidents and public opinion's biased perception of risk; the effect of liberalizing electricity markets on nuclear investments; the consequences of industrial nationalism on reactor exports. My aim is to use economics to analyse effects, not to dictate the decisions that public and private-sector policy-makers should make, less still teaching people the *right* way to think and behave.

Secondly, I believe much the best way to throw light on the individual and collective decisions before us is to gain an understanding of the many uncertainties weighing on the cost, risks, regulation and politics of nuclear power. A possible sub-heading for this book might be 'In the Light of Uncertainty'. It is vital to set aside a whole series of categorical claims such as the notion that there is a single *true* cost of nuclear power, be it high or low; that a major nuclear accident will certainly occur somewhere in the world over the next twenty years, or alternatively that a disaster is impossible in Europe; that safety regulation in the US is above reproach, or on the contrary in the hands of the nuclear lobby; that a national nuclear industry is a sure-fire asset for France's future balance of payments, or a complete waste of resources. Such assertions only serve to rubber-stamp decisions that have already been taken. Any attempt to settle the many questions raised by nuclear power must allow for such uncertainties, and to do so they need to be circumscribed. The present work shows how the theory of decision under uncertainty can throw light on nuclear debate, how probabilistic assessment can prompt a reappraisal of beliefs, and how the median voter theorem and the theory of political marketing can explain some public decisions.

The book is in four parts, addressing costs, risks, regulation and politics. Each one provides a wealth of detail on its subject, providing the facts of the matter and their theoretical basis, backed by references to academic literature. I am convinced that a degree of immersion gives the reader proportionately many more insights than a brief summary of ideas and arguments.

## Estimating the costs of nuclear power: points of reference, sources of uncertainty

Predictably the first part of a book by an economist is given over entirely to the competitiveness of nuclear power. Does it cost less or more than electricity generated using coal, gas or wind? Does it make financial sense for electricity utilities to invest in nuclear technology? The cost of nuclear power has escalated since the first plants were built. How could it break out of this vicious circle and prevent a further drop in its relative competitive advantage? Financial and economic factors now furnish anti-nuclear campaigners with compelling arguments.

## The risk of a major nuclear accident: calculation and perception of probabilities

Measures to enhance safety are among the factors which are making nuclear power increasingly expensive. The second part focuses on the risks of an accident and efforts to limit them. On the one hand, although it is still difficult to assess nuclear risk precisely, it can be analysed dispassionately using a whole series of instruments and methods. As an overall trend, nuclear risk is declining. On the other hand, risk as perceived by the general public, in the wake of major disasters such as Chernobyl or Fukushima Daiichi, is on the rise. So is the public attitude to such hazards irrational? Should government decisions be based on risk as assessed by experts or the general public? Is it possible to narrow the gap between calculated and perceived risk? This part explores in detail the biased perception of probabilities brought to light by experimental psychology, a discipline which now significantly influences economic analysis. It shows how modern probabilistic analysis enables us to reconcile our prior perception of a hazard with input from material knowledge.

## Safety regulation: an analysis of the American, French and Japanese cases

The more effective nuclear-safety regulation is, the fewer accidents there will be. How can this technology be expected to inspire confidence among the general public if reactor-safety standards are badly designed by the authorities or improperly applied by operators? But how is effective regulation to be achieved? This is not a simple matter, safety regulation being dogged by imperfections and uncertainty. The third part of the book analyses several examples closely. The Fukushima Daiichi accident has shown that regulation as practised in Japan is an example to be avoided. Regulation operates along very different lines in France and the US, yet both are exemplary on account of the transparency, independence and competence of their respective nuclear safety authorities. If the same criteria were enforced worldwide the risk of an accident would be much lower.

## National policies and international governance

The fourth and last part deals with politics, to which decision-making under uncertainty devolves. This process plays a considerable part in nuclear power: witness the diversity of choices made by individual nations. Some countries have embraced the atom for military and economic reasons, whereas others – the majority – have only developed civil nuclear applications. Some countries are now phasing out their nuclear power plants, others are keen to adopt the technology. Why? Over and above national policies, mechanisms of international governance are trying to contain the risk of proliferation, and improve the safety of reactors and their operation, for the good reason that both safety and security have a planetary dimension. But these efforts to institute supranational governance must come to terms with the sovereignty of states. The economic and commercial interests of countries which export nuclear technology are also at stake here. Clearly, political and strategic considerations still weigh heavily on the world market for reactors.

This book aims to adopt a non-partisan stance, neither pro- nor anti-nuclear. But it does not claim to be objective. As in any essay, the choice of issues, facts and perspective reflects the author's personality and situation. I teach at Mines ParisTech, Paris, one of France's top engineering schools. Many of those who have gone on to build and operate France's nuclear industry were educated here.

Many of my former students still work in the industry. Furthermore, the research carried out by my laboratory is funded by EDF, which also numbers among the clients of my consultancy company, alongside other electricity-generating companies elsewhere in the world. This record may prompt some readers to query the independence of the views expressed in the book, perhaps even suspecting the author of having 'sold out' to the international nuclear lobby. Others, familiar with the intellectual freedom which prevails in the academic world and the open minds of energy engineers, will soon set aside such suspicions. Others still will conclude that the author's links with the 'nucleocrats' are after all a guarantee of the validity of the information contained in these pages.

PART I

# Estimating the costs of nuclear power

## Points of reference, sources of uncertainty

The debate on this topic is fairly confusing. Some present electricity production using nuclear power as an affordable solution, others maintain it is too expensive. These widely divergent views prompt fears among consumers and voters that they are being manipulated: each side is just defending its own interests and the true cost of nuclear power is being concealed.

Companies and non-governmental organizations certainly adopt whatever position suits them best. But at the same time, the notion of just one 'true' cost is misleading. As we shall see in this section there is no such thing as *the* cost of nuclear power: we must reason in terms of costs and draw a distinction between a private cost and a social cost. The private cost is what an operator examines before deciding whether it is opportune to build a new nuclear power plant. This cost varies between different investors, particularly as a function of their attitude to risks. On the other hand the social cost weighs on society, which may take into account the risk of proliferation, or the benefits of avoiding carbon dioxide emissions, among others. The cost of actually building new plant differs from one country to the next. So deciding whether nuclear power is profitable or not, a benefit for society or not, does not involve determining the real cost, but rather compiling data, developing methods and formulating hypotheses. It is not as easy as inundating the general public with contradictory figures, but it is a more effective way of casting light on economic decisions made by industry and government.

Without evaluating the costs it is impossible to establish the cost price, required to compare electricity production using nuclear power and rival technologies. Would it be preferable to build a gas-powered plant, a nuclear reactor or a wind farm? Which technology yields the lowest cost per kWh? Under what conditions – financial terms, regulatory framework, carbon pricing – will private investors see an adequate return on nuclear power? In terms of the general interest, how does taking account of the cost of decommissioning and storing waste affect the competitiveness of nuclear power?

This part answers these questions in three chapters. We shall start in Chapter 1 by taking a close look at the various items of cost associated with nuclear power.[1] We shall look at how sensitive they are to various factors (among others the discount rate and price of fuel) in order to understand the substantial variations they display. Chapter 2 reviews changes in the cost dynamic. From a historical perspective nuclear technology has been characterized by rising costs and it seems most likely that this trend will continue, being largely related to concerns about safety. Finally in Chapter 3 we shall analyse the poor cost-competitiveness of nuclear power, which provides critics of this technology with a compelling argument.

# ONE

# Adding up costs

Is the cost per MWh generated by existing French nuclear power plants €32 or €49? Does building a next-generation EPR reactor represent an investment of about €2,000 per kW, or twice that amount?

The controversy about the cost price borne by EDF resurfaced when a new law on electricity was passed in 2010,[1] requiring France's largest operator to sell part of the output from its nuclear power plants to downstream competitors. Under this law the sale price is set by the authorities and must reflect the production costs of existing facilities. GDF Suez, EDF's main competitor, put these costs at about €32 per MWh, whereas the operator reckoned its costs were almost €20 higher. How can such a large difference be justified? Is it just a matter of a buyer and a seller tossing numbers in the air, their sole concern being to influence the government in order to obtain the most favourable terms? Or is one of the figures right, the other wrong?

The figures for investments in new nuclear power plants are just as contradictory. Take for example the European Pressurized Reactor, the third-generation reactor built by the French company Areva. It was sold in Finland on the basis of a construction cost of €3 billion, equivalent to about €2,000 per kW of installed capacity. Ultimately the real cost is likely to be twice that amount. At Taishan, in China, where two EPRs are being built, the bill should amount to about €4 billion, or roughly €2,400 per kW of installed capacity. How can

9

the cost of building the same plant vary so much, simply due to a change in its geographical location or timeframe?

## The notion of cost

The disparity between these figures upsets the idea, firmly rooted in our minds, that cost corresponds to a single, somehow objective value. Surely if one asks an economist to value a good, he or she will pinpoint its cost like any good land surveyor? Unfortunately it does not work like that. Unlike physical magnitudes, cost is not an objective given. It is not a distance which can be assessed with a certain margin of error due to the poor accuracy of measuring instruments, however sophisticated they may be; nor is it comparable to the invariant and intrinsic mass of a body. Cost is more like weight. Any object, subject to the force of gravity, will weigh less at a certain elevation than at sea level, and more at either Pole than at the Equator. In the same way cost depends on where you stand. It will differ depending on whether you adopt the position of a private investor or a public authority, on whether the operator is subject to local competition or enjoys a monopoly, and so on. Change the frame of reference and the cost will vary.

In economics opportunity plays the same role as gravity in physics. Faced with two mutually exclusive options, an economic agent loses the opportunity to carry out one if he or she chooses the other. If I go to the movies this evening I shall miss a concert or dinner with friends. The cost of forgoing one of the options is known as the opportunity cost. As economic agents must generally cope with non-binary options, the opportunity cost refers more precisely to the value of second-best option forgone. As preferences are variable (Peter would rather see a movie than spend the evening with friends; for John it is the opposite), the opportunity cost depends on which economic agent is being considered. As a result it is eminently vari-able. Ultimately there may be as many costs as there are consumers or producers. Regarding our present concern, the cost of building nuclear power plants in Russia, which exports gas, will be different from the cost borne by another state. Investing in nuclear plants to generate electricity, rather than combined-cycle gas turbines, enables locally produced gas to be directed to a more profitable outlet. The economic concept of opportunity cost puts an end, once and for all, to any idea that cost might be an objective, invariant magnitude.

Moreover, it should be borne in mind that cost relates, not to a good or service, but to a decision or action. The opportunity cost is not the cost of something, rather the cost of *doing* something. This of course applies to the cost of production, which is defined by economists using an equation, the production function. This function expresses the relation, for a particular technology, between the quantity produced – a kWh for instance – and the minimum production factors required to obtain such output: labour, capital, natural resources. The production function enables us to determine the cost of an additional unit of the good, or the marginal cost, the opportunity of this additional production being measured against the decision not to produce. The production function also allows us to determine the fixed cost of production, this time compared with the alternative option of producing nothing at all. The fixed cost is not zero, because before producing the first unit, it was necessary to invest in buildings and machines. So, even if the infrastructure is not used, it must be paid for.

To assess officially the cost of a good or service, it is advisable to ask an accountant, using the appropriate methods. An accountant will calculate direct costs, in other words the costs directly related to the product (e.g. steel purchases in car manufacturing), and indirect costs (R&D expenditure, overheads, etc.) depending on the prevailing rules on cost allocation. Accountants will distinguish between operating and maintenance costs, and between capital expenditure drawing on shareholders' equity or on borrowing in order to make investments. For 2010 France's Court of Auditors estimated that the accounting cost, not including decommissioning, of electricity production by EDF's nuclear fleet amounted to €32.30 per MWh.[2] This figure corresponds to annual operating and maintenance expenditure of nearly €12 billion, to produce 408 TWh, and €1.3 billion annual capital costs, restricted to provision for depreciation. Obviously the production cost found by an accountant depends on the method used. Using the full cost accounting method for production the Court of Auditors found a total cost of €39.80 per MWh. This figure is higher than the previous one, because the first method, cited above, only includes depreciation in the capital costs, but does not allow for the fact that the fleet would cost more, in constant euros, to build now than it did in the past. With the full-cost accounting method for production, assets not yet depreciated are remunerated and the initial investment is paid back in constant currency.

To conclude, neither an approach based on accountancy nor on economics yields a single cost. For one kWh of nuclear electricity, much as for any other good or action, the idea of a true or intrinsic cost for which accountants or economists can suggest an approximate solution is misleading. On the other hand, as we shall see, their methods do help to understand variations in costs, identify the factors which determine costs, compare such costs for different technologies, and also observe the efficiency of operators. All these data are valuable, indeed necessary, to deciding whether or not to invest in one or other electricity-generating technology.

## Social, external and private costs

So cost is not invariant. Moreover, sometimes it is quite simply impossible to put a figure on it. This additional complication concerns the external effects of using nuclear power generation, be they negative – such as the unavoidable production of radioactive waste and damage in the event of accident – or positive: avoiding $CO_2$ emissions and reducing energy dependency. Such external effects (or 'externalities' in the jargon of economics) explain the disparity between the private cost, borne by producers or consumers, and the social cost, borne by society as a whole.

Economic theory requires us to fill the gap. Because of this disparity, the decisions taken by households and businesses are no longer optimal in terms of the general interest; their decisions no longer maximize wealth for the whole of society. For example, if it costs €10 less per MWh to generate electricity using coal rather than gas, but the cost of the damage caused by emissions from coal is €11 higher, it would be better to replace coal with gas. Otherwise society loses €1 for every MWh generated. But in the absence of a tax or some other instrument charging for carbon emissions, private investors will opt to build coal-fired power plants. Hence the economic concept of internalizing external effects.

How, then, are externalities to be valued in order to determine the social cost of nuclear power? How much does it cost to decommission reactors and store long-term waste? What price should be set for releasing one tonne of carbon into the atmosphere? How can the cost of a major nuclear accident be estimated? What method should be used to calculate the external effects of nuclear power generation on security with respect to energy independence or the risk of proliferation?

We shall see that the answers to these questions raise not so much theoretical or conceptual issues as practical difficulties posed by the lack of data and information. As a result, the positive and negative external effects of nuclear power are only partly internalized. But then the same is true of other sources of energy.

## External effects relating to independence and security

We shall start with the trickiest question: putting a figure on the effects of national independence. This is such a complicated task that no one has ever attempted it. Analysis so far has only been

qualitative. We often hear that nuclear electricity production contributes to the energy independence of the country developing it. It
purportedly yields greater energy security. Many political initiatives
are justified by such allegations (see Chapter 10), but the terms of the
debate are muddled. Conventionally energy dependence refers to the
supply of oil products. The latter weigh down the balance of trade of
importing countries and subject them to price shocks and the risk of
shortages in the event of international conflict. Nuclear electricity
production only replaces oil and its derivatives in a marginal way.
Only 5 per cent of the electricity generated worldwide is produced
using oil derivatives.

In fact it would make more sense to look at gas, to justify the
claim that nuclear power contributes to energy independence and
security. In this respect Europe, for example, is dependent on a small
number of exporting countries. The European Union imports two-
thirds of the gas it requires and the Russian Federation is its main
supplier. One may remember the disruption of Russian gas transit
through Ukraine in the winter of 2008–9. As a knock-on effect, gas
deliveries to Europe were held up for almost three weeks. Millions
of Poles, Hungarians and Bulgarians were deprived of heating and
hundreds of factories ground to a halt. There is no doubt that Poland's
determination sooner or later to start nuclear electricity production
is partly due to the need to reduce its dependence on Russia. On the
other hand we have heard no mention of calculations putting a figure
on the expected benefit: a calculation resulting in acceptance, for
instance, of nuclear power costing €5 or €10 per MWh more than that
of electricity generated using imported gas. The concepts of energy
independence and security are too fuzzy to measure. The best one can
do is estimate the cost of the shortfall for the Polish economy per day
of disrupted supply. But to calculate the gain in independence, this
cost would have to be multiplied by the probability of such disruption.
However, in forty-one years Russia has only failed to honour its
commitments twice, with one interruption lasting two days, and

the other twenty. It would be difficult, on the basis of such a small number of events, to extrapolate a probability for the future.

To take full account of security issues, allowance must be made for the risk of military or terrorist attacks, and the risk of proliferation (see Chapter 13). In this case the externality is negative and could counterbalance nuclear power's advantage in terms of energy independence. A nuclear power plant is vulnerable to hostile action. For example, during the Iran–Iraq war in 1980–88 the nuclear plant being built at Bushehr in Iran was bombed several times by Iraqi forces. All other things being equal, the higher the number of nuclear plants in a country, the larger the number of targets available to enemy action.

The development of civilian applications for the atom may entail the additional risk of facilitating proliferation of nuclear weapons. Nuclear weapons can be manufactured using plutonium or highly enriched uranium. The latter may be obtained by using and stepping up enrichment capacity that already exists for producing fuel for nuclear reactors. Such fuel must contain about 5 per cent of uranium-235, whereas the concentration of this fissile isotope must exceed 80 per cent in order to produce a bomb. Plutonium is obtained from reprocessing spent fuel.

No country has so far used fissile material from commercial reactors to produce weapons (see Chapter 13). Reactors used – purportedly at least – for civilian research have however been used to produce plutonium which can be used in weapons. India and North Korea are two instances of this diversion. Iran's nuclear programme also substantiates the claim that civilian nuclear materials may be diverted toward military purposes.

One way of reducing the risk of proliferation would be to guarantee countries launching programmes to develop nuclear power a supply of fuel for their reactors. In this way they would no longer need their own enrichment capacity. This measure would restrict the spread of enrichment technology, which can be diverted from its original civilian purpose. The United Arab Emirates has, for example, made a commitment not to produce its own fuel. The UAE will have to import it. Similarly the Russians will guarantee a supply of fuel for the nuclear plants they are due to build in Turkey. However, agreements of this sort are contrary to the goal of reducing energy dependence often associated with the decision to resort to nuclear power, there being only a limited number of potential fuel suppliers. It seems likely that in the long run more countries will want to have their own enrichment units, at least once they have a sufficient number of reactors.

There are no firm figures for the external effects of nuclear power on national and international security, any more than there are for energy independence. Any attempt to calculate such figures is hampered by the scope of these concepts, both too broad and too fuzzy. It would probably be wiser to leave it up to the diplomats and military strategists to persuade their governments – using qualitative arguments – to revise, upwards or downwards, the cost of resorting to nuclear power in their country.

## The price of carbon

How are we to assess nuclear power's contribution to combating global warming? Stated in these terms the question is too general to allow an economist to provide an accurate answer. There are too many uncertainties regarding the goal being sought and the consequences if no action is taken. How is global warming to be defined? How large a share should be attributed to human activity? Which greenhouse gases should be taken into account? We are back to the previous problem. On the other hand, values may be suggested for the benefit of nuclear power in relation to reducing $CO_2$ emissions to the atmosphere. To calculate this benefit we need to know the price per (metric) tonne of carbon emissions, which can then be combined with the emissions avoided for each MWh generated.

At first sight this seems straightforward. At a theoretical level, all the textbooks on environmental economics explain how to determine the optimal price of a pollutant. In practical terms trade in $CO_2$ emissions credits provides an indication of the price of carbon. But in fact, the problem is still a thorny one: we lack the data to apply the theory and the carbon markets produce the wrong price signals.

The theory for determining the optimal price of a pollutant emission is simple enough in principle. The optimal price is found at the point where the curve plotting the marginal cost of pollution abatement intersects the curve for the marginal benefit of the avoided damage. The general idea is that the level of pollution which is economically satisfactory for society is the point beyond which further abatement costs more than the benefit from avoiding additional damage. Or put the other way round, it is the point below which the situation would not serve the public interest, the cost of additional abatement being lower than the benefit it would yield: abatement is consequently worth carrying out. This coincides with a basic economic principle according to which all actions for which the social cost is lower than the social benefit should be carried out. As is the case with any equilibrium, the optimal price corresponds to the optimal amount of pollution. Normative economics does not prescribe zero pollution. The economically optimal amount of waste or effluents is only equal to zero in the rare event of it being less expensive to eliminate pollution, down to the last gram, rather than suffering the damage it entails.

Applying this theory is another matter. The data required to plot curves for $CO_2$ emissions abatement and avoided damage do not exist. Obviously there are estimates of the cost of various actions such as insulating homes or recycling waste, which in turn limit carbon emissions to the atmosphere. But economists need future costs, not just current costs. The former are unknown, because technological innovation – such as carbon capture and storage – has yet to yield conclusive results. It would also be necessary to know the cost of measures to adapt to global warming. It may be more economical, at least for part of the temperature increase, to adapt to the situation rather than combating it. But it is future generations which will have to adapt. How can we know how much it will cost them? We cannot ask them. The same applies to the damages suffered by our descendants. How could they be calculated without an exact idea of their extent and without questioning those who might be exposed to them? For example the cost of migration to escape changing geographical conditions depends on the individuals concerned, in particular how much value they attach to the loss of their land. The last, but by no means the smallest, obstacle to assessing damages is the lack of a robust formula for converting the concentration of $CO_2$ in the atmosphere into temperature increase. It is not the amount of carbon which causes the economic loss but the climate change it may bring about. In this situation economists are dependent on the scientific knowledge of climatologists. Unfortunately analysis of the exact consequences for climate change of a rise in the amount of greenhouse gas stored in the atmosphere is still tentative.

But does looking at the markets make it any easier to find the price of carbon? At first sight, yes. Since 2005 Europe has had a market for tradable emissions permits. On this market the price of a tonne of $CO_2$ fluctuated on either side of €15 in 2009–10 (equivalent to €55 a tonne of carbon, a tonne of $CO_2$ containing 272 kilograms of this element). Given that generating one MWh using coal releases roughly a tonne of $CO_2$, the operators of coal-fired plants had to pay an average of €15 per MWh for their emissions in 2009–10. In other words, all other things being equal, if in the course of this period, 1 MWh generated by a coal-fired power plant had been replaced by 1 MWh generated by a nuclear plant, €15 would have been saved. Projecting ourselves into the future and anticipating that the market price of carbon will double, switching electricity production from coal-fired to nuclear power plants would save €30 per MWh.

So far, so good. On the basis of the market price we can obtain the opportunity cost we sought, be it past or future. In addition the private and social costs seem to have been reconciled: private operators are forced to make allowance for the price of carbon emissions when choosing to invest in coal-fired or nuclear plants. Thanks to the market, the external effect has been internalized.

In fact, nothing has been settled. For two reasons. Firstly, the European Emissions Trading Scheme is not a market for polluters and polluted, but an exchange for companies at the source of emissions. It reflects the abatement performance of the various players, but in no way the damage done. Secondly, the market was badly calibrated. The prices it reveals are not sufficient to achieve the targets set by the EU for reducing $CO_2$ emissions. We shall now take a closer look at these two reasons.

Economic theory explains that externalities occur in the absence of a market, so the answer is to design one. With no market, there is no price, hence no purchasing cost and no accountable expenditure. When manufacturers discharge harmful emissions into the atmosphere, they are using the latter as a huge tip, access to which is free of charge. Some polluters are not against the principle of a toll system, particularly for the sake of their image. Similarly, to improve the market value of their home, some residents would be prepared to pay polluters to restrict their emissions. But there being no marketplace where polluters and polluted can meet, pollution is free; it appears in no accounting system and remains an external cost.

The European market for tradable emissions permits is not a place of exchange between polluters and polluted. On the contrary it brings together companies to which individual $CO_2$ emissions quotas have been distributed, but for a total amount capped below the level of industry's overall emissions. Let us suppose that, for example, 100,000 permits, each for a tonne of $CO_2$, are allocated, whereas emissions from polluting companies amount to 120,000 tonnes. In this fictitious case, the companies would have to reduce emissions by 20,000 tonnes. In some companies the cost of cutting $CO_2$ emissions is low, for others it is higher. The first group will become sellers and carry out more abatement, the second group will buy permits and abate less. At equilibrium, the price will be equal to the marginal cost of eliminating the last tonne required to meet the limit. The advantage of this market is that expenditure by industry on cutting emissions is minimized. Economic theory demonstrates that a tax yields a similar benefit. With a tax on each unit of pollution, companies with low costs for cutting emissions will abate more to reduce their liability for taxation; on the other hand companies with high abatement costs will pay proportionately more tax and do less to cut emissions. The main difference between a permit and a tax is the initial variable selected by the competent authority. In the case of a tax, the price is set in advance and deploying the instrument will show *ex post* the corresponding cut in emissions. For example, a tax pegged at €20 a tonne will lead to a 20,000 tonne drop in emissions. For a permit, the amount is decided first and the market then reveals the price per tonne of emissions avoided. If an upper limit of 100,000 tonnes is set to reduce emissions by 20,000 tonnes the market will balance out at a price of €20 a tonne.

The decision to base the system on price or quantity is closely related to the political consensus underpinning the action. In the first instance agreement was reached on the level of the acceptable surcharge per unit, in particular for consumers and business. Here the unknown factor was the amount of abatement; it might be too low, but in any case the economic conditions were such that a higher surcharge could not be applied. In the second instance agreement was reached on the level of a significant reduction that needed to be achieved, in particular according to scientific experts.

The unknown was the price to be paid for such a reduction, but in any case setting a lower target for pollution abatement would certainly not have achieved the desired environmental effect. This is obviously an oversimplification. In the absence of accurate data on the cost of abatement, orders of magnitude may sometimes be posited. When the initial level of taxation is announced, business and government may be able to estimate how much pollution will be abated within a certain range. In the other case, when the initial abatement target is published, the various players can estimate an approximate price. Once the first variable has been set, the second is not usually completely unpredictable. However, as the European example (see box) shows, the initial calibration may be faulty.

The preceding discussion of the price of carbon is important, as we shall see, for it is one of the determining factors in the competitiveness of nuclear power: without taxes on $CO_2$ emissions or in the absence of an emissions trading scheme, nuclear power cannot compete with coal or even gas. Furthermore, a consideration of how the cost of carbon is assessed highlights the dual role played by economic analysis. In the world of perfect information posited by economic theory, such analysis would enable us to set the optimal level of abatement, at the intersection between the cost of the damage done by an additional tonne of emissions and the cost of reducing pollution by an additional tonne. The role of government would be simply to plot curves and enforce the resulting target price or quantity. Economic analysis would dictate its prescriptions to policy-makers. In the real world of limited information in which we live, economics occupies a humbler position and the roles are reversed. Political decisions, through voting, debate or consultation lead to the definition of an acceptable level of either damages or expenditure. Economic analysis mainly intervenes to minimize the cost of achieving the degree of damage decided by government or to maximize the quantity produced corresponding to the level of expenditure set by government.

## The failures of the EU Emissions Trading Scheme

By mid-2013 the price of $CO_2$ had dropped to less than €5 a tonne. Five years earlier the European Commission predicted that by this point in time it would be worth €30. The financial crisis and the drop in industrial output obviously explain part of the difference. But in 2006 the price had already fallen below €15 a tonne. The main structural reason behind the persistently low price of $CO_2$ is the failure to create scarcity, too many permits having been distributed. The resulting downward pressure on prices has been exacerbated by lower than expected emission-abatement costs. The European $CO_2$ trading system does not fulfil its purpose: it does not send a reliable signal enabling industry to curtail long-term investments, in particular enabling electricity utilities to choose between various generating technologies according to their $CO_2$ emissions performance.

The case of the United Kingdom is a perfect illustration of this failure. The UK has undertaken to halve $CO_2$ emissions by 2030. To achieve this target it plans to set a carbon-price floor at €40 a tonne by 2020 and €87 by 2030. The price in the EU Emissions Trading Scheme (ETS) will only exert an influence if it exceeds these price-floors, which is unlikely to happen very often unless the ETS is reformed in the meantime. France offers another example. In 2009 the government was planning to introduce a carbon tax as an incentive to reduce the use of oil products. The rates recommended to achieve a fourfold cut in emissions by 2050 were €32 a tonne in 2010, rising to €100 a tonne in 2030, and twice that amount by mid-century.[3] The ETS price for carbon is far below the value recommended by experts to achieve long-term targets for reducing emissions.

## Decommissioning and waste: setting the right discount rate

Nuclear-power companies are responsible for the waste and by-products they produce. In this field, much as elsewhere, the polluter-pays principle applies. This principle is not disputed by the operators of nuclear plants, nor yet by opponents of nuclear power. So the controversy does not centre on the need to internalize the costs of decommissioning reactors and storing waste (spent fuel, decommissioning debris), but on the amount to be set aside now to cover these costs, thus ensuring that these back-end activities can be carried out tomorrow.

Worldwide we have almost no experience of dismantling power plants and burying radioactive waste. Nowhere in the world has anyone built a permanent storage facility for burying long-term waste. In France not a single nuclear power plant has been completely decommissioned. Work decommissioning the Chooz A reactor, in the Ardennes, is only scheduled to end in 2019. The reactor was commissioned in 1967 and shut down twenty-four years later. Worldwide less than twenty commercial reactors have been completely dismantled.

The lack of references makes appraisal very uncertain. We cannot rule out the possibility that the technical costs of dismantling and the costs of waste-management may prove very high. However, even if this were the case, it would have little effect on the return on investment from a new nuclear power plant. The return is not very sensitive to this parameter because the costs at the end of a nuclear plant's service life are very remote in time, and a euro tomorrow is worth less than a euro today, and even less the day after tomorrow. Future costs or benefits are wiped out by the rate of exchange used to convert present funds into future funds (or vice versa). For example, at an annual rate of 8%, €1 million would only be worth €455 in a century. This amount drops to €0.20 after two centuries and in

500 years it would have dwindled to almost nothing. Taking the same rate of 8% and supposing that decommissioning would cost 15% of the total cost of a new reactor, this share only represents 0.7% of the total cost if decommissioning is carried out forty years after construction. This rate, known as the discount rate, plays a decisive part in assessing the costs of decommissioning plant and managing waste. To avoid wiping out such costs, a discount rate close to zero would need to be used. Certain environmental conservation groups advocate this position, but there is little support among economists. We shall now look in greater detail at how the discount rate works.

To avoid confusion, we should start by explaining what this rate is not. Firstly, the discount rate bears no relation to inflation. The latter, whether its origin is monetary or results from indexing wages, is a phenomenon which raises prices. Consumers will buy less tomorrow because the same shopping basket will cost more. Secondly, the discount rate does not reflect the risks associated with the investment project being assessed. Such risks cast doubt on income and expenditure and change the way they are estimated, but not due to the discount rate.

To convert current euros into future euros we must start from existing knowledge. Despite the limited experience mentioned above, we do have preliminary orders of magnitude.

In 2012 France's Court of Auditors used the EDF estimate of how much it would cost to decommission its fifty-eight reactors. The cost entered in the company's accounts amounts to €18 billion, equivalent to €300 per kW of installed capacity. In comparison with assessments in other countries, and consequently relating to reactors and conditions which may be very different, this figure is near the lower end of the range. The management consultants Arthur D. Little estimated that the upper value in Germany would be close to €1,000 per kW. In the United States estimates of the cost of decommissioning the Maine Yankee plant, completed in 2005, are in the region of €500 per kW.

As for waste destined to be buried in deep geological repositories, only very preliminary estimates have been made. Work is still focusing on pilot schemes or has barely started. The only site currently operating stores radioactive waste of military origin at Carlsbad, New Mexico. This waste is easier to manage because it does not release any heat. To store the amount of long-term waste produced by a reactor in one year, the order of magnitude currently cited is €20 million. This figure is based on various British, Japanese and French estimates.[4] The amount is likely to change with progress by research and technical know-how. In 2005 France's Nuclear Waste Authority (Andra) estimated that it would cost a little under €20 billion to build and operate a deep geological repository. Five years later the price was adjusted to €36 billion. The second amount makes allowance for additional parameters, integrating return-on-experience from excavating underground galleries, requirements for greater capacity and tougher safety constraints, among others.

The timescales we are dealing with here are very long. Some categories of nuclear waste will go on emitting radiation for several hundred thousand years. For example, plutonium-239 has a half-life – the time required for half the radioactive atoms to disintegrate – of 24,000 years. For technetium-99 it rises to 211,000 years and for iodine-129 it is 15.7 million. Such time spans are stupendous when compared to the scale of human life. Our most distant ancestors, *Australopithecus*, appeared on Earth 4 million years ago and modern humans (*homo sapiens*) only emerged about 200,000 years ago. Of course there are no plans for the storage facilities to operate for such long periods. For example the deep geological repository projected by Andra is expected to last for 120 years, from the start of construction to final closure. If a decision were taken now to invest in a new reactor, the plant would be commissioned in 2020 and operate for sixty years. Only in 2100 would decommissioning be complete, with the last tonnes of waste finally being buried in 2200. These economic deadlines are short compared with the half-life of certain waste products, but nevertheless dizzying. A century is a very long time in the life of an economy, with its multiple crises. Government bonds, the investments with the longest time span, spread over periods of twenty or sometimes thirty years, stretching to fifty in exceptional cases.

With such long timeframes, how can we account for this expenditure now? The reference to government bonds suggests a preliminary approach to the discount rate and its basis. If someone offers to give you €100 today or in twenty years' time, there is no need to think twice. If you take the €100 now you can make a very sound investment in US Treasury bonds. Thanks to the interest you will have more than €100 in twenty years. You will thus be able to consume more than if you had agreed to wait before receiving the funds. The decision, based on a simple trade-off, justifies the use of discounting and the long-term interest rate may be used to find the future value of today's euros. With 4% interest, €100 today will be worth €220 in twenty years or, inversely, €100 from twenty years ahead would be worth only €45.60 at present. However, using this interest rate to discount the value does not solve the problem, if the aim is to determine the value of a euro in a century. As the business weekly *The Economist* amusingly observed,[5] 'At a modest 2% rate [...] a single cent rendered unto Caesar in Jesus' time is equivalent to [...] thirty times the value of the entire world economy today.'

Furthermore, interest rates only partly justify discounting. According to economic theory discounting is necessary for two reasons: people are impatient and future generations will be better off. Economic agents display a pure time preference for the present. Instead of taking an interchangeable value, let us suppose that the choice concerns the possibility of attending the performance of an opera in the course of the coming year, or the same performance in five years' time. Which ticket would you choose? The interest rate argument does not hold because you can neither lend nor resell the ticket. If you do not use it, it will be wasted. It is highly likely that you will opt for the performance in the coming year, rather than waiting five years. This impatience is reflected in a pure time preference for the present, which crushes future consumption to give it less weight.

It is more difficult to illustrate the notion that future generations will be better off. The discount rate depends on the growth rate of the economy and a barbaric term, the elasticity of inter-temporal substitution in consumption. The overall idea is that the richer you are, the less satisfaction an additional euro will yield. If you give €100 to someone with a low income, you will be making them a present worth much more than if you give the same amount to a millionaire. Marginal utility decreases with income. Consequently, if society is €1 billion richer tomorrow, it will respond less to this gain than now. With an ordinary utility function, society ten times better off than at present, and elasticity equal to 1, contemporary society would see its well-being increase ten times more for each marginal unit of consumption (€1 billion) than tomorrow's society. So it would be advisable to limit our efforts to provide benefits for future generations.

Economists thus provide the following key, forged by Frank Ramsey in 1928,[6] for calculating the discount rate: the discount rate (d) is equal to the sum of the pure preference rate for the present (p) and the product of the elasticity of the marginal utility of consumption (e) multiplied by the growth rate of per capita GDP (g), in other words $d = p + eg$ (see box). With the three values often used [2; 2; 2], the discount rate is 6 per cent.

The discount rate cuts both ways, exerting a decisive influence on decisions regarding public and private investment, but it is impaired by numerous unknowns. One recent attempt to reduce this tension has involved using a rate that varies over time – rather than being constant – declining as it advances into the future. The per capita growth rate can be used to illustrate the intuition behind this idea: the more remote the future, the greater the uncertainty regarding economic and technological progress; so, the greater our caution

## Interpreting and selecting p, e and g

The pure preference rate for the present may be seen as an equity parameter, its value depending on how fairly we wish to treat future generations. Let us suppose that the output from a new nuclear plant entails a waste-management cost of 100 in a century, but yields a present gain because nuclear technology is cheaper. If we want to treat future generations even-handedly, we should only commit ourselves to the investment if its present benefit for us is greater than 100. In this case we would apply a pure preference rate equal to zero. This position in favour of equality between generations is defended by some economists, including Frank Ramsey. It rejects the idea of discrimination depending on the date of birth and involves treating all generations on the same footing, even if they are more prosperous. If we want to treat our descendants slightly less favourably (if, for example, we are convinced they will find smarter means of storage or recycling), we need to use a slightly positive preference rate. A gain of 14 will then suffice (equal to 100 discounted at 2% over 100 years). If, on the other hand, we are feeling selfish and have no concern for what comes after, the rate will be very high: even if nuclear power only yields a unit gain, it is worth taking, its value exceeding 100 in the future (0.40 today is enough to get 100 in a century with an 8% discount rate). We may also interpret the pure preference rate in terms of our chances of survival. With a one-in-ten chance of mankind not surviving for 100 years (following, for example, collision with a meteorite), the value of the preference rate is 0.1; it rises to 1 if we assume the likelihood of survival is 0.6 (a 4-in-10 chance of the end of the world).

The elasticity of the marginal utility of consumption also measures equity. The greater the difference in utility for a marginal unit of consumption between low- and high-income households, the more justification there is for high levels of transfer, through taxation for instance, from rich to poor. Such transfers raise the utility of the whole of society. In other words, this parameter reflects our attitude to unequal levels of consumption, between different people in the present day, or between them and their descendants. Unlike the previous variable, the difference in treatment is not related to time. The more egalitarian we are, the more we favour redistribution from rich to poor and the higher the value we need to use for elasticity when calculating the discount rate. If we assume that future generations will be richer than today, it is legitimate to limit our efforts to improve their welfare. Looked at another way, elasticity equal to 1 would be unfair. Given a constant population it would justify spending 1% of today's GDP to give future generations the benefit of an additional 1% of GDP, even if they are incomparably more prosperous. Per capita fractions of GDP can therefore be traded between generations on equal terms. The elasticity of marginal utility may also be linked to risk. According to economic theory risk aversion is proportional to elasticity. The higher the elasticity, the more a person is prepared to pay for the certainty of consuming 100, rather than a random outcome (for example, a one-in-two chance of consuming 200, or zero). Taking a higher value for this parameter, which is then multiplied by the growth rate of per capita GDP, is tantamount to assuming that the present generation is averse to risk.

Setting the three values which make up the discount rate is no easy matter; but they will play a decisive role in how we act now. An instance of this point is the controversy prompted by the publication of the Stern Review in 2006.[7] This report caused quite a stir because it concluded that substantial, immediate expenditure (about 1% of GDP) was needed to reduce greenhouse gas emissions. This recommendation contradicted the conclusions of most climate-change economists, who suggested a more gradual increase in expenditure. The work of the US economist William D. Nordhaus,[8] for example, recommends a carbon tax of $13 a tonne over an initial period in order to internalize the damage done by global warming. Nicholas Stern prescribed $310 a tonne. Half of this difference is simply due to the discount rate used by the two parties: 4% for the former, 1.4% for the latter.

In his estimate Stern uses a preference rate for the present of 0.1 and elasticity of 1. These two values represent the lower limits of the ranges economists generally accept. His choice is open to criticism because it raises a logical contradiction. A low preference rate for the present should go hand-in-hand with high elasticity or, on the other hand, low elasticity should match a high preference for the present. It would be mistaken to suppose that one of these two parameters reflects equity between generations, the other solidarity within a single generation. A low value for the elasticity of the marginal utility due to consumption can be justified on the grounds of reducing inequality between rich and poor, regardless of when they were born. This choice coincides with a high preference rate for the present, which endorses the idea that the present generation should only make limited sacrifices for future generations (given that the latter will be better off, as Stern posits with a positive growth rate for per capita GDP). Using a simplified economic model the Cambridge economist, Partha Dasgupta,[9] has demonstrated that the parameters used by Stern would lead to inconceivably high saving ratios. With a preference rate for the present of 0.1, elasticity of 1, and a world with neither technological progress nor population growth, we should be investing 97.5% of our current output in boosting the standard of living of future generations.

The Stern Review asks whether it makes economic sense to spend 1% of today's GDP to prevent damage amounting to 5% of GDP in a century. The three values it uses, [0.1; 1; 1.3], would lead to a discounted benefit five times *greater* than the cost. But if we use [2; 2; 2] the discounted benefit would be ten times *smaller* than the cost! In other words, with a 1.4% discount rate it is entirely justifiable to spend 1% of GDP on reducing greenhouse gas emissions, whereas with a 6% discount rate it would be quite out of the question.

We have so far set aside the question of the third parameter, the future growth rate of per capita GDP. Its value is just as uncertain as the others, but setting it does not raise equity-related issues. Looking back in time, the annual growth rate of per capita GDP was 1.4% in the UK from 1870 to 2000, and 1.9% in France. However these averages conceal significant variations. In the UK the growth rate was 1% in 1870–1913, 0.9% in 1913–50, 2.4% in 1950–73, and 1.8% from 1973 to 2000. Over a very long period of time – 1500 to 1820 – it is estimated to have been 0.6%. Which of these different rates should we use? The growth rates for the next century or two may be very different. Nor can we rule out a negative growth rate, though it is not very likely. However, global warming in excess of 6°C in 200 years could have precisely that effect.

regarding action that might jeopardize the well-being of future generations, the lower the discount rate should be. Christian Gollier recommends using a 5% annual discount rate for costs borne over the next thirty years, dropping to 2% for subsequent costs.[10] It is also possible to set the discount rate on a downward path, with either several steps or a steady decline. In a report submitted to the British government in 2002,[11] Oxera Consulting Ltd suggested adopting a 3.5% rate from 0 to 30 years, 3% from 31 to 75 years, 2.5% from 76 to 125 years, and so on, with the rate ultimately bottoming out at 1% after 300 years. In France the Lebègue report,[12] on a review of discount rates in public investment, recommended a 4% rate for the first 30 years, then a rate that would steadily decrease to reach 3% after 100 years, tending towards 2% for a time horizon of over 300 years.

A varying rate also seems to represent a compromise between two demands: on the one hand taking account of our preference for the present and our contribution through technical progress to the prosperity of future generations; and on the other hand allowing for the potentially very negative consequences of our action, or inaction, with regard to future generations.[13]

It is obviously up to the relevant authorities to ensure that adequate provision is made for the projected costs of decommissioning and waste management, in accordance with the discount rate they have decided. In both the United States and France the government took such measures long ago. Left to themselves, utilities would stand to gain by underestimating future expenditure on this work and by opting for high discount rates in order to minimize projected costs. In the US and France – but also in many other countries – today's consumers are paying for tomorrow's expenditure. There are no hidden costs for decommissioning and waste which once internalized would make the cost of nuclear electricity production prohibitive (see box).

So back-end activities have no significant impact on the cost competitiveness of existing or new nuclear power. Unless of course one adopts a very low, or even zero, discount rate for very distant time horizons – as is the case in the Stern Review's calculations for climate change (see the last but one box above). In my opinion, this stance – which its advocates justify by the hazardous nature of nuclear waste and its very long life – boils down to using inconsistent economic reasoning to endorse a legitimate argument. In this part we have not allowed for the possibility that such waste might represent a risk for future generations. The only waste-related costs taken into account are the preventive costs built into the quality of repositories and their supervision. These costs vary depending on the safety standards set

## Taking into account the costs of decommissioning and waste

In France, regulation is based on special discounted provisions imposed on EDF. They appear on its balance sheet and the utility is required to secure them with specific cover assets. The law sets an upper limit for the discount rate pegged to 30-year government bonds, currently close to 3%. With this rate EDF's provisions for decommissioning and waste amount to €28 billion. They would increase by 21% with a 2% discount rate, adding 0.8% to the overall cost of a MWh. Furthermore, if just the cost of decommissioning were to rise by 50% (amounting to €30 billion as opposed to €20 billion), the cost of electricity would increase by 2.5%.[14] If the cost of deep geological repositories were to double, it would result in a 1% increase. In the US a special fund has been set up to cope with the future expense of deep repositories for spent fuel. Utilities pay a fee into the fund equal to $1 per MWh they generate. The Department of Energy checks at regular intervals that the fee is sufficient. For this purpose it has developed about thirty cash-flow models designed to balance out by 2133.[15] These scenarios depend on a large number of parameters, including the discount rate. The lowest rate considered is 2.24% per year. Two out of the four scenarios based on this rate result in a deficit, whereas the proportion is only one in four for the scenarios using a higher rate. The figures above are valid for existing reactors in the US and France. For new nuclear plants, the time horizon for expenditure would be longer, so decommissioning and waste-management costs would have even less impact on the present value of projects. A Massachusetts Institute of Technology study on the future of nuclear power puts the overnight cost of building a reactor at $4,000 per kW, and the cost of decommissioning it at $700 per kW, or 17.5%.[16] Spreading decommissioning expenditure out between the 41st and 110th year after the reactor is commissioned, and assuming a 6% discount rate, would bring the present value of decommissioning down to $11 per kW. This value would be five times higher ($52) if the rate was almost halved (3.5%). But as before, this cost is negligible compared with construction costs. With the above discount rates, the 17.5% shrinks to 0.2% or 1.3%, respectively.

by government for decommissioning and storage. To take a trivial example, the cost of a repository increases in relation to the length and depth of its tunnels. On the other hand, we have made no allowance for the cost of possible accidents, despite the fact that the risk does exist. Protecting future generations against a disaster of this sort poses the problem of assessing the uncertain damages associated with events with a very low probability and a very high cost. The release of radioactive substances into the atmosphere following the meltdown of a reactor core raises the same question: how is one to estimate the costs without knowing how the risks are distributed? We shall address this key question in the following part. The discount rate is of only limited value for finding an answer. The wrong solution would be to give the matter no further thought and select a very low, or zero, value to allow for the hazards of waste. Making allowance for a possible disaster caused by downstream activities may mean opposing the construction of new nuclear reactors without it being necessary to hide behind a very low or zero discount rate.

## Liability in the event of accident

The parts of this book devoted to risks and regulation deal in detail with the cost of major accidents and the legal framework for the civil liability of nuclear power. But we need to mention the matter briefly here, many authors having suggested that estimates of the cost of nuclear power fail to make allowance for the risk of disaster.

The operators of nuclear power plants are liable in the event of accident, but it is true that in most cases an upper limit is placed on such liability. The amount of compensation they must pay in the event of massive radioactive emissions is less than the value of the damage. In France, for instance, the limit is €91.5 million. It will soon be raised to €700 million. Such caps on liability raise the question of whether the costs of major accidents are sufficiently internalized. According to the opponents of nuclear power, limited liability is equivalent to a hidden subsidy. A Swedish study estimated that the Chernobyl disaster cost nearly $400 billion.[17] There is no way of settling the matter without a detailed review of the expected and observed frequencies of accidents and the uncertainty surrounding the level of damage. Here we shall make do with a much simplified examination of the risk involved, in order to determine how much impact it has on the full cost of existing nuclear plants and new reactors.

Risk is classically defined as the result of multiplying the probability of an accident by the severity of the outcome. For the sake of argument, we shall take the highest values cited in the literature for these parameters. We shall suppose that there is one chance in 100,000 that a disaster may occur during one year of a reactor's service life, a probability 100 times higher than the figure cited by Areva for the EPR. We shall then suppose that the massive release of radioactivity causes damage to public health and the environment worth €1 trillion, ten times higher than the provisional estimates for Fukushima. So the risk is equal to 0.00001 × 1,000,000,000,000, or €10 million a year. Supposing that the reactor's annual production amounts to 10 million MWh, the risk would be equivalent to €1 per MWh, or between 1 and 2 per cent of the estimates of the average cost of new nuclear. This scratch calculation shows that, under much simplified conditions – in particular with no allowance for uncertainty – internalizing the full cost of an accident has only a very slight impact on the cost of nuclear electricity.

We should however point out that, on the basis of these hypothetical data, the upper limits on liability currently in force mean that only a relatively small share of costs is internalized. If we take the case of the €91.5 million limit in France, it only amounts to 0.4 per cent of the full cost of an accident we took as an illustration.[18] Raising the limit to €700 million would still leave 97 per cent unaccounted for. In other words internalization is indeed partial, but internalizing the full cost would only result in a slight increase in the cost of nuclear electricity.

## Technical and financial production costs

Here at last we may venture onto more solid ground. Engineers do not base their decisions on externalities which are so difficult to grasp and estimate. On the contrary they work on data relating in particular to the costs of reactors built in the past and current operating costs. They can use proven, widely accepted methods for calculating costs, in particular for project funding. They juggle with concrete, steel, enriched uranium, man-months, assets and deadlines.

To come to grips with the subject we shall start with construction. This involves an overnight cost and a capital cost. The overnight cost refers to a hypothetical construction project completed in an instant, or 'overnight'. Spending on material, machines and wages is entered into the accounts at the prices in force when construction starts. This does not overlook financial costs; they are simply processed separately. It takes from five to ten years to build a power plant, from initial preparation of the site to the moment it is connected to the grid. During this time there is no return on investment. On the contrary, it represents a cost. If the operator borrows half the amount it needs from banks, at a 4 per cent real interest rate (allowing for inflation), and funds the rest out of its own resources at 6 per cent, for instance, the average cost of capital is 5 per cent. This cost must be added to the overnight cost to obtain the cost of investment, or installed cost.

The overnight cost is useful if we want to make an abstraction from the variability of construction lead-times. It makes comparisons easier, because construction times vary depending on the reactor model and size, but also due to non-technical causes, particularly changes in the prevailing regulatory framework or local opposition. In the US, for example, the shortest construction project lasted less than four years, but the longest one took twenty-five years.

Although it overlooks such factors the overnight cost can vary a great deal. Firstly, over time. On a per kW basis the first reactors were much cheaper than at present. We shall examine this dynamic in the following chapter. The overnight cost also varies geographically. In its 2010 study of electricity production costs the OECD noted a difference of one to three between the overnight costs, expressed in $ per MW, for building a reactor in South Korea and Switzerland.[19] The size, model and country (cost of labour, regulatory framework, etc.) are not the same, but such a large difference may nevertheless come as a surprise. However, it is not specific to nuclear power. The OECD observed a similar disparity for gas, with South Korea and Switzerland once again at the two extremes.

The overnight construction cost is one of the three main factors affecting the cost of generating nuclear electricity. The other two are the load factor and the financing cost (see box).

Once construction of the plant is complete, expenditure concerns fuel and other operating and maintenance costs. Roughly speaking fuel costs represent between 5% and 10% of the cost of generating electricity, with the other costs totalling between 20% and 25%. The cost of fuel varies depending on the amount of electricity generated, because it is depleted as the chain reaction proceeds. It is this chain reaction which releases heat, used in turn to generate electricity. The level of production has little impact on the other operating costs, which may be treated as relatively fixed, at least as long as the reactor is in service. When the nuclear plant is finally shut down, most of these costs disappear.

## Load factor and cost of financing

Nuclear power plants are characterized by very long construction times and a very high fixed investment cost compared to a variable operating cost, particularly with respect to fuel expenses.

As a result, if a reactor does not operate at full capacity, once it has been built, the fixed cost must be paid off by a smaller amount of electricity production, which in turn makes each MWh more expensive. Over the past decade the load factor of existing nuclear plants was about 95% in South Korea, 90% in the US and 70% in Japan. To illustrate the weight of this factor, we may use an example from the book by Bertel and Naudet:[20] improving the load factor from 75% to 85% boosts output by 13% and cuts the cost of a MWh by 10%.

The financing cost or interest during construction (IDC) depends on how long it takes to build the plant, but also on the choice of discount rate. As the overnight construction cost is spread over several years, expenditure must be discounted. The calculation uses the date on which the plant was commissioned as its baseline and a discount rate decided by the operator. The difference between this discounted expenditure and the overnight cost measures the financing cost. For a private-sector operator the discount rate may range from 5% to 10%. With construction lasting six years, the overnight cost must be multiplied by 1.16 with a 5% discount rate, and by 1.31 with a 10% rate. Obviously the sooner construction is complete, the sooner income will start to flow in, with interim interest reduced accordingly. In the example borrowed from Bertel and Naudet, shortening the construction time to five years reduces the cost of financing by 27%, with a 10% discount rate, and by 13% with a 5% discount.

## Adding up the costs: the levelized cost method

The technical and financial costs of building and operating a nuclear power plant, the downstream costs of decommissioning and processing waste, and the external costs (avoided carbon emissions, accidents) must all be added up to obtain the full cost of nuclear power. It will then be possible to monitor variations in this cost over time and to compare it with the cost of electricity generated using other technologies. To do so, we need to convert the euros at different points in time into constant euros and MWs into MWhs. The discount rate is used for the first conversion. The second operation is required in order to add up fixed costs – expressed as value per unit of power, for example in € per MW – and variable costs – expressed as value per unit of energy, for example in € per MWh. By definition, one MWh is the amount of electricity generated by one MW of power in one hour. A 1,000 MW nuclear plant operating at full capacity round the clock will generate 8,760,000 MWh a year. To allocate investment costs we need to know or anticipate the plant's load factor and its projected service life.

The full cost is worth knowing, but what is really important is whether it is greater or less than the revenue, in order to determine whether there is a net gain for the utility or any other company venturing into nuclear power. So far we seem to have disregarded revenue. Nor have we addressed the price of electricity and how it is sold. However, in conceptual terms, there is no difference between a cost and a benefit. One switches back and forth between them just by changing the sign. They are two sides of the same coin: a purchasing cost for a producer is a source of income for its supplier; an avoided carbon emission cost is a benefit for the environment.

Cost–benefit analysis, which compares discounted costs and benefits, is the canonical method used by economists to estimate the private or social merits of a project or decision. However, a variant is used in the field of electricity, the levelized cost. It is used to determine the price of electricity required to balance income and outgoings throughout a power plant's service life. In a way it takes the opposite route to the economic canon: instead of calculating a project's rate of return as a function of assumptions on the future price of electricity, this variant sets a zero profit rate from which to deduce a price for electricity which balances discounted income and outgoings. For example, taking €75 per MWh as the levelized cost of the EPR plant at Flamanville, in western France, means it will break even if the average price recorded reaches this level during the plant's operational service life for the projected number of hours' operation. But bear in mind that zero profit does not mean that there is no return on capital. The outgoings accounted for by this method include the cost of bankers' loans and raising funds from investors.

The levelized cost method goes back to before liberalization of the electricity sector and the creation of wholesale electricity markets. It enabled a regulator to determine the sale price of a monopolistic operator on the basis of the latter's costs. It also allowed the two parties to identify, by comparison, the cheapest generating technology in which to invest in order to meet rising demand. For the economics of today's electricity markets, only the comparison is of any interest. In principle private operators, not government, take decisions on investment. Operators tend to base such decisions on forecasts of future electricity prices and consequently on cost–benefit analysis. On the other hand, to decide whether it is preferable to add coal or gas-fired or nuclear plant to existing capacity, they will use the levelized cost variant, because it makes it easier to compare technologies. In practice, even after liberalization of the electricity market, government has continued to have a say in the choice of generating technology. At the very least it plays a part in some European countries setting long-term targets for decarbonizing electricity generation in line with policy on emissions abatement. In this case the average discounted social cost will be used. Technical and financial costs, including back-end costs (site remediation and waste management), are added to estimates of external effects (such as accidents, pollutant emissions), unless they have already been fully integrated in private costs due to regulatory or legal constraints (liability, carbon tax, safety standards). Applying this method in the general interest also involves discounting future factors differently. The authorities' choice of discount rate is based on notions of equity discussed above, not on bank interest rates and investors' demands regarding the rate of return.

Predictably, the disparities between levelized cost estimates are even greater than those observed between estimates of overnight construction costs, the latter being just one component of the former. According to the OECD the cost of construction varied by a factor of one to three between South Korea and Switzerland. In the case of the levelized cost these two countries still occupy the upper and lower extremities of the range, but with a one-to-five variation in estimates: $29 MWh for South Korea; $136.5 MWh for Switzerland.[21]

The values taken into account for the overnight cost of construction and its duration, the load factor and discount rate explain much of the disparity between the various estimates of nuclear costs. The cost may be multiplied by four if only extreme, yet realistic, values are taken into account. Take for example the base case in the 2009 MIT study,[22] an update of the 2003 MIT study.[23] The estimated cost of $84 per MWh is based on four parameters [$4,000 per kWe; 5 years; 85%; 10%]. Taking the extreme values [$2,000 per kWe; 4 years; 95%; 5%], on the one hand, and [$5,000 per kWe; 6 years; 75%; 12%] on the other, we obtain, respectively, a levelized cost of $34 per MWh and $162 per MWh. The operating costs, including the cost of fuel, weigh less heavily in the balance, whereas decommissioning and waste-management costs weigh more heavily. Of course we are referring here to the cost of next-generation nuclear plants. For ageing reactors nearing the end of their service life, operation accounts for the lion's share of costs. Furthermore, decommissioning expenditure being imminent, it adds substantially to costs unless the operator has already made sufficient provision.

Allowing for external effects does not significantly change the ranking of cost determinants. According to the simplistic estimate discussed earlier, at the most the risk of an accident only adds one euro to the average cost per MWh. This is negligible compared with the cost of a new facility, and low even compared to the cost of operating existing plant. However, it is still only partly internalized, the liability of operators being capped at low levels in the event of an accident. Nuclear power's advantage with regard to $CO_2$ emissions could certainly be taken into account as a social benefit. It could have a substantial impact on the levelized cost of nuclear power if the price for $CO_2$ emissions was in the upper range (€50 to €100 per tonne). However, it makes more sense to integrate the price of carbon in the levelized cost of technologies responsible for emissions: indeed, it is integrated through taxes or emissions permits which directly affect these technologies. We shall consequently examine its impact when discussing the relative competitiveness of nuclear power (see Chapter 3).

TWO

# The curse of rising costs

It is well known that the cost of a technology drops as it is deployed and becomes more widely used. We have all noticed that we pay less for using a telephone, computer or airplane than our parents did, simply because the cost of these goods has been substantially reduced since the first products rolled off factory production lines. Economic theory cites two causes to explain this phenomenon: the scale effect and the learning effect. The first one is both familiar and intuitive. The bigger the factory, the less each unit costs to produce. In other words, the unit cost of large production runs is lower than for smaller volumes. At the start of a technology cycle the capacity of each production unit is relatively small, in particular because demand is still limited. Subsequently the size of factories gradually increases, stabilizing when diseconomies of scale start to appear (due, for instance, to time spent moving from one workshop to another). The learning effect in manufacturing is linked to the know-how which accumulates over time. The most intuitive example to illustrate this point is the repetition of a single task. You may spend more than ten minutes folding your first paper hen, but barely a minute after making a thousand or so. Manufacturing an airliner, steam turbine or solar panel is much the same. The learning effect is generally measured by the learning rate which corresponds to the reduction in cost when cumulative production doubles. The cost per kWh of wind power, for example, drops by about 10 per cent each time installed capacity doubles.[1]

Nuclear technology displays the opposite trend. The per-kW construction cost of the most recent reactors, in constant (inflation-adjusted) euros or dollars, is higher than that of the first reactors. A technology with rising costs is a very strange beast, which requires closer study, particularly as this feature distinguishes it from several competing technologies, such as wind or solar. If nuclear engineering firms fail to find a solution in the near future, the cost of nuclear power will continue to rise, undermining its competitiveness.

## The rising costs of nuclear power

The rising cost of building nuclear reactors is a well-established fact. In particular, it has been studied in depth for installed capacity in the US. The overnight cost of the first reactors, built in the early 1970s, was about $\$1{,}000_{2008}$ per kW. It has increased steadily ever since, reaching $\$5{,}000_{2008}$ per kW for the most recent reactors, built in the early 1990s. In other words, a one-to-five difference in constant dollars. The increase in the installed cost is even more striking. The average construction time has increased with time, so interim interest has increased too. The time taken to build a nuclear power plant has risen from between five and six years for the first plants to be connected to the grid to more than twice as long for the most recent units. The average total cost per kWh displays the same upward trend. Maintenance and operating costs have dropped and the load factor has improved with time, but these two factors are not enough to counteract the very large increase in the fixed cost of construction.[2]

In France the overnight construction cost reported by EDF for its various plants was made public for the first time in a 2010 report by France's Court of Auditors.[3] It amounted to $€860_{2010}$ per kW for the first four reactors at Fessenheim and Bugey, commissioned in the late 1970s, and $€1{,}440_{2010}$ per kW for the last four reactors, at Chooz and Civaux, which came online in the early 2000s.[4] Although it is less than twice the initial amount, the increase is nevertheless substantial.

Nuclear power consequently has a record of rising costs. But what is the explanation for this anomaly? A great many factors may have come into play, such as the rising cost of materials and machinery, or the lack of economies of scale. The figures cited above are the result of several forces, invisible to the naked eye, which may conceal causes exerting an opposite force, with varying degrees of influence. To highlight all these factors we need to use a statistical method known as econometrics. This tool enables us to isolate each of the factors determining a phenomenon and to measure their respective influence. As early as 1975 econometrics was used to scrutinize the costs of nuclear power in the US.[5] Other work using the same method has been done since, yielding very interesting results.

Firstly, these studies show the absence of any significant economies of scale. The cost per MW of installed capacity is no lower for the construction of the largest reactors. Why? Because they are not just scaled-up replicas of their predecessors. They are more complex, fitted with more parts and components, often of a different design. Some research even shows diseconomies of scale. For instance, Robin Cantor and James Hewlett, in a paper based on Geoffrey Rothwell's work,[6] calculated that a 1% increase in the size of a reactor resulted in a 0.13% rise in the overnight cost per kW.[7] They demonstrated in a first step that, other things being equal – in other words, maintaining the other factors they examined at a constant level – the construction cost was significantly less with higher reactor power (a 1% increase in capacity cut the cost by 0.65%). However another key factor, construction time, also varied with size. Increasing the size by 1% added 0.6% to construction time, entailing in turn a 0.78% increase in cost. The net effect was therefore 0.78 – 0.65, making a 0.13% increase in cost. Large reactors would have been more economical had they been built as quickly as their smaller counterparts.

Secondly, there were few if any learning effects. This result concerns possible savings for the nuclear vendor. For example, according to Martin Zimmerman,[8] if the experience accumulated by a firm rises from four to eight units, it reduces the overnight cost by 4%. Taking the US nuclear industry as a whole it is difficult to isolate the learning effect specifically. The figures show that the cost increases with the overall volume of installed capacity in the US. However, this correlation is not due to diseconomies of learning but rather, as we shall see below, to regulation, which, with passing time, has increased the construction cost of all reactors. It is important to remember that a correlation does not necessarily mean there is a relation of cause and effect. There is a correlation between sales of ice cream and suntan lotion, but one does not drive the other. The correlation is due to a single hidden variable, the weather, which affects sales of both products.

Thirdly, learning effects appear or are simply greater when utilities act as the prime contractors on projects, rather than simply purchasing a turnkey plant. There is less incentive for engineering firms to cut costs. But diminished economies of learning may also be due to their market power and a better understanding of costs. Firms may take advantage of their experience to boost profits, to the detriment of their customers. This conceals learning effects.

Lastly the rising costs were not the result of the accident in 1979 at Three Mile Island, though it did contribute to the trend.[9] The partial reactor meltdown which occurred there delayed some ongoing construction projects, but the rising costs also concern the overnight cost, which is not directly impacted by the duration of the project. Furthermore, the slowdown in the US nuclear programme started before the accident. In 1977 the volume of capacity ordered but subsequently cancelled exceeded built and commissioned capacity. The two curves crossed over. The already visible rise in costs partly explains the slowdown in the US programme.

One variable is missing yet omnipresent: safety regulation. But this variable is hard to measure, unlike reactor capacity or construction time. The number of texts and their length is not much use as an indicator, making no distinction between major and minor regulations. As a result, safety regulation is rarely taken into account as a variable in econometric equations. In 1979 two authors, Soon Paik and William Schriver,[10] invented an ad hoc index in an attempt to integrate regulation. They listed all the regulations issued by the US Nuclear Regulatory Commission (NRC) and sorted them into four categories, depending on their supposed importance. They were thus able to calculate that between 1967 and 1974 regulation had caused a 70 per cent increase in the investment cost per kW. In most other publications economists have used a temporal milestone (start or end of construction, issue of building permit) as an approximation for regulation. The work of the NRC continued at a steady rate all the way through the period during which nuclear plants were being built in the US; every year it published new standards, rules and measures. The regulation variable may thus be correlated with time. Any simple variable representing the passing of time, such as the year when a nuclear plant is connected to the grid, is just as useful as a complex indicator based on compiling and analysing NRC publications. Using temporal milestones to inform the regulation variable, US economists estimate that it is responsible for a 10% to 25% annual increase in construction costs.

The inflation in safety regulation is by far the largest factor in the escalating costs observed in the US. Stricter regulations require larger numbers of safety devices and systems, thicker containment walls, and completely isolated control rooms. In response to these tougher requirements engineers design increasingly complex facilities and systems. Only at the end of the 1990s did it occur to anyone that a possible solution might be to make things simpler, leading to Westinghouse's AP1000, which is based on a passive safety system. Rather than increasing the number of back-up pumps, for instance, a gravity-fed flow would be maintained if the cooling system failed. In the meantime safety was reflected in higher construction inputs and overall a more cumbersome framework for coordinating the construction of plants. The frequent changes in regulations also had a direct impact on the duration of construction projects. Work on a large number of US power plants had to be stopped in order to make allowance for new rules introduced since the start of work. Longer lead times meant higher financial costs, which of course added to the cost of investment. When new rules required additional inputs, this also impacted indirectly on the overnight cost. And, despite it being based on the assumption that a plant was built in one night, longer lead times pushed up overnight costs in the US.

At first sight, analysis of the escalating costs of nuclear power in the US might suggest that stricter safety requirements imposed by the regulator are to blame. But several factors contradict such a simplistic conclusion. It is not so much the severity of regulation as its defects that cost US nuclear power so dearly. Fluctuating rules and shifting priorities, excessive delays in decision-making and an inadequate understanding of the fundamental technical issues may generate excess costs for utilities, which far outstrip the impact of rising safety requirements. It seems more probable that, up to the end of the 1970s, the regulations did not so much attempt to raise the original safety level as simply to achieve it. It is far from easy to assess the safety level of a nuclear power plant, particularly before the fact, simply on the basis of drawings. Building and operating a plant may ultimately reveal that it does not meet the safety targets set by the regulator, and the operator, at the design stage. So the regulator intervenes to ensure that the original safety targets are fulfilled. This may remedy defective quality but does not raise its level. Some authors, such as Mark Cooper,[11] assert that early US reactors were quite simply defective in safety terms and that regulation imposed a form of making good. Lastly, if we read between the lines of escalating US costs we may detect serious shortcomings in industrial organization. Divided into a large number of utilities, often small and limited in territorial reach, and with a host of engineering firms,

the industrial organization failed to achieve sufficient standardization of procedures, reactor models and construction practices. Apart
from Bechtel, which built twenty-four reactors, the experience of
engineering firms and operators was limited to building just a few
nuclear plants. In short, unlike many other fields of technology in
which the US led the way, the development of nuclear power on an
industrial scale was not a great success.

The picture in France was very different, whatever its critics may
have maintained (see box). It has now been firmly established that
the escalation in costs was far less spectacular, with overnight costs
rising by 1.7 per cent a year, compared with 9.2 per cent in the US.

## International comparisons

Econometrics is unfortunately not much help here. On the one hand,
only a small amount of work has focused on France's nuclear reactors; on the other, the sample itself is small. In all we only have
twenty-nine records of costs. France has a total of fifty-eight reactors, but they were built in pairs and the EDF accounting system did
not itemize them separately. With such a small sample, few variables
can be tested. With respect to economies of scale, there is no sign of a
positive effect, quite the opposite. The nameplate capacity of French
reactors increased in three steps, rising from 900 MW for the first
reactors, through 1,300 MW for the majority of them, culminating at
1,450 MW for the last four. It is immediately apparent that the cost
per kW went up with each *palier*, or step, with a particularly spectacular leap at the end. The overnight cost reached €$1,442_{2010}$ per kW
for the last four reactors, compared with an average of €$1,242_{2010}$
per kW for the twenty second-step reactors, or €$1,121_{2010}$ per kW
for the first fifty-four overall. Econometric analysis yields no further information on this point; the diseconomies of scale persist.
Here again the explanation is to be found in the relation between
size and complexity. Not only did the reactors on each step differ

## A dizzy rise in costs based on mistaken analysis

In 2010 an academic journal published an article which attracted considerable attention.[12] For the first time the construction costs of French reactors were detailed and tracked over time. But contrary to what everyone imagined, the figures showed that France, despite its assets, had also suffered a steep escalation in costs: the cost of building France's last four reactors was allegedly 4.4 times higher than that of the first four. Worse still, the last reactor to be completed (Civaux 2) purportedly cost 7.5 times more than its cheapest counterpart (Bugey 4). It seemed that through some intrinsic fault nuclear technology was incapable of controlling costs and impervious to learning effects. The large scale of the construction projects, the limited unit count, the need to adapt to different sites, and the task of managing such a complex undertaking all contributed to cancelling out the cost-cutting mechanisms observed elsewhere: standardization, production runs comprising several thousand units and the repetition of almost identical processes.

This diagnosis would have been justified, had it not been founded on a mistaken estimate. In the absence of publicly available data on the construction costs of each French reactor, the author of the article, Arnulf Grubler, extrapolated the cost of plants from EDF's annual report on investments. Work had been carried out on several reactors – often of different sizes – in the course of the same year, so Grubler had broken down annual investment, using a theoretical model of expenditure to estimate the cost of each plant. Unfortunately this extrapolation yielded figures which subsequently proved to be at odds with reality. Far from a more than fourfold increase in the construction cost of reactors, from start to finish, the data later published by the Court of Auditors revealed a slightly less than twofold increase, in no way comparable to what had happened in the US.

in size, they varied in other ways. Each step brought technological advances. For example, the second-step plants were equipped with a completely updated control room and system. The design of the last four was almost completely different. When it comes to learning effects, econometric analysis is more helpful, revealing that the overnight cost of a reactor fell depending on the number of reactors already built on a given *palier*. Each additional reactor brought a 0.5% drop in cost. On the other hand, the effect is no longer visible if we look at the total number of reactors previously built. Apparently the experience gained building one model of reactor did not benefit a different model.

It is essential to grasp the step-related learning effect, because it throws light on a recent controversy. The French nuclear programme offered the best possible conditions for powerful learning effects. The power plants were built by a single operator, EDF, which was able to appropriate all the experience accumulated with each new project. The plants were built in a steady stream over a short period of time. In the space of just thirteen years, from late 1971 to the end of 1984, work started on construction of the first fifty-five reactors. The programme as a whole only slowed down at the end, with work on the last three units starting between late 1985 and mid-1991. The average construction time was consistent, only increasing slightly over time. Unlike what happened in the US, the regulatory framework did not upset construction of nuclear plants. The fleet expanded gradually thanks to cooperation between all the players (EDF, Commissariat à l'Energie Atomique, Framatome, Ministry of Industry), well out of sight of non-specialist outsiders.

So, despite the fact that France enjoyed the most favourable conditions for a gradual drop in the cost of building nuclear power plants, this did not materialize. What went wrong? We may suggest a series of specific explanations: the easiest sites were chosen first; quality assurance was gradually tightened up; the rising price of energy impacted on the price of machinery; project ownership expenses increased.[13]

At a more fundamental level, the French nuclear programme was over-ambitious and nationalist. The standardization and learning effects it made possible were cancelled out by changes in reactor models. The two capacity increases, from 900 MW to 1,300 MW, and then from 1,300 MW to 1,450 MW, coincided with substantial, expensive changes in technology. Some were adopted to make the technology French. In an effort to achieve greater independence and improve its chances of exporting its own reactors, France was determined to break free from the US technology used in the first pressurized-water reactors built there (see Chapter 10). The first stage in this process involved the design of the P'4 variant of the first-step 900 MW reactor. This dispensed with the need to pay licence fees to Westinghouse. The second stage brought the original design of a 1,450 MW reactor, but ultimately only four units were built. This model proved more expensive than its predecessor, due to its greater technological complexity and the exclusive use of components and machinery made in France.[14] In addition, construction times grew longer, reaching an average of 126 months for the last four plants, half as much again as for the plants built during the previous step. The French nuclear programme was nearing its end, indeed rather sooner than expected, because growth in demand for electricity, with a corresponding increase in capacity, had been overestimated. Completion of the last reactors was deliberately spread out in time, to adjust to demand and cope with the gradual winding down of the workforce caused by the end of the construction programme. Things are always clearer with the benefit of hindsight, but it does look as though France could have done without the last four reactors, which would have yielded a substantial saving.

Together the US and France have a total of 162 reactors, equivalent to just under a third of global capacity. What is known about the costs of other reactors? Nothing! There is no public source of data for all the nuclear capacity deployed in the former Soviet Union, Japan, India, South Korea or the People's Republic of China. No figures are available to say whether costs escalated there too, less still at what rate. We can only resort to qualitative reasoning. Apart from South Korea and China, it is hard to imagine costs rising less than in France. South Korea enjoys similar conditions, which should have enabled costs to be contained: swift pace of construction; reasonably similar reactor design and layout; well integrated industry and a single operator; nationalist fervour. In fact it may have done better than France. The picture in China is much more disparate, featuring all types of technology – boiling water, pressurized water, heavy water – and many sources: Canada, Russia, France and even the US. However, less than ten years ago China decided to give priority to building large numbers of its own CPR-1000 reactor, derived from the French 900 MW model. The speed of construction has been stupendous, great efforts have been made to standardize processes and the industry is very well organized. The cost of building this reactor has probably dropped with each new unit.

On the other hand the former Soviet Union and India would be plausible candidates for notching up escalating costs even worse than in the US. In the first case because costs under the socialist system were never a key issue when deciding to invest in infrastructure. Politics had more say than economics in the siting of plants, in the choice of model and the speed of construction. India is well placed too, no country having witnessed such a chaotic civil nuclear programme.

## Is there no limit to escalating costs?

Will what happened yesterday hold true tomorrow? We are confronted with a classic case of inductive reasoning. We have seen that the second reactor costs more than the first one, the third one more than the second... and that reactor $n$ costs more than $n - 1$. So can we conclude that the same progression will hold true for $n + 1$ and $n + 2$? The immediate answer is affirmative. If you have only seen black cats in the past, you will be quite prepared to bet they are all black. In the past nuclear power has reported rising costs, so nuclear technology is synonymous with rising costs. It is tempting to generalize, particularly as new next-generation reactors – the ones following the nth reactor such as the EPR – are again more expensive than their predecessors. However, we shall see that it is possible to upset this progression, even if it is much less likely than the previous trend continuing. Research would also need to explore new routes, with industry finding ways of standardizing models and developing modular machinery. If no spell is found to lift the curse of escalating costs, nuclear power will be gradually sidelined.

At the beginning of the 2000s costs seemed to stop escalating. Next-generation reactors were expected to bring improved safety, but they would also be cheaper than their forebears (see box). On paper the outlook for nuclear costs was rosy, on both sides of the Atlantic.

Barely ten years later, the first construction projects soon showed that the de-escalation everyone hoped to see had not yet started. The next-generation reactors were even more expensive. Present trends are after all entirely consistent with those of the past.

In 2009 the MIT published a second report,[15] updating the findings of the initial study six years earlier. The increase in the overnight cost was spectacular: expressed in current dollars it doubled, rising from \$2,000 to \$4,000 per kW.[16] In particular this figure took into account the estimated costs of eleven projected plants in the US, for which the relevant utilities had applied to the regulatory bodies for reactor licensing. Meanwhile, the University of Chicago investigated applications for construction licences for the Westinghouse AP1000. On average, the overnight cost quoted in applications was $4,210_{2010}$ per kW, multiplied by a factor of 2.3, in constant dollars, compared with a study seven years earlier.[17]

Unlike what occurred in the US, where next-generation reactors went no further than the drawing board, construction projects in

## Costs at renaissance

After a long, sluggish period in western countries, nuclear power woke up again in the early 2000s. New construction projects were tabled in the US and Europe. Many countries with no previous experience of nuclear power were also eager to enter the technological fray. This, it seemed, marked the so-called renaissance of nuclear power. The International Energy Agency forecast the construction of several hundred new plants by 2030. The outlook on costs was naturally just as upbeat. In 2003 the MIT published a study estimating the cost of building a plant with a next-generation reactor. In its base case it assumed an overnight cost of about $2,000 per kW, which yielded a levelized cost of $67 per MWh (with an 11.5% discount rate). To situate the latter cost in relation to the past,[18] let us imagine a scale of 1 to 100 ranking existing US plants by rising cost (calculated in constant dollars, adjusted for inflation and with a uniform 6% interest rate[19]). The MIT's projected plant would be ranked nineteenth, in the top 25% least expensive plants ever built, reaching back to the 1970s. In an even rosier scenario, positing a swifter, more flexible response by administrative bodies for the issue of construction permits, the cost would be lower than any plant previously built in the US. A year later the University of Chicago carried out a similar study, reaching comparable conclusions. On the supply side Westinghouse announced an overnight cost for its AP1000 of $1,400 per kW and a levelized cost of $27 per kWh.[20] Predictably, this estimate was more optimistic than the ones produced by university research laboratories.

In France the baseline costs were published by the Ministry of Energy. In 2003 the costs for third-generation nuclear plants were estimated at €1,300 per kW for the overnight cost and €28.4 per MWh for the levelized cost (with an 8% discount rate).[21] With these values the EPR bettered, in terms of cost, the reactors on the last step built in France. Industry was slightly less optimistic, with EDF suggesting an overnight cost of between €1,540 and €1,740 per kW and a levelized cost of €33 per MWh.[22]

Europe got off the ground. Work started on two EPRs, one at Olk-iluoto, Finland, the other at Flamanville, France. Here the increase in costs has been even more spectacular. In Finland the initial cost of the project when work started was €3 billion, or €1,850 per kW.[23] It has since been revised upwards on several occasions; delays have accumulated too. The final cost is now estimated at €6.6 billion, or €4,125 per kW. The job was supposed to last four and a half years, with grid connection in mid-2009. In the end, production will not start before 2015, at best: say ten years, to be on the safe side. Work at Flamanville started two years later and took the same unhappy route as its elder sister. The initial cost of €3.3 billion[24] has soared to €8.5 billion[25] and the original construction time of under five years will probably stretch to nine years. In the UK, where EDF is considering construction of two EPRs at Hinkley Point, the reported cost is between €17.2 billion and €19.7 billion.[26] So the first EPRs cost much more than the preceding 1,450 MW reactor model, on which they are based.

The changes in academic studies and industrial quotes are so large that it would be easy to make fun of them, or even to suspect decep-tion. But it would be a mistake. It is only natural that the initial estimates of experts and vendors should be a little optimistic. But for new nuclear neither experience nor facts were available to temper initial optimism. After a long period without any new plant being built, a large share of American and French expertise had vanished. Most of the engineers and senior executives who had taken part in the golden age of nuclear power had either moved to another sector or retired. Furthermore, the first cost estimates were drafted when design of the next-generation reactors was still in its early stages. Millions of man-hours were still needed to finalize detailed plans,[27] which inevitably revealed additional costs. Then it was time to obtain quotes from suppliers and to sign contracts for parts and machinery, a process which took the true understanding of costs one step further. The last set of estimates generally focuses on indexed values, in particular the price of raw materials and building materials. This brought additional price increases, the first decade of the 2000s having seen substantial upward pressure on these commodities. The overnight cost of gas- and coal-fired power plants also increased steeply over this period.[28] The difference with nuclear power was that the initial estimates for the fossil-fuel plants were more accurate. They were based on a building process which had never stopped, nor yet slowed down, all over the world, with hundreds of examples on which to draw.

Optimism may also be dictated by self-interest. Utilities in favour of nuclear power and reactor engineering firms stand to gain by reporting low costs in their initial estimates, by only publishing values at the lower end of their spread estimates. On the other hand, much as any trader selling goods to a small number of buyers, on whose custom the business depends, it is not in the interest of reactor vendors and turnkey plant integrators to announce miraculous figures. Making promises which they know they cannot keep permanently saps their credibility in the eyes of customers, bankers and governments. If there was any deceit regarding costs at the renaissance of nuclear power, it was industry which fooled itself.

To put an end to any notion of across-the-board deceit, it should also be borne in mind that the baseline academic studies did not only work on a set of assumptions favourable to nuclear power. The reason why the first MIT study caused such a stir in 2003 was that it made the iconoclastic choice of a high discount rate, which was unfavourable to nuclear power. The MIT highlighted the high financial risk associated with this investment in liberalized electricity markets. As a result, the assumptions regarding the structure and cost of nuclear capital were less attractive than for gas or coal. Nuclear power involved higher capital outlay, less debt and a 15 per cent return on assets, rather than 12 per cent. Without these assumptions the MIT study would have concluded that the excess cost of nuclear power, compared to gas, was only half as large.[29]

There is no escaping the facts and they are particularly stubborn: nuclear power now is much more expensive than before. For the time being third-generation reactors are still plagued by rising costs, and new reactor models bring additional costs. What does the future hold?

With the same design, costs should certainly drop, but by how much? It is impossible to say whether there will be a slight reduction or a huge one. Take the EPR. Its cost is bound to drop, but how far? First-of-a-kind costs are known to be higher, generally by about 20 to 30 per cent,[30] but it is not known how the excess cost is amortized. Does the full burden fall on the first unit, or is it spread over the first five or ten reactors? For obvious reasons – the first customers do not like teething problems – data of this sort is confidential. Furthermore, there has been a loss of experience on the construction side, following a long period without any new projects. Lastly, the first two EPRs are not being built by the same company.

Seen from abroad, the French nuclear industry may look like a homogeneous block: EDF and Areva, both publicly owned companies, seem barely distinguishable. But in fact they have been keen rivals in recent years. Areva went it alone in Finland, operating as a turnkey plant vendor, rather than just selling a reactor, which is its core business. EDF has long-standing experience as both the prime contractor and project owner of nuclear plants. It sees Areva as an original equipment manufacturer, or even – rather disparagingly – as a boiler manufacturer. So learning effects between Olkiluoto and Flamanville are limited. The two firms have been at loggerheads, rather than confidently pooling their experience. The opposite seems to have happened at Taishan, in China, where two EPR-powered plants are being built. EDF and Areva are working together with the prime owner, the China Guangdong Nuclear Power Group, the utility in Guangdong province. For the time being Taishan-1 is on target for both construction time (five years) and cost (€3 billion). Areva management say this is thanks to the return on experience from the Finnish and French jobs.[31] Certainly, between Olkiluoto and Taishan, the supply deadlines have improved by 65%, engineering man-hours for the nuclear steam-supply system are down by 60% and the time taken to build the main components has been cut by 25% to 40%. So the third reactor seems poised to finish first. Work on Taishan-1 started in 2009, after the other two, but it should be connected to the grid before the end of 2014, several years ahead of Olkiluoto and Flamanville. However, return on experience is not the only reason for the impressive performance in China regarding costs and deadlines. The PRC boasts top-notch civil engineering contractors, can count on a seasoned nuclear industry, is deploying a massive programme (with twenty-eight reactors under construction in 2013), and has the advantage of a cheap, well qualified workforce and a well organized site where work continues round the clock, even at weekends.

The last unknown regarding the scale of the drop in the cost of the EPR relates to the number of units ultimately built worldwide. Four, ten, twenty or more? All other things being equal, the more reactors sold, the lower the cost and vice versa. The serpent eats its tail. Potential buyers are price-sensitive – though we do not know whether this effect is very slight or substantial – and learning effects cut costs, though here again we cannot say by how much.

From a technical point of view the key to lower costs is to be found in standardization and modularity. Standardization requires every unit of a particular reactor model to be identical, which is not always the case, due to specific changes demanded by customers or safety authorities. As mentioned above, standardization allows learning effects; we may add that it also facilitates competition between suppliers, another powerful mechanism pushing costs down. Modularity means construction in modules, in other words component parts which are relatively independent one from another, making it easy to separate them and simply assemble them on-site (structural elements, but also cable ducts, reinforced concrete mats, etc.).[32] A good example of modular building is factory-assembly of the roof timbers of a detached house, rather than erecting them piece by piece on-site. Pre-assembly is advantageous because a factory is a sheltered environment and such operations lend themselves to automation, yielding productivity gains. Pre-assembly also reduces the amount of clutter on a building site, streamlining its organization. So modularity has the potential for substantial gains.

So far, our reasoning has been based on an unchanging technological framework. What happens to the costs entailed by nuclear power if we take into account innovation, and the design and development of new reactors? Past form is far from encouraging. We have seen that in France, where conditions were most favourable, each new model led to an increase in the construction cost per kW of installed capacity. Two insurmountable obstacles seem to be preventing a reduction in the cost of new models. The first relates to the increasingly strict rules on safety. It is hard to imagine the authorities certifying a new model with lower safety performance than its predecessors. As time passes experience gained from building and operating plants reveals defects; progress in science and technology provides solutions to correct them. Furthermore, with time, new political risks may emerge (terrorist hijacking of an aircraft to target a power plant, for instance) and, in general, public opinion is increasingly averse to technological risks. The above is true for countries already equipped with nuclear power. For new players safety requirements may be less stringent and they may not require the latest generation of reactors. But, keen to develop their science and technology, such countries are unlikely to resist the appeal of modernity for long.

So the question is whether it is possible to build reactors which are similar to the current generation, but safer and cheaper. Very probably not, but as it is still too soon to pass judgement on the AP1000, we should allow for a positive outcome. Westinghouse designed this reactor with two aims: to provide a mechanical solution to some of the safety problems; and to simplify the overall design. For example, water tanks are positioned on the roof in order to cool the reactor vessel should the need arise, fed by gravity and the pressure inside the system. This more or less halves the need for pumps, valves and pipework. Four AP1000s are currently under construction in China. It will be interesting to see, in a few years' time, whether they cost substantially less to build than the EPR. If the concept is a success, it could lead to the development of improved versions, using the new design rules, but at even lower cost. Nuclear power may finally cast off the curse of rising costs.

The second, apparently insurmountable obstacle concerns on-site construction and short production runs. Much like other large civil engineering projects – bridges, airports or dams – nuclear power plants are mainly built on-site. Progress may be made towards greater modularity, but there is little hope of a 1,000-MW plant one day being put together like a flat-pack kitchen. Civil nuclear power differs from other electricity-generation technologies in that only a small number of units are built. Whereas hundreds or thousands of wind-farms or coal- or gas-fired plants are ordered worldwide every year, there are just a few dozen new nuclear construction projects. One of the reasons is the trend towards building increasingly large reactors. The scale of fixed costs justifies this option, because they can be recouped on a larger volume of electricity output. But there is nevertheless a downside. All other things being equal, the more powerful the reactor, the smaller the number of identical units built. So production runs are short and only a few similar parts and components are manufactured. The trade-off between economies of scale per unit and manufacturing economies of scale has so far tipped in favour of the former.[33] Giving fresh impetus to small-reactor projects would break with this approach.

The example of small reactors is worth looking at, because it demonstrates the scope for radical innovation, which in my opinion offers the only lasting antidote to the curse of rising costs. People have been developing low-power nuclear reactors for many years. They are used to drive nuclear submarines, drawing on work and trials going back to the 1950s. What is new is the sudden emergence of futurist projects. Take for instance the best-known example, funded by Microsoft founder Bill Gates. The project is being developed by TerraPower, in which Gates is the main shareholder. The initial aim has been to produce a mini-reactor several metres high, running on natural uranium and cooled by liquid sodium. It is based on the travelling-wave principle, with the reaction slowly spreading outwards from the core of a block of uranium. Picture a candle with a flame inside gradually advancing as it consumes the surrounding wax. For the reactor itself, imagine a cylinder less than one metre high, which requires no outside intervention once the reaction has started and which shuts down on its own after several tens of years. We may also cite the project for an underwater nuclear power plant being developed by France's naval defence firm DCNS. In this case the cylinder is 100 metres long and 15 metres in diameter, containing a reactor and remote-controlled electricity generating plant. With several tens of MWs' capacity, it would be located out to sea, several kilometres from the coastline, anchored to the seabed. The cylinders would be modular units, several of which could be placed side by side in the case of higher output requirements. The units would be taken back to a shipyard for maintenance and replaced by other units, much like bottles with a refundable deposit.

These projects, which sound even more fantastic when described so succinctly, will very probably never see the light of day. Either they will founder completely or change so much that the final application bears no resemblance to the initial concept. It matters little to our current concerns. That is how radical innovation works: projects pursuing a large number of original ideas are launched; very few give rise to pilot schemes; an even smaller number lead to commercial projects; and in each case the ongoing redefinition process will shift pilot schemes and commercial goods further and further away from the original idea. Obviously there is no way of knowing in advance whether, out of the hundreds of current and future projects to develop modular small or mini-reactors similar to those discussed above, at least one could reach fruition and enter industrial production. But unless nuclear research moves away from the present model of large, non-modular plants and gigantic construction projects, the costs of nuclear technology will likely continue to rise, which is a serious drawback in the competition between nuclear power and other electricity-generating technologies.

# THREE

# Nuclear power and its alternatives

We cannot do without oil but we may, on the other hand, stop using the atom. We should never lose sight of the fact that there are several means of generating electricity, using among others coal, gas, oil, biomass, solar radiation and wind. At the scale of a whole country these generating technologies are generally combined to form an energy mix, which may or may not include nuclear power, much as it may or may not include thermal coal or gas, wind or solar.

The various technologies are both competitors and complementary. Conventionally a distinction is made between baseload generating technologies, coal-fired power stations for example, which operate round the clock all year long, and peak generating technologies, such as oil-fired power plants, which only operate at times of peak demand. With a finer mesh, further categories can be distinguished, of semi-baseload and extreme-peak generation. The overall idea is to classify production resources in such a way that the ones with high fixed costs and low variable costs are used for as many hours a year as possible, while on the other hand those with low fixed costs and high variable costs are only used for a few hours a year.

Two categories of baseload technology – coal and nuclear – are in competition, whereas oil-fired technology is complementary. However, in situations where they overlap this ranking may change. For example gas, which tends to be seen as a semi-baseload resource, may play a primary role as a baseload resource; nuclear power may lend itself to load-balancing (as in France, for example) and is consequently suitable as a semi-baseload resource.[1] Renewable energy sources also upset the ranking. Hydro-electric power from dams is generally seen as a peaking resource, despite its extremely high fixed cost and variable operating cost close to zero, the explanation being that its variable cost should in fact be treated as a marginal opportunity cost. It is preferable to hold back a cubic metre of water for peak hours with correspondingly high prices, rather than wasting it by generating electricity at times when demand drops and the price is low. Regarding wind and solar, production is intermittent because it depends on the force of the wind or the amount of sunlight, which vary in the course of a day, and from one day to the next, quite beyond our control. Here again variable technical costs are close to zero, but the irregular nature of output makes it impossible to classify these technologies among baseload resources. At the same time, the lack of any way of controlling them means they cannot be treated as peaking resources. If intermittent renewable energy sources play a significant part in the energy mix, back-up capacity must be available – generally gas-fired plants – to take over in the absence of sunlight and wind. Under these circumstances gas and the renewable energy are complementary. However the growth of intermittent energy sources pushes the market price of electricity down and baseload and semi-baseload sources operate for shorter periods. This creates competition between nuclear power and gas, on the one hand, and renewable energy sources, on the other.

Lastly, nuclear power and renewables have one characteristic in common: they produce no $CO_2$ emissions. They may consequently be seen as rivals for achieving the targets set for reducing greenhouse gas emissions, or alternatively, as it seems difficult to rely exclusively on just one of these sources, they may be seen as complementary, with a view to completely carbon-free electricity generation. To simplify matters, any comparison of nuclear electricity should make allowance for two factors: on the one hand its competitive or complementary position in relation to coal or gas, for baseload electricity production; and on the other hand its competitive or complementary position in relation to other carbon-free energy sources.

## The relative competitive advantage of nuclear power over gas or coal

The levelized cost enables us to classify the various generating technologies. Which one, out of coal, gas or nuclear power, offers the lowest cost? How do these forms of energy rate in the overall cost ranking? In fact, our obsession with rank prompts us to ask the wrong questions, which only yield contingent answers.

There is no single ranking system because the costs depend on different locations and hypotheses on future outcomes. With regard to nuclear power we have seen that the cost varies from one site to another, from one country to the next, and that it above all depends on the discount rate. The cost of fuel is the key parameter for coal and gas. But the price of energy resources depends on geography. The cost of transporting coal or gas being high, building a fossil-fuel power plant in one place or another yields different results. Furthermore, market prices fluctuate a great deal, particularly for gas, often indexed to the price of oil. The rate of return on an investment in a new fossil-fuel plant depends on assumptions as to how fuel prices will behave over the next ten or twenty years. Consequently it is only possible to use the levelized cost to rank coal, gas or nuclear power on the basis of a very specific set of conditions, valid at the geographical scale of a country and in line with the expectations of specific operators. For example, taking a broad-brush approach to the current position in the US, gas enjoys a comfortable lead, followed by coal, with nuclear power in third place. This ranking may vary between US states depending on the proximity of coal-mining resources and unconventional gas reserves.

We should nevertheless bear in mind a few, almost universal trends and shifts, which also happen to explain to a large extent the current US ranking of baseload generating technologies: before and after climate-change policy; before and after shale gas; before and after deregulation of the electricity market.

In a world with no pollution-abatement measures, coal would lead the pack with the cheapest MWh almost all over the world. But using it to generate electricity causes local pollution (release of dust, soot, sulphur and nitrogen oxides) and $CO_2$ emissions. The first group is by far the most costly, unless a very high price is set for $CO_2$ (in excess of \$100 per tonne). In ExternE, the major European study of the externalities of generating electricity, the damage caused by coal, setting aside that linked to $CO_2$ emissions, was estimated at between \$27$_{2010}$ and \$202$_{2010}$ per MWh. The lower value in this range is about the same as the one reported by William Nordhaus and other authors in a conservative assessment dating from 2011.[2] As for the upper value, it is comparable to that found in a maximalist study by Professor Paul Epstein, at Harvard, published the same year.[3] Taking the values which the experts consider to be the 'best estimates', we may note that the cost of a coal-generated MWh doubles when we include its externalities.

The large divergence between the upper and lower values in the estimates can be partly explained by the different types of plant under consideration and the prevailing environmental standards. In OECD countries the regulatory framework for local emissions from coal is very strict. Part of the externalities is internalized by emissions standards, which raise the overnight cost of coal-fired thermal plants, and consequently the levelized cost of energy for the utility. Similarly, some OECD countries have introduced a carbon price, or are planning to do so. Depending on their level, such taxes and tradable emissions permits internalize, to a greater or lesser extent, a share of $CO_2$ externalities and add to the variable cost borne by the utility. On the other hand, in most developing or emerging countries, the cost of a coal-generated MWh is still low because neither investors nor utilities pay for any part of the environmental damage entailed, in the absence of both regulations on local pollution and a carbon price. This lack of symmetry explains why it is now almost out of the question to build coal-fired power plants in the US, the UK or Japan, whereas such facilities are springing up in China, Malaysia, Senegal and South Africa. In terms of new electricity-generating capacity being installed, coal is the technology which has enjoyed by far the strongest growth worldwide since 2000. In the long term, the cost of a coal-generated MWh in non-OECD countries is expected to rise, reducing the gap. The localized pollution and damage this technology entails for public health exert pressure which encourages a shift towards other, more expensive technologies which cause less pollution. In OECD countries it is more difficult to predict future developments. The application of R&D work on clean coal, particularly for carbon capture and storage technology, is uncertain. Future trends for the price of $CO_2$ emissions are equally uncertain.

Gas has a very different environmental profile from coal, with little or no local pollution, and half the volume of $CO_2$ emissions. This explains its growth in OECD countries, at the expense of coal. The price of gas delivered to the generating plant is generally higher than for coal, but this competitive disadvantage is counterbalanced by incomparably lower environmental costs.[4] There is certainly a before and after unconventional gas here, because this advantage is now being enhanced by lower costs due to new gas-exploitation techniques (horizontal drilling and hydraulic fracturing), and the resulting extension of reserves. In the US, where shale gas was first exploited (alongside Canada), this change means that nuclear power is durably losing its status as a baseload generating technology. Gas is now in first place and is likely to stay there for a long while.

However, it should be borne in mind that unconventional gas currently enjoys a novelty effect, which means its social cost is underestimated. It took decades to estimate the economic externalities of coal, conventional natural gas and nuclear power. They took shape as science advanced in its understanding of the effects of pollution and on-site measurements. The dissemination of scientific advances and the results of metrology, beyond the confines of laboratories and a small number of experts, works on a specific time scale. None of this applies to shale gas, yet. The measurements and studies have barely started, particularly to estimate greenhouse gas emissions and possible damage to aquifers. It is plausible to suppose that what has so far been gained through lower exploitation costs may tomorrow be lost to rising environmental costs. Lastly it is worth noting that the decision by some markets to delink oil and gas prices gives the latter an advantage which is likely to last. Until now, in many countries gas prices were driven up by the rising price of oil. Oil-indexed gas supply contracts were encouraged by various factors: comparable extraction conditions; joint production in some cases; and markets offering imperfect competition, due to the dominant position of monopsonists. In places where the exploitation of conventional gases has developed, this arrangement has been destroyed.

Liberalization of the gas and electricity markets is the third key shift which changes the relative competitiveness of baseload generating technologies. Here too nuclear power has lost ground on the whole. For many years the gas and electricity markets were organized as municipal, regional or national monopolies subject to regulated tariff schemes. Regardless of whether generating companies belonged to the public or private sector, the investments they made were exposed to little risk, being paid back by captive consumers. Dependent on the authorities, these companies often acted as cogs in the implementation of energy policies based on factors related to cost, but also to national independence, scientific prestige, job creation and such.

Instigated by some US states and the UK, privatization and the opening up of the gas and electricity markets to competition upset this model. In its place, or alongside it, another model was established in which the link between production and captive consumption was broken, and in which investment was decided by shareholders and bankers. From being utilities – public service providers – the electricity generating companies became operators at the head of merchant plants, power stations selling electricity to the wholesale market. The risks here were not of the same order. Much like football teams that compete on the same playing field, be it muddy or too hard, one might suppose that liberalization would affect all the electricity generating technologies in the same way. Accordingly the new deal should not alter their competitive positions in relation to one another. In practice this did not prove to be the case for nuclear power, which, as far as the financiers were concerned, involved greater, more serious risks:[5] higher risks of budget overruns and missed deadlines, in the course of construction and during operation (e.g. safety defects leading to unpredictable reactor shutdowns and consequently lost output); a long period over which to recover investment, increasing the risk due to uncertainty in prices on the wholesale electricity markets; higher regulatory and political risks due to the opposition of part of public opinion and some political parties to atomic energy. In the face of these additional risks, the MIT 2003 study cited above set a weighted average capital cost 25 per cent higher than for gas and coal, which pushed up the cost per MWh of nuclear power by 33 per cent.

We may observe that it makes little sense to rely on the levelized cost method in an economy with liberalized energy markets. The rationale used to establish the price of electricity, which balances income and expenditure, including the remuneration of capital, is more in keeping with regulated electricity tariffs set by the authorities. In a market economy, electricity prices fluctuate; they are uncertain, just like the price of fuel consumed by generating plants, or the price of tradable emissions permits. The solution is to use the conventional method for calculating the return on a project in terms of net present value, while taking into account the uncertainties.

The price of electricity can thus be treated as a variable, which is associated with a distribution function (e.g. a bell curve, on which the peak represents the most probable expected value and the extremities the lowest and highest values, of low probability). Similarly various values with a range of probabilities are allocated to the other variables affecting income or outgoings. Then we shake up all these data, carrying out repeated random sampling, thousands of times – using the Monte Carlo method, in reference to roulette. We thus obtain the risk profile for the investment, in other words a curve showing the losses and gains it may produce, each level of loss and gain being associated with a probability. If the curve is relatively flat the risk is high, because the probability is more or less the same for low or high rates of return, both positive and negative. If the curve rises sharply, the risk is low, with a substantial probability that the rate of return will be centred near the peak, be it positive or negative.

The merit of this probabilistic approach is that it yields a mean value (obviously essential to knowing whether the return will be positive or negative, low or high), but also an indication of the possible variances on either side of the mean. Assisted by other authors, Fabien Roques has used this approach to obtain a better comparison of baseload electricity generating technologies.[6] With a whole series of possible hypotheses – in particular a 10 per cent discount rate – their research shows that gas yields higher profits than nuclear power, at a lower risk, the latter point being due to the gains achieved by more flexible plant operation. The load factor, instead of being constant throughout the service life of plants, varies according to the market price of electricity. If the price results in a loss, production stops, starting again when the net present value is once again positive. A second interesting outcome of this work is that it puts figures on the complementary relation between gas and electricity. A portfolio of assets, with gas-fired plants making up 80% of capacity and nuclear power the remainder, yields a lower average return than an exclusively gas-fired portfolio, but entails less risk. Investors may prefer this combination, which offers better protection, particularly from high but unlikely losses incurred if gas and carbon prices are high, a situation which has no effect on nuclear power.

## The competitive advantages of nuclear power and renewable energies

In suitable locations onshore wind farms display levelized costs comparable to those of nuclear plants. Neither technology releases $CO_2$ emissions and both are characterized by high fixed costs. However, although nuclear power has a low marginal cost (about €6 per MWh for fuel[7]), for wind the cost is zero. (The same is true of solar but, except under extremely favourable conditions, its levelized cost is way above that of nuclear.) From an economic point of view this difference is of fundamental importance, because in an electricity market the optimal price is equal to the marginal cost of the marginal unit, in other words the unit that needs to be generated to meet instantaneous demand. When instantaneous demand is at its lowest, generally in the middle of the night, only baseload plants are used. If massive wind capacity were to be installed, the night breeze would blow away gas and coal (perhaps even nuclear) during off-peak hours, reducing their load factor and raising their respective costs per MWh. In fact the loss would be even greater. Coal-fired or nuclear power plants do not ramp up to full capacity or shut down instantaneously. So slowing down or stopping output at night would reduce the power available in the early morning. To sell more electricity at times when prices are higher, it may be in the interest of baseload plant operators to bid negative prices in order to keep their plants running all night. So at certain times of the day, large scale wind capacity would result in a market price equal to its marginal cost, in other words zero, and even, at other times, in a lower market price, equal to the opportunity cost of baseload operators forgoing a reduction in output.

Not taking into account variations in demand distorts the results when calculating the levelized cost. Paul Joskow, at MIT, has shown that this method is unsuitable for intermittent renewables.[8] Only exceptionally are intermittent energies in sync with demand. The wind does not blow harder at the beginning or end of the day, nor yet during the five working days of the week, which is when power demand is highest. To simplify matters we shall suppose that peak and off-peak hours are evenly distributed throughout the year. We shall then suppose that an intermittent renewable plant produces two-thirds of its output at off-peak hours, the remaining third at peak hours, and that its levelized cost per MWh is the same as that of a baseload plant. If the country as a whole needs one additional MWh of power, the levelized cost method tells us that it makes no difference whether we invest in wind power or a baseload technology. Yet the second option is more useful because it will produce proportionately more at peak hours: with all-year-round output it operates half of the time at peak hours, the other half off-peak. So the levelized cost method is biased against investment in baseload technology. It is also worth noting that it distorts the ranking of intermittent renewable energies. As the sun does not shine at night, a solar plant generally responds in a larger proportion to peak demand in summer than a wind farm. To compare investment projects in various generating technologies, it is consequently wiser to use the net present value method to estimate income on the basis of the hourly generation profiles of plants and the electricity prices expected at different times of the day.

A third form of distortion which handicaps nuclear power is specific to Europe. The EU has set targets for renewable energies. By 2020 renewables are slated to account for 20 per cent of final energy consumption. As applied to electricity this target means that renewables should supply 35 per cent of all electricity. Measures of this sort requiring a share of renewables in the overall energy mix are commonplace. Most US states apply similar measures. But the EU is unusual in that the measure operates in parallel with a carbon price. The EU system of tradable emissions permits already adds to the cost of fossil-fuel-generated electricity, compared to nuclear power or renewables, changing their relative competitiveness in the same way as a carbon tax. For example, with a permit costing €30 per tonne of $CO_2$, it costs about €30 per MWh more to generate 1 MWh using coal. Adding a target for renewables to this scheme pushes the price of carbon down. The 20 per cent target for 2020 was set without adjusting the cap on $CO_2$ emissions decided when the Emissions Trading Scheme was originally set up. As a result the cut in emissions, made compulsory by the renewables quota, restricts demand for permits. So their price drops. David Newbery has estimated that the price of permits will be driven down by €10 per tonne by 2020, from €60 to €50.[9] To avoid this downward pressure, the cap on emissions should have been lowered to allow for the volume of $CO_2$ recently avoided, in such a way as to achieve the target of 35 per cent electricity from renewable sources. In conclusion, the quota for renewable energies in the EU energy mix has a dual effect. It deprives nuclear power of part of its potential market, despite its also being carbon-free, and makes it less competitive by doing less to increase the price of competing baseload technologies, due to a lower carbon price.

We need to see the electricity system as a whole in order to grasp the relative competitive advantages of nuclear power and renewables. As wind, solar and wave are intermittent energy sources, and storing electricity is very expensive, large-scale development of renewables involves building back-up capacity to make up for the lack of wind, sunlight or tide at certain times. Such capacity is far from negligible. For Ireland to meet its target for the 2020 renewables quota, it will have to install 30 GW more renewable capacity, while providing a further 15–20 GW of non-intermittent capacity as a back-up.[10] To enable such supplementary capacity to be built, the country must either agree to stupendous electricity prices (several thousands of euros per MWh) at certain times of day, or set up capacity markets to pay utilities even when they are not producing anything. Otherwise the plant will simply not be built, because investors will anticipate difficulties covering fixed costs due to the insufficient load factor. Nuclear power, dogged by higher fixed costs than gas and less flexible production, is ill suited to catering for this new demand. All other things being equal, the more intermittent energies develop, the more the competitiveness of nuclear power with regard to gas will be undermined.

Looking beyond 2020 we see no sign of a possible improvement in the competitiveness of nuclear power compared with renewables: quite the opposite. The development of storage technologies and ongoing learning effects for wind and solar represent serious threats. Using batteries to store electricity is still outrageously expensive. So far the only alternative solution to have been developed is pumped-storage hydroelectricity. This involves using electrical pumps to raise water from one reservoir to another at a higher elevation. Meanwhile research is focusing on countless other possibilities. What results and applications will research yield over the next twenty years? Without an answer it is hard to see whether electricity storage will one day be sufficiently affordable to be deployed on a very large scale. Realizing such a possibility would remedy the main shortcoming of intermittent renewable energy and substantially increase its economic value, at least for renewables for which the cost is currently close to that of more traditional technologies. This is the case for onshore wind, setting aside the sources of distortion cited above. In the future it might also be the case for offshore wind, and photovoltaic or concentrated solar. The costs of these technologies have dropped substantially, with scope for powerful learning effects. But we shall once again concentrate on terrestrial wind power.

The levelized cost per MWh of onshore wind power was divided by three, allowing for inflation, between the early 1980s and the late 2000s.[11] Estimates indicate learning effects between 10 and 20 per cent.[12] However, a closer look reveals that the reduction in the levelized cost in constant dollars stopped in 2005, and that the levelized cost has actually risen since. Is this a sign that the technology has reached maturity, with an end to diminishing costs? Very probably not, as shown by the report by the US National Renewable Energy Laboratory. The rise in costs towards the end of the 2000s is due to the increase in the price of materials and machinery, and a flattening out of performance gains. But since then performance gains have started to improve and the cost per MW of installed capacity is steady. The levelized cost across all wind speeds started dropping again in 2012, down on 2009. The NREL has also compared a dozen prospective studies looking ahead to 2030, covering eighteen scenarios in all. Most of them predict a 20–30 per cent reduction in the levelized cost. Only one forecasts that it will remain steady. These results obviously concern specific wind classes. The average performance of wind capacity in a country or region may decline over time, due to the less favourable characteristics of more recent locations, the first wind farms having occupied the spots with the best conditions. The issue of siting is the only factor driving costs upwards. However, in the future it would be more than offset by the gains derived from mass production and higher performance fed by R&D.

So nuclear has been caught in a pincer movement, so to speak. In OECD countries its high cost, particularly with regard to capital, is a handicap compared to gas. It is only competitive if a carbon price is introduced – a fairly high one at that. On the other hand, setting aside onshore wind power, it is still more cost-effective than intermittent renewables. So in principle there is every reason why it should feature in a mix of carbon-free generating technologies. But only in principle, because in practice it is sidelined and hampered by quotas for renewable energies. In other countries nuclear power is at a disadvantage when compared to cheap, polluting coal, but at least the prospects are a little better. Demand for energy is often so great that all technologies are considered. Large countries such as

China and India can plausibly hope to reduce costs through large-scale production and learning effects. Smaller nations may count on the advantage derived from keen competition between vendors of turnkey solutions (see Chapter 10).

On reaching the end of the first part of this book, readers may feel slightly bereft, having lost any sense of certainty regarding costs. There is no such thing as a 'true' cost for nuclear power, which economists may discover after much trial and error. Nor yet are there any hidden external costs, such as those related to managing waste or the risk of serious accidents, which might completely change the picture if they were taken into account. Far from reducing the cost of nuclear power, technical progress has actually contributed to its increase. It makes no sense to assert that it is currently more or less expensive, in terms of euros per MWh, to build a wind farm or a nuclear power plant. There can be no universally valid ranking order for coal, gas and the atom based on the cost of generating electricity.

But the loss of such illusions should not leave readers in a vacuum. The first part has also provided a firm basis for assessing the costs of electricity, which depend on location and various hypotheses on future developments. Consequently such costs can only be properly calculated with a clear understanding of both factors.[13] The construction cost of a nuclear power plant is not the same in Finland, China or the United States. Overall expenditure may vary a great deal depending on the influence of the safety regulator, scale effects and the cost of capital. Regarding wagers, the future prices of gas, coal and carbon dioxide will be largely decisive in the ranking of coal, gas and nuclear power. These same prices will also affect the profit margins of nuclear plants, their revenue depending on the number of hours per year during which they operate, and whether the prices per kWh during those hours are decided by a marginal generating plant burning coal or gas, or one powered by sunlight or wind. Confronted by the risky long-term wagers which investors must make to calculate costs and take decisions, even the most laissez-faire public authority will feel obliged to intervene. Concerned by the general interest, it must set a discount rate, the parameter with the greatest impact on the cost of nuclear power. This particular wager hinges on how prosperous future generations may be: the richer they are, the lower the discount rate will be, making nuclear power that much cheaper. Furthermore, there is a political choice to be made, in order to maintain a certain degree of equity between rich and poor, and between generations, a choice which influences the rate set for converting present dollars into future dollars.

What is more, analysing trends for past costs throws light on their future behaviour. Historically, nuclear technology has been characterized by rising costs. Today's third-generation reactors are no exception to this iron rule. They are safer than earlier counterparts, but also more expensive. The escalation of costs may stop, but only on two conditions: through a massive scale effect – if China chooses one type of reactor and sticks to it, it may achieve this effect – or through a fundamental change in direction of innovation, giving priority to modular design and small reactors, for instance. Failing this, nuclear technology seems doomed to suffer a steady decline in its competitiveness compared with any thermal technologies spared by taxes and renewable energies boosted by high learning effects.

Setting aside any consideration of possible accidents, it would be an economically risky choice for an operator to invest in building new nuclear power plants or for a state to facilitate such projects.

# PART II

# The risk of a major nuclear accident

## Calculation and perception of probabilities

The accident at Fukushima Daiichi, Japan, occurred on 11 March 2011. This nuclear disaster left a lasting mark in the minds of hundreds of millions of people. Much as Three Mile Island or Chernobyl, yet another place will be permanently associated with a nuclear power plant which went out of control. Fukushima Daiichi revived the issue of the hazards of civil nuclear power, stirring up all the associated passion and emotion.

The whole of Part II is devoted to the risk of a major nuclear accident. By this we mean a failure initiating core meltdown, a situation in which the fuel rods melt and mix with their metal cladding. Such accidents are classified as at least level five on the International Nuclear Event Scale. The Three Mile Island accident, which occurred in 1979 in the United States, reached this level of severity. The explosion of reactor four at the Chernobyl plant in Ukraine in 1986 and the recent accident in Japan were classified as level seven, the highest grade on this logarithmic scale.[1] The main difference between the top two levels and level five relates to a significant or major release of radioactive material to the environment. In the event of a level-five accident, damage is restricted to the inside of the plant, whereas, in the case of level-seven accidents, huge areas of land, above or below the surface, and/or sea may be contaminated.[2]

Before the meltdown of reactors one, two and three at Fukushima Daiichi, eight major accidents affecting nuclear power plants had occurred worldwide.[3] This is a high figure compared with the one calculated by the experts. Observations in the field do not appear to fit the results of the probabilistic models of nuclear accidents produced since the 1970s. Oddly enough, the number of major accidents is closer to the risk as perceived by the general public. In general we tend to overestimate any risk relating to rare, fearsome accidents. What are we to make of this divergence? How are we to reconcile observations of the real world, the objective probability of an accident and the subjective assessment of risks? Did the experts err on the side of optimism? Is public opinion irrational when it comes to the hazards of nuclear power? How should risk and its perception be measured?

# FOUR

# Calculating risk

In Part I we made the tentative suggestion that the cost of accidents was low, indeed negligible, when compared with the value of the electricity generated. This introductory conclusion was based on a scratch calculation drawing on upper-case assumptions, multiplying the (ill) chance of a disaster of 1 in 100,000 years of a reactor's operation by damage costing €1,000 billion. The cost of such an accident amounts to €1 per MWh generated. Setting aside the back of an envelope, which lends itself to quick, simple calculations, we shall now look at the matter in greater detail, from both a theoretical and an empirical standpoint.

## Calculating the cost of major accidents

Is the cost per MWh of an accident negligible, at less than 10 euro cents? Or is it just visible, at about €1, or rather a significant fraction of the several tens of euros that a MWh costs to generate? How does our scratch calculation, in Part I, compare with existing detailed assessments and how were the latter obtained?

Risk is the combination of a random event and a consequence. To calculate the risk of an accident, the probability of its occurrence is multiplied by the damage it causes.[1] Much like many other forms of disaster, a major nuclear accident is characterized by an infinitesimal probability and huge damage. A frequently used short-cut likens the former to zero, the latter to infinity. As we all know, multiplying zero by infinity results in an indeterminate quantity. The short-cut is easy but idiotic. The probability of an accident is not zero; unfortunately, some have already occurred. Nor yet is the damage infinite. Even the worst-case accident on a nuclear reactor cannot lead to the destruction of the planet and humankind. (Would the latter outcome, following a collision with an asteroid 10 kilometres in diameter, for example, or quarks going out of control in a particle accelerator, count as infinite damage?[2]) Mathematically, multiplying a very small number by a very large one always produces a determinate value. So nuclear risk assessments seek to approximate the two numbers and then multiply them.

In its worst-case scenario ExternE, the major European study of the external effects of various energy sources published in 1995, estimated the cost of an accident at €83 billion. This estimate was based on the hypothetical case of a core melt on a 1,250 MW reactor, followed two hours later by emissions lasting only an hour containing 10 per cent of the most volatile elements (caesium, iodine) of the core. The population was exposed to a collective dose of 291,000 person-sieverts.[3] This contamination ultimately caused about 50,000 cancers, one-third of which were fatal, and 3,000 severe hereditary effects. In a few days it caused 138 diseases and 9 fatalities. The impact on public health accounted for about two-thirds of the accident's cost. The study also assessed the cost of restrictions on farming (lost production, agricultural capital, etc.), and the cost of evacuating and re-housing local residents. This string of figures gives only a tiny idea of all the data required to estimate the cost of a nuclear accident. It merely lists some of the main parameters, in other words those that may double, or indeed multiply by ten, the total cost of economic damage. Let us now take a closer look.

In theory the extent of emissions may reach the release of the entire contents of the reactor core. The explosion at Chernobyl released the equivalent of 30 per cent of the radioactive material in the reactor, a huge, unprecedented proportion. Emissions from

Fukushima Daiichi's three damaged reactors are estimated to have amounted to ten times less than the amount released in Ukraine. The collective dose is the radiation dose measured in person-sieverts and the sum of the radiation absorbed by groups of people subject to varying levels of exposure. The collective dose depends on emissions, but also the weather conditions and population density. Depending on whether radioactivity is deposited by rain on an area of woodland or a city, the number of people exposed will obviously vary. The person-Sv unit is used because it is generally assumed that the biological effects of radiation follow a linear trend: the health effect of exposure of 20,000 people to 1 millisievert or 20 people to 1 sievert is consequently taken as being the same. This approach is based on the assumption that even the lowest level of exposure is sufficient to increase the number of fatalities and diseases, in particular cancer. It is controversial because it implies that the natural radioactivity that exists in some areas, such as Brittany in France, exposes local residents to a specific hazard. For our present purposes we shall treat it as an upper-case hypothesis, in comparison with the one setting a threshold below which ionizing radiation has no effect.

Translating the collective dose into figures for fatalities and diseases then depends on which effect one decides to use. For example, positing a 5 per cent risk factor of a fatal cancer to 100 people who have accumulated an equivalent-dose of 1 Sv means that five will be affected by the disease under consideration. The final step in assessing health effects involves choosing a monetary value for human life. Without that it is impossible to add the damage to public health to the other consequences, such as population displacement or soil decontamination. There are several methods for calculating the value of human life, based for example on the amount allocated to reducing road accidents or the average contribution of a single individual to their country's economy, in terms of gross domestic product.[4]

The assumptions used in studies of the quantity and dispersion of emissions, the collective dose received, the risk factor, the value of human life, the number of people displaced or indeed the amount of farmland left sterile all contribute to creating substantial disparities in estimates.[5] Just looking at two indicators – the number of additional cancers and the total cost of an accident – is sufficient to grasp the scale of variations. In the ExternE study cited above, the scenario corresponding to the largest volume of emissions led to 49,739 cancers and cost €83.2 billion$_{1995}$, whereas the scenario with the lowest emissions led to 2,380 cancers and cost €3.3 billion$_{1995}$. In a recent German study,[6] the low-case values calculated were 255,528 cancers and €199 billion$_{2011}$ – which corresponds to frequently quoted orders of magnitude. But the high-case figures reported by the study are far larger, with 5.3 million cancers and €5,566 billion$_{2011}$. It is unusual for experts to produce such a high estimate, with a single accident leading to millions of cancers and total damage amounting to thousands of billions of euros. However, it is close to the orders of magnitude reported in the first studies carried out in the 1980s after the Chernobyl disaster.[7] Allowance must nevertheless be made for such extreme figures, which correspond to worst-case scenarios. For example, in the German study just mentioned, the weather conditions were strong wind, changing in direction, and light rain (1 mm per hour). The rain and wind severely contaminated an area of 22,900 square kilometres (a circle about 85 km in diameter), occupied by millions of people who had to be evacuated. The most catastrophic scenarios obviously correspond to accidents at power plants in densely populated areas. Some 8.3 million people live inside a 30-kilometre radius round the nuclear power plant at Karachi, Pakistan. Worldwide there are twenty-one nuclear plants with more than a million people living within a 30-kilometre radius around them.[8]

Rather than picking a random number from the various damage assessments, the right approach would be to take these uncertainties into account, particularly the ones which affect the collective dose and risk factor. Let us suppose an accident has occurred with a given quantity of emissions. We need to plot a curve indicating, for example, that there is a 1% probability of the event causing economic losses in excess of €1,000 billion, a 10% probability of losses ranging from €500 billion and €999 billion, or indeed a 5% probability of losses below €1 billion. In conceptual terms an exercise of this sort is easy to carry out. But in practice the problem is obtaining sufficient data on variations in the determinants of damage. This is the case for weather parameters: the wind and rainfall conditions have been statistically established for each plant. But for many other factors, the scale of their variation is unknown, in which case it must be modelled on the basis of purely theoretical considerations.

There have been too few accidents in the past with significant emissions to allow observation of the statistical variations affecting their impacts, such as the frequency of cancers. We do not even know exactly how many fatal cancers followed the Chernobyl disaster. This is not so much due to a lack of epidemiologic studies or monitoring of the local population – several studies have been carried out since the accident – nor yet to their manipulation inspired by some conspiracy theory. On the contrary, the uncertainty is due to the fact that cancer is a very common cause of death and cancers caused by ionizing radiation are difficult to isolate. According to the Chernobyl Forum – a group of international organizations, including the World Health Organization – of the 600,000 people who received the highest radiation doses (emergency recovery workers, or liquidators, and residents of the most severely contaminated areas), 4,000 fatal cancers caused by radiation are likely to be added to the 100,000 or so cancers normally expected for a population group of this size. Among the 5 million people exposed to less severe contamination and living in the three most affected countries (Ukraine, Belarus and the Russian Federation), the Forum forecasts that the number of additional fatal cancers will amount to several thousand, ten thousand at the very most. This estimate is one of the lowest in the literature. For the whole population of the contaminated areas in the three countries, estimates vary between 4,000 and 22,000 additional deaths.[9] It should be noted that these figures do not take into account emissions outside the officially contaminated areas, nor in other parts of these countries, nor yet in Europe or the rest of the world. The second set of estimates is less reliable, more controversial, as radiation exposure per person is very low. Only the hypothesis of a linear relationship between dose and effect leads to additional fatal cancers outside the three areas, estimated for example at 24,700 by Lisbeth Gronlund, a senior scientist at the US Union of Concerned Scientists.[10]

If we accept the rough estimate suggested in Part I for the purposes of illustration, with a €1,000 billion loss for a major nuclear accident, we would not only be somewhere in the upper range of estimates but substantially exceeding the estimated cost of actual accidents in the past: Three Mile Island cost an estimated $1 billion[11] and Chernobyl several hundreds of billions of dollars.[12]

## Calculating the frequency of major accidents

We shall now turn to the task of putting figures on the probability of a major nuclear accident. The ExternE study reports a probability of a core melt at $5 \times 10^{-5}$ per reactor-year,[13] in other words a 0.00005 chance of an accident on a reactor operating for one year; or alternatively, due to the selected unit, a frequency of five accidents for 100,000 years of reactor operation, or indeed a frequency of five accidents a year if the planet boasted a fleet of 100,000 reactors in operation. Following a core melt, two possibilities are considered: either an eight-in-ten chance that radiation will remain confined inside the reactor containment; or a two-in-ten chance that part of the radiation will be released into the environment. In the first case damage is estimated at €431 million, in the second case at

€83.252 billion. As we do not know which of the two scenarios will actually happen, the forecast damage is calculated using its expected value, in other words: $0.8 \times 431 + 0.2 \times 83{,}252$, which equals roughly €17 billion. This simple example illustrates two connected concepts, which are essential to understanding probabilistic analysis of accidents: conditional probability and event trees (see box).

The main purpose of probabilistic safety assessments is not to estimate the probability of an accident for a specific plant or reactor, but rather to detect exactly what may go wrong, to identify the weakest links in the process and to understand the failures which most contribute to the risk of an accident. In short, such studies are a powerful instrument for preventing risks and ranking priorities, focusing attention on the points where efforts are required to improve safety. But our obsession with single numbers has pushed this goal into the background and all we remember of these studies is the final probability they calculate, namely *core-melt frequency*.

This bias is all the more unfortunate because the overall result is rarely weighted by any measure of uncertainty. If no confidence interval is indicated, we do not know whether the figure reported – for example 1 accident per 100,000 reactor-years – is very close to the mean value, for example an eight-in-ten likelihood that the accident frequency ranges from 0.9 to 1.1 accidents per 100,000 reactor-years, or more widely spread, with an eight-in-ten likelihood that the frequency ranges between 0.1 and 10 accidents per 100,000 reactor-years. Intuitively it is not the same risk, but in both cases the mean frequency is the same. In the second case there is some likelihood that more than ten accidents will occur per 100,000 reactor-years, whereas in the first case this risk is almost non-existent. Of the studies which have estimated the uncertainty, it is worth noting an NRC appraisal carried out in 1990 on five plants. It found, for example, that for the two pressurized water reactors at Surry (Virginia) the confidence interval for the mean frequency of $4.5 \times 10^{-5}$ ranged from $1.3 \times 10^{-4}$ at its upper limit to $6.8 \times 10^{-6}$ at the lower end. The

## Conditional probability, event trees and probabilistic safety assessment

The probability that in the event of core melt the radiation will remain confined inside the reactor containment is 0.8 (eight-in-ten chance). This is a conditional probability. It is commonly denoted using a vertical bar: p(release | melt). In a general way, A and B being two events, it is written as p(A | B), which reads as 'the probability of A given B'. Conditional probability is a key concept. In Chapter 6 we shall see that it gave rise to a fundamental mathematical formula, known as Bayes' rule, or theorem. It enables us to update our appraisals on the basis of new information. In the present case conditional probability is used as a tool for estimating the probability of various sequences of events, and for estimating the cost of the event among their number which leads to a major accident. For example, p(release | melt | cooling system failure | loss of emergency power source | protective wall round plant breached by wave | 7.0 magnitude quake). All the possible sequences form an 'event' tree, with a series of forks, each branch being assigned a probability, for example p for one branch and therefore (1–p) for the other.[14] Try to picture an apple tree growing on a trellis, and the route taken by sap to convey water to each of its extremities, one of which is diseased. The event tree maps the route taken by a water molecule which reaches the diseased part rather than taking any of the other possible routes.

Probabilistic assessment of nuclear accidents is based on this type of tree structure. It seeks to identify all the possible technical paths leading to an accident, then assigns a probability to the faulty branch at each fork. The starting point is given by the probability of a factor triggering an accident, for example a 1-in-1,000 chance per year of a quake resulting in peak ground acceleration of 0.5g at the site of the plant. The outcome is the occurrence of core melt, or the massive release of radiation into the environment following meltdown. There may be many intermediate forks, concerning both technical (1-in-10,000 chance that a pump will break down for each year of operation) and human failures (1-in-5,000 chance that an operator disregards a flashing light on the control panel).

The first large-scale probabilistic assessment was carried out in the US in the 1970s.[15] It was led by Norman Rasmussen, then head of the nuclear engineering department at the Massachusetts Institute of Technology. The study was commissioned by the Atomic Energy Commission, which was keen to reassure public opinion by demonstrating that the risks, albeit real, were actually infinitesimal. To this end it circulated a misleading summary of the study, which over-simplified its findings. To impress readers the document made dubious comparisons – not contained in the Rasmussen report – with other risks. For example, it asserted that the likelihood of a person dying as a result of a nuclear accident was about the same as being hit by a meteorite. Such distortion of the report and some of the errors it contained prompted a major controversy.[16] The Nuclear Regulatory Commission, which was set up to separate nuclear safety from the other missions allocated to the AEC, rejected the contents of the report summary in 1979. But ultimately the Rasmussen study is remembered for the work done in establishing a detailed method, rather than the values it calculated for probabilities and damages. Since then probabilistic assessment has become more rigorous and an increasing number of such safety studies have been carried out.

The first probabilistic safety assessments changed several deeply rooted beliefs. They highlighted the possible input of operators, capable of either interrupting a sequence of material faults, or in some cases making it worse. Accidents and preventive measures do not only have a technical dimension. Rasmussen and his fellow scientists showed that the loss of liquid through small breaks in the primary cooling system could also be a frequent cause of accidents, largely disregarded until then. Several years later the Three Mile Island disaster prompted new interest in probabilistic safety assessment. Since then its scope has broadened and it has grown more complex. It has taken into account additional factors which may initiate an accident, both natural and human (for example the risk of falling aircraft). Probabilistic safety assessments have now been carried out on all the nuclear plants in the US and many others worldwide. Similarly reactor vendors carry out such studies for each reactor model while it is still in the design stage.

first figure indicates that there is a 5% chance that the value of the frequency may be even greater; the second that there is a 5% chance it may be lower. In other words, there is a 90% chance that the core-melt frequency is somewhere between the two limits. However, this interval was not calculated. It was based on the judgement of various experts who were questioned.

The widespread lack of any mention of the distribution on either side of the mean may be explained by the method employed. The probabilities assigned to the various branches of the tree used to calculate the overall core-melt frequency are selected as best estimates. The final number is single, because it is the sum of a succession of single numbers. Safety specialists naturally know how to use statistics and calculate uncertainties. They do not make do with averages. But they are concerned with the details, because it is here that there is scope for improving safety, for example using a probability density function to model the failure of a particular type of pump. At this scale, the error and distribution parameters – with barbaric names such as standard deviation, mode, variance or kurtosis – are generally entered. So why are they not systematically used to obtain an overall core-melt probability expressed as more than just a single number? The first reason is that uncertainty propagation in an event tree is far from trivial, compared to addition or multiplication. It is not just a matter of adding up or combining the standard deviations at each fork to obtain the one governing core-melt frequency. The second reason is that the prime aim of probabilistic safety assessments is not to obtain a final result. Rather they focus on the details of each branch. Only in recent years have the specialists started paying sustained, systematic attention to presenting the uncertainty affecting the aggregate probability of an accident.

Non-specialists are consequently inclined to think that probabilistic safety assessments reveal the true value for accident frequency for a given reactor, whereas in fact this value is subject to uncertainty. At best it is possible to offer a confidence interval within which the probability of an accident will fall.

It is worth noting that with advances in reactor technology the probability of a core melt has dropped. For example, on its 1,300 MW *palier*, or step, EDF estimated the core-melt frequency as $7.2 \times 10^{-6}$. On the following generation, represented by the EPR, the results of safety studies carried out by Areva and vetted by the British regulator show a core-melt frequency of $2.7 \times 10^{-7}$ per reactor-year,[17] lower by a factor of more than 25. In the US an Electric Power Research Institute study found that the mean core-melt frequency of the US fleet had dropped by a factor of five since the early 1990s.[18]

## Divergence between real-world observation of accidents and their frequency as predicted by the models

The Fukushima Daiichi disaster revealed an order-of-magnitude difference between the accident frequencies forecast by probabilistic safety assessments and observed frequencies.[19] Since the early 1960s and grid-connection of the first nuclear reactor, 14,400 reactor-years have passed worldwide. This figure is obtained by adding up all the years of operation of all the reactors ever built that generated kWhs, whether or not they are still in operation, were shut down earlier than planned or not. In other words the depth of observation currently at our disposal is equivalent to a single reactor operating for 14,400 years. Given that the global fleet currently numbers about 500 reactors, it may make more sense to say that this depth is equivalent to 28.8 years for 500 reactors. At the same time, since grid-connection of the first civil reactor, eleven failures initiating a core-melt have occurred, of which three were at Fukushima Daiichi. So the recorded core-melt frequency is 11 over 14,400, or $7.6 \times 10^{-4}$, or an accident for every 1,300 reactor-years. Yet the order of magnitude reported by probabilistic safety studies ranges from $10^{-4}$ to $10^{-5}$, or an accident every 10,000 to 100,000 reactor years. Compared to 1,300 that means a ten- to hundred-fold divergence between calculated and observed probabilities.

How can such a large divergence be explained? The reasons are good or bad, trivial or complicated.

The first possible reason is simply bad luck. Just because you score 6 five times running with the same die, it does not mean it is loaded. There is a 1-in-7,776 chance of this sequence with a perfectly balanced die. So experts firmly convinced of the accuracy of their models or passionate advocates of nuclear power may set aside the suggestion that the calculated frequencies are erroneous, despite being much lower. Much as with the die, 14,400 reactor-years are not sufficient to obtain an accurate picture. This reason is legitimate in principle but it does mean ignoring observations if there are only a limited number. All in all it is not very different from the opposite standpoint, which consists in discarding probabilistic safety assessments and only accepting observations. The right approach is to base our reasoning on data obtained from both observation and modelling. Faced with uncertainty, all data should be considered, whether obtained from the field or from laboratories. We shall examine this approach in greater depth in Chapter 6.

A variation on the bad luck theory is to point out that the observed frequency actually falls within the range predicted by probabilistic assessments. As we saw above, in their forecast for the Surry plant, the experts estimated a 5 per cent likelihood that the core-melt frequency would exceed $1.3 \times 10^{-4}$, in other words an accident every 769 reactor-years. The observed value of an accident every 1,300 reactor-years is actually lower than this limit value. So convinced experts have no reason to review their position: there is no divergence between observations and the model. This stance might carry some weight if core-melt frequencies were reported with a confidence interval, but this is not the case. Moreover, the previous comment still holds true: the upper and lower limits on uncertainty must move to accommodate fresh observations.

The second possible reason is that the probabilistic assessments are not exhaustive. The event trees they examine and assess do not cover all the possible scenarios. The first safety studies only focused on internal initiating events, such as a device failure. The sequences of faults initiated by an earthquake or flood which might lead to core-melt have only gradually been taken into account. The validity of calculated frequencies is restricted to the perimeter under study. If no allowance is made for the risk of a falling aircraft the frequency is lower. The studies which do take it into account estimate that it is lower than $10^{-7}$ per reactor-year.[20] On its own this figure is too small to significantly change core-melt frequency, which is much higher. All this example shows is that by adding scenarios the frequency gradually increases. Little streams make big rivers. The Fukushima Daiichi accident is a concrete illustration of missing scenarios. It made people realize that spent-fuel pools could cause a massive release of radioactive material into the atmosphere. Probabilistic safety assessments do not usually register a break in the water supply to these pools as a possible initiating event. At Fukushima Daiichi, much as at other Japanese nuclear plants, the possibility that two risk factors – an earthquake and a tsunami – might coincide was apparently not studied either. Readers may be surprised by this oversight. Tidal waves and quakes are frequent in Japan and the two events are connected: one triggers the other. In fact the scenario which was not considered (nor its probability assessed) was the failure of the regional electricity network, knocked out by the quake, combined with flooding of the plant, due to the tsunami. With the surrounding area devastated, the backup diesel pumps underwater and the grid down, the power plant was left without an electricity supply for eleven days. Safety assessments generally assume power will be restored within twenty-four hours.

But is it possible for probabilistic studies to take into account all possible scenarios? Obviously not. It is impossible to imagine the unimaginable or to conceive the inconceivable. Joking apart, there is an intrinsic limit to probabilistic analysis: it can be applied to risk and uncertainty, not to situations of incompleteness (see box).

The third possible reason is that every event is unique, so probability theory cannot apply. The divergence between the observed frequency of accidents and their calculated probability is not a matter of bad luck, but results from the impossibility of applying probability theory to exceptional or one-off events. This reason is intuitive but must nevertheless be discarded.

In our minds the concept of probability is associated with that of frequency, and the law of large numbers too. We all learned at school that probability is the ratio of favourable or unfavourable outcomes to the number of possible cases. We all remember having to apply this definition to the observation of rolling dice or drawing cards. We also recall that calculating a probability requires the operation to be repeated a large number of times. We need to toss a coin several dozen times to grasp that it will land on one or the other side roughly the same number of times. This frequency-based approach to calculating probabilities is the best known and it does not work without data. But there are other ways of – or theories for – analysing

## Risk, uncertainty and incompleteness

These basic concepts may be explained using the example of an urn containing different coloured balls. We shall start from what is a *certainty*. It may be described using the case of an urn only containing balls of the same colour, red for instance. We know that if we pick one ball out of the urn, it will inevitably be red. The outcome is a foregone conclusion. *Risk* corresponds to an urn of which the content is known: thirty red balls and sixty white, for instance. We can no longer be sure of picking out a red ball, but we do know that we have a one-in-three chance of picking a red, a two-in-three chance of picking a white. In theoretical jargon, we would say that all the states of the world (or indeed the universe of events) are known with certainty, and for each state or event there is a corresponding probability, also known with certainty. *Uncertainty* may be represented by an urn known to contain thirty red balls, where we do not know whether the sixty others are black or white. So only the probability of picking a red ball (one-in-three) is known. On the other hand all the states of the world (picking a red, black or white ball) are known. There is no possibility of a surprise, such as picking a blue ball. Lastly, *incompleteness* corresponds to an urn full of balls of unspecified colours. We may pick a white or a purple ball, perhaps even a multicoloured one. Unlike risk and uncertainty, in a situation of incompleteness all the states of the world are not known. So probability theory cannot apply. A probability cannot be assigned to an unknown event.[21]

This presentation makes a distinction between uncertainty and risk. However, this vocabulary is not universally accepted and must be handled with care. The term 'uncertainty' is often used with a broader sense, which encompasses the notion of risk. The part of uncertainty which is not covered by the term 'risk' is then referred to as 'ambiguity' or 'non-specific uncertainty'.[22] But such quibbles are of secondary importance, the priority being to draw a line between situations in which we have probabilities for all the events under consideration, and situations in which we do not. In the second case it is necessary to make assumptions in order to assign probabilities to events for which they are unknown.

Returning to the urn containing thirty red and sixty black or white balls. A simple way of assigning a probability to picking a black ball and a white ball would be to posit that the two events are of equal likelihood, namely a one-in-two chance of picking a black or white ball from among the sixty which are not red.[23] The probability of picking a black, white or red ball, from among the ninety balls in the urn, is one-in-three (30/90). In other words, ignorance is treated by assuming equiprobability: if $n$ outcomes are possible, and if we have no idea of their chances of occurring, we consider them to be equiprobable (equally probable) and equal to $1/n$. This approach provides a way of treating non-specific uncertainty as a risk, thus making it possible to apply the calculation of probabilities to situations of uncertainty in general.

probabilities, which do away with the need for repeated experiments, and consequently a large number of observations, to calculate frequencies. It is quite possible to carry out probabilistic analysis without any observation at all. The reader may recall the wager made by Pascal. The French thinker was puzzled about the right approach to adopt regarding the uncertainty of God's existence. Here there could be no repeated events. The existence of a Supreme Being was a singular proposition, which Pascal subjected to probabilistic reasoning.

So probabilistic logic can be applied to one-off events. The concept of probability refers to a degree of certainty regarding the veracity of a proposition. It applies for instance to the reasoning of a court judge who takes a different view of the guilt of the accused if it is known that the latter lacks an alibi or that traces of his or her DNA have been found on the body of the victim. According to John Maynard Keynes the theory of probability 'is concerned with [the] part of our knowledge we obtain [...] by argument, and it treats of the different degrees in which the results so obtained are conclusive or inconclusive'. This reference to the author of *The General Theory of Employment, Interest and Money* may surprise readers unfamiliar with the work of the Cambridge economist. A *Treatise on Probability* was one of the first works published by Keynes, but is still well worth reading even now.[24]

The theory of subjective probability opens up a second approach to probability, not based on frequency. It has been advocated and developed by three leading figures in economic science: Britain's Frank Plumpton Ramsey, already cited with reference to discount rates; Bruno de Finetti, from Italy; and an American, Leonard Jimmie Savage. These authors understand the concept of probability as the belief which an individual invests in an event, regardless of whether or not the event recurs. According to De Finetti, probability is the degree of confidence of a given subject in the realization of an event, at a given time and with a given set of data.[25] So probability is not an objective measurement, because it depends on the observer and his or her knowledge at that time. Probability is thus assimilated to the odds a gambler will accept when betting on a given outcome, odds at which he or she will neither lose nor win money. Imagine, for example, that two nuclear experts are asked to bet on the likelihood of a nuclear accident occurring in Europe in the course of the next thirty years. One agrees to odds of 100-to-1, the other 120-to-1, or 200-to-1. So according to subjective probability theory anyone can bet on anything. However, we should not be misled by the terms 'belief' and 'subjective'. The odds accepted do not depend on a person's mood or state of mind, but are supposed to be based on their knowledge. Furthermore, the person making the wager is deemed to be rational, being subject to the rules governing the calculation of probabilities. For example, he or she cannot bet three-to-one for and one-to-two against a given outcome. The theory of decision under uncertainty, a monument elaborated by Savage in the middle of the last century, assumes that economic agents comply with all the axioms for calculating probabilities: they must behave as perfect statisticians.

The fourth reason for the divergence between observations and forecasts is that the models may be faulty or use the wrong parameters. It is vital to avoid mistakes when assigning a probability to the known states of the world, and consequently when measuring the probability and associated uncertainty. Returning to what happened at Fukushima Daiichi, the six reactors at the plant were commissioned in the 1970s. We do not have access to the probabilistic studies carried out by the plant operator, Tokyo Electric Power (Tepco), at the time of the plant's construction or afterwards. But we do know some of the values taken into account by the generating company or the regulator for the risk of an earthquake or tsunami. The figures were largely underestimated. The plant was designed to withstand a magnitude 7.9 earthquake and tidal wave of several metres. On 11 March 2011 at 14:46 JST it was subjected to a magnitude 9 tremor, then swamped by a wave more than 10 metres high. So, much like Tepco, the nuclear industry is purportedly inclined to play down risks by picking values or models favourable to growth. Unless one is an adept of conspiracy theories, this explanation for the divergence between observations and forecasts is barely convincing. For many years safety authorities and independent experts have scrutinized such probabilistic studies.

Two points are of particular note in this long list of possible reasons for the divergence between the observed frequency and calculated probability of an accident. Firstly, with regard to method, we should make use of all the available instruments, combining empirical and theoretical data. Assessments of the risk of a major nuclear accident based exclusively on either data from past observations or theoretical

## The wrong values taken into account for Fukushima Daiichi

When the nuclear power plant at Fukushima Daiichi was built, the risk of an earthquake exceeding magnitude 8 on the Richter scale was estimated at less than $2 \times 10^{-5}$ per year.[26] This value was taken from the work of modelling and numerical simulation carried out for each plant in Japan by the National Research Institute for Earth Science and Disaster Prevention (NIED). However, historical research has identified six major quakes which have occurred on the Sanriku coast since 869. That was the year of the Jogan undersea earthquake, probably the most devastating ever known on this stretch of coastline until March 2011. The various pieces of evidence which have been gathered suggest that these quakes all exceeded magnitude 8. This means the observed annual frequency should be about $5 \times 10^{-3}$, in other words more than 200 times higher than the results calculated by the NIED.[27]

The protective seawall at Fukushima Daiichi was built in 1966. It was six metres high, a dimension decided in line with a classical deterministic principle: take the most severe historic event and add a twofold safety margin. The three-metre wave which struck the coast of Chile in 1960 was taken as a baseline value. This was a surprising choice, the Jogan quake having triggered a four-metre wave locally, a fact that was already known when the plant was built.[28] Be that as it may, forty years later, our understanding of past tsunamis had obviously made significant progress, but the initial height of the wall was thought to comply with the 2002 guidelines for tsunami assessment set by the regulatory authority.[29] Such compliance was based on an annual probability of less than $10^{-4}$. Yet historical studies have established that waves exceeding eight metres have been recorded on the Sanriku coast: witness the stones set into the ground marking the points furthest inland reached by the flood. Some of these stones are more than 400 years old. The inscriptions on them urge residents not to build homes lower down the slope. Moreover, core samples have revealed sedimentary deposits left by previous tsunamis. At the Onagawa plant, on the basis of remains found in the hills one kilometre inland, it was estimated that the 1611 quake had caused a six-to-eight-metre wave.[30]

On the evidence of old records Woody Epstein estimated that the average frequency of a tsunami of eight metres or more on the Sendai plain, behind the Fukushima Daiichi plant, was about 1 every 1,000 years. He estimated as $8.1 \times 10^{-4}$ the probability of an earthquake of a magnitude equivalent to or greater than a seismic intensity of 6 or more,[31] followed by a tidal wave of over eight metres. Due to the layout of the plant, this dual shock would very likely lead to flooding of the turbine building, destruction of the diesel generators, loss of battery power and station blackout lasting at least eight hours. This entire sequence of events would correspond, according to Epstein, to a probability of core-melt of about $5 \times 10^{-4}$, five times higher than the authorized limit. Tepco based its various probabilistic studies on very low values, very probably taken from a badly designed historical database or unsuitable simulation models.[32]

probabilistic simulations lead to a dead end. Secondly, in practical terms, the limitations of probabilistic assessment lead to the deployment of deterministic safety measures. As we do not know all the states of the world, nor yet the probability of all those we do know, it is vitally important to install successive, redundant lines of defence as a means of protection against unforeseeable or ill-appraised events. In short, we must protect ourselves against the unknown. We could raise protective walls or quake-resistant structures able to withstand events twice as severe as the worst recorded case, install large numbers of well-protected back-up diesel generators and build nesting containments – the first one in the fuel cladding itself, forming an initial barrier between radioactive elements and the environment; the largest one, made of reinforced concrete, encasing the reactor and steam-supply system.

# FIVE

# Perceived probabilities and aversion to disaster

The concept of subjective probability can be a misleading introduction to individuals' perception of probabilities. Perceived probability, much like its subjective counterpart, varies from one person
to the next but it does not take the same route in our brain. The
former expresses itself rapidly, effortlessly, in some sense as a reflex
response; the latter demands time, effort and supervision. Perceived
probability is based on experience and routine, whereas subjective
probability is rooted in reason and optimization. In response to the
question, 'Which is the most dangerous reactor: the one for which
the probability of an accident over the next year is 0.0001; or the
one which has a 1 in 10,000 chance of having an accident over
the next year?', most people will spontaneously pick the second
answer. Yet reason tells us that the two reactors are equally dangerous (i.e. 1:10,000 = 0.0001). So how do we perceive risks? Are
individuals poor statisticians, or perhaps not statistically minded at
all? In which case, what use is the theory of decision under uncertainty, given that it requires us to calculate probabilities accurately?
How do individuals make a choice if they do not optimize their
decisions? These questions are central to forty years' work by experimental cognitive psychologists on how individuals assess the probability of events. Understanding this work is essential, for almost
all its findings contribute to amplifying the perception of nuclear
risk.

## Biases in our perception of probabilities

What is the connection between cognitive psychology – which is an experimental science – and economics? In fact, there is a significant link. In 2002, in Stockholm, Daniel Kahneman was awarded the most coveted distinction to which an economist can aspire: the Sveriges Riksbank prize in Economic Sciences in memory of Alfred Nobel. His work was distinguished for 'having established laboratory experiments as a tool in empirical economic analysis, especially in the study of alternative market mechanisms'. That year, the winner of the Nobel prize for economics was not an economist, but a psychologist!

Economic analysis focuses on many subjects and lends itself to many definitions: decision theory is among these subjects; it may, among others, be defined as the science of human behaviour. Economics investigates how humans seek the best means of achieving their aims. When an individual needs to take a decision under uncertainty,[1] he or she assigns more or less weight to the probabilities associated with each choice. A rational being faced with an alternative between an action yielding a satisfaction of 100 associated with a probability of 0.3, and an action yielding a satisfaction of 105 associated with a probability of 0.29, will choose the second option because its mathematical expectation is greater ($100 \times 0.3 < 105 \times 0.29$). The expected utility of the outcome is thus optimized.

The Swiss mathematician Daniel Bernoulli laid the foundations of expected utility theory. In his 1738 essay he raised the question of how to give formal expression to the intuition according to which the rich were prepared to sell insurance to the poor, and the latter were prepared to buy it. He also sought to resolve an enigma of great interest at the time, subsequently known as the Saint Petersburg paradox: why does a gamble which holds an infinite expectation of gain not attract players prepared to bet all they own?[2] The answer to both these questions is to be found in our aversion to risk. This psychological trait means that we prefer a certain gain of 100 to an expected gain of 110. This also explains why a rich person attaches less value to a sum of €100 than a poor person does, which translates into contemporary economic parlance as the declining marginal utility of income. In mathematical terms it is represented by the concave form of the utility function. (The curve, which expresses our satisfaction depending on the money we own, gradually flattens out. Bernoulli thus used a logarithmic function to represent utility.)

The connection between risk aversion and the form of the utility function may not be immediately apparent to the reader. In which case, illustrating the point with figures should help. Let us assume that €100 yields a satisfaction of one; €200 yields a proportionately lower satisfaction, 1.5 for example; €220 less still, say 1.58. I offer you €100 which you either accept or we toss a coin with the following rule: heads, I keep my €100; tails, you take the €100 and I add €120 more. Which option would you choose? The certainty of pocketing €100, or a one in two chance of winning €220? In the first case your gain is €100, whereas in the second case there is the expectation of €110 (€220 × 1/2). But what matters to you is not the money, but the satisfaction – or utility – it yields. The first case yields a utility of one, the second an expected utility of 0.79 (1.58 × 1/2). So you choose the first option, which does not involve any risk. With the resolution of the Saint Petersburg paradox, by altering the form of the utility function, Bernoulli opened the way for progress towards decision theory, which carried on to Kahneman. This was achieved through a back-and-forth exchange between economic modelling and psychological experimentation. The latter would, for instance, pick up an anomaly – in a particular instance people's behaviour did not conform to what the theory predicted – and the former would repair it, altering the mathematical properties of the utility function or the weighting of probabilities. The paradoxes identified by Allais and Ellsberg were two key moments in this achievement (see the box).

### Allais, Keynes, Ellsberg and the *homo probabilis*

During an academic lecture in 1952 Maurice Allais, a Professor at the Ecole des Mines in Paris, handed out a questionnaire to those present, asking them to choose between various simple gambles, arranged in pairs. He then collected their answers and demonstrated that they had contradicted expected utility theory in what was then its most advanced form, as developed by Leonard Jimmie Savage. Their responses violated one of the axioms of the theory, which was supposed to dictate the rational decision under uncertainty. In simple terms the Allais paradox may be expressed as follows: the first gamble offers the choice between (A) the certainty of winning €100 million, and (B) winning €500m with a probability of 0.98, or otherwise nothing. The second gamble offers the choice between (C) receiving €100 million with a probability of 0.01, or otherwise nothing, and (D) €500 million with a probability of 0.0098, or otherwise nothing. The paired gambles are therefore the same, but with a probability 100 times smaller. Most of the students chose A, not B, but D rather than C. It seemed to them that the probability of winning €500 million with D (0.0098) was roughly the same as that of winning €100 million with C (0.01), whereas in the previous case the same difference of 2% between probabilities seemed greater. Yet, according to Savage, rational behaviour would have dictated that if they chose A rather than B, they should also pick C, not D. Ironically, Savage also attended the lecture and handed in answers to the questionnaire which contradicted his own theory.

A common solution to the Allais paradox is to weigh the probabilities depending on their value, with high coefficients for low probabilities, and vice versa. Putting it another way, the preferences assigned to the probabilities are not linear. This is more than just a technical response. It makes allowance for a psychological trait, which has since been confirmed by a large body of experimental study: people overestimate low probabilities and underestimate high probabilities. In other words they tend to see rare events as more frequent than they really are, and very common events as less frequent than is actually the case.

To our preference for certainty, rather than risk, and our varying perception of probabilities depending on their value, we must add another phenomenon well known to economists: our aversion to ambiguity. This characteristic was suggested by Keynes, and later demonstrated by Daniel Ellsberg in the form of a paradox. In his treatise on probabilities, the Cambridge economist posited that greater weight is given to a probability that is certain than to one that is imprecise. He illustrated his point by comparing the preferences for a draw from two urns containing 100 balls. One contains black and white balls, in a known, half-and-half proportion; the other urn also contains black and white balls but in an unknown proportion. In the second urn, all distributions are possible (0 black, 100 white; 1 black, 99 white; . . . ; 100 black, 0 white) with equal probability (p = 1/101). So the expected probability of drawing a white ball is one-in-two,[3] the same value as for the probability of drawing a white ball from the first urn. But we would rather win by drawing white (or black) balls from the first urn.

In 1961 Ellsberg revisited Keynes' example, developing it and making experiments. We shall now look at his experiment with an urn containing balls of three different colours. In all it contains ninety balls, of which thirty are red and the remaining sixty either black or yellow. So all we accurately know is the probability of drawing a red (one-in-three) and that of drawing a black or yellow – in other words one that is not red (two-in-three). You are presented with two pairs of gambles. In the first pair, (A) you win €100 if you draw a red ball from the urn, (B) you win €100 if you draw a black ball from the urn. Which option do you choose? If, like most people, you are averse to ambiguity, you choose A. The other pair is more complicated: (C) you win €100 if you draw a red or a yellow ball from the urn, (D) you win €100 if you draw a black or a yellow ball from the urn. The question, however, is the same: which option do you choose? The answer is also the same: aversion to ambiguity prompts most people to pick D rather than C. You know very well that you have a two-in-three chance of drawing a black or yellow ball (one that is not red). The paradox resides in the fact that the preference for A and D is inconsistent on the part of a rational person, as modelled in the classical theory of expected utility. Preferring A to B implies that the player reckons subjectively that the probability of drawing a black ball is lower than one-in-three, whereas for a red ball it is higher than one-in-three. Knowing that there is a two-in-three probability of drawing a black or yellow ball, the player deduces that the probability of drawing a yellow ball is higher than one-in-three. As the probability of drawing a yellow or black ball is higher than one-in-three in both cases, their sum must exceed two-in-three. According to Savage, the player picks C. But in the course of experiments, most players who choose A also choose D, which suggests that there is an anomaly somewhere, which the aversion to ambiguity corrects.

Just as there is a premium for taking risks, some compensation must be awarded to players for them to become indifferent to gain (or loss) with a one-in-two probability or an unknown probability with an expected value of one-in-two. Technically speaking, there are several solutions for this problem, in particular by using the utility function, yet again. It is worth understanding Ellsberg's paradox because ambiguity aversion with regard to a potential gain, has its counterpart with regard to a loss: with the choice between a hazard associated with a clearly defined probability – because the experts are in agreement – and a hazard of the same expected value – because the experts disagree – people are more inclined to agree to exposure to the first rather than the second hazard. Putting it another way, in the second instance people side with the expert predicting the worst-case scenario. Simple intuition enables us to better understand this result. If players prefer (A) the prospect of drawing a red ball associated with the certainty of a one-in-three probability, rather than that (B) of drawing a black ball, it is because they are afraid that in the latter case the person operating the experiment may be cheating. The latter may have put more yellow balls in the urn than black ones. The experimental proposition is suspect. Out of pessimism, the players adjust their behaviour to suit the least favourable case, namely the absence of black balls in the urn, comparable to the worst-case scenario among those proposed by the experts.

Kahneman's work followed on that of Bernoulli, Allais, and Ellsberg, but it also diverged in two respects.

He and his fellow author, Amos Tversky,[4] introduced two changes to the theory of expected utility.[5] Firstly, individuals no longer base their reasoning on their absolute wealth, but in relative terms with regard to a point of reference. For example, if your boss gives you a smaller rise than your fellow workers, you will perceive this change as a loss of utility, rather than a gain. The value function does not start from zero, corresponding to no wealth, subsequently rising at a diminishing rate as Bernoulli indicated with his model. A whole new part of the curve, located to the left of the zero, represents the value of losses in relation to the status quo. This part is convex (see Figure 5.1): just as we derive less satisfaction from our wealth rising from 1,000 to 1,100, than when it rises from 100 to 200, so our perception of the loss between –1,000 and –1,100 is less acute than between –100 and –200.

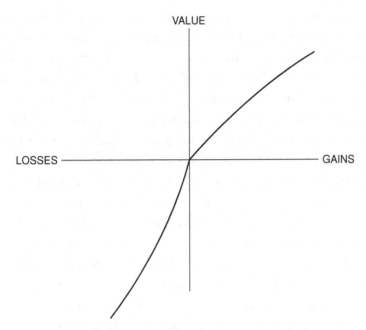

**Figure 5.1** The Kahneman–Tversky asymmetric value function

Secondly, individuals are more affected by loss than gain. For example, if a professor randomly hands out mugs marked with their university shield to half the students in a lecture theatre, giving the others nothing, the recipients, when asked to sell their mugs, will set a price twice as high as the one bid by those who did not receive a gift.[6] Reflecting such loss aversion, the slope of the value function is steeper on the loss side than on the gain side. Regarding the perception of probabilities, Kahneman and Tversky revisit the idea of distortion, in particular for extreme values (overweighting of low probabilities, underweighting of high probabilities). Armed with these value functions and probability weighting, the decision-maker posited by the two researchers assesses the best option out of all those available and carries on optimizing the outcome.

Kahneman has also experimented widely and published on heuristics, the short-cuts and routines underpinning our decisions. In so doing he distances himself from previous work on optimal decision-making under uncertainty. The decision-maker no longer optimizes nor maximizes the outcome. Anomalies in behaviour with regard to expected utility theory are no longer detailed to enhance this theory and enlarge its scope, but rather to detect the ways we react and think. Here the goal of research is exclusively to describe and explain: in the words of the philosophers of science it is positive. The aim is no longer to propose a normative theoretical framework indicating how humans or society should behave. Observing the distortion of probabilities becomes a way of understanding how our brain works. This line of research is comparable to subjecting participants to optical illusions to gain a better understanding of sight. For example, a 0.0001 probability of loss will be perceived as lower than a 1/10,000 probability. Our brain seems to be misled by the presentation of figures, much as our eyes are confused by an optical effect which distorts an object's size or perspective. This bias seems to suggest that the brain takes a short-cut and disregards the denominator, focusing only on the numerator.

Psychology has also done a great deal of experimental work revealing a multitude of micro-reasoning processes, which sometimes overlap, their denomination changing from one author to the next. Economists may find this confusing, much as psychologists are often thrown by the mathematical formalism of decision theory. Readers may feel they are faced with a choice between two unsatisfactory options. On the one hand, an economic model has become increasingly complex as it has been built up with the addition of *post hoc* hypotheses, the theory of expected utility having gradually taken onboard aversion to risk, ambiguity, loss, while nevertheless remaining a schematic, relatively unrealistic representation. On the other hand, a host of behavioural regularities, identified through observation, throw light on the countless facets of the decision-making process but, lacking a theoretical basis, simply stand side by side. In one case we have a discipline which is still basically normative, in the other a science obsessed by detailed description. Building bridges between economics and psychology does not eliminate their differences. However, in a recent book targeting the general public, *Thinking, Fast and Slow*,[7] Kahneman sets out to reconcile the two disciplines. In a synthesis which rises above doctrinal differences he draws a distinction between two modes of reasoning which direct our thoughts and decisions, one automatic, the other deliberate. The first mode is swift, and requires no effort on our part nor supervision, being mainly based on association and short-cuts. The second one is quite the opposite, slow, demanding effort and supervision, largely underpinned by deduction and rules we have learnt. Supposing we submit the following problem to a group of mathematically adroit students:[8] a baseball bat and ball cost €110, the bat costs €100 more than the ball, so how much does the ball cost? The spontaneous answer is €10. But in that case the bat would cost €110, and the whole kit €120! Leave the students to think for a while or to write out the equations on paper and they will find the right answer.

By revisiting and extending the classic psychological distinction between the two cognitive systems Kahneman stops seeing heuristics and calculation as mutually exclusive. It is no longer a matter of deciding whether humans are rational or irrational: they are both.

## Perception biases working against nuclear power

The overall biases in our perception of probabilities, discussed above, amplify the risk of a nuclear accident in our minds. To this we must add other forms of bias, with which economists are less familiar, but which contribute to the same trend.

A major nuclear accident is a rare event. Its probability is consequently overestimated. Much like smallpox or botulism, the general public thinks its frequency is higher than it really is. Nuclear technology is thus perceived as entailing a greater risk than other technologies. Paul Slovic, an American psychologist who started his career working on perception of the dangers of nuclear power, asked students and experts to rank thirty activities and technologies according to the risk they represented.[9] The students put nuclear power at the top of the list (and swimming last), whereas the experts ranked it in twentieth position (predictably putting road accidents first). Faced with a low probability individuals are inclined to over-reassure themselves and demand higher protection or compensation. Protecting the personnel of nuclear power plants against workplace accidents costs more than in any other field.[10] In a general way we over-invest in protection against events with a low probability, the incremental improvement being seen as more beneficial than it really is.

The risk of a nuclear accident is ambiguous. Expert appraisals diverge, depending on whether they are based on probabilistic assessments or the observation of accidents. Furthermore, probabilistic assessments produce different figures depending on the reactors or initiating events they consider; the same applies to observations of accidents, for which the definitions and consequently lists differ. Lastly there is a divergence of views between industry experts working for operators and engineering firms, and scientists opposed to the atom, such as the members of Global Change in France or the American Concerned Scientists Association. The effect suspected by Keynes and demonstrated by Ellsberg is clearly at work here. In the face of scientific uncertainty we are inclined to opt for the worst-case scenario. The highest accident probability prevails. The same phenomenon affects the controversial estimation of damages in the event of a major accident. The highest estimates tend to gain the most widespread credence.

The asymmetry between loss and gain highlighted by Kahneman and Tversky is behind a widespread effect which is less decisive for nuclear power and not as specific.[11] The main consequence of this commonplace, third bias is that it favours the status quo, the drawbacks of change being seen as greater than the benefits (there is a bend in the curve of the value function shown in Figure 5.1 near the point of reference separating the feeling of gain from that of loss). As a result local residents will tend to oppose the construction of a nuclear power plant in their vicinity, but on the other hand they will block plans to close a plant which has been operating for several years. Obviously what holds true for nuclear power is equally valid for other new facilities, be they gas-fired plants or wind farms.

The risks of a nuclear accident are also distorted by the dimension of potential damages and their impact on public opinion. With low frequency and a high impact they are among the various 'dread' risks, along with plane crashes and terrorist attacks targeting markets, hotels or buses, or indeed cyclones. The perception of the consequences of such events is such that their probability is distorted. It is as if the denominator had been forgotten. Rather than acknowledging the true scale of the accident, attention seems to focus exclusively on the accident itself. Disregard for the denominator, mentioned very briefly above, is connected to several common routines of varying similarity which have been identified by risk psychologists. Let us look at the main routines at work.

*The availability heuristic* is a short-cut which prompts us to answer questions on probabilities on the basis of examples which spontaneously spring to mind. The frequency of murders is generally perceived, quite wrongly, to be higher than that of suicides. Violent or disastrous events leave a clearly defined, lasting mark, in particular due to the media attention they attract. It is consequently easy to latch onto them. Chernobyl reminds us of the accident at Three Mile Island, and in turn Fukushima Daiichi brings Chernobyl to mind. It matters little that each case is different, with regard to the established causes, the course of events or the consequences in terms of exposure of the population to radiation. The spontaneous mode of thought works by analogy and does not discriminate.

*The representativeness heuristic* is based on similarity to stereotypes. It originated with a well-known experiment by Kahneman and

Tversky. Test participants were told that Linda was thirty-one, single, outspoken and very bright. She had majored in philosophy. As a student, she was deeply concerned with issues of discrimination and social justice, and also took part in anti-nuclear demonstrations. Then they were asked to pick a description of Linda, in descending order of likelihood: elementary schoolteacher; bookstore salesperson, attending yoga classes; active feminist; psychiatric social worker; member of the League of Women Voters; bank teller; insurance salesperson; bank teller and an active feminist. Most of the participants placed the last answer ahead of the antepenultimate one. Yet logic tells us that there is a higher probability of being a bank teller, than being a bank teller and a militant feminist, the latter category being a subset of the female bank-teller population. The representativeness heuristic prompts us to confuse frequency and plausibility. We are consequently able to find regularities and trends where they do not exist. Two successive accidents separated by a short lapse of time will be interpreted as a substantial deterioration in nuclear safety, whereas their close occurrence is merely a matter of chance. This heuristic resembles another bias which often misleads us, namely our tendency to generalize on the basis of low numbers. Rather than waiting to see how a low-probability event repeats itself, tending towards infinity or even with 100 recurrences, we immediately deduce the probability of its recurrence.

To round off this list of biases in our perception of frequencies unfavourable to nuclear power, a word on *very low probabilities*. It is difficult to grasp values such as 0.00001, or even smaller. Much as it is hard to appreciate huge figures, in the billion billions, often due to the lack of tangible points of reference to which to connect them, dividing one by several hundred thousand makes little sense to the average person. Cass Sunstein, a Harvard law professor, carried out an interesting experiment. He asked his students how much they were prepared to pay, at the most, to eliminate a cancer risk of 1-in-100,000 or 1-in-1,000,000. He also presented the problem in slightly different terms, adding that it 'produces gruesome and painful death as the cancer eats away at the internal organs of the body'. In the first case his students were prepared to spend more on saving one individual in 100,000 than on one in a million. On the other hand, playing on their emotions narrowed the difference. They forgot the denominator, despite having a basic grasp of fractions and probabilities! To an economist, the biased perception of frequencies is very odd. The way individuals react is light years away from the marginalist reasoning which underpins economic models. As an economist it is disconcerting to discover that cutting a risk from $5 \times 10^{-4}$ to $10^{-4}$ can pass unnoticed, even if it took time and effort to achieve this shift.

In practice, if the authorities want to reassure citizens worried by nuclear risk, they will have difficulty basing their arguments on progress in reactor safety. It would apparently be more effective to emphasize the benefits. Slovic and a fellow author have noted a negative correlation between perception of risks and benefits.[12] If the benefits of a given activity are thought to be high, the corresponding risks are perceived as being low and vice versa; on the other hand, if the risks are thought to be high, the associated benefits will be underestimated. In more prosaic terms, if we like something, we will downplay the risks involved; if we dislike something, we will minimize the benefits.

Unlike very low probabilities, everyone understands zero. By playing on people's emotions it seems possible to persuade them to spend large amounts to eradicate risk completely. In the field of nuclear power, the decision by Germany to shut down its reactors following the Fukushima Daiichi disaster is a case in point (see Chapter 11). Economists have estimated the loss resulting from the decision to retire the plants earlier than planned at between €42 billion and €75 billion.

In short, however psychologists address perception biases, the latter go against nuclear power. Their effect is cumulative, one consolidating another: in this way they amplify the perceived probability of an accident. At a practical level this has two major consequences. Firstly, there is a risk of over-investing in safety. For this to happen all that is required is for the authorities to follow the trend, either because policy-makers have an interest in giving way to the electorate's demands, or because their perception of probabilities is no different from that of private individuals. Secondly, the options for alternative investment are distorted. On account of its dread nature, the perceived risk of a nuclear accident has been much more exaggerated than the risk inherent in other generating technologies. In 2010 the Organisation for Economic Cooperation and Development published a study on the comparative assessment of serious accidents (causing more than five deaths).[13] The survey covered the whole world, from 1969 to 2000. In terms of fatalities and accidents coal came top of the list, a very long way ahead of nuclear power.[14] Does this correspond to your perception, given that the list of accidents during this period of time includes the Chernobyl disaster? Very probably not, much like most other people. If the authorities base their action on perceived probabilities, their decisions may ultimately detract from the overall goal of reducing risks and loss of human life.

The decisions taken following the terrorist attacks on 11 September 2001 provide a simple illustration of the dual effect of our aversion to dread risk. During the three months following the disaster Americans travelled less by plane, more by car. The well-known German psychologist Gerd Gigerenzer has demonstrated that this behavioural shift resulted in an increase in the number of road deaths, exceeding the 265 passengers who perished in the hijacked planes.[15] As the fearsome pictures of the collapse of the twin towers of the World Trade Center receded into the past, the sudden distortion of probabilities must have been partly dissipated. On the other hand we are all affected by the long-term impact of 9/11 whenever we travel by plane. Air passengers all over the world undergo a series of trials – queuing to be checked before entering the boarding area, taking off belts and shoes, emptying pockets, taking electronic devices out of hand luggage, trashing water bottles, putting other liquids into plastic bags, forgetting keys and setting off the alarm, and being handled by complete strangers. Not to mention the rise in airport taxes to cover the extra security costs. In view of the infinitely low risk actually involved, these measures seem to have been taken too far.[16]

In conclusion, there is good reason to suppose that the discrepancy between the perceived probability of a nuclear accident and the figures advanced by the experts will persist. The work by experimental psychologists on the amplification of risks for low-probability events seems convincing. The distorting effect of the Fukushima Daiichi disaster is likely to be a durable feature, particularly as we receive regular reminders, if only when progress is achieved or shortcomings are observed in the resolution of the many outstanding problems (soil decontamination, public health monitoring, dismantling of reactors, waste processing, among others). It will also require great political determination to avoid pitfalls similar to those entailed by tightening up checks on airline passengers.

# SIX

# The magic of Bayesian analysis

An English Presbyterian minister and a French mathematician forged a magic key which enables us to update our probabilistic judgements, use probability as a basis for reasoning without being a statistician, reconcile objective and subjective probability, combine observed frequency and calculated probability, and predict the probability of the next event.

## The Bayes–Laplace rule

Thomas Bayes was the first person to use the concept of conditional probability, for which Pierre-Simon Laplace found a more widespread application. (See the box.) We presented this concept briefly in Chapter 4 on the escalating severity of nuclear accidents, more specifically in relation to the probability of massive release of radioactive material in the event of a core melt, or in other words given that core meltdown has occurred. This probability is denoted using a vertical bar: $p(release|melt)$. In a general way, A and B being two events, the conditional probability is written as $p(A|B)$, which reads as 'the probability of A given B'. Or, to be more precise, we should refer to two conditional probabilities; as we are focusing on two events, we may also formulate the conditional probability $p(B|A)$, the probability of B given A. In the case of a core melt and the release of radioactivity, $p(melt|release)$ is almost equal to one: with just a few exceptions (water loss from a spent-fuel pool) a massive release is impossible unless the core has melted.

## Bayes' rule and conditional probabilities

The Bayes–Laplace rule, more commonly known as Bayes' rule, enables us to write an equation linking two conditional probabilities. It is written as:

$$p(A|B) = p(A)[p(B|A)/p(B)] \qquad (1)$$

It reads: 'the probability of A given B is equal to the probability of A multiplied by the probability of B given A divided by the probability of B'.

It may also be written as:

$$p(A|B) = [p(A)p(B|A)]/[p(A)p(B|A)$$
$$+ p(notA)p(B|notA)] \qquad (2)$$

This formula is typically illustrated with examples from medicine. We shall use the following data for the problem:

The probability that a patient undergoing an X-ray examination has a cancer is 0.01, so p(cancer) = 0.01

If a patient has a cancer, the probability of a positive examination is 0.8, so p(positive|cancer) = 0.8

If a patient does not have a cancer, the probability of a positive examination is 0.1, so p(positive|no cancer) = 0.1

What is the probability that a patient has cancer if the examination is positive, so p(cancer|positive)?

Using equation (2), we find 0.075.[1]

Bayes' rule, set forth in the box, is the key to inductive reasoning. Let us look at how it works.

As the reader is probably more familiar with the concept of deduction rather than induction we shall start by contrasting the two approaches. Deduction generally proceeds from the general to the specific: 'All dogs have four legs; Fido, Rover, Lump and Snowy are dogs; so they have four legs.' Induction generalizes on the basis of facts or empirical data: 'Fido, Rover, Lump and Snowy are dogs; they have four legs; so all dogs have four legs.'

A simple equation enables us to grasp the inductive power of Bayes' rule. Let us suppose that event A is a hypothesis, H, and event B is an observed data, d. So p(H|d) is the probability that H is true, given that d is observed. Bayes' rule is written thus:

$$p(H|d) = p(H)[p(d|H)/p(d)]$$

The conditional probability p(d|H) encapsulates the essence of deductive reasoning. It indicates the probability of datum d being observed when hypothesis H is true. The reasoning starts from a hypothesis, rooted in theory, to arrive at the probability of an observation. The conditional probability p(H|d) expressing inductive reasoning takes the same route but in the opposite direction. It indicates the veracity of a hypothesis, if datum d has been observed. Here, we start from the observation to infer the degree of certainty of a hypothesis.

The equation above also shows how Bayes' rule provides a means of reviewing an appraisal in the light of new information.

At the outset only the general, or prior, probability p(H) is known. It may be based on objective data from the past, on scientific theory or indeed on personal belief. New data, which changes the picture, then comes to notice. Thanks to Bayes' rule we can calculate a new, posterior probability p(H|d), namely that H is true now that data d has been brought to light. It depends on the prior probability p(H) and the multiplier [p(d|H)/p(d)]. The inverse probability, p(d|H), is the probability of observing data d given that H is true; it is often referred to as the likelihood. p(d) is the prior probability of observing data d. The multiplier determines how much the prior probability is updated. It is worth noting, for example, that if the observation yields no additional information, then p(d|H) equals p(d) and the multiplier equals one, so there is no reason to update the prior probability. Intuition would suggest that if there is no relation between d and H, observing d should not change the degree of veracity assigned to H. On the other hand, when the observation is more probable if the hypothesis is true, $p(d|H)$ being greater than $p(d)$ and the multiplier greater than one, the probability will be revised upwards. In other words, if d and H are linked I must upgrade my appreciation of the hypothesis. But in the opposite case, when the observation is less probable if the hypothesis is true, $p(d|H)$ being smaller than $p(d)$ and the multiplier less than one, the probability will be revised downwards. Or in other words, if d and H conflict, I must downgrade my appreciation of the hypothesis. Lastly, it should be borne in mind that a posterior probability calculated on the basis of a new item of information may in turn be used as a prior probability for calculating another posterior probability taking into account yet another new item of information. And so on. As the new inputs accumulate, the initial prior probability exerts less and less influence over the posterior probability.

Bayes' rule thus makes it possible to calculate probabilities with only a limited number of observations. It even provides for a complete lack of data, in which case the prior probability is unchanged. For example you did not toss the coin your adversary offered, or perhaps you have never tossed a coin. If you do not know whether the coin is loaded, you will choose one-in-two as the subjective initial odds and you will stick to these odds if the game is cut short before it even starts.

## Are we naturally good at statistics?

In the above we presented Bayesian analysis as a relatively intuitive mechanism for inductive reasoning and updating earlier appraisals. However, applying the rule on which it is based is tricky. In the previous section we explained that Ivy League students sometimes came up with the wrong answers to much simpler probability problems. Disregard for the denominator, apparently so common, is such that one may wonder whether the average person can really grasp the concept of probability. Probability is commonly defined by a frequency, the ratio of observed cases to possible cases. Omitting the denominator renders a probability completely worthless.

Thanks to experimental psychology the focus of the classical debate on whether decision-making by humans has a rational or irrational basis has shifted to the statistical mind, raising the question of whether we are capable of reasoning in terms of probabilities.[2] The response by most experimental psychologists is categorical. Kahneman and Tversky sum up their position thus: 'People do not follow the principles of probability theory in judging the likelihood of uncertain events [...]. Apparently, people replace the laws of chance by heuristics which sometimes yield reasonable estimates and quite often do not.' According to Slovic our lack of probabilistic skills is a proven fact: humans have not developed a mind capable of grasping uncertainty conceptually. This view was endorsed by the well-known evolutionary biologist Stephen Jay Gould. After examining the work of the winner of the 2002 Nobel prize in economics and his main co-author, Gould concluded that: 'Tversky and Kahneman argue, correctly I think, that our minds are not built (for whatever reason) to work by the rules of probability, though these rules clearly govern our universe.'

Under these circumstances Bayesian logic seems even further beyond our reach. The case of the medical examination described in the previous chapter was presented to American physicians and medical students. An overwhelming majority answered that the chances of being ill if the examination was positive ranged from 70% to 80%, 10 times higher than the right answer, 7.5%. A survey in Germany revealed that HIV-Aids counsellors were just as confused by Bayesian reasoning. After being given the data to calculate the probability that a person who had tested positive was actually a carrier of the virus, test participants produced answers very close to one, whereas the right answer was closer to 50%. If you think you are ill and decide to have an examination, make sure you choose a doctor who is trained in statistical analysis!

In short it seems to be an open-and-shut case. As Kahneman and Tversky assert: '[In] his evaluation of evidence, man is apparently not a conservative Bayesian: he is not Bayesian at all.'[3] Participants in tests go so far as to resist attempts by experimenters to correct and instruct them. This reaction is comparable to the impact of optical effects: although we know they are optical illusions we persist in taking a bump for a hole, or in seeing an area as smaller or larger than it really is.

In the 1990s Gerd Gigerenzer firmly opposed this standpoint. According to the German psychologist humans are not deaf to statistical reasoning. Unlike Gould he upholds the idea that the human mind evolved, while integrating Bayesian algorithms. However, to be of any use such algorithms must be expressed in practical terms, close to their internal format, as is apparently the case with frequencies in their natural form, but not probabilities. (See the box.)

Cognitive science has made further progress, since the experimental work of Kahneman, Tversky and Gigerenzer, in particular with regard to language learning. Recent work seems to favour the idea of a Bayesian brain,[4] endorsing the conviction expressed by Laplace two centuries earlier in the introduction to his philosophical treatise on probabilities: 'the theory of probabilities is basically just common sense reduced to calculus; it makes one appreciate precisely something that informed minds feel with a sort of instinct, often without being able to account for it'.

## Choosing the right prior probability

We explained above that the degree of prior certainty we choose to assign to a hypothesis may be based on objective data from the past, scientific theory or indeed subjective belief. It is now time to

## Presenting Bayes' rule in other terms

People make mistakes in Bayesian calculations because the problem is not clearly stated. They make far fewer mistakes if data are set forth in a 'natural frequency format', instead of being given as frequencies expressed as percentages or values ranging from 0 to 1.

This approach presents the medical examination in the following way:

(a) 10 out of 1,000 patients who undergo an X-ray examination have cancer;
(b) 8 out of 10 patients with cancer test positive;
(c) 99 out of the 990 patients who do not have cancer test positive.

If we take another group of patients, who have also tested positive, how many would you expect to have cancer?

Presented in these terms one may easily reason as follows: 8 patients who tested positive have cancer and 99 patients who tested positive do not, so in all 107 patients tested positive; the proportion of patients with cancer among those who tested positive is therefore 8/107, in other words a probability of 0.075.

This problem and numerous variants were employed by Gigerenzer and his fellows in experiments on groups comprising doctors, students and ordinary people. They also developed a method for learning Bayes' rule in under two hours, which is now widely used in Germany to train medical staff.

tighten up that statement and show how this prior choice influences the degree of posterior certainty.

Bayes' rule applies equally well to statistical reasoning as to probabilistic logic. In the first case the prior value we take is based on observation. For example, a die has been rolled ten times, the six has come up twice, and we are about to roll the same die, starting a new series. The prior probability of the six coming up again (or, more simply, the prior) is two-in-ten. Before rolling any dice at all, we may also choose a theory such as the one which posits the equiprobability of any of the faces coming up. This can no longer be treated as an observation, rather a belief. It assumes that the die is perfectly balanced, which is never the case. The strict equiprobability of one-in-six is rarely observed, even if a die is rolled several thousand times. Some people have taken the trouble to check this. So by choosing the prior, we are engaging in probabilistic logic. Equiprobability is the most plausible initial hypothesis, lacking any knowledge of the properties of the die on the board. Adopting a more subjective stance we may also opt to assign a prior probability of one-in-ten to the six, because we know we are unlucky or having a bad day.

The effect of the prior is described visually in the next box. It shows that only in two cases does it have no effect: either there are an infinite number of observations or measurements, or the beliefs are vague and all the outcomes equiprobable. Between the two the prior carries a varying amount of weight.

In this example Bayes' rule worked one of its many magic tricks, reconciling perceived probability with its objective counterpart. It combines an initial perception regarding the plausibility of a hypothesis or the probability of an event with a calculated probability based on new knowledge.

## Predicting the probability of the next event

Will the sun rise tomorrow? According to Laplace the answer is uncertain, having estimated the probability of it not rising to be 1/1,826,215! In this example – which earned him much mockery and criticism – he applied a general formula which he had worked out himself. His rule of succession is based on the same principles as Bayes' rule.

## Illustrating Bayesian revision

Figure 6.1 describes a probability density function.[5] It expresses your prior judgement on the proportion of your compatriots who hold the same opinion as you do on nuclear power. According to you, the proportion is probably (95% chance of being right) between 5% and 55%, and its most likely value is 20%. Of course, someone else may be more sure than you are, in which case the curve would be more pointed; on the other hand a third party might only have a very vague prior conviction, making the curve very flat.

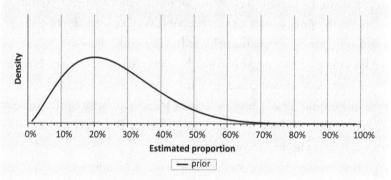

**Figure 6.1** Probability density function

Let us now assume that one of your friends has carried out a mini-survey of acquaintances. Of the twenty people she has questioned, eight (40%) shared your point of view. With some knowledge of statistics you will be able to define what is known as the likelihood function, summarizing the survey's findings. The light grey curve in Figure 6.2 plots its density. Had a higher number of people been surveyed, the curve would have had a steeper peak. As we all know, the larger the number of people polled, the greater the certainty associated with poll results.

You are now ready to update your initial judgement by applying Bayes' rule. Just multiply the two previous functions. The dark grey curve in Figure 6.2 plots the new density function which expresses the proportion of your compatriots who hold the same opinion as you do on nuclear power, given the results of your friend's mini-survey.

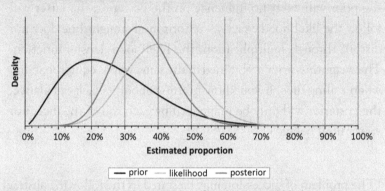

Figure 6.2 Prior distribution, posterior distribution and likelihood function

It is immediately apparent that the posterior is more certain than the prior – the dark grey curve is not as flat as the black one. This is hardly surprising, because new data result in a better judgement. It is also worth noting that observation carries more weight than the prior: the posterior peaks at 34%, a value closer to 40% than to 20%. Finally, you should bear in mind that if the survey had polled a larger number of people, the dark grey posterior curve would have been even closer to the light grey curve, which would – others things being equal – also rise to a steeper peak. The likelihood would have weighed even more heavily on the prior. Ultimately, if observations are carried out on very large samples, the event is repeated a large number of times, or a quasi-exhaustive survey is conducted, the choice of the prior no longer has any bearing on the posterior. It makes no difference.

On the other hand, what happens if the prior is changed? If your prior is subject to greater uncertainty (a flatter black curve), it will weigh less heavily on the posterior, which will resemble more closely the likelihood. The prior exerts less influence. If, at the outset, you have absolutely no idea of how many people share your views (a horizontal line from one end to the other), the prior will exert no influence on the posterior; the latter will follow the likelihood exactly – a horizontal straight line does not distort, through multiplication, the likelihood density function. The weightless prior is obtained in the same way as equiprobability when rolling dice. If you know nothing about its lack of balance, the posterior will only be influenced by each roll of the dice, not your initial belief.

The problem of succession may be stated in the following abstract terms: what is the probability of picking a red ball out of an urn at the $(n + 1)$th attempt, given that for the $n$ previous attempts, $k$ red balls were picked? Or alternatively, posing a more concrete question addressed in the following section, given that the global reactor fleet has experienced eleven core-melt events in 14,400 reactor-years, what is the probability of another core-melt occurring tomorrow?

Your intuitive answer to the question on the next draw is very probably $k/n$. Previously $k$ red balls have been picked from the urn in the course of $n$ draws, so you conclude that the same proportion will also be valid next time. It is not a bad solution, but there is a better one. From a mathematical point of view your answer does not work if $n = 0$.[6] So you cannot use this formula to determine the probability of the first event, for example a core-melt before an accident of this type ever happened. Using the formula means you can only base your judgement on observation. Much like Doubting Thomas, you only believe what you can see. You may, however, have carried out previous observations (though this is fairly unlikely in the case of picking balls out of an urn), or perhaps you fear that the person who set the urn on the table wants to cheat you. In short, you are not basing your decision on your prior; you are not using Bayesian reasoning.

To solve the succession problem Laplace resorted to Bayes' rule, though it was not yet known as such. Indeed Laplace reinvented it, very probably never having heard of the work done by the British minister. He started from an *a priori* judgement, assuming that only two outcomes were possible: picking a red ball or a ball that was not red. Each had an equal chance of occurring, so the probability was one-in-two. He thus obtained a formula $(k + 1)/(n + 2)$. For example, if five draws are made and a red ball is only picked once, the probability predicted by the formula for the sixth draw is $(1 + 1)/(5 + 2)$, or $2/7$, or indeed 0.286. Alternatively, if we return to the example of the sun,[7] Laplace went back to the earliest time in history, 5,000 years or 1,826,214 days earlier, making 1,826,213 successful sunrises out of 1,826,213. The probability that the sun would also rise on the following day was consequently $(1,826,213 + 1)/(1,826,213 + 2)$ and the probability of the sun not rising was $1-(1,826,213 + 1)/(1,826,213 + 2)$, or one chance in 1.8 million.

Laplace proceeded as if two virtual draws had been made, in addition to the real ones, one yielding a red ball, the other a non-red ball. Hence the simple, intuitive explanation of the origin of his formula: he adds two to $n$ in the denominator, because there are two virtual draws; and adds one to $k$ in the numerator, to account for the virtual draw of a red ball. From this point of view the choice of the sun is unfortunate. The merit of the prior depends on it being selected advisedly. Laplace was a gifted astronomer, yet he chose a prior as if he knew nothing of celestial mechanics, as if he lacked any understanding of the movement of the sun apart from the number of its appearances.

The choice of the prior to predict picking a red ball at the next draw is more successful. It anticipates Keynes' principle of indifference:

when there is no *a priori* reason to suppose that one outcome is more probable than another, equiprobability is the only option. If there is nothing to indicate a bias, the prior probability that a coin will land heads up is one-in-two, the prior probability of scoring a three with a roll of a die is one-in-six, and so on. A $1/n$ probability makes complete sense if no prior knowledge or expertise suggests that a specific outcome among the $n$ possible outcomes is more probable than another.

Thanks to progress in probability theory we now have a better instrument than the Laplace formula for solving the succession problem. The solution found by the French mathematician, $(k+1)/(n+2)$, has become the particular case in a more general formula. This formula is still the result of Bayesian reasoning, but it has the key advantage of introducing a parameter which expresses the strength of the prior in relation to new observations and knowledge. For the two extreme values of this parameter, either your confidence in the prior is so strong that nothing will change your mind – so the posterior cannot diverge from the prior – or your confidence in the prior is so weak that the tiniest scrap of new data will demolish it – leaving the posterior exclusively dependent on the new data.

In this modern version the expected probability of picking a red ball in the $(n + 1)$th draw is written as $(k + st)/(n + s)$, where $t$ is the prior expected probability – for example 0.5 for equiprobability – and $s$ is the parameter measuring the strength of the prior. The two virtual draws, of which one is successful, added by Laplace, are replaced here by $st$ virtual draws (where $t < 1$) of which $s$ are successful. In other words we return to Laplace's formula if $s = 2$ and $t = 1/2$. Bear in mind that if $s = 0$, the posterior, in other words the probability of drawing a red ball in the $(n + 1)$th draw, becomes $k/n$. Only the observations count. On the other hand, if $s$ tends towards infinity, the posterior tends towards the prior.[8] Only the prior counts.

The general formula, $(k + st)/(n + s)$, corresponds to the summit of a curve similar to the light grey curve in Figure 6.2. Similarly, the value $t$ marks the top of the black curve in the same figure. So Bayes' rule does apply to functions. The prior probability and the observed probability are both random variables for which only the distribution and classical parameters (expectation, variance) are known.[9] The exact value is not known, but simply estimated by statements such as 'there is a 95 per cent chance that the target value is between 0.3 and 0.6'. The parameter $s$ which expresses the strength of the prior may therefore be interpreted as the uncertainty of the prior. The greater the value of $s$, the steeper the curve; the lower the value of $s$, the flatter the curve (in mathematical terms $1/s$ is proportional to the variance, which measures the distribution on either side of the mean value of the prior function). Expert opinions can be finely quantified, thanks to this property. They can be queried too, for the most probable prior measurement of a phenomenon ($t$), but also the strength of their opinion ($s$). In this way two experts can state that they estimate the probability of radioactive emissions in the event of a core melt as 0.1, but one may be confident, assigning a value of 10 to $s$, whereas the other is much less certain, only crediting $s$ with a value of 0.5.

An interesting feature of this way of calculating the probability of the next event is that it enables us to grasp the connection between perceived probability and acquired objective probability. At the outset all you have is your subjective prior (the expected probability, $t$). Then you receive the objective information that $k$ red balls have been picked in $n$ draws (an expected probability of $k/n$). So your perception of the next draw is given by the posterior (the expected probability of $(k + st)/(n + s)$), which combines the two previous elements. In other words the perceived probability is seen as the updated prior probability taking into account the acquired objective probability. This interpretation is fruitful because it provides a theoretical explanation for our biased perception of low and high probabilities.[10] There is a linear relation between the expectation of the perceived probability $(k + st)/(n + s)$ and the expectation of the acquired objective probability $k/n$. If $k/n < t$, the perceived probability is less than the objective probability.

## What is the global probability of a core melt tomorrow?

In the wake of the Fukushima Daiichi disaster numerous scratch calculations appeared in the press to determine the observed frequency of nuclear accidents and estimate the probability of future accidents.[11] They started from the fact that four core melt events followed by a massive release of radioactive material (Chernobyl 4, Fukushima Daiichi 1, 2 and 3) had occurred with a global nuclear fleet which had operated for a total of 14,000 reactor-years. So the observed accident frequency was 4/14,000, or about 0.0003. If we restrict ourselves to Europe and its 143 reactors, we should therefore expect 1.29 accidents in the course of the next thirty years $(30 \times 143 \times 0.0003)$. Of course, this number is not a probability because it is higher than one. Confusing it with a probability is tantamount to saying that in a family with four children the probability of having a daughter is 2 $(0.5 \times 4)$, whereas in fact it is 15/16 $(1-1/(2 \times 2 \times 2 \times 2))$! To calculate a probability, one needs to adopt a different approach. The most straightforward explanation could be presented in the form of an exercise for a student starting a course in statistics. 'If 14,000 reactor-years were asked: "Have you had a major accident?", and four answered yes, then what is the probability of an accident over the next 30 years for a fleet of 143 units?" The answer is 0.72. This result is calculated by modelling the probability of a major accident using a binomial distribution function with the following parameters [30 × 143; 0.0003].[12]

This result is frightening, because it means that it is almost certain that an accident will occur in Europe over the next thirty years, a probability of 0.72 being almost equivalent to a three-in-four chance. Fortunately, this figure can be discarded, because it is based on assumptions and scientific method which make no sense. It assumes that all reactors are identical, representing the same degree of risk regardless of their design, their exposure to natural hazards, the precautions taken by their operators or indeed the ability of their regulatory authority. It also assumes that safety does not vary over time. What was true in the past will hold true tomorrow, so no provision may be made for improved safety. According to this approach, more effective preventive measures, stricter standards, better reactor designs or even the lessons learned from past accidents have no impact on performance. Nor would it make any difference if efforts and expenditure to improve safety were dropped. In short, it supposes that the entire European fleet is in an imaginary state of permanence, perfectly homogeneous and immutable, with each reactor being replaced by an identical unit and all of them exposed to the same risk of failure regardless of their geographical situation and operations.

Technically speaking, the method outlined above is open to criticism because of the very low number of observations on which it is based. It is hazardous, to say the least, to estimate the value of a parameter statistically on the basis of just four observations!

But, in our view, the main methodological shortcoming of a scratch calculation of this sort is the assumption that our knowledge of the subject is based entirely on these observations. It proceeds as if the probability of another accident tomorrow could only be elucidated and parameterized by the past observed frequency, $k/n$. This reasoning focuses exclusively on observation, completely disregarding all the other learning which has accumulated on the subject. Yet tens of thousands of engineers, researchers, technicians and regulators have been working on precisely this topic for the past sixty years. It is as if the only hope of salvation was in major accident data, despite their number being extremely limited, unlike other forms of knowledge. It considers that the parameter indicating the strength of the prior (previously designated as $s$) has a value of zero. Such an approach is as extremist as its opposite: refusing to attach any credit to observations of past accidents to predict the probability of an accident in the future, in other words, leaving $s$ to tend towards infinity.

A simple example may be used to illustrate how observation and other knowledge can be combined to predict future nuclear risks. There are no particular difficulties regarding observations, apart from their very limited number. This constraint may, however, be slightly reduced if we focus on the risk of a core-melt accident, of which there are eleven cases,[13] rather than the risk of core melt leading to release, of which there are only four instances. As ever, choosing the prior is tricky, but we may take advantage of the availability of probabilistic safety assessments, the results of which summarize all the existing knowledge on accidents, apart from observations. On the basis of assessments carried out in the 1990s on US reactors, we may choose to add to the data on 11 accidents in 14,000 reactor-years, a virtual observation of 1.6 accidents in 25,000 reactor-years. This virtual observation is derived from the expected value of the probability of core melt, estimated by the experts as $6.5 \times 10^{-5}$ per reactor-year, associated with an uncertainty measured by a prior strength of 25,000.[14] We thus obtain values for $t$ and $s$ and can apply the formula $(k + st)/(n + s)$, or $[11 + (25{,}000 \times 6.5 \times 10^{-5})]/(14{,}000 + 25{,}000)$, which equals $3.2 \times 10^{-4}$. In other words the experts propose a mean probability of core-melt per reactor-year of $6.5 \times 10^{-5}$; and observed evidence leads to $7.8 \times 10^{-4}$. By advisedly combining the two sources of knowledge we obtain a probability of $3.2 \times 10^{-4}$. To make this figure easier to grasp, we may translate it by calculating that the probability of a core-melt accident next year in Europe is 4.4 in 1,000. This probability is less than half the value obtained from the observed frequency of core-melt accidents on its own.

It is nevertheless very high, because the approach used in this example is based on the same oversimplifying hypotheses criticized above, especially the assumption that the accident probability we want to measure has remained constant over time. But progress has been made on safety and these gains are reflected in the risk of a core melt. A large number of the accidents taken into account happened during the early years of the development of civil nuclear power.[15]

Nor is the above approach ideal, because it assumes that events are unconnected. Yet the European, or global, nuclear fleet in 2012 is fairly similar to the one that existed in 2011, in 2010, and so on. In other words, we are not dealing with a situation akin to rolling dice, in which the result of one roll has absolutely no bearing on the following one. On the contrary the occurrence of one major accident may point to others, because we are still dealing with the same reactors.[16] For example, an accident may indicate that a similar problem affects other reactors of the same design or subject to comparable natural risks. A recent study sought to remedy these two shortcomings.[17] The model it used weights accident observations according to the date when they occurred. It calculates a sort of discount rate which, with time, tends to reduce the influence of old accidents over the probability of an accident now. With this model the probability of a core-melt accident next year in Europe is 0.7 in 1,000, less than one-sixth of that calculated with the previous approach.

How may we conclude this second part devoted to risk? By emphasizing the pointless opposition between experts and the general public, with both parties accusing the other of rank stupidity. The experts, with their sophisticated calculations, have allegedly made massive mistakes estimating the frequency of major accidents and the associated damage. Locked up in their laboratories, their certainty has purportedly blinded them to events that are glaringly obvious, namely the recurrence of disasters since Three Mile Island. On the other hand the general public is supposedly guilty of yielding exclusively to its fear of disaster, stubbornly clinging to its rejection of calculated probabilities. Brainwashed by a stream of disaster scenes and incapable of basing the slightest decision on statistical evidence, people's reactions are ruled by emotion and nothing else; they are ostensibly quite unable to see that coal or hydroelectric dams constitute risks every bit as serious as nuclear power plants. So you may choose sides, denigrating either the experts or the general public! However, there is no contradiction between analysing on paper sequences of events that may potentially lead to a core melt, and observing accidents and their causes. Furthermore, theory provides ways of combining and associating such knowledge. It enables us to find rational explanations for the irrational, which becomes much less or not at all irrational. We have a better grasp of how probabilities are distorted, either through simple heuristics or gradual acquisition of Bayesian methods, depending on one's school of thought. What was once perceived as irrational loses that connotation once it is seen to conform to identified mechanisms. A fuzzy border now separates rational decision-making from other forms of choice. Is it irrational to take decisions under uncertainty without allowing for risk aversion? Surely it would in fact be irrational to go on denying the existence of such aversion? When Maurice Allais presented Leonard Jimmie Savage with a puzzle, the latter's answer contradicted the results of his own theory. When he appears as a guest-speaker in Israel, Daniel Kahneman avoids travelling on Tel Aviv buses for fear of an attack. In so doing he is fully aware that his actions are dictated by perceived, rather than calculated, probability.

The loss of bearings which fuels discord between experts and the general public, opposing the rational and irrational, is most uncomfortable. It raises a terrifying question: should government base its decisions on perceived probabilities or on those calculated by experts? In the case of nuclear power, the former are largely overestimated, a situation which is likely to last. It would be foolish to treat the attitude of the general public as the expression of fleeting fears which can quickly be allayed, through calls to reason or the re-assuring communication of the 'true' facts and figures. If government and the nuclear industry did share an illusion of this sort, it would only lead to serious pitfalls. The reality test, in the form of hostile demonstrations or electoral reversals, may substantially add to the cost for society of going back on past decisions based exclusively on expert calculations. On the other hand, the airline security syndrome will inevitably spread to nuclear power if the authorities only pay attention to public perception of probabilities. If the propensity to invest in nuclear safety is not brought under control, many expensive new protective measures will accumulate, without necessarily having any effect.

# PART III

# Safety regulation

## An analysis of the American, French and Japanese cases

The accident at Fukushima Daiichi starkly revealed the deficiencies of safety regulation. How is public confidence in nuclear power to be maintained if safety standards are badly designed or simply not enforced? The disaster in Japan resulted from a conjunction of natural forces, but it probably would not have caused so much damage had there been a clearer division between the regulatory authorities and the nuclear power companies they were supposed to be supervising. How then is safety to be effectively regulated? In the absence of powerful, independent, transparent and competent regulatory authorities, nuclear power will not be able to progress worldwide.

Nuclear power is regulated long before work starts on building plants. Investors must obtain a licence from the government or regulatory authority tasked with safety. The design of reactors is subject to a whole series of standards and principles laid down by the public authorities. For construction of the plant to be authorized it must comply with these 'specifications'. During the construction project the regulator must check that the relevant standards are being implemented. It must subsequently intervene all the way through the operating life of the plant, from initial grid-connection to final decommissioning. It monitors the operator's compliance with the safety regulations it has established. So a safety failure with the potential to cause an accident is not necessarily only the failure of the organization which built or is operating the plant, which may not have complied with regulatory safety requirements. It may also be due to shortcomings in the regulatory framework itself, its loopholes or fuzzy definitions, lack of supervision and inadequate sanctions. But in the nuclear power industry it is not easy to draw a line between good and bad regulation, between proper and improper enforcement. A nuclear reactor is a complex system. It is difficult to define and assess its safety level. This depends on the design of reactors, their maintenance, the way they are operated and their geographical location. It is impossible to reduce the safety level to a single variable. But supposing for a moment that a probabilistic criterion – for example a core-melt frequency of less than 0.000001 per year of operation – could fulfil this role, there would be no way of observing it directly. Nor would it be easy to define precise technical guidelines to ensure this safety level was actually maintained. The regulation of nuclear safety is necessarily based on a large number of criteria and standards.

It is much easier to perform an economic analysis of safety regulation.[1] Such analysis highlights the overall problems that need to be solved, advocates regulatory, legal and institutional solutions, checks that the measures already in force are satisfactory, and shows why public intervention is necessary and the form it should take. In this way it goes far beyond simply estimating the impact of safety measures on costs. Why then are self-regulation and civil liability not sufficient to guarantee an adequate safety level? Why must the safety regulator be independent from both industry and government? Is it necessary to set quantitative safety goals, such as risk thresholds? Would it be preferable to establish detailed standards to which engineering firms and operators must conform, or rather to set overall goals for performance? Economic analysis of regulation shows that safety is not an exclusively technical concern. There are other institutional and organizational solutions – often less expensive than adding certain redundant devices – which may substantially increase the level of nuclear safety.

# SEVEN

# Does nuclear safety need to be regulated?

What an odd question! Surely everyone accepts the need to regulate nuclear safety? Only economists would think of querying something so obvious! Naturally we have no intention of claiming that the need for state regulation in this field is a recent discovery. The value of economic theory resides in the analytical process it offers for this purpose. In terms of methodology it always starts from opposition to state intervention per se. Such intervention is always open to question and the first thing an economist will ask is why the market is not sufficient in itself to solve the problem. This does not necessarily reflect an ideological preference, on his or her part, for the invisible hand or outright rejection of public intervention, but simply the knowledge that economic theory has identified specific conditions in which the market is not an effective means of securing the general economic interest, in other words maximizing wealth for the whole of society.[1] Externalities, such as pollution, or a public good, such as national security, are the main impediments to market efficiency. If there is no market failure, there is no justification for public intervention. But economists also know that the state – be it embodied by planners, regulators or legislators – has weaknesses of its own, and that public intervention is not perfect. It is easy to summarize what economic theory prescribes with regard to regulation: resorting to the visible hand of the state is justified if, and only if, the defects of public intervention are less than those of the market it sets out to correct. In short, the benefit of public intervention must exceed its cost, otherwise *laissez-faire* is the only option and one must make do with an inefficient market. (Unless, of course, the means can promptly be found to reduce the cost and inefficiency of the visible hand of the state.)

When it comes to nuclear safety, the market does not provide operators with sufficient incentives to make the necessary efforts. The signals it sends are too weak. On the demand side of the electricity market, consumers are exposed to a relatively low average risk of accident. They are not spontaneously inclined to pay for all the necessary safety efforts, in particular the protection of local residents, who are exposed to much greater risk. Consumers are primarily concerned about the price of electricity, which is seen as a commodity. On the supply side, it is in the collective interest of both operators and reactor vendors to invest in safety. Another disaster would damage their image and future prospects. But individually each company stands to gain by doing nothing (or very little, as we shall see below). It can maximize its gains by pocketing the collective benefits of a lower accident-risk without additional expenditure. In keeping with this rationale all the companies would tend to behave like stowaways, but if so there would no longer be a ship to carry them. To deal with this two-sided failure, the most cost-effective form of regulation would be to establish rules on civil liability. But full insurance against accidents must be possible for this to work. Unfortunately, as we shall see, this is not the case for nuclear power.

## Inadequate private incentives

The economically optimal level for nuclear safety is determined in the same way as efforts to mitigate pollution. The same principles apply as those presented in Chapter 1 with regard to reducing carbon-dioxide emissions. The level of safety efforts is optimal when the marginal social cost of protection is equal to the marginal social benefit of the damage avoided. Beyond this level, additional efforts to improve safety cost more than the additional benefit they yield. Below this level additional efforts can still be made, their cost being lower than the benefit they procure.

Left to itself a market balances private costs and private gain. An operator will invest in safety as long as the resulting profit exceeds the cost entailed. This cost roughly corresponds to the social cost. Setting aside a few emergency services devoted to nuclear risk, the operator shoulders the full cost. In France the government even charges operators for the police officers tasked with supervising nuclear power plants and fuel transport.[2] On the other hand, an operator only receives part of the benefit from the safety measures it deploys, for example in terms of the industry's image. Its efforts to improve safety being trimmed to suit the limited benefit it receives, they will necessarily be inadequate.

Though inadequate, it will still make some effort on safety. The private incentives for operators are far from negligible. Improving safety contributes to reducing the number of breakdowns; in turn, fewer breakdowns mean higher electricity production, so more revenue. As we saw in Chapter 1, the load factor is critical for the profitability of a nuclear power plant, due to its very high fixed costs. Drops in output or temporary reactor shutdowns are very expensive for an operator. There is every incentive to restrict them as much as possible. Furthermore, in the event of a core melt, even without the release of radioactive material to the environment, the generating company loses its means of production and the value of the asset is wiped out. An operator which only owns one reactor will go bankrupt. Even with several in its fleet it may well suffer the same fate, having lost the confidence of consumers, banks, small investors and even its workforce. The risk of disappearing is a powerful incentive for an operator to concern itself with safety. So it would be foolish to imagine that nuclear operators are actively hostile to prevention and regulation. This point is important for two reasons. Regulation would be much more expensive and a great deal less effective if nuclear operators stood to gain by blocking initiatives and getting round measures introduced by the regulator. In addition, government sometimes fails in its duty to protect the general public: when confronted with a toothless regulator, operators are not exposed only to temptations that jeopardize safety.

However, it would also be a mistake to conclude that, for lack of anything better, self-regulation would be an acceptable solution. Our concern is merely to emphasize that market forces and safety are not wholly antagonistic. Self-regulation may set in motion a virtuous circle in order to reduce what is known as 'free-riding'; witness the case of the Institute of Nuclear Power Operations (see box).

So the question is not whether self-regulation can take the place of state intervention, but rather the extent to which it may play a complementary role. Self-regulation streamlines the flow of information and the dissemination of good practice among operators, the latter no longer restricting themselves to the single channel connecting them to the safety authority. The larger the number of operators, the greater this advantage: they currently number more than twenty-five in the US. Self-regulation can partly fill a vacuum too. National regulators are trapped inside their respective borders; they can neither inspect nor stigmatize operators outside their jurisdiction, nor yet act against black sheep behaving in a way that threatens the global industry as a whole. To do that would require an international federation of operators, playing the same role as the INPO but worldwide. A World Association of Nuclear Operators, dedicated to the safety of nuclear power plants, does now exist, but its operating rules are a far cry from the US equivalent. The peer inspections it implements are neither systematic nor mandatory; results are only reported to the operator being inspected; there is no scoring system or ranking, nor any sense of stigma associated with shortcomings. This pale replica of the INPO was set up after the Chernobyl accident. The Fukushima Daiichi disaster could have given rise to an increase in its powers,

## The Institute of Nuclear Power Operations: a case of successful collective regulation

After the accident at Three Mile Island on 28 March 1979 the reputation and credibility of all the nuclear operators in the United States were seriously damaged. They realized that the image of the nuclear electricity industry was a common good and that the existence of just one black sheep constituted an economic threat to them all. On the strength of this realization they decided to set up a trade federation to promote progress and excellence in safety. The INPO came into existence seven months after the accident. Marking a new departure, it relied on peer pressure to oblige operators, in particular those with a poor record, to improve their safety performance. At two-year intervals mixed teams, comprising INPO employees (currently about 400) and safety experts working for operators other than the power plant being inspected, assessed the safety of each plant. Since the INPO was first started, some 13,000 peer-assessors have contributed to the work of these teams. Their appraisals are reported and discussed at an annual gathering, devoted exclusively to this topic, for the CEOs of nuclear operators. Poor performers are singled out and must remedy any detected deficiencies. If they fail to improve safety conditions they run the risk of being ostracized by the rest of the trade and barred from the INPO.

The INPO is a success.[3] It has contributed to continuously and significantly improving the safety of the US nuclear fleet, while reducing the gap between the overall average and poor performers. This is not a cover-up, with cosmetic safety measures acting as a front for communication and lobbying objectives. At first sight an economist is inclined to view such an achievement as an anomaly. Economic theory explains that corporate coalitions are by nature unstable. As the free-rider phenomenon also operates over time, the INPO should supposedly have folded once the shockwave of Three Mile Island had passed. An operator which persists in substandard performance avoids expenditure while benefiting from the positive collective image built up by its fellows, and from the lower probability of another accident in the US. Coalitions can only be sustained if there is an effective mechanism offering incentives and imposing sanctions on deviants. Peer pressure, a force identified by sociologists, does not seem to carry enough weight. In fact it is underpinned by economic mechanisms. Firstly, a form of financial solidarity comes into play among US nuclear operators if one of their number suffers an accident. It is in their interest to institute mutual supervision, as they are under a legal obligation to contribute to compensation paid out for damage caused by a fellow operator. Secondly, INPO assessments affect the level of compulsory civil liability insurance premiums. The better an operator performs, the less it contributes to Nuclear Electric Insurance Ltd, which is jointly owned by nuclear generating companies. NEIL requires the organizations it insures to be INPO members, giving added weight to the threat of exclusion: an operator excluded by its peers will struggle to obtain insurance or will have to pay a much higher premium. Lastly, the INPO is supported by the US Nuclear Regulatory Commission. It submits inspection reports, annual updates on assessments and its ranking of operators to the NRC. If poor performers repeatedly fail to remedy shortcomings, it may threaten intervention by the regulatory authority. The INPO is not authorized to impose fines, nor yet to order a reactor to be shut down, but on the other hand it can persuade the NRC to sanction those who fail to toe the line. As with most self-regulation mechanisms, the effectiveness of the INPO is dependent on the presence of public regulation in the background.

but no one seized the opportunity. Nuclear operators which adopt an uncompromising approach to safety consequently remain at the mercy of black sheep operating plants tens of thousands of kilometres away.

## Civil liability

From a legal standpoint the rules on civil liability are designed to compensate victims for damages. Inspired by considerations related to morality and fairness, such rules come into play after an accident. But an economist sees the rules on civil liability as an incentive. They encourage potential wrong-doers to take preventive action. In theory they bring about an optimal level of effort. A potentially liable economic agent seeks to minimize spending on *ex ante* prevention and *ex post* compensation for damage. It is therefore in its interest to stop preventive action if an additional measure costs more than compensating for any potential damage. The law of civil liability thus appears to offer an alternative to regulation. Rather than some authoritarian regulator laying down standards of behaviour and monitoring compliance, an economic agent has an incentive to behave virtuously. Public intervention can thus be restricted to the work of the courts which assess the value and order the payment of compensation to victims. As those liable for the damages may be insolvent, civil liability is often backed by a legal obligation to take out insurance. But even as cumbersome a system as this usually costs less than public regulation, which is why economists prefer it and recommend it whenever possible.

A system of this sort cannot take the place of regulation to guard against nuclear accidents: much like self-regulation, it can only play a complementary role. The financial limit on operators makes it impossible for civil liability to play an exclusive role. Liability must be unlimited, in order to provide an incentive for optimal preventive efforts. Unfortunately, in practice, the cost of the damage caused by a major accident can exceed the value of generating companies' assets. Disaster leads to bankruptcy, leaving third parties only partly compensated. This means that the operator is not being given sufficient incentive to take the necessary preventive measures in relation to the economically optimal level.

The problem of solvency can be overcome by a mandatory insurance mechanism, but this raises another obstacle: the impossibility of insuring the full risk of a major nuclear accident.[4] Major accidents are too infrequent for the average risk to be determined with sufficient accuracy. Nor is there sufficient certainty that the limited number of accidents and incidents will compensate an insurer for the occasions when it will have to pay out very high compensation. Confronted with a major accident, insurers and re-insurers may in turn be made insolvent. The annual insurance premium, not including margin and overheads, must be equal to the probability of an accident in the course of the coming year multiplied by the cost of the damage. As the probability of an accident occurring next year is equal to that of it occurring in any other reactor-year, the insurer must build up a reserve ahead of the event. However, if the accident occurs next year, the insurer will not have received a sufficient number of premiums to be able to compensate victims. The initial premiums, paid during the first few years of coverage, would therefore have to be high enough to constitute a reserve equal to the cost of average damages, raising the cost of the premium for the insured. Lastly, premiums will be too expensive for operators to pay if the size of the reserve is based on worst-case, rather than average, damages.

Given these various obstacles, civil liability can only play a secondary role. It is imperative for it to be associated with public safety regulation, to which it serves as a useful complement, acting as an incentive on operators to make up for the gaps and mistakes of regulation. The regulator cannot dictate every aspect of the behaviour and action of an operator and its workforce. Even if the regulator could do so, it would be unable to check compliance with its many requirements. In economics the fact that the regulator is relatively powerless is referred to as information asymmetry. The regulator does not possess all the information it needs – particularly regarding costs and performance – to regulate a firm. Nor is it able to check how much effort a firm makes to comply with its orders. The firm is better informed, but it is not in its interest to be entirely open towards the regulator without something in return. Telling all would be tantamount to shooting itself in the foot, the immediate consequence being more costly regulation. So regulation is necessarily imperfect when compared to the ideal situation of an all-knowing authority. The regulator inevitably makes mistakes. Practically speaking, a safety regulator may not establish a standard for an important safety criterion, for example if it can neither observe nor check this criterion. Lacking adequate information the regulator may also set a technical standard at too low a level. If the operator is subject to rules on civil liability, it is in its interest to make up for at least part of the regulator's shortcomings, introducing an in-house rule on the missing safety criterion or raising, on its own accord, the severity of the standard applied to the whole power-plant workforce. The higher its civil liability and the greater its assets, the more a firm will feel the incentive to make up for such shortcomings.

## Civil liability for nuclear damage, in practice

In practice civil liability for nuclear-generation companies is usually capped. The extent of liability is generally slight in comparison to the cost of a major accident. In France, for instance, EDF's total liability amounts to €91.5 million. In the United Kingdom the upper limit is £140 million. Unlimited liability is exceptional. In the European Union it only applies in four Member States. But three of them – Austria, Ireland and Luxembourg – have no nuclear power plants.

Only German operators are required by law fully to compensate accident victims regardless of the extent of damage.

The low limit on civil liability is mainly a legacy of the past. It only makes sense in the light of the other characteristics of this regime, which is very particular compared to other sectors. In the event of an accident, the operator of a nuclear power plant is deemed liable regardless of how it has behaved. It is liable even if it has not committed any form of negligence, and has complied with all the regulatory requirements. Under this regime of 'strict' liability, there is no burden of proof. Furthermore, operator liability is exclusive: it must compensate for any damage caused by the negligence of a supplier present on-site or due to faulty design on the part of an equipment manufacturer. Responsibility for an accident may be shared, in terms of what actually happened, but only the operator is legally liable. These two characteristics, which are unusual in industry, offer an additional incentive for operators to maintain a high level of safety, entailing greater expenditure. Furthermore, this system secures quicker, more extensive compensation for victims.[5] This was in fact the goal pursued by early legislators framing nuclear law, when these terms were imposed on operators in the 1960s. In return, operators obtained a provision setting limits on their liability. They also pointed out that there was in any case an upper limit on the mandatory insurance associated with their liability, simply because insurers at that time were unable to offer more extensive coverage. In economic terms the system negotiated at this point increased incentives on the one hand – through strict, exclusive liability – while reducing them on the other by capping liability.

This limited-liability regime is obviously related to governments' determination to encourage the development of the nuclear industry. In the late 1950s initial debate on nuclear legislation in the US emphasized that it was critical to its vital national interests to develop nuclear energy and protect the industry against as yet unforeseeable demands for compensation.[6] Subsequently the 1963 Vienna

Convention on Civil Liability for Nuclear Damage underlined that 'unlimited liability [ . . . ] could easily lead to the ruin of the operator'.

Today the nuclear industry is no longer in its infancy, nor yet nuclear risk insurance. It is time to raise the upper limit on civil liability, or perhaps remove it altogether. Indeed the process has already started. The UK government, for example, plans to raise the cap on operator liability to £1.2 billion, eight times its current level. In the next few years France will probably set a new maximum limit at €700 million, a more than sevenfold increase. The highest limit applies to all US generating companies. The 1957 Price-Anderson Act, which has been revised several times, sets the maximum limit at $12.6 billion. This amount includes an operator's financial liability for its own plants ($375 million per site), which must be insured by a pool of private insurers. What is unique about the system is that if damage exceeds this amount, the other operators must pay. As mentioned above, each operator is collectively liable in the event of an accident on any reactor in the fleet. This liability takes the form of a payment, after the accident, which may be as much as $117.5 million for each reactor it owns. With 104 nuclear power plants currently operating, this collective mechanism thus provides for third-party compensation of up to $12.2 billion.

## A subsidy in disguise?

If a major accident does occur in the US, leading to damages in excess of $12.2 billion, the US taxpayer will have to pay, with Congress, the insurer of last resort, drawing on the federal budget to supplement third-party compensation. Can this be construed as a form of subsidy and on what scale?

Opponents of nuclear energy are adamant that the upper limit on liability constitutes a subsidy:[7] the insurance premiums paid by generating companies would be higher if the cap were removed. This is indeed the case. Nuclear advocates counter that there is no subsidy unless an accident leading to damage exceeding the cap really occurs. Prior to such an eventuality taxpayers pay nothing. This too is true. Rather than fighting over how to define a subsidy, let us return to the basic economic principles outlined in Chapter 1. Theory prescribes that the decision-maker should pay the full cost of his or her action, and pocket the entire gain. When there is a gap between the private and public values of these two amounts, it must be filled by a series of taxes or subsidies which internalize external effects. The question is consequently whether the external cost of the accident should be partly or completely internalized. Economists favour the second solution, so that optimal quantities of goods are produced and consumed. For example, if the price of electricity generated using coal does not reflect the cost of carbon-dioxide emissions, too much electricity will be consumed and too much will be invested in this technology. Similarly, if the price of nuclear electricity makes inadequate, or no, provision for the cost of accidents, the incentives for consumers and investors in nuclear power will be too great.

How large is the residual externality, in other words the externality covered by neither the operator nor the consumer of nuclear electricity because of the cap on liability (see Chapter 1)? Or to use a term which is more straightforward but laden with connotations, how big is the hidden subsidy? Naturally the answer depends on the cap itself. If it is very high, exceeding for example the worst conceivable damage, there is no difference between limited and unlimited liability. A note of caution, though: there is no question of comparing maximum liability and the actual cost of an accident. Even the US limit falls far short of the damage caused by the Chernobyl or Fukushima Daiichi disasters. An accident in the US entailing $100 billion in damages would cost the taxpayer $87.4 billion. But this amount does not measure the residual externality. Allowance must be made for an accident's probability too. The subsidy is equal to the gap between the cost of the insurance required to cover the entire risk, and the cost of insurance only covering the maximum liability. To calculate this gap involves making assumptions about the damage distribution function. Unfortunately very few full estimates have been made: two in the US, one in France. In the US the gap is $2.3 million per reactor-year, according to old studies by academic economists,[8] or $600,000 per reactor-year, according to the Congressional Budget Office in its 2008 report on nuclear power.[9] Allowing for the size of the US fleet, these amounts correspond to an annual subsidy about a hundred times larger to the industry as a whole. In the case of France,[10] the disparity in the annual cost of insurance between the liability of EDF, limited to €91.5 million, and unlimited liability ranges from €140,000 to €3.3 million per reactor-year, depending on various assumptions as to accident probability and costs (making an upper-range annual subsidy worth €500 million for the whole French fleet). If we compare these amounts to the price per MWh of nuclear electricity, they are very low, about $0.07/MWh, or 0.1% of the levelized cost,[11] for the Congressional estimate, and €0.45 per MWh, or 0.75% of the levelized cost, for the highest French estimate.[12]

The low level of the preceding figures should not come as a surprise. To illustrate a particular point in Chapter 1 we made a scratch calculation, which found a cost of €1 per MWh. This value necessarily marks the upper limit of the subsidy provided by limited liability. It is equivalent to assuming that operators would not compensate any victims and that the cap on liability is set at zero. The gap between fully and partly internalizing the cost of accidents is always lower than the first term of this disparity.

Of course the residual externality can be inflated by making extreme assumptions on the probability of an accident and the cost of damages. In Part II we cited a study, commissioned by Germany's Renewable Energy Federation,[13] which reported an extremely high risk. It found, for example, that the probability of a terrorist attack was one-in-a-thousand per reactor-year, with upper-range damages costing several thousands of billions of euros. Under such conditions the subsidy would soar, exceeding even the full cost of generating electricity. Liability in Germany is unlimited, but the mandatory financial guarantee underwritten by private insurance or comparable means is limited to €2.5 billion per plant. The authors of the study claim that if the financial guarantee was unlimited, it would cost – depending on the various scenarios and assumptions they studied – from €139 per MWh to €6,730 per MWh, between twice and a hundred times the price of electricity![14]

By setting an upper limit on the operator's civil liability, the legislator is more or less certain to make a mistake, as it does not know the exact probability of the worst-case accident, less still the cost of the resulting damage. Would it not be preferable to put an end to this confusing situation by decreeing that the liability of nuclear operators is unlimited? There would be no need for any more calculations. Moreover, from an economic point of view, this seems the best solution. After all academics specializing in insurance and nuclear law have advocated doing just that. We shall rapidly demonstrate that a reform of this nature would be appropriate but very expensive to implement. To start with, we need to return to the two approaches to analysing liability mentioned in the introduction: one, economic, based on incentives for operators to improve safety; the other, legal, hinging on compensation to victims in the event of an accident. In an incentive-driven system, unlimited liability must be recommended, for it is always beneficial. If there is no safety authority or it is too lax, removing the cap on liability will prompt the operator to increase safety efforts. Such efforts will not attain their optimal level due to the finite value of its assets, but they will come much closer. Whenever regulation is too strict, the regulator imposes safety measures the marginal cost of which exceeds their marginal benefit. At first sight unlimited liability makes no difference to this situation. Regulation alone dictates the safety efforts which exceed the optimal level. Here too, reform can in fact be beneficial, bearing in mind that the regulator – which neither sees nor knows all things – may overlook some aspects of safety. Liability corrects this shortcoming: with unlimited liability efforts will come much closer to the optimal level than if it is limited.

Regarding compensation of victims, removing the cap on liability only has an effect if, at the same time, the legislator requires the operator, through insurance or a comparable mechanism, to obtain a financial guarantee which is also unlimited. Without this provision, victims will only receive compensation in the event of an accident up to the value of the operator's assets. In Germany, as we saw above, liability is unlimited but the financial guarantee is capped at €2.5 billion. Above this amount victims are compensated out of the assets owned by the bankrupt operator. If that is not sufficient, the taxpayer must pay. Implementation of the financial guarantee is impeded by the constraints described above regarding the degree to which nuclear risk can really be insured. The operator needs a market with insurers of adequate size and sufficient number. In practice few insurance companies specialize in nuclear accidents. Furthermore, they are generally organized on a strictly national basis, which limits the pooling of risk, competition and, ultimately, supply. So implementing an unlimited guarantee would fail due to insufficient capacity or excessive cost. It would reveal the shortcomings of the market for insurance and its derivatives; or, failing that, with help from creative financiers,[15] it would lead to premiums so expensive that operators could not afford them (at levels close to those estimated in the German study cited above). Predictably, unlimited liability underwritten by an unlimited guarantee is advocated by the opponents of nuclear power.

To conclude, reform of nuclear civil liability is needed. It should rest on three pillars: unlimited liability, capped financial obligations and pooling of risk at a European level. The first pillar is essential due to its positive effect on incentives, forcing them to rise to meet the glass ceiling constituted by liability, at a height determined by the value of the assets belonging to the person liable for damage. The cap on financial obligations is justified by the catastrophic nature of major nuclear accidents and the structures of the associated insurance market. The first two pillars correspond to the German model. The recommendation to introduce Europe-wide pooling of risk is inspired by the US example.

EIGHT

# The basic rules of regulation

## An engineer's view of safety regulation

Nuclear safety is primarily a matter for engineers, and that is just as well. A reactor operates at a crossroads between many disciplines, including chemistry, mechanical engineering, physics, automation, computing, neutronics and thermodynamics. To design and safely operate these technological monsters requires theoretical and practical knowledge of which engineers have a better grasp than managers or administrators. An accident may occur when a chain reaction goes out of control or a break appears in the cooling system. So the parameters which must be monitored are the fuel and moderator temperature coefficients, excess radioactivity, the neutron life cycle, capture cross-sections for neutrons of varying energies, calorific capacity and thermal conductivity, and radiation resistance of cladding. In short, a series of values with which accountants and financiers are wholly unfamiliar.

If you open a book on nuclear power written by an engineer, you will find the key principles and full details of how a reactor works, with a section on safety. This describes two possible approaches: a deterministic approach based on the defence-in-depth concept; or a probabilistic approach based on risk calculation. The first one dates from the 1950s, the second took shape in the late 1960s. For many years they were presented as rivals, but the quarrel of the ancients and moderns is now largely behind us. The probabilistic approach, discussed in Chapter 4, is based on identifying fault sequences which have the potential to initiate a major accident. The deterministic approach considers a whole list of plausible incidents and accidents, but without assigning them a probability. It frames rules, either to avoid such events purely and simply, or to mitigate them, but without trying to calculate the scale of the reduction. For example it is a basic rule that a failure on a single device must not be able to cause an accident on its own. Defence in depth is closely linked to the deterministic approach, because it involves raising a succession of barriers to prevent the release of radioactive materials into the environment. Furthermore, the dimensions of each barrier include a safety margin.

Engineers have designed and specified three main lines of defence: the cladding on fuel rods, the structure containing the primary cooling system and reactor vessel, and the concrete containment building which houses the entire system. Up to the mid-1960s this final barrier was thought to be impassable, unless it was hit by a missile.[1] A major accident leading to the massive release of radioactive material was not considered plausible. But the first theoretical calculations showed that it was possible: meltdown in the reactor core, following a break in the cooling system, would open a break in the pressure vessel and rupture the final barrier formed by the outer containment building. This was the so-called China syndrome, popularized by the eponymous film. Unfortunately the term is misleading: China is not at the antipodes in relation to the US; the nuclear reaction, if it carried on into the depths of the Earth, would stop when it reached the central core. The purpose of the first large-scale probabilistic assessment in the US was to calculate the frequency of such an event (see Chapter 4).

For some time the probabilistic and deterministic approaches were seen as alternative options, one taking the place of the other. Once it became possible to estimate probabilities, an approach which sought to determine the right dimensions of system components on the basis of constraints such as maximum pressure, corrosion, temperature or indeed seismic intensity looked obsolete. Criticism also focused on the effectiveness of a deterministic approach, because it operates blindly. It provides no indication of what should be prioritized, whether, for instance, it is more important to double the thickness of the containment shell or the number of back-up pumps on the primary cooling system. However, it gradually emerged that probabilistic safety assessments were not a universal panacea either, being unable to encompass all the parameters and causes of an accident. In particular, it is very difficult to assign probabilities to human error or building faults. As we saw in Chapter 4, a probability cannot be calculated in a situation of incompleteness, in other words when all the states of the world are not known. A deterministic approach is the only way of countering 'unknown unknowns', in other words when you do not know what is unknown.

A deterministic approach is also a key component in non-preventive defence in depth. As human and material weaknesses cannot all be eliminated, safety measures must detect them in good time and stop them. They must be prevented from cascading into a core melt. Lastly, as the worst cannot always be avoided, safety must plan and organize action to contain the accident and limit its impact, for example by setting up rapid-action forces or taking measures to evacuate residents. For thirty or forty years nuclear engineers have debated, designed and worked out in detail these three aspects of defence in depth – prevention, supervision and action.

That safety is primarily the concern of engineers does not mean it should be their exclusive preserve. On the contrary, safety should involve many other trades: physicians, psychologists, public-safety specialists, experts in crisis-management and logistics, among others. Technicians, team leaders and managers play a key role in accident prevention inside nuclear power plants. There is now broad consensus regarding the crucial importance of organization and the human factor in risk management. Engineers may lack the necessary competencies in these fields, much like economists, so we shall not venture any further down this path. On the other hand, safety comes at a cost and must be regulated, two domains on which economic analysis has focused for many years. It seeks to clarify the ends and means of implementing regulation, and to characterize good and bad regulatory practices.

## Japanese regulation, an example to avoid

I must start by apologizing to my Japanese friends for being so blunt, but the serious shortcomings of nuclear-safety regulation in Japan reveal several essential basic principles.[2] In a word, the regulator must be independent, competent, transparent and powerful. These four characteristics have in the past all been cruelly lacking in Japan's Nuclear Industrial Safety Agency.

The shortcomings of nuclear regulation in Japan go back a long way, but it took the accident at Fukushima Daiichi to bring them to public attention. Specialists in nuclear safety and reactor operations were fully aware of these weaknesses. A scientific journalist recalls the following exchange, in the course of a conversation long ago with the head of a foreign safety agency.[3] 'So, where do you think the next nuclear accident will happen?' 'In Japan.' 'Why?' 'Because their system of control and supervision of nuclear safety is not up to scratch, and their safety authority isn't independent enough. On the contrary it is governed, to too great an extent, by political, industrial and financial priorities. I don't trust Japan's nuclear-safety system.'

In June 2007 a mission comprising a dozen foreign experts in safety regulation visited Nisa to carry out a peer review of its activities. Such missions are frequent. They are organized by the International Atomic Energy Agency at the request of governments and national authorities. Specialists refer to the process as the International Regulatory Review Service. The visit to Japan gave rise to a public report which gave the impression that the Japanese regulator was doing a pretty good job. Good marks were awarded, substantial progress was underlined. There was however some criticism, barely perceptible to anyone not familiar with IRRS reports on other countries. It noted, for instance, that some safety incidents were not followed by corrective measures due to 'limited use of resources for evaluating operating experience, and lack of systematic inspection and enforcement of licensees' activities by Nisa'. Those who took part in the mission complied with the rules of Japanese courtesy, which make it difficult to voice criticism, both face-to-face and in a public report.

The Fukushima Daiichi disaster uncovered a whole series of recurrent, serious misdemeanours committed by various operators. This information, which had previously escaped notice, revealed gaping holes in the supervision carried out by the safety regulator. Seven out of Japan's ten generating companies admitted to having knowingly deceived the regulator, falsifying the results of safety tests and reports on maintenance and repairs on their nuclear power plants.[4] Among their number, Tepco, the operator of the Fukushima Daiichi plant, was the most frequent offender.[5] In 2002 Nisa revealed that Tepco had omitted to report a whole series of faults observed on its containment buildings.[6] To begin with, the operator admitted to twenty-nine falsifications, finally owning up to about 200 between 1977 and 2002. In 2006 it admitted having censored primary-cooling-system temperature recordings. In 2007, when an accident occurred at the Kashiwazaki-Kariwa plant, Tepco initially denied that any radioactivity had been released, subsequently admitting that several thousand litres of contaminated water had been discharged into the sea. The same year the firm acknowledged having failed to report six emergency shutdowns at the Fukushima Daiichi plant. The *Japan Times* has alleged that just ten days before the disaster on 11 March 2011 Tepco sent yet another falsified report to the regulator.

Fukushima Daiichi also revealed the scale of conflicts of interest in Japan's nuclear village, highlighting shortcomings, inconsistencies and failures to implement safety regulations.

## Revolving doors and failures

Revolving doors are common practice in the Japanese administration. For officials who transfer to industry, this process is known as *amakudari*, or descent from heaven. Much as the Shinto gods coming down to Earth, senior public-sector officials at the end of their career are invited to take up honorary or executive positions in business. In the past nearly seventy high-ranking Ministry of Economy, Trade and Industry (Meti) officials have joined the senior management of generating companies.[7] In 2011 thirteen of them were on the boards of various power operators. The *New York Times* reported that there was a position as vice-president at Tepco reserved as a retirement sinecure for a top-grade bureaucrat.[8] The head of the Agency for Natural Resources and Energy was the last official who was supposed to enjoy this benefit. The agency is a branch of Meti; the Japanese safety authority, Nisa, is affiliated to it, its budget and staff being dependent on it. ANRE Director-General Toru Ishida moved to Tepco in January 2011, initially working as an advisor before obtaining the reserved position as vice-president. The Fukushima Daiichi disaster derailed this plan, bringing the practice to public notice. Mr Ishida was forced to resign a month later.[9] Amakudari helps utilities stay on good terms with officials in charge of energy and obtain favourable decisions. The prospect of a prestigious, well-paid job is a sufficient incentive to convince some people to do the occasional favour. Here is just one example. In 2000, as part of Tepco's in-house inspections, a General Electric engineer, Kei Sugaoka, discovered that Tepco had falsified data. He reported this to the ministry in charge of energy and nuclear safety. For two years he received no response, and his warning was not followed up. All that happened was that Meti passed on his name to Tepco.

Fukushima Daiichi brought to light a large number of failures in safety regulation. They were detailed in the very thorough reports on two inquiries, one conducted by the government,[10] the other by the Diet (parliament).[11] Published after the accident, they are most instructive. Their analysis of the disaster draws attention to the mistakes, negligence and lack of preparation of all parties. None of the government, the civil service, or Tepco are spared. Point by point the two reports show how their respective shortcomings led to loss of control over the nuclear power plant and disastrous mismanagement of the ensuing crisis. Both reports also focus on the underlying causes: backtracking over the main decisions on safety regulation, particularly measures to protect against earthquakes, flooding and station blackout. For safety regulation in Japan it was typical for changes in the regulations either to be scrapped altogether, amended to reduce their impact, delayed or never implemented. Three brief illustrations will suffice.

Shortly after it was set up, Nisa was supposed to frame a baseline document charting long-term changes in regulatory practice. In the end the report was only written and published ten years later.[12] The Guideline for the Reactor Site Evaluation, which was established in 1964, is still in place regarding construction permits for nuclear power plants.[13] In the US and most other countries nuclear power plants must be capable of coping with an eight-hour outage. This threshold was set following numerous incidents on the US fleet in the 1980s. In Japan the long-standing requirement for plant designers and operators was thirty minutes. Various working groups addressed and reported on the issue, ultimately concluding that there was no need for it to be changed.[14]

The anecdotes in the box on 'revolving doors' are appalling, but they are all due to the way safety regulation is organized in Japan. Even on paper the institutional system there cannot work, for the simple reason that Nisa lacks the full attributes of a nuclear safety authority. It only carries out part of the necessary regulation work, enjoys no real independence from policy-makers, and also supervises other hazardous industrial facilities. In this respect it cannot be compared to the modern model embodied by the NRC in the US, or the Autorité de Sûreté Nucléaire in France. Safety regulation in Japan is spread over several bodies. Nisa works alongside the Nuclear Safety Commission.[15] The latter reports directly to the Prime Minister and is supposed to supervise and check up on Nisa. But the NSC is not really empowered to investigate or sanction Nisa. The NSC is also supposed to duplicate some of its work, in particular carrying out inquiries before issuing operating licences, and framing regulatory directives and guidelines. However, it has no mandate to investigate operators. This 'reserve' regulator is headed by five commissioners, assisted by several hundred officials. Inspections, which are a key component of safety, are partly carried out, by legal provision, by a third body, the Japan Nuclear Energy Safety Organization (JNES). Nisa entrusts work to this technical support department, but has no authority over it. As the IRRS mission to Japan briefly noted, all the functions and responsibilities expected of a safety authority are present in the Japanese regulatory framework, but they are unfortunately dispersed. Their responsibilities are consequently diluted and pooling of information is inefficient.

Nisa is not strictly speaking an independent authority, rather an administrative body that is part of ANRE, which in turn is a branch of Meti, a powerful ministry. Nisa does not have its own budget, nor even autonomous personnel. There is no *de facto* separation from the organization tasked with energy and the economy. Indeed, most of its staff belong to ANRE. It is commonplace for personnel to move from an office supervising safety regulation to Meti departments tasked with nuclear affairs. Others transfer to or from the Ministry of Education, Science and Technology. Exactly the same situation prevails at the NSC, where staff are managed in a similar way. The NSC reports to the Cabinet, which decides policy on energy and trade. This institutional set-up does not guarantee an airtight seal between safety regulation and the promotion of nuclear power. Moreover, the lack of any legislation requiring complete transparency in the work of the three safety organizations – NSC, Nisa and JNES – exacerbates the problem. Very few of the discussions, decisions, reports, minutes, inspections and projects are brought to the attention of the general public, the media and local authorities. There is no way for them to monitor progress on regulatory work and its results, or to determine whether efforts are influenced by political imperatives or interest groups. It all goes on behind closed doors.

No legislative or judicial power counterbalances the pressure on safety exerted by the executive. In Japan the government appoints Supreme Court judges. In turn the Supreme Court supervises the office in charge of the team tasked with assessing, promoting and sanctioning less senior judges. The same office, itself made up of judges, vets all job allocations. According to J. Mark Ramseyer, Professor of Japanese legal studies at Harvard Law School, the appointment system in Japan ensures that sensitive cases will be settled, even in courts of first instance, in line with the political priorities set by the government.[16] As a former judge at the district court explained to the *New York Times*,[17] 'Judges are less likely to invite criticism by siding and erring with the government than by sympathizing and erring with a small group of experts.' District courts have received about fifteen complaints about safety defects, due in particular to the proximity of nuclear power plants to seismic fault lines and non-compliance with prevailing standards. The cases were lodged by local opponents to projects to build new reactors or restart existing ones. In only two cases did the courts rule in favour of the plaintiffs, but their decision was overturned by the appeal court.

The Diet does not counterbalance government influence over regulation either. Japan is a parliamentary democracy and the prime minister is appointed by the lower chamber, of which he or she is a member. In practice the leader of the party holding a majority is elected prime minister. As the largest party in many majorities, the Liberal Democrats (LDP), held power almost without interruption from 1955 to 2009. It dominated the various assemblies and ruling coalitions. A separation exists between executive and legislative powers, but in political terms, they have marched in step for decades. In a democratic regime, parliament controls the authorities tasked with regulating network industries and public health. It is an essential characteristic of nuclear safety agencies, because their work extends into the long term, a responsibility which is generally better suited to law-makers than the executive. Japan is an exception in this respect, the Diet exerting no control over Nisa's work. On the contrary, the latter body reports to the NSC, itself answerable to the prime minister.

## The Japanese regulator is a captive to industry

There are two sides to the independence enjoyed by a safety authority. We have just examined its position in relation to the executive. Our concern here is to put regulation out of the reach of the many, shifting goals which may interfere with nuclear safety, priorities such as maintaining living standards thanks to the price of electricity, or sustaining the trade balance by exporting nuclear materials and technology. The other, even more crucial, side hinges on independence from industry, primarily engineering firms and operators. The regulator must not adjust its actions to suit the interests of the companies it regulates. This would produce the same level of safety as with self-regulation, which is inadequate even under an unlimited-liability regime.

Since the work of George J. Stigler in the 1960s, economic theory has taken an interest in the risk of the regulator acting in the interest of those it regulates. This University of Chicago economist discovered an astonishing phenomenon. The price of electricity in US states where it was set by the regulator was no lower than in states where utilities were free to decide for themselves. Stigler then asked two questions, previously disregarded by economists, blinded by a conciliatory view of regulation in the general interest: who benefits from regulation? And why? His answer was that 'regulation is acquired by the industry and is designed and operated primarily for its benefits'.[18] To explain this bias he cited the dominant position enjoyed by industry in relation to other interest groups. Taxpayers, consumers or indeed people who suffer damages are more numerous than companies, yet they have fewer resources. It is more difficult for them to form a collective group. The theory of regulatory capture took shape with these preliminary ideas, but it has since progressed. We shall return to this point later, but in the meantime this sketchy account should be sufficient to understand relations between Japan's safety regulator and the operators of nuclear power plants, relations which the report on the Diet investigation rightly referred to as 'incestuous'.[19]

Questioned by members of the Diet, the NSC head Haruki Madarame said:

> The nuclear safety regulations until today have been based on a convoy system of the regulatory authority and utilities. The utilities proposed the least expensive safety standards, which in turn were approved by the authorities. This led to a vicious cycle in which the utilities did nothing and justified their inaction by claiming that the government had approved the safety standards.

He admitted that there are consequently many shortcomings in safety guidelines. 'Though global safety standards kept on improving, we wasted our time coming up with excuses for why Japan didn't need to bother meeting them', Mr. Madarame said.[20]

The Japanese regulator adjusted its behaviour to suit operators' concerns: witness the lamentable discussions between Nisa and the industry on preventive action on major accidents. They started in 2007, following the IRRS mission. The aim was to consider a new regulation specifying preventive measures required to meet international standards. The position of the Federation of Electric Power Companies was perfectly clear. It may be summed up in three points: avoid work to upgrade plants which would interrupt their operation; prevent stricter requirements which would force reactors to be shut down; and ensure that the regulation would not have a negative impact on current or subsequent litigation. The regulator accepted these imperatives and concluded that safety levels were adequate. The head of Nisa wrapped up the talks by saying that he shared industry's concern about litigation and its impact on output. He added that it might be necessary to convene a formal committee, but that first he wanted to settle the matter in hand. Ultimately these talks led to no changes until the accident at Fukushima Daiichi, and without the ensuing investigation no one would have been any the wiser.

Obviously there is nothing inevitable about regulatory capture by industry. As we have a better understanding of this process than in Stigler's day, we now have more effective solutions to counter it. The predominant influence of industry, compared to other interest groups, is due to unequal access to regulation. The financial resources at industry's disposal naturally spring to mind. But setting aside means, economic theory shows that the representation of interests is systematically distorted in favour of small homogeneous groups comprising potential losers.[21] Their homogeneity makes it easier to form a cohesive group; their small number is a powerful incentive for each member to act, and also makes it easier to identify those shirking their obligations; losers are more pugnacious than winners due to the endowment effect. You may recall the example, cited in Chapter 5, of mugs being randomly handed out to half the students in a class. The lucky ones set a higher selling price on their mugs, than their unfortunate counterparts were prepared to pay. But the authorities are empowered to deal with this disparity between interest groups: they can broadcast information widely, appeal for expressions of interest, invite stakeholders to make themselves known, officially recognize certain groups, make consultation procedures compulsory, encourage independent expert appraisals. In France there is a legal requirement for a local information committee to be set up for each nuclear power plant, with guidelines for its composition. It comprises policy-makers and the representatives of various organizations: environmental watchdogs, trade federations, chambers of commerce, public health bodies, among others. Officials and the operator take part in meetings and committee work, but only in an advisory capacity. France's Autorité de Sûreté Nucléaire supports the committees financially and ensures that the information they receive is as comprehensive as possible. It also encourages committee members to take part in plant inspections and to resort to outside expertise, distinct from advice offered by itself or the operator.[22]

Regarding regulation, industry draws its strength from the information advantage it enjoys, a situation encapsulated in the information asymmetry concept, which we introduced in Chapter 7. Regarding economic regulation, such as setting the tariff for using a transport or telecommunications network, the solution is to pay a company to reveal its real costs. Similarly to bonuses given to CIOs to ensure that they genuinely uphold shareholder interests, it is necessary to award a bonus to the most successful regulated companies. This principle – which is complex to implement and rarely ideal – is designed to prevent companies from inflating their costs, which saps productivity. On the one hand the regulator loses, by awarding the company a rent in exchange for the information it discloses, but on the other hand it has the benefit of a lower tariff. Overall the net gain for society is positive. In the case of nuclear regulation, the regulator also tries to avoid excessively high safety costs. But this is a secondary consideration. Above all it seeks to prevent operators from lying about their shortcomings. The nuclear safety authority must know whether incidents occur, whether their cause has been identified, and whether corrective measures have been taken. It must be in a position to assess a plant's safety performance and check compliance with its requirements. Far from rewarding regulated companies, it is simply a matter of monitoring and sanctioning them in the event of deficiencies. To know, to inspect and to sanction are the three priorities for safety regulation. Which is why the safety authority must have competent staff and the means to enforce the dissuasive sanctions at its disposal. A nuclear safety regulator which has neither top-notch technical and scientific expertise nor real powers will not achieve a level of safety significantly better than self-regulation.

To achieve effective safety regulation, powers of inspection and sanction, underpinned by the requisite skills, are certainly necessary but not sufficient in themselves. The regulator must be prepared to exercise such powers too. In modern economic theory, the regulator is an economic agent much as any other: it is moved by its own interests and pursues an individual goal – which is not spontaneously in the general interest, contrary to the model presented by classical public economics. It will act in the general interest – in more concrete terms it will try to fulfil its legally defined mandate – providing it is in someone's personal interest to do so: for example, to advance one's career, for fear of sanction, to enjoy the benefits of a larger or more highly qualified team, to occupy more comfortable premises, among others. In short, a regulator must also be supervised and given incentives. Its own interests must be brought into line with that of the law-maker, distancing it from that of the regulated. To this end a whole range of instruments may be deployed to combat corruption and conflicts of interest (banning revolving doors in regulated companies, operating as a collegial body, making it compulsory to report any meeting with any company or its management).

In conclusion, Japanese regulation failed because it was trapped by industry in a pincer movement. On the one hand this was a classic case of a regulator allowing itself to be captured by the utilities it was supposed to be regulating. But in another respect it was more unusual, because the authority's lack of independence from government also brought it under the influence of industry. The political system in Japan is tightly meshed with the interests of utilities, energy-intensive industries and nuclear engineering firms. So written between the lines of the Nisa mandate was the imperative neither to hinder electricity production nor yet to impede promotion of nuclear power. Safety was never its prime concern, less still the main focus of its action. It acted as a fully paid-up member of the *genshiryoku mura*, or nuclear village, an expression hatched to describe the select club formed by the stakeholders in Japan's nuclear industry: businesses, but also the relevant administrative bodies and scientists, and of course policy-makers. The village concept here reflects the loyalty owed by members to their common cause. Nisa was built in the village, not outside.

NINE

# What goal should be set for safety and how is it to be attained?

The above review of safety regulation in Japan illustrates the recommendations of economic analysis on how an authority should be organized, and how such prescriptions are justified. But economic analysis also focuses on regulatory goals and instruments. In this section we shall draw on the examples of regulation in the United States and France. Naturally they both have their shortcomings and they are very different, yet both may be taken as good models.

Let us start with a reminder of the economic principles guiding the setting of goals and the choice of instruments. Economics is useful for discarding hollow slogans such as 'nuclear safety is our absolute priority' or 'nuclear safety is priceless'. Catchphrases of this sort suggest that there is not, or there should not be, any limit to safety efforts.

The first limit established by economic theory relates to determination of the optimal level of safety efforts on the basis of marginal costs. The incomplete nature of data reduces the practical value of this response to the question of when safe is safe enough. Very little is known about the marginal costs of safety efforts and even greater uncertainty affects the marginal costs of avoided damages. However, this is not specific to nuclear accidents. Economists have difficulty determining the optimal level of carbon-dioxide emissions or safety on oil-rigs. It is equally tricky to estimate the right discount rate, or put a value on loss of human life or damage to the natural environment. In practice economic calculation very rarely has any say in setting health and safety goals, generally decided by scientific and technical debate, and political process.

The second limit relates to the maximum level of safety efforts, which results from the law of substitution. From an economic standpoint there can be no absolute measurement of the value of a technology; it can only be estimated in relation to competing technologies. In the case, for example, of two rival technologies, the value of the better one is equal to the gain it procures for society compared to the less efficient technology. Over the long term the market achieves a form of Darwinian selection, discarding less successful technologies. The natural economic limit on safety efforts is the one which forces nuclear power out of the market because it has become too expensive compared to alternative technologies. This limit becomes apparent when attempts are made to extend the service life of ageing reactors. If the additional safety work required by the regulator to extend its licence exceeds a certain threshold, operators stand to gain from investing in new means of production.

Having set a goal for regulation, economic analysis then looks at how it can be achieved at the lowest cost. The value of this approach is obvious: savings are made by minimizing expenditure to achieve a given level of polluting emissions or a given safety objective. The money thus saved can be used to mitigate even more pollution or increase safety further – doing more with the same amount – or it can be used for something else.

To minimize costs, economic analysis recommends the use of standards based on performance rather than technology. The latter type of standard focuses on the characteristics of an installation, such as the thickness of concrete, or the number and start-up time of emergency pumps. The regulator prescribes standards and checks compliance. With performance-based standards the regulator simply defines criteria and levels, for example the resistance of a reinforced concrete containment to the start of a core melt, or the longest acceptable duration of a failure of the secondary cooling system. It is up to the regulated party to choose the right measures to achieve the necessary performance level. This provides an incentive for the nuclear operator to select and fine-tune effective, less costly measures, reducing expenditure and increasing margin. From the point of view of the general interest, performance-based standards have the dual advantage of reducing information costs and protecting innovation. By delegating the choice of the appropriate component or device to the operator, the regulator spares itself the effort of collecting information and expertise. But it is important to understand that it is not simply transferring costs to the operator. The operator incurs lower costs because it already has the information. Furthermore if it can see ways of achieving performance goals by developing new processes, it will invest in R&D work, which may ultimately benefit the community.

The story of nuclear regulation started without any explicit goal. It has gradually taken shape, alongside the design and construction of nuclear power plants. Means having preceded ends, we shall start by looking at the issue of standards before that of goals, proceeding in chronological order.

## Technology versus performance-based standards in the United States

Performance-based standards, guided by probabilistic safety assessments, now play a central role in nuclear regulation in the US, the goal being to manage and contain costs. But for a long while the balance tipped in the opposite direction and nuclear safety in the US was beset by mountains of prescriptive regulations. Between 1967 and 1977 the pace of annual publication of regulations, implementation guidelines and other documents setting forth regulatory doctrine increased by a factor of twenty-five. In the early 1970s there were less than ten such publications; by the end of the decade the full catalogue contained almost 150 items.[1] Indeed regulatory inflation and its exclusively prescriptive focus, in terms of technical options and details, were instrumental in slowing down the construction process (see Chapter 1).

Slow progress is a prime characteristic of regulation by technology-based standards. The pitfalls associated with the application of the Nuclear Regulatory Commission's fire-protection standards are symptomatic of this approach. In 2010 there were still almost fifty reactors which did not comply with these regulations. The standards were drawn up after a fire at Browns Ferry nuclear power plant, which rapidly went out of control, putting reactor 1 out of service for more than a year. Setting aside Three Mile Island, this failure was certainly the most severe yet to occur in the US. It is well known because the event that triggered it seems so absurd only a newspaper cartoonist could have imagined such a tale: a candle flame set light to polyurethane foam sealing material, propagating from there to the adjoining electrical cables. This occurred in the cable-spreading room, below the control room, through which control cables passed on their way to the reactor building. Two electricians were applying inflammable foam to stop air leaks from the adjoining room. To check the seal round the cables one of the electricians lit a candle and placed it beside the stopped holes. Coming too close it set light to the foam. The fire then spread into the cables and from there all the way down the line. A sequence of errors followed. Many safety systems were damaged, including the cooling system, and a core-melt accident was narrowly avoided.

To prevent similar accidents the NRC set out to establish new standards. One required operators to separate the circuit supplying the primary cooling system from its secondary counterpart by six metres (at Browns Ferry they were almost side-by-side). Failing this, operators must fit automatic fire-suppression systems, fire barriers or fireproof cladding that would protect cables for three hours. The operators thought these measures were excessive and extremely expensive. A series of legal proceedings started, accompanied by protracted hostilities between operators and regulator. In 1982 the Connecticut Light and Power Company went on the offensive, claiming that the NRC had failed to provide adequate technical justification for its fire-protection regulations. The District of Columbia Court of Appeals ruled in favour of the safety authority. Some operators opted to apply only part of the new regulations, others not at all. In the face of this refusal to comply with its demands the NRC finally announced at the start of the 2000s that it would no longer tolerate any violation of its standards. However the industry managed to have the deadline postponed, until new regulations came into force in 2004. They have still not been applied. Deadlines have been extended several times. The most recent one requires compliance by 2016. If all the parties fulfil this requirement it will have taken almost forty years for the fire-protection regulations to be deployed on the entire US fleet.

My grasp of the technicalities of nuclear safety are such that I cannot pass judgement on the validity of the 1980 fire-protection standards. Were the measures impracticable, at least in some cases? Was their cost excessive compared to the avoided risk? Were all the delays the fault of a few operators, determined to limit costs, which forced the regulator to retreat? Whatever the answer, this example is emblematic of the dead end to which nuclear safety regulation leads when it is based exclusively on detailed technical prescriptions. The complexity and diversity of reactors is so great that it is necessary to draw up huge catalogues of standards. The cost of their *ex post* verification comes on top of the expense of framing the standards. The cost of a single check is low, but as they pile up, it soars. Take for example the building industry, with its myriad standards, which confuse both operators and regulator: it is impossible for all the rules to be applied, let alone checked. And even if they were, there would still be a major problem: technology-based standards induce a passive attitude to safety in the industry, dispelling any sense of responsibility. In other words, compliance with regulations is no guarantee of safety. This was one of the key lessons learnt from Three Mile Island: full compliance with the detailed technical rules would have done nothing to prevent the accident.

Awareness of the many shortcomings of technology-based standards made it easier to switch to a new approach to regulation based on performance. Its development was fuelled by the boom in probabilistic safety methods and the adoption, in 1986, of the first quantitative safety goals. Risk-informed and performance-based regulation, better known by its acronym RIPBR, has since taken over as the dominant approach.[2]

A regulatory guide published in 2002 perfectly illustrates the new approach, describing how the NRC assesses 'licensee requests for changes to a plant's licensing basis'.[3] The guide starts by recounting the US authority's convictions regarding the merits of probabilistic risk assessments. Such studies enable the authority to improve safety, but also to better direct its efforts and reduce pointless regulatory constraints on operators. As Richard A. Meserve, NRC Chairman when the guide was published, explains, 'The aim, of course, is to use risk as the tool for dissecting and reforming our regulatory system so that the NRC focuses on risk-significant activities, thereby both enhancing safety and reducing needless regulatory burden.'[4] The guide then explains the authority's attitude to requests by operators for changes, according to the reactor's overall level of safety performance and how any changes may alter such performance. Core-damage frequency and large-early-release frequency values are used to define risks. But they are also used to characterize various situations when the licensing-basis change requested by an operator may entail 'safety degradation'.[5] The key idea is that the NRC may accept a change leading to a slight incremental risk providing the plant's overall safety performance is high, but not if it is poor. For example, if the change increases the core-damage frequency by less than $10^{-6}$ per reactor-year, and the overall frequency is lower than $10^{-5}$ per reactor-year, the licensing-basis change is acceptable. On the other hand, it is not acceptable if the total core-damage frequency is close to or higher than $10^{-4}$ per reactor-year. Several areas of acceptance or refusal are thus defined depending on how the results of the probabilistic risk assessments compare with the various thresholds.

George E. Apostolakis was sworn in as an NRC Commissioner in 2010, for a four-year term. Professor of Nuclear Science and Engineering at the Massachusetts Institute of Technology and a keen advocate of quantitative risk assessment, Apostolakis maintains that this guide constitutes a breakthrough. Probabilistic risk assessments are no longer used simply to pinpoint negative features, revealing previously unidentified failures. They also 'pay attention to the "positive" insights, i.e. that some of the previously imposed safety requirements can be relaxed because either they do not contribute to safety or they contribute a very small amount that cannot be justified when it is compared to the corresponding cost'.[6]

However, performance-based nuclear-safety regulation is by no means a panacea. Much of the performance can only be observed indirectly, by probabilistic means. It may not actually be achieved and it is not always possible to measure exactly and rigorously the uncertainty affecting it. Of course, verification becomes easier with time. It will be possible to confirm or invalidate some aspects of safety performance measured during the reactor-design phase, once it starts operating. Much as with forensic medicine, dismantling reactors and exploring their entrails will also provide new evidence as to the weakness or solidity of certain components. But even this ultimate phase will never be sufficient to reconstruct the whole picture.

The shortcomings in the verification of performance mean that regulation based exclusively on performance must be ruled out. Even in the US, which has gone furthest down this path, regulation still has a dual basis. Let us return to the most recent episode in the fire-prevention saga, when a new regulation was adopted in 2004. Rather than revising its technology-based standards, the authority proposed to take a risk-informed, performance-based approach. The operators of the forty-seven reactors which are still not compliant must either opt to apply the standards or to take ad hoc measures on the basis of fire-propagation models and probabilistic safety assessments. The latter option enables them to concentrate on the parts of the plant which display the highest risks. The vast majority of the operators have indeed taken the second route. As a result some US nuclear power plants comply with certain fire-prevention rules, whereas others disregard them. Some operators are actively committed to risk-informed regulations, whereas others are more 'passive' in their attitude to the new approach.[7] Technology and performance-based standards now coexist in the US and are sometimes combined. Indeed the prevalent model is hybrid. As we shall now see, the same applies to its objectives, with US nuclear safety regulation pursuing both qualitative and quantitative goals.

## Choosing a goal for overall safety: words and figures

On 11 July 1977 the French Industry Minister, tasked with nuclear safety, sent a letter to EDF mentioning for the first time a quantitative safety goal: the overall probability of a reactor initiating unacceptable consequences must not exceed $10^{-6}$ per reactor-year.[8] On 21 August 1986, after talks lasting more than five years, the NRC issued a statement setting an upper limit on any additional health risk to members of the community in the vicinity of a nuclear power plant. The additional risk should not exceed one-in-a-thousand of the fatality risks resulting from other causes. A comparison of these two dates would suggest that France was quick to adopt quantitative safety goals. In fact the minister's letter has still not come into effect, whereas the US has steadily increased its reliance on quantitative goals for safety regulation.

The 1986 NRC statement followed the recommendations of the Commission on the Accident at Three Mile Island, also known as the Kennedy Commission. It set four objectives, two qualitative, two quantitative. We shall cite all four, as they illustrate key principles:

> Individual members of the public should be provided a level of protection from the consequences of nuclear power plant operation such that individuals bear no significant additional risk to life and health;
>
> Societal risks to life and health from nuclear power plant operation should be comparable to or less than the risks of generating electricity by viable competing technologies and should not be a significant addition to other societal risks;
>
> The risk to an average individual in the vicinity of a nuclear power plant of prompt fatalities that might result from reactor accidents should not exceed one-tenth of one percent (0.1%) of the sum of prompt fatality risks resulting from other accidents to which members of the US population are generally exposed;
>
> The risk to the population in the area near a nuclear power plant of cancer fatalities that might result from nuclear power plant operation should not exceed one-tenth of one percent (0.1%) of the sum of cancer fatality risks resulting from all other causes.[9]

These principles are based on good common sense. In a word, nuclear risk must not diverge from the norm. To start with, this may be compared with other health hazards in general: for people living close to power plants, who are obviously more exposed, the acceptable risk – with a ratio of one-in-a-thousand (one-tenth of 1%) – scarcely exceeds the normal level of risk. It may then be compared with the hazards involved in alternative electricity-generating technologies, in particular coal: the safety level required of nuclear power should neither favour nor handicap this technology.

Looking more closely, this list of goals also raises many questions: why is there no mention of costs or risks other than those affecting public health? Why are quantitative and qualitative goals mixed up?

Are these goals in any sense legally binding? Why were they issued by the regulator, not the legislator? We shall start by providing precise answers to these questions, with contextual details reflecting the specificities of US safety regulation. Then we shall offer some more general, analytical answers.

The four goals introduced in 1986 are still top priorities for safety regulation in the US. As fundamental principles they answer the question of what safe enough really means. Yet they only address the nuclear hazards affecting public health and are a poor translation of the NRC mission: 'to licence and regulate the Nation's civilian use of radioactive materials to protect public health and [...] protect the environment'.[10] In a press conference in March 2012 NRC Chairman Gregory Jaczko regretted the fact that safety goals did not deal with 'significant land contamination and displacement, perhaps permanently, of people from their homes and their livelihoods'.[11] He expressed the hope that they would be changed to ensure that, under the US regulatory framework, an accident such as the one at Fukushima Daiichi would be unacceptable, even if, as he emphasized, radiation had not had any immediate health impact. At present such impacts are, in a sense, acceptable, because the four goals established by the NRC focus exclusively on risks to public health. While on the matter of oversights, it should be noted that this answer to the question of what safe enough really means completely overlooks the concept of cost or savings. The choice of the one-in-a-thousand ratio is arbitrary. It is not the result of any cost assessment and less still of any cost–benefit analysis. Furthermore, the ratio is the same for all the nuclear power plants, identical for Palo Verde, in the Arizona desert, and Zion, close to Chicago. No allowance is made for population density around reactors, whereas the economic benefit of safety efforts substantially increases with the size of the local population protected.

Could the NRC have restricted its list to just the first two basic goals? Residents living close to nuclear power plants are a subset of the US population. As such their health is protected too. Of course, there is no question of them enjoying less protection than other Americans. It would be a mistake to suppose that the specific mortality thresholds concerning them are appreciable. These thresholds in fact translate into figures the idea of not being 'a significant addition' to existing risks, asserted in the qualitative goals. Without some firm basis, any attempt to gauge the extent of a phenomenon, be it significant or not, is either quite impossible or simply subjective. In the present case the figures provide a firm basis. Given the annual rate of accidental death and the annual rate of death by cancer in the US, the limits placed on additional deaths among local residents due to nuclear power plant operations are, respectively, $5 \times 10^{-7}$ and $2 \times 10^{-6}$, or 5 and 20 persons per 10 million residents. According to the NRC, if the figures were any higher they would constitute a significant addition to the risks affecting local residents and as such would not be acceptable.

Introducing quantitative goals also reflects the concern, following Three Mile Island, to make the goals pursued by the regulator more explicit. The 1986 statement was preceded by a large number of reports and workshops devoted to safety goals, focusing in particular on choosing quantitative criteria to specify them. These criteria seemed necessary to clarify terms used to qualify risks, such as 'unreasonable', 'undue', 'intolerable' or indeed 'unacceptable'.[12] They were consequently seen as supplementing, rather than replacing, qualitative goals, it not being possible to quantify all aspects of safety. When it was possible, safety was still subject to great uncertainty. The NRC took care to stipulate that selected quantitative goals could not form the exclusive, nor even prime, basis for decisions on granting licences or on regulations. Moreover, it soon emerged that the goals must be underpinned by subsidiary benchmarks. The additional mortality thresholds proved difficult to use as a basis for standards and practical rules implemented by operators. In a memorandum dating from 1990 the NRC introduced two new limits: a core-damage probability lower than $10^{-4}$ per reactor-year and an overall mean frequency of a large release of radioactive materials to the environment of less than $10^{-6}$.[13]

The NRC set the subsidiary benchmarks, just as it did for the fundamental goals. They were adopted by the five commissioners (who make up the Commission). So the goals were not prescribed by Congress. The fundamental goals set in 1986, in particular the requirement that there be no significant additional risk to life, did not reflect any legal obligation imposed by Congress. It is also worth noting that the NRC decided not to make its fundamental and subsidiary goals binding. It cannot sanction operators which fail to comply with them. It sees the goals as guidelines, enabling it to frame tighter standards, which may on the contrary be binding.

The reader may be surprised by this approach, but Congress lacks the necessary technical expertise to set a quantitative goal. It would have to allocate substantial resources to determining the right thresholds. Moreover, this effort would have to be repeated at each update, the thresholds being likely to vary over time in line with changes in the understanding of accident risks. Qualitative goals, such as a 'tolerable' or 'reasonable' risk, are obviously of a more permanent nature. The legislator may also stipulate how the regulator should go about setting quantitative goals, according to what procedure, striking a balance in this way between contradictory factors. It may also check *ex post* that the regulator's goals do really correspond to its wishes. In a further example of savings on information costs and labour division, the regulator may ask its administrative services to draft proposals for possible quantitative goals, selecting the ones which best reflect the legislator's concerns and wishes; then ask its staff to transform the goals into operational rules; finally checking that they are really being implemented.

The combination of quantitative and qualitative goals is justified by their respective shortcomings and advantages. Without quantitative goals the regulator cannot separately assess how efficiently an operator achieves a particular goal, and at the same time how effectively it minimizes the cost entailed. If such and such a safety expense is pinpointed by an operator, is it trying to achieve the goal or just concealing its inability to identify the most cost-effective means? The regulator needs to know whether achieving a goal is very expensive, or whether the operator is artificially inflating the cost by wasting resources! This is clearly impossible if the stated goal is a 'tolerable' or 'reasonable' level of safety. Without quantitative goals the regulator enjoys greater discretionary powers. Obliging it to set a threshold for a particular risk is tantamount to tying its hands. It will be less at liberty to demand over-ambitious safety efforts, yielding to pressure from environmental pressure groups, or alternatively efforts that are too feeble to satisfy industry. There is less danger of regulatory capture by interest groups. Lastly, without quantitative goals it is difficult to put technologies on an equal footing. Thanks to quantitative risk thresholds safety performance in the nuclear industry can be compared to that of other hazardous activities, in particular coal- and gas-powered energy.

But setting quantitative goals does not solve all the problems either, so by necessity they must be backed up by their qualitative counterparts. Defining a risk threshold leads to a focusing effect. Concerned with a particular figure, the administrative services of both regulator and operator may disregard other goals, despite their importance for safety, such as staff training or new knowledge production. It is impossible to quantify many of the factors determining nuclear risk and the corresponding mitigating measures. So there is a danger of making insufficient allowance for them. Above all, verification problems are a further obstacle to meeting goals regarding additional mortality or the probability of large-scale contamination. It is not just a question of checking the speed of motor vehicles on a road. There are no radar speed-guns, with a low margin of error, for measuring nuclear risk. The limit here is a probability and it cannot be observed. Nor is it always easy to interpret. For example, a threshold corresponding to a core-melt probability of less than, say, one-in-ten-thousand per reactor-year seems at first sight to be clearly defined. But what does core melt mean? Slight damage to the zirconium-based cladding round the fuel rods? Partial meltdown? The formation of corium, a molten mixture of parts of the nuclear-reactor core? And interpreting the threshold itself is no easy task. Does it correspond to a mathematical expectation? Or is it an interval, such as the core-melt probability must be, with a nine-in-ten likelihood, lower than the limit of one accident in 10,000 reactor-years? The issue of whether it is possible to verify the goal persists even after an accident. The date of a failure within the relevant interval cannot be anticipated on the basis of the calculated core-melt frequency. If an accident occurs during the first few years of a fleet's operation, this does not mean that the limit has been passed, no more than if the accident had occurred much later.

Much as for performance-based standards, the limited scope for verifying quantitative goals explains why they are not legally binding. The regulator does not require the operator to prove that it has achieved a goal and is not in a position to sanction it in the event of non-compliance. Quantitative goals, in particular core-melt frequencies, which are the most commonly used goals of this sort, are used as guidelines rather than as benchmarks. The regulator uses them to focus the action of its administrative services, which in turn use them to direct operators' safety efforts.

The only binding quantitative goals concern personnel exposure to ionizing radiation under normal plant-operating conditions. This exception confirms the previous rule. The radiation dose to which operatives are exposed, on-site or in its vicinity, is relatively easy to monitor thanks to dosimeters. In the United Kingdom, just as in the rest of the European Union, 'employees working with ionizing radiation', through access to certain parts of the plant, must not receive an annual dose exceeding 20 millisieverts.[14]

## A French approach in complete contrast to the American model

France is well known for its long, diverse tradition of codifying rules and its over-zealous officials. Almost everything must be put down in writing as law, the Conseil d'Etat supervises the enforcement of a very extensive body of public law, and officials draw up and implement large numbers of regulations. Yet this stereotype does not apply to nuclear safety. The first major item of legislation dates from the mid-2000s.[15] There are only a limited number of ministerial orders and rulings. Surprisingly enough, French regulation has been very sparing in its use of explicit goals and technical standards.

The Autorité de Sûreté Nucléaire, France's nuclear safety authority, is an administrative body which reports to the Industry and Environment Ministries. It was set up in 2006. The Transparency and Nuclear Safety Act[16] stipulates its mission: ASN is tasked with defining technical requirements for the design, construction and operation of civil nuclear power plants, necessary to uphold safety, and protect public health and the environment. Before issuing an operating licence ASN checks that the operator has taken the necessary organizational measures 'to prevent or limit *sufficiently* the risks or drawbacks which the installation presents'.[17] The regulator has not subsequently indicated what 'sufficiently' means. The only explanation provided by ASN regarding its objective is that it 'ensures that the safety of French civil nuclear facilities is continuously improved'. This improvement concerns both the safety of existing reactors and that of subsequent generations. ASN justifies this approach by the fact that safety can never be taken for granted and that scientific and technical knowledge are constantly evolving. Incidents and accidents on reactors in France and elsewhere throw light on further shortcomings too. Analysed in detail they give rise to a valuable return on experience, contributing to improved safety in existing and new plants. Continuous improvement is facilitated by mandatory safety visits to nuclear power plants every ten years. This periodic review of reactor safety, imposed on the operator by the 2006 act, enables ASN to introduce new technical provisions, which in particular make allowance for advances in knowledge. The ten-year check-up is extremely thorough, shutting down the reactor for several months.

The goals set forth in the technical directives adopted in 2000 for the European Pressurized Reactor, the fruit of Franco-German cooperation, are not much more explicit.[18] They simply stipulate that the core-melt frequency must be lower than for existing reactors and accidents leading to the release of material must be 'practically eliminated'. Clearly this wording is open to interpretation. Is the aim to eliminate almost all accidents? What unspoken goals does the word 'practically' conceal? Are accidents to be made impossible, by raising a series of physical barriers to prevent any radioactive material from escaping into the environment?[19]

The only officially stated quantitative safety goal dates back to the previously cited administrative document of 1977. On 11 July the Industry Minister, tasked with nuclear safety, sent a letter to EDF specifying that the overall probability of a reactor initiating unacceptable consequences must not exceed $10^{-6}$ per reactor-year.[20] This letter had no effect at the time, and it has had even less effect since, because it was soon followed by a sort of correction. In a second letter the minister specified that the threshold carried no regulatory weight and that EDF was under no obligation to demonstrate compliance using probabilistic assessments.[21] The threshold was in fact no more than a back-of-the-envelope calculation.[22] The following reasoning underpinned the letters penned by officials at the Service Central des Installations Nucléaires:[23] France planned to build a fleet of fifty reactors scheduled to operate for forty years; if the probability of a serious accident was $5 \times 10^{-4}$ per reactor-year, an accident during the service life of the fleet was foreseeable: this was out of the question. The expected probability of an acceptable accident must therefore be about one-in-hundred or one-in-thousand, or to steer a middle course, a probability of $10^{-6}$ per reactor-year. Much as for its US counterpart, the 1977 letter constituted a general guideline. However, unlike the NRC, regulation in France did not adopt it. The safety authority has no intention, no more now than in the past, of defining a benchmark indicating a good enough level of safety. The safe-enough concept is entirely foreign to the French conception of safety. The explicit goal set by ASN is continuous progress.[24]

The idea of ongoing progress is difficult to reconcile with specific, detailed, technology-based standards. Indeed, less than thirty-five guidelines and rules have been established since 1980, mostly only a few pages long.[25] In contrast, the goal of continuous progress is reflected in an overall performance-based standard, enshrined in the 'baseline' concept. This is an essential part of French regulation. In practical terms it involves compiling all sorts of safety requirements (technical standards and specifications, general operating rules, good practice), from a wide range of sources (ministerial orders establishing safety rules, ASN guides, operator practices), for each reactor category. So there is a baseline for each *palier* (900 MW, 1,300 MW and 1,450 MW). The baseline changes to reflect the few new ministerial regulations, extensive return on experience and a large number of scientific and technical studies. The relevant baseline is used during the periodic safety reviews carried out by ASN, to detect non-conformity and identify areas in which progress is required, even if current compliance is demonstrated. Whatever happens, a review ends with a list of items for which performance must be improved. The following review will measure progress as a function of the previous and existing baselines. This system is obviously made easier by the relative homogeneity of the French nuclear fleet and the presence of a single operator, EDF, which centralizes data on the safety of each reactor and can compare performance between individual reactors, plants and *paliers*. The company can thus oversee the overall trend towards improvement, between various types of reactor and plants, in order to ensure that progress towards greater safety is evenly spread throughout the fleet.

Setting large numbers of detailed standards is also ill suited to good quality technical dialogue between operator and regulator, the latter acting as both supervisor and expert. In this second capacity the regulator addresses its counterpart on an equal footing. Technical exchanges resemble academic discussions and confrontations. The expert seeks to make his or her conclusions as accurate as possible, convincing the other party of their sound basis. The aim is to find an original solution not previously considered or, more modestly, to improve a solution others have suggested. Exchanges between safety experts – regardless of whether they work for ASN, EDF or other bodies – take the form of debate between peers. As this other characteristic of regulation in France is part of a long-standing tradition, some historical background is needed to explain and appraise it.

Until work started on building the first *palier* of its fleet of reactors, nuclear safety regulation in France played a very minor role. Only in 1973, with the launch of a large-scale programme to build the existing fleet, was the first centralized body established to administer nuclear safety. Reporting to the Industry Ministry, it consisted of five engineers and drew largely on the expertise of the Commissariat à l'Energie Atomique (CEA). This science and technology research organization, set up by General de Gaulle in 1945, designed France's first prototype reactors and had a complete department, with staff of about a hundred, specializing in safety. Ministry officials referred to this department the safety reports submitted by EDF as part of its application for licences to build and operate each of its new nuclear power plants. These reports were then studied by a committee known as the Groupe Permanent Réacteurs. It consisted of a dozen highly qualified specialists from EDF, CEA and the Ministry. Even allowing for these two additional inputs, there were very few people regulating the safety of projects and licensing them. All the more so given the rapid rate of procedures, with two to six licensing applications filed every year, each one described in a nuclear safety report long enough to reach the office ceiling if stacked.

But the prime concern of this proto-safety regulator was to avoid any delays to the nuclear programme.[26] It realized that detailed regulations were beyond its reach and counter-productive. To draw up standards required some practical grasp of building and operating a plant. The nuclear safety administrators only acquired these skills gradually in the process of carrying out technical reviews of installations. Reflecting its scanty resources or deep-rooted convictions, France's first nuclear authority adopted a critical stance regarding regulation based on written standards and conformance. It espoused the doctrinal views of F. R. Farmer, an official working for the UK Atomic Energy Authority's Safety and Reliability Directorate. In 1967 he called for greater care to avoid being guided by the pressures of administrative or regulatory commodity to produce formulae or rules in the light of which safety would be tested.[27] He was afraid that designers and operators would only concern themselves with conformance and neglect more searching thought on the effects of their efforts to achieve fundamental safety.

In its early days France's safety regulator benefited from US experience in two ways. The reactors built at this stage by EDF had been developed by Westinghouse, so the technology had already been tested on an industrial scale. Several units had been built in the US and their safety examined, assessed and codified in standards by the NRC. French experts could thus draw on this knowledge base, with its mass of existing data. France's regulator also learnt from the difficulties encountered by its US counterpart. It was consequently keen to avoid the instability of this regulatory framework and its standards, and its negative impact on construction deadlines. The French saw US regulation as an example to avoid, enabling it to sidestep similar teething problems with nuclear safety. But things could easily have been very different. France initially tried to develop its own gas-cooled, graphite-moderated reactors. The decision to shelve this home-grown technology in favour of another, under foreign licence, was a difficult and lengthy process (see Chapter 10),[28] which delayed the launch of France's large-scale nuclear programme. Had it persisted with research on gas-cooled reactors, it would have saved time, perhaps leading to a very different safety setup.

As for the organization of technical exchange and expertise, the situation in the 1970s foreshadowed the present arrangement. The CEA department specializing in nuclear safety has been hived off to form the Institut de Radioprotection et de Sûreté Nucléaire (IRSN). It now fields a team 2,000 strong. IRSN is a research institute but it also provides technical support for ASN, carrying out in particular the safety assessments of installations on which the regulator bases its decisions. ASN itself is a remote relation of the centralized safety department established in 1973. It still draws on the work of permanent expert committees, partly manned by EDF representatives. The Permanent Reactor Group still exists. It is very active and much in demand at ASN. It remains an outstanding body, distinguished by the skills of its members and the quality of technical debate.

The big changes have happened elsewhere. Nowadays technical dialogue no longer occurs behind closed doors. The general public can obtain and consult IRSN expert appraisals. ASN publishes a statement each time it refers an issue to one of the permanent expert committees. Committee reports are published too. The make-up of the expert committees has been extended to include views other than those of specialists working for the nuclear industry. Scientists from abroad and university researchers may now contribute to expert appraisals. Such transparency and diversity are completely at odds with the form of expertise, out of the public eye and among people with the same background – or more bluntly nucleocrats – which characterized the period when France's nuclear programme was developed.

A word in passing on the term 'nucleocrat', because it is an opportunity to dismiss two widespread pre-conceptions. Originally hatched by Philippe Simonnot, who writes for the French daily *Le Monde*, as the title of a book published in 1978,[29] this portmanteau word refers to a small clique of bureaucrats and leaders of the nuclear industry in cahoots to make France a great nuclear power. They belong to the same world, because 'for the most part [they] are graduates of Ecole Polytechnique and belong to the same elite: the Corps des Mines, Corps des Ponts et Chaussées, in other words two career paths which attract the best brains from Ecole Polytechnique [...] CEA is packed with people belonging to the Corps des Mines, EDF with their counterparts from the Corps des Ponts'. The first misconception is not that engineers belonging to these two *corps* played a decisive role. They were indeed prime movers in France's technical and economic expertise on nuclear power, as well as actively promoting its development with the government. Rather, the mistake lies in the idea that the government had no say in such developments nor in their overall direction. The nucleocrats allegedly usurped the powers of policy-makers, the latter becoming mere puppets in their hands. In fact, throughout the second half of the twentieth century, the two arms of French government – the Presidency and the Cabinet – were in the front line, taking full responsibility for the nuclear programme.[30] The second misconception is that the nuclear lobby and central government faced little opposition. On the contrary, in the 1970s France's anti-nuclear movement was one of the most powerful in Europe.[31] Simonnot's book bears this out, much like all the other literature combating nuclear power. Opposition took many forms, backed by trade unions, politicians, environmental watchdogs and leftwing extremists. It led to various acts of violence, such as the 1977 attack on the home of Marcel Boiteux, then head of EDF, or the death of Vincent Michalon, an anti-nuclear campaigner, during a demonstration at Creys Malville which was violently repressed.

## Regulator and regulated: enemies or peers?

The ongoing technical dialogue between EDF and the safety authority, now ASN, should not be interpreted as a form of collusion or conniving between regulator and regulated. Such a simplistic view of the theory of regulatory capture should be discarded. The world is neither all white – no capture – nor all black, with the regulator enmeshed in industry's net. Modern economic theory on interest groups sees collusion between the authority and one of the stakeholders as a risk.[32] It has established, for example, that the risk is greater with specialist authorities supervising a small number of companies. A sectoral regulator – overseeing sectors such as telecommunications or air transport – remains in contact with the same companies, and it is hard for officials wishing to transfer to the private sector to capitalize on their acquired expertise anywhere else than in the firms they once regulated. The phenomenon is all the more acute when there are only a few players. Furthermore, with fewer companies the regulator has more limited sources of information. In contrast, competition authorities intervene in all types of market, generally only once. They supervise a merger or sanction abuse of a dominant market position, one day in the fine chemicals industry, the next day in office furniture. They can easily draw on information provided by competitors too. So the risk of capture is potentially higher for a nuclear safety authority, greater still when faced with just one company operating nuclear power plants, as in France. But just because the risk is potentially higher does not mean that it is realized. We may conclude that greater vigilance is required, with more substantial safeguards.

Modern economic theory on interest groups also provides for a trade-off between regulatory capture and information asymmetry. One way of limiting the first factor is to distance the regulator from regulated companies. But limiting contacts and connections between regulator and regulated also reduces the flow of vital information to the former. The regulator may lock itself up in an ivory tower, but it will be struck dumb by ignorance. The knowledge published in books or reports is not enough to work out standards. Similarly, undercover inspections are not up to the task of checking conformance. In the US at least, NRC inspectors are present at all times in power plants. They live on the spot, with an office on-site. Resident inspectors can provide first-hand information on reactor-safety conditions and performance. At the same time, rubbing shoulders on a daily basis with plant personnel and other families living nearby increases the risk of a loss of independence with regard to the operator. The two go hand-in-hand.

It would also be a mistake to over-simplify the aims and behaviour of the regulated company. It is commonplace to highlight cheating and resistance to the application of standards. After all, any company aims to maximize its profits. Reducing expenditure on safety imposed by the regulator by delaying or falsifying compliance is one way of cutting costs and boosting margin. No one is in any doubt that as a short-term strategy this may pay off. The story of the Millstone nuclear power plant in Connecticut, which belonged to Northeast Utilities, amply illustrates this point. It was studied in minute detail by a researcher and a professor at Yale School of Management,[33] demonstrating that in the mid-1980s the management deliberately adopted a low-cost, low-safety strategy. Spending on safety, and other items, was reined in. Safety performance was gradually degraded, but for some time, during which the reactors were operating at full capacity, the regulator was fooled. In 1996, after several incidents and warnings, the NRC forced Northeast Utilities to shut down its three plants. The share price plummeted, losing 80 per cent of its value. The oldest reactor was never restarted, the other two started operations again but after more than two years lying idle. The company itself failed and its assets were sold off. Shareholders lost substantial amounts of money, but management earnings soared. In the course of several years of excellent financial performance, achieved by cost-cutting, managers received several million dollars in bonuses. Unlike the shareholders, management suffered no losses when the reactors were shut down and the company failed; some of them simply had to retire early. This example illustrates the concept of moral hazard, brought to public awareness by the financial crisis of the late-2000s. Management takes greater risks, contrary to the interests of shareholders which it is supposed to uphold, not having to take full responsibility for the consequences. In modern economic theory large companies do not maximize their long-term profit, they simply tend to do so. On the one hand we have shareholders who are naturally concerned about the long term – a realistic assumption in the case of industrial and infrastructure assets – and on the other, management, with other aims and preferences: increasing their personal wealth in a few years or all the way through their career, more or less pronounced risk aversion, quest for social status and utility. Between the two parties, a contract seeks to reconcile management's interests with those of shareholders, for example through bonuses, but with only limited success.

One does, therefore, come across members of the management of electrical utilities who, unlike their counterparts at Northeast Utilities, care about safety just as much as their shareholders, if not more. Witness the Onagawa nuclear power plant, managed by Tohoku Electric Power, a regional utility. One of its former vice-presidents, Yanosuke Hirai, has become a legend in Japan. Onagawa is about 100 kilometres north of Fukushima Daiichi. Located closer to the epicentre of the quake on 11 March 2011, the force was even greater though the wave was the same height. The reactors were shut down in the normal way and no damage was registered. The plant withstood the shock, whereas the two nearby towns were devastated. It even served as a refuge for several hundred local people who had been made homeless. This was by no means a matter of luck, rather the result of the conviction and obstinacy of one man, Hirai san, who died in 1986. In the account that former colleagues gave to the press, he instigated the construction of a 14.8-metre dyke designed to protect the plant against a tsunami. Sure enough, it proved most effective against the Sendai tidal wave. Hirai also convinced the board of his company to install special pumps for the cooling system, to cope with the drop in sea level immediately before a tsunami. Yet no regulations required the company to take these measures. According to his colleagues, Hirai was driven by a powerful sense of duty.[34] He thought it essential to take responsibility for the consequences of one's acts. He was reportedly the sort of person who did not believe that everything would be all right as long as one complied with standards. His aim was more than just compliance. He was the sort of engineer who goes beyond regulations and makes the necessary checks to get to the root of the problem.

Much has been published in economic literature on the behaviour of companies doing more than the legal requirements, particularly with regard to the environment. Many reasons are cited. As in the case of Onagawa, management may have a strong sense of social or ethical responsibility. More commonly, and setting aside any considerations about the psychological make-up of management, companies may do more than just comply with standards and regulations because it is in their interest to do so. There are plenty of powerful incentives such as the threat of stricter legislation, the need to improve their image, the better to recruit more highly qualified young executives or to raise funds more easily. In the case of nuclear power, the most powerful stimulus is undoubtedly the operator's responsibility with regard to its workforce and the neighbouring community. The operations manager at a nuclear power plant reports to his or her superiors at head office, but is also accountable to fellow workers, whose families live nearby. Asked why he spent so much more on safety than was required by the regulations, the head of a Swiss plant would answer: 'Because we live here!'

The same mechanisms and factors explain why the regulator and a regulated company can sustain a dialogue and cooperate without straying into collusion. There is more to a company's behaviour than just bribing or duping the authority, rejecting its requests or achieving minimal conformance. When it comes to technical dialogue there is an additional reason: the need to share the cost of producing information and knowledge. The plant manager at the Diablo Canyon nuclear power plant, questioned by researchers from the University of California, Berkeley,[35] explained that he always welcomed NRC inspectors in the same manner. 'They [NRC inspectors] seem to think we invariably cover-up. They come in here with that idea. When they come in [to inspect], this is what I do.' Pulling out a pad and a pen, he leaned forward and began to write. 'I say to 'em, "Here's a list of our four or five most serious problems," and hand it to 'em. "Go see if we aren't right. And let me know if our solutions aren't working." They don't expect this. And they go away and look. We try to be better than they are at finding and fixing problems.'

This attitude is at odds with the conventional US approach to regulation. In the past, and to a lesser extent now, whether it is a matter of policy on safety, the environment or telecommunications, regulator and regulated have regarded one another as enemies, not peers. There is little mutual trust between them. Such wariness is due in particular to the weight and nature of the US judicial system. Industry often disputes regulatory decisions and measures. Up to the 2000s one-third of the regulations issued by the Environmental Protection Agency were disputed in the courts. Before judge and jury, the two parties, assisted by an army of lawyers and experts, vigorously defended their respective positions, fighting every inch with questions to friendly and hostile witnesses. Anything said or written before the complaint could be used by the opposite party. The best course of action was great caution at all times, particularly before the event. A climate of this sort is obviously not conducive to technical dialogue between NRC administrative staff and the management and safety experts of utilities.

In a more general way, political scientists attribute the antagonistic relations in the US between regulator and regulated to a deep-rooted distrust of the state: citizens need to be protected by the courts against its power of coercion. Conversely, political scientists ascribe the consensus-based regulation found in France and mainland Europe to the model of an informed elite and a benevolent state. Study of national systems and different modes of regulation gave rise to abundant literature in the last two decades of the twentieth century. The stream seems to have dried up since, the differences having become less noticeable. The European and US models have moved closer together.[36] Cooperation between industry and the authorities in the framing and application of new regulations has developed in the US, whereas on the other side of the Atlantic a more judicial approach to regulation has gained ground.

## Pros and cons of American and French regulation

Let us start by briefly summarizing the similarities and differences, discussed in the preceding sections, between safety regulation in France and the US.

Basic principles apply on both sides of the Atlantic: the nuclear safety authorities are powerful and competent, independent and transparent. Yet the institutional environment is very different: the US regulator acts under the close supervision of Congress and the courts, whereas its French counterpart is comparatively free in this respect. ASN submits an annual report on its activity to the lower chamber of the French Parliament. Debate remains courteous and the hearing attracts only limited attendance. Very few MPs take an interest in the topic, let alone understand its finer points. Civil or administrative courts have almost no say on the matter. The executive, in contrast, is much more actively involved. The government enjoys far-reaching legal powers for intervention, though in practice it makes little use of them.

The institutional balance of power, with its respective strengths and weaknesses, applies to nuclear safety regulation in the US and in France. We have also seen that the two industries and the installations under supervision are very different too. The NRC must monitor a large variety of reactors and a fair number of operators. Collective self-regulation, through the good offices of the INPO, lightens this burden. In France a more uniform fleet and a single operator facilitate safety checks and performance assessments. But the regulator must take greater care to avoid capture. Overall, regulation has learnt to cope with the particular features of the industry it supervises.

The issue on which there is a major divergence between regulation in the US and France concerns allowance for costs and the role of economic factors. On the one hand we find quantitative goals, regular use of probabilistic assessments, concern about containing costs and effective allocation of efforts; on the other, poorly defined goals and few regulatory prescriptions,[37] with priority given to technical dialogue and no reference to cost-benefit analysis. Clearly we are dealing with two approaches to regulation founded on very different principles, with the US seeking the level at which safety is safe enough, whereas in France the aim is to achieve continuous progress.

Despite such significant differences, the safety regulation systems in both countries are frequently cited as examples. They constitute a baseline for specialists. The regulatory community pays just as much attention to statements by the NRC as it does to ASN. They both influence the course of debate in the various international forums on nuclear safety. But despite this peer recognition we must not overlook several intrinsic defects which have become increasingly visible in recent years.

We shall start with the shortcomings of US regulation. It is exemplary in its management of safety costs, both with respect to the limits it places on the assigned goals and its efforts to define effective measures for achieving those goals. On the other hand, the cost of drafting regulations is very high, which hinders change, delays implementation and reduces responsiveness.[38] The attempt to impose fire-protection standards, described above, is an extreme case, but it is a good illustration of an overall pattern of regulation, litigation and more regulation. To start with, the NRC sets forth new requirements. Then a stakeholder – either an operator or non-government organization – lodges a complaint before a court. In a third step, several years later, a federal judge issues a ruling, at which point the NRC must adjust its requirements. This syndrome is particularly frequent in the case of measures to upgrade existing installations to allow for return on experience following incidents or accidents. Since a decision in 1987 by the District of Columbia Court of Appeals, US regulation operates with two separate categories. The first one corresponds to the standard of 'adequate protection' of public health and safety, which is a legal requirement.[39] The safety requirements enshrined in the initial licensing process for the construction and operation of nuclear power plants, for example, belong in this category. This first set of obligations applies to all plants and the NRC enjoys full discretionary powers for their prescription. The second category comprises all the measures the NRC deems reasonable to limit damage. It must justify them by cost–benefit analysis, demonstrating a substantial gain in safety. So this category covers all the regulatory measures which go beyond an adequate level of protection. They are mandatory, but there is scope for operators to negotiate and adapt the manner in which they are deployed. Each time the NRC frames a new requirement it must choose the right category, a constant source of conflict. Operators accuse the NRC of arbitrarily choosing the first category when costs are high, whereas parties opposed to or critical of nuclear power condemn use of the second category, because making allowance for costs boils down to limiting everyone's safety.

Disputes over the appropriate category oppose the administrative services and the Commission, and also take place between the commissioners themselves. Witness the efforts to introduce stricter safety regulations following the Fukushima Daiichi disaster. The NRC set up a taskforce to highlight the lessons learned which were relevant to the safety of the US nuclear fleet. It recommended a whole range of actions, in particular three urgent new measures: hardened vent designs to prevent hydrogen explosions; instrumentation for spent-fuel pools; and a series of strategic actions to cope with extreme events (portable emergency diesel pumps, emergency intervention team). The staff thought these three mandatory requirements were necessary to secure adequate protection of public health and safety. The then NRC Chairman, Gregory Jaczko, endorsed this move, voting in favour of it when it was tabled before the Commission. However, Commissioner Kristine Svinicki, well known for her pro-industry stance, voted in favour of downgrading all these measures to the second category. The three other commissioners expressed mixed preferences depending on individual measures. Ultimately, a majority decided to put instrumentation on spent-fuel pools in the second category, delaying implementation of this measure till cost–benefit analysis could be completed. On several occasions Jaczko found himself in the minority, alone against the four other members of the Commission. Advocating a hard line requiring additional safety improvements without further delay, he was confronted by the relative inertia of his fellows, often irritated by his style of management. Jaczko finally resigned in May 2012.

In comparison to the US regulator, its French counterpart is more like light cavalry. It travels fast, when necessary, and is neither encumbered by the heavy armour of legal or procedural constraint, nor hemmed in by countless goals and standards. The principle of continuous progress governs its actions without being linked to any specific safety goal, particularly a quantitative one, or to cost–benefit analysis. There are, however, some disadvantages to the extensive discretionary powers invested in the French regulator, raising a series of problems for the years to come.

First of all, this regulatory model requires an enlightened regulator. Until recently ASN has been represented by André-Claude Lacoste, a man of fine character. A graduate of Ecole Polytechnique and the Ecole des Mines, he spent a large part of his career in industrial safety. He took over as head of the Direction de la Sûreté des Installations Nucléaires in 1993 and was in charge of nuclear safety for twenty years, initially under the aegis of the Industry Ministry, then at ASN. Indeed this body was to some extent his brainchild. He managed to convince the policy-makers of its necessity and worked tirelessly to affirm its independence and make its workings transparent. Lacoste's successor, appointed by the French President, is one of his former deputies, a graduate of Polytechnique and member of the Corps des Mines. On taking office in 2012, Pierre-Franck Chevet said that he intended to follow in his predecessor's footsteps. And there is every reason to suppose he will do so. But there is no guarantee that at the end of each six-year term the incumbent president will choose another enlightened regulator.

Secondly, pressure on safety costs is bound to increase. By 2015 two-thirds of the reactors currently operating in France will be at least thirty years old. Spending on the necessary safety and upgrade work will rise, because to continue operations parts will have to be replaced. Some of them, such as steam generators, are large and very costly. At the same time, French policy-makers are firmly opposed to any increase in administered electricity tariffs, keen to maintain living standards and industrial competitiveness. Several times in the past ten years the government has refused to allow electricity prices to make allowance for the full increase in costs, whether caused by rising gas prices, spending on improvements to the grid or subsidies for renewable energies. In principle the pricing system is supposed to reflect the full cost, but in practice this rule does not apply. EDF cannot count on being able to pass on to the consumer the full cost of spending on safety and upgrades. The regulator will have to make allowance for this new factor and methodically weigh up the economic consequences of its prescriptions.

Thirdly, the lack of quantitative goals adds yet another unknown to the tricky equation of plant retirement. The date for the final shut-down of reactors needs to be anticipated as accurately as possible in order to ensure that new capacity – nuclear or not – can take over, satisfying demand without undue disruption. The process of deter-mining this date is already fraught with much uncertainty. Should calculations be based on a forty-, fifty- or sixty-year service life? The whole picture could change almost overnight: the government may suddenly change its mind on the pros and cons of nuclear power; safety inspections may discover serious defects; or a technological breakthrough may substantially reduce the cost of alternative energy sources. With no specific safety goals, the degree of improvement demanded by ASN constitutes a sizeable unknown. All one can say is that it lies somewhere between two extremes. ASN may make do with barely perceptible, yet continuous improvement: for exam-ple, a reactor must be slightly safer as it enters its thirties or forties than it was in its late twenties; alternatively, ASN might require the safety performance of ageing reactors to be brought into line with the European Pressurized Reactor. In the latter case, this would entail substantial compulsory investment. Let us suppose, for example, that the reactors belonging to the 900 MW *palier* had to be fitted with as many redundant systems as their more recently designed 1,300 MW counterparts. The cost of work might even exceed the gain yielded by additional electricity sales from extended operation. The opera-tor would have no option but to shut down the plant completely. Raising the level of safety necessarily stops when the gain expected from electrical revenue is absorbed. This state of affairs benefits nei-ther the consumer nor EDF shareholders. Continuous progress, with no limit set by a specific goal or cost–benefit analysis, leaves wide open the question of how the economic surplus should be shared out between consumers, the generator and potential beneficiaries of the lower probability of an accident. The regulator alone decides how far it should go one way or the other. But a decision of this importance on the distribution of wealth should not be its responsibility, but that of policy-makers.[40]

The principle of unbounded, continuous progress may ultimately create the impression that safety is never adequate. Inevitably, various parties will try to up the odds. Hearing an operator assert that safety is its absolute priority or the regulator maintain that safety is 'never safe enough'[41] makes us forget that the resources available for safety are necessarily limited. Public opinion and policy-makers may end up losing sight of the fact that there comes a time when raising the safety level means closing existing plants which are fully operational. Anti-nuclear campaigners have realized this and make a point of insisting that safety is never adequate. Having no specific safety goals leaves the door wide open for ill-conceived political decisions dictated by the imperatives of fleeting alliances, and subject to local or national haggling between parties. The political decision to close the Fessenheim nuclear power plant – which we shall analyse in Chapter 11 – perfectly illustrates this danger.

On paper the respective shortcomings of nuclear safety regulation in the US and France could be remedied by borrowing from the other's strong points. The US authority lacks freedom of movement, its French counterpart enjoys too much latitude: on the one hand continuous progress is impeded, on the other it lacks direction. But the ideal balance for either organization cannot be achieved at the drop of a hat by importing ready-made solutions. Our case study shows that in both cases national institutions and the lessons of history exert considerable influence on both the form and content of nuclear safety regulation. Change can only be slow and gradual.

In conclusion, regulating the safety of nuclear power is a trade that deals in uncertainties. The better to tirelessly eliminate its defects, we must understand and accept that it is necessarily imperfect.

Identifying the rules to set for operators is no easy task: their effects are hypothetical and it is often impossible to verify their enforcement. Ideally, we should be completely sure about the measures to be taken, their impact on safety and their implementation, but that is beyond our reach. Much like any other public intervention, nuclear safety regulation is, and will remain, imperfect. The perfect regulator does not exist, no more than zero risk. But this is not a reason to down tools. On the contrary it should be seen as an incentive for greater vigilance and an ongoing quest for improvement.

Economic analysis helps us to understand why some forms of regulation are less imperfect than others; it highlights the root causes on which attention must focus to identify potential hazards and find remedies. Four essential, universal pillars are needed to contain the risk of regulatory capture: giving the regulator extensive powers to inspect and sanction; endowing it with adequate resources to secure and consolidate its skills; guaranteeing its independence from government and industry; imposing the transparency of its actions. The example of Japan is a reminder of the pitfalls to which regulatory capture may lead.

Substantial progress towards improved safety can also be achieved by changing the legal framework and balance of power. Shareholder structure, governance and management incentives all feature among information that must be taken into account by safety authorities, prompting them to exercise closer or more remote supervision. Stakeholders in the enterprise can provide the regulator with valuable support, counterbalancing shareholders and management, who may be too concerned with short-term goals. Staff may act as whistleblowers and the local community can bring pressure to bear if the operator slackens its safety efforts, providing the regulator makes it easy for residents to access information. In short, through the overarching design of regulation and its fine-tuning, economics provides an essential complement to the scientific and technical prescriptions of engineers. Moreover, in financial terms, these additional solutions cost almost nothing. Those who stand to gain from bad governance will of course lose out. Circumscribing their resistance sometimes requires a massive political effort, but no money, unlike a thicker concrete containment.

Our certainty about the best way of regulating the nuclear industry is founded on economic analysis applied to the United States, France and Japan. But what of the rest of the world? Setting aside the majority of OECD countries, we are dealing with an unknown quantity. How does nuclear safety work in China, India, South Korea, Russia or Pakistan? Not very well, in all probability. Regulation is almost certainly far from perfect in these countries and regulatory capture extremely widespread. In the absence of public studies it is difficult to be more specific, but there seems good reason to fear that there may be many other cases similar to Japan.

# PART IV

# National policies and international governance

Politics is a possible route for taking decisions under uncertainty. Faced with the controversial findings of experts, it is up to public policy-makers to choose and act. It should consequently not come as a surprise to discover that politics plays a predominant role in decisions on nuclear power. Considerable uncertainty weighs on its competitive advantage, risks and regulation. Politics is also a way of providing and managing common goods. Only through mechanisms of public governance can safety and non-proliferation be secured. But, in contrast with this concept of strategically minded governments boldly advancing into the fog or labouring for the general interest, we are sometimes confronted with a more prosaic picture of public decision-making, a process not so much guided by scientific experts and legal advisors, as by the support of lobbies and vote-seeking. The pressure brought to bear on governments by 'green' parties or industrialists also explains many nuclear-policy decisions.

Regardless of where it originates, the decisive weight of politics is illustrated by a trivial observation: countries have adopted and are still pursuing very different courses on the atom. Germany, which invested in nuclear power at an early stage, is now attempting a rapid exit; in neighbouring Austria not a single reactor has ever generated any electricity and the constitution bans nuclear power. France, on the other hand, is still banking on nuclear power, but wants gradually to reduce its share in the energy mix. Meanwhile, Switzerland and Belgium aim to gradually phase out nuclear power altogether. The former Soviet Union, the United States and the People's Republic of China have developed civilian and military uses for the atom. In contrast, Japan and South Korea have both built large nuclear fleets but possess no weapons. Israel, however, probably does, but has built no power plants. Whereas some countries are phasing out nuclear power, others, such as Turkey and Vietnam, are signing up, driven by various motives ranging from energy independence to national defence, through climate change mitigation and international prestige.

This diversity highlights the fact that the decision to develop or retire nuclear power is taken by national governments. Research centres, electricity generating companies and equipment suppliers in each country obviously attempt to influence the political decision. But they do not hold all the necessary cards. After the accident at Fukushima Daiichi, Germany's nuclear industry had to comply with the decision by Chancellor Angela Merkel speedily to shut down the country's nuclear power plants. Despite the heavy losses this decision imposed on them, Germany's three electricity utilities – Eon, EnBW and RWE – could not oppose this choice. Countries may be subjected to diplomatic and commercial pressure from abroad, urging them to adopt or shelve nuclear power, but ultimately it is their decision. In the early 2010s the Czech Republic decided to add two new reactors to the existing plant at Temelin, disregarding vigorous protests by its Austrian neighbour.

Why have some countries decided in the past, and are still deciding now, to develop nuclear power? Why are others phasing out this technology? How do they go about adopting nuclear power and subsequently retiring it? The first two chapters of this final part aim to explore these questions.

The subsequent two chapters focus on supranational governance. The European Union, the inception of which was closely connected to atomic energy, has become a patchwork of energy mixes and policies. It is a classic illustration of the difficulties of framing and implementing policy on nuclear power, which without necessarily being common to all Member States is at least compatible between them.

We then turn to nuclear governance by treaty, and the question of proliferation linked to trade in nuclear raw materials and equipment. The Treaty on the Non-Proliferation of Nuclear Weapons came into force in 1970. It was based on a bargain between nuclear and non-nuclear countries, exchanging access to civilian applications for a commitment not to pursue military goals. The International Atomic Energy Agency is tasked with promoting nuclear power and monitoring installations for possible abuses. Its dual role is subject to great tension: the more international trade in nuclear goods develops, the greater the risk of proliferation, which in turn requires stricter governance of action to monitor possible diversion to military applications.

# TEN

# Adopting nuclear power

About thirty countries worldwide have built one or more nuclear power plants. Some fifty others have expressed an interest in the technology and asked the International Atomic Energy Agency (IAEA) for practical assistance developing a project. What prompts them to start generating nuclear electricity? Are their motives the same as those of their predecessors?

We shall start by presenting the countries which currently generate electricity using nuclear technology. About ten former soviet socialist republics belong to this category. They include Armenia, Lithuania and Ukraine, and satellites such as Bulgaria, Hungary and the Czech Republic. The choice to invest in nuclear power was above all made by Moscow. Three other major nuclear countries – the United States, the United Kingdom and France – belong in the same category as the Soviet Union. They all adopted nuclear power at an early stage, combining civilian and military applications. India and China joined the game later, but have also developed civilian and military applications. In contrast, Canada, Japan, South Korea and Germany have large nuclear fleets too, but no weapons.[1] There follow a large number of countries with only a small number of reactors. Did you know that Mexico, Argentina, Brazil, Taiwan, Belgium, Switzerland, Sweden, Spain, Finland, the Netherlands and South Africa all own nuclear power plants? To this list of small-scale players we should add Pakistan and Iran. Unlike the others, the development of civilian nuclear power here is closely linked to military ambitions. Pakistan used the development of civilian nuclear power as a blind and Iran has developed an uranium enrichment programme far in excess of its needs for power generation.[2] Counting the former Soviet Union as a single unit, there are currently twenty-one countries which have successfully developed nuclear power.

The above list shows there is no systematic link between possessing atomic bombs and building nuclear power plants. Indeed, most of the countries generating nuclear electricity have not developed such weapons. The military dimension of the atom is nevertheless essential to understanding the background to this technology's development and deployment.

## Atoms for peace

Atomic energy was born of science and warfare: science because nuclear physicists played a key part in finding applications for nuclear fission; warfare because, after the atrocities which scarred Europe and Japan, atomic energy contributed to hopes of peace.

A letter from Albert Einstein to Franklin D. Roosevelt in August 1939 is testimony to the contribution of scientists. The physicist drew the President's attention to the possibility that uranium might be used to produce bombs of incomparable force. He implied that Germany, which had just invaded Czechoslovakia, had already understood the possible military significance of nuclear fission, which had recently been discovered. His warning gave rise to the Manhattan Project, the code-name for the US programme to develop nuclear weapons during the second world war, with assistance from Canada and the UK. Under the leadership of the physicist Robert Oppenheimer, the project brought together a host of US scientists and distinguished refugees from Europe. Thanks to its work, the US became the first country to deploy atom bombs, dropped on Hiroshima and Nagasaki in 1945. Through its civilian applications, this military programme also meant that the US was the first country to connect a nuclear reactor to the power grid and to generate electricity.[3]

The end of hostilities heralded a period of collective enthusiasm about the potential of civilian nuclear power, high hopes which now look unfounded. At the time it seemed to have two key assets: being abundant and cheap, this new source of energy could fuel economic growth in industrialized nations and the developing world; its expansion, thanks to scientific cooperation between advanced countries and international supervision of fissile materials, could be achieved while preventing widespread nuclear proliferation.

A famous speech by Dwight D. Eisenhower, entitled Atoms for Peace, reflected this conviction. Addressing the United Nations General Assembly on 8 December 1953, the 34th US President drew on ideas and proposals which had been under discussion for several years.[4] He urged the UN to pursue two goals: reducing the destructive potential of the nuclear stockpile which was starting to build up; and expanding the atom's civilian applications. By this time the US had already lost its monopoly of the bomb. The UK and the Soviet Union were now in the running, with France not far behind. Several tens of tonnes of enriched uranium and plutonium suitable for use in weapons had already been accumulated, whereas just a few kilos was enough to produce a bomb. Eisenhower realized that many, perhaps all, countries would ultimately acquire the scientific and technical knowledge needed to build their own weapons. With the prospect of global nuclear armament, it was time to find and apply remedies to counter the coming arms race. The solution he advocated was to divert efforts away from the atom's military applications and to concentrate on civilian uses. Today a shift of this nature seems barely credible, because just the opposite occurred: some states, such as India, used development of this energy source as a basis for obtaining nuclear weapons. But that was the conviction underpinning the Atoms for Peace initiative and the proposals it set forth. Eisenhower advocated setting up a sort of international bank, on the one hand receiving fissile material to prevent its destructive uses, and on the other distributing it for peaceful purposes, in particular as a source of abundant energy. Hence the close connection between efforts to combat proliferation and to promote nuclear power.

Were such contradictory aims the result of cool calculation or utopian thinking? Many commentators have claimed that Eisenhower's proposals were no more than realpolitik. In compensation for shifting attention from military uses countries would enjoy economic growth fuelled by the civilian applications of the atom in medicine, farming and above all energy. Cooperation on science and technology by countries which had already tamed nuclear fission, primarily the US and USSR, would speed up development for those just starting out. By opting not to compete with the superpowers for control of nuclear weapons, these countries would enjoy the benefits of nuclear power sooner. That was the deal and it was certainly a little idealistic. As the US President emphasized:

> The United States knows that peaceful power from atomic energy is no dream of the future. That capability, already proved, is here – now – today. Who can doubt, if the entire body of the world's scientists and engineers had adequate amounts of fissionable material with which to test and develop their ideas, that this capability would rapidly be transformed into universal, efficient, and economic usage.

This idyllic vision is the other side of the nuclear threat, that 'dark chamber of horrors'. Convinced of their natural symmetry, Eisenhower argued that the benefits the atom could bring to humankind were just as great as its destructive power. The apocalyptic description of nuclear ills in his speech served to highlight the promise of peaceful applications.[5]

Ultimately it makes little difference whether this offer was a cynical ploy or entirely sincere. Either way the linkage between the struggle to prevent global nuclear armament and promotion of nuclear power produced lasting effects. An immediate outcome of

Eisenhower's initiative was the setting up of the IAEA, which subsequently played a key part in the spread of nuclear power worldwide. The Agency must strive 'to accelerate and enlarge the contribution of atomic energy to peace', as well as being tasked with limiting the development of military applications by monitoring installations. The Treaty on the Non-Proliferation of Nuclear Weapons, signed eleven years later (see Chapter 13), had the same goal, to contain the spread of nuclear weapons while encouraging that of nuclear power.

It was not until India's first nuclear test that the Atoms for Peace doctrine began to be questioned. The underground explosion in the Thar desert in 1974 was carried out using plutonium from a research reactor built with Canadian assistance and technology. The test made Pakistan, its hostile neighbour, all the more determined to obtain nuclear weaponry. It achieved its aims thanks to solid scientific and technological skills gained through the international cooperation advocated by Eisenhower. Starting in the late 1950s, Pakistan had received scientific and technical assistance from many countries, including the US, Canada and the UK.[6] China's backing, in the form of military technology under the pretence of an agreement on civilian nuclear cooperation, was decisive too.

This brief summary of the early days of nuclear power should help the reader to understand why a single international body, the IAEA, should be tasked with monitoring nuclear installations while at the same time promoting them, and why a treaty on non-proliferation should encourage exchanges between nuclear scientists and engineers. What now seems a contradiction in terms made perfect sense at the end of the second world war, in the aftermath of Hiroshima and Nagasaki and then during the cold war.

This period is also a reminder of the two-sided nature of nuclear research and technology. To design and produce atom bombs or nuclear power plants involves building large-scale research facilities and training hundreds of physicists, chemists and engineers. Similar resources and skills are needed, at least in part, whether the end result is civilian or military. Top scientists worked on both, either in parallel or alternately. Major atomic institutes, such as the US Atomic Energy Commission, developed civilian and military applications. Of course, advocates of Atoms for Peace were aware of this duality, but they were convinced that the pursuit of military goals could be effectively prevented by control of fissile materials and installations.

The international cooperation on science and technology established and promoted by Washington did speed up and extend adoption of nuclear power by many countries. By 1960 the US had already signed cooperation agreements in this field with forty-four countries, ranging from Algeria to Yugoslavia. The Soviet Union was almost as active, reaching bilateral agreements with seventeen countries, mainly within its sphere of influence (e.g. Bulgaria, German Democratic Republic, Romania). At the beginning of the 1960s there were almost 200 research reactors worldwide, most of which had been imported, wholly or in part, from the US or the USSR.

## Pioneers and followers

The countries that pioneered development of nuclear power were also the ones that led the way on research into the atom bomb. The first to master the use of nuclear power to generate electricity were the three nations involved in the Manhattan Project, the US, Canada and the UK, soon followed by the USSR, which tested its first atom bomb in 1949. France only caught up with this group some time later, after resurrecting its research capability, which had been mothballed during the German Occupation.

In terms of technology the pioneering countries each took a separate route, but all of them made allowance for military factors. In the US the uranium-enrichment plants, built to produce weapons, facilitated the development of light water reactors for generating electricity. Experience using this type of reactor to power nuclear submarines also helped. Meanwhile, other countries initially chose technologies using natural uranium as fuel. They saw this as a way of by-passing Washington's monopoly of enrichment. Natural-uranium technology, used in conjunction with graphite or heavy water as a moderator to reduce the speed of neutrons sufficiently to sustain a nuclear chain-reaction, could meet military demand for plutonium while producing energy. In fact, for some time, electricity was seen as a by-product. The priority for the UK and France, which had opted for this technology, was to design a reactor which would produce spent fuel with a high plutonium content, rather than a high output in kWh per tonne of uranium consumed. Military considerations influenced even the technology chosen by Canada,[7] which gave up the idea of nuclear weapons at an early stage. It developed a heavy-water reactor, having gained experience of this type of moderator during the second world war. As part of the Manhattan Project, Canada produced one of the very first prototype heavy-water reactors, manufacturing the moderator to supply US arms factories.

The pioneering countries were not the only ones to develop civilian nuclear power as a sideline to military research. China, for instance, carried out its first nuclear test in 1961, only starting to build its first nuclear power plant twenty-three years later. It is worth noting that all the nuclear-weapon states have also developed atomic energy.[8] Their decision to develop civilian applications was largely determined by military capability.

The case of such countries also throws light on the three main reasons for adopting nuclear power: national prestige, energy independence and the development of science and technology. Indeed, nuclear ambitions may exacerbate such motives. Take France, for instance. In 1945 General de Gaulle established France's Commissariat à l'Energie Atomique (CEA) to develop research and technology related to the atom. This decision, much like many others he took, was intended to restore France's lost grandeur. He firmly believed that scientific excellence and technological prowess were one of the main sources of the nation's prestige. 'France cannot be

France without grandeur', he wrote in his wartime memoirs, and this meant contributing to technical progress, for 'a state does not count if it does not bring something to the world that contributes to the technological progress of the world'. By the time he returned to power in 1958, CEA's investment in equipment and research was beginning to produce results. De Gaulle was clearly determined that France should bolster its independence with nuclear weapons. In the context of the escalating cold war he sought to steer a separate course from that taken by the US or the USSR. He approved plans to build an enrichment plant at Pierrelatte, in the Rhône valley, opening the way for the subsequent construction of a fleet of light-water reactors.

Developing nuclear power was seen as essential to France's continuing growth. It would supply electricity to industry and contribute to exports of equipment. As De Gaulle put it, 'Being the French people, we must achieve the rank of a great industrial state or resign ourselves to decline.' Making up for France's lack of energy resources was also a priority. It had little oil or gas, and coal was expensive to mine, with only limited reserves. The oil crisis in 1973 underlined the importance of energy independence as a justification for developing nuclear power. Indeed, it prompted measures to speed up the French programme: most of the existing fleet was built between 1970 and 1985.

Advances in science and technology are essential for energy independence too. Though less frequently cited as one of the factors prompting a country to adopt nuclear power, science and technology weighs heavily in the balance. In economic terms technological innovation is a response to both supply and demand. Reducing a country's dependence on oil, gas or coal imports responds to a public, political imperative, rather than consumer demand. But the pressure is nevertheless still on the demand side, and it must be met by supply. Innovation comes out of research institutes and university laboratories, not factories like ordinary goods. In the case of nuclear power the supply side involves colossal human and material resources. Research in nuclear physics is the archetypal big science, which came to the fore after the second world war. All its demands – in terms of budget, machinery, labour and research – are huge. France's Court of Auditors has estimated that CEA spent nearly €10 billion$_{2010}$ on research into civilian nuclear electricity in 1957–69, a sum equivalent to the cost of building the first fourteen reactors in the French fleet operated by EDF. In relation to the whole period from 1957 to 2010, and all the reactors, CEA's research spending amounted to half EDF's outlay building the entire fleet.[9] A very large proportion. Furthermore, nuclear research requires highly qualified staff and specific equipment only produced in small amounts. As a result it is not easily redeployed in other fields of innovation. Taken as a whole, these factors have turned nuclear scientists, engineers and technicians into a force driving the growth of nuclear power (and resisting moves to phase it out), regardless of any variations in demand. To this must be added the massive resources at EDF's disposal – in terms of both engineering skills and operating staff – and of course the industrial assets deployed by the boiler-maker Framatome and electrical engineering firm CGE (later Alstom), tasked with developing giant turbo-generators for the 1,300 MW reactors. Anti-nuclear campaigners are fully aware of the pressure exerted by science and technology, when they condemn the allocation of funds to nuclear research and seek to reduce the budgets of bodies such as CEA.

France is an interesting case in another respect, for it may be seen as the last of the countries to pioneer civilian nuclear power, and the first of the large followers. Much like the US, Canada and the USSR, it started research early on and set its own course, developing gas-cooled, graphite-moderated reactors.[10] However, in keeping with other follower countries, France ultimately opted to import foreign technology, when it reached the stage of large-scale industrial deployment (see box).[11] Its fleet consists of pressurized water reactors, the first of which were based on a model developed by Westinghouse. France went on paying licence fees to the US firm until 1991, by which point the technology in use had become entirely French.

## Why did France give up gas-cooled, graphite-moderated technology?

If De Gaulle was so concerned about grandeur and standing up to US hegemony, what prompted France to adopt US nuclear technology?

We should start by pointing out that the decision to scrap gas-cooled, graphite-moderated technology was taken by Georges Pompidou, who took over from De Gaulle as President in 1969. It is unlikely that his predecessor would have agreed to this move.[12] Both were keen to promote French industry and foster industrial champions, but Pompidou was more open-minded when it came to private enterprise, more concerned about bringing his country's industry into line with international competition. The fact that the nuclear engineering company Framatome was not publicly owned was not a critical obstacle for Pompidou. Moreover, in the late 1960s, pressurized-water reactors were seen as *the* technology of the future for winning overseas contracts. In 1969 Siemens and AEG were awarded an initial contract to build a pressurized-water reactor in the Netherlands. Shortly afterwards the authors of the CEA-EDF nuclear action plan noted: 'It is sad to see [German companies] building a plant in Holland, despite the fact that Germany's nuclear industry started much later than in France.'[13]

It should also be borne in mind that collaboration with West-inghouse was already well established. National pride blinded observers to the fact that Framatome (short for Société Franco-Américaine de Constructions Atomiques) had been set up in 1959. Westinghouse was one of its initial shareholders, hold-ing 15 per cent of its stock, alongside Schneider, Merlin Gerin and the Empain group.[14] Prior to winning the contract to build the nuclear power plant at Fessenheim, the starting point of the French nuclear programme, Framatome had equipped the Franco-Belgian plant at Chooz and another one at Tihange in

Belgium. Before it finally sold out in 1981, Westinghouse's share in the company had risen to 45 per cent. Framatome went on paying licence fees to the US firm for another 10 years.

Lastly, the gas–graphite technology was thought to be more expensive than its light-water rival. Many comparative studies were carried out, in particular by EDF, more concerned than CEA about obtaining the best cost per kWh. There was heated dispute between the two organizations over which technology to choose. CEA wanted to seize the opportunity to capitalize on its skills and past research, whereas EDF contested the former's growing importance in civilian applications of nuclear power and its industrial development. Naturally both parties did their best to convince successive policy-makers to choose their preferred technology. Ultimately the pressurized-water option supported by EDF prevailed.[15]

Many countries benefited from transfers of technology and know-how from pioneer countries, subsequently graduating to greater technological independence. Germany built its first industrial reactors under licence from Westinghouse and General Electric, but, just as in France, its equipment industry gradually rose to stand on its own two feet. The same was true of Japan and India. China has built, on home ground and under licence, almost all the various types of reactor that exist. It has put its own stamp on Franco-American technology, working upwards from its first nuclear power plants at Daya Bay. It seems likely that it will appropriate and transform the third-generation technology it has imported more recently, with the Areva EPR and Westinghouse AP1000. South Korea has gained its independence too, with its own pressurized water model, after building its early reactors under licence from Canadian, French and US firms. Taking its cue from France, Germany and Japan, South Korea has evolved from importing technology to exporting its own nuclear equipment. The Doosan conglomerate supplied the steam-supply systems for two AP1000s built in China; the Korea Electric Power Corporation won the contract to supply four reactors to the United Arab Emirates. China is progressing along similar lines and has stated on several occasions that it wishes to take part in international tenders for nuclear power plants.[16]

As for the smaller countries, among the new entrants, such as Switzerland, Brazil, South Africa and Taiwan, their first steps in nuclear power were not preceded by massive investments in research nor followed by the launch of a national nuclear industry. These countries have nevertheless built up solid teams of physicists and established high-quality laboratories. They have also taken care to combine domestic content with orders for foreign reactors, in order to support local firms and create jobs. With regard to their motives, these players all want to reduce their energy dependence and boost their prestige, particularly in science and technology.

## Aspiring nuclear powers

The countries which now hope to adopt nuclear power are very similar to small past entrants. Their arrival will speed up a trend in the acquisition of nuclear power by non-OECD countries which started with Argentina and South Africa. Their motives are a combination of energy independence, desire for grandeur, and in some cases hidden military ambitions. They do not plan to make nuclear power the key component in their energy mix, nor yet to develop their own technology. Indeed they often buy turnkey plants. The only new development is the appearance of an additional motive, the quest for carbon-free generating technology.

A review of the aspiring nuclear-power countries will enable us to look at these characteristics in greater detail. The IAEA has listed about fifty countries, on the basis of expressions of interest for the advice it offers newcomers.

Most of the aspiring nuclear powers are countries which do not stand a chance of joining the club in the next twenty years. For example, the IAEA list includes the Republic of Haiti, Jamaica, Bahrain, Bangladesh, Tanzania and Sudan. A substantial budget and a powerful electricity grid are pre-conditions for building a nuclear power plant. If we follow the advice of José Goldemberg and exclude all those with gross domestic product lower than $50 billion and grid capacity of less than 10 GW,[17] the list of candidates only contains fifteen names: Algeria, Belarus, Chile, Egypt, Indonesia, Kazakhstan, Kenya, Malaysia, Philippines, Poland, Saudi Arabia, Thailand, Turkey, Venezuela and Vietnam.[18]

What distinguishes them from past entrants is the large number of countries with significant oil and gas reserves.[19] The recent arrival of the UAE is emblematic in this respect. Following a call for tenders completed in 2009 and awarded to a South Korean consortium, four reactors have been ordered. Oil states draw attention to rapidly increasing electricity demand, particularly to power sea-water desalinization plants. However, such demand could very well be met by gas, which is plentiful in many cases. The energy-independence argument rings hollow. In fact, these candidates see nuclear power as a means of coping, in the very long term, with dwindling hydrocarbon reserves. Furthermore, they often have difficulty bringing domestic oil and gas prices into line with international rates. Iran and Nigeria are examples of the popular outcry prompted by attempts to reduce subsidies. Using nuclear power to generate electricity is consequently a way of preserving export revenue.

The final point made by countries with rich hydrocarbon reserves relates to climate change. This too may seem something of a paradox, but should be taken as reflecting a determination to improve their international image and reputation rather than any powerful commitment to combat global warming. In a general way it seems odd that doubtful or even serious aspiring nuclear powers should cite climate change as a motive. Some countries, such as Chile and

Turkey, are far from having exploited the full potential of their hydraulic resources. Others, such as Kenya, have substantial biomass potential. Above all, almost none of them have made any ambitious commitments to cut greenhouse gas emissions. Poland is an exception to this rule. As a member of the European Union it is a party to Community policy on climate change and must meet its targets for reducing carbon emissions. At present 92 per cent of Polish electricity is generated using coal. The option of importing large amounts of Russian gas to diversify its energy mix is not on the cards as the country has no desire to depend on its powerful neighbour. In 2005 this combination of factors led the government to adopt nuclear power.[20] A call for tenders for the first two reactors was due to be issued in 2012, but was postponed when it emerged that Poland may have large shale-gas reserves. The priority is currently to explore that possibility.

Of course, in the case of Poland and other candidates, nuclear power is a less effective way of achieving energy independence than local resources, be they hydrocarbons or renewables. The countries will have to import the fuel. In the past some new entrants – Germany, the Netherlands, Japan, Brazil, China, India and Pakistan – have invested in uranium enrichment plants, reducing their dependence on imports of raw uranium. This is a relatively minor constraint, ore reserves being fairly evenly spread between developed countries, such as Australia and Canada, and less politically stable sources, such as Kazakhstan. Countries now planning to build nuclear power plants will find it almost impossible to enrich their own uranium. Global installed capacity already far outstrips demand and for a new plant to break even in a single country it would need to serve about twenty reactors. Under the circumstances, plans to build such a plant may be taken as signalling military ambitions, or at least interpreted as such by the international community, with the risk of outside pressure, followed by sanctions.

To round off this review of aspiring nuclear-power countries let us look at Turkey, the most advanced nation after the UAE to seek to join the nuclear club. Work on building the first power plant at Akkuyu, in the south, is due to start in 2014. It will mark the culmination of a process which started in the early-1970s. Turkey first expressed a wish to embark on nuclear power years ago. A whole succession of development plans foundered. Calls to tender were issued, but were cancelled or never completed. With strong economic growth since the early 2000s, enhancing its financial position and boosting demand for electricity, the various obstacles have been overcome. Access to nuclear power is also testimony to the political determination to restore the nation's glorious past, a desire borne out in international relations, with Turkey adopting an increasingly assertive stance as a key regional player. Nuclear power will help boost its prestige, dispelling the image of a backward country. In terms of energy independence the motives are less clear. On the one hand, nuclear power will enable it to reduce imports of Russian gas, which currently account for 30 per cent of electricity output. But at the same time, the first nuclear power plants in the Turkish fleet will be funded, built, owned and operated by Atomstroyexport, a subsidiary of Russia's nuclear giant Rosatom.[21]

ELEVEN

# Nuclear exit

Several dozen countries are eager to adopt nuclear power, yet others, such as Germany or Switzerland, are trying to phase out this energy source. Why are nations which once banked on the atom now pulling out? It seems to make no sense, particularly as others, following the example set by the United Kingdom, are still determined to renew their fleet. Even France seems to be in a quandary: just a few years ago it started building a new reactor; now it has decided to reduce the share of nuclear power in its energy mix.

For countries already equipped with nuclear power plants, the first step towards phasing out nuclear technology is to stop further construction. A small number of countries have taken the plunge: Germany, Switzerland, Belgium, Spain and Sweden.[1] The next question is what to do with existing plants. Should they be shut down as soon as possible or is it preferable to wait till the safety authority or utility decides they no longer fulfil the conditions for safe or cost-effective operation? In short, a choice between swift or gradual retirement. Germany took a decision on both issues in just a few months, making it a good example for study. France is equally interesting, on account of its decision to trim its sails. We shall analyse both cases in detail. What is interesting is that some of the political motives are similar.

## German hesitation over a swift or gradual exit

Germany started to consider phasing out nuclear power in the mid-1980s. Part of public opinion lost confidence after the Chernobyl disaster. The country's last nuclear power plant to be built, which was nearing completion in 1986, was subsequently shut down after three years' operation. Environmental campaigners in Rhineland-Palatinate took the case to court, which ultimately cancelled the plant's licence. The Social Democrats (SPD), once advocates of nuclear power, changed sides, adopting the same position as the Green party. The latter gained a lasting toe-hold in the Bundestag and some regional assemblies. Meanwhile, the realists prevailed over the fundamentalists. The battle against the state, which had been one of the Green party's priorities, was gradually sidelined in favour of mobilization for environmental conservation and against nuclear power. Well organized and electorally effective, the Greens evolved into the world's most powerful environmental party.

At the end of the 1990s, after sixteen years during which the federal government had been dominated by the Christian Democrats (CDU), the SPD won the general election and formed a coalition to govern with the Greens. It was no longer a question of promising the end of nuclear power; it was time for action. There were many obstacles, particularly of a legal nature, because the validity of the licence contracts issued to the operators was not limited in time and expropriation was out of the question. Ultimately, an agreement was negotiated with the utilities providing for the scheduled shutdown of plants.[2] It was ratified in 2002 by an amendment to the Atomic Energy Act. This established a quota for residual production by each plant, which once used up would lead to final shutdown. The last nuclear reactor was due to be turned off in about 2022. In exchange the government made various commitments to the operators, undertaking not to introduce any new taxes and only to prevent the use or transport of waste for technical reasons.

The CDU opposition condemned the scheduled phase-out of nuclear power. It promised to extend the operating life of reactors on returning to power in order to limit any increase in electricity prices and to leave a reasonable amount of time to deploy a new energy plan based on renewables. The German Right saw nuclear power as a transition technology that was necessary until such time as wind farms, solar panels, biogas units, energy-efficiency measures and high-voltage transmission networks could be rolled out. The aim was not to promote the construction of new reactors, simply to delay the closure of existing ones. German advocates of nuclear power had few illusions regarding a possible resurgence of this technology in their country. Given the influence of the Greens, new projects would never be able to overcome local and regional opposition. Indeed, a majority of German public opinion opposed a return to this source of energy.

Following the general election in 2009,[3] the CDU and its Liberal (FDP) allies set about implementing the projected nuclear transition. The Atomic Energy Act was accordingly amended once more in December 2010, three months before the accident at Fukushima Daiichi. It fitted into the larger framework of an Energykonzept designed to halve national electricity output by 2050 and triple the amount of power generated by renewables. The final phase-out of nuclear power was set for 2036, with the shutdown of Neckarwestheim 2, the last plant still operating. The closure date of each plant was determined in the same way as in the 2002 amendment: operators would be awarded a residual production quota, only this time it was larger, amounting to about 1,800 TWh.

German utilities grudgingly acknowledged the new law, which further encroached on their profits. They complained that the extension was too short. Every year they would have to pay a special tax on the uranium and plutonium they consumed, amounting to just over €2 billion. They would also have to pay several million euros a year into an energy-transition fund. The overall cost of about €30 billion was high but ultimately made sense. Extending the operating life of a nuclear power plant would boost the operator's profits, once most of the initial investment had been paid off. Of course, the extension entailed additional costs to maintain the level of safety and reliability, but they were generally far less than the revenue from additional operation. Allowing for work amounting to €500 million on each reactor, a variable cost of €12 per MWh, and assuming that the cost of building the fleet had been fully depreciated, J. H. Keppler estimated the total cost of generating the residual amount authorized by the quota to be about €30 billion (roughly €17 per MWh).[4] This sum must be set against revenue which could amount to €90 billion, on the assumption of an average market price of €50 per MWh over the relevant period. So the €30 billion in taxation would halve the gain the operators derived from their reactors' extended operating life. The utilities' lack of enthusiasm was all the more understandable given that the initial operating life of the reactors, used to calculate depreciation in their accounts, had been forty years. The new timetable for nuclear exit corresponded to an average of forty-four years, adding only four extra years. Of course, if one took as a baseline the 2002 agreement, which involved shutting down plants after an average of only thirty-two years, the new deal was more acceptable for the operators. Even allowing for the special taxes – theoretically non-existent in the case of the 2002 agreement, €30 billion with the new deal – it was more attractive.

The Fukushima Daiichi disaster upset this arrangement. The extension, which Parliament had just approved, was cancelled, and the timetable for phasing out nuclear power that had been negotiated ten years earlier was reinstated. All this happened very quickly. Three days after the disaster a moratorium on the extension of the operating life of reactors was decreed. The next day, it was decided immediately to shut down seven of the oldest reactors for three months. Plans to restart another reactor (Krümmel) were shelved. A week after Fukushima Daiichi, Chancellor Angela Merkel set up an Ethics Committee for a Safe Energy Supply, tasked with reassessing nuclear risk. At the end of May the Committee recommended phasing out nuclear power over the next decade. In August Parliament approved yet another amendment to the Atomic Energy Act.

It reinstated the shutdown timetable from the 2002 plan for the nine plants still operating and indefinitely extended the moratorium on the eight others. This was a serious blow for the utilities, taking them back to square one but with the additional burden of the special tax on fuel. They filed a complaint with the Constitutional Court and sued the Federal Government for damages, demanding several tens of billions of euros in compensation.

Why did Germany choose such a hasty nuclear exit? Foreign observers were stunned by the speed of this move. In a country committed to consensus-based politics decision-making is reputedly slow. Moreover, Mrs Merkel is not one to change her mind easily. The political context influenced the process. The Fukushima Daiichi accident occurred in the middle of an election campaign, with three regional elections set for late March in Germany. The CDU was determined not to lose Baden-Württemberg, which it had governed for the previous sixty-two years. By announcing a moratorium just a few days after the nuclear disaster, party leaders hoped they might contain the rising tide of Green voters. It certainly did some good, but not enough to prevent a Green–SPD coalition winning a narrow majority and electing a Green President to head the regional assembly. The CDU's share of the poll dropped by 5 per cent, whereas the Greens gained 12 per cent, compared with the previous election. Without the moratorium the swing would probably have been even greater. Another election was held in Bremen in May 2011, returning the incumbent SPD–Green majority to power. Here the Greens finished in second place, polling more votes than the CDU.

The cost of fast-tracking the nuclear phase-out also came as a shock abroad. In economic terms there is nothing to be gained from a swift shutdown, rather than a more gradual process. But the Germans, who habitually avoid unnecessary expense, did just the opposite. Early closure of safe, fully operational nuclear reactors represented a dead loss for the operator: the power no longer generated had to be replaced by drawing on other, more expensive sources.[5] The price of electricity for consumers went up, leading to unfavourable macro-economic impacts. Keppler, cited above, estimated the excess cost of replacing 20.5 GW of phased-out nuclear capacity with alternative sources at €45 billion, equivalent to an increase of €25 per MWh. The purpose of this estimate was to measure precisely the difference in cost between the two exit scenarios, before and after Fukushima Daiichi. In the first case generating 1,800 TWh would have cost €30 billion (see above). In the second case the author estimated that it would cost about €75 billion.[6] Other microeconomic studies of the cost of fast-tracking the shutdown of reactors have yielded comparable results, amounting to several tens of billions of euros.[7]

The economic loss entailed in hastening nuclear exit underlines the fact that the motives for the German decision were not economic. It was the Ethics Committee which recommended taking the shortest route. Chaired by Klaus Töpfer, a former Environment Minister, the committee achieved a consensus in a couple of months: nuclear power should be phased out as soon as possible. Its report cites 'the three pillars of sustainability: an intact environment, social justice and healthy economic strength', noting that the 'eternal burden' of nuclear-waste pollution runs counter to these goals. The Committee found that Fukushima Daiichi had changed public awareness of risk. Firstly, the fact that a major accident had occurred in a technically advanced country such as Japan undermined the conviction that a similar event could not happen in Germany. Secondly, this previously inconceivable disaster had revealed the inadequacy of probabilistic risk assessments. Lastly, the Committee emphasized that the damages could not be assessed, even approximately, before an accident, or even afterwards. Several weeks after the meltdown of the reactors at Fukushima Daiichi, there was still considerable uncertainty as to the final extent of pollution and its effects. None of the hazards were under control. It was not even possible to circumscribe a worst-case scenario. So there was no way of comparing the risks associated with nuclear power with those of other energy sources. Yet there were alternative means of generating electricity, in particular renewables, which were not a danger for humans or the environment. They should consequently be deployed as soon as possible in the place of nuclear power, within a decade according to the authors of the report. This brief summary outlines the arguments set forth by the Ethics Committee. We may note in passing that its diagnosis of the Japanese disaster was mistaken. The accident at Fukushima Daiichi could have been foreseen; it was not a 'black swan', as risk specialists put it. Furthermore, the real cause of the accident was an institutional breakdown, due to industry's capture of the regulatory authorities (see Chapter 8). Being technologically advanced, or not, had no bearing on the matter.

What prompted the decision to make the transition over a decade, rather than five or fifteen years? The report offers no explanation, simply suggesting that a shorter period, though ideally desirable, would jeopardize German competitiveness. No evidence was cited to substantiate the claim that ten years would be sufficient to avoid this risk. The ten-year timeframe was probably taken from the previous exit plan, which stipulated that the last power plant should be shut down in 2022. The same milestone was adopted without further calculation or debate. Whether it reflects lack of thought or political determination, the deadline has now been signed into law and Germany must lose no time in implementing its *Energiewende* (energy transition).

We lack sufficient historical hindsight fully to appreciate the consequences of speeding up transition. Apart from the dead loss discussed above and the compensation which the Federal Government – and consequently the taxpayer – may have to pay to the utilities, the change of timetable will probably have a negative impact on employment, inflation and economic growth. Such impacts are hard to assess, for all we can see at present are effects such as job cuts at nuclear power companies, higher electricity prices for households and a surge in investment in other energy sources.

They are difficult to untangle from the effects of the financial crisis, which has led to a drop in demand for energy, and those of the shift to renewables which started well before the decision to speed up nuclear exit was announced. Moreover, the impact of price shocks and investment bubbles must be seen in a long-term perspective, which may mean they have the opposite effect: an immediate rise in the number of jobs may in fact lead to long-term losses.

We cannot estimate the scale of the macroeconomic cost of accelerated transition, but it will probably be negative. It is hard to see any tangible economic benefit from such haste. In terms of labour, we shall see more rapid destruction of jobs in the nuclear industry, with less scope for reconversion than with a more gradual exit timetable, because training takes time. The renewable energy industry should benefit from the sudden increase in demand, but the inertia caused by training will also come into play here, preventing firms from taking full advantage of the situation. As for infrastructure, the massive investment in the power grid which will be needed is a major obstacle for transition. High-voltage power lines, thousands of kilometres long, will have to be built across Germany, in order to connect the wind farms in the north to major industrial centres in the south, where two-thirds of nuclear capacity is currently located. Here again there is considerable inertia. It takes years to complete power transmission infrastructure projects, due to the powerful local opposition they encounter. The plan for a gradual transition had already run into problems in this respect; the decision to speed up the process will only make things worse.

The only hope of any benefit depends on a sudden change of heart. Speeding up the nuclear phase-out, by accentuating the challenge posed by the energy transition, could galvanize the German people, stirring up its enthusiasm; greater collective effort and higher productivity could make the country more competitive in all the trades and industrial sectors involved in next-generation energy technology. The report by the Ethics Committee suggests a scenario of this sort. It is certainly the only one which would reconcile economic interest with the decision to fast-track nuclear exit. But such hopes are faint, perhaps completely vain. In economic terms, speeding up a transition process leaves no option but to make greater use of existing technology, even if it has to be imported, while investing less in research and development to improve future competitiveness. Since 2011 the change in plan seems, above all, to have played into the hands of coal and lignite, which also benefit from lower prices on foreign markets and on the EU Emission Trading Scheme.

In conclusion, on the basis of the German case we may suggest a scheme for explaining the decision to phase out nuclear power, and why the process was speeded up. It is based on two conditions: a political party which makes phasing out this technology a central plank of its platform; and government intervention on risk led by perceived probabilities, not expert calculations (see Chapter 5). Let us imagine then that the entire electorate is spread along a straight line of finite length, with at one end the person most hostile to nuclear power, who advocates the immediate shutdown of all reactors, and at the other the person most in favour of nuclear power, who would endorse the construction of dozens of new reactors. Moving in from the two extremities, towards the median voter – leaving the same number of people on either side – the views will become increasingly moderate. To be elected, the candidate must determine the median voter's preferences, in order to target that person and gain his or her support. If this can be achieved the candidate takes half the poll plus one vote, thus winning a majority.

Obviously an election is much more complex, with voters expressing a view on a large number of different topics such as living standards, jobs and education. Furthermore, there are alliances between parties specializing in various segments of the electorate. But this simple model taken from the theory of political marketing is sufficient for a basic understanding. If nuclear power ranks as a key issue and there has been a shift in public opinion, following a nuclear accident for instance, the median voter will view this technology less favourably. Unless the candidate changes his or her platform, votes will be lost with the risk of election defeat. A choice must be made between personal conviction, prompting the candidate to advocate keeping nuclear power, and gradual phase-out, if the latter option corresponds to the results of the expert appraisal; economic calculation; and his interest in winning the election by keeping step with the electorate's changing perception. On the other hand, if anti-nuclear feeling ebbed, a Green candidate would have to choose between keeping faith with party militants and combating nuclear power, and the risk of losing votes.

## In France early plant closure and nuclear cutbacks

Unlike Germany, France has not decided to phase out nuclear power. But it plans to close the Fessenheim nuclear power plant, in Alsace, ahead of schedule. It has also made a commitment to reduce this technology's share in its future energy mix. Underpinning these decisions we find the same factors as in Germany: a party built around combating nuclear power; a more acute perception of nuclear risk in the aftermath of the Fukushima Daiichi disaster; electoral competition; and political alliances which make allowance for risks as perceived by the general public, not as calculated by experts. Just as in Germany, we shall see that decisions on targets and the time schedule have been based mainly on approximation, without paying much attention to economic factors.

The Fessenheim plant, located close to Germany and Switzerland – west of Freiburg, north of Basel – has been a bone of contention ever since it was built. Its two reactors were connected to the grid in April and October 1977. Due to its position it is subject to specific safety measures. The seismic risk, though low, is higher than average due to the proximity of a fault line in the Rhine plain. The plant must also be able to withstand flooding, in the event of a break in the dyke on the Grand Canal d'Alsace. Lastly, it is built on top of a large aquifer; contamination of this source of freshwater would be a major disaster. In July 2011 France's Autorité de Sûreté Nucléaire reported favourably on plans for Fessenheim's number 1 reactor to continue operating for a further ten years. The decision was conditional on work estimated at several tens of millions of euros being carried out, most importantly measures to consolidate the concrete slab below the reactor in order to limit the risk of polluting the aquifer. The second, similar reactor had also qualified for an extension, subject to comparable requirements for improved safety. In short, the regulator concluded that, subject to the prescribed work being carried out, in compliance with the law, the risks would be 'adequately' limited (see Chapter 9).

Does the French President have a different opinion? Opening an environmental conference on 14 September 2012, Mr Hollande announced that the Fessenheim plant would close in 2016.[8] Justifying this decision, he cited its age – it is indeed France's oldest nuclear power plant – and position in a seismically active area, subject to flooding.[9] At first sight, there is little one can say: the democratically elected government supervises the independent regulator in the exercise of its mission. The former is empowered to intervene if it considers that safety targets are not being reached, or because the regulator has wrongly interpreted, or opted not to apply, the necessary measures. There is nothing surprising about policy-makers exercising their power over safety on national territory and setting stricter safety requirements to make allowance, for example, for new data.

Unfortunately this explanation does not add up. Firstly, however counter-intuitive this may seem, age is not a very useful criterion for determining how dangerous reactors are. Think for a second and you will soon see why. Contrary forces compensate for the dual effect of wear and technological obsolescence on the safety of old reactors. Due to their age, they are the focus of greater attention on the part of operators and safety authorities. Frequent, thorough inspections are carried out. In France, for example, the thirty-year inspection, prior to a possible extension of its operating life, involves a complete overhaul. Furthermore, the older a reactor is, the more new parts and equipment it boasts, for instance the new steam generators EDF has fitted to its oldest reactors. Above all, the regime implemented by the safety authority is not a two-tier process, with lower requirements for old reactors and much higher standards demanded of their more recent fellows. All the reactors, young and old, are subject to the same rules, the same level of safety with regard to maintenance and operation. It is consequently misleading to assert categorically that the oldest reactors are not as safe (see box).

Secondly, the French President's decision has not been endorsed by any new study or review of existing data by his advisors, the government, the administration or even an ad hoc group of experts. No technical input on the risks entailed by reactors informed the choice made by Mr Hollande. He was no better informed as a presidential candidate, when retiring Fessenheim featured among his campaign commitments. In fact closure of the plant is entirely due to a political compromise. The French Socialist party (PS) is a long-standing ally of the Green party (EELV), which has opposed nuclear power since its inception. Six months before the presidential election in 2012 an agreement was signed by the two parties. It included Fessenheim's immediate closure and a cut in nuclear power's share of the energy mix, bringing it down to 50 per cent by 2025. At the same time agreement was reached in preparation for the parliamentary election, following the presidential poll. It allocated about sixty constituencies to the Greens with no opposing Socialist candidate in the first round, in order to strengthen their local powerbase. Agreement on their political platform also opened the way for Green support for the Socialist candidate in the second round of the presidential contest. On the campaign trail, a few months later, Hollande endorsed the target of a 50 per cent nuclear share in the energy mix. But he

### Age is a misleading criterion for determining how dangerous reactors are

'Start by shutting down the old nuclear plants, as they're the most dangerous' would appear to be a slogan rooted in common sense. The older a reactor, the more worn-out its parts; they must consequently be more fragile, making the reactor more accident-prone. QED. Furthermore the oldest reactors were built longer ago, at a time when safety standards were less strict than nowadays and safety technology more rudimentary. Unfortunately, the facts do not confirm this intuitive reasoning.

A simple method involves checking whether there is a relation between the age of reactors and the number of times an emergency shutdown is caused by a human or material failure jeopardizing safety. Unplanned reactor shutdowns, not connected with scheduled maintenance work, refuelling or inspections, are a widely used safety performance indicator. It – the Unplanned Unavailability Factor, as it is technically known – is listed in the International Atomic Energy Agency database, which is easily accessible online. We used it to analyse the 143 reactors in the European fleet, in order to work on a larger number of observations than for just the French fleet.[10] This method shows that reactors do, on average, suffer more technical problems giving rise to unplanned shutdowns between the age of 30 and 40 years – 16 per cent more to be exact. But if we compare this age group with the performance of reactors in the first years of their operating life, the balance tips in favour of the oldest plants. More technical problems occur during the first years of operation. This may come as a surprise, but it reflects the difficulties encountered on reactors while they are being commissioned and during the first few years of operation. Care should also be taken with reasoning based on an average. There are few old reactors. Only two reactors in Europe are more than 40 years old: Oldbury 1 and 2 in the UK. As a result, the safety-indicator average is based on too small a number, reducing its statistical value. With this exercise there is no way of knowing whether the curve plotting safety against age flattens out after the first five years, or whether great age translates into declining safety. Indeed, there is no statistical evidence against the hypothesis that safety actually improves with age.

A more complicated method, based on data which though available to the general public takes a long time to obtain, involves looking for a correlation between the age of reactors, their operating characteristics and the number of incidents (level 1 or 2 on the INES scale: see introduction to Part II). The number is about ten per reactor per year. Here the results are more conclusive. The first observation is that incidents occur most frequently during reactor shutdown. There are two reasons: it is easier to observe failures, a much larger part of the reactor being accessible to inspectors; and work is mainly carried out during reactor shutdown, with intervention by machines or humans, which in turn entails a higher risk of incidents. Consequently, if the aim is to analyse the number of incidents on two nuclear reactors, one cannot compare a unit which has been shut down for some while for refuelling or a ten-year inspection with one which has been operating non-stop. After correcting this bias, we found that the oldest reactors suffered fewer incidents than the more recent ones. Those familiar with the EDF fleet will not be surprised. The fleet consists of various reactor models, known as *paliers*. Reactors belonging to the first *paliers* are characterized, on average, by a lower number of incidents than those in the last group.

To distinguish the age effect from the 'reactor model' effect, we must compare reactors in the same *palier*, depending on whether they were built first or last. Statistical processing shows that there are fewer incidents on the oldest reactors in any one *palier*. This finding may reflect a learning effect. The first unit in a series certainly displays more defects than the following one, due to the management and running of construction; commissioning is fraught by teething problems. Similarly there is more trouble on the second unit than the third. Only four reactors belonging to the N4 *palier* were actually built, which partly explains why this model, despite being the most recent and the most technologically advanced, has also notched up an above-average number of safety incidents. So if the construction date were to be taken as the only criterion for retiring a reactor, attention should focus on Civaux and Chooz, France's most recent operational reactors! This quip should serve to underline how misleading it is only to use age as a basis for ranking reactors, deciding in which order they should be retired, and reducing the share of nuclear power in a nation's energy mix.

It is consequently inaccurate to state categorically that the oldest reactors are not as safe.

only made a firm commitment to close one nuclear power plant – Fessenheim – during his first term of office. So his statement in the environmental conference in September 2012 merely confirmed an earlier decision.

Clearly a concern for political and electoral balance prevailed. The PS-EELV agreement is a compromise between a party which wants to keep nuclear power and another which advocates its rapid phasing-out. At first sight agreement seems impossible. In fact by adding an adjective both sides could accept the accord, for it promised 'la sortie du tout nucléaire', an end to [the] all-nuclear [energy mix]. France's nuclear-electricity fleet currently covers almost all of semi-base and baseload demand, amounting to three-quarters of the nation's electricity consumption. Of all the large countries which developed nuclear power, France is the only one where this technology is so predominant. Expressed in figures, ending the all-nuclear energy mix means halving the share of electricity generated by nuclear power plants. But cutting back its share of the mix from 75 to 50 per cent by 2025 is little more than a game with round numbers which are easy to remember and communicate. No technical factors underpinned this choice, nor yet an age limit requiring the closure of all reactors more than 30 or 40 years old.[11] The 2025 deadline echoes the 25 per cent difference between three-quarters and a half![12]

In terms of electoral balance Hollande seems to have made the right choice, winning the Socialist primary and then the presidential election. Let us return to the median voter model, which posits that in order to be elected a candidate must guess the preference of this hypothetical voter. Countering his main adversary in the primary, Martine Aubry, Hollande refused to support completely phasing out nuclear power, even in the long term. The number of Green supporters this stance lost him was no doubt fewer than the number of Socialist voters that he gained from it, the latter taking a much more moderate view of the role of nuclear power. Subsequently, in the presidential campaign, Hollande came up against Nicolas Sarkozy, who was in favour of an 'almost all' nuclear energy mix and criticized plans to close Fessenheim. Here again, Hollande was probably closer to the preferences of the median voter.

So, if any calculations preceded the decision to close Fessenheim, they focused on vote projections, not nuclear safety.

Nor did the calculations focus on projected costs. Debate on how to assess the economic cost of closing Fessenheim only started once the decision had been reiterated.[13] Newspapers asserted that several billions would go up in smoke.[14] The plant's early retirement would entail particularly damaging asset destruction for the operator: large components, such as the steam generators, have recently been replaced. Moreover, as we mentioned above, ASN demanded additional work costing several tens of millions of euros as a condition for issuing a licence for another ten years' operation. All these investments may be considered a dead loss and EDF might demand compensation. The fact that the company is publicly owned does not necessarily make a difference. Private minority shareholders are unlikely to allow themselves to be cheated. In addition, Swiss and German utilities hold special production rights on almost a third of the output from the two reactors. They will probably expect compensation, which might help EDF press its case. In terms of the general interest, early retirement of Fessenheim is not an ideal solution. It entails a microeconomic loss which depends, as we saw previously with the German plants, on a whole series of assumptions, such as electricity prices in the future, the cost of alternative resources, investment in further refurbishment, the level of the load factor and the 'normal' operating life of nuclear power plants. This loss could amount to several billions or even more.[15]

I wish to make it clear that I see no problem with policy-makers opting to prematurely retire nuclear power plants, even if their motives are exclusively electoral or driven by a political alliance. That is how democracy works. My concern is that at no point was the debate properly informed, documented or debated in technical and economic terms. Decisions were taken on the early retirement of reactors, despite their having received a clean bill of health from the safety authority and with total disregard for the consequences, other than shifts in voting patterns and the balance within and between political parties. It is quite possible that making allowance for these technical and economical factors would not have made any difference to the final outcome, but at least the political decision would have been taken with full knowledge of the facts and consequences.

In conclusion, the issue of phasing out nuclear power needs to be uncoupled from the question of the length of a nuclear fleet's operating life. This separates the decision not to build any more nuclear reactors from the other one regarding the early or late retirement of existing plants. Deciding to keep nuclear power as part of the nuclear mix is a major gamble. As we have seen in preceding chapters, the future competitive advantage of nuclear power is uncertain, much like the frequency and scale of accidents and the effectiveness of safety regulation. The decision to extend the life of existing reactors once the safety authority has given the green light is in no way comparable. We have a clear idea of how much it costs to generate electricity using existing nuclear power plants, and each plant has been examined in painstaking detail. When policy-makers decide in favour of fast-track rather than gradual exit, they are not settling an uncertain wager but engaging in massive asset destruction contrary to the general interest.

# TWELVE

# Supranational governance

## Learning from Europe

Let us step back now from domestic politics to look at international nuclear governance. We shall start by examining cooperation to prevent nuclear accidents, with a review of work in Europe. In this respect the European Union is a remarkable test-bench: the political and economic integration of Member States is far-reaching and long established; cooperation on safety operates within the framework of supranational bodies such as the European Commission and Parliament; there is even a specific treaty, which set up the European Atomic Energy Community. Moreover, the EU possesses an impressive nuclear fleet, almost a third of global capacity, occupying a relatively small space, divided by national borders: one in four nuclear power plants is located less than 30 kilometres from another Member State. A major accident is therefore likely to affect the population of several countries at the same time. Lastly, almost all of Europe's centres of electricity consumption and production are connected by a power grid. Shutting down power plants in one country impacts on the others; consumers in nuclear-free nations are partly supplied by nuclear-generated electricity. In short, with this level of political and energy interdependence, Europe is an ideal place to observe supranational nuclear governance, its advantages and limitations.

## Why is the national level not sufficient?

The national level is not sufficient, quite simply because radioactive clouds do not stop at borders, and nor do electrons. Chernobyl made the transnational nature of damage caused by a nuclear accident painfully clear to Europeans. Starting from Ukraine, the radioactive pollution mainly contaminated neighbouring Belarus. Driven by southeasterly winds, material from the reactor core also soon reached Scandinavian countries. Poland, Germany, France and the United Kingdom were also affected. The pollution had no impact on public health in the EU but caused serious concern, fuelling a lively controversy between the authorities and environmental groups.[1] As for electrons, the only border they encounter is the one formed by the power grid. The European grid interconnects all Member States,[2] reaching from Finland to Greece, from Portugal to Poland. More precisely, it is the electrical load which does not encounter any borders, as electrons only travel very slowly. It is a bit like water in a garden hose, which immediately spills out of the end when you open the tap but bears no relation to the liquid entering the other end of the main. It is the pressure which is transmitted so quickly, not the water.

The concept of a collective – or public – good casts light on the problem posed by permeable borders. As far as economic theory is concerned, it is at the root of the need for international governance. Public goods are collectively consumed products or services: they are accessible to all and each user consumes the same good. In other words, it is impossible to exclude users, and consumption by one user does not reduce availability for others. Street lighting in towns is a simple example often cited in the literature on such goods. It is there for everyone – residents, passers-by and even thieves who would probably prefer the cloak of obscurity. None of them deprive others of light by consuming it. These characteristics justify public intervention in supply of these goods, in particular to provide funding and determine both the quantity and quality available for all users.

Protecting the population against nuclear accidents counts as a collective good. Everyone benefits from this protection and enjoys, for better or worse, the same level of quality, which they may consider excessive or inadequate. Compared with street lighting, this service is more abstract, particularly as it is universal, not local, in its coverage. So it is more difficult to grasp its collective dimension. It is nevertheless of prime importance, the same problem arising as for street lighting with the need to take everyone's preferences into account, while ultimately imposing a uniform standard. Some people, because they live close to a power plant or have a heightened perception of risk, may demand a higher level of protection; others would happily make do with less. Were this an 'ordinary' good, it would not be a problem. The market supplies apples of various sizes and colours, and shoppers are free to choose whatever suits their taste. Not everyone is obliged to accept the same product.

To some extent local authorities and national governments can cope with such divergent preferences regarding collective goods. They can base their decisions on the institutions and rules which govern local and national democracy. But what is to be done when a collective good concerns several foreign jurisdictions? There is no alternative but to set up mechanisms of international governance. In the absence of common elections or rules, each state will decide on its own. On one side of the border residents might, for example, benefit from safety targets for nuclear installations twice as strict as the standards in force on the other side. Similarly, in the event of an accident, with liability regimes varying from one state to the next, people living on either side of the border would not receive the same compensation. Moreover, without common rules, if damages exceed the upper limit on liability, taxpayers in the country least affected will not contribute to compensating their neighbours in the most severely affected country.

Without supranational governance, tension is potentially highest between neighbouring countries which have adopted opposite courses of action on nuclear power. It is further exacerbated when their electrical systems are interconnected, as is the case in Europe.

The security of the power transmission network, balancing input and output, is a collective good in itself. The Germans and Italians, both now opposed to nuclear power, enjoy the same quality of service – in terms of voltage and security of supply – as the French. The collective dimension of the network is more clearly apparent in the compulsory nature of consumption which characterizes public goods. Just as thieves would gladly dispense with street lighting, those opposed to nuclear power would rather not be supplied by reactors in their own or neighbouring countries. However, this is impossible, unless they refuse to connect their dwelling to the power line. Europe's interconnected network means that consumers cannot forgo nuclear power. Short of disconnecting their homes, anti-nuclear campaigners determined not to use this energy source have no option but to leave the EU.[3] The same applies to militant nuclear advocates opposed to renewable energy sources: there is no escaping green electricity.

How, at a practical level, does the EU go about reducing such tension? With regard to its success achieving a relatively high level of safety, it may serve as an example; on the other hand, when it comes to rules on liability, it is in no position to lecture other countries. We shall see in detail how and why this is the case.

## European safety standards

Surprising as it may seem, there is no economic reason why all EU Member States should apply the same safety standards. In other words, there is nothing odd about a reactor being retired in one country after fifty years' service, whereas a unit registering exactly the same safety performance could be authorized to operate for a further ten years in another country. European safety standards need to be harmonized, but not equalized. This is due to the variation in costs and benefits of nuclear energy, depending on the population and where they reside.

Either they live in the immediate proximity of the nuclear power plant, or in a much larger surrounding area. Damage in the event of an accident mainly concerns local residents. The latter also enjoy a share in the benefits, in the form of jobs and local taxes. From an economic standpoint the safety target set for the plant must make allowance for local interests. These may vary a great deal from one region to the next. Some local authorities are very keen to keep nuclear power, others would rather it was banned. Looking at the bigger picture, damage may affect a much larger area and even remote residents may benefit from nuclear power, in particular with regard to electricity prices. Safety targets – and the standards required to achieve them – must consequently make allowance for the preferences of both local residents and people living well beyond the immediate proximity of the plant. As ever, a balance must be struck.

We may now apply the same local–global dichotomy to the separation between Member States and the EU. This is not a misleading over-simplification, because many EU countries are small and a large number of reactors are located close to an intra-Community border. Furthermore, Member States display widely divergent attitudes to nuclear power. An average of 35 per cent of EU citizens think the advantages of nuclear power counterbalance its risks, but the share rises to 59 per cent in the Czech Republic and plunges to 24 per cent in neighbouring Austria.[4] It is consequently impossible to set the same safety goals for the whole of the EU, in line with the highest, lowest or even median level. On the other hand, no country can behave as if the rest of the Community did not exist. In particular, low safety goals, which one Member State might accept, must not be possible if its immediate neighbours set higher targets. In short, neither the national nor the supranational safety level should be unilaterally imposed.

Europe seems to have found a satisfactory way of achieving a local–global balance in this respect (see box). Obviously, given the

## Europe's successful balancing act

The Commission succeeded in reconciling the Czech and Austrian positions on starting up the nuclear power plant at Temelin. The village is located less than 60 kilometres from the border with Bavaria and Upper Austria. In 1986 work started here on two reactors of Russian design, but things did not go smoothly. The original command and control system – the reactor's nervous system – proved defective and had to be replaced. At the end of the 1990s Austria opposed commissioning of the plant, on the grounds of inadequate safety. The Czech Republic, which had applied to join the EU, said it was prepared to continue upgrading the reactors to meet the strictest safety standards. Bilateral negotiations between the two countries ran into stalemate and the Commission stepped in to act as a mediator.[5] In the early 2000s the two parties signed an agreement which entitled Austria to monitor safety at the Temelin plant, in particular regarding compliance with the new standards the Czech government had promised to enforce. In those days there was no relevant EU directive. Had it existed, the previous agreement and the arbitration process carried out by the Commission would not have been necessary. On joining the EU the Czech Republic would have been required to raise the level of safety at Temelin to comply with the directive, and Austria, in the event of non-compliance, would have been able to take the matter to the Commission and the European Court of Justice. On the other hand, had the Czech Republic not wanted to join the EU it could have exercised its sovereign right to decide the appropriate level of safety at its plant. So, despite pressure from Austria, Temelin did not close, and thanks to the Czech Republic's determination to join the EU, its safety was substantially upgraded prior to commissioning. Europe did its job.

It was equally effective in imposing the shutdown of eight first-generation Soviet reactors in Bulgaria, Slovakia and Lithuania. It would have been impossible, without exorbitant expenditure, to upgrade these reactors to meet prevailing European standards.[6] The EU made their closure a condition for allowing the three countries to join the Community. To substantiate its case it drew on the expertise of the heads of national safety authorities, who belong to the Western European Regulators Association. In so doing, it succeeded where international cooperation had previously failed. When the Soviet bloc collapsed, the Group of Seven countries set up a programme to improve reactor safety in Central and Eastern Europe. But it was never really implemented. It only led to one closure: reactor number 3 at Chernobyl. Reactor number 2, at the Metsamor nuclear power plant in Armenia, is one of the most dangerous in the world but it is still running.

difficulties in setting targets and criteria for assessing safety performance, the balance does not take the form of quantitative risk thresholds, more a set of rules and standards. The EU requires for instance that safety in each Member State should be regulated by an independent authority. More broadly, all the standards recommended by the International Atomic Energy Agency are binding throughout the EU. They are mandatory and non-compliance is sanctioned. This constraint is enshrined in the European Directive, of 25 June 2009, 'establishing a Community framework for the nuclear safety of nuclear installations'. As with most European legislation, drafting this directive was a long, difficult process, spread over several years and involving many versions. Member States disagreed over its necessity and aims; the Commission advocated a more federal project than the Council.[7] But the result is there for all to see, the first binding supranational legal framework ever to apply to nuclear safety.

## Europe's patchwork of liability

European Union citizens do not all enjoy the same compensation rights in the event of a major nuclear accident. The amount varies depending on the country in which the plant is located. This discrimination is exacerbated by the many discrepancies from one state to the next regarding the legal definition of damages, the time limit for filing claims and their subsequent investigation, the rules prioritizing the allocation of funds, and so on. As a result, if an accident happens in a power plant close to the border between two Member States, residents will not receive the same compensation, depending on which side they live. Exaggerating slightly, there are as many civil liability regimes for nuclear damage as there are Member States. Yet, in a broader perspective, international nuclear law should bring about a certain uniformity. There are only two Conventions – one signed in Paris, the other in Vienna – to which countries may be a party, and the Euratom Treaty provides a basis for joint legislation on liability. Unfortunately, in practice changes and revisions to the two conventions have created an impenetrable legal jungle, and Euratom is, to all intents and purposes, still an empty shell or almost (see boxes).

There is nothing inevitable about the patchwork of liability which blankets Europe. A legal framework exists, in particular in the Euratom Treaty, within which Community legislation on liability and insurance could be established. The Commission plans to table a draft directive on the matter in 2015. Member States hostile to nuclear power will probably press for very high levels of liability and insurance coverage, whereas advocates of further nuclear growth will certainly oppose such measures. A compromise is quite possible, even if coexistence between countries for and against nuclear is increasingly fraught.

## The Paris and Vienna Conventions

The Paris Convention of 1960 on Nuclear Third-Party Liability was adopted under the auspices of the Organisation for Economic Cooperation and Development, and its Nuclear Energy Agency. The Vienna Convention, which is slightly more recent, was the work of the IAEA. Each of these liability regimes has established its own rules, capping compensation, determining the scope of damage covered and investigating complaints. So no country can ratify both conventions. For historical reasons, the first nations to join the European Community signed up to the Paris Convention, whereas more recent Member States such as Bulgaria and Romania, and all former eastern-bloc countries, subscribed to the Vienna Convention. This initial separation was followed by a multitude of other divisions. An additional Brussels Convention, already amended once, supplemented its Paris counterpart, which has itself been amended twice. The Vienna Convention has been revised once. But signatory countries have not necessarily kept pace with these changes. Some have stuck to the original versions, others have signed the latest versions, but not ratified the texts. Similarly some countries have signed or ratified a protocol bridging the gap between the two conventions, others have not. The protocol establishes mutual recognition between the two international regimes: victims in a third country will be compensated in line with the regime in force in the country where the accident occurred. To complicate matters further, a third convention was drawn up in 1997. It commits countries – whether they are bound by the Vienna or Paris–Brussels regime – to pay additional compensation to victims. Only three EU countries have signed it. Lastly, any Member State is at liberty to set ceilings and levels of liability higher than those established in the various conventions. All in all, these disparities make for substantial differences between individual countries in the EU. Take for example three neighbours: in France the upper limit on EDF's liability is about €100 million, to which the state undertakes to add the same amount; in Belgium the ceiling for GDF-Suez is €300 million, but with no commitment by the state; in Germany there is no limit to the liability of operators, with a state commitment amounting to €2.5 billion.

## Euratom, a pointless treaty?

In 1957 two treaties were signed in Rome. One, as we all know, established the European Economic Community. It brought into existence the Common Market, evolving after a series of revisions into the European Union as we know it today. The second treaty, now largely forgotten, established the European Atomic Energy Community. It aimed to encourage and assist the development of joint nuclear activities. But Euratom, as it is known, made little contribution to European integration and is still a largely empty shell.

Of the two Communities, Euratom seemed at first sight the more promising. Jean Monnet, one of the prime movers behind European integration and an instigator of this treaty, thought it was more likely to succeed, due to its focus on a specific sector. Nuclear power followed on from coal and steel in the gradual advance towards economic integration. Once its benefits had been demonstrated in a particular sector, it would be easier to overcome resistance in other sectors. Pro-Europeans all subscribed to this belief. In contrast, the customs union affected all industrial activities and could be construed as impinging on national sovereignty. Euratom was expected to be all the more effective as nuclear power was a brand-new technology, purportedly without deep-rooted national interests. The benefits of nuclear integration were perfectly obvious: the Six were individually too small to cope with the massive investment required to develop this energy source; they all faced rising demand for energy and were fully aware of their dependence. It should be borne in mind that in the late 1950s civilian nuclear power was seen as the solution to the lack of cheap, abundant energy. The preamble to the Treaty asserted that 'nuclear energy represents an essential resource for the development [...] of industry'. The aim was 'to create the conditions necessary for the development of a powerful nuclear industry which will provide extensive energy resources, lead to the modernization of technical processes and contribute, through its many other applications, to the prosperity of their peoples'. Germany was strongly in favour of the customs union, but France less so. On the other hand, it advocated nuclear integration. So to win over both parties it was decided to join the two projects.

Euratom was based on mistaken assumptions and never developed. Nuclear power was not brand new. Some states, including France, had already taken the lead, whereas others, such as Germany, reckoned their industry could catch up, particularly with American help. There was no sense of an equal footing, and there were already powerful national interests, both industrial and diplomatic. It proved hard to achieve the smooth progress posited by the advocates of an atomic confederation. Civilian nuclear power, much like its military cousin, had become a national affair, a field in which to assert nationalist ambitions. Unlike the Common Market, negotiations as part of Euratom failed to contain the rival interests of Member States through trade-offs between sectors.[8] Furthermore, the energy environment had changed. The blockade of the Suez Canal in 1956 had prompted fears of lasting oil shortages and high prices, but these were soon allayed, with oil prices dropping again.

The missions set forth in the treaty included establishing joint companies, to develop a European nuclear industry; securing a centralized supply of uranium ore and nuclear fuel; harmonizing rules on liability; and facilitating insurance contracts. But in practical terms these, and other objectives, came to little, due to the lack of large-scale concrete undertakings. Indeed, there was even talk of disbanding Euratom altogether. However, the treaty has survived, with its original aims. This other Community is still a legal person, separate from the EU, though they share various institutions such as the Commission and the Court of Justice.

Euratom has nevertheless progressed in some areas, particularly research, public health and safety. It funds a joint research centre and has engaged in an ambitious project to develop nuclear fusion, with the construction of an international experimental 500 MW reactor, known as ITER. It has also established common standards to combat ionizing radiation, particularly to protect people exposed at work (staff at nuclear power plants, radiologists, among others). It contributes to efforts to contain proliferation and, thanks to its regional reach, carries more weight than individual Member States. Lastly, with regard to safety, Euratom was behind the 2009 Directive on the safety of nuclear installations. It also set conditions for former eastern-bloc countries joining the EU, requiring them to upgrade safety on existing nuclear power plants. Ironically, the treaty made no provision for this particular advance. None of its 294 articles, despite all the detail, make any explicit reference to the Community's competence over the safety of ionizing-radiation sources, either to establish standards or to monitor compliance. A ruling by the Court of Justice was needed to give a broad interpretation of Euratom's initial competence for protecting public health, extending its remit from setting standards on the admissible effects of ionizing radiation to include equipment likely to cause radiation.

In view of its initial ambitions the Euratom Treaty has not achieved a great deal. But compared with the vacuum its absence would have left, it has not been entirely useless.

## Coexistence fraught by conflicting views on nuclear power

Once perceived as a force for European integration, nuclear power has become a source of discord between Member States. They all share the same network, but some are building new reactors (Finland, France) or plan to do so (UK, Czech Republic), whereas others have outlawed the technology (Austria) or decided to phase it out of their energy mix (Germany, Belgium). Tension between the two sides was palpable after the Fukushima Daüchi accident, when stress tests were carried out on the European fleet. The nuclear divide is preventing the EU from framing a joint energy policy.

In Europe, just as elsewhere, the disaster in Japan prompted national governments and regulators to review the safety of nuclear installations. With the financial crisis still looming large in people's minds, the term 'stress test', originally hatched to assess the resilience of banks, was recycled to assess the response of nuclear power plants to various hazards, in particular flooding and earthquakes. Unlike elsewhere, the tests were coordinated between neighbouring countries under the aegis of the European Commission and the European Nuclear Safety Regulation Group (Ensreg), which advises on safety.[9] This group brings together heads of administrative bodies or independent nuclear safety authorities in all Member States.

Coordinating the tests caused some tension. During the weeks following the accident at Fukushima Daiichi the European Energy Commissioner Günther Oettinger adopted an extreme position. Referring to an 'apocalypse', he claimed that some reactors in Europe did not meet the standards set by the Directive.[10] He asked for the stress tests to be extended to include terror risks, such as the possibility of an airliner being hijacked and targeted at a nuclear power plant. Most national regulators dismissed this idea. The Council of Ministers had not given the Commission a mandate to assess such extreme situations, quite unconnected with the causes of the Japanese accident. The head of France's Autorité de Sûreté Nucléaire, André-Claude Lacoste, suggested that this move by the Commission reflected its determination to shift the balance of power,[11] calling into question the existing national and supranational distribution of competence and power, laboriously hammered out during negotiations for the 2009 Directive. Ultimately, a compromise was reached. Ensreg would check certain issues at the interface between safety and security, such as falling planes, and an ad hoc group would be set up to address the security threat posed by acts of terrorism.

Another skirmish followed when the Commission presented the results of the stress tests. In a communication to the Council of Ministers and Parliament, it returned to the Ensreg results published six months earlier.[12] The summary report by the regulators concluded that given the level of safety there was no need to shut down any reactors. It also proposed a series of recommendations designed to increase reactor resilience in extreme situations. The detailed reports highlighted the shortcomings which needed to be corrected for individual countries and plants. But the summary of Ensreg results which the Commission included in its communication was much more alarmist. It criticized France, where allegedly only three installations made allowance for seismic risks. It also addressed seven recommendations for improvements to France, and only two to Germany. In a press release the ASN regretted 'the working methodology used by the Commission for drawing its conclusions' which 'ignore some important recommendations of the stress-tests final report'.[13] Lacoste subsequently explained that this statement was a much toned-down version of what he really thought. He accused Oettinger of intervening unduly in the stress-test process, of using Ensreg, and sapping national regulators' confidence in the Commission, which would make future cooperation between the two echelons more difficult.[14] In referring to national regulators, Lacoste meant active and experienced regulators. It should be remembered that he instigated the Western European Nuclear Regulators Association. Unlike Ensreg, Wenra only brings together the authorities of countries which possess nuclear power plants. States such as Ireland or Austria, which condemn nuclear power and have no experience in its supervision, are not represented. At Ensreg each country can vote and have its say, advocating safety requirements so strict as to be unattainable. The two organizations are emblematic of Europe's divisions over energy, with on the one hand the EU's twenty-eight Member States and on the other its fifteen nuclear-power countries. The Commission seeks to reconcile the interests of all parties, but the balance varies over time. The nuclear states – Wenra members – gained ground in the 2000s with the arrival of former eastern-bloc countries, most of which were open to this technology. But Germany's sudden determination to phase out nuclear power quickly was a blow, as was Italy's decision to shelve construction plans.

The nuclear divide is an insurmountable obstacle in the path of a common energy policy. It limits the effectiveness of EU policies to combat climate change, and establish a single gas and electricity market. A common energy policy, as advocated by Jacques Delors, former President of the European Commission,[15] would require Member States to give up, or at least share, sovereignty over their energy mix. But as long as the nuclear divide persists, a transfer of this sort is out of the question. It is comparable to the problem posed by the euro single currency. The euro crisis confirmed the views of theoretical economists who demonstrated that a single currency was not viable without coordinated or common macroeconomic, budgetary and fiscal policies. The Member States which make up the Eurozone do not all hold the same views on inflation, unemployment or growth; in the absence of coordinated budgets and economic targets, conflicting national strategies prevail, undermining trust in the currency and its overall resilience. In a similar way, a common energy policy cannot be viable without close coordination of decisions on the energy mix.

The Treaty of Rome, which established the EEC, left major decisions on energy policy in the hands of Member States. The exploitation of energy resources (the conditions of access and rate of extraction, for example), the choice of energy sources (in the case of electricity the mix between coal, uranium, gas, biomass, wind, water and so on), and the modalities of energy procurement (in particular its geographical origin) fell within their exclusive remit. The most recent update, the Lisbon Treaty, reiterated this national prerogative. But, for the first time, a specific article was devoted to European policy on energy. It is reproduced in the box here, almost in its entirety, there being no better way of summarizing it or high-lighting the tension between the national and supranational levels.

A country's choice to include nuclear power in its energy mix, or not, is certainly the most crucial decision for its EU neighbours. As we explained above, neither radioactive clouds nor electrons stop at borders between Member States. Furthermore, as individual EU countries display highly contrasting preferences, borders may separate – as with Austria and the Czech Republic – a Member State fiercely opposed to nuclear power from a keen advocate of the technology. Under the circumstances, it is hard to see supranational rules being framed which would require one state to phase out nuclear power, or on the other hand oblige another to adopt it. There is no likelihood of a unanimous vote on such a motion. However, setting aside nuclear power, national energy mixes in the EU are gradually converging. The overall trend suggests that coal is on the way out, whereas renewable energy sources are gaining ground. With increasing similarity in the mix among Member States, national sovereignty with regard to its make-up should be less of an obstacle for engaging and developing a common energy policy. As nuclear power exists, and is here to stay for a long time, it makes convergence impossible,

## Article 194 of the Treaty on the Functioning of the European Union

1. In the context of the establishment and functioning of the internal market and with regard for the need to preserve and improve the environment, Union policy on energy shall aim, in a spirit of solidarity between Member States, to:
   (a) ensure the functioning of the energy market;
   (b) ensure security of energy supply in the Union;
   (c) promote energy efficiency and energy saving and the development of new and renewable forms of energy; and
   (d) promote the interconnection of energy networks.
2. Without prejudice to the application of other provisions of the Treaties, the European Parliament and the Council, acting in accordance with the ordinary legislative procedure, shall establish the measures necessary to achieve the objectives in paragraph 1. Such measures shall be adopted after consultation of the Economic and Social Committee and the Committee of the Regions.

   Such measures shall not affect a Member State's right to determine the conditions for exploiting its energy resources, its choice between different energy sources and the general structure of its energy supply [ ... ].

The four goals cited in Article 194 endorse previous decisions by the European Council, Commission and Parliament. The decision to set up a large internal market for gas and electricity goes back to the late 1990s. Its aim was to open these sectors to competition and reduce barriers inside the Community, particularly regulatory obstacles to energy trade. Since then a great deal of legislation has been introduced, in the form of directives and regulations. The market is nearing completion. But to be a real success it must develop infrastructures connecting Member States for gas and electricity transmission. Progress has been made building interconnections, but much remains to be done. The goal for energy efficiency and renewables also corresponds to earlier commitments, which have been calculated precisely and established as mandatory targets for 2020. As for the security of supply, this above all concerns the gas imports on which Europe is dependent. This goal became a priority following the stoppage, during two successive winters in the late 2000s, of gas deliveries between Russia and Ukraine. The interruption in supply had a knock-on effect on several Member States. Hundreds of thousands of people in Bulgaria, Hungary and Poland were left without heating. The EU struggled to cope, hampered by poorly coordinated action by individual countries, non-existent exchange of information and the lack of a single spokesperson to deal with its Russian and Ukrainian neighbours.

The list of goals set forth in the Treaty is now closed, so new objectives can only be added if all the Member States agree. A unanimous vote is also required for measures to achieve the four targets. The end of Article 194 emphasizes that these measures must not impinge on Member States' right to exercise the sovereignty they have enjoyed since the Rome Treaty, that right being backed by a veto.

The spirit of solidarity invoked at the beginning of Article 194 offers the only opening for closer cooperation between EU countries. But the instruments, outlines and extent of this spirit remain vague.[16] Solidarity tends to affect relations with the outside world. Even in this respect it is of limited value. The EU as a whole was unable to agree on the routes to be taken by pipelines transporting gas from Russia and competing countries on its borders. Member States funded rival projects displaying a complete lack of organization, going so far as to short-circuit one another's schemes. Germany, for instance, gave preference to a pipeline along the Baltic seabed, rather than an intra-European land route via Poland.

With regard to EU energy policy, whether it concerns the single market, energy efficiency or development of renewables, the spirit of solidarity is even hazier. Does it involve compensatory measures in favour of countries such as Poland, where decarbonizing electricity generation is particularly costly for geographical or historical reasons? With regard to the choice of conventional means of generating electricity, does the spirit of solidarity require countries to announce their decisions in advance and coordinate their application in order to limit the resulting disruption in other Member States? Germany has decided to phase out nuclear power by 2022. It did so without consulting any of its partners. France is thinking about halving this technology's share in its energy mix by 2025. It has not discussed the matter with its neighbours either. It seems therefore that the solidarity cited in Article 194, already in force before these decisions, does not require national decisions on the energy mix to be coordinated nor yet information to be exchanged. Perhaps one day the solidarity proclaimed by Article 194 will exert leverage on a common energy policy, but for the time being there is no sign of any shift in that direction.

ruling out a European Energy Community. This divide has become all the more impossible to bridge since France and Germany have adopted different stances on the issue. So the team formed by the two countries, acting as a driving force for their partners in other fields, will not work for energy.

We should point out in passing that national sovereignty over the energy mix limits the effectiveness of two key policies: measures to combat global warming and the development of the internal market. Take, for instance, the case of renewables. To achieve the target of a 20 per cent share by 2020, it is 'every man for himself', with no overall organization. Germany is well ahead of Spain in terms of solar installed capacity, despite enjoying fewer hours of sunlight. Spain itself is far ahead of the UK for wind power, despite being less windy. In other words, installations are decided by national policy – in practice in the places where the subsidies for their development are highest – not according to resource availability. This makes the 2020 target, for the whole of the EU, all the more costly to achieve. Regarding the internal gas and electricity market, the effectiveness of European policy is undermined by the industrial nationalism which drives sovereignty over the energy mix. To achieve energy equilibrium, a country needs equipment and generating companies which are pursuing that goal. With certain energy sources being subsidized – by consumers or taxpayers – government and policy-makers are tempted to offer local jobs in exchange. A national mix often goes hand-in-hand with national industry. Stiffer competition in the Community gas and electricity markets does not necessarily suit some Member States – certainly as long as they lack a national champion capable of taking a share of neighbouring markets. The setting up of a single gas and electricity market consequently has been and still is hindered by the creation and consolidation of national energy industries.

In short, the nuclear issue makes it impossible for the EU to transfer sovereignty over the energy mix to the supranational level, which in turn rules out a common energy policy. Yet the lack of such a policy jeopardizes policies to consolidate the internal market and to combat global warming. Nuclear power is the root of the problem holding up a European Energy Community and it is hard to see how a solution may be found.

Reaching the end of this tour of Europe, what may we conclude about international nuclear governance? It seems fair to say that the EU has achieved a degree of cooperation between states on nuclear safety which it would be difficult to surpass. It certainly represents the upper limit of what the IAEA might one day achieve in terms of cooperation. The EU's legal framework provides precisely what the IAEA lacks, namely the authority necessary to make its safety standards and recommendations binding. The European Commission can enforce the requirements of the 2009 Directive approved by Member States, sanctioning those which fail to comply. It can impose fines and refer cases to the European Court of Justice. The only form of coercion available to the IAEA is peer pressure. If, for instance, an exceptional event occurred – a disaster causing a shock sufficient to galvanize all the countries defending this technology – and more extensive powers for monitoring safety were invested in the Agency, it would still not be able to do any better than the Commission. The latter body succeeded in having some hazardous reactors shut down, but only because it could play on the determination of the relevant countries to gain EU membership. In the aftermath of the Fukushima Daiichi disaster, had the result of a stress test been sufficiently bad to justify a plant being closed, the Commission would not have been empowered to take the decision on its own. The decision to commission or shut down a nuclear power plant is still within the remit of national governments. This sovereign power limits the scope of European nuclear governance, but it shows little sign of waning.

# THIRTEEN

# International governance to combat proliferation

## Politics and trade

The proliferation of nuclear weapons is a collective security issue, but microeconomics does not throw much light on efforts to manage this particular global public good, and the associated pitfalls. Political analysis and the application of game theory seem more helpful for understanding rivalry between states and their various strategies. Much has been written on global politics and strategy, and more specifically on the arms race between the Soviet Union and the United States, followed by de-escalation. We shall consequently not address the rising number of warheads deployed by nuclear powers, referred to as vertical proliferation. Its only connection with the economics of nuclear power is historical.

On the other hand, horizontal proliferation – the acquisition of nuclear weapons by countries not previously armed – is linked to the development of this energy. We have already cited the example of Pakistan, which has used civilian nuclear power as cover. We have also noted the utopian attitude of advocates of the Atoms for Peace doctrine, who believed that help with developing nuclear power could contribute to preventing widespread nuclear armament. This form of proliferation, by the gradual spread of nuclearized territory all over the planet, is linked to trade. The larger the number of new countries adopting nuclear power, the greater the risk of proliferation. The vast majority of the countries which currently own nuclear power plants have certainly refrained from developing military applications, but there is no technical obstacle to them doing so in the future. They already have the fissile materials, and the necessary scientific and engineering skills. Japan, Brazil or indeed South Korea could very rapidly become nuclear powers. Nations on the verge of deploying nuclear power, such as Turkey or the United Arab Emirates, could start by abstaining, then change their minds. It is also interesting to note that many of today's aspiring nuclear-power nations display features which justify fears of further proliferation. Compared with the countries already exploiting this technology, they are, on average, characterized by more widespread corruption and greater political instability, weaker democratic institutions, higher crime rates and more frequent terrorist attacks.[1] All these factors potentially increase the risk of nuclear power being diverted for military ends.

Stronger international governance is needed to contain the growing risk of proliferation. This is the purpose of the Treaty on the Non-Proliferation of Nuclear Weapons and the mission of the International Atomic Energy Agency. It is also the purpose of export controls over sensitive technologies, and the mission of a little known, but effective organization, the Nuclear Suppliers Group.

## The IAEA and the NPT: strengths and weaknesses

The International Atomic Energy Agency (IAEA) is one of many agencies established by the United Nations, much like the World Health Organization or the International Civil Aviation Organization. It was launched in 1957, largely inspired by the ideas proclaimed by President Eisenhower in his Atoms for Peace speech. It was consequently tasked with promoting the development of nuclear power worldwide and ensuring it was not diverted for military purposes. The Agency subsequently played an increasingly important part in nuclear safety, framing standards and guidelines. Its mission as an atomic police force was substantially reinforced with the signature of the Non-Proliferation Treaty (NPT), which came into force in 1970.[2] It was tasked with establishing supervisory measures to prevent nuclear technology being diverted. All the parties to the NPT were encouraged to accept technical safeguards applied by the IAEA.[3] The Treaty was initially ratified by 70 states. They now number 159, more than four-fifths of UN membership.

The best-known countries that have never signed up to the Treaty are the ones who acquired nuclear weapons after 1 January 1967: India, Pakistan and Israel. In the jargon of proliferation experts they are the 'non-NPT states', as opposed to the authorized nuclear-weapon states – France, the United States, the United Kingdom, China and Russia – which developed nuclear weapons before 1967. These five states have made a commitment not to help other countries to acquire nuclear weapons and to 'undertake effective measures in the direction of nuclear disarmament'. This obligation goes hand-in-hand with that placed on the other signatories never to receive, manufacture or acquire nuclear weapons. It is the source of considerable tension because the authorized nuclear-weapon states consider that the requirements of the Treaty regarding disarmament only constitute a limited constraint. It simply states that they must 'pursue negotiations in good faith on [ . . . ] a treaty on general and complete disarmament'. Most other signatories dispute this interpretation, demanding the destruction of the 20,000 or so nuclear warheads the 'big five' have accumulated. In the words of the former head of the IAEA, Mohamed el-Baradei, it is like 'some who have... continued to dangle a cigarette from their mouth and tell everybody else not to smoke'.[4]

Over and above disarmament and proliferation the Treaty encourages the spread of nuclear power through scientific cooperation and technology transfer. It restates the terms of the deal outlined by Eisenhower and set forth in the IAEA statute, granting access to civilian technologies in exchange for giving up military aims.

At first sight the Non-Proliferation Treaty and the IAEA are totally ineffective. This system obviously does not prevent non-signatories from acquiring nuclear weapons, as was the case with India; but above all it does not prevent signatories from attaining the threshold for producing nuclear weapons, while remaining a party to the Treaty: Iran attempted to do just this before international sanctions finally dissuaded it; North Korea simply withdrew.[5] But such a categorical judgement would be out of place, because a system's effectiveness cannot be assessed on a binary basis. Rules and organizations are necessarily imperfect and neither the Agency nor the Treaty are an exception. So the question is rather how effective they are. To find out, we need to make allowances for their failures and successes – some countries such as South Africa having shelved their nuclear weapons programme. We also need to consider what might have happened in the absence of this system, or if a different one had existed. But to judge by the literature, the jury is still out on the matter of feeble or strong effectiveness. In the view of some authors,[6] the Treaty – and the various forms of pressure it can exert – has done little to prevent proliferation. Non-proliferation is primarily the result of the determination and capacity of individual states. Some countries which could have produced nuclear weapons decided against the idea, typically Germany and Japan. Others, such as Egypt or Iraq, wanted to become nuclear powers but were unable to do so. Lastly, a small number took the decision and ultimately succeeded in producing weapons, witness North Korea. But other analysts contend that the Agency and Treaty definitely curbed proliferation of nuclear weapons, the last two countries cited above being particular cases.

Be that as it may, we must underline two shortcomings in efforts to stop proliferation. The first is the IAEA's lack of resources, both in terms of its budget allocation and the tools at its disposal for intervention. The Agency's budget has remained low and for a long time its interventions consisted merely of limited checks on installations, subject to the authorization of the host country. This left very little scope for inspecting and more importantly detecting diversion of civilian technology. The second relates to the linkage between non-proliferation and the promotion of nuclear power. Scientific cooperation has turned out to be the main vector for proliferation. To develop nuclear weapons, it is not even necessary to possess nuclear power plants; a research reactor is quite sufficient.[7] The IAEA and the NPT have greatly contributed to disseminating them all over the world. In this respect the deal set out by Atoms for Peace was an illusion.

Relations between international nuclear trade and global efforts to combat proliferation are complicated. There would certainly be less proliferation if there was no trade, but there would also be less nuclear trade with no international governance combating proliferation. We shall now look at how markets, politics and governance are entangled.

## Nuclear trade

International trade in nuclear goods is a small market. This may seem surprising, as public attention often focuses on large contracts for the sale of power plants costing tens of billions of euros. Understandably, the news that the South Korean consortium led by the Korea Electric Power Corporation had won the $20 billion tender to build four reactors in the United Arab Emirates made a powerful impression. But the payments for such contracts are spread out over about ten years – the time it takes to build the plant – and, above all, there are very few large contracts of this nature. In 2000–10 the global export market amounted to orders for two new reactors a year, some awarded following a call for tenders, others by mutual agreement. Furthermore, a nuclear power plant is a complex assembly comprising a pressure vessel, steam generators, piping and a control room, associated with equipment for generating electricity with steam. As with any thermal power plant, it is necessary to install turbines, alternators, capacitors and such. The nuclear island – the specifically nuclear part of a plant – accounts for roughly half its cost, with the conventional part making up the rest. Thus reduced to its essentials, the global reactor market is worth less than €5 billion a year.

However, to this relatively low figure for annual sales of new reactors must be added trade in uranium, fuel, maintenance services, spare parts, reprocessing of spent fuel and waste management. These specifically nuclear up- and downstream activities multiply by three the value of the world market. For integrated companies, such as Areva or Rosatom, which cover the whole cycle, business connected to building new reactors represents, at the most, only one-fifth of total revenue. Upstream and downstream activities are crucial for such companies, for they are recurrent and profitable. They are less erratic than orders for new equipment, less subject to intense competition. The operators and owners of nuclear power plants are to a large extent tied by their inputs to the company which built the reactor. For reasons of compatibility, know-how and technical information, the vendor has a competitive advantage over other suppliers of enriched fuel, spare parts and maintenance services. It enjoys market power, allowing it to increase prices and thus profits. Industrial economists who focus on markets for complementary goods (printers and ink cartridges, razors and blades, coffee machines and pods) are familiar with this mechanism. The first item is sold at cost price, or perhaps even subsidized, but the supplier makes up the initial loss on sales of recurring products.[8]

The US dominated the international market for many years. Until the mid-1970s three-quarters of the plants built elsewhere were either built by US firms or under licence.[9] But this dominant position subsequently crumbled. Its share has dropped to less than a quarter of all the reactors built in the past twenty years. By value, the US accounts for less than 10 per cent of global exports of nuclear equipment (reactors, large components and small parts), and about 10 per cent of materials (natural uranium and plutonium).[10] Indeed, it has become a net importer, amounting to $15 billion a year.[11] Meanwhile Canada, Russia and France have increased their market share. The first two export their proprietary reactor technology; France too,[12] though its technology was originally derived from US imports. South Korea recently joined the nuclear exporters' club, taking a similar route to achieve technological independence gradually. Japan is poised to do likewise. Its nuclear engineering companies have responded to calls to tender by new entrants such as Turkey and Vietnam. Former General Electric and Westinghouse licensees, they have gone further than their French and Korean counterparts, purchasing the nuclear assets of the two US companies.[13]

The decline in US nuclear sales abroad is not due to the arrival of more powerful competitors, rather to the collapse of domestic demand. From the mid-1970s to the end of the 2000s not a single contract was signed for a new nuclear power plant in the US. Engineering firms have had to weather more than thirty years without any domestic demand, preceded by years of uncertain profits sapped by the vicissitudes of regulatory pressure (see Chapter 9), with a major accident in 1979 (Three Mile Island) to crown it all. Enough to floor any industrial operation. The decline at home led to a massive loss of industrial capacity and skills in enrichment, the manufacture of large forged parts and construction engineering. On the other hand, reactor R&D and design has survived and is still of first-class quality. The first new reactors to be built on US soil in the 2010s are Westinghouse AP1000s. In 2006 China ordered four of these innovative next-generation reactors. General Electric has developed advanced boiling-water reactors too. So the US still features in new nuclear and the international market thanks to innovation and technology transfer. Westinghouse now sells brainpower rather than equipment. It is still earning money thanks to licence fees. For example, it received its share of fees on the construction of the four Kepco reactors in the UAE. The export version of this reactor still contains parts which belong to Westinghouse, including the software which controls the nuclear chain reaction in the reactor core. In short, the nuclear industry still operating in the US no longer comprises many factories and is partly controlled by Japanese firms, but it is still alive and profitable.

In general, the state of a nuclear engineering firm's domestic market and its export potential are very closely connected. Just as with the US, it is difficult to enjoy a significant share of the export market, without at the same time building nuclear power plants at home. The industrial fabric is not responsive enough, highly trained staff are not available in sufficient numbers and the technical skills are lacking. Oddly enough, when a domestic market is enjoying powerful growth, there is also less scope for exports: all efforts are focused on success on the home front, meeting deadlines, coordinating production and construction. China is a good example of this point. It is currently building twenty-eight reactors at various places in the People's Republic. Bearing in mind that the manufacture of heavy engineering parts is scheduled several years ahead of the construction project, winning foreign contracts would mean reallocating manufacturing output originally intended for the home market. This would slow down the national programme and delay the projected supply of additional electricity. So the ideal situation for exports is somewhere between non-existent and booming domestic demand. This is the case in Russia, which has exported the largest number of reactors in the past fifteen years. Unlike the US, work building new capacity never stopped, though new orders were temporarily shelved in the aftermath of Chernobyl and the collapse of the Soviet Union. Exports to Iran, China and India helped compensate for the momentary drop in demand at home. Ten new reactors are at present being built in the Russian Federation, and since the end of the 2000s additional contracts have been signed with India and China, but also Turkey and Vietnam. Another case in point is South Korea. Much as in many other sectors this country has succeeded in developing a top-grade national nuclear industry in a very short time. Initially output catered exclusively for the domestic market, building a fleet of reactors which now numbers twenty-three and supplies a third of the country's electricity. The aim is to reach 60 per cent by 2030. There is little likelihood of the fleet growing any more. South Korea is a small island, in electrical terms, with no scope for selling surplus output to either its neighbour in the north, or Japan. So without exports there is no room for South Korea's nuclear engineering industry to expand further. Just as it has done in shipbuilding, car manufacture or consumer electronics, it must export or die. It scored its first success with the UAE and it very much hopes others will follow soon.

## Which model: the armament industry, or oil and gas supplies and services?

The nuclear industry is very similar to defence procurement in many respects, but in the future it could resemble oil and gas supplies and services. In addition to the reasons cited above, the international nuclear market is small because individual states keep their orders for national industry, just as for arms. They give priority to technology that is either indigenous (Canada, Russia) or was originally licensed but has subsequently been developed locally (France, South Korea). In the immediate future, it is hard to imagine Russian or South Korean utilities issuing an international call to tender for the construction of a nuclear power plant on their home ground. China today and India tomorrow – if the latter launches an ambitious construction programme – also rely primarily on national firms and their own reactor models. The international market is thus restricted to the delivery of the first plants to be built by one of the main new entrants – in other words, countries which will subsequently develop their own fleet – and to supplies to countries which will never possess more than a few units. In both cases, a certain proportion of local content is one of the factors determining the outcome of a tender. The market is more open for large engineering components. EDF recently purchased forty-five steam generators, worth an estimated €1.5 billion, to refurbish its plants, entrusting a quarter of the order to Westinghouse and the rest to its traditional French supplier Areva. As there is only limited global capacity for producing the pressure vessels fitted to the largest reactors, two of China's AP1000s are fitted with boilers manufactured by Doosan,[14] of South Korea, whereas the French EPRs being built in France, Finland and China will use pressure vessels made by Japan Steel Works.

Another feature reminiscent of the arms industry is the active involvement of government in export contracts. Ministers and even heads of state intervene at a diplomatic level, but also meddle in finance, strategy and even organization. The UAE tender is emblematic in this respect. By 2009 two national consortiums remained in the running for the contract to build the Barakah nuclear power plant. One, led by Kepco, brought together Doosan for the steam-supply system, Hyundai for civil engineering, Korea Hydro and Nuclear Power for system engineering and Korea Power Engineering for design. The rival team, led by Areva, consisted of the utility GDF-Suez, turbine manufacturer Alstom, oil company Total and civil engineering specialist Vinci. Each consortium enjoyed the political and diplomatic backing of their respective head of state. Mr Sarkozy and Mr Lee Myung Bak visited Abu Dhabi several times to persuade UAE President Sheikh Khalifa bin Zayed al-Nahyan to choose their champion. To clinch the deal they offered financing facilities to the buyer, a move that seems almost laughable given the UAE's ample liquidity. The Import-Export Bank of Korea subsequently took out an international loan to fund half the project, no doubt borrowing at a higher rate than UAE banks would have obtained. The two political leaders very probably offered additional sweeteners. It is commonplace for large nuclear contracts to be associated with offers of military assistance, arms sales or infrastructure development projects, but such information is seldom made public. Regarding the Korean bid, the only detail which leaked to the press was that a battalion had been promised by Seoul to train Emirati armed forces.[15] More surprisingly, heads of state may even become involved in details of organization. In the run-up to the final decision Mr Sarkozy intervened to bring EDF into the French consortium, alongside Areva and GDF-Suez. On the Korean side, Mr Lee behaved like a commander-in-chief hectoring and encouraging his troops.[16] He intervened repeatedly in the preparatory stages of the project and negotiations to seal the contract.

State involvement at the highest level in nuclear export contracts results in companies rallying round the flag. Consortiums are national. Unlike major gas and oil infrastructure projects, they do not field companies from all over the world, which choose to make a joint bid for a tender on the basis of their respective affinities and complementary assets. This inevitably means nuclear consortiums are less competitive. Despite being less effective in terms of costs or know-how, a civil engineering firm or own-equipment manufacturer may nevertheless be co-opted because, like the other members of the team, it is French, South Korean, Japanese or Russian. State intervention is not necessarily an advantage for the vendors either. It forces them to reduce their margin, sometimes excessively. As a large nuclear contract attracts considerable media attention, any head of state is very keen it should be awarded to his or her country, in the hope of basking in the glory of successful national firms, particularly as an election campaign approaches. To clinch the deal a head of state may push the national consortium to offer the buyer more favourable terms and prices. It is particularly easy to exert such pressure when, as is often the case, nuclear companies are wholly or partly state-owned. The shareholder in person orders senior management to make do with a pitiful margin, or even to sell at a loss. The winner of a tender is often the biggest loser!

So on the one hand, the consortium needs the diplomatic, financial and strategic support of its state apparatus, but on the other hand, this may come at a high price. When it comes to political intervention, Russia leads the pack (see box).

## Russia: exports under Kremlin's control

Civilian nuclear exports are a priority for this country. Like most gas and oil exporting countries, it has very little industry which can compete on the export market. It must rely on raw materials. There is no manufacturing sector but arms which can compete in global markets with top international firms. The only exception is nuclear power. In this field Russia possesses considerable scientific and technical skills, a range of recently designed, powerful reactor models and dense industrial fabric. Russian leaders see the export of nuclear technology as a matter of national pride and a source of great prestige. Above all, it helps to achieve their diplomatic and strategic goals. The Russian reactors sold abroad are pieces on a global chessboard. In 1995 Russia carried on the job started by Siemens, building a reactor at Bushehr in Iran, on the shore of the Persian Gulf. After a whole series of setbacks, it was finally connected to the grid in 2013. The Russians lost a great deal of money on this scheme. In the same part of the world they won a contract with Turkey in the late 2000s for the construction of four reactors. Moscow is funding the whole project, drawing to a large extent, if not wholly, on the federal budget.[17] This subsidy is understandable when seen in the larger context of Russian gas interests. Turkey agreed to allow the projected South Stream gas pipeline to run through its territorial waters. In so doing it changed sides, withdrawing its earlier support from the rival Nabucco project, backed by the European Union. The sale of Russian reactors to Vietnam at the end of the 2010s was also sweetened by advantageous financial terms, this time in the form of export credits. Vietnam has long been a Russian ally in southeast Asia, particularly on the military front. For nuclear vendors from France, Japan or South Korea, Russia is a particularly tough competitor, the authorities being prepared to invest massively to facilitate reactor exports. Thanks to its gas rent, Russia's pockets are well lined. French, Japanese and South Korean heads of state are keen to help their nuclear companies win large contracts abroad, but they do not enjoy the same latitude as Vladimir Putin, nor are their arms so heavily laden with gifts.

State intervention obviously plays a key role in importing countries. Reactor vendors have two customers: the utility which will be operating the nuclear power plant, and the state. It is often more important to win over the latter, particularly if it is the former's only shareholder. The political dimension which dominates the importer's choice of a reactor vendor is manifest in bilateral agreements. For example, China and Vietnam did not organize an open call for tenders prior to choosing the plants they purchased from Rosatom in the late 2000s. An opaque selection process enables government to exercise its political and strategic preferences more freely and obtain often unavowable forms of compensation. Allowing an electric utility to organize an international call for tenders substantially reduces government's discretionary powers. However, in some cases the transparency and competitive openness of the tendering process is merely a front. What really counts is not the score awarded by the expert committee which analyses the bids, but the opinion of government. It may choose the losing party. Or the loser may be brought back into the running for equally political reasons. China opted to base its first four third-generation reactors on Westinghouse's AP1000, not Areva's EPR. But it nevertheless ordered two EPRs from the firm shortly afterwards.

Taking the oil and gas supplies and service industry as its model would make the nuclear industry truly international. This industry comprises the firms which supply the infrastructure for oil and gas exploration and production. It is cited here as an example of the engineering, procurement and construction industry, commonly known as EPC. It covers the whole supply chain that contributes to delivery of an oil rig or refinery, but in a broader context refers to any major industrial installation. So it can equally well apply to nuclear power plant construction projects, which involve drawing up an overall plan, adapting to a given site, purchasing hundreds of thousands of parts and the corresponding services to implement the project, and of course its overall completion. Who does what in this huge puzzle depends on the specific contracts, customers and suppliers. Some buyers just want to take possession of a turnkey project. In this case, either it will be delivered by a single contractor, as is the case with the plant supplied by Areva at Olkiluoto in Finland, or it will be the work of a consortium comprising various contractors, as with the Barakah plant in the UAE. Other buyers want separate contracts for the various parts of the job, for instance making a distinction between the nuclear island and the conventional generating units. In this case the customer must take charge of, or delegate to a design office, the coordination of the various contractors and their respective work packages. The utility may also opt to draw up a large number of supply contracts, acting as its own architect and engineer, as EDF has done at Flamanville, France. Or alternatively it may hire an outside service provider.

The diversity of approaches to project management is no different in oil engineering, procurement and construction. Firms such as Technip, Halliburton or Schlumberger organize themselves in much the same way, depending on the circumstances and the demands of the oil companies for which they are working. What is strikingly different in the nuclear industry is the uniformity of the national colours flown by individual companies making up a consortium, and most of their main customers. Russian firms work primarily for Russian utilities; the same is true in Japan; and so on, and so forth. Basically, the nuclear industry is not a global industry selling to customers all over the world, working with similarly diverse partners and suppliers. In oil engineering, procurement and construction, issues related to national politics are only apparent on the demand side, as the world's largest oil companies are publicly owned, from Saudi Arabia to Venezuela, through Norway and Russia.

So could the supply side of the nuclear industry become a multinational undertaking? Could the various companies open up to foreign, private capital, form alliances which disregard their nationality, and cast off the yoke of domestic politics? Or in other words, could the nuclear industry take its cue from oil engineering, procurement and construction, rather than mimicking defence procurement? Two examples (see box) suggest this may be possible.

But one swallow does not make a summer. These alliances, between US and South Korean, or French and Japanese companies, are not the first signs of a shift towards multinational consortiums exporting nuclear power plants. At least not in the immediate future. A change of this nature is not on the cards, for it would have to satisfy several improbable conditions. The national character of most bids is due to the diplomatic and geopolitical stakes for nuclear power. We have seen how government meddles in these contracts, on both sides of the bargaining table. If only with respect to the risk of proliferation the stakes will remain just as high. For nuclear companies keen to export their technology, collaboration with government – and the collective game they must consequently play – is all the more critical, given that the firms are dependent on domestic orders. It would make no sense to take the risk of undermining their position at home in exchange for a few sales abroad, as part of multinational consortiums disapproved of by government. Only companies confronted with a moribund domestic market have sufficient latitude to break loose. Substantial growth of the international market, driven by widespread adoption of nuclear power or the opening up of protected home markets, would certainly encourage the formation of multinational consortiums. But hopes of a nuclear renaissance are fading and the main domestic markets, in China, Russia and even South Korea, are closed. As long as the export market for nuclear power plants remains so restricted, there is little likelihood of the industry developing in the same way as its oil counterpart.

## Korean–US alliance in the UAE, Franco-Japanese partnership in Turkey

The consortium which won the UAE tender was not exclusively South Korean. Westinghouse, headquartered in Pittsburgh, PA, is part of the team, supplying parts, technical and engineering services, and licensing its intellectual property. Toshiba, which holds a majority share in the US firm, is also involved. The contract does not explain its role, but it will be supplying equipment as a subcontractor for Doosan. The South Korean companies did not have much option but to accept the presence of these two partners. As they are not yet fully independent regarding technology, they needed Westinghouse's agreement. But in turn the latter needed to be authorized by both the US Administration, which controls exports of nuclear equipment, and by its Japanese shareholder. Without the agreement of these two parties, South Korea stood no chance of honouring an export contract. But the US and Toshiba used this bargaining power to their economic advantage. Over and above such legal considerations, US involvement also brought the South Korean bid a valuable political endorsement to counter the French offer. Indeed, the project to build the nuclear power plant started life as a mutual agreement between France and the UAE. The latter's decision to issue an international call for tenders came as a surprise, marking the beginning of the end for the French consortium led by Areva. The US purportedly had a hand in this volte-face.[18] It was bound to take an interest in a nuclear project in the Gulf, opposite Iran. The US has a very strong presence in the UAE, with about 2,000 military and 30,000 residents, some of whom hold key positions in civilian nuclear power.[19] The tender reshuffled the cards and brought General Electric into the game, through its joint venture with Hitachi, GE Hitachi Nuclear Energy. Washington would no doubt have preferred General Electric to have been awarded the contract, but it soon emerged that its bid was more expensive than the others, at which point US support switched to the Korean option.

The order, which has yet to be finalized by Turkey at time of writing, is for a medium-sized Atmea reactor, designed by Areva and Mitsubishi Heavy Industries. The consortium is led by the Japanese firm and also comprises Itochu, a Japanese fuel supplier, and the French energy company GDF-Suez. The latter, which operates Belgium's nuclear fleet and boasts a highly qualified team of nuclear engineers, took an early interest in the new reactor. Acting as both architect and engineer, it hoped to build and operate one in France, but it ran into opposition from the government and trade unions at EDF. As well as being a potential buyer, it has also positioned itself as a partner in future consortiums, when the customer wants plant operation to be entrusted to an experienced nuclear generating company, at least for the first few years. GDF-Suez has taken on this role for the Turkish project. Japan has several companies with experience in this field, but never overseas. Furthermore, their financial situation has been very challenging since the Fukushima Daiichi accident, not to mention the stain on their reputation. Parliamentary inquiries have shown that Tokyo Electric Power was not the only Japanese operator to cut corners on safety in the past. The bid for the project in Turkey was, however, largely Japanese, not Franco-Japanese as the French media rather hastily claimed. Indeed, the agreement was sealed by the President Tayyip Erdogan and Prime Minister Shinzo Abe, with no French representatives to be seen. This project is nevertheless much closer to oil engineering, procurement and construction than its UAE counterpart, the political dimension having played a much smaller part on both sides of the deal. All the firms in the consortium are privately owned[20] and the two-nation alliance is the result of a strategy based on industrial cooperation on a new reactor model, not some legal obligation.

Finally, a change of paradigm of this sort would require massive industrial reorganization. The global nuclear industry is still dominated by vertically integrated companies, spanning the entire cycle, such as Areva, Rosatom, China Nuclear Power and even Korea Electric Power.[21] Furthermore, these companies are wholly or partly under state control. It seems unlikely that a subsidiary specializing in fuel preparation, reactor design, engineering or operation would go it alone and join a consortium. Such a move would mean joining forces with foreign competitors of its fellow subsidiaries. Why should the mother company's management and shareholders encourage behaviour of this sort? You can count the potential candidates for creating international consortiums on the fingers of one hand. They are all private companies, with limited vertical integration and an almost non-existent home market: General Electric and Westinghouse in the US, GDF-Suez, a company with international interests and the incumbent operator of Belgium's nuclear fleet, and maybe one or two Japanese firms.

## Export controls

Controls over the export of nuclear material, equipment and technology are primarily the preserve of international cooperation. That is not to say that individual countries do not enforce their own policies, but being uncoordinated they are inevitably less effective. Countries introducing strict legislation are likely to see their industry ousted from international contracts, in favour of companies subject to fewer constraints at home. Buyers prefer to deal with the latter to avoid waiting too long, cumbersome obligations regarding enrichment and reprocessing, or indeed bans on re-exports. Without international governance the threat of losing a contract reduces national policies to the lowest common denominator. 'If our industry isn't chosen, the job will go to another, less law-abiding country', some might argue. 'All in all, it would be better if our industry wins the contract;

## The US control on exports

The US is an excellent example of how national policy works in this area. As long as US firms had few competitors, it was relatively easy to apply effective export controls, only one country really being involved. Things are very different now. Of all the nuclear countries exporting materials or technology, the US has so far been the one most concerned about the risk of proliferation. Witness, as a by-product, the voluminous output of US academics on this topic. Dozens of researchers have specialized in the field, exploring it in the light of games theory, law or political science. US policy on nuclear export controls is the strictest, and by consequence it places the greatest constraints on its industry.[22] Companies can only sell goods abroad if the importing country has signed a nuclear cooperation agreement with Washington. So-called 123 Agreements (a reference to Section 123 of the US Atomic Energy Act which requires their use) place various obligations on signatories. The agreement negotiated with the UAE – commonly referred to now as the 'gold standard' – is one of the strictest ever seen. It expressly bans the construction of uranium enrichment units, to produce fuel, and facilities for reprocessing spent fuel. Twenty-two 123 agreements have so far been signed, but most of them date from the time when US technology and industry monopolized the market. These days prospective purchasers can choose from various suppliers. They can place an order with Rosatom, Areva or Mitsubishi without having to wait while a lengthy agreement is hammered out with the US authorities. With the rising threat in the 2000s from Iran and North Korea, Washington tightened up its policy on export controls. A bill[23] sponsored by a two-party committee of the House of Representatives aims to make it virtually compulsory[24] for purchasing states to undertake never to develop enrichment or reprocessing facilities. But an amendment of this nature could be completely counterproductive, ultimately increasing the risk of proliferation. If Washington sets even stricter rules it will drive potential buyers away from US firms. In other words, if US firms, hampered by excessive anti-proliferation legislation, no longer export their goods, the leverage exerted through export controls will be lost, depriving US policy-makers of the key component in their policy for combating proliferation.

The advocates of a tougher line are adamant that this risk is entirely theoretical. The House of Representatives Foreign Affairs Committee 'believes that a serious effort to convince potential recipients of US nuclear exports to adopt some form of the UAE example would have significant proliferation benefits without sacrificing legitimate commerce'.[25] In their view, there is no evidence to justify concerns that higher requirements in Section 123 agreements would have a negative impact on the competitive position of US industry.[26] According to the hard-liners, many prospective buyers, such as Taiwan or the UAE, are prepared to accept more stringent rules because they are concerned about the quality of their relations with the US. Its seal of approval is still sought after. Furthermore, other exporting countries, yielding to Washington's friendly pressure, will follow its example and set higher, non-proliferation standards.

Opponents of the bill, led by US nuclear companies, claim that none of these arguments hold water. Turkey, which signed a 123 agreement in 2008, would probably not have accepted it if it had been obliged to rule out subsequent development of certain parts of the nuclear cycle. Though an ally of the US, Turkey thinks it is entitled to develop its own fuel enrichment capability, this being a basic right enshrined in the Non-Proliferation Treaty. Any attempt by Washington to deprive it of this right would have been interpreted as a hypocritical, discriminatory measure.[27] Turkey's leaders have not forgotten the favourable terms India obtained from Washington. This country, which is not a party to the Non-Proliferation Treaty, was allowed to reprocess its spent fuel. In addition, it is most unlikely that other exporting countries will toe the line on tougher rules. France, for instance, has in the past displayed much less concern about the risk of proliferation entailed by nuclear trade.[28] The authorities attach great importance to the nation's nuclear industry. They are more inclined to assist exports than to toughen up legislation on export controls, and have actively canvassed many countries in the Middle East and North Africa promoting French nuclear know-how. In the past ten years, Paris has signed cooperation agreements with Saudi Arabia, the UAE, Algeria, Libya, Jordan and Tunisia. Finally, opponents of tougher controls maintain that illicit trafficking is the real problem, not above-board sales of technology. With respect to enrichment and reprocessing, trade has only ever concerned countries which were already equipped,[29] for instance between Australia and France, or France and the US. Up to now, not a single previously unequipped country has built an enrichment or reprocessing unit without involving illicit networks. The best known is the network headed by Abdul Qadeer Khan, the father of Pakistan's atom bomb. In addition to plans for producing nuclear weapons he supplied designs for an enrichment unit to Libya.[30] The same network was also behind the transfer of gas centrifuge technology to Iran and supply of parts to North Korea in order to build centrifuges.[31]

The criticism of stricter export controls voiced by the US nuclear industry is well founded and should convince Congress not to adopt the draft amendment – unless the senators and representatives consider its effect on industry will be negligible. They may be convinced that their country's nuclear industry is doomed, due to the lack of domestic outlets. The exploitation of unconventional gas has lastingly undermined its competitive position, not export controls. So it makes no difference whether they are tightened up or not. With increasingly negligible economic stakes at home, US foreign policy to combat proliferation on security grounds could become even stricter.

so we'll set lower standards.' In keeping with this rationale, export controls would be kept to an absolute minimum, a stance reminiscent of the attitude, outlined above, to collective goods. Why should people act for the common good if it is contrary to their interests? Why should people contribute to funding a common good if they can consume it free of charge? The reader will recall that if everyone takes to free-riding, the good will not be produced at all. Here it is much the same, except that we are dealing with a collective ill. Self-interest means nothing is done to reduce it. So too many ills are produced. For lack of cooperation between parties the battle against proliferation is doomed to failure.

We shall now turn our attention to collective action on export controls, the main player in this field being the Nuclear Suppliers Group. This multilateral organization was formed in 1975 by the seven countries which pioneered nuclear power: the US, the UK, France, the Federal Republic of Germany, Canada, Japan and the Soviet Union. It now represents almost fifty states with varying profiles – emerging economies such as China, states with no real nuclear industry such as Switzerland, or even no equipment whatsoever, as in the case of Austria. One of the main rules established by the NSG is the obligation to publish any refusal to allow restricted material or information to be exported to a buyer country. In this way the state publishing the information has an assurance that other NSG members will behave likewise, strengthening its resolution because no one else will take advantage of its refusal.

The NSG came into existence to act on proliferation at source, in contrast to the IAEA, which devotes most of its energy to monitoring nuclear installations after the fact in the countries where they have been deployed. The NSG thus supplements supervision of compliance with the safeguards imposed by the Non-Proliferation Treaty on countries developing nuclear power. Many cases – Iran and Iraq immediately spring to mind – have highlighted the inability of IAEA inspections to detect civilian nuclear investments being diverted for military purposes. The NSG also stops the gaps left by countries which are not a party to the Treaty and are consequently beyond the Agency's remit. For these countries action at source, through export controls, is the only way of halting proliferation. Indeed, the Group was instigated with precisely this in mind, following India's first nuclear test. A non-party to the Treaty, India used imports of equipment from the US and Canada to develop its bomb. The last aspect of the complementary relationship between the NSG and the IAEA is the obligation imposed by supplier states on their customers to sign up to the Treaty and consequently to accept its constraints, in particular inspection.

Thanks to this two-pronged action, the NSG is a prime instrument for combating proliferation, and certainly a great deal more effective than stacks of national policy on export controls. We have already cited the theoretical mechanism of the collective ill which, in the absence of coordination, reduces national policies to the lowest common denominator. At a practical level, the NSG has succeeded in gradually implementing increasingly ambitious, strict policy on controls. In its early days supplier states concentrated on nuclear goods properly speaking. They drew up a list of equipment for reactors and enrichment facilities, and collectively established guidelines setting conditions for the export of such goods. For example, products on the list could only be sold to countries where the nuclear installations fully comply with IAEA safety rules. In the 1990s the Group extended controls to include 'dual-use' goods and technologies such as certain lasers or machine tools, the purchase of which could contribute indirectly to manufacturing arms. The two lists are updated at regular intervals, with slight delays and, inevitably, wording that is sometimes open to interpretation. Given these imperfections NSG countries have committed themselves not to sell equipment – even if it is not mentioned on either list – to countries developing nuclear weapons. To conclude, we should note the rule adopted in 2011 which makes exports related to enrichment and reprocessing conditional on beneficiaries being a party to the IAEA additional protocol, which establishes stricter, more extensive obligations for inspections.

In addition to stricter safeguards, the NSG has gained in effectiveness as growing numbers of countries have introduced export controls. To qualify for membership states must sign into national law and then deploy controls reflecting NSG guidelines. Forty other states have now joined the Group's seven founder members, extending the impact of their action. Of course they are not all leading supplier countries and some, like Cyprus, produce nothing vaguely resembling exportable nuclear goods. But their participation is nevertheless useful, because illicit trafficking no longer goes from A to B – for instance from North Korea to Iran – but passes through a host of transit countries. Policies on export controls adopted by countries such as Cyprus limit the number of platforms through which contraband can transit.

The effectiveness of the NSG may also be measured with respect to the cooperation permitted by the Non-Proliferation Treaty. We suggested above that without the NSG there would be no more than the sum of uncoordinated national policies. A few years before the Group was established an IAEA committee was convened to review the wording of an article in the Treaty imposing safeguards on exports.[32] But the committee soon ran into two obstacles. Firstly, the conditions and lists drawn up by this body could only apply to importing countries party to the Treaty. It obviously had no control over other countries. Secondly, the Treaty already contained an article which tended to encourage exports. This article stipulates that parties have an 'inalienable right' to develop the production of nuclear energy for peaceful purposes and that they undertake to facilitate the 'fullest possible exchange of equipment, materials and scientific and technological information'.[33] The committee still exists, but in a marginal capacity. It has fewer members than the NSG and simply adopts the latter's lists each time they are updated.

Citing their inalienable right, many countries have accused the NSG of constituting a cartel which holds back technology transfers and prevents developing countries from obtaining nuclear power. In their view, the spirit in which the IAEA was set up – access to civilian technology in exchange for giving up military goals – has been totally disregarded. The notion of fair exchange has vanished. Over and above this political and ideological criticism voiced by non-aligned countries for many years, the comparison to a cartel is relevant. But it is a 'good' cartel, in so far as it is a way of reducing a collective ill. Of course, one must subscribe to the idea that combating nuclear proliferation in general and more specifically its application to export controls contributes to greater global security; that it is not simply the expression of the hegemonic position of a small number of nuclear-weapon states, determined to prevent others from obtaining these weapons. I am convinced by this idea, but in defending it we would stray too far from the field of economics. What matters is that the NSG displays the characteristics of a cartel of states.

Firstly, it restricts supply and competition. The volume of nuclear material, equipment and technology exported is lower than it would be in the absence of multilateral controls. Prices are also higher – in the short term due to a certain scarcity, a mechanical effect of the law of supply and demand; in the long term, too, because the cartel bars the way to newcomers, with enrichment and reprocessing remaining in the hands of the same states and companies.

Secondly, as a cartel, the NSG has difficulty making its members uphold their commitments. Establishing a cartel does not necessarily solve the problem of free-riders. Measures must be taken to ensure that each member abides by their commitments, despite the fact that

self-interest dictates that they do not comply or at least only pretend
to do so. By cheating, an NSG member could advantage its national
industry while reaping the collective benefit of less global proliferation. Take, for instance, Chinese exports to Pakistan.[34] The People's
Republic of China joined the NSG in 2004. In so doing it under-
took only to engage in trade if the recipient state had enforced an
agreement with the IAEA applying safeguards to all its installations.
China had helped Pakistan develop its civilian nuclear programme
in the 1970s. It wanted to continue along these lines, as did Pakistan,
but the latter was against extensive inspection of its nuclear installations, which included weapon stockpiles. China claimed a right
of precedence on joining the NSG, arguing that its agreement with
Pakistan predated its joining the Group and that it was consequently
entitled to continue supplying fuel and various services to the nuclear
power plant at Chasma in Pakistan. At the end of the 2000s, when it
emerged that Beijing planned to supply two new reactors, it was more
difficult to persuade other parties that the deal had again been 'grand-
fathered' by the original understanding. Washington thought it was
a trick, China having made no mention of such plans when it joined
the NSG. The members of the Group remain divided on the issue,
but also on how to respond to Beijing's opportunistic behaviour.
Either way the NSG, as an organization operating on a purely voluntary basis, only has limited margin for manoeuvre on sanctions.
There is no treaty binding member states and consequently no scope
for litigation against members which fail to fulfil their obligations.
NSG commitments are not binding. The only sanction available to
its members – at least those with sufficiently persuasive arguments
at their disposal, such as the US[35] – is political pressure. Lastly, the
NSG has only limited means of checking that its members are keeping their commitments. It is easy enough to check that they have
transposed into national law the export safeguards it advocates, but,
in the absence of an appropriate administrative body, it is much more
difficult to ensure that safeguards are strictly enforced. The NSG is

a lightweight organization. It has neither a headquarters nor a secretariat, the latter service being provided by the Japanese delegation to the IAEA. The lack of any means of control is particularly bothersome when monitoring members' obligation to declare exports. The US has complained several times that Russia has failed to provide full details of trade with Iran.[36]

Thirdly, the NSG is prone to instability and paralysis. One way of taking advantage of a cartel's action without paying its price is either not to join, or to leave: witness the Organization of Petroleum Exporting Countries. Countries which are not Opec-members, such as Norway or Russia, sell their oil at a higher price thanks to the organization's work. Former members, such as Gabon or Ecuador, have left; Iraq has threatened on several occasions to do so. Oddly enough, the NSG's stability is less affected by the risk of members leaving, more by the possibility that newcomers might join. Group membership is very appealing because decisions are taken by consensus. Newcomers must bring their export policy into line with NSG guidelines, but in exchange they gain influence over the Group's future decisions, enabling them to block proceedings. Moreover, NSG membership boosts a country's credibility in the eyes of potential buyers. There is no danger of orders being cancelled or partnership agreements scrapped because the international community has outlawed the supplier. Lastly, membership brings knowledge and information which can help a newcomer bolster its market position.

At present, having almost fifty members is in fact a weakness. The interests the NSG represents are too disparate, slowing down or completely stopping new initiatives. For example, it took seven years to make it mandatory for countries importing parts and equipment related to enrichment or reprocessing to sign up to the IAEA additional protocol. There is a great deal of internal tension over how to deal with nuclear-weapon states other than those designated by the Non-Proliferation Treaty. The NSG almost fell apart over the decision to authorize exports to India, but in the end the US, backed by France and Russia, succeeded in overcoming opposition. In 2008 the NSG granted a waiver for the supply of nuclear materials to this country, despite its refusing to allow IAEA inspectors into all its installations. Some members, notably the US, maintained this was an exception, whereas others, led by China, countered that Pakistan should be treated in the same way. The current controversy hinges on whether a country which is not a party to the Non-Proliferation Treaty can join the Group. India has not filed an official application, but it has expressed the wish to do so.[37] Here again the US, France and Russia are thought to support Indian membership, for reasons related to bilateral trade agreements. But its powerful neighbour China is against the idea – unless, of course, it opens the way for Pakistan, another non-designated nuclear-weapon country, to join too. It is a tricky situation, particularly because if India gains membership it will promptly block Pakistan's bid to join. Moreover, a solution based on new members also signing up to the Non-Proliferation Treaty is out of the question, because for that to work both India and Pakistan would have to scrap their nuclear weapons.

Many observers believe the NSG has reached a crossroads.[38] It is time to choose between continuing its political role as a part of efforts to combat nuclear proliferation or changing to become an open club for countries capable of exporting nuclear goods and above all concerned about their collective economic interest: not encouraging proliferation, because the diversion of exports by importing countries for military ends can only strengthen public opposition to the growth of nuclear power. In the first case the NSG would continue to serve as a lever for stiffer upstream controls. For example, it succeeded in forcing recipient countries to adopt the additional protocol of the Non-Proliferation Treaty, an addition which had only been signed by less than a quarter of countries party to the Treaty. In the second case the NSG would cast off the safeguards imposed by the IAEA and start treating nuclear-weapon countries other than the original big five as importers and potential members.

International trade in nuclear goods is thus subject to a delicate balance between contrary forces, torn between promoting and restricting exports. At a national level, there is a political will to encourage exports by an industrial sector often plagued by sluggish domestic demand. Foreign contracts help keep it afloat, perhaps even creating jobs, and boost the political credit and prestige of heads of state. The latter are consequently eager to support their national team and quick to offer sweeteners, particularly if the potential buyer is an ally – or future ally – in the international arena. But the same governments must make allowance for security imperatives, ensuring that exports of nuclear material, equipment and technology do not fuel proliferation.

The extent to which security concerns counterbalance industrial and commercial ambition varies. Security is a more pressing issue in the US than in France, for example. But whether it features to some extent or not at all at home, this concern is imported from abroad via international governance of non-proliferation. We have seen how this governance is organized: the Non-Proliferation Treaty, which bans nuclear weapons; the IAEA, which controls installations downstream; and the NSG, which intervenes upstream. But this governance is also subject to the imperatives of promotion. In pursuit of their mission to develop atomic energy, particularly scientific and technological cooperation, the Treaty and Agency encourage nuclear trade.

International governance thus reproduces the same balance of contrary forces observed at a national level. However, there is an essential difference in the way the balance is achieved. At a national level the legislative and the executive both have a say in reconciling promotion and nuclear controls, in other words the same people who are in charge of economic, diplomatic and defence policy. At the multilateral level, on the other hand, the contradictory forces are balanced by a specialized administrative agency with no head. The focus of IAEA's efforts and the allocation of its budgets, between promotion and non-proliferation, varies mainly according to contingent events: the personality of its Director-General; the ideological stance of its respective Board members; even the centres of interest of its administrative staff. The only constant, combating proliferation, is reduced to after-the-fact control, due to the need to continue promoting nuclear power. This severely limits the effectiveness of its action, mercifully remedied by the NSG. This situation strongly suggests that the two missions of the IAEA should be separated. All the supervisory work should be entrusted to a single body, whereas promotion should be hived off. The only problem with what might at first sight appear to be simple common sense is that this reform would spell the end of the deal set forth by Eisenhower – access to nuclear power in exchange for giving up nuclear weapons – the deal on which nuclear governance has been founded for over fifty years. Reform of this sort would be difficult.

As we explained in the introduction to this fourth and final part of the book, politics steps in to take decisions held up by uncertainty about the cost, risk and regulation of nuclear power. Policy-makers decide whether or not to adopt nuclear power, prolong its use or phase it out. Such choices, which are the prerogative of sovereign states, have repercussions on safety and security well beyond national borders. They necessitate multilateral coordination, the only approach which can create global collective goods and eliminate global collective ills. Over and above the diversity of multilateral arrangements and responses to uncertainty, some recurrent trends emerge. Firstly, the motives for adopting nuclear power now and in the past are largely similar: prestige, energy independence, growth of science and technology. The only novelty is that sometimes these goals are coupled with the determination to combat global warming. Secondly, decisions to reduce the share of nuclear power in the energy mix more quickly or phase it out altogether are based on approximation, with little real consideration for economic factors. The choice of goals and the timetable for shutting down nuclear power plants are dictated by assumptions about shifts in voter preference and the balance between and within political parties, not by forecasts of the economic costs and benefits for society. Thirdly, little progress has been made towards reducing nuclear power's transboundary impact on safety and security. The multilateral model embodied by the European Union, despite having gone furthest down this path with some notable successes, highlights the handicaps which inevitably plague international governance, due to the diversity and number of parties. Opposition by EU Member States to the continued exploitation or immediate shutdown of atomic energy drastically reduces scope for joint action. In the struggle to limit the deployment of nuclear weapons, the divergent interests of parties to the Non-Proliferation Treaty and members of the Nuclear Suppliers Group are also a major hindrance for effective collective action.

# Conclusion

Let us summarize the main points in this book.

With respect to costs, we have seen why the idea of a single, universally valid cost for nuclear power is misleading. Not because figures can be manipulated or assessments influenced by specific interests, but because there is no such thing as the *true* cost. Nuclear costs depend on the options available to economic decision-makers, public or private, and on a range of factors subject to considerable future uncertainty. So it does not cost the same amount to build a new plant in Finland, China or the United States.

The curse of rising costs has dogged nuclear technology from the outset, particularly in the US. Far from falling as more plants have been built, costs have soared, making new nuclear even more expensive. If there is no change in the immediate future, particularly if the price of carbon stays low, the competitiveness of nuclear power compared with other electricity-generating technologies will vanish. The outcome of any decision by a generating company to invest in the construction of a new nuclear power plant, or by a government advocating such a project, remains uncertain.

Regarding risk, we have noted and explained the divergence between expert calculations of the probability of a nuclear accident and the public perception of such a disaster. Nuclear technology is increasingly safe, thanks to the knowledge accumulated on operating reactors and the probabilistic safety assessments developed by scientists and engineers over the past thirty-five years. But this progress makes no impression on public opinion, petrified by pictures of Fukushima Daiichi. At a subjective level, people tend to overestimate risk when it relates to rare, terrifying events. Our understanding of what drives such perception is improving, thanks to the work of experimental psychology, which is gradually shedding the light of reason on irrational behaviour. We have also learnt that we constantly update our prior appraisals of probability, drawing on new, objective data and knowledge. It is time to rise above the pointless opposition between experts and the general public, each accusing the other of rank stupidity. The former are purportedly out of touch, locked up in their laboratories and theories, while the latter are supposedly far too sensitive to disasters and incapable of probabilistic reasoning. We must reconcile subjective and objective probability.

On the matter of regulation, we have presented the economic theory of safety regulation and compared the approaches adopted in the US, Japan and France. There are incentives for nuclear power plant operators to prevent failures and compensate for damage under an unlimited-liability regime. But the resulting safety efforts and investments are limited. Intervention by a regulatory authority is necessary to define an appropriate safety level, prescribe standards and monitor their application.

Regulation is inevitably imperfect – the regulator is not omniscient – but to a varying extent. Among the cases which are known and well documented, nuclear safety regulation as practised in Japan is an example to be avoided. The safety authority has allowed itself to be captured by the nuclear operators it was supposed to supervise and lacks the necessary independence with regard to government, itself under the influence of industry. The immediate cause of the Fukushima Daiichi disaster was natural, an earthquake followed by a very large tidal wave. But its underlying cause was the lack of transparent, independent, competent safety regulation. In contrast, safety regulation in the US and France, though strikingly different and not without fault (with, respectively, too many stan-dards or vague safety targets), offers valuable examples.

With regard to policy-makers, we have shown how they fill the vacuum created by uncertainty about costs, risk and regulation, and how they may agree to provide global collective goods, or to reduce equally pervasive ills. National policies on nuclear power are widely divergent. Various countries decided to develop nuclear power sev-eral decades ago, some are taking that step now, while others are opting to retire their nuclear capacity. The will to power, the need to achieve energy independence, the growth of science and tech-nology and more recently the desire to combat climate change have all played a part in such decisions. To this we must add efforts to acquire nuclear weapons, although this only concerns a few cases: the vast majority of countries operating nuclear power plants have not taken this route. The Treaty on the Non-Proliferation of Nuclear Weapons contributed to this state of affairs, reducing the collective ill of proliferation threatening the planet. But in some cases it made it worse, by promoting scientific and technological cooperation on nuclear power: access to this technology by a growing number of countries increases the risk of proliferation.

Multilateral cooperation has also attempted to improve nuclear safety. The European Union has gone furthest towards achieving this collective good. Underpinned by the Treaty on the European Atomic Energy Community and specific legislation, European safety governance has proved an undeniable success, notably shutting down hazardous reactors on its borders. Much as with other forms of safety cooperation, its effectiveness is hampered by the number of its mem-bers and their divergent interests, some in favour of nuclear power, others against.

Let us now recapitulate the key determinants of the future of nuclear power worldwide, as analysed in the preceding pages.

The first challenge is to lift the curse of rising costs. Standard-ization and innovation may be the answer. France has in the past benefited slightly from large-scale production and learning effects. China could in the future achieve even greater standardization in its reactors. So far, innovation has above all involved building larger, more complex systems. Other routes must be explored, for instance giving priority to modular plants and small reactors.

Secondly, it is essential to pool our understanding of risk. All too often perceived and calculated risk have been at odds, leaving us no option but to side with the experts or the general public. When such opposition is sustained and exploited by leaders of industry and policy-makers, we may fear the worst possible economic outcome. Billions of euros are wasted when nuclear projects, founded on calculated risk and blind to public fears, are shelved at the last moment. Similar amounts are squandered when a government impervious to all but perceived probability orders the premature shutdown of reactors. Experts need to be made aware of how risk perception may be distorted, and the general public educated to recognize this bias and understand probabilistic reasoning.

Thirdly, it is imperative that safety authorities should be powerful and competent, independent and transparent. Giving the regulator extensive powers and a mandate to sanction non-compliance; securing and consolidating its skills with adequate resources; guaranteeing its independence with regard to government and industry; requiring complete transparency in its actions: these are the essential pillars to make regulation as near perfect as possible. Safety can also be much improved by changing the legal framework and balance of power. Many countries need to make progress on safety regulation. Unlike building multiple physical barriers, the financial cost of such gains is almost negligible.

The future of nuclear power is also conditional on the adoption of measures to separate promotion of this form of energy from global governance of security and safety. A throwback to a world divided into three parts – Communist, developed and non-aligned countries – the same international bodies and laws still have a three-pronged mission: to disseminate electronuclear technology; to establish safety principles and standards; and to combat the proliferation of nuclear weapons. This confusion, present even at the heart of the International Atomic Energy Agency, undermines the effectiveness of its work to improve safety, monitor exports and inspect installations.

Lastly, the nuclear industry and corresponding trade must become international. Individual states currently play a key role, by protecting their home market – giving priority to national firms and 'home-grown' technology – and by the active involvement of political leaders in negotiating export contracts. There are few openings for foreign capital, and foreign markets only see competition between national teams. To sustain the growth of both trade and industry, international regrouping is needed, organized along lines closer to oil and gas supply and service than to defence procurement.

Let us end with an inevitable question: for or against nuclear power.

The reader will have gathered that nuclear power is durably established all over the world. Every year new nuclear power plants are commissioned and built, though some will quibble over whether there are too many or too few. The priority is therefore to ensure that the system operates properly.

It should be clear by now that a polemical approach to new nuclear makes no sense. As for existing plants, nothing can do more harm to the general economic interest than the early retirement of reactors which are still perfectly serviceable and have been licensed to operate by the safety operator.

It should also be apparent that the author is in favour of new nuclear and its growth, on condition that costs are contained, risks controlled and safety regulation properly implemented, and that international governance on safety and security is stronger and more effective. On the other hand, the author opposes further growth if costs continue to soar, if some plants go on being operated in an irresponsible manner, if regulation is non-existent in practice and the state omnipresent.

The above observation is not an attempt to dodge the issue. To refuse to take a stand for or against nuclear power is simply to acknowledge the many uncertainties relating to costs, risk, regulation and governance which surround this technology. Economic appraisal can only give categorical answers when the most probable hypotheses have been singled out. Too much uncertainty still besets nuclear power for the general economic interest to serve as a basis for decisions for or against.

# Notes

## Part I   Introduction

1 For a full, accurate account of the various items of cost, see W. D. D'Haeseleer, Synthesis on the Economics of Nuclear Energy, Study for the European Commission (DG Energy, 2013), available at: ec.europa.eu/energy/nuclear/forum/doc/final_report_dhaeseleer/synthesis_economics_nuclear_20131127-0.pdf.

## 1   Adding up costs

1 Loi *no.* 2010–1448 du 7 Décembre 2010 portant organisation du marché de l'électricité.
2 Cour des Comptes, *Les Coûts de la Filière Nucléaire* (La Documentation Française, Paris, 2012).
3 Centre d'Analyse Stratégique, *La Valeur Tutélaire du Carbone.* Report by the committee chaired by Alain Quinet (La Documentation Française, 2009).
4 Cour des Comptes, *Les Coûts de la Filière Nucléaire*, p. 150.
5 'Is it worth it?', *The Economist* (5 December 2009).
6 F. P. Ramsey, A mathematical theory of saving, *Economic Journal*, 38 (1928), pp. 543–59.
7 The Stern Review of the Economics of Climate Change (2006), available online: http://webarchive.nationalarchives.gov.uk/+/http:/www.hm-treasury.gov.uk/sternreview_index.htm.
8 W. Nordhaus, Global warming economics, *Science*, 294:5545 (2001), pp. 1283–4.
9 P. Dasgupta, Comments on the Stern Review's Economics of Climate Change, *National Institute Economics Review*, 199 (2006), pp. 4–7.
10 C. Gollier, Discounting an uncertain future, *Journal of Public Economics*, 85 (2002), pp. 149–66.

11 Oxera Consulting Ltd, *A Social Time Preference Rate for Use in Long-Term Discounting* (2002).

12 Report by the group of experts led by D. Lebègue, *Révision du Taux d'Actualisation des Investissements Publics* (Commissariat Général du Plan, 2005).

13 The intuitive, commonsense solution of a declining, variable discount rate now has a solid base in economic theory: see C. Gollier, *Pricing the Planet's Future: The Economics of Discounting in an Uncertain World* (Princeton University Press, 2013).

14 Cour des Comptes, *Les Coûts de la Filière Nucléaire*, p. 282.

15 US Department of Energy, *Civilian Radioactive Waste Management Fee Adequacy Assessment Report* (2008), RW-0593.

16 Massachusetts Institute of Technology, *Update of the MIT 2003 Future of Nuclear Power* (2009).

17 Economic losses estimated at $148 billion for Ukraine and $235 billion for Belarus. Figures cited by Z. Jaworowski, *The Chernobyl Disaster and How it has been Understood* (2010), available online: www.world-nuclear.org/uploadedFiles/org/info/Safety_and_Security/Safety_of_Plants/jaworowski_chernobyl.pdf.

18 A fleet of fifty-eight reactors with a forty-year service life, or an expected number of accidents of $2{,}320 \times 10^{-5} = 0.0232$ and damages of $0.023 \times €1{,}000$ billion, in other words €23.2 billion.

19 Data compiled from fourteen countries, of which three non-OECD, but not the US: see p. 59 of the joint OECD-IEA report, *Projected Costs of Generating Electricity* (2010).

20 E. Bertel and G. Naudet, *L'Economie du Nucléaire* (EDP Sciences, Paris, 2004).

21 A small part of the difference results from the discounting rates: 5% for South Korea and 10% for Switzerland.

22 Y. Du and J. Parsons, *Update on the Cost of Nuclear Power* (Center for Energy and Environmental Policy Research (CEEPR) 09-004, 2009).

23 Massachusetts Institute of Technology, *The Future of Nuclear Power: An Interdisciplinary MIT Study* (2003).

## 2   The curse of rising costs

1 A. Lindman and P. Söderholm, Wind power learning rates: a conceptual review and meta-analysis, *Energy Economics*, 34 (2012), pp. 754–61.

2 N. E. Hultman and J. Koomey, A reactor-level analysis of busbar costs for US nuclear power plants 1970–2005, *Energy Policy*, 35 (2007), pp. 5630–42.

3 The overnight construction cost for the French fleet, cited in Cour des Comptes, *Les Coûts de la Filière Nucléaire* (La Documentation Française, Paris, 2012), amounted to about €83 billion$_{2010}$. The report also published the construction cost of each pair of reactors, but these detailed figures do not correspond to the overnight cost, strictly speaking, as they omit engineering expenses and pre-operating costs.

4 Averages based on figures in the Cour des Comptes, *Les Coûts de la Filière Nucléaire*. The difference between Chooz 1 and 2 (€1,635 per kW), and Civaux (€1,250 per kW) vanishes. The high figure for Chooz is due to the fact that these were the first units of the N4 series, a new reactor model.

5 I. C. Bupp, J.-C. Derian, M. P. Donsimoni and R. Treitel, The economics of nuclear power, *Technology Review* 77:4 (1975), pp. 14–25, quoted by Hultman and Koomey, A reactor-level analysis of busbar costs.

6 G. S. Rothwell, Steam-electric scale economies and construction lead times, *Social Science Working Paper*, 627 (California Institute of Technology, 1986).

7 R. Cantor and J. Hewlett, The economics of nuclear power: further evidence on learning, economies of scale and regulatory effects, *Resources and Energy*, 10:4 (December 1988), pp. 315–35.

8 M. B. Zimmerman, Learning effects and the commercialization of new technologies: the case of nuclear power, *The Bell Journal of Economics*, 13 (1988), pp. 297–310.

9 L. W. Davis, Prospects for nuclear power, *Journal of Economic Perspectives*, 26:1 (2012), pp. 49–66.

10 S. Paik and W. R. Schriver, The effects of increased regulation on capital costs and manual labor requirements of nuclear power plants, *The Engineering Economist*, 26 (1979), pp. 223–44.

11 M. Cooper, *Policy Challenges of Nuclear Reactor Construction: Cost Escalation and Crowding Out Alternatives* (Institute for Energy and the Environment, Vermont Law School, 2010).

12 A. Grubler, The costs of the French nuclear scale-up: a case of negative learning by doing, *Energy Policy*, 38 (2010), pp. 5174–88.

13 E. Bertel and G. Naudet, *L'Economie du Nucléaire* (EDP Sciences, Paris, 2004); G. Moynet, Evaluation du coût de l'électricité nucléaire en France au cours des dix dernières années, *Revue Générale Nucléaire*, 2 (1984), pp. 141–52.

14 Grubler, The costs of the French nuclear scale-up.

15 Massachusetts Institute of Technology, *Update of the MIT 2003 Future of Nuclear Power* (2009).

16 Expressed in $\$_{2007}$; the levelized cost increases by 25%.

17 The lower end of the overnight cost for the 2004 study was $1,413_{2010}$ to $2,120_{2010}$ with a mean value of $1,765_{2010}$, hence the increase by a factor of 2.3. Focusing just on the AP1000, the 2004 range was $1,554_{2010}$ to $2,331_{2010}$ with a mid-point at $1,943_{2010}$. So the study used $2,000 per kW and compared this to $4,210 per kW, an increase by a factor of 2.1.

18 Hultman and Koomey, A reactor-level analysis of busbar costs.

19 With a 6% discount rate, the levelized cost in the MIT study is $42 per MWh.

20 Hultman and Koomey, A reactor-level analysis of busbar costs.

21 J.-M. Glachant, Generation technology mix in competitive electricity markets, in F. Lévêque (ed.), *Electricity Reform in Europe: Towards a Single Energy Market* (Edward Elgar Publishing, London, 2009).

22 B. Dupraz and L. Joudon, Le développement de l'EPR dans le marché électrique européen, *Revue Générale Nucléaire*, 6 (2004); P. C. Zaleski and S. Méritet, Point de vue sur l'EPR, *Contrôle*, 164 (2005), pp. 51–6.

23 According to French MP Marc Goua, tasked with reviewing the accounts of Areva and EDF: www.enerzine.com/2/12796+lepr-finlandais-couterait-au-final-6-6-mds-deuros+.html (2011).

24 *Le Monde*, www.lemonde.fr/planete/article/2011/11/10/sur-le-chantier-de-l-epr-a-flamanville-edf-est-a-la-moitie-du-chemin_1602181_3244.html.

25 EDF communiqué, cited in the Cour des Comptes, *Les Coûts de la Filière Nucléaire*.

26 See the EDF Energy press release dated 21 October 2013, 'Agreement reached on commercial terms for the planned Hinkley Point C nuclear power plant'. The figure of €17.2 billion – or €8.6 billion per reactor – can be directly compared with the cost of building the EPR at Flamanville, France. The sum of €19.7 billion includes additional expenditure such as purchase of the land and building a facility for storing fuel. Such expenditure is not included in the cost of Flamanville 3.

27 See second study by the University of Chicago: R. Rosner and S. Goldberg, *Analysis of GW-scale Overnight Cost, Technical Report* (University of Chicago, 2011).

28 For example, in its updated study the MIT revalues the overnight cost per kW of a gas plant, resulting in a 70% increase, and a 130% increase for a coal-fired thermal plant.

29 The levelized cost for the base case was $67 per MWh for nuclear, $43 per MWh for coal and $41 per MWh for gas. With the same financial conditions the cost of nuclear power drops to $51 per MWh. See Table 1, Yangbo Due and J. E. Parsons, *Update on the Cost of Nuclear Power* (Massachusetts Institute of Technology, 2009).

30 The 2004 University of Chicago study suggests that first-of-a-kind costs may amount to as much as 35% of the overnight cost. Dupraz and Joudon, Le développement de l'EPR, estimate that such costs add 20% to the levelized cost of the first in a series, for a series of ten units (€41$_{2004}$ per MWh, rather than €33$_{2004}$ per MWh).

31 Areva défend son EPR, *Energies Actu* (2012), available online at: www. energiesactu.fr/production/areva-defend-son-epr.

32 Organisation for Economic Cooperation and Development, *Reduction of Capital Costs of Nuclear Power Plants* (2000).

33 G. F. Nemet, Demand-pull energy technology policies, diffusion and improvements in California wind power, in T. Foxon, J. Koehler and C. Oughton (eds.), *Innovations for a Low Carbon Economy: Economic, Institutional and Management Approaches* (Edward Elgar Publishing, Cheltenham, 2007).

## 3   Nuclear power and its alternatives

1 L. Pouret and W. J. Nuttall, Is nuclear power inflexible?, *Nuclear Future*, 5:6 (2009), pp. 333–41.

2 N. Z. Muller, R. Mendelsohn and W. Nordhaus, Environmental accounting for pollution in the United States economy, *American Economic Review*, 1001:5 (2011), pp. 1649–75. Value reported as $28.3 per MWh.

3 P. Epstein et al., Full cost accounting for the life cycle of coal, *Annals of the New York Academy of Sciences*, 1219 (2011), pp. 73–98. Here the value was $269 per MWh, of which $44 per MWh corresponds to the impact on public health.

4 European Commission, Directorate General XII, Externalities of Energy, (2009). The ExternE study estimates the external costs of electricity generated using natural gas (excluding carbon emissions) at between $13.4$_{2010}$ and $53.8$_{2010}$ per MWh, as against $27$_{2010}$ to $202$_{2010}$ for coal. The latter figures were reported by S. Grausz, The Social Cost of Coal, *Climate Advisers* (2011). Natural gas produces half as much carbon emissions as coal.

5 Organisation for Economic Cooperation and Development, Nuclear Energy Agency, *The Financing of Nuclear Power Plants* (2009).

6 F. Roques, D. M. Newbery and W. J. Nuttall, Fuel mix diversification incentives in liberalized electricity markets: a mean-variance portfolio theory approach, *Energy Economics*, 30 (2008).

7 Cour des Comptes, *Les Coûts de la Filière Nucléaire* (La Documentation Française, Paris, 2012), p. 51.

8 P. L. Joskow, Comparing the costs of intermittent and dispatchable electricity generating technologies, *American Economic Review*, 101:3 (2011), pp. 238–41.

9 D. M. Newbery, Reforming competitive electricity markets to meet environmental targets, *Economics of Energy and Environmental Policy*, 1:1 (2012), pp. 69–82.

10 Quoted by S. Ambec and C. Crampes, *Electricity Production with Intermittent Source of Energy*, Lerna Working Paper, University of Toulouse 10.07.313 (2010).

11 E. Lantz, R. Wiser and M. Hand, *IEA Wind Task 26: The Past and Future Cost of Wind Energy*, Work Package 2 (National Renewable Energy Laboratory, 2012).

12 In a meta-study carried out for the International Panel on Climate Change, bearing on eighteen estimates, R. Wiser et al., Wind Energy, in *IPCC Special Report on Renewable Energy Sources and Climate Change* (Cambridge University Press, 2011) suggest a 4% to 32% range. But the gap narrows to a 9% to 19% range, if only post-2004 estimates are considered.

13 W. D. D'Haeseleer, Synthesis on the economics of nuclear energy: Study for the European Commission (DG Energy, 2013), available at: ec.europa. eu/energy/nuclear/forum/doc/final_report_dhaeseleer/synthesis_economics_nuclear_20131127-0.pdf.

## Part II  Introduction

1 The severity of an event increases tenfold, at each level: accordingly level two is ten times worse than level one, level three ten times worse than level two, and so on.

2 The term 'major accident' used in this part is more general than its usage in the terminology established by the INES scale. In this ranking only level-seven accidents rate as 'major'; a level-five core-melt counts as an 'accident with wider consequences', whereas its level-six counterpart counts as a 'serious accident'.

3 SER (1959), Enrico Fermi 1 (1961), Chapelcross 2 (1967), Saint-Laurent A1 (1969) and A2 (1980), Three Mile Island 2 (1979), Chernobyl 4 (1986), Greifswald 5 (1989). List taken from T. B. Cochran, Fukushima Nuclear Disaster and Implications for US Nuclear Power Reactors (2011), at www.nrdc.org/nuclear/tcochran_110412.asp. This figure only includes grid-connected reactors. It does not include accidents on research reactors such as the SL-1 meltdown in Idaho, USA, in 1961. Note that this list includes cases of very limited core damage. For these cases, the term 'major accident' could be viewed by specialists as an overstatement.

## 4 Calculating risk

1 Other risk dimensions such as the duration and reversibility of damage are sometimes taken into account too.

2 R. A. Posner, *Catastrophe: Risk and Response* (Oxford University Press, 2004), estimates it at $600 trillion ($6.10^{14}$).

3 The Sievert is an International System (SI) unit used to measure the biological effects of the absorption of ionizing radiation.

4 These methods are open to dispute and actively debated. Readers wishing to pursue the matter may refer to Chapter 4 of W. K. Viscusi, *Rational Risk Policy* (Oxford University Press, 1998).

5 Some recent estimates have also sought to estimate direct and indirect macroeconomic consequences, as well as the impact on society as a whole and on private individuals, through the emotional and psychological disturbance caused by an accident. See L. Pascucci-Cahen and P. Momal, *Les Rejets Radiologiques Massifs Different Profondément des Rejets Contrôlés* (Institut de Radioprotection et de Sûreté Nucléaire, 2012).

6 Versicherungsforen Leipzig, *Calculating a Risk-Appropriate Insurance Premium to Cover Third-Party Liability Risks that Result from Operation of Nuclear Power Plants*, commissioned by the German Renewable Energy Federation (2011).

7 Versicherungsforen Leipzig, *Calculating a Risk-Appropriate Insurance Premium.*

8 Reactors, Residents and Risk, *Nature News* (2011), doi: 10.1038/472400a.

9 The Chernobyl Forum, *Chernobyl's Legacy: Health, Environmental and Socio-Economic Impacts*, second revised version (2005). The estimate is based on the linear no-threshold model.

10 Lisbeth Gronlund, How many cancers did Chernobyl really cause? *Monthly Review Magazine*, available at: http://mrzine.monthlyreview.org/2011/gronlund070411.html.

11 Estimated, for example, at $3.4 billion by B. Sovacool, The costs of failure: a preliminary assessment of major energy accidents, *Energy Policy* (2008), pp. 1802–20, of which $1 billion just for the cost of cleaning up. The remainder is an estimate of the cost of loss of property.

12 Belarus, for instance, has estimated its losses over thirty years at $235 billion: Chernobyl Forum, *Chernobyl's Legacy*, p. 33, n. 6.

13 This figure is taken from the very first studies carried out by EDF in the late 1980s on its 900 MW nuclear reactors. In its subsequent, more detailed studies, the figure was ten times lower, at about $5 \times 10^{-6}$. But in both cases the relevant studies made no allowance for external initiating factors such as earthquakes or floods.

14 If there are only two options – such as drawing a red ball or a black ball from an urn only containing balls of these two colours – then once you

know the probability of one option, the other can be deduced, the sum of the two being equal to one. So if the probability of drawing a red ball is one-in-three, the probability of not drawing a red ball is two-in-three; as all the non-red balls are black, the probability of drawing a black ball is two-in-three.

15 US National Regulatory Commission, *Reactor Safety Study: An Assessment of Accident Risks in US Commercial Nuclear Power Plants* (NUREG-75/014) (1975).

16 See in particular the criticism voiced by the Union of Concerned Scientists, published in its report: H. W. Kendall, R. B. Hubbard and C. M. Gregory (eds.), *Risks of Nuclear Power Reactors: a Review of the NRC Reactor Safety Study* (1977).

17 Health and Safety Executive, Nuclear Directorate, *Generic Design Assessment – New Civil Reactors Build* (Step 3, Probabilistic Safety Analysis of the EDF and Areva UK EPR, Division 6, Assessment report no. AR 09/027-P) (2011).

18 Electric Power Research Institute, *Safety and Operational Benefits of Risk-Informed Initiatives* (White paper, 2008).

19 Cochran, Fukushima Nuclear Disaster.

20 EDF studied the risk of falling aircraft for the Flamanville 3 reactor. The probability of an attack on one of its safety functions is $6.6 \times 10^{-8}$ per reactor-year. See EDF Probabilistic Safety Analysis, available at: www.edf.com/html/epr/rps/chap18/chap18.pdf.

21 Nor is it possible to assign probabilities to the share of known events in an incomplete whole. The sum of all probabilities must be one, but as the probability associated with the subset of disregarded events is not known, it cannot be subtracted from one to obtain the probability assigned to known events.

22 In an even more comprehensive definition of uncertainty, this term also encompasses incompleteness. The term 'radical uncertainty' may be used to distinguish this form of uncertainty from non-specific uncertainty.

23 The intuition underpinning this hypothesis is that all the distributions of black and white balls (one white, fifty-nine black; two white, fifty-eight black; ... ; thirty white, thirty black; ... fifty-nine white, one black) are possible, and that there is nothing to suggest that one is more probable than another. It is no more likely that the distribution contains more black than white balls, than the opposite.

24 J. M. Keynes, *A Treatise on Probability* (Macmillan, London, 1921).

25 Quoted by S. Morini, Bernard de Finetti: l'origine de son subjectivisme, *Journal Electronique d'Histoire des Probabilités et de la Statistique*, 3:2 (2007), pp. 1–16.

26 W. Epstein, *A PRA Practitioner Looks at the Great East Japan Earthquake and Tsunami* (Ninokata Laboratory White Paper, 2011).

310 • Notes to pages 99–108

27 The estimated frequencies for the other nuclear power plants on the same coastline do not display such a large difference. At Fukushima Daini the estimated frequency is ten times lower than the historic frequency; at Onagawa the two frequencies converge. The differences are due to the choice of too fine a mesh to differentiate seismic risks in the Sanriku area. See R. J. Geller, Shake-up time for Japanese seismology, *Nature*, 472 (2011), pp. 407–9.

28 Epstein, *A PRA Practitioner*, p. 52.

29 Japan Society of Civil Engineers, Nuclear Civil Engineering Committee, Tsunami Evaluation Subcommittee, *Tsunami Assessment Method for Nuclear Power Plants in Japan* (2002 guidelines): 'We have assessed and confirmed the safety of the nuclear plants [at Daiichi] based on the JSCE method published in 2002' (Tsunami Study for Fukushima 1 and 2, p. 14); written statement by Tepco submitted to the regulator, cited by Epstein, *A PRA Practitioner*, p. 24.

30 Epstein, *A PRA Practitioner*, p. 23.

31 On Japan's Shindo scale of seismic intensity. Seismic scales not being strictly speaking comparable with one another, seismic intensity six on the Shindo scale is roughly equivalent to between magnitude six and seven on the Richter scale.

32 It is of little concern here whether this was done knowingly or not. This point is addressed in Chapter 8.

## 5 Perceived probabilities and aversion to disaster

1 I. Gilboa, *Theory of Decision under Uncertainty* (Cambridge University Press, 2009). This work presents the classical and recent axiomatic theories, and discusses their conceptual and philosophical aspects.

2 Imagine the following game in a casino: a coin is tossed a number of times, with an initial kitty of $1, which doubles each time the coin comes up tails. The game stops and the player pockets the kitty when the coin lands heads up. At the first toss, the winner pockets $1 if the coin lands heads up; if it is tails, the game continues, with a second toss. If the coin then lands heads up, the player pockets $2, otherwise it is tossed a third time and so on. The expectation of gain equals $1 \times 1/2 + 2 \times 1/4 + 4 \times 1/8 + 8 \times 1/16 + \cdots$ or $1/2 + 1/2 + 1/2 + 1/2 + \cdots$, making an infinite amount provided the casino has unlimited resources.

3 Equal to $(1/101)(100/100 + 99/100 + \cdots + 1/100 + 0/100)$.

4 Tversky and Kahneman co-authored a great many academic papers, in particular those awarded the economics prize by Sweden's central bank. But Tversky died at the age of 59 and was consequently not awarded the Nobel prize for economics alongside Kahneman.

5 D. Kahneman and A. Tversky, Prospect Theory: An Analysis of Decision under Risk, *Econometrica*, 47 (1979), pp. 263–91; and D. Kahneman and A. Tversky, Advances in prospect theory: cumulative representation of uncertainty, *Journal of Risk and Uncertainty*, 5:4 (1992), pp. 297–323.

6 This type of experiment, simplified here, is presented in detail in D. Kahneman, J. L. Knetsch and R. H. Thaler, Anomalies: the endowment effect, loss aversion, and status quo bias, *Journal of Economic Perspectives*, 5:1 (1991), pp. 193–206.

7 D. Kahneman, *Thinking, Fast and Slow* (Farrar, Straus and Giroux, New York, 2011).

8 Kahneman submitted this problem to his students at Princeton.

9 P. Slovic, B. Fischhoff and S. Lichenstein, Facts and fears: societal perception of risk, *Advances in Consumer Research*, 8 (Association for Consumer Research, 1981), pp. 497–502.

10 W. K. Viscusi, *Rational Risk Policy* (Oxford University Press, 1998).

11 Moreover it is not strictly speaking a biased perception of probability. Rather the utility function is distorted, depending on whether there is a loss or gain.

12 A. A. Siddik and P. Slovic, A psychological study of the inverse relationship between perceived risk and perceived benefit, *Risk Analysis*, 14:6 (1994), pp. 1085–96.

13 Organisation for Economic Cooperation and Development, Nuclear Energy Agency, *Comparing Nuclear Accident Risks with Those from Other Energy Sources* (OECD, 2010).

14 In all, 80,250 people lost their lives in 1,870 accidents. Which accident caused the largest number of fatalities? The Chernobyl disaster? No! The failure of the Banquiao and Shimantan dams in China in 1975. This accident, disregarded by all but the local population, claimed 30,000 lives. As we saw in Chapter 4, the number of fatalities at Chernobyl did not exceed 60 and 20,000 was an upper-range estimate of subsequent early deaths. Even if the OECD study had considered these deaths, serious accidents in the coal industry during this period caused more deaths. Moreover, in all fairness, the early deaths caused by coal should also be taken into account in any comparison with nuclear power.

15 G. Gigerenzer, Dread Risks, September 11 and fatal traffic accidents, *Psychological Science*, 15:4 (2004), pp. 286–7.

16 In addition many observers maintain that they have not been very effective. See, for instance, K. Harley, Why airport security is broken, *The Wall Street Journal* (15 April 2012).

## 6  The magic of Bayesian analysis

1  p(cancer|positive) = [p(cancer) × p(positive|cancer)]/[p(positive|
cancer)p(no cancer) + p(positive|no cancer)p(no cancer)] = [0.01 × 0.8]/
[0.01 × 0.8 + 0.99 × 0.1] = 0.075.

2  This is not a matter of innumeracy, a shortcoming similar to illiteracy but
for numbers. The experiments described above involved educated
participants, students or graduates.

3  D. Kahneman and A. Tversky, Subjective probability: a judgment of
representativeness, in D. Kahneman, P. Slovic and A. Tversky (eds.),
*Judgements under Uncertainty: Heuristics and Biases* (Cambridge University
Press, 1982), p. 46.

4  S. Dehaene, Le cerveau statisticien: la révolution Bayésienne en sciences
cognitives (a course of lectures at Collège de France, Paris, 2011–12); C.
Kemp and J. B. Tenenbaum, Structural statistical models of inductive
reasoning, *Psychological Review*, 116 (2009), pp. 20–58.

5  Figure taken from E. Grenier, Introduction à la démarche Bayésienne sans
formule mathématique, *Modulab*, 43 (2010–11).

6  If $n$ equals zero, $k$ equals zero; zero divided by zero is indeterminate.

7  'Thus we find that an event having occurred successively any number of
times, the probability that it will happen again the next time is equal to
this number increased by unity divided by the same number, increased by
two units. Placing the most ancient epoch of history at 5,000 years ago, or
at 1,826,213 days, and the sun having risen constantly in the interval at
each revolution of 24 hours, it is a bet of 1,826,214 to 1 that it will rise
again tomorrow.' – *Essai Philosophique sur les Probabilités* (1825), p. 23.

8  If $s$ is very high, the denominator $n$ is negligible compared to $s$, while the
numerator $k$ becomes negligible compared to $st$, and so the ratio is close to
$t$, the initial expectation of probability.

9  The general formula is obtained by choosing a beta distribution with
parameters $[st, s(1 - t)]$ for the prior function, and a binomial distribution
for the likelihood function. In this case a beta distribution with parameters
$[st + k, s(1 - t) + n - k]$ is used for the posterior function.

10  W. K. Viscusi, *Rational Risk Policy* (Oxford University Press, 1998).

11  See B. Dessus and B. Laponche, Accident nucléaire: une certitude
statistique, *Libération*, 5 June 2011; and the correction made to their
calculations by E. Ghys, Accident nucléaire: une certitude statistique,
*Images des Mathématiques* (Centre National de Recherche Scientifique,
Paris, 2011).

12  To simplify matters, we have assumed here that the number of accidents
follows a binomial distribution. The probability of there not being a major
accident in Europe over the next thirty years is 1–0.0003 per reactor-year,

or $(1-0.0003)^{30 \times 143}$, or about 0.28. The probability of there being a major accident in Europe over the next thirty years is therefore 0.72. The choice of a thirty-year period is largely arbitrary; the calculation can be reduced to the probability of an accident in Europe next year, for which we find 0.042, or nearly a one-in-twenty-three chance.

13 The eight core-melt accidents previous to Fukushima Daiichi (see introduction of Part II), plus the meltdown of reactors 1, 2 and 3 at this nuclear power plant.

14 This value is derived from an American study encompassing seventy-five probabilistic risk assessments: US Nuclear Regulatory Commission, *Individual Plant Examination Program: Perspectives on Reactor Safety and Plant Performance (NUREG-1560)*, Vol. 1–3 (1997). This study indicates a mean core-melt frequency of $6.5 \times 10^{-5}$. It also gives the 95% confidence interval from which $s$ could be derived. For more details, see F. Lévêque and L. Escobar, How to predict the probability of a major nuclear accident after Fukushima Daiichi, Communication at the International Days of the Chair Modeling for Sustainable Development 2012 (MINES ParisTech, Sophia-Antipolis, 2012), available at: www.modelisation-prospective.org/Days-Chair_2012.html.

15 See also the changes in US probabilistic safety studies highlighted by the Electric Power Research Institute.

16 At least with respect to their location and type. Improvements to safety through changes to equipment and operating procedures mean that a reactor is not quite the same over time.

17 L. Escobar-Rangel and F. Lévêque, How Fukushima Daiichi core meltdown changed the probability of nuclear accidents, *Safety Science*, 64 (2014), pp. 90–98.

## Part III  Introduction

1 At least when an economic study is based, as is the case here, on publicly available documents (articles in scientific journals, parliamentary reports, information published by companies or administrative bodies, and so on), rather than detailed interviews with nuclear-safety professionals and practitioners the world over. Field investigations of regulation reveal the distance between stated principles and written rules, on the one hand, and their practical application, on the other.

## 7  Does nuclear safety need to be regulated?

1 F. Lévêque, *Economie de la Réglementation* (Collection Repères, Editions La Découverte, Paris, 2004).

2 Cour des Comptes, *Les Coûts de la Filière Nucléaire* (La Documentation Française, Paris, 2012), pp. 62–70.

3 J. V. Rees, *Hostages of Each Other: The Transformation of the Nuclear Industry Since 1980* (University of Chicago Press, 1985); N. Gunningham and J. Rees, Industry self-regulation, *Law and Policy*, 4:19 (1997), pp. 363–414.

4 See the detailed work of debate and analysis done by K. Fiore, Industrie nucléaire et gestion du risque d'accident en Europe, PhD thesis in economic science (Aix-Marseille 3 University, 2007), pp. 85–8.

5 For full analysis of the pros and cons of the strict, exclusive liability enshrined in the nuclear civil liability regime, see R. A. Winter and M. Trebilcock, The economics of nuclear accident law, *International Review of Law and Economics*, 17 (1997), pp. 225–43.

6 T. Vanden Borre, Shifts in governance in compensation for nuclear damage, 20 years after Chernobyl, in M. Faure and A. Verheij (eds.), *Shifts in Compensation for Environmental Damage* (Tort and Insurance Law, Vol. 21, Springer, 2007).

7 D. Koplow, Nuclear power: still not viable without subsidies (Union of Concerned Scientists, 2011); R. Bell, The biggest nuclear subsidy: pathetically inadequate insurance for a colossal liability (www.planetworkshops.org, 2012).

8 The first study dates from 1990 and was done by J. A. Dubin and G. S. Rothwell, Subsidy to nuclear power through the Price-Anderson liability limit, *Contemporary Policy Issues* (1990), pp. 73–8. It yielded an estimate of $21.7 million per reactor-year, compared with the then cap of $7 billion. This study contained a mistake which was corrected by A. G. Heyes and C. Liston-Heyes, Subsidy to nuclear power through the Price-Anderson liability limit: comment, *Contemporary Economic Policy*, 16 (1998), pp. 122–4. The estimate was cut to $2.32 million per reactor-year.

9 Congressional Budget Office, *Nuclear Power's Role in Generating Electricity*, (2008), pp. 28–9.

10 M. G. Faure and K. Fiore, An economic analysis of the nuclear liability subsidy, *Pace Environmental Law Review*, 26 (2009), pp. 419–47.

11 The 2008 Congressional Budget Office report adopted a levelized cost of $72 per MWh for its baseline scenario for an advanced generation of nuclear reactors.

12 Faure and Fiore, An economic analysis of the nuclear liability subsidy, took €0.457 per MWh as the value of the highest risk. They compared this amount to a levelized cost for nuclear power of €30 per MWh, which is not very realistic. We have adopted a levelized cost twice as high, at €60 per MWh.

13 Versicherungsforen Leipzig, *Calculating a Risk-Appropriate Insurance Premium to Cover Third-Party Liability Risks that Result from Operation of Nuclear Power Plants*, commissioned by the German Renewable Energy Federation (2011).
14 Also of note is the study by B. A. Leurs and R. C. N. Wit, *Environmentally Harmful Support Measures in EU Member States* (CE Delft, January 2003), for the Environment Directorate General of the European Commission, on energy subsidies and how they harm the environment. In the part on nuclear power, the subsidy through limited liability is estimated, at its upper limit, at €50 per MWh, in other words with a baseline production cost for nuclear electricity of €25, the final cost would be tripled.
15 For example the risk coverage provided by pension funds or cover funds, in the form of bonds. See M. Radetzki and M. Radetzki, Private arrangements to cover large-scale liabilities caused by nuclear and other industrial catastrophes, *The Geneva Papers on Risks and Insurance*, 25 (2000), pp. 180–95.

## 8 The basic rules of regulation

1 For a detailed account of changes in the conceptions of safety in the US and France, see the PhD thesis by C. Foasso, Histoire de la sûreté de l'énergie nucléaire en France (1945–2000) (Université Lumière-Lyon II, 2003).
2 We have chosen regulation in Japan as an example to avoid because its shortcomings have been largely documented. This does not mean it comes last in any global ranking. Safety regulation in the Russian Federation or Pakistan may be even more defective, but these countries do not publish the information needed to assess their authorities.
3 S. Huet, *Nucléaire: Quel Scénario pour le Futur?* (Collection 360, Editions La Ville Brûle, Montreuil, 2012), p. 132.
4 J. Kingston, Power politics: Japan's resilient nuclear village, *The Asia-Pacific Journal*, 10 (2012), available at: www.japanfocus.org/-Jeff-Kingston/3847.
5 A. Gundersen, The echo chamber: regulatory capture and the Fukushima Daiichi disaster, in *Lessons from Fukushima* (Greenpeace, 2012), pp. 41–9.
6 Citizens Information Center, Revelation of endless N-damage cover-ups, *Nuke Info Tokyo*, 92 (2002).
7 *Japan Times*, 4 May 2011.
8 Culture of Complicity Tied to Stricken Nuclear Plant, *New York Times*, 26 April 2011.
9 *Asahi Shimbun*, 3 May 2011.

10 Final report, Investigation Committee on the Accident at the Fukushima Nuclear Power Stations of Tokyo Electric Power Company (2012).

11 Official Report of the Fukushima Nuclear Accident Independent Investigation Commission, National Diet of Japan (2012).

12 Final report, Investigation Committee on the Accident at the Fukushima Nuclear Power Stations, pp. 414–15.

13 Official Report of the Fukushima NAICC (2012), Executive Summary, p. 73.

14 Official Report of the Fukushima NAICC (2012), Section 5, p. 12.

15 For a definition of its mission, see the Official Report of the Fukushima NAICC (2012), Chapter 5, p. 52.

16 J. M. Ramseyer and E. B. Rasmusen, Why are Japanese judges so conservative in politically charged cases?, *American Political Science*, 95:2 (2001), pp. 331–44.

17 Japanese officials ignored or concealed dangers, *New York Times*, 16 May 2011.

18 G. J. Stigler, The theory of economic regulation, *Bell Journal of Economic and Management Science*, 2:1 (1971), pp. 3–21.

19 'The incestuous relationships that existed between regulators and business entities must not be allowed to develop again.' Official Report of the Fukushima NAICC (2012), Executive Summary, p. 44.

20 Japan ignored nuclear risks, official says, *New York Times*, 15 February 2012.

21 F. Lévêque, *Economie de la Réglementation* (Collection Repères, Editions La Découverte, Paris, 2004), p. 14.

22 Commissions locales d'information (local information committees): see ASN website, www.asn.fr/L-ASN/Les-autres-acteurs-du-controle/CLI.

## 9   What goal should be set for safety and how is it to be attained?

1 S. Paik and W. R. Schriver, The effect of increased regulation on capital costs and manual labor requirements of nuclear power plants, *The Engineering Economist*, 1 (1980), pp. 223–44.

2 M. W. Golay, Improved nuclear power plant operations and safety through performance-based safety regulation, *Journal of Hazardous Materials*, 1–3:71 (2000), pp. 219–37.

3 Regulatory Guide 1.174, *An Approach for Using Probabilistic Risk Assessment in Risk-Informed Decisions on Plant-Specific Changes to the Licensing Basis* (Nuclear Regulatory Commission, 2002).

4 R. A. Meserve, The evolution of safety goals and their connections to safety culture, *NRC News*, S-01–013 (2001).

5 Author's emphasis.

6 G. E. Apostolakis, How useful is quantitative risk assessment?, *Risk Analysis*, 24:3 (2004), pp. 515–20.

7 A. C. Kadak and T. Matsuo, The nuclear industry's transition to risk-informed regulation and operation in the United States, *Reliability Engineering and System Safety*, 92 (2007), pp. 609–18.

8 Lettre d'orientation (guideline) SIN 1076/77 (1977), on the main safety options for plants operating one or more pressurized water reactors, sent by the minister in charge of industry to EDF's general manager.

9 Nuclear Regulatory Commission Policy Statement, 4 August 1986.

10 Nuclear Regulatory Commission Strategic Plan, fiscal years 2008–13, p. 5. For an older reference on the NRC mandate issued by Congress to *protect the environment* (author's emphasis), see Atomic Energy Act (1954), p. 149.

11 Regulatory Information Conference, 13 March 2012, Rockville, Maryland.

12 Workshops on Frameworks for Developing a Safety Goal, Palo Alto (NUREG/CP-0018) (1981).

13 Nuclear Regulatory Commission, Staff Requirements Memorandum on SECY-89-102, Implementation of the Safety Goals (1990).

14 UK Health and Safety Executive, Office of Nuclear Regulation, *Numerical Targets and Legal Limits in Safety Assessment Principles for Nuclear Facilities* (2006).

15 Whereas most nuclear nations were quick to legislate on the procedures for monitoring nuclear installations, France added a few, very general provisions to an act on atmospheric pollution and odours, passed in 1961, adding more specific rules in a brief ministerial order in 1963. Prior to that, nuclear power plants were subject to a 1917 act on hazardous, inconvenient and insalubrious facilities. See the PhD thesis by A.-S. Vallet, quoted by G. Rolina, in *Sûreté Nucléaire et Facteurs Humains, La Fabrique Française de l'Expertise* (Presse des Mines, Paris, 2009), p. 37. Until 2006 a substantial part of the administration's work monitoring nuclear safety was based on a pragmatic approach that was not legally binding.

16 Law *no.* 2006–686, on the transparency and safety of nuclear facilities, commonly known as the 'Loi TSN'.

17 Article 29, author's emphasis.

18 Groupe Permanent d'Experts pour les Réacteurs Nucléaires, *Directives Techniques pour la Conception et la Construction de la Nouvelle Génération de Réacteurs Nucléaires à Eau Sous Pression* (2000).

19 In 2012 the wording became more precise: Article 3.9 of the ministerial order dated 7 February 2012 establishing General rules relative to basic nuclear installations stipulates that: 'The demonstration of nuclear safety must prove that accidents that could lead to large releases of hazardous substances or to hazardous effects off the site that develop too rapidly to allow timely deployment of the necessary population protection measures

are physically impossible or, if physical impossibility cannot be demonstrated, that the measures taken on or for the installation render such accidents extremely improbable with a high level of confidence.'

20 Lettre d'orientation SIN 1076/77 (see n. 8).

21 Lettre d'orientation SIN 576/78 (1978).

22 C. Foasso, Histoire de la sûreté de l'énergie nucléaire en France (1945–2000), PhD thesis (Université Lumière-Lyon II, 2003), p. 379.

23 A forerunner of ASN, the Central Department for Nuclear Plant Safety (SCSIN) was set up in 1973.

24 The 2006 Act does not mention the principle of continuous progress, but it does appear in the previously cited ministerial order (February 2012), with a whole chapter under this title.

25 A list of basic rules and safety guides can be found at: www.asn.fr/ Reglementer/Regles-fondamentales-de-surete-et-guides-ASN.

26 Foasso, PhD thesis, p. 655.

27 F. Farmer, Siting criteria, a new approach, in *Containment and Siting of Nuclear Power Plants* (International Atomic Energy Agency, Vienna, 1967), pp. 303–29.

28 Giving rise to serious dispute, in particular between CEA and EDF: see G. Hecht, *Le Rayonnement de la France* (Collection Textes à l'appui, Editions La Découverte, Paris, 2004).

29 P. Simonnot, *Les Nucléocrates* (Collection Capitalisme et Survie, Presses Universitaires de Grenoble, 1978).

30 See A. Beltran, J.-F. Picard and M. Bungener, *Histoire(s) de l'EDF, comment se sont prises les decisions de 1946 à nos jours* (Editions Dunod, Paris, 1985) and Pierre Guillaumat, *La Passion des Grands Projets Industriels* (Editions Rive Droite, Paris, 1995).

31 S. Topçu, L'agir contestataire à l'épreuve de l'atome. Critique et gouvernement de la critique dans l'histoire de l'énergie en France (1968–2008), PhD thesis (Ecole des Hautes Etudes en Sciences Sociales, Paris, 2010).

32 G. M. Grossman and E. Helpman, *Special Interest Politics* (Massachusetts Institute of Technology Press, Cambridge, MA, 2001).

33 P. W. MacAvoy and J. W. Rosenthal, *Corporate Profit and Nuclear Safety: Strategy at Northeast Utilities in the 1990s* (Princeton University Press, Princeton, NJ, 2005).

34 *Mainichi Shinbun*, 7 March 2012.

35 T. R. La Porte and C. W. Thomas, Regulatory compliance and the ethos of quality enhancement: surprises in nuclear power plant operations, *Journal of Public and Administration Research and Theory*, 5:1 (1995), pp. 109–37.

36 E. Löfstedt and D. Vogel, The changing of regulations: a comparison of Europe and the United States, *Risk Analysis*, 21:3 (2001), pp. 399–405.

37 However, the reappraisal of the basic rules on nuclear installations that started in 2006 has led to a significant increase in general regulations, affecting all plants, and individual prescriptions specific to each case.
38 The report on the Integrated Regulatory Review Service Mission to the US notes for example that 'licensees have not been as proactive in making voluntary measures to upgrade systems, structures, and components with improved technology as many foreign countries have done to enhance safety'.
39 Article 182 of the US Atomic Energy Act of 1954.
40 With the obvious risk that political decisions may be based on perceived risk rather than calculations by experts: see Chapter 11.
41 R. Meserve, *International Atomic Energy Agency Bulletin*, 49/1 (2007).

## 10 Adopting nuclear power

1 Canada is not such a clear-cut example, because its policy on nuclear weapons has varied over time. As a member of Nato, much like Germany, it allowed US nuclear warheads to be stationed on Canadian soil for some time.
2 North Korea has also tried to develop a civilian and military nuclear programme. However, unlike Pakistan and Iran it has not yet succeeded in producing nuclear power.
3 Some authors have also presented the UK as the first country to have built a nuclear reactor and connected it to the grid.
4 M-H. Labbé, *Le Nucléaire à la Croisée des Chemins* (La Documentation française, Paris, 1999), ch. 6, discusses in detail the studies, reports and debate preceding Eisenhower's speech.
5 The reference to the Apocalypse highlights the religious side to the speech to the UN General Assembly. Humankind must be saved, while redeeming the scientists and engineers who laboured to build the bomb. Civilian nuclear power would earn them absolution and wash away the original sin of the atom bomb. The longest analysis (162 pages) devoted to the Atoms for Peace speech was produced by a little known specialist in religious studies at the University of Colorado at Boulder: Ira Chernus, *Eisenhower's Atoms for Peace* (Texas A&M University Press, 2002).
6 Between 1955 and 1970 Pakistan signed thirteen bilateral agreements on nuclear power with the countries mentioned in addition to France, Belgium, Denmark, the USSR, Spain and the Federal Republic of Germany.
7 S. D. Thomas, *The Realities of Nuclear Power: International Economic and Regulatory Experience* (Cambridge University Press, 1998). Chapter 7 is a monograph devoted to Canada.

8 Israel is the only exception. This country has nuclear weapons but not a single grid-connected nuclear power plant. This unusual choice is no doubt due to the secrecy surrounding its nuclear armament, now and in the past. Israel has not signed the non-proliferation treaty and has never officially acknowledged possessing nuclear weapons. Moreover, the limited size of its power grid and the target which a nuclear power plant would constitute for enemy attacks make it most unlikely that such a plant will ever be built.

9 It should, however, be noted that during this time a significant part of CEA research, and therefore expenditure, targeted technology and applications with no direct or indirect connection to the demands of the nuclear fleet.

10 The UK also adopted graphite-moderated, gas-cooled technology, but unlike France it went on to use it for its own fleet of reactors.

11 For discussion of this decision, in its technical and institutional context, see D. Finon and C. Staropoli, Institutional and technological co-evolution in the French electronuclear industry, *Industry and Innovation*, 8 (2001), pp. 179–99.

12 According to Robert Schuman, had De Gaulle stayed on as president, the gas-cooled graphite-moderated programme would have continued. Quoted by G. Hecht, *Le Rayonnement de la France* (Collection Textes à l'Appui, Editions La Découverte, Paris, 2004), p. 261.

13 Quoted by Hecht, *Le Rayonnement de la France*, p. 269.

14 Christian Bataille and Robert Galley, *Rapport sur l'aval du cycle nucléaire*, volume 2: *Les coûts de production de l'électricité*, ch. 1, sect. II (Rapport de l'office parlementaire d'évaluation des choix scientifiques et technologiques, 1999), available at: http://www.assemblee-nationale.fr/rap-oecst/nucleaire/r1359.asp.

15 Hecht, *Le Rayonnement de la France*, in particular ch. 7, La guerre des filières.

16 China has already exported equipment to Pakistan, but now hopes to take part in international tenders, rather than just making do with mutual agreements governed exclusively by strategic considerations.

17 J. Goldemberg, Nuclear energy in developing countries, *Daedalus* (2009), pp. 71–80.

18 OECD countries still feature largely in this selective list, but this is misleading, because not many of them now want to extend their nuclear-power capacity; their preponderance is due to the filter used to eliminate doubtful applicants. Turkey, Chile and Poland are the only OECD members out of a total of fifty, but all three are on the short list. Disregarding their real potential and chances of remaining solvent over the next twenty years, developing countries account for most applications to adopt nuclear power.

19 So far only Mexico and Iran have gained a foothold in nuclear power.

20 Poland is not strictly speaking a newcomer to nuclear power. In the 1980s work started at Zarnowiec to build four Russian reactors, but the project was shelved in 1991.

21 This is the nuclear industry's first build, own and operate contract. It goes much further than the sale of a turnkey plant, for the foreign investor is not only taking a risk as the main shareholder, but also as the operator of the plant. The only source of revenue is future sales of electricity, the price of which is fixed in advance. A transaction of this sort, by which an installation is supplied in exchange for the purchase of electricity shows that competition between vendors is very stiff, playing into the hands of buyers. All other things being equal, this situation is an incentive for aspiring countries to take the plunge.

## 11 Nuclear exit

1 The Swedish Parliament banned construction of additional reactors in 1980, but overturned the decision thirty years later.

2 The deal was negotiated in 2000 with Germany's four nuclear operators (Eon, RWE, Vattenfall and EnBW).

3 The CDU returned to power in 2005, but as the senior partner in a broad coalition with the SPD and CSU. So Angela Merkel could not go back on the 2000 agreement between the Federal Government and the utilities.

4 J. H. Keppler, The economic costs of the nuclear phase-out in Germany, NEA News, 30 (2012), pp. 8–14.

5 This amount corresponds to the cost of building and operating alternative capacity. The author has, however, allowed for the fact that part of the 20.5 GW capacity which was retired early did not need to be replaced. In anticipation of nuclear exit, investment shifted to other generating technologies from 2000 onwards, mainly coal-fired and to a lesser extent gas-fired plants. This outlay should consequently not be treated as an additional cost.

6 Or by electricity not consumed, given the expenditure on boosting energy efficiency, or the costs due to network outages.

7 A study by various research institutes for the German Economy and Technology Ministry put the cost of early exit at €48 billion (Energiezenarien 2011, Projekt 12/10, available at: www.prognos.com/fileadmin/pdf/publikationsdatenbank/11_08_12_Energieszenarien_2011.pdf); another study, by the Institute for Energy Economics at Stuttgart University, put the cost at €42 billion.

8 'The nuclear power plant at Fessenheim, which is the oldest in our fleet, will be closed at the end of 2016', speech by President Hollande at the opening of the Environmental Conference.

9 In the course of a television debate, prior to the second round of the presidential election, between François Hollande and the incumbent Nicolas Sarkozy, the former said: 'People ask: "Why [close] Fessenheim?" [I answer] it is the oldest one in France. It is in a seismically active area, beside the Canal d'Alsace. There is considerable pressure, from all sides, to have it closed.' – Full transcript of the debate, *Le Monde*, 3 May 2012.

10 M. Berthélémy and F. Lévêque, Don't close nuclear power plants merely because they are old!, energypolicyblog.com (March 2011).

11 Roughly speaking and making a few assumptions, nuclear electricity will account for half of France's energy mix in 2025, with all the reactors aged over 40 having been shut down. But age is not a relevant technical or economic criterion when deciding to retire a plant. The 40-year age limit corresponds to a rule governing the period of depreciation used in EDF accounts.

12 The lack of anything other than political and communication considerations in the choice of a 25% cut by 2025 is also apparent in the possibility of this slogan making the wrong impression on advocates of nuclear exit. With sustained growth in domestic demand, no change in foreign trade and a mediocre load factor, retiring reactors before the age of 40 would make it necessary to build several new EPRs in order to attain the requisite 50%. See 50% d'électricité nucléaire implique la construction d'au moins 13 EPR, rebellyon.info/50-d-electricite-nucleaire-en-2025.html.

13 In a radio interview just a few days after the speech by Mr Hollande at the Environmental Conference, the Minister of the Economy and Finances said, 'We are putting figures on the table with which I'm not familiar, regarding a decision announced by the President the day before yesterday' – Arrêt de Fessenheim: le paiement de lourdes indemnités en question, *Le Nouvel Observateur* (16 September 2012).

14 *Le Journal du Dimanche* predicted €2 billion costs on 16 September 2012; eight months later it led with the headline, 'Five to eight billion [euros] to close Fessenheim' (5 May 2013).

15 French MP Hervé Mariton, Rapport Législatif 251, Annexe 13 (2013), estimates the loss entailed by shutdown in 2016 rather than 2022 at €2.4 billion. To justify this figure, he suggests a net profit on the Fessenheim plant of €400 million multiplied by the number of operating years lost. According to an anonymous blogger, Olivier68, Quel impact économique si on arrêtait Fessenheim? (AgoraVox.fr, 2013), the loss due to early shutdown would amount to about €10 billion. He tested several assumptions for additional operating life (ten or twenty years), average cost of replacement electricity (€110 or €130 per MWh) and discount rate (4% and 8%).

## 12 Supranational governance

1 In France the head of the Central Department for Protection against Ionizing Radiation (SCPRI), Pierre Pellerin, was accused of minimizing the effects of Chernobyl on French territory and concealing information. A journalist credited him with the apocryphal claim that 'The cloud from Ukraine stopped at our border.' On the legal front, the handling of the crisis and the impact of the disaster in France gave rise to much litigation. This ended in 2012 when the Cour de Cassation (final court of appeal) dismissed charges that Pellerin had endangered people who subsequently suffered cancers due to radiation. The court acknowledged that 'with the current state of scientific knowledge it is impossible to establish a certain relation of cause and effect between observed pathologies and the fallout from the radioactive cloud from Chernobyl'.

2 Except for islands such as Cyprus and Malta, and the Baltic states. The electrical isolation of the latter from the rest of the EU should end before 2020, thanks to an undersea connection between Estonia and Finland and a land connection between Lithuania and Poland.

3 Unless, of course, one opts for one of Europe's many islands, most of which are not on the grid.

4 Europeans and Nuclear Safety (2010), http://ec.europa.eu/energy/nuclear/safety/doc/2010_eurobarometer_safety.pdf, p. 42.

5 For a detailed examination of the conflict between the Czech Republic and Austria, and the part played by the EU, see R. S. Axelrod, Nuclear power and EU enlargement: the case of Temelin, *Environmental Politics*, 13 (2004), pp. 153–72.

6 World Nuclear Association, Early soviet reactors and EU accession, www.world-nuclear.org/info/Safety-and-Security/Safety-of-Plants/Appendices/Early-Soviet-Reactors-and-EU-Accession/.

7 R. S. Axelrod, The European Commission and Member States: conflict over nuclear safety, *Perspectives*, 13 (2004), pp. 153–72.

8 L. Scheinman, *Euratom: Nuclear Integration in Europe* (Carnegie Endowment for International Peace, New York, 1967).

9 Often anticipating the European Commission, national governments also asked their own nuclear safety authorities to carry out stress tests.

10 Désaccord européen sur les stress tests nucléaires, *Le Monde* (12 May 2011).

11 Les relations se tendent entre Bruxelles et l'Autorité de sûreté du nucléaire, Euractiv.fr (2012).

12 Communication from the Commission to the Council and the European Parliament on the comprehensive risk and safety assessments ('stress tests') of nuclear power plants in the Union and related activities (Brussels, 2012).

13 ASN has reservations about the communication of the European Commission (press release, Paris, 5 October 2012).

14 French nuclear regulator slams EC, Oettinger over EU stress tests, www. platts.com (2012).

15 J. Delors, Pour une Communauté européenne de l'énergie, in S. Andoura, L. Hancher, and M. Van Der Woude, *Vers une Communauté européenne de l'énergie: un projet politique* (Notre Europe, Etudes et Recherches 76, 2010); J. Delors et al., L'Europe de l'énergie c'est maintenant, *Le Monde* (25 May 2013).

16 Z. Laïidi and R. Montes Torralba, *Les Conditions de Mise en Oeuvre de la Solidarité Energétique Européenne* (funded by Conseil Français de l'Energie, Centre d'études européennes de Sciences Po, Paris, 2012).

## 13    International governance to combat proliferation

1 S. E. Miller and S. D. Sagan, Nuclear power without nuclear proliferation? *Daedalus: Journal of the American Academy of Arts and Sciences*, 138:4 (2009), pp. 7–18, based on measurements by the World Bank of state performance on good governance, and terrorism data from the US Counterterrorism Center.

2 For more details on the connection between the NPT and the IAEA, see La Commission des affaires étrangères, Les Enjeux Géostratégiques des Proliférations (Assemblée Nationale, Paris, 2009), available at: www. assemblee-nationale.fr/13/rap-info/i2085.asp.

3 These safeguards apply to all parties to the NPT, except a few African countries (Guinea Bissau, Eritrea) and small island states (Micronesia, East Timor, Sao Tome and Principe).

4 Quoted by Professor Ole Danbololt Mjos during the award ceremony of the 2005 Nobel Peace Prize to the IAEA and its Director General Mohamed El-Baradei.

5 North Korea signed the NPT in 1985, then left in 2003. The Treaty allows parties to resign membership after nine months. So a country can start by signing up to benefit from the science and technology cooperation on nuclear power provided by the IAEA then, having gained its expertise, withdraw and use it for military purposes in complete impunity.

6 M. Kroenig, E. Gartzke and R. Rauchhaus (eds.), *Causes and Consequences of Nuclear Proliferation* (Routledge Global Security Studies, Routledge, Abingdon, UK, 2011).

7 B. L. Cohen, *The Nuclear Energy Option* (Plenum Press, New York, 1990), ch. 13.

8 Anne Lauvergeon, former Areva CEO and well known for being plain-spoken, cited the pods invented by Nestlé in 2008: 'Our model is

Nespresso, we sell coffee machines and the coffee to go with them. And coffee is very profitable.' – *Le Point* (10 December 2010): www.lepoint.fr/economie/areva-un-geant-de-l-atome-de-la-mine-d-uranium-au-traitement-des-dechets-10-12-2010-1273598_28.php.

9 K. Piram, Stratégies gouvernementales pour le développement de l'énergie nucléaire civile: pratiques françaises et américaines sur le marché des centrales nucléaires, *Cahier Thucydide*, 8 (2009).

10 US Government Accountability Office, Nuclear Commerce (GAO-11–36, 2010).

11 US Government Accountability Office, Nuclear Commerce.

12 For a review of the French export offering, see D. Finon, *La Recomposition de l'Industrie Nucléaire Française est-elle Nécessaire?* (Document de travail du CIRED, Paris, 2011).

13 Hitachi holds an 80% share of GE-Hitachi Nuclear Energy, a company resulting from the merger of the two companies' nuclear interests; Toshiba acquired Westinghouse in 2006, and has enjoyed full control over it since 2012.

14 Half the AP1000s and EPRs ordered by China will be fitted with pressure vessels manufactured by Chinese firms.

15 South Korean elite forces arrive in UAE, *The National* (13 January 2011): www.thenational.ae/news/uae-news/south-korean-elite-forces-arrive-in-uae.

16 F. Chevalier and K. Park, The winning strategy of the late-comer: how Korea was awarded the UAE nuclear power contract, *International Review of Business Research Papers*, 6 (2010), pp. 221–38.

17 In 2012 State-owned Rosatom received an initial payment of $750 million. This sum was taken out of the national budget, as a share in the assets of the company specially created to build, own and operate the nuclear power plant at Akkuyu.

18 M. Berthélémy and F. Lévêque, Korea nuclear exports: Why did the Koreans win the UAE tender? Will Korea achieve its goal of exporting 80 nuclear reactors by 2030? (Cerna Working Papers Series, 4, 2011).

19 Before the tender, former NRC Director-General William Travers was the Executive Director of the Emirati authority. He contributed to framing Abu Dhabi's nuclear strategy. David F. Scott is on the board of the Emirates Nuclear Energy Corporation, which will be operating the plants.

20 The French state owns a 36.7% stake in GDF-Suez but in this capacity has no say in the firm's international development policy, unlike EDF in which it holds an 84.5% share.

21 The Korean case is unusual in that the long-standing publicly owned monopoly was split into various units, including KHNP, which operates

hydraulic and nuclear power plants. It was due to be privatized, but a change of government derailed this plan and all the subsidiaries, including numerous firms involved in nuclear power, are still wholly owned by Kepco, in which the state holds a majority stake.

22 J. A. Glasgow, E. Teplinsky and S. L. Markus, *Nuclear Export Control: A Comparative Analysis of National Regimes for the Control of Nuclear Materials, Components and Technology* (Pillsbury Winthrop Shaw Pittman LLP, Washington DC, 2012). This study compares export controls imposed by the US, Japan, France and Russia.

23 House of Representatives bill 1280: 'To amend the Atomic Energy Act of 1954 to require congressional approval of agreements for peaceful nuclear cooperation with foreign countries, and for other purposes', 31 March 2011.

24 Congress has ninety days to oppose a draft 123 agreement submitted by the US President. To do so, a majority of both chambers must approve a motion rejecting the agreement. In which case the President would very probably use his veto, which would mean that Congress would have to vote down the agreement by a two-thirds majority. To get round this almost insurmountable obstacle, the bill provides that Congress should approve, before it comes into force, any agreement which does not include a permanent ban on a potential customer ever acquiring enrichment and spent-fuel reprocessing facilities. The lack of a majority in favour of the agreement in one or other chamber would be sufficient to block the project.

25 Report to amend the Atomic Energy Act of 1954 to require congressional approval of agreements for peaceful nuclear cooperation with foreign countries, and other purposes (House of Representatives, Washington DC, 2012).

26 There is no evidence to support the concern 'that a wider application of the "gold standard" might have a potentially negative impact on commercial sales by the US nuclear industry, which asserted that efforts to secure commitments from countries not to engage in ENR activities would place US companies at a competitive disadvantage vis-a-vis their foreign competitors' (Report to amend the Atomic Energy Act of 1954).

27 J. C. Varnum, US nuclear cooperation as nonproliferation: reforms, or the devil you know?, *The Nuclear Threat Initiative* (2012): www.nti.org/analysis/articles/us-nuclear-cooperation-nonproliferation-reforms-or-devil-you-know/.

28 La Commission des affaires étrangères, Les Enjeux Géostratégiques.

29 Only the US, France, Russia and, as part of a joint venture, the Netherlands, the UK and Germany, possess large-scale enrichment capacity. Similarly, only France, Japan, Russia and the UK have industrial reprocessing units.

30 D. Albright, Holding Khan accountable, an Isis statement accompanying Release of Libya: a Major Sale at Last, *Institute for Science and International Security* (2010).

31 A. Rose, interview with M. Hibbs, How North Korea built its nuclear program, *The Atlantic* (10 April 2013).

32 Treaty on the Non-Proliferation of Nuclear Weapons, Article III, section 2.

33 Article IV of the Treaty states that 'Nothing in this Treaty shall be interpreted as affecting the inalienable right of all the Parties to the Treaty to develop research, production and use of nuclear energy for peaceful purposes without discrimination and in conformity with Articles I and II of this Treaty' and that 'All the Parties to the Treaty undertake to facilitate, and have the right to participate in the fullest possible exchange of equipment, materials and scientific and technological information for the peaceful uses of nuclear energy.'

34 M. Hibbs, *The future of the Nuclear Suppliers Group* (Carnegie Endowment for International Peace, 2011).

35 The same process also worked in the opposite direction, when the US had to persuade other NSG members to agree to the waiver granted to India.

36 F. McGoldrick, *Limiting Transfers of Enrichment and Reprocessing Technology: Issues, Constraints, Options* (Belfer Center, Harvard Kennedy School, 2011).

37 Oliver Thränert and Matthias Bieri, *The Nuclear Suppliers Group at the Crossroads* (CSS Analysis, 127, Center for Security Studies, 2013).

38 M. Hibbs, I. Anthony, C. Ahlström and V. Fedchenko, *Reforming Nuclear Export Controls: The Future of the Nuclear Suppliers Group* (SIPRI Research Report 22, Oxford University Press, 2007).

# Index

AEC (Atomic Energy Commission) 89–90
AEG 220
Algeria, as aspiring nuclear power 222–3
Allais, Maurice 104–8, 136
Allais paradox 104–8
Alstom 274–5
Andra (Nuclear Waste Authority) 25
ANRE (Agency for Natural Resources and Energy) 162, 164–5
Apostolakis, George E. 178
Areva 9–10, 58–9, 221–2, 270, 274–5, 281
    dominant position in nuclear industry 282
Argentina, as nuclear power user 211–12
Armenia, as nuclear power user 211–12
ASN (Autorité de Sûreté Nucléaire) 169, 186–7, 191, 198, 199, 203, 236, 241–2, 257
aspiring nuclear powers 222–5
Atomic Energy Act 227, 228, 229, 279–82
    Section 123 283–4
Atomic Energy Authority Safety and Reliability Directory 190
Atoms for Peace 213–15, 265, 266
Austria, political decisions 207–9, 256

Bahrain, as aspiring nuclear power 222–3
Bangladesh, as aspiring nuclear power 222–3
el-Baradei, Mohamed 267
Barakah nuclear power plant 274–5
baseload generating technology 64–6
baseline standards 188–9
Bayes' rule 88, 118–25
Bayes–Laplace rule see Bayes' rule
Bayesian algorithms 123
Bayesian revision 126–8
    and conditional probabilities 119–21
    Laplace reinvention of 125–32
    probability density function 126–8
    and rule of succession 125–32
    and statistical reasoning 124–5
    see also nuclear accidents; probability
Bayes, Thomas 118
Bechtel nuclear plant 49
Belarus
    as aspiring nuclear power 222–3
    Chernobyl pollution 244–5
    Chernobyl-related cancers 85–6
Belgium
    as nuclear power user 211–12
    phasing out nuclear power 226, 256
Bernoulli, Daniel 103–4

Bertel, E. 38
Brazil, as nuclear power user 211–12, 221–2
Browns Ferry reactor fire 175–6
Brussels Convention 252
Bugey reactors 44–5, 50
Bulgaria
  as nuclear power user 211–12
  reactor safety 249
Bushehr reactor 276

Canada
  heavy-water reactor 216–17
  in international nuclear market 271
  as nuclear power user 211–12
Cantor, Robin 45
carbon emissions 13, 73–4, 223–4
  coal 67–8
  cost of 16–21, 42
  markets and cost 18–20
  optimal level 172
  pollution abatement targets 20–1
  see also climate change; European Union Emissions Trading Scheme
carbon tax 2–3, 20, 30
Carlsbad, New Mexico waste storage 25
CEA (Commissariat à l'Energie Atomique) 189, 192, 217–18, 219
CEA-EDF nuclear action plan 220
Chernobyl Forum 85–6
Chernobyl nuclear accident 1–2, 4, 79–80
  areas affected 244–5
  cancer incidence 85–6
  cost of 86
  liability 1–2, 4
  radioactive emission 82–3, 84

Chevet, Pierre-Franck 202
Chile as aspiring nuclear power 222–3
China
  as authorized nuclear-weapon state 267
  early nuclear development 217
  and NSG 289–91
  nuclear aid to Pakistan 290
  as nuclear power user 211–12
  political decisions 207–9
  reactors 1–2, 52–3, 61, 271, 272, 274
    EPR (European Pressurized Reactor) 9–10, 58–9, 61
    technology transfer 221–2
    and Westinghouse reactors 271, 272
China syndrome 158
Chooz A reactor 23
Civaux 2 reactor 50
civil liability see nuclear accidents
climate change 223–4, 298
coal
  compared to nuclear power 66–72
  emissions costs 67–8
  fatalities/accidents incidence 116
  levelized costs 66
coal baseload technology 64–6
cognitive science, and probability 121–3
cognitive systems 110–11
Commission on the Accident at Three Mile Island 179–85
conditional probability see probability
Connecticut Light and Power Company 176
Conseil d'Etat 186
Conventions, nuclear 251–6
Cooper, Mark 48
core-damage frequency 177
core-melt see nuclear accidents
cost–benefit analysis 39–40

costs
  accountancy methods 11–13
  capital cost 36–7
  carbon emissions 16–21
  civil liability *see* nuclear accidents
  competitiveness 3–4, 8
  dynamic of 8
  economies of scale 43–4,
    45
  EPRs 58–9
  evaluation 7–8
  external 11–13
  external effects 13–16, 19
  financing 37
  fuel 8
  IDC (construction financing/interest
    cost) 38
  interest rates and discounting
    26–7
  investment 39
  levelized cost method 39–42, 70,
    72–3
  load factor 37, 143
  maintenance 44
  modular building 59–60, 61–2
  Monte Carlo method 70–2
  nuclear accident liability *see* nuclear
    accidents
  operating 44
  and opportunity 10–11
  overnight construction cost 36–7,
    41, 44–5, 46, 57
  per unit of power 39
  private 11–13
  rising *see* rising costs
  safety goals 172–4
  social costs 11–13
  standardization 59–60
  technical production costs
    36–7
  and technical progress 77
  variations 10–12, 77–8, 296
CPR-1000 reactor 52–3

Czech Republic
  as nuclear power user 207, 208,
    211–12
  reactor building 256
  Temelin reactor safety 249

Daya Bay nuclear power plant 221–2
DCNS 62–3
De Finetti, Bruno 97–8
decision theory 103, 121–3
decommissioning *see* reactors,
    decommissioning
Delors, Jacques 258
discount rate 8, 23–32
  calculation 27, 33–4, 41, 172
  and euro conversion 39
  Stern Review (2006) 29–31
District of Columbia Court of Appeals
    176, 200
dominant companies in nuclear
    industry 282
Doosan 221–2, 274–5, 280–1
dual-use goods 287

earthquake risk 93–4, 98, 195–6,
    297–8
economics, and psychology 110–11
EDF
  and Areva 58–9
  civil liability costs/limits 149–51,
    152–3, 253
  costs 11–13
  decommissioning/waste costs 24, 33,
    241–2
  international trade 274
  reactor building 51
  safety goals 179, 187–8
Egypt, as aspiring nuclear power 222–3,
    268
Einstein, Albert 212–13
Eisenhower, Dwight D. 213–15, 266
electricity market liberalization 2–3,
    40–1, 57–8, 69–70

electricity storage 75–6
Ellsberg, Daniel 104–8, 112
emissions *see* carbon emissions;
    nuclear accidents
energy independence 13–15, 218, 223,
    224, 298
Ensreg (European Nuclear Safety
    Regulation Group) 256–8
EPC (engineering, procurement and
    construction) industry 277–9
EPR (European Pressurized Reactor)
    9–10, 54–6, 58–9, 61, 187
Epstein, Paul 67
Erdogan, Tayyip 281
Ethics Committee for a Safe Energy
    Supply 229, 231–2, 233
Euratom Treaty 253–6
European Atomic Energy Community
    244
European Commission, enforcement
    powers 264
European common energy policy 258–9
European power grid 244, 245
European safety standards 247–51
European Union
    carbon costs 18–20
    civil liability
        discrimination/variations 251–6
        limits 149–51
    coexistence and conflicting views
        256–9
    gas imports 13–15
    international governance issues
        295
    nuclear safety 1–2, 298
    internal standards variations 247–51
    international cooperation 244–7
    local–global balance 248–50
    reactor stress safety tests 256–8
    renewable energy targets 73–4
European Union Emissions Trading
    Scheme 19, 73–4
    failure of 22

expected utility theory 103–4, 108–9
export controls *see* nuclear trade
ExternE study 82, 84, 86–7

Farmer, F. R. 190
Fessenheim nuclear plant 44–5, 204,
    220–1, 235–6, 237–40, 241–2
Finland
    as nuclear power user 211–12
    reactor building 256
Flamanville, France reactor 40, 54–6,
    58–9
Framatome 220–1
France
    ASN (Autorité de Sûreté Nucléaire)
        *see* ASN
    as authorized nuclear-weapon state
        267
    carbon emissions 22
    civil liability costs/limits 149–51,
        152–3
    Conseil d'Etat 186
    consensus-based regulation 198
    early reactor development 190–1
    EDF costs 10–11
    energy independence 218
    and gas-cooled, graphite-moderated
        technology 220–1
    in international nuclear market
        271
    national prestige 217–18
    nuclear accident liability 35, 36
    nuclear exports 285
    nuclear industry 2–3
    as nuclear power user 211–12
    nuclear research 219
    nuclear trade 285
    overnight construction cost 44–5
    plant closure 235–43
    political party policies 237–40,
        241–2
    reactor building 256
    reactor risk probability 187–8

France (*cont.*)
  reasons for adopting nuclear power
    217–19
  regulatory model 201–4
  rising reactor costs 49–50, 52, 55–6
  safety costs 202–3
  safety regulation 186–92, 297–8
    comparisons with US 198–9
    early years 189–91
  safety supervisory bodies 189–90
  safety systems 4–5
  technology transfer 221–2
  and US experience 190–1
  and US technology adoption 220–1
  waste management 32
  and Westinghouse 190–1, 220–1
  *see also* EDF
free-riding 144, 145
fuel cost 8
fuel rod cladding 158
Fukushima Daiichi nuclear accident
    1–2, 4–5, 296–7
  and European policy 256–8
  and frequency predictions 91–2
  and German policy 207, 208
  prediction failures 93–4, 98–100
  safety regulation 139, 160–71, 297–8
  seawall design 99–100
  and US safety regulation 201
  *see also* Japan; safety regulation

gas
  compared to nuclear power 66–72
  emissions costs 68–9
  levelized costs 66
  market liberalization 69–70
  reserves 223
  semi-baseload technology 64–6
  shale gas 1–2, 68–9
gas-cooled, graphite-moderated
    technology 220–1
Gates, Bill 62

Gaulle, General Charles de 217–18
GDF-Suez
  liability limit 253
  price controversy 9–10
  reactor building 274–5, 281
GE Hitachi Nuclear Energy 280–1
General Electric 271, 272, 280–1
Germany
  Chernobyl pollution 244–5
  civil liability limits 149–51, 153–4
  compensation rules 155
  and Fukushima Daiichi disaster
    229–32
  as nuclear power user 211–12
  phasing out nuclear power 226
    economic cost 228–9, 230–1
    political party policies 226–30,
      234–5
    reason for decision 232, 256
    speed of exit 226–35
  reactor closure/decommissioning
    1–3, 207–8, 209
  technology transfer 221–2
  and Westinghouse/General Electric
    221–2
Gigerenzer, Gerd 116, 123
Gollier, Christian 31
Gould, Stephen Jay 122–3
governance
  internal differences 259–64
  international export controls
    282–6
  international governance, and
    collective good 245–7
  international tensions 246–7,
    259–62
  national sovereignty, and policies
    262–3
  nuclear security issues 293–4, 299
  and nuclear weapon proliferation
    266
  supranational 244–64

greenhouse gases *see* carbon
Gronlund, Lisbeth 86
Grubler, Arnulf 50
Guangdong Nuclear Power Group 58–9

harmonization of European safety
    standards 247–51
heuristics 109–10
    availability heuristic 113
    representativeness heuristic 113
Hewlett, James 45
Hinkley Point reactor plans 56
Hirai, Yanosuke 195–6
Hiroshima bomb 212–13
Hollande, François 236, 237–40, 241–2
hostile attacks 15
House of Representatives Foreign
    Affairs Committee 284
Hungary, as nuclear power user 211–12
hydro-electric peaking technology 64–6
hydrocarbon reserves 223
hydrocarbon-producing countries, as
    aspiring nuclear powers 223
Hyundai 274–5

IAEA (International Atomic Energy
    Agency) 160–1, 209, 211,
    214–15, 250–1
    budget/resources 268–9
    cooperation between states 263–4
    ineffective inspections by 287
    list of aspiring nuclear powers 222–3
    and Non-Proliferation Treaty 266–9
    strengths and weaknesses of 266–9,
        293–4
    and weapon proliferation 266, 287
IDC (construction financing/interest
    cost) 38
Import-Export Bank of Korea 274
India
    fuel reprocessing 284–5
    IAEA inspection issues 291–2

as non-NPT state 267, 284–5, 291–2
    NSG dispute over exports 291–2
    as nuclear power user 211–12
    and nuclear weaponry 215
    reactors 52–3
    technology transfer 221–2
indifference principle 129
Indonesia, as aspiring nuclear power
    222–3
industrial nationalism, and reactor
    exports 2–3
INPO (Institute of Nuclear Power
    Operations) 145–6, 199
intermittent energy sources 64–6, 73,
    74–5
international borders, and collective
    good 245
international governance, and
    collective good 245–7
international regrouping, of nuclear
    industry 300
inverse probability 120–1
investment cost *see* cost
Iran
    non-proliferation treaty 1–2
    as nuclear power user 211–12
    nuclear threat from 283–4
    reactor construction 276
    uranium enrichment 1–2
Iraq, as aspiring nuclear power 222–3,
    268
IRRS (International Regulatory
    Review Service) 160–1
IRSN (Institut de Radioprotection et
    de Sûreté Nucléaire) 191
Israel, as non-NPT state 267
ITER reactor 255
Itochu 281

Jaczko, Gregory 181, 201
Jamaica, as aspiring nuclear power
    222–3

Japan
 Diet 165–6
 government influences 165–6
 Nisa 163–5, 167–8
 as nuclear power user 211–12
 reactors 52–3
 regulatory capture by industry
  166–71
 revolving door (*amakudari*) policy
  162–3
 safety regulation 139, 160–71,
  195–6, 297–8
 safety systems 4–5, 160–71
 Supreme Court judges 165
 technology transfer 221–2
 *see also* Fukushima Daiichi; safety
  regulation
Japan Steel Works 274
JNES (Japan Nuclear Energy Safety
 Organization) 164
Joskow, Paul 73

Kahneman, Daniel 103, 108–11,
 112–13, 122–3, 136
Kazakhstan, as aspiring nuclear power
 222–3
Kennedy Commission 179–85
Kenya, as aspiring nuclear power
 222–3
Kepco 274–5
Keppler, J. H. 229, 231
Keynes, John Maynard 96–7, 104–8,
 112, 129
Khan, Abdul Qadeer 285
Korea Electric Power Corporation
 221–2, 269–70
Korea Hydro and Nuclear Power 274–5
Krümmel reactor 229

Lacoste, André-Claude 202, 256–8
Laplace, Pierre-Simon 118, 125–32; *see
 also* Bayes' rule

Lebègue report 31
Lee Myung Bak 274
levelized cost method *see* cost
liability *see* nuclear accidents
Lisbon Treaty 259
Lithuania
 as nuclear power user 211–12
 reactor safety 249
load factor *see* cost
load-balancing 64–6

Madarame, Haruki 167
Maine Yankee nuclear plant 24
Malaysia, as aspiring nuclear power
 222–3
Manhattan Project 212–13, 216–17
Merkel, Angela 207, 208, 229, 230
Meserve, Richard A. 177
Mexico, as nuclear power user
 211–12
Millstone nuclear plant 194–5
MIT studies 54, 55, 57–8, 70, 89–90
Mitsubishi Heavy Industries 281
Monnet, Jean 253

Nagasaki bomb 212–13
al-Nahyan, Khalifa bin Zayed 274
national independence 13–15
national policies 5, 207–9, 298
national prestige 217–18
national sovereignty 262–3
natural frequency format 124
Naudet, G. 38
Neckarwestheim nuclear plant 228
Netherlands, as nuclear power user
 211–12
Newbery, David 74
NIED (National Research Institute for
 Earth Science and Disaster
 Prevention) 99
Nisa 163–5, 167–8
Nordhaus, William D. 30

North Korea
  and NPT 268
  as nuclear power 268
  nuclear threat from 283–4
Northeast Utilities 194–5
NPT *see* Treaty on the Non-
    Proliferation of Nuclear
    Weapons
NRC (Nuclear Regulatory
    Commission) 89–90, 175–6,
    177–8, 179, 182–3, 199–201
NREL (National Renewable Energy
    Laboratory) 76
NSC (Nuclear Safety Commission)
    164–5
NSG (Nuclear Suppliers Group) 266,
    286–95
  and China 289–91
  constituent states 286, 288
  disparate interests 291–2
  export controls 288
  and exports to India 291–2
  future role 292
  instability of 291
  lack of control/enforcement 289–91
  membership benefits 291
  membership size 291–2
  and Non-Proliferation Treaty 288–9
  roles of 286–7
  seen as restrictive cartel 289–91
  supply/competition restriction by
    289
  and weapon proliferation 286–9
nuclear accidents 1–3, 4
  Bayes' rule *see* Bayes' rule
  Bayesian algorithms 123
  cancer incidence from 82–6
  certainty 95–6
  civil liability 34–6, 42, 113, 142,
    146, 147–54, 156, 253
    hidden subsidy 151–6
    within Europe 251–6

and collective good 245–7
core-melt frequency 87–90, 91, 92,
    93, 132–7
core-melt probability 132–7, 185
costs of 31, 86–7
disaster aversion 102–17
earthquake risk 93–4, 98
emissions from 82–6
European safety standards 247–51
event trees 88–91
ExternE study 82, 84, 86–7
fires 175–6
frequency calculation 37
frequency predicted/actual 91–101
heuristics *see* heuristics
international cooperation 244
levels 79
person-Sv unit 83
predicting next event 125–32
probability *see* probability
reactor risk probability 187–8
risk ambiguity 112
risk assessment 4
solvency and accident liability
    147–8
spent-fuel risk 93–4
tsunami/tidal wave risk 93–4, 98,
    99–100
uncertainty 95–6
  *see also* risk; safety regulation
nuclear baseload technology 64–6
nuclear fuel 15–16
nuclear power
  *Atoms for Peace* speech 213–15
  compared to gas/coal 66–72
  competitiveness 72–8
  current users 211–12
  and European discord 256–9
  international cooperation
    agreements 215–16
  military use 207, 208
  national policies 5, 207–9, 298

nuclear power (*cont.*)
  perception biases against 111–17
  pioneering countries 216–17
  post-war policy 213–15
  reasons for adoption 217–19
nuclear power plants *see* reactors
nuclear research 219
  power/weaponry duality 215–16
nuclear risk *see* risk
nuclear safety *see* safety regulation
nuclear trade 269–73
  comparisons with armament and
      oil/gas industries 273–82
  consortiums 275
  and EPC (engineering, procurement
      and construction) industry
      277–9
  export controls 282–95
    collective action on 286–95
    dual-use goods 287
    and IAEA *see* IAEA
    international cooperation 282–6
    and NSG *see* NSG
    political support for 293
    security issues 293–4
    United States 283–6
  government involvement 274–5
  government/companies
      collaboration 279–82
  heads of state involvement 274–5
  state intervention and imports 277
  tendering 277
  *see also* IAEA; NSG; risk; safety
      regulation
nuclear waste *see* waste
nuclear weapon proliferation
  and governance 266
  horizontal proliferation 265
  and NSG 286–9
  and nuclear exports security 293–4
  risk 15–16, 283–4
  and scientific cooperation 269

  Treaty *see* Treaty on the
      Non-Proliferation of Nuclear
      Weapons
  vertical proliferation 265

Oettinger, Günther 256–8
oil reserves 223
oil-fired baseload technology 64–6
oil/gas industry model 277–9
Olkiluoto, Finland reactor 54–6, 58–9
Onagawa nuclear plant 195–6
  123 agreements 283–4
OPEC (Organization of Petroleum
      Exporting Countries) 291
Oppenheimer, Robert 212–13
overnight costs *see* cost

Paik, Soon 47
Pakistan
  as non-NPT state 267
  nuclear aid from China 290
  as nuclear power user 211–12
  and nuclear weaponry 215, 285
Paris Convention on Nuclear
      Third-Party Liability 252–3
peaking generating technology 64–6
person-Sv unit 83
Philippines, as aspiring nuclear power
      222–3
photovoltaic *see* solar energy
Pierrelatte enrichment plant 218
Poland
  as aspiring nuclear power 222–3, 224
  Chernobyl pollution 244–5
  as gas consumer 224
political consequences 2–3
pollution *see* carbon emissions
pollution abatement 20–1
Pompidou, Georges 220
posterior probability *see* probability
power demand variations 73
PRC *see* China

Price-Anderson Act (1957) 151
prior probability *see* probability
probability
  conditional probability 38, 88–91,
    119–21
  core-melt *see* nuclear accidents
  Keynes on 96–7, 104–8, 112, 129
  perceived probabilities 102–17,
    234
  posterior probability 120–1,
    126–8
  predicting next nuclear accident
    125–32
  prior probability 120–1, 123–5,
    126–8, 296–7
  probabilistic safety assessment 81–7,
    88–101, 157–8
  results divergence causes 88–101,
    296–7
probabilities perception
  biases 103–11
    against nuclear power 111–17
    psychology of 121–3
  subjective probability 97–8
  very low probabilities 114–15
  *see also* nuclear accidents; risk
proliferation *see* nuclear weapon
    proliferation
psychology
  and decision-making 121–3
  and economics 110–11

Ramsey, Frank Plumpton 27, 28, 97–8
Ramseyer, J. Mark 165
Rasmussen, Norman 89–90
Rasmussen report 89–90
reactor model effect 239
reactors
  accidents *see* nuclear accidents
  age and safety 236–40
  construction 1–3, 77–8, 256,
    269–70, 271, 272; *see also* costs

construction consortiums 274–5,
    280–1
construction/design regulations
    139–40
construction times 44, 45, 269–70
containment building 158
decommissioning 1–3, 23–32, 178,
    256
  costs 33–4
financing 274
learning effects 46
and military ambitions 224
modular building 59–60, 61–2
*paliers* 239
R&D 271
reactor age 236–40
reactor structure 158
rising costs 49–50, 52–3, 55–63,
    77–8
safety 46–9, 236–40
shutdowns and age 238–9
simplification 47–8
small/mini reactors 62–3, 298–9
standardization 59–60
underwater 62–3
Unplanned Unavailability Factor
    238
regulation *see* safety regulation
renewable energy
  competitive advantages 72–8
  EU targets 73–4
  sources 64–6
representativeness heuristic 113
Republic of Haiti, as aspiring nuclear
    power 222–3
RIPBR (risk-informed and
    performance-based regulation)
    177
rising costs 44–63, 296, 298–9
  learning effects 46
  limits to 53–4
  overnight cost 44, 46

rising costs (*cont.*)
    reactors 49–50, 52–3, 298–9
    21st century construction 55–63
       and safety regulation 46–9, 60
    and Three Mile Island 46
risk
    Allais paradox 104–8
    aversion 103–4, 116–17
    cooperation on 299
    definition 35
    economic costs 299
    nuclear accidents *see* nuclear
       accidents
    nuclear weapon proliferation risk
       15–16
    objectives/principles 179–80
    perception 2–3
    political decision-making 294–5
    population density around reactors
       181–2
    tidal wave/tsunami risk 93–4, 98,
       99–100, 297–8
    *see also* probability
Roosevelt, Franklin D. 212–13
Roques, Fabien 71
Rosatom 270
Rothwell, Geoffrey 45
rule of succession 125–32
Russia
    as authorized nuclear-weapon state
       267
    Chernobyl-related cancers 85–6
    early nuclear development 216–17
    fuel supply 15–16
    gas exports 13–15, 224
    in international nuclear market 271
    nuclear accident liability 35
    nuclear exports 276–7
    political decisions 207–9
    reactor construction 271, 272
    reactor exports 271, 272
    reactors 52–3

safety degradation 177
safety goals 179–85
    quantitative/qualitative objectives
       179–85
    subsidiary benchmarks 182–3
safety investment 116
safety regulation 46–9, 60, 299
    and civil liability *see* nuclear
       accidents
    collective regulation 145–6
    companies exceeding legal
       requirements 195–6
    competition authorities 193
    consensus-based 198
    containment building 158
    and cost-cutting 194–5
    countries exceeding legal
       requirements 195–6
    defence in depth 158
    determinist approach 157–9
    economic costs 172–4, 199, 205
    economic limits on 173
    engineers' 157–60
    Ensreg (European Nuclear Safety
       Regulation Group) 256–8
    Euratom Treaty 255
    European Directive 250–1
    European safety standards 247–51
       local–global balance 248–50
    European stress safety tests 256–8
    fire-protection 175–6, 178–9, 200
    free-riding 144, 145
    fuel rod cladding 158
    Fukushima Daiichi 139
    incentive-driven 142–7, 154–5
    influences on 205–6
    INPO (Institute of Nuclear Power
       Operations) 145–6
    inspection 146, 170, 194
    international tensions 246–7
    and law of substitution 173
    legal framework 205–6

main lines of defence 158
market forces 142
need for 141–56
operator rules 205
performance-based standards 173–9
political decision-making 294–5
probabilistic approach 81–7, 88–101,
    157–8, 159
quantitative/qualitative objectives
    179–85
reactor age 236–40
reactor/cooling structure 158
regulator independence 299
regulator powers 299
regulators/regulated relationships
    193–8
regulatory capture by industry
    166–71
RIPBR (risk-informed and
    performance-based regulation)
    177
sectoral regulators 193
self-regulation 142–7
state/public intervention 141–2, 147
technical dialogue 197
and technology-based standards
    173–9
undercover inspections 194
see also France; IAEA (International
    Atomic Energy Agency); Japan;
    NSG (Nuclear Suppliers
    Group); nuclear accidents;
    United States
safety systems 4–5
Saint Petersburg paradox 103–4
Sarkozy, Nicolas 241, 274, 275
Saudi Arabia, as aspiring nuclear power
    222–3
Savage, Leonard Jimmie 97–8, 105, 136
Schriver, William 47
security issues 15–16, 293–4, 299
semi-baseload technology 64–6

Service Central des Installations
    Nucléaires 187
shale gas 1–2, 68–9
Siemens 220
Simonnot, Philippe 192
Slovakia, reactor safety 249
Slovic, Paul 111
solar energy 73, 74–6
solvency, and accident liability 147–8
South Africa, as nuclear power user
    211–12, 221–2
South Korea
    in international nuclear market 271
    as nuclear power 266
    as nuclear power user 211–12
    nuclear trade alliance with US
        280–1
    reactor construction 221–2, 269–70,
        271–3
    reactors 52–3
    technology transfer 221–2
sovereignty 262–3
Spain
    as nuclear power user 211–12
    phasing out nuclear power 226
spent-fuel pools 93–4, 201
Stern Review (2006) 29–31, 32
Stigler, George J. 166–7
subjective probability see probability
succession, rule of 125–32
Sudan, as aspiring nuclear power 222–3
Sugaoka, Kei 162
Sunstein, Cass 114
supranational governance 244–64
Svinicki, Kristine 201
Sweden
    as nuclear power user 211–12
    phasing out nuclear power 226
Switzerland
    as nuclear power user 211–12, 221–2
    phasing out nuclear power 226
    political decisions 207–9

Taishan-1 nuclear plant 58–9
Taiwan, as nuclear power user 211–12,
    221–2
Tanzania, as aspiring nuclear power
    222–3
technological advances 218–19
technology transfer 221–2, 267
Temelin nuclear power plant
    249
Tepco 161, 162, 163
Terrapower 62
Thailand, as aspiring nuclear power
    222–3
Three Mile Island reactor accident 46,
    79–80, 271
  Commission report/
    recommendations 179–85
  cost of 86
  and INPO 145
  and probabilistic safety assessment
    90
  quantitative/qualitative objectives
    179–85
  technical rules 177
  see also safety regulation; United
    States
tidal wave/tsunami risk 93–4, 98,
    99–100, 297–8
Tohoku Electric Power 195–6
Tokyo Electric Power 98,
    281
Töpfer, Klaus 231
Toshiba, in US/South Korea
    consortium 280–1
Total 274–5
Transparency and Nuclear Safety Act
    186
Treaty on the European Atomic Energy
    Community 298
Treaty on the Functioning of the
    European Union (Article 194)
    260–2

Treaty on the Non-Proliferation of
    Nuclear Weapons 1–2, 209,
    215, 266, 298
  authorized nuclear weapon states
    267
  cooperation/technology transfer 267
  and disarmament 267
  and the IAEA 266–9
  and India 291–2
  non-NPT states 267
  Turkey's rights within 284–5
  see also nuclear weapons
    proliferation
tsunami/tidal wave risk 93–4, 98,
    99–100, 297–8
Turkey
  as aspiring nuclear power 222–3,
    225, 266
  independent nuclear development
    284–5
  political decisions 207–9
  reactor building 276
    international consortium 281
Tversky, Amos 108–11, 112–13, 122–3

Ukraine
  Chernobyl pollution 244–5
  Chernobyl-related cancers 85–6
  as nuclear power user 211–12
uncertainty, and nuclear accident
    probability 95–6
United Arab Emirates
  as aspiring nuclear power 222–3, 266
  reactor building 274–5
United Kingdom
  as authorized nuclear-weapon state
    267
  carbon emissions 22
  Chernobyl pollution 244–5
  civil liability limits 149–51
  as nuclear power user 211–12
  reactor building 1–2, 256

United States
  as authorized nuclear-weapon state
    267
  baseload technology rankings 67
  civil liability costs/limits 149–51,
    152–3
  decommissioning/waste costs 33–4
  early nuclear development 216–17
  INPO (Institute of Nuclear Power
    Operations) 145–6
  in international nuclear market
    270–2
  nuclear export controls 283–6
  as nuclear power user 211–12
  nuclear trade alliance with South
    Korea 280–1
    123 agreements 283–4
  post-war nuclear policy 213–15
  regulation shortcomings 199–201
  regulatory model 199–201
  rising reactor costs 49–52
  safety regulation 2–3, 297–8
    compared with France 198–9
  safety systems 4–5
  shale gas 1–2
  technology versus performance-based
    safety standards 174–9
  waste management 32
  see also Three Mile Island
Unplanned Unavailability Factor 238
uranium enrichment 216–17

Venezuela, as aspiring nuclear power
  222–3
vertically integrated companies
  282
very low probabilities see probability
Vienna Convention on Civil Liability
  for Nuclear Damage (1963)
  150, 252–3
Vietnam
  as aspiring nuclear power
    222–3
  reactor construction 276
Vinci 274–5

waste lifespan 25–6
waste management 23–32, 33–4
Westinghouse
  in nuclear trade 271–3, 274
  in US/South Korea consortium
    280–1
Westinghouse reactors
  AP1000 47, 54, 55, 61, 221–2, 271,
    272
  in China 271, 272
  in France 190–1, 220–1
  in Germany 221–2
wind farms 72–8
World Association of Nuclear
  Operators 144–7

Zimmerman, Mark 46

Printed in the United States
By Bookmasters